IMMUNOASSAY
AND OTHER BIOANALYTICAL TECHNIQUES

IMMUNOASSAY
AND OTHER BIOANALYTICAL TECHNIQUES

Edited by
Jeanette M. Van Emon

CRC Press
Taylor & Francis Group
Boca Raton   London   New York

CRC Press is an imprint of the
Taylor & Francis Group, an informa business

CRC Press
Taylor & Francis Group
6000 Broken Sound Parkway NW, Suite 300
Boca Raton, FL 33487-2742

International Standard Book Number-10: 0-8493-3942-1 (Hardcover)
International Standard Book Number-13: 978-0-8493-3942-4 (Hardcover)

Library of Congress Cataloging-in-Publication Data

Immunoassay and other bioanalytical techniques / edited by Jeanette M. van Emon.
    p. ; cm.
    Includes bibliographical references and index.
    ISBN-13: 978-0-8493-3942-4 (alk. paper)
    ISBN-10: 0-8493-3942-1 (alk. paper)
    1. Immunoassay. 2. Biosensors. 3. Environmental chemistry. I. Van Emon, Jeanette M., 1956- .
    [DNLM: 1. Immunoassay--methods. 2. Biosensing Techniques--methods. 3. Chemistry,
Analytical--methods. 4. Environmental Monitoring--methods. 5. Toxicity Tests--methods. QW
525.5.I3 I3295 2007]

    QP519.9.I42I426 2007
    616.07'56--dc22
                           2006030215

Visit the Taylor & Francis Web site at
http://www.taylorandfrancis.com

and the CRC Press Web site at
http://www.crcpress.com

Foreword

This excellent reference text informs analytical chemists of the strengths and special applications of bioanalytical technologies. It describes several trends in the development and implementation of immunoassays and biosensor methods. The text illustrates that scientists involved in environmental diagnostics are working on better, faster, smaller, and more cost-effective technologies. As instrumental techniques and new materials become available, immunoassays and other biosensors are quick to adopt these new technologies. Certainly, if the field of immunodiagnostics is to remain a major analytical tool, the technologies upon which it is based must continually improve. It is interesting to note that material science is reaching into the nano realm where biological molecules have always resided. Innovative interfaces between new materials and biological molecules will be a driver in methods development for years to come. The choice of binding molecules is increasing as polyclonal and monoclonal antibodies are supplemented with aptamers, phages, peptides, imprints, and other receptors. Binding molecules are being modified with recombinant technologies and also by such techniques as sol–gel stabilization. Detection technologies using optical phosphors, quantum, dots and other reporters complement electrochemical and other transduction systems.

Of course the targets to which these analytical tools are being applied are expanding. Part of this trend is due to the changing nature of agriculture as we add genetically modified organisms to our more classical targets of pesticides, environmental contaminants, and food adulterants. Food safety is becoming an important issue with the need to detect analytes ranging from metals and bioterrorism agents to prions. Certainly, human monitoring is a need well served by immunodiagnostics and other biosensor technologies. This text also clearly illustrates the strength of biosensors in their application to homeland security measures and human safety.

Biosensors lend themselves to arrays in which multiplexed systems provide the opportunity to "taste" a sample. Such information-rich assays, of course, require newer computational approaches to address sophisticated data sets. In the final analysis we must do more than just develop assays; rather, we need to integrate them into analytical laboratories where numerous techniques are needed to solve complex problems. This text introduces the analytical chemist to the subject of sensor and immunoassay technology, presents new and developing technologies, and provides insight to implementing these technologies in a functional analytical laboratory. It should serve as a useful reference as well as a catalyst to further research.

Bruce Hammock
Distinguished Professor of Entomology &
Cancer Research Center

Preface

There is a continuing and pressing need for economical analytical methods that can detect trace levels of hazardous compounds in complex environmental and biological media. The availability of rapid and reliable methods for diagnostic indicators of human disease and exposure is also of key importance. Immunochemical methods are adaptable to a variety of matrices, are capable of achieving very low detection levels, and do not require heavy or expensive instrumentation. Significant time and cost savings can be achieved for studies requiring repetitive analysis or having high sample loads when specific immunoassays are employed.

Today, immunochemical methods are part of a growing cadre of bioanalytical methods for compounds of environmental and biological significance such as pesticides, biomarkers, genetically modified organisms, prions, and heavy metals. These methods are being applied to environmental monitoring and human exposure assessment studies and homeland security measures. This reference work is intended to inform analysts about the strength and versatility of immunochemical and related bioanalytical methods for a wide spectrum of environmental and biological measurement applications. Chapters describe the wide base of immunoassay methods illustrating the maturation of the technology. Readers who are already experienced analysts employing immunoassay methods will find a selection of readings describing other areas of applied bioanalytical research. The authors present established and innovative uses of bioanalytical methods as well as data analysis and quality assurance guidelines. Insights into recently developed technologies and procedures can be found in later chapters. The field is rapidly expanding and updates on breakthrough research are also presented, including the impact of nanotechnology.

As described in this reference work, bioanalysis is rapidly advancing in many areas such as engineering and generating specific antibodies and the development of antibody mimics and other receptors, microarrays, improved detection systems, immunoaffinity sample preparations, multiplexed immunoassays, and sensors for unattended and real-time analysis. It is hoped that this book will recruit additional practitioners of traditional analytical chemistry into the diverse field of bioanalytical method research. Bioanalytical methods are not a panacea, but they should be considered when deemed to provide appropriate data in the required time frame.

This reference work provides both a basic understanding of immunochemical and other bioanalytical methods and an update on important technological advances such as new platforms and detection systems. Therefore, the book should be a reference for students as well as practicing analysts and an aid for bioanalytical methods development and application.

Acknowledgments

The editor expresses her gratitude to family, colleagues, and friends for their unwavering support and enthusiasm during the many stages of this project. Special thanks to Gary Coates (my husband) for the cover art, Andrew Coates (my son) for helping to proof read, Bob Hope for his tireless efforts in checking details, and Guohua Xiong for keeping the research flowing in the lab while I was otherwise engaged. I am grateful for each of the authors who devoted their time in preparing chapters and who made this book a reality. I am indebted to those who have contributed greatly to the field of bioanalytical methods, in particular Rosalyn Yalow, Helen Van Vunakis, Ralph Mumma, and Bruce Hammock.

While this text is edited, and to some extent authored, by an employee of the U.S. Environmental Protection Agency (EPA), the views and opinions expressed herein are those of the respective authors and editor and do not represent those of the U.S. EPA.

Mention of trade names or commercial products does not constitute endorsement or recommendations for use.

Editor

Jeanette M. Van Emon is the director for the immunochemistry program at the US EPA, National Exposure Research Laboratory, Human Exposure and Atmospheric Sciences Division, Methods Development and Applications Branch in Las Vegas, Nevada. She received her BS in environmental studies from California State University, Hayward, and her PhD in agricultural and environmental chemistry from the University of California, Davis. During her postdoctoral research appointment at the University of California's, Lawrence Livermore National Laboratory, she developed monoclonal antibodies and immunoassays for environmental contaminants. Her research interests include development of tandem bioanalytical and instrumental methods, metabolomics, and nanotechnology. Her current research is focused on bioanalytical methods development for environmental contaminants and biomarkers of exposure.

Dr. Van Emon received the Award of Distinction from the College of Agricultural & Environmental Sciences, University of California, Davis, in 2001 and a Citation for Excellence award by the University Alumni Association. She received the US EPA Office of Research and Development Statesmanship Award for building partnerships and facilitating the exchange of information in immunochemistry within and outside the EPA. She is the recipient of several other agency awards including two Bronze Medals. She was twice awarded the US EPA and American Chemical Society joint Science Achievement Award in Chemistry (1992 and 1996) for the development and sponsorship of annual international Immunochemistry Summit Meetings. She has served as chairman and program chairman of the Agrochemicals Division of the American Chemical Society (ACS) as well as chairman of her local ACS Section. She is a former member of the Chemical Sciences Roundtable, the National Science Foundation, and the National Research Council. She has served as an advisor in the area of biotechnology to the United Nations International Atomic Energy Agency, Vienna, Austria. Dr. Van Emon was featured as an Environmental Pioneer on a US EPA Women in Science and Engineering Poster.

Dr. Van Emon has numerous publications (e.g., 50 journal articles, 26 US EPA reports, 13 proceedings, 18 book chapters, and 100 abstracts). She has edited two books on immunochemical methods and has developed and organized several national and international symposia. She is frequently sought as a speaker and lecturer.

Contributors

Elizabeth R. Abboud
Department of Biochemistry
Tulane University School of Medicine
New Orleans, Louisiana

Farid E. Ahmed
Leo W. Jenkins Cancer Center
The Brody School of Medicine
East Carolina University
Greenville, North Carolina

Miriam Altstein
Department of Entomology
Institute of Plant Protection
The Volcani Center
Bet Dagan, Israel

Lars I. Andersson
Bioscience, AstraZeneca R&D Sodertalje
    Research
Sodertalje, Sweden

Raymond E. Biagini
Biomonitoring & Health Assessment Branch
Robert A. Taft Laboratories
Cincinnati, Ohio

Diane A. Blake
Department of Biochemistry
Tulane University School of Medicine
New Orleans, Louisiana

Robert C. Blake II
College of Pharmacy
Xavier University of Louisiana
New Orleans, Louisiana

James F. Brady
Health Assessment & Environmental Sciences
    Department
Syngenta Crop Protection, Inc.
Greensboro, North Carolina

Jennifer R. Brigati
Division of Natural Sciences
Maryville College
Maryville, Tennessee

Alisa Bronshtein
Department of Entomology
Institute of Plant Protection
The Volcani Center
Bet Dagan, Israel

Franco Cardone
Department of Cellular Biology
    and Neuroscience
Institute Superiore di Sanita
Viale Regina Elena, Rome

Wilfred Chen
Department of Chemical & Environmental
    Engineering
University of California
Riverside, California

Jane C. Chuang
Battelle
Columbus, Ohio

Zoe Cobb
Bioscience, AstraZeneca R&D
    Sodertalje Research
Sodertalje, Sweden

David E. Cooper
SRI International
Menlo Park, California

Annalisa D'Andrea
SRI International
Menlo Park, California

Ibrahim A. Darwish
College of Pharmacy
King Saud University
Riyahd, Saudi Arabia

Kilian Dill
Antara Biosciences
Mountain View, California

Joyce Durnford
Battelle
Columbus, Ohio

Per the licence

Gregory W. Faris
SRI International
Menlo Park, California

H. Sho Fuji
CombiMatrix Corporation
Mukilteo, Washington

Xiaohu Gao
Winship Cancer Institute
Emory School of Medicine
Emory University
Atlanta, Georgia

H. Geue
Technical University of Munich
Freising-Weihenstephan, Germany

Andrey L. Ghindilis
CombiMatrix Corporation
Mukilteo, Washington

David S. Hage
Chemistry Department
University of Nebraska
Lincoln, Nebraska

H. Brian Halsall
Department of Chemistry
University of Cincinnati
Cincinnati, Ohio

William R. Heineman
Department of Chemistry
University of Cincinnati
Cincinnati, Ohio

B. Hock
Technical University of Munich
Freising-Weihenstephan, Germany

Loredana Ingrosso
Department of Cellular Biology
and Neuroscience
Institute Superiore di Sanita
Viale Regina Elena, Rome

Joany Jackman
Applied Physics Laboratory
Johns Hopkins University
Laurel, Maryland

Kanchan A. Joshi
Department of Chemical & Environmental
Engineering
University of California
Riverside, California

Mehraban Khosraviani
Amgen, Inc.
Thousand Oaks, California

K. Kramer
Technical University of Munich
Freising-Weihenstephan, Germany

Alison M. Kriegel
Department of Biochemistry
Tulane University School of Medicine
New Orleans, Louisiana

Kalle Levon
Motorola, Inc.
Tempe, Arizona

Xia Li
Beijing Vegetable Research Center
Beijing, China

Robin Liu
CombiMatrix Corporation
Mukilteo, Washington

Brent MacQueen
SRI International
Menlo Park, California

Marco Mascini
Chemistry Department
Universita degli Studi Di Firenze
Sesto Fiorentino, Italy

Maria Minunni
Chemistry Department
Universita degli Studi Di Firenze
Sesto Fiorentino, Italy

Annette Moser
Chemistry Department
University of Nebraska
Lincoln, Nebraska

Ashok Mulchandani
Department of Chemical & Environmental
  Engineering
University of California
Riverside, California

Mary Anne Nelson
Chemistry Department
University of Nebraska
Lincoln, Nebraska

Shuming Nie
Winship Cancer Institute
Emory School of Medicine
Emory University
Atlanta, Georgia

Ruth M. O'Regan
Winship Cancer Institute
Emory School of Medicine
Emory University
Atlanta, Georgia

Valery A. Petrenko
Department of Pathobiology
College of Veterinary Medicine
Auburn University
Auburn, Alabama

Maurizio Pocchiari
Department of Cellular Biology
  and Neuroscience
Institute Superiore di Sanita
Viale Regina Elena, Rome

Lon A. Porter Jr.
Department of Chemistry
Wabash College
Crawfordsville, Indiana

Niina J. Ronkainen-Matsuno
Department of Chemistry and Biochemistry
Benedictine University
Lisle, Illinois

Michael J. Schöning
University of Applied Sciences, Aachen
Jülich, Germany

Kevin R. Schwarzkopf
CombiMatrix Corporation
Mukilteo, Washington

John E. Snawder
Biomonitoring & Health Assessment Branch
Robert A. Taft Laboratories
Cincinnati, Ohio

Cynthia A. F. Striley
Biomonitoring & Health Assessment Branch
Robert A. Taft Laboratories
Cincinnati, Ohio

Raquel M. Trejo
Battelle
Columbus, Ohio

Sara Tombelli
Chemistry Department
Universita degli Studi Di Firenze
Sesto Fiorentino, Italy

Jeanette M. Van Emon
National Exposure Research Laboratory
U.S. Environmental Protection Agency
Las Vegas, Nevada

Joseph Wang
Departments of Chemical Engineering
  and Chemistry & Biochemistry
Arizona State University
Tempe, Arizona

Frances Weis-Garcia
Memorial Sloan-Kettering
  Cancer Center
New York, New York

William H. Wright
SRI International
Menlo Park, California

Yun Xing
Winship Cancer Institute
Emory School of Medicine
Emory University
Atlanta, Georgia

Maksym Yezhelyev
Winship Cancer Institute
Emory School of Medicine
Emory University
Atlanta, Georgia

Haini Yu
Department of Biochemistry
Tulane University School of Medicine
New Orleans, Louisiana

Bin Yu
Motorola, Inc.
Tempe, Arizona

Yanxiu Zhou
Motorola, Inc.
Tempe, Arizona

Table of Contents

1 Integrating Bioanalytical Capability in an Environmental Analytical Laboratory

Jeanette M. Van Emon, Jane C. Chuang, Raquel M. Trejo, and Joyce Durnford

CONTENTS

1.1 INTRODUCTION*

Bioanalytical methods prepare samples for analyses, detect residue levels of pesticides in food and the environment, follow products of genetic engineering, sense changes of pollutant levels in lakes and streams, and provide diagnostic medical information in real-time. Antibody-based bioanalytical measurement and separation techniques have been routinely used in medical and clinical settings for the quantitative determination of large target analytes (e.g., hormones, drugs) with molecular mass of more than several thousand Daltons. Immunochemical bioanalytical methods have also been used to detect and trace genetically modified organisms in the food production chain by measuring novel proteins [1]. These types of methods are also an important tool in detecting and studying prion disease or transmissible spongiform encephalopathies [2]. Antibodies for small molecules and heavy metals have been formulated into specific and sensitive bioanalytical methods such as immunoassays, biosensors, and immunoaffinity chromatography separations [3–11]. Quantitative immunoassay methods for pesticides and their metabolites and environmental degradation products, dioxins, polycyclic aromatic hydrocarbons (PAHs), polychlorinated biphenyls (PCBs), microbial products, and biomarkers of exposure have been useful for environmental monitoring and exposure assessment [12–20]. The application of immunoassay testing kits for field screening environmental contaminants and biosensors for real-time field monitoring has also been reported [21–27]. Flow injection techniques enable automated immunoassay and flow immunosensor formats [28,29]. Immunoaffinity chromatography has been used as a separation technique for analyte enrichment and/or cleanup for subsequent on-line or off-line analysis for environmental pollutants [30–41].

Analysts interested in applying bioanalytical methods should be familiar with how easily the capability can be integrated into an existing analytical facility. One approach to easily introduce bioanalytical techniques is to use specific immunoassays as screening methods to determine appropriate dilution factors for instrumental analysis. Used in this manner, immunoassays can minimize instrument down time by protecting sensitive components such as detectors and columns while making more efficient use of expensive instrumentation. As another example, the powerful resolving capability of high-performance liquid chromatography (HPLC) can be coupled to the sensitive detection of immunoassay. Leveraging the benefits of both technologies provides a synergistic effect. For real-time monitoring, biosensors are firmly established for the measurement of blood analytes such as glucose, urea, and other diagnostic indicators. This bioanalytical capability has useful application for environmental monitoring and homeland security. Analysts will undoubtedly find other ways to include these types of techniques into their repertoire of methods.

* Notice: The US Environmental Protection Agency (EPA), through its Office of Research and Development, funded and collaborated in the research described here under Contract 68-D-99-011. It has been subjected to agency review and approved for publication. Mention of trade names or commercial products does not constitute endorsement or recommendation for use.

TABLE 1.1
A Glossary of Commonly Used Terminology of Immunochemical Methods

Term	Glossary
Antibody	A class of serum proteins that are induced by exposing to an immunogene (sometimes referred to as an antigen) and will bind specifically to the antigen/analyte forming an antibody-antigen complex
Antigen	A molecule capable of selectively binding to an antibody. Antigens are not necessarily immunogenic, sometimes used synonymously with immunogens
Antiserum	Serum from an animal containing specific antibodies
Chromophore	A chemical that changes color, resulting from enzyme interaction with substrate
Conjugate	Formed by covalently coupling two molecules together such as a hapten with a protein or antibody with an enzyme
Cross-reactivity	The ability of antibodies to react with other compounds (i.e., target analyte) in addition to the immunogen/antigen that induced their production
ELISA	Enzyme-linked immunosorbent assay (ELISA) is a commonly used immunoassay format for environmental monitoring
Enzyme	A substance that can react at a low concentration as a catalyst to promote a reaction
Enzyme conjugate	An enzyme that is covalently coupled to a protein such as an antibody or to a target analyte/hapten
Hapten	A small antigenic molecule that can bind to an antibody. It is not immunogenic and cannot generate an antibody response.
IgG	A divalent immunoglobulin-G antibody molecule, consisting of two identical H-chains and two identical Y-chains held together with disulfide bonds
Immunogen	A substance that can generate a strong immune response. Sometimes used synonymously with antigen
Monoclonal antibody	A homogeneous antibody population, possessing identical selectivity and affinity produced by a single clone of antibody-producing cells
Polyclonal antibody	An antibody population of various selectivities and affinities produced by many clones of antibody-producing cells
Substrate	A chemical that specifically reacts with an enzyme
Titer	A dilution of a substance that produces a desired result such as the working dilution of an antibody in an immunoassay

This chapter presents general concepts and uses of immunochemical bioanalytical methods for sample preparation and detection of pesticides and other small molecules. A discussion of the general approaches to equipping a bioanalytical laboratory is also provided. Instrumentation and supply lists and laboratory space considerations are given as guidance for integrating bioanalytical capability into an analytical facility. Guidance for preparing quality assurance documents and standard operating procedures are also presented. Selected immunoassay and immunoaffinity chromatography procedures are presented as illustrations of the required expertise, reagents, supplies, and laboratory equipment. A glossary of terms commonly used in immunochemical methods is given in Table 1.1.

This chapter does not describe the laboratory facilities, equipment, and instrumentation for antibody production or for developing biosensors. These issues are discussed in other chapters. Chapter 2 and Chapter 3 discuss the development of monoclonal antibodies and Chapter 16 and Chapter 17 describe biosensor development.

1.1.1 ANTIBODY REAGENTS

Vertebrate animals produce antibodies as part of their defensive immune response when exposed to a foreign organism or toxic substance. Antibodies (Abs) are immunoglobulins with specific reactive binding sites for a corresponding antigen (Ag). The immunoglobulin typically employed in analytical methods is the divalent IgG that has two identical binding sites for the intended target. Antigens (Ags) have particular reactive sites called epitopes or antigenic determinants for binding to specific Abs. An Ab binds to an Ag (analyte) and forms an immune complex (Ab–Ag) that is the basis for immunochemical analytical procedures.

A key factor in immunochemical techniques is the response of the specific Ab for the target analyte in a sample matrix or extract. The specificity of the immunochemical method is largely determined by the specificity of the Ab employed. Many environmental pollutants such as insecticides, herbicides, PAHs, PCBs, and dioxins are too small (mw < 10,000 Daltons) to elicit Ab production by themselves. Although these small molecules may be antigenic (i.e., can react with an Ab), they are not immunogenic (i.e., cannot elicit Ab production). Low molecular weight analytes must be converted to immunogenic compounds to induce a specific Ab response. This typically involves three steps: (1) designing a hapten that preserves important structural features of the analyte, (2) synthesis of the hapten, and (3) coupling the hapten with a macromolecule such as a protein [42–44]. Figure 1.1 gives an example of a target analyte, hapten, hapten-protein conjugate, and generalized antibody structure [12].

Paraquat dichloride

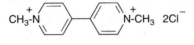

Paraquat hapten

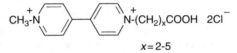

$x = 2\text{-}5$

Paraquat hapten-protein conjugate

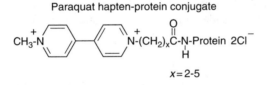

$x = 2\text{-}5$

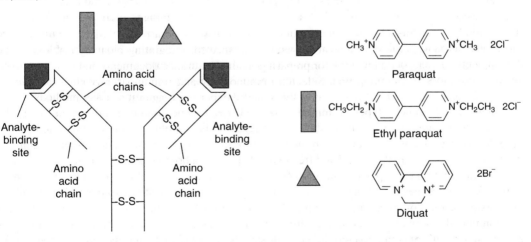

FIGURE 1.1 Example of a target analyte, hapten, protein conjugate, and generalized antibody structure. (From Van Emon, J. M. and Lopez-Avila, V., *Anal. Chem.*, 64, 79A, 1992.)

A hapten is a derivative of the small molecule analyte that is not immunogenic by itself but can stimulate the production of specific Abs when attached to a large carrier molecule such as a protein (e.g., keyhole limpet hemocyanin, bovine serum albumin) and can bind with the antibody. A cornerstone of immunoassay development for small molecules was Landsteiner's research on immunologic specificity using qualitative hapten inhibition techniques that suggested that Abs could bind to small molecules [45]. The design of a hapten must allow important structural features of the analyte to be preserved after attachment of a functional group (i.e., carboxyl acid, amino, hydroxyl, sufhydryl) as a side arm for covalent binding with the carrier. The linker or side arm should be introduced distal to the determinants of interest for Ab recognition. Thus, the hapten design greatly affects the selectivity and sensitivity of the resulting Ab. The newly introduced functional groups are used to conjugate the hapten to the carrier molecule using various reaction schemes (42-44). The resulting hapten-carrier conjugate is an immunogenic compound used to stimulate the immune system for specific Ab production. Some of the Abs will be directed toward the carrier protein, others to the linkage, and some toward the hapten or target portion of the conjugate. This mixture of Abs can be used for developing immunochemical methods providing there is a strong overall response to the hapten. The mixture is termed polyclonal because the resulting Ab population is a product of more than one clone responding to the immunogen. Alternatively, clonal selection may be performed to produce an Ab preparation responding from a single clone. Monoclonal antibodies (mAbs) are a uniform reagent of a single Ab population having the same selectivity and affinity [46,47]. Further refinements using genetic antibody engineering or directed evolution can provide reagents with improved affinity and selectivity as well as increased thermal stability and tolerance to organic solvents (see Chapter 2).

Large-scale in vitro Ab production can provide a highly concentrated supply of reagent mAb. These in vitro production techniques are very attractive as they can be performed on a large scale in culture with minimal labor and cost (see Chapter 3). Both polyclonal Abs (pAb) and monoclonal Abs (mAb) have been successfully employed in clinical and environmental bioanalytical methods. Each type of Ab preparation has advantages and disadvantages (e.g., cost, development time, affinity, titer) that need to be considered prior to Ab production [48].

1.1.2 IMMUNOASSAY PRIMER

An immunoassay is based on the reaction of an analyte/antigen (Ag) with a selective antibody (Ab) forming an Ab–Ag complex. The immunochemical reaction may be visualized with enzyme labels such as alkaline phosphatase and horseradish peroxidase, providing a colored end product. Radio-immunoassays use labels such as I^{125} and may offer the advantages of better precision and sensitivity. Other labels and techniques such as fluorescence polarization [49], chemiluminescence [50], and electrochemical detection [51] as well as more novel labels have also been employed. The development of electrochemical immunoassays is presented in Chapter 16. A discussion of upconverting phosphors as emerging sensitive labels for immunoassays [52] is presented in Chapter 9. Table 1.2 presents various enzymes, substrates, and labels used for immunoassay methods.

There are two general types of immunoassays: (1) heterogeneous and (2) homogeneous. In the heterogeneous immunoassay, one of the immunoreagents is immobilized on a solid support (e.g., latex beads, microplates, magnetic particles, test tubes, microspheres, filter paper) and requires a washing step to separate the bound and free reagents [53]. In a homogeneous immunoassay format, the immunoreaction takes place in solution and does not require a separation step but is more prone to matrix interferences. Heterogeneous immunoassays are used in most environmental applications as a result of the complex samples typically encountered. A commonly used assay format for environmental pollutants is the competitive enzyme-linked immunosorbent assay (ELISA). A particular strength of immunoassay is the high sample throughput capability. Polystyrene microplates containing 96-wells in an 8-row by 12-column format enable the analysis of several quality control (QC) samples and controls appropriate for a large sample load. The 96-well microplate

TABLE 1.2
Example of Detection Methods Used in Immunoassays

Detection Method	Label	Substrate
Colorimetric (end point or kinetic)	horseradish peroxidase	Diaminobenzidine ortho-phenylenediamine tetramethylbenzidine 2, 2′-azino-di-(3-ethylbenzthiazoline sulfate) 2,2′azino-bis-(3-ethylbenzothiazoline-6-sulfuonic acid)
	alkaline phosphatase	p-nitrophenylphosphate 3,3′5,5′-tetramethylbenzidine
Fluorescence	alkaline phosphatase, β-galactosidase	4-methylumbelliferyl phosphate β-galactosylumbelliferone
Chemiluminescence	Peroxidase	Luminol
Liminescence	Luciferase	Luciferin
Radioactivity	I^{125}	—
Electroactive (oxidation current)	Ferocenes	—

format can accommodate 0.225 mL of reagent per well. Microplates containing a larger number of smaller wells are also available. For example, microplates containing 384- or 1536-wells are routinely used in the pharmaceutical industry. Figure 1.2 shows examples of ELISA methods in both indirect and direct competition formats.

In one indirect format, an Ag is immobilized onto a solid surface such as the wells of a microtiter plate. The analyte and a known amount of specific Ab are mixed and react to form solution phase Ab–Ag complexes. This solution is added to the Ag-coated microtiter plate where the immobilized coating Ag competes with remaining free analyte in solution for available Ab binding sites. A buffer rinse removes reagents not bound to the solid surface. A secondary species-specific Ab covalently bound to an enzyme such as alkaline phosphatase (i.e., Ab-enzyme conjugate) is then added and binds with the specific Ab bound to the immobilized Ag (Figure 1.2). Unbound material is washed away with another buffer rinse. Enzyme substrate is added to produce a color, and the absorbance is read, or the kinetics of color development is monitored. This ELISA is also called an inhibition assay as the analyte in the samples or standards inhibits the specific Ab from binding to the immobilized coating Ag on the solid phase. The absorbance value of each sample is compared with the absorbance values obtained from a standard curve. A decrease in the signal is directly related to the amount of target analyte present in the sample. This inverse relationship between the target analyte concentration and the measured absorbance is the basis for quantitation. Chapter 10 provides a discussion on the statistical analysis of immunoassay data.

In a typical direct format, the specific antibody is immobilized on the wells of a microtiter plate. The analyte and an enzyme-labeled hapten (analyte mimic) are added to the prepared microtiter plate. The analyte in the sample competes with a known amount of the enzyme-labeled hapten for binding sites on the immobilized Ab. A washing step removes all unbound reagents, and substrate is added for color development. The direct format is commonly employed in immunoassay testing kits.

A guidance document developed by the EPA described the basics of environmental immunoassay techniques as an aid for analytical chemists to apply this alternative technology as methods became more widespread [4,54]. Quality control/quality assurance (QA/QC) guidance documents helped in the early evaluations of field testing kits. Immunoassay methods have since been included in an EPA methods compendium [55]. Immunoassay methods and conventional gas chromatographic (GC) and GC/mass spectrometry (GC/MS) methods have been studied in tandem to document the overall accuracy and precision of ELISA methods [27,56–60]. One consideration when comparing methods is that the immunochemical method may respond to compounds related

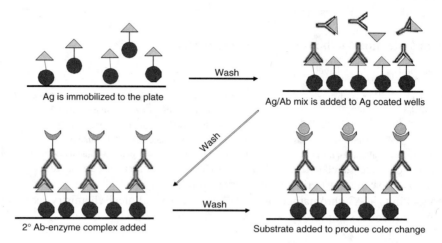

Indirect competitive ELISA

Antibodies are immobilized to the plate

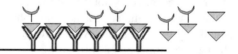

Analyte and enzyme labeled hapten compete for antibody sites

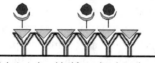

Wash removes unbound analyte and labeled hapten

Substrate is added for color detection

Direct competitive ELISA

FIGURE 1.2 Example of indirect and direct ELISA formats.

to the target analyte, whereas the instrumental method may not be able or optimized to do so. However, the compounds detected in the immunochemical procedure may be of toxicological significance and may help in identifying metabolites or environmental breakdown products previously undetected. An immunoassay that responds to a broader suite of related compounds may also be used prior to instrumental analysis in an analytical triage approach. An ELISA for determining PAHs in residential and contaminated soil samples was compared with a GC/MS procedure for application to a monitoring study [57,58]. The ELISA data were significantly different and higher than the GC/MS data, but they were highly correlated for 52 contaminated soil samples. This was due to the GC/MS not measuring all of the PAHs and alkylated PAHs present that responded in the assay and are typically found in environmental samples. Only the 16 EPA priority pollutant

PAHs are typically measured by GC/MS because of cost limitations. However, the ELISA provides a boarder indication of the contaminants present that may be useful for toxicological or exposure studies.

1.1.3 IMMUNOAFFINITY CHROMATOGRAPHY

Residue analyses of trace levels of contaminants (e.g., pesticides, PCBs) in environmental samples generally require a cleanup prior to detection by GC or HPLC. The cleanup is often accomplished by SPE techniques with the stationary phase (e.g., silica, C_{18}) determined by the chemical properties of the target analyte. Even when cleanup procedures are employed, the quantitative determination of trace levels of contaminants in complex matrices can be difficult. Thus, the cleanup step is often the most critical step in the overall analysis.

Immunoaffinity (IA) chromatography has been used in clinical chemistry as an extraction method for aqueous samples as well as a cleanup method for various sample media [30–41,61–65]. An IA column employs a solid phase extraction (SPE) cartridge. However, the conventional sorbent material is replaced with a sorbent of specific antibody that is immobilized onto an inert solid support forming an immunosorbent column resin (i.e., stationary phase).

The process is based on the reversible reaction of a specific Ab and its target analyte. A liquid extract containing the analyte is passed over the IA column under conditions favoring the binding of Ab to the analyte. The column containing the bound analyte is washed to remove matrix contaminants and nontarget analytes. The mobile phase conditions are then altered such as a change in pH or ionic strength, allowing the Ab to release the analyte. The analyte is eluted from the column in a small volume of liquid for detection by immunoassay or off-line or on-line instrumental analysis. The Ab remains immobilized on the column for another use. The immunologic reaction imparts selective extraction and enrichment of the target analyte. The selective separation of the target analyte from other contaminants in the sample through Ab–Ag interactions is particularly suited to the separation challenge in complex sample media.

IA columns as cleanup methods have been developed for carbendazim [39], triasulfuron [64], triazinic biocides [35,36], PCBs, dioxin/furans [40,65] and ivermectin [31,32] in various sample media, including water, soil, serum, and biological tissue. Immunoaffinity capillary electrochromatography has been employed for selective trace enrichment prior to capillary electrophoresis with detection by laser induced fluorescence [41]. Automated IA columns and columns linked with HPLC methods have also been reported [36–39]. Chapter 14 and Chapter 15 provide detailed examples of immunoaffinity chromatography procedures and applications.

1.1.4 IMMUNOSENSORS

Biosensors are analytic instruments composed of two parts: a biological recognition element such as an Ab, enzyme, receptor, DNA, or cell and a transducer that converts the biorecognition process into a measurable physical signal. Biosensors can be classified based on transducer type as optical, electrochemical, piezoelectric, or thermometric. An immunosensor is an affinity-based biosensor designed to detect the direct binding of an Ab or an Ag, forming an immunocomplex at the transducer surface. Immunosensors can provide continuous, in situ, and rapid measurements. Biosensors utilizing several different transducer types have been developed and applied to pesticide monitoring [21–26, 66–67]. Chapter 4, Chapter 16, and Chapter 17 describe the development and application of biosensors for environmental monitoring and homeland security.

1.1.5 MICROARRAYS

Microarray technology enables the parallel analysis of hundreds or thousands of biological molecules such as DNA on a single chip or glass slide. The molecules are separately arrayed on a chip so that individual, but simultaneous, analyses can occur. The analysis is based on the reaction of a labeled target molecule to the molecular array on the chip. Thus, information regarding the reaction of a single molecule with several other molecules is obtained. A tremendous amount of data regarding biological interactions can be obtained from a single chip [68,69]. These data aid in the study of functional relationships and biological processes, including disease mechanisms. Microarray data also have a tremendous research impact for genomics and proteomics and environmental monitoring. Chapter 18 and Chapter 19 provide detailed discussions on microarray technology and applications.

1.1.6 ADVANTAGES AND LIMITATIONS OF BIOANALYTICAL METHODS

Immunochemical bioanalytical methods generally offer high selectivity and lower detection limits when compared to instrumental methods. Other advantages are high sample throughput and relative low costs. These attributes are necessary for large-scale biological monitoring to support epidemiologic studies (see Chapter 11). Bioanalytical methods are applicable to a wide range of analytes, including small molecular environment contaminants, genetically modified organisms (see Chapter 12), and medically important prions (see Chapter 13). Field portable immunoassay testing kits enable screening tests to be performed on site with results obtained close to real-time by field technicians. Biosensors can provide monitoring data in real time with networking capability and remote access.

The performance of immunoassays and immunoaffinity chromatography depends strongly on the properties of the specific Abs. However, the greatest limitation is Abs specific for environmental pollutants are not readily available commercially. It can be time-consuming and costly for analytical laboratories to synthesize haptens as well as generate and characterize Abs. Recently, researchers have used recombinant DNA technology to produce recombinant Abs, but the processes can be expensive, time-consuming, and labor-intensive. However, through recombinant DNA technology, phage-antibody libraries have been developed as sources of selective phage probes for use in ELISA-type formats [70]. Chapter 8 presents the development and use of phage as biospecific probes. Molecularly imprinted polymers (MIPs) may be used as Ab mimics in traditional antibody-based method formats such as immunoassay and immunoaffinity chromatography [71,72]. A primary advantage of MIPs is the potential for a large supply of a uniform reagent. However, MIPs may have a greater potential for organic solvent tolerance and greater selectivity than biological reagents. Because of the imprinted binding site, the physical and chemical attributes of MIPs (see Chapter 5 and Chapter 7) may be advantageous for the development of sensors for chemical warfare agents [67]. Another alternative is the application of aptamers or nucleic acid ligands as reagents. Reagent grade aptamers can be obtained in large quantity and have been applied to biosensors and other bioanalytical methods [73]. These alternative receptor molecules and their application to bioanalytical methods are discussed in Chapter 6.

Bioanalytical procedures can be used in conjunction with conventional analysis for enhanced method performance and quality assurance. ELISAs can be used to analyze fractions from a HPLC separation as a selective detector in contrast to the typical non-selective detectors typically employed for HPLC [74]. Immunoaffinity chromatography columns for sample enrichment and purification can be coupled with either on-line or off-line HPLC or HPLC/MS analysis [37–39]. GC or GC/MS detection can also be used for off-line analysis after immunoaffinity cleanup [36,40].

Because Abs function in aqueous environments under physiological conditions (pH 7.0, 0.15 M NaCl), most immunochemical methods have a restricted tolerance for organic solvents. For compounds having a high lipophilicity such as the dioxins/furans, a challenging task in assay development may be determining an optimal solvent system for both analyte solubility and method performance. Through the use of mixed solvents and surfactant, a suitable solvent system can be found.

In order to enhance the advantages of simplicity and high throughput, simple sample preparation procedures without any cleanup steps are desirable. Environmental matrices range from simple (e.g., drinking water) to difficult (e.g., contaminated soils or sediments). Sample matrices used to assess human exposure include biological fluids and complex dietary samples. Antibodies can often specifically bind the target analyte even in the presence of a complex matrix by diluting the sample within practical detection limits. Alternatively, similar cleanup procedures employed in conventional methods such as SPE or liquid–liquid partitioning can also be used. However, complicated extraction and cleanup procedures prior to use of immunochemical methods may offset the advantage of a high sample throughput.

1.2 ESTABLISHING BIOANALYTICAL METHOD CAPABILITY

A bioanalytical laboratory typically includes bench space, fume hoods, refrigerators, freezers, analytical balances, and other commonly used analytical equipment and supplies. Although other equipment such as a 96-well microplate reader, small incubator, and orbital shaker may be required, bioanalytical capability can easily be integrated into an existing analytical facility. A typical analytical chemistry facility requires laboratory space of approximately 2000 ft.^{2} with multiple rooms. The majority of space is necessary for sample receipt and handling, preparing standards and reagents, sample extraction and cleanup, and performing detection procedures such as GC/MS. The space requirements for performing immunoassay detection and immunoaffinity chromatography are typically quite small, and these procedures are often performed next to GC/MS or HPLC instrumentation.

1.2.1 EQUIPMENT AND INSTRUMENTATION

1.2.1.1 General Equipment and Instrumentation

Table 1.3 summarizes the general equipment and instrumentation for sample handling, storage, and preparation for immunoassay and immunoaffinity techniques. Most of this equipment and instrumentation is also required for operating a conventional analytical chemistry laboratory and can be used for several different types of analyses.

TABLE 1.3
Summary of General Equipment and Instrumentation Used in a Bioanalytical Laboratory

Function	Instrument/Equipment
Storage for solvents, reagents, standards, and samples	Solvent storage cabinets, refrigerators, and freezers (-80°C desirable)
Cleaning glassware	Muffle oven
Buffer and reagent preparation	Reagent water system (i.e., ASTM Type 1 reagent water system or equivalent)
Sample preparation and processing	Chemical fume hood, analytical balance, centrifuge, homogenizer, incubator, and Vortex mixer
Sample extraction	Pressurized solvent extraction system, Soxhlet extraction apparatus, solid phase extraction columns, ultrasonic bath, orbit shaker, microwave oven
Organic solvent contration/evaporation	Nitrogen evaporator or Kuderna-Danish evaporator, turbo-evaporator
Analyzing standards and samples, confirmatory analyses	GC, GC/MS, HPLC, LC/MS

1.2.1.2   Special Immunoassay Equipment and Instrumentation

Some of the required equipment and instrumentation used for bioanalytical methods that may not normally be found in an analytical chemistry laboratory include

- Microplate washer: Plate washers range from manual (a simple wash bottle) to fully automated. Prices for these washers vary significantly depending upon the features.
- Magnetic separation rack: This apparatus is required for performing washing and separation steps in a magnetic particle-based immunoassay.
- Spectrophotometer: A spectrophotometer specially designed to either read 96-microwell plates, test tubes, or other formats to measure the signal output of the particular method format for quantitative analysis. Software to obtain and quantify the data, change wavelengths, control temperature, and perform statistical analysis needs to be considered before purchasing the equipment. Instrumentation for other detection systems such as fluorescence, chemiluminescence, luminescence, radioactivity, and electroactive may be required.

1.2.2   SUPPLIES AND REAGENTS

1.2.2.1   General Supplies

Common supplies used in a bioanalytical laboratory are basic to most conventional analytical laboratories and include:

- Assorted disposable gloves
- Assorted glassware (i.e., beakers, volumetric flasks, graduated cylinders, etc.)
- Assorted test tubes
- Boiling chips
- Felt-tipped markers
- Label tape
- Laboratory plastic film (parafilm)
- Laboratory wipes such as Kimwipes
- Magnetic stirrers and stir bars
- Organic solvents (e.g., acetone, acetonitrile, methanol)
- Paper towels
- pH meters
- Spatulas
- Syringes
- Timers
- Ultrasonic water bath
- Various sizes of polypropylene test tubes
- Vortex mixers
- Weigh boats/paper

1.2.2.2   Special Supplies for Immunoassays and Immunoaffinity Chromatography

In addition to using general laboratory reagents, supplies, and equipment, immunoassay and immunoaffinity chromatography methods may also require supplies not typically found in an analytical chemistry laboratory. Supplemental items may include:

- Adhesive acetate plate sealers
- Antibodies (e.g., compound-specific/primary antibody, antibody-enzyme conjugates)
- Antigens (e.g., analyte standards, coating antigen conjugates)
- Buffer salts (e.g., phosphate-buffered saline, mono- and dibasic sodium phosphate, sodium chloride, mono- and dibasic potassium phosphate, carbonate-bicarbonate)
- Carboys (10–20 L) for storage of large amounts of buffer
- Detection labels (e.g., enzymes, fluorophores, isotopes, electroactive and/or chemiluminescent compounds)
- Enzymes (e.g., alkaline phosphatase, β-galactosidase, horseradish peroxidase, luciferase)
- Enzyme-linked secondary antibodies, dependent on the source of the primary antibody (i.e., goat anti-rabbit IgG antibody conjugated to alkaline phosphatase or horseradish peroxidase, etc.)
- Enzyme substrates (i.e., p-nitrophenylphosphate; 3,3'5,5'-tetramethylbenzidine; diaminobenzidine; ortho-phenyldiamine; 4-methylumbelliferyl phosphate; luminol; luciferin)
- Microtiter plates, 96-well format (e.g., Nunc Nalgene or equivalent) with protein binding capacities
- Multichannel and single channel pipettors
- Orbital shaker
- Pipet guns
- Pipet tips for multi- and single channel pipettors
- Pre-packed columns with various types of support materials used to prepare immunoaffinity chromatography columns as well as a control column (e.g., HiTrap NHS-activated HP, 1 mL)
- Reagents such as bovine serum albumin (Fraction V), skim milk, or fetal bovine serum used as blocking agents to limit non-specific binding to microtiter plates
- Reagent troughs for multichannel pipettors
- Resin materials (e.g., HiTrap NHS activated and CDI-activated Sepharose CL 4B) used as support for packing an IA column (or pre-packed columns)
- Serological pipettes
- Sodium azide used as a buffer preservative
- Thimerosal used as a buffer preservative
- Tween 20 used as a surfactant in buffer solutions to prevent nonspecific binding

Some immunoassays are commercially available as self-contained testing kits and may contain supplies such as coated microplates, magnetic particles, test tubes, and buffers. These kits provide detailed protocols using the supplied reagents to obtain a specified assay performance. Results can be compromised if the protocols are not strictly followed or if reagents other than those supplied or specified in the kit are substituted.

1.2.3 REAGENTS AND BUFFERS FOR IMMUNOCHEMICAL METHODS

Antibodies may be produced in-house or obtained through various sources (commercial vendor, academic, or government laboratories). Antibodies are proteins, and appropriate measures should be taken to prevent their degradation. Upon receipt of the Ab, it should be stored at the temperature recommended by the source. It may be appropriate to store the stock solution of Ab in small aliquots for single use portions. If the required amount for the immunoassay is very small ($<$ 20 μL), larger volumes may be made into aliquots and stored refrigerated for multiple uses over a short time period to minimize freeze-thaw cycles. Repeated freeze-thaw cycles can lead to partial denaturation of the Ab that may result in unwanted protein-protein aggregates and diminished Ab performance [75]. The expiration date designated on commercial immunoassay testing kits should be followed. Most Abs are viable for at least a year at $-20°C$ or even longer with minimal freeze-thaw cycles.

Antibody-enzyme conjugates may be obtained from the same types of sources that provide specific Abs, or they can be produced as part of the method development process. Immunoassays commonly employ species-specific secondary Abs that bind with the primary or analyte-specific Ab. Secondary Abs labeled with an enzyme (e.g., goat anti-rabbit alkaline phosphatase) are readily available commercially. Either primary or secondary Abs can be labeled with biotin, horseradish peroxidase, alkaline phosphatase, fluorochromes, I^{125} or other labels, depending on the desired method. Hapten-protein or antigen-protein conjugates may be required, depending on the assay format. These conjugates can be prepared during methods development or prepared by commercial sources. Conjugates are usually stored refrigerated with a preservative such as sodium azide rather than frozen. Enzymes, substrates, and other labels are available from a variety of commercial sources. The storage conditions and expiration dates provided for individual reagents should be followed.

A variety of assay buffers are used in immunochemical procedures. Coating buffers (i.e., carbonate-bicarbonate) are designed to maximize the immobilization efficiency of the coating Ag or Ab to the solid support (i.e., test tubes, microtiter plate wells). Blocking buffers may contain BSA or other agents that bind to unreacted spots on the solid support without reacting with the coating reagent. Phosphate buffers are typically used to dilute standards and samples to enhance the performance of the Ab and minimize non-specific binding. Wash buffers (e.g., phosphate with Tween 20) are designed to minimize interferences and remove un-reacted materials without affecting or damaging the reagents bound to the solid support. Substrate buffers (e.g., diethanolamine) maximize the color reaction between the substrate and enzyme. Sodium bicarbonate and phosphate buffers are also employed in immunochromatography procedures for Ab immobilization and sample elution. Buffers must not interfere with the reactants or contribute to the color end-point signal. Buffers can be easily prepared using standard protocols (Appendix 1.A), or pre-formulated buffers in liquid or solid form are available that provide the desired buffer composition.

1.3   ESTABLISHING AND MAINTAINING IMMUNOASSAY CAPABILITY

In this section, it is assumed that the immunochemical procedure is introduced into the user laboratory with standard operating protocols (SOPs) and appropriate immunologic reagents. Under these conditions, the used laboratory does not need the capability to develop immunochemical reagents that includes such tasks as hapten synthesis, antibody production, and characterization in addition to method development [42–44]. To establish immunoassay capability, the user laboratory must demonstrate the ability to validate an established procedure and apply the method for analytical research or other application.

1.3.1   ASSAY VALIDATION

Initially, the assay is conducted using the SOPs established by the source laboratory. To validate method performance, the assay should be assessed on at least three different days. The assay results must indicate that the quality assurance/quality control (QA/QC) parameters are acceptable. Each analysis should include a standard curve, QC calibrators (low, middle, high), and other controls necessary for documenting the performance of the assay. If initial assay results indicate a sub-optimal performance, then new assay conditions must be established. For solid-phase assays such as those performed in 96-well microplates, a checkerboard titration can determine the optimal concentrations of specific antibody and coating antigen [4]. The reported concentrations from the source laboratory are typically used as a basis for determining optimal concentration values of reagents for the new laboratory. It is a good practice to also check all buffers for proper pH prior to analysis. Commercial reagents such as enzyme conjugates should initially be used at the dilution specified by the manufacturer.

Whenever new lots of reference standards or immunochemical reagents are obtained, they should be comparatively tested with the lot currently used in the immunochemical method. It is essential that the performance of new lots of immunoreagents (i.e., antibodies, enzyme conjugates) be evaluated prior to use as reagents produced at different times could have different activities that may influence assay performance. The composition of replacement lots should closely match the critical characteristics of those previously used for consistent assay performance, particularly in regard to Ab selectivity. If activities should differ, then SOPs can be adjusted to resetablish the required assay performance.

1.3.2 SAMPLE PREPARATION

The key steps in preparing environmental samples are extraction, enrichment or concentration, cleanup, and detection. The extraction technique removes the target analyte from the sample medium for subsequent analysis. Extraction methods such as Soxhlet, sonication, mechanical or manual shaking, or solid phase procedures can be used for both conventional instrumental detection methods and for immunoasssays. An established method for extracting the analyte of interest is a good starting point for quantitative modifications can be made for immunoassay compatibility. The simple manual shaking procedure used for most field-testing kits usually does not quantitatively remove the target analyte from the sample matrix. Evaluation and comparison studies must be performed to establish the efficiency of the extraction method to accurately interpret data generated from immunoassays. This is particularly important for field portable methods or testing kits that typically employ the less efficient extraction procedures.

Immunoassays are traditionally performed in an aqueous phase at or near physiologic conditions where antibodies are most effective. Antibody-based reactions cannot occur in high levels of the organic solvents typically used to extract many environmental pollutants (i.e., pesticides, PAHs, PCBs, and dioxins). The tolerance of a particular assay for various water miscible organic solvents must be determined to establish an optimal assay solvent/buffer system (e.g., 20% methanol in 80% phosphate buffer) where the target analyte remains soluble and the antibody active. This is easily accomplished by generating a series of standard curves using various concentrations of organic solvent in buffer. The solvent/buffer system that shows minimal or no undesired effect on the antibody performance is selected. All calibration and standard solutions, controls, and samples should be prepared in the same solvent/buffer system to normalize any effects on quantification in the assay.

The choice of extraction solvents for conventional and immunoassay methods may differ. The organic solvent most commonly used in extractions prior to immunoassay detection is methanol because of its water solubility and compatibility with phosphate buffer that is used throughout the immunoassay procedure. Other extraction solvents such as dichloromethane or hexane can either be evaporated or solvent exchanged into methanol for immunoassay analysis.

As antibodies can be highly selective, a cleanup step is rarely required to remove extraneous materials for ELISA. In contrast, instrumental detection methods frequently require a multi-step cleanup to protect chromatography columns and expensive instrumentation. For complicated sample matrices, cleanup methods may be required prior to ELISA to remove sample interferences.

1.3.3 MATRIX EFFECTS

A sample matrix effect is often encountered in environmental analysis for both instrumental and immunochemical methods. Interferences in the analysis may be the result of nonspecific binding of the analyte to the matrix. For antibody-based methods, additional matrix effects may be encountered from nonspecific binding of the matrix to the Ab or enzyme or denaturation of the Ab or enzyme. Effects of the sample matrix can be determined by analyzing a number of samples at different dilutions. Typically, results from different dilutions should be within $\pm 30\%$. Larger variations in the data suggest a matrix interference problem, indicating cleanup procedures may be necessary. For immunoassays based on highly sensitive antibodies, the matrix effect may be overcome by simply diluting the sample if a practical detection level can still be achieved. For more complex sample matrices, conventional cleanup methods (e.g., solid phase extraction or accelerated solvent extraction) or immunoaffinity column chromatography may be applied to sample extracts prior to ELISA.

1.3.4 DATA PROCESSING

A calibration curve consisting of a series of standard solutions is generated for each specific immunoassay. These curves are sigmoidal in shape, suggesting that the best fit curve is either a four-parameter fit or a log-logit fit [76]. Four parameter fit curves are commonly used in the 96-microwell plate assays, and the log-logit fit is built into most commercial testing kits and data analysis software [77]. The four parameter fit calibration curve employs the following formula $y = (A - D)(1 + (X/C)B + D$ where:

- y is the ELISA response in absorbance
- A is the y-value corresponding to the asymptote at the lowest value of X
- B is the slope of the curve
- C is the central point of the linear portion of the curve (i.e., the X-value midpoint between A and D, frequently termed the concentration yielding the 50% inhibition point or IC_{50}
- D is the y-value corresponding to the asymptote at the highest value of X
- X is the concentration of the analyte

Specifications are determined from each calibration curve for a mid-point on the curve at 50% inhibition (IC_{50}), a maximum absorbance for the lower asymptote, and a minimum absorbance for the upper asymptote. An established ELISA method usually has well documented historical data for the specifications of the curve fit constants (Table 1.4). The specific curve fit constants may vary from day to day, and the accepted ranges of such variations must be determined and documented. Triplicate analyses of each standard, control, and sample are generally performed for 96-microwell plate assays. The acceptable percent relative standard deviation (%RSD) of each triplicate analysis may be as low as $\pm 10\%$, depending on the specific assay and required data quality objectives. Recoveries of positive controls typically range from 70 to 130%, similar to instrumental methods. The mean value of each triplicate measurement is used for calculating the sample concentration. If the results of the samples are outside the calibration range, the sample is diluted and reanalyzed. A more detail discussion of immunoassay data analysis is given in Chapter 10.

1.3.5 QUALITY ASSURANCE/QUALITY CONTROL

Every analytical laboratory must maintain a QA program to ensure that all activities affecting the quality and integrity of the data produced are planned, coordinated, and implemented in a reliable manner to provide data of known quality. An established and reliable QA program ensures that individual study requirements are met for complete, accurate, precise, and traceable data. Specific QC protocols are designed to assess data precision and accuracy as well as incorporate established procedures for verifying prescribed assay performance in the measurement process. Generally, a study-specific Quality Assurance Project Plan (QAPP) is prepared and followed throughout the

TABLE 1.4
Historical Data for Curve Fit Parameters for a PCB ELISA Calibration Curve

Assay Date	Plate No.	A	B	C	D	R
2/26/2003	1	0.982	0.864	357	0.163	0.991
	2	0.905	1.036	357	0.188	0.997
2/28/2003	1	1.022	1.007	437	0.223	0.998
	2	0.976	1.042	530	0.209	0.995
10/16/2003	1	1.268	0.914	302	0.239	0.998
	2	1.190	1.356	248	0.359	0.996
10/20/2003	1	1.182	1.06	421	0.299	0.996
	2	1.185	1.152	384	0.327	0.995
6/30/2004	1	1.175	0.865	250	0.292	0.998
7/7/2004	1	1.338	0.84	306	0.171	0.998
	2	1.265	1.116	436	0.242	0.992
7/8/2004	1	1.427	0.897	306	0.216	0.999
7/12/2004	1	1.669	0.796	400	0.265	0.999
	2	1.489	0.859	378	0.281	0.997
7/14/2004	1	1.568	0.883	399	0.269	0.999
	2	1.451	1.041	340	0.339	0.998
7/19/2004	1	1.63	0.995	326	0.348	0.999
	Mean	1.278	0.984	363	0.261	0.997
	Std Dev	0.235	0.142	72.3	0.062	0.002
	%CV	18.37	14.4	19.9	23.6	0.239

Calibration curve formula: $y = (A - D)/(1 + (X/C)^B) + D$. A, the y-value corresponding to the asymptote at the lowest value of the X axis; B, the curve slope; C, the X-value corresponding to the y value at the midpoint between A and D, showing 50% inhibition (IC_{50}); D, the y-value corresponding to the asymptote at the highest value of the X axis; and R, correlation coefficient of the curve fit.

study. Appendix 1.B has an example of an outline for developing a QAPP for a sampling and analysis study of environmental pollutants.

QC protocols for ELISA methods are the same as those for instrumental methods and may include:

- Establishing method acceptance criteria (e.g., acceptable false positive and false negative rates)
- Establishing the performance of QC samples, including negative and positive controls, laboratory and field blanks, matrix spikes, and duplicate samples
- Establishing the frequency for QC sample analysis
- Establishing a reporting system for QC samples

QA protocols to be considered for ELISA methods are also the same as those appropriate for other analytical methods and include:

- Establishing data quality objectives (DQOs), data acceptance criteria, and appropriate corrective actions for data that do not meet the DQOs
- Assuring the training or experience of staff prior to assignment of specific tasks
- Establishing appropriate storage and handling procedures for samples and reagents

Once the above criteria have been established, standard operating procedures (SOPs) for sample preparations and the ELISA analysis are written. A guidance outline for developing SOPs for analytical methods is given in Appendix 1.C.

Because immunoassay methods are high throughput methods, it is easy to incorporate good QA/QC practices. However, as samples are analyzed in parallel, problems are usually not detected until complete analysis of the entire sample set. The throughput of instrumental methods are typically lower than immunoassay and do not easily lend to the analysis of several QA/QC samples per analysis run. However, as samples are processed in sequence rather than in parallel, problems with the analysis can be detected earlier.

1.3.5.1 Quality Control Protocols

Table 1.5 summarizes the QC measures generally expected in analyzing environmental samples and the frequency with which these measures should be employed. The table also includes the QA acceptance criteria and corrective actions associated with each QC measure. QC samples include blanks, matrix spikes, and replicate samples. Typically, for each set of field samples, at least one field blank, one laboratory method blank, one matrix spike, and one duplicate sample are included and processed through the extraction, cleanup (if required), and detection processes in the same manner as the field samples. In addition, QC samples are also generated for the detection method to verify that the detection technique is under control. Each 96-microwell plate assay generally includes blank wells, blank controls, positive controls, calibration standards, and samples in triplicate. The 96-microwell format enables the analysis of several QC samples, an advantage over instrumental methods.

Assay performance is monitored by characterization of the calibration curve and QC data. Parameters to characterize the calibration curve (i.e., lower and upper asymptote values, IC_{50} and linear range) in tandem with the QC data enable the performance of each assay to be monitored.

TABLE 1.5
Summary of Typical Quality Assurance and Quality Control Measures

QC Measure	QC Frequency	QA-Acceptance Criteria	Corrective Actions
Calibration curve	One per each assay plate (or assay analysis sequence)[a]	Multi-point curve; establish specifications for curve fit parameters; $r^2 > 0.99$	Check for error; remake standards; recalibrate
Assay blank	One or more per each assay plate (or assay analysis sequence)[a]	Less than detection limit	Check for contamination source; correct; reassay
Assay positive control	One or more per each assay plate (or assay analysis sequence)[a]	Recovery 70%–130%	Check for source; correct; reassay
Triplicate analyses	All standards, controls, and samples per each assay (or assay analysis sequence)[a]	%RSD within $\pm20\%$	Check for source; flag data; reassay
Laboratory and field blanks	One for each sample set	Less than quantification limit	Check for source; correct; flag data set
Matrix spike samples	One for each sample set	Recovery 50%–130%	Check for source; correct; flag data set
Replicate samples	One for each sample set	%RSD within $\pm30\%$ or %D within $\pm40\%$	Check for source; correct; flag data set

[a] Assay sequence is for test tube format analyzed as one group.

Statistics for the calibration curve data are recorded to provide historical assay data. The mean, standard deviation, and percent coefficient of variance for all QC samples is determined for each data group. For example, Table 1.4 shows historical values for the four-parameter fit data and correlation coefficient data over a period of time for an ELISA method for PCBs. A trend analysis allows the analyst to determine the stability of the assay over time and to set QC ranges for each parameter. Immunoassay QC data may include %RSD of triplicate assays, percentage recovery of positive controls, and measurements of sample blanks. Overall method QC results may include %RSD or percent difference (%D) of replicate samples, percent recovery of matrix spikes, and measurements of laboratory and field blanks. All these QC results can be documented in electronic spreadsheet formats that fulfill the needs of individual studies. The acceptance criteria for commercial testing kits should be stated by the manufacturer; however, the user may perform additional QC criteria. If the kits are modified for a particular need, the user must establish additional acceptance criteria.

1.3.5.2   Quality Assurance Protocols

For most analytical measurement studies, the QA objectives address overall method precision and accuracy as well as method detection and quantification limits. Overall method accuracy is based on the recoveries of the matrix spiked samples. Precision is determined by the %RSD of triplicate samples or %D of duplicate samples. For most sample media, the analytical measurement of target compounds involves extraction and detection. Thus, the overall accuracy and precision measurements must include both the extraction and detection techniques. In general, the ELISA detection method performed in a 96-microwell plate format provides a tighter range of accuracy (e.g., $> 90\%$) and assay precision (e.g., within $\pm 10\%$) than rapid testing kits. Accuracy of the assay is based on analysis of post-spiked sample extracts (that does not include any variation introduced by the extraction/cleanup steps). Assay precision is based on %RSD of triplicate analyses on the same plate. For many field testing kits, duplicate and/or triplicate analyses are performed for each sample within a sample set. If duplicate analyses are performed, assay precision is then based on percent difference (%D) of the duplicate measurements.

The analysts who perform the sample preparations and ELISA analysis should have proper training for their assigned tasks. Staff should be trained in preparation techniques that are commonly employed in a conventional analytical laboratory such as preparation of standards, accurate use of balances, ultrasonic baths, accelerated solvent extraction (ASE), Soxhlet extraction, SPE cartridges, and concentration apparatuses. ELISA methods often require additional techniques that staff should be trained in such as the use of pipettors as well as the operation of plate washers, mechanical shakers, incubators, and 96-well microplate spectrophotometers. Proper procedures for storage and handling of standards, immunoreagents, and samples should be established and recorded in SOPs for staff to follow.

1.3.6   TROUBLESHOOTING

The most common problems encountered in ELISA methods utilizing 96-microwell plate assays are poor precision, uneven color development, and no or low color development [4,43]. Problems and solutions for them are discussed in the paragraphs below.

The performance of an antibody-based method is largely determined by the selectivity and concentration of the specific Ab. During method development, the optimal concentration of the specific Ab is determined. However, if a decrease in assay performance is observed, the working titer of the Ab should be checked in relation to the other reagents. Checkerboard titrations of the specific Ab and coating antigen [4] are often used to obtain the optimal reagent concentrations for an indirect 96-well assay format.

Any particles present in the sample extract may interfere with the assay performance. Particles should be removed with an Acrodisc PTFE (or equivalent) 25 mm, 0.45 micro syringe filter. The filtrate can then be analyzed by an ELISA analysis. A solvent spike sample is typically prepared for Acrodisc filtration to assess any loss occurring during the filtration step.

Laboratory staff should be trained to properly use mechanical pipettors [4]. Poor pipetting techniques can cause a serious decline in assay precision, and it can also affect the accuracy of the method. ELISA testing kits typically have a small dynamic optical density range (i.e., 1.0 OD–0.35 OD), and small changes in OD correlate to large changes in derived concentrations. The differences between absorbance values from duplicate assays are generally small and are well within the acceptance requirement (%CV <10%) for the calibration standard solutions. However, the %D of the derived concentrations of the standard solution from duplicate assays can sometime exceed 30%. The greater %D values obtained for some of the measured concentrations for the standards and samples may be due to a small volume of standard or sample retained in the pipette tip during the transfer step. Extreme care should be taken when measuring small volumes of standard or sample. A trace amount of liquid not delivered may result in a large variation in the data from duplicate assays. Laboratory pipettors must also be routinely calibrated and properly maintained for accuracy (Appendix 1.D).

Inadequate plate washing can also cause poor precision. For most immunoassays, washing 3–6 times after each incubation step is sufficient to remove unbound material from the microtiter wells. A tight sealing plate cover should be used to avoid any evaporation during the incubation steps. Small changes in volume may affect the precision and accuracy for a quantitative immunoassay. For commercial testing kits, the washing instructions given by the manufacturer should be strictly followed. A variation in the binding capacity of the solid phase (i.e., test tubes, microtiter plates) can also cause poor precision. It is important to purchase microtiter plates from reliable manufacturers to avoid this problem.

It may be important to warm the liquid reagents to room temperature before measuring or conducting the immunochemical procedure. Immunoassay incubations are usually equilibrium reactions and are subject to temperature fluctuations that can cause uneven color development in the assay resulting in poor precision. Use of an incubator at the proper temperature (usually 37°C) may be warranted. Assay performance may be affected if the reagents are not at the proper temperature. Oftentimes, a cool laboratory temperature will result in a slower and lower assay response. It is a good practice to record the laboratory temperature when conducting an immunochemical procedure. A constant laboratory temperature will reduce inter-assay variability throughout a study.

As in an instrumental quantitative analysis, if the result of a sample extract is above the calibration range, the sample extract should be diluted and reanalyzed. To assess if there is any effect of the sample matrix in the immunoassay, selected sample extracts are generally diluted (within the calibration range) and assayed. Similar results should be obtained for both the diluted and non-diluted samples demonstrating the lack of matrix interference.

Problems with color development such as little or no color or too intense color may be due to errors in performing dilutions, reagent degradation, sub-optimal incubation conditions, or matrix interference. The overall procedure must be carefully analyzed stepwise to determine the cause of the problem and eliminate error probability.

There are other areas not discussed in this section that can cause problems in an immunoassay. Many vendors of immunochemical reagents and testing kits offer free on-line comprehensive troubleshooting guidance through their websites. Other resources that discuss problems and give potential solutions are also available for analysts [4,43,78]. Communications should be established with the assay developer or manufacturer to solve assay-specific issues.

1.3.7   Safety Considerations and Waste Disposal

Safety and waste disposal issues need to be addressed for each ELISA method. A benefit of immunoassay methods is the minimal use of hazardous organic solvents. However, even minimal amounts of these reagents should be carefully handled and disposed of properly. Enzyme substrates, organic solvents, and sample residues may be hazardous. All hazardous materials should be properly disposed of according to individual facility regulations for safety and waste disposal.

1.4   EXAMPLES OF ELISA METHODS FOR DETERMINING ENVIRONMENTAL POLLUTANTS AND METABOLITES

Immunoassays such as ELISAs have been developed for determining environmental contaminants and metabolites in various sample media [8]. The performance of many immunoassay methods has been determined with real-world environmental samples. The comparison of ELISA data with GC/MS data is often favorable, suggesting that ELISA methods may be suitable as either quantitative or qualitative monitoring tools for target pollutants. The use of bioanalytical methods such as ELISA in tandem with instrumental procedures often provides advantages such as cost reduction and additional data may be obtained using a tiered analytical approach.

Environmental monitoring and human exposure assessment studies generate a large sample load that must be analyzed in a timely manner. This section gives examples of how ELISA methods have been used to monitor environmental contaminants and metabolites in various sample media.

1.4.1   ELISA Methods for Measuring 3,5,6-TCP in Dust and Soil

1.4.1.1   Summary of Results

Chlorpyrifos [O,O-diethyl O-3,5,6-trichloro-2-pyridyl phosphorothioate] is a broad-spectrum organophosphorus pesticide. It has been used on agricultural crops, in and around residential buildings, and on pets. The related compounds chlorpyrifos-methyl and trichlorpyr are also commonly used in agriculture. 3,5,6-Trichloropyridin-2-ol (3,5,6-TCP) is a major urinary biomarker of exposure for chlorpyrifos, trichlorpyr and chlorpyrifos-methyl. 3,5,6-TCP is also an environmental marker and has been detected in samples of air, dust, and soil [79]. An ELISA method to measure 3,5,6-TCP in dust and soil samples collected for a population-based exposure study [80,81] was evaluated using a conventional GC/MS method [27]. The assay precision in the 38 dust and 38 soil samples was within ±50% for sample extracts in the concentration range of 0.25 to 1.0 ng/mL and within ±25% for sample extracts in a higher concentration range (1–6 ng/mL). The estimated detection limit for a 1 g dust or soil sample was 0.25 ng/g based on a final extract of 1 mL. The assay accuracy was greater than 90%, and the estimated assay detection limit was 0.25 ng/mL. The overall ELISA method accuracy (sample extraction and detection) was greater than 80% based on recoveries from fortified exposure dust and soil samples. Linear and positive relationships were observed between the ELISA and GC/MS data with a correlation coefficient of 0.982 for dust and 0.980 for soil. Slopes of the regression lines were 1.079 for dust and 0.999 for soil, confirming that the ELISA was a reliable tool for the quantitative measurement of 3,5,6-TCP in soil and dust matrices.

1.4.1.2   Summary of Methods

Establishing analytical method performance prior to a large-scale exposure study can assist in the development of realistic monitoring criteria. Figure 1.3 illustrates the overall approach for the analysis of dust and soil samples by the ELISA and GC/MS methods. Briefly, the dust and soil samples were extracted with either acetone or methanol using an accelerated solvent extraction

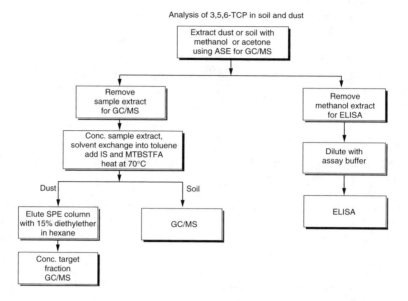

FIGURE 1.3 Analytical procedures for determining 3,5,6-TCP in dust and soil samples.

(ASE) technique (2200 psi, 120°C for 3 cycles). The ASE extracts were dried over Na_2SO_4, filtered, and analyzed by both ELISA and GC/MS. For ELISA, an aliquot of the ASE methanol extract was diluted with a phosphate buffer prior to analysis. Acetone was a better solvent than methanol for GC/MS to minimize sample matrix interferences. The acetone extract was solvent-exchanged into isooctane and derivatized with an aliquot (100 µL) of N-(tert-butyldimethylsilyl)-N-methyltrifluoroacetamide (MTBSTFA). A Florisil SPE column cleanup was needed for the dust but not for the soil samples prior to GC/MS. Differences in sample extraction and preparation often occur when evaluating immunoassay and instrumental methods and may complicate the comparison.

The study acceptance criteria for the 3,5,6-TCP magnetic particle were that (1) the %CV of the absorbance values of the standard solution must be less than 10%; and (2) the correlation coefficient for the calibration curve must be greater than 0.99. All of the assay results met the acceptance criteria. The differences between absorbance values were within the acceptance criteria (%CV < 10%) for all standard solutions. However, the %D of derived concentrations from the duplicate analyses of the lowest level standard (0.05 ng/mL) exceeded ±25%. The observed greater %D for the low level standard may have been due to a small volume of standard retained in the pipette tip. Satisfactory results (3.12±0.08 ng/mL) were obtained for the analyses of the control sample that agreed well with the expected value (3.0 ng/mL). Each method blank yielded a non-detectable value for 3,5,6-TCP.

Duplicate analyses were performed for all dust and soil samples yielding a %D value within ±50%, for the derived concentrations of the duplicate analyses, based on an assay detection limit of 0.25 ng/mL. The %D values improved to within 25% when a detection level of 1 ng/mL was used. Because of the small working range of the assay, most dust samples were diluted and reanalyzed. Satisfactory recoveries were obtained (99±6.7% for dust and 92±8.6% for soil) for post-spiked sample extract. Approximately 10% of the sample extracts were analyzed at different dilution levels in the ELISA as a QC check. Similar results were obtained for these extracts, suggesting that neither matrix interfered with the assay performance. Quantitative recoveries were achieved averaging 81±18% for dust and 91±5.2% for soil samples. In general, the ELISA and GC/MS data were in good agreement (Table 1.6). A regression analysis showed that the data were highly correlated, demonstrating that the ELISA was generating quantitative data.

TABLE 1.6
Summary Statistics for ELISA and GC/MS Data for 3,5,6-TCP

Summary Statistics[a]	Dust		Soil		Food		Urine	
	ELISA	GC/MS	ELISA	GC/MS	ELISA	GC/MS	ELISA	GC/MS
Sample Size	38	38	38	38	18	18	60	60
Unit	ng/g	ng/g	ng/g	ng/g	ng/g	ng/g	ng/mL	ng/mL
Mean	688	535	5.11	4.77	4.56	4.32	15.4	13.0
Standard Deviation	1332	1304	9.39	9.57	2.80	2.99	11.5	11.0
Maximum	6033	7267	39.5	40.0	11.2	11.0	53.4	49.9
Minimum	27.3	12.4	0.08	Non-detect[b]	Non-detect[b]	Non-detect[b]	1.70	0.80

[a] Sample size is the total number of samples analyzed. Mean, standard deviation, maximum, and minimum are from the measured concentrations of 3,5,6-TCP in each respective sample set (dust, soil, food, and urine).

[b] Estimated detection for soil is 0.2 ng/g by GC/MS, and the estimated detection for food is 0.1 ng/g by either GC/MS or ELISA [27]

1.4.2 ELISA Methods for Measuring 3,5,6-TCP in Food

1.4.2.1 Summary of Results

Duplicate diet food samples obtained from a non-occupational exposure study were analyzed using a 96-microwell plate immunoassay format [27]. The %RSD of the 96-microwell plate triplicate analyses was within $\pm 15\%$ for all samples. The estimated assay detection limit was 0.1 ng/mL. Quantitative recoveries ($>90\%$) were obtained for post-spiked food sample extracts as well as for fortified food samples ($87 \pm 7.0\%$). Overall method precision was with $\pm 15\%$, and method accuracy was greater than 85%. The ELISA and GC/MS results were well correlated for 18 duplicate diet solid food samples.

1.4.2.2 Summary of Methods

Figure 1.4 and Figure 1.5 outline the different required approaches for ELISA and GC/MS analyses of the food samples. Duplicate diet food samples collected from an exposure study were homogenized using a Waring blender. Individual aliquots of the homogenates were thoroughly mixed with a 50% weight amount of Extrelut. The food homogenates were extracted with methanol using ASE at 2000 psi and 110°C for 3 cycles of 5 min each with 100% flush. The extract was filtered, concentrated, and transferred into a silylated separatory funnel for subsequent liquid–liquid partitioning to remove the fat content. The resulting dichloromethane (DCM) extracts were concentrated on a water bath under a stream of nitrogen. One portion of the DCM extract was solvent exchanged into methanol and diluted with phosphate buffer for ELISA. Another portion of the extract was derivatized with MTBSTFA prior to GC/MS detection.

The above sample preparation procedure gave unsatisfactory ELISA results, possibly because of the residual fatty acids and fatty acid esters remaining from the liquid–liquid partitioning steps. Different sample preparation procedures were then developed for the ELISA that actually resulted in a more simplified approach. An aliquot of a food sample was mixed with Celite 545 and sonicated with acidic methanol (72% methanol, 26% water, and 2% acetic acid) for 30 min. The mixture was centrifuged at 2500 rpm at 4°C for 20 min. An aliquot of the supernatant located beneath the fat layer was removed and diluted with

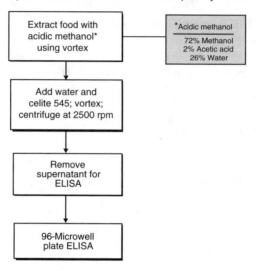

FIGURE 1.4 Analytical procedures for determining 3,5,6-TCP in solid food samples by ELISA.

800µL of phosphate buffered saline with 0.05% Tween 20 (PBST). The ELISA was performed as previously described in detail [59].

Data acceptance criteria for the 96-microwell plate assay were established based on study requirements and used as guidance throughout the analyses. The %RSD of the triplicate analyses was less than ± 15% for all samples and standard solutions. The %D of the measured values of the

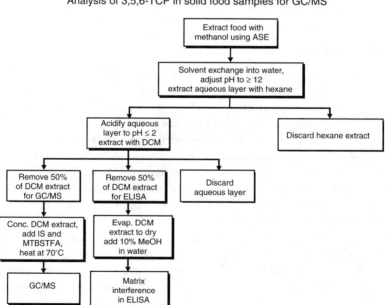

FIGURE 1.5 Analytical procedures for determining 3,5,6-TCP in solid food samples by GC/MS.

standard solutions and the expected values were less than 10%. The means of the triplicate analyses from each assay were used to calculate the final concentrations of 3,5,6-TCP in the food samples.

Although different sample preparation procedures were used for the ELISA and GC/MS analysis, the data obtained from the two methods were generally in good agreement (Table 1.6). The ELISA data were correlated with the GC/MS data over a concentration range of 2.28 to 11.2 ppb. The Pearson correlation coefficient was 0.930 with a slope value of 0.996 for the linear regression line. Composite food samples are an analytical challenge, but they play a key part in determining dietary exposures of pesticides. Improvements in analytical capability such as more rapid and cost-effective methods may often be provided by bioanalytical procedures such as immunoassay.

1.4.3 ELISA Methods for Measuring 3,5,6-TCP in Urine

1.4.3.1 Summary of Results

The same ELISA format (96-microwell plate assay) that was used for the food samples (Section 4.2) was applied to urine samples from an observational field study to determine levels of the biomarker 3,5,6-TCP. Assay precision was within $\pm 10\%$ for the urine samples with an accuracy greater than 90% and a detection limit of 0.1 ng/mL. The overall method accuracy was greater than 85% (87%–91%) based on recoveries from the fortified samples. Recoveries for post-spiked urine sample extracts were greater than 90%. Sixty human exposure urine samples were analyzed by the ELISA and compared with data from a GC/MS method. The data compared well with a Pearson correlation coefficient of 0.983 and a slope of 0.936.

1.4.3.2 Summary of Methods

Figure 1.6 shows the analytical procedures for determining 3,5,6-TCP in urine samples. Briefly, an aliquot of each urine sample was hydrolyzed with concentrated HCl at 80°C for one hour then extracted with 20% NaCl and chlorobutane. The sample was centrifuged for 10 min at 2500 rpm.

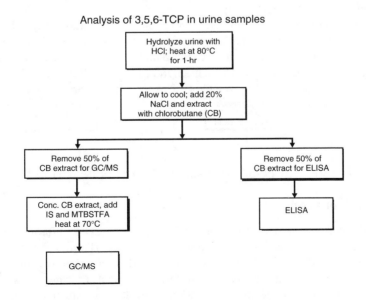

Analysis of 3,5,6-TCP in urine samples

FIGURE 1.6 Analytical procedures for determining 3,5,6-TCP in urine samples.

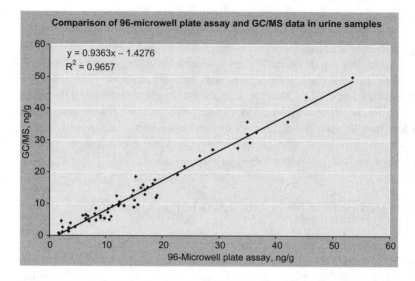

FIGURE 1.7 Comparison of ELISA and GC/MS data for 3,5,6-TCP in urine samples.

For ELISA, an aliquot of the supernatant was removed and evaporated to dryness under a stream of nitrogen and re-dissolved in phosphate buffer prior to analysis. For GC/MS analysis, an aliquot of the supernatant was derivatized with MTBSTFA. No cleanup procedures were required for the urine samples prior to GC/MS.

The four-parameter curve-fit values obtained for the urine samples were similar to the food samples except that better assay precision was achieved for the urine. The %RSD values of the triplicate analyses were less than $\pm 10\%$ for urine samples as opposed to $\pm 15\%$ for food samples. The means of triplicate measured values were used to calculate the final biomarker concentrations in the samples. Quantitative recoveries ($>90\%$) were obtained for post-spiked sample extracts. Recoveries for the fortified urine samples were $87 \pm 5.2\%$, $89 \pm 3.7\%$, and $91 \pm 4.6\%$ at 3,5,6-TCP levels of 1, 5, and 10 ng/mL, respectively.

The human exposure samples underwent similar sample preparation procedures for both the ELISA and GC/MS methods except that a derivatization step was required for the GC/MS. Thus, the discrepancies between the ELISA and GC/MS data were mainly attributable to the detection techniques. In general, the ELISA data were highly correlated with the GC/MS data (Table 1.6). The results suggested (Figure 1.7) that both the ELISA and GC/MS are reliable analytical methods for measuring 3,5,6-TCP. The ELISA had a higher throughput and provided a more cost-effective analysis.

1.4.4 A High Throughput ELISA Method for Measuring 2,4-D in Urine

1.4.4.1 Summary of Results

An ELISA method was developed to quantitatively measure 2,4-dichlorophenyoxyacetic acid (2,4-D) in human urine to support a monitoring study [60,82]. The method consisted of diluting the urine samples with PBST (1:5) prior to analysis with a 96-microwell plate immunoassay. Diluting the samples minimized the matrix interference so that a cleanup was not required for the ELISA. The assay precision, %RSD of the 96-microwell plate triplicate analyses ranged from 1.2 to 22% for the exposure samples. The overall method precision and day-to-day variation of real-world exposure samples was within $\pm20\%$ with method accuracy greater than 70%. The experimental quantification limit for 2,4-D in urine was 30 ng/mL. A multi-step sample preparation and cleanup procedure was required for the GC/MS. Urine samples were extracted with acidic DCM, methylated with diazomethane and processed through a SPE column prior to GC/MS detection. A more streamlined approach of sample dilution was used for the immunoassay. The ELISA and GC/MS data in 50 human urine samples were highly correlated with a correlation coefficient of 0.9438 and a slope of 1.0008. These results indicate that the ELISA method is suitable as a high throughput, quantitative monitoring tool for identifying individuals with exposure to 2,4-D above the typical background levels (>30 ng/mL). The method could be applied in a cost-effective manner for obtaining large-scale epidemiologic information.

1.4.4.2 Summary of Methods

For initial ELISA development, the urine samples were hydrolyzed with chlorobutane and concentrated HCl. The resulting hydrolysate was extracted with DCM (2×5 mL) and 20% NaCl. The DCM extracts were concentrated, solvent exchanged into methanol, and diluted with PBST (1:10) for ELISA analysis. The GC/MS analysis also required an acid hydrolysis and extraction with DCM. The DCM extracts were then methylated, passed through a SPE column, and analyzed by GC/MS. The final ELISA method included a 10-point calibration curve (0.78–400 ng/mL), with controls, and exposures, and QC samples analyzed in triplicate. The solvent used for preparing the standard solutions and diluting (1:5) all samples was 80%PBST with 20% drug-free urine (DFU).

The 2,4-D ELISA could tolerate 10% methanol in PBST and 10% DFU in PBST. The urine matrix effect in the assay was investigated to develop a more simplified sample preparation in comparison to the GC/MS preparation. Based on these results, a streamlined sample preparation approach was developed where urine samples were simply diluted prior to the ELISA.

A constant amount of 2,4-D standard was added to a series of dilutions of a 2.5 DFU concentrate to determine the maximum amount of urine that could be analyzed in the assay without a cleanup step. The results showed that a 1:8 dilution of the concentrate (31.2% urine) could be analyzed without a cleanup step and not adversely effect method performance. Three levels of the urine matrix (31.2%, 15.6%, and 7.8%) were further evaluated for assay performance. Standard curves of 2,4D (6.25–400 ng/mL) were constructed using solutions containing 31.2%, 15.6%, 7.8%, and 0% urine in 10% methanol/PBST. The standard curves with these urine concentrations were superimposable, suggesting that at these concentrations, (7.8%–31.2% urine) the matrix does not interfere with the assay performance. Because of the high variability in the biological composition of urine samples, a dilution factor of 1:5 (20% urine) was chosen for the final ELISA method to reduce the potential for matrix interference.

Fifty urine samples obtained from an exposure field study [83] were analyzed for 2,4-D by the ELISA and GC/MS. The %RSD of triplicate values was within $\pm30\%$ for all urine samples, QC samples, and standard solutions. Results of the post-spiked sample extracts were within $\pm30\%$ of the expected values for 2,4-D concentrations, ranging from <30 to 2480 ng/mL. Eighteen urine samples were reanalyzed over several different days, and the %RSD of these replicate analyses (day-to-day) variation) was within $\pm20\%$, indicating good inter-assay reproducibility. Quantitative recoveries (70%–124%) were obtained for the fortified urine samples. The overall accuracy for the final ELISA method was greater than 70%, and the overall assay precision was within $\pm20\%$ for the exposure samples. The ELISA and GC/MS data were in good agreement despite the different sample preparation and detection techniques employed. In general, there was a strong and positive relationship between the ELISA and GC/MS data (correlation coefficient of 0.9438).

1.4.5 A Qualitative Screening ELISA Method for Dioxins in Sediments

1.4.5.1 Summary of Results

Polychlorinated dibenzo-p-dioxins (PCDDs) and polychlorinated dibenzofurans (PCDFs) are typically found as mixtures of congeners in the environment. Of the 210 individual congeners, 17 of the most toxic have been assigned a toxic equivalent factor (TEF) by the World Health Organization based on toxicity [84]. The most toxic of these compounds is 2,3,7,8-tetrachlorodibenzo-p-dioxin (2,3,7,8-TCDD), a probable human carcinogen. A Toxic Equivalents (TEQ) value is calculated for a mixture of dioxins/furans by multiplying the concentration of each individual dioxin or furan by its respective TEF.

A sensitive 96-microwell plate ELISA method was developed to determine 2,3,7,8-tetrachlorodibenzo-p-dioxin (TCDD) equivalent concentrations in soil and sediment samples. The cross-reactivity of the less toxic TMDD in the TCDD assay was 130%. This response enabled the less toxic TMDD to be used as a surrogate standard for assay calibration and quantitation. The assay had an IC_{50} value of approximately 100 pg/mL TMDD and a detection limit of 30 pg/mL TMDD in assay buffer (50% DMSO-Triton X-100). The assay precision from triplicate analyses of standards and samples was within $\pm 35\%$, and assay accuracy was greater than 70%. The overall method accuracy (including extraction, cleanup, and detection) was greater than 60% (60%–113%) in fortified sediment samples. There was good correlation between the measured TMDD equivalent concentrations and the TEQ values derived from the GC/high resolution mass spectrometry (HRMS) methods. For this set of samples, there were no false positives for TMDD equivalent levels greater than 100 pg/g.

1.4.5.2 Summary of Methods

Sediment samples collected from an EPA superfund site. For ELISA, the sediments were dried, ground, and extracted with hexane. The hexane extract was cleaned by a multilayered acid/base silica gel column followed by a carbon column with a toluene elution. The target fraction was solvent exchanged into DMSO-Triton X-100 [17]. Different aliquots of the sediment samples were prepared and analyzed by GC/HRMS based on EPA method 1613 [85]. The sediment samples were extracted with DCM using ASE and prepared for GC/HRMS analysis using a gel permeation column, an acid-base silica column, and a carbon column.

Various cleanup steps were evaluated to develop a simple, high throughput cleanup approach able to remove the interferences prior to ELISA. The most effective cleanup procedure consisted of a multi-layered silica column followed by a carbon column. The overall method accuracy and precision were established with various sediment samples fortified with different amounts of TCDD (50, 200, 600, 2500, 7500 pg/g). These samples were processed through the cleanup method and analyzed by ELISA. The recoveries of the fortified sediment samples ranged from 70 to 113%, with two exceptions at the 50 and 200 pg/g level where 63% and 60% of the spiked amounts were detected. The %RSD of triplicate spiked samples ranged from 5 to 34%.

Most of the dioxins and furans (PCDDs and PCDFs) with high TEF values (>0.1) also exhibited a strong or moderate cross-reactivity in the ELISA. The TEQ value for each sediment sample was calculated using the GC/HRMS data and based on WHO TEF criteria. There was a strong and positive relationship (correlation coefficient of 0.987, slope of 1.06) between the ELISA-derived TCDD concentrations and the GC/HRMS-derived TEQ values. No false positives were detected for levels higher than 100 pg/g, but the data did indicate some false positives for samples with levels lower than 100 pg/g. Thus, this ELISA may be a good indicator of dioxin toxicity in contaminated sediment samples.

1.5 IMMUNOAFFINITY (IA) CHROMATOGRAPHY

There is a growing interest in the application of immunoaffinity (IA) chromatography to the analysis of environmental pollutants such as pesticides, PCBs, dioxins, and non-conventional pollutants, including pharmaceutical and personal care products. The impact of immunochemical methods for the environmental monitoring of these non-conventional types of pollutants can be significant. Particularly, as a supply of Abs for these larger molecules already exists for clinical purposes, facilitating methods development for environmental applications. A key issue for IA purification research and routine environmental analysis is the availability of appropriate Abs. IA chromatography consumes a larger amount of Ab than do immunochemical detection techniques. However, once an IA column is prepared, it may be used for several times without a loss in activity. Table 1.7 summarizes some of the methods employing IA columns as a cleanup method prior to immunoassay or instrumental (i.e., GC, GC/MS, HPLC, LC/MS) detection for environmental pollutants.

1.5.1 GENERAL APPROACH TO IA COLUMN CLEANUP

The key factors to consider in establishing an IA column cleanup method are (1) selection of the suitable IA sorbent (e.g., HiTrap NHS-activated Sepharose) for antibody coupling and efficient loading; (2) evaluation of a suitable loading and washing solvent/buffer system; (3) evaluation of an optimal elution solvent/buffer system; and (4) regeneration and storage of the IA column.

An IA sorbent is prepared by immobilizing specific Ab on the surface of a rigid or semi-rigid support (stationary phase). The IA column is activated through the functional groups of the absorbent in preparation for reaction with the Ab. Proteins A and G are often used as a bridge between the solid support and the specific Ab. The protein bridge helps to provide optimum orientation of the Ab binding site for reaction with the sample extract. An ideal immobilization condition maintains the selective binding activity of the specific Ab during the activation and coupling procedures. In general, there are two types of supports: semi-rigid, non-pressure resistant support and rigid, pressure-resistant support. An IA column using the rigid support can be directly interfaced to an HPLC system for on-line analysis (see Chapter 15).

Once the IA column is prepared with the immobilized Ab, the sample extract is loaded onto the column for the Ab to bind with the target analyte. The solvent/buffer system for sample loading varies among analytes and is designed to enhance the specific binding of the analyte with the Ab while minimizing nonspecific binding for the support. The pH of the solvent/buffer system is typically neutral with low to moderate ionic strength and low organic solvent content. The breakthrough volume is the maximum sample volume that can be applied without loss in recovery of the analyte. Accurate determination of the breakthrough volume is critical as this is an important parameter related to the loading step. A large breakthrough volume enables more sample extract to be loaded on to the column, resulting in a better detection limit. The column flow rate also affects the interaction between Ab and analyte. The binding reaction is less efficient with faster flow rates. Typical flow rates are generally between 0.5 and 5 mL/min.

A washing step removes weakly bound and nonspecifically bound materials (unwanted substances) from the column. The same solvent/buffer system for sample loading is typically used for washing. If the Ab has cross-reactivity for other structurally similar compounds or isomers, the solvent strength of the wash solvent/buffer system can be increased to remove the weakly bound chemicals. To extract related compounds and isomers, the solvent strength is not increased, and the solvent/buffer system used for sample loading is also used for washing. This allows the relatively weakly bound, but related, compounds to be retained during the washing cycle.

An elution solvent/buffer system must elute the analyte as quickly as possible and avoid any irreversible damage to the immobilized Ab. Common elution solvent/buffer systems include acidic or basic buffers, high ionic strength solutions, and organic solvents (e.g., methanol or ethanol).

TABLE 1.7
Examples of Immunoaffinity Methods Used in Environmental Analysis

Analyte	Antibody[a]	IA Support	Analytical Method	Sample Matrix	LOD[b]	Reference
Trisulfuron	pAb to triasulfuron; 5 mg/mL	Reacti-gel 6X	Extract soil with buffer (NH_4HCO_3 and $CaCl_2$, pH 8.2), elute with 100% methanol on IA column, off-line ELISA	Soil samples	0.1 ng/g	64
Avermectin	pAb	CDI activated Sepharose CL-4B	Homogenize and extract sample with methanol, elute with PBS + 0.5 M NaCl/methanol (9:1) then water/methanol (9:1), off-line HPLC/MS	Swine liver samples	5 ng/g	31
Dioxins, coplanar PCBs	pAb	Epoxy silica beads	IA extraction for coplanar PCBs and dioxins in solutions	PBS/methanol (9:1) buffer	Not reported	34
Irgarol	pAb	Hi Trap, NHS activated Sepharose	IA extraction for water samples and elute with ethanol/water (7:3), then off-line GC analysis	Sea water samples	2.5 ng/L	35
Triazine	mAb	Beaded cellulose material, ONB carbonate	IA extraction for water samples and on-line GC analysis	Water samples	15–25 ng/mL	36
Triazine and phenylurea	Anti-atrazine and anti-chlortoluron mAbs	Silica	IA purification for water samples and on-line LC/MS. Soxhlet extraction, IA purification and on-line LC/MS for sediment samples	Water and sediment samples	1–5 ng/L	37
Atrazine	Not specified	Diol-bonded silica	IA extraction with water sample and on-line HPLC analysis	Water samples	0.1 ng/mL	38

[a] pAb and mAb denote polyclonal antibody and monoclonal antibody, respectively.
[b] LOD denotes limit of detection.

The last step is regeneration of the column that removes the elution solvent/buffer system and reconditions the column with a loading solvent/buffer system for the next sample injection. Phosphate buffer with a preserving agent such as sodium azide may be added to store the column between uses.

A performance check is conducted before reusing the column for sample analysis. This check is done by passing a standard solution through the column to determine the reactivity of the support for the target analyte. Columns should be replaced when the QC check sample does not meet the recovery requirement of the study that generally ranges from 80% to120%.

1.5.2   PURIFICATION OF ATRAZINE IN SOIL AND SEDIMENT

1.5.2.1   Summary of Methods

Four IA columns were made by immobilizing rabbit anti-atrazine polyclonal Ab (1 mL of 4.25 ng/mL) onto HiTrap NHS-activated Sepharose column resin. Atrazine specifically bound to the IA columns, but it did not bind non-specifically to the control protein or the Sepharose resin. Three sample loading solvents were evaluated: 2% ACN in water, 100% water, and 10% methanol in PBS. Three elution solvents to quantitatively release the antibody-bound atrazine from the IA column were evaluated: 70% ethanol in water, 70% methanol in water, and 100% methanol. The amount of Ab coupled to the column was consistent when making multiple columns using the same procedures (93%–97% of the Ab was bound to each of the four columns). The maximum binding capability of atrazine on the IA column was approximately 700 ng per 1 mL of the resin bed (0.16 μg of atrazine per mg Ab). The elution profile for atrazine was reproducible each time the column was used. The column-to-column variability was within $\pm 12\%$. The IA columns were robust, and they could be regenerated and reused. The binding efficiency did not decrease even after repeated use (>50 times). The IA columns were challenged with real-world soil and sediment samples with the resulting extracts analyzed by ELISA and GC/MS. Quantitative recoveries were achieved for the fortified samples. The ELISA and GC/MS data were in good agreement, supporting the application of IA chromatography as a cleanup method.

1.5.2.2   Analytical Method for IA Purification of Atrazine in Soil and Sediment Samples

Soil and sediment samples were extracted with DCM using ASE at 2000 psi and 125°C for 3 cycles of 10 min each with 100% flush. The collected DCM extracts were treated with Na_2SO_4, filtered through a muffled quartz fiber filter, and concentrated to a final volume of 10 mL. Each sample extract was divided into two portions: portion I was either used for GC/MS without a prior cleanup step or reserved for future analysis, and portion II underwent an IA column cleanup with subsequent analysis by both ELISA and GC/MS. Prior to IA column cleanup, the DCM sample extract was solvent exchanged into 100% methanol. The methanol sample extract was diluted to 10% methanol in PBS before loading on to the IA column.

Four Sepharose IA columns were prepared using a polyclonal Ab specific for atrazine. A control column was prepared with a nonspecific rabbit IgG antibody following the same procedure. The coupling efficiency of the four IA columns ranged from 93 to 97% with an average of $96 \pm 2\%$. A control column was prepared using an Ab not specific for atrazine that bound to the resin with 98% efficiency. Various sample loading and elution buffers/solvents were evaluated. An elution solvent of 100% methanol was selected for the final procedures, and the elution profile was refined to minimize the water content in the target fraction for subsequent GC/MS analysis.

1.5.2.3   IA Column Performance

Atrazine was effectively retained by the IA column when the loading solvents were either 2% ACN in water, 100% water, or 10% methanol in PBS. However, atrazine was not retained by the control column. These findings indicated that the specific binding of atrazine to the IA column was significant and that atrazine did not bind indiscriminately to the protein or the Sepharose support. Column-to-column variability ($\pm 12\%$) was determined by the %RSD of recoveries of fortified atrazine solutions (5, 50, and 500 ng/mL) from the four IA columns.

The four IA columns were challenged with 37 real-world sediment and soil samples. QC samples (20 ng/mL of atrazine in 10% methanol in PBS) were processed through each of the IA columns at the beginning and at the end of each day and analyzed by ELISA. Quantitative recoveries ($114 \pm 17\%$) were obtained for the QC samples, indicating that the IA columns continued to function properly after processing the real-world sample extracts.

Selected sample extracts were analyzed by GC/MS with and without IA column cleanup. A co-eluting interference component misidentified by the GC/MS as atrazine was present in some of the samples without the IA column cleanup step. After the IA column cleanup, the interference was removed from the sample extracts. These results demonstrate that IA column cleanup effectively removed the interfering compound from the soil and sediment samples. Sample processing through the IA columns was approximately 30 min using a syringe to manually push the liquids through the column. However, the IA column could easily be adapted to an automated system by using an HPLC system equipped with a fraction collector.

For ELISA, duplicate measurements were performed for all soil and sediment samples. The %D values of the duplicate measurements were within $\pm 30\%$. Selected sample extracts were also spiked with a known amount of atrazine prior to the IA column cleanup and analyzed by both ELISA and GC/MS yielding satisfactory recoveries ($93 \pm 17\%$, and $96 \pm 21\%$, respectively). Good agreement between the two different detection methods throughout the study illustrates the utility of IA chromatography as an effective cleanup procedure.

1.6 FUTURE OUTLOOK

Although not a panacea, bioanalytical methods provide powerful complementary analytical capability and may, indeed, be the method of choice for many analytes. The breadth of the technology as discussed in this volume includes rapid screening tests, high throughput immunoassays, immunoaffinity sample preparations, multi-analyte sensors, and microarrays among other techniques. New formats such as chip type flow immunoassays will continue to be introduced [86]. Many of the rapid easy-to-use formats cannot compete with GC/MS for accuracy, but they are appropriate for rapid screening, thereby filling an important analytical void and justifying inclusion in an analytical laboratory.

Immunochemical technologies have gained increased recognition and acceptance for supporting environmental monitoring and human exposure assessment studies. Immunoassay methods have been applied to the detection of pesticides, industrial contaminants (e.g., PAHs, PCBs, and dioxins), and biomarkers of human exposure. Immunoaffinity chromatography has been used for sample cleanup with subsequent off-line or on-line analysis for environmental and biological samples.

Multianalyte microarrays enable the simultaneous analysis of several different compounds that may not be analyzed in a single chromatographic run. Microarrays can effectively monitor viruses, determining antigenic drift to guide vaccine development in a timely manner [87]. This bioanalytical capability can significantly decrease the time and cost for identification of potential lethal viral strains. The impact of this capability on human health is immeasurable. Multianalyte biosensors can provide rapid and specific environmental monitoring surveillance to support the regulatory monitoring of pollutants or help protect municipal water supply facilities for national security [88]. A network of automated and unattended monitoring systems with remote control, allowing communications between the measurement and control stations, could support early warning systems [89,90].

It is relatively easy to integrate bioanalytical capability into an existing analytical chemistry facility as general laboratory needs are the same. A bioanalytical laboratory needs only a few additional items such as a microplate washer, magnetic separation rack, spectrophotometer for microtiter plates, and immunoreagents. The skills, knowledge, and professionalism of the laboratory staff remain the key components to maintaining high quality analytical capability regardless of the methods employed.

As the need for rapid, cost-effective, and high throughput analytical methods for environmental monitoring and human exposure research increases, the development and use of immunochemical and other bioanalytical methods will continue to expand. Researchers have been building upon the elegant research performed by Yalow and Berson whose pioneering research in the development of immunoassays for small molecules resulted in the Nobel Prize in Medicine [91]. Their work was followed by Engvall and Perlman describing the first enzyme-linked immunosorbent assay [92]. Detailed procedures for reagent synthesis and methods development were compiled in excellent references by Van Vunakis and Langone that assisted researchers in developing immunoassays for a variety of applications [93]. A primary reference on pesticide immunoassays was written by Hammock and Mumma describing the potential of immunoassay technology for agrochemicals [44]. The first American Chemical Society symposium dedicated to immunochemical methods research and applications followed in 1992 [94]. As analytical needs change and technological advancements continue, new bioanalytical methods will appear. Developments in nanotechnology (see Chapter 20 and Chapter 21) will be a compelling driving force in the research and development of bioanalytical methods, providing new formats, labels, and, ultimately, new users [95].

APPENDIX 1.A   PROCEDURES FOR PREPARING BUFFERS FOR ELISA

Buffer solutions can be prepared from pre-made tablets or capsules or from starting materials from common commercial sources. Sodium azide is generally used as a preservative. If the buffer solutions are used rapidly or stored in the refrigerator, sodium azide may be omitted. If utilizing horseradish peroxidase as the enzyme label, it is best to omit the sodium azide because it is inhibitory to the enzyme. In this case, sodium ethylmercurithiosalicylate may be substituted (0.005%). Phosphate buffered saline-Tween (PBS-Tween) is commonly used as the wash buffer solution and for making all dilutions involving antibody or sample. Tween-20 in the buffer solution is used to minimize nonspecific binding. Generally, buffer solutions are brought to room temperature prior to use. Procedures for preparing common buffer solutions using individual starting materials are described as follows:

a. Substrate Buffer for Alkaline Phosphatase
   1. Remove 97 mL of diethanolamine, 0.2 g of NaN_3, and 0.1 g of $MgCl_2$ to a clean container
   2. Bring to 800 mL with reagent grade water; adjust pH to 9.8 with 6N HCl.
   3. Bring final volume to 1 L
   4. Store the solution at refrigerated temperature; check pH after prolonged storage
b. Tween/Azide
   1. Remove 10 g of NaN_3 (2%), 25 mL of Tween 20 (5%) to a clean container
   2. Bring final volume to 500 mL with reagent grade water (add water slowly to limit foaming)
   3. Store the solutions at room temperature
c. Coating Buffer (store at refrigerated temperature)
   1. Remove 0.795 g of Na_2CO_3, 1.465 g of $NaHCO_3$, and 0.1 g NaN_3 to a clean container
   2. Dilute to almost 500 mL with reagent grade water; adjust pH to 9.6 and bring to final volume of 500 mL

    3. Store the solution at refrigerated temperature; check pH after prolonged storage (> 2 wks)
  d. 10X PBS (store at room temperature)
    1. Add slowly 640 g of NaCl, 16 g of KH_2PO_4, and 91.96 g of Na_2HPO_4; to the water while stirring to prevent salts from clumping
    2. Bring to approximately 7L with reagent grade water. Stir well until all salts dissolve. Adjust pH to 7.4 and bring to final volume of 8L
    3. Store the solution at room temperature
  e. 1X PBS (store at room temperature)
    1. Remove 800 mL of 10X PBS (#d above) to a clean container
    2. Bring to approximately 7L with reagent grade water
    3. Add 80 mL of 100X Tween/Azide (#b above), after most of the water has been added to avoid foaming due to Tween 20, adjust pH to 7.4, if necessary, then bring to final volume of 8 L
  f. Substrate Buffer for Horseradish Peroxidase
    1. Prepare citrate-acetate buffer by using 13.61 g of sodium citrate (100mM); bring to approximately 1 L with reagent grade water; adjust pH to 5.5 with acetic acid
    2. Prepare 1% hydrogen peroxide by using 1 mL of 30% H_2O_2 in 29 mL reagent grade water; store in a plastic container in the refrigerator
    3. Prepare 0.6% 3,3'5,5'-Tetramethylbenzidine (TMB) by using 60 mg of TMB in 10 mL dimethylsulfoxide (DMSO); store at room temperature in the dark
    4. Just prior to use prepare the final substrate buffer by mixing 0.4 mL of 0.6% TMB in DMSO and 0.1 mL of 1% hydrogen peroxide in 25 mL of citrate-acetate buffer
    5. Store buffer and TMB solutions at room temperature before mixing to avoid precipitation

APPENDIX 1.B   GUIDELINES FOR PREPARING A QUALITY ASSURANCE PROJECT PLAN AND STANDARD OPERATING PROCEDURES

GUIDELINES FOR PREPARING A QAPP

There are different requirements for preparing specific QAPP. The requirements are listed in EPA Requirements for Quality Assurance Project Plans for Environmental Data Operations, EPA QA/R-5, EPA/240/B-01/003, March 2001. Guidance for Quality Assurance Project Plans, EPA QA/G5, EPA/600R-98/018, February 1998 may be used to help address these requirements. The documents are available at http://www.epa.gov/quality. The following requirements should be addressed as applicable for a study involving analysis of environmental samples for pollutants using an existing analytical method and/or a modified analytical method.

Section 1.0 Project Description and Organization

  1.1 The purpose of the study shall be stated in the QAPP
  1.2 Responsibilities of all project participants shall be identified, meaning that key personnel and their organizations shall be identified along with the designation of responsibilities for planning, coordination, sample collection, measurements (i.e., analytical, physical, and process), data reduction, data validation (independent of data generation), data analysis, report preparation, and quality assurance.

Section 2.0 Sampling

If sampling is involved in the study, sampling-related issues such as sampling points, sampling frequency, and sample types should be addressed.

Section 3.0 Testing and Measurement Protocols

3.1 Each analytical method to be used shall be referenced.
3.2 If applicable, modifications to EPA-approved methods or other validated methods shall also be described.

Section 4.0 QA/QC Checks

4.1 All calibrations and QC checks and/or procedures used in the field and laboratory shall be listed and defined. QC checks may include spikes, replicates, blanks, controls, surrogates, etc.
4.2 For each specified calibration, QC check, or procedure, required frequencies and acceptance criteria shall be included. Generally, for chemical methods, quality control procedures to determine the precision, accuracy, and method detection limit should be described. For microbiological methods, positive and negative control procedures should be described.

Section 5.0 Data Reduction and Reporting

5.1 Data reduction procedures to be used shall be described.
5.2 The reporting requirements (e.g., units) for each measurement and matrix shall be identified.

Section 6.0 Reporting Requirements

The report will be a method written in a format appropriate for the application with supporting method performance data appended.

Section 7.0 References

References shall be provided either in the body of the text as footnotes or in a separate section.

APPENDIX 1.C   GENERAL OUTLINE FOR PREPARING STANDARD OPERATING PROCEDURES FOR ANALYTICAL METHODS

Technical SOPs generally include a wide variety of activities and/or procedures. For examples, SOPs may include procedures on how to perform a specific bioanalytical method to be followed in the laboratory or field. Technical SOPs can also cover activities on sample preparation and cleanup procedures for subsequent bioanalytical method analysis. The following sections are generally included in the technical SOP.

TITLE PAGE

The first page or cover page of each SOP generally contains the following information: a title that identifies the procedures, a SOP identification number, date of issue and/or revision, and the signatures and signature dates of those individuals who prepared and approved the SOP.

Section 1.0 Scope and Applicability

A brief description of the purpose of the process or procedures should be included in this section.

Section 2.0 Summary of Methods

Procedures used in the SOP should be summarized in this section.

Section 3.0 Definitions

Acronyms, abbreviations, or specialized terms used in the SOPs should be described in this section.

Section 4.0 Cautions

This section include activities and/or procedures that are critical to the overall procedures regarding safety, equipment damage, degradation of sample, or possible invalidation of results.

Section 5.0 Responsibilities

This section describes the tasks and responsibilities of the personnel involved in the SOP.

Section 6.0 Interferences

This section describes any components that may interfere with the accuracy of the overall method.

Section 7.0 Reagents, Materials, and Apparatus

This section lists the necessary equipment, materials, reagents, chemical standards, and biological specimens.

Section 8.0 Procedures

This section describes all pertinent steps in order to accomplish the SOP such as

- Sample preparation and extraction
- Instrument parameters
- Instrument Calibration
- Analysis sequence
- Data processing

Section 9.0 Records

Forms to be used, data and record storage information, and how to document experimental activities are included in this section.

Section 10.0 Quality Control and Quality Assurance

This section includes the preparation of appropriate QC procedures, QC materials, and specific QC criteria as well as the frequency of required calibration and QC checks and remediation actions if the QC criteria do not meet.

Section 11.0 Extract Storage

This section describes the sample extract storage conditions.

Section 12.0 References

This section lists all references related to the SOP.

APPENDIX 1.D   STANDARD OPERATION PROCEDURE FOR THE OPERATION, CALIBRATION, AND MAINTENANCE OF FIXED AND ADJUSTABLE VOLUME PIPETTES

SIGN OFF PAGE (APPROVAL OF APPROPRIATE PERSON)

Section 1.0 Scope and Applicability

This Standard Operating Procedure (SOP) describes the general procedures for the operation, calibration, and maintenance of fixed-volume and adjustable-volume pipettes.

Section 2.0 Summary of Methods

The method describes procedures to ensure that fixed-volume and adjustable-volume pipettes used in the laboratory are dispensing the correct volume of liquids. Calibration will be performed at least every six months. Calibration will be done gravimetrically by weighing aliquots of purified water (e.g., deionized water). Adjustable-volume pipettes will be calibrated at a minimum of two delivery volumes, preferably mid-range and upper range.

Section 3.0 Definitions

None.

Section 4.0 Cautions

Standard laboratory protective clothing, gloves, and eye protection is required.

Section 5.0 Responsibilities

A designate qualified laboratory technician is responsible for regularly performing the calibration tests and for making sure all pipettes in the laboratory are calibrated up to date. The laboratory manager will perform routine laboratory inspection to ensure that all pipettes used are calibrated.

Section 6.0 Apparatus and Materials

6.1 Analytical balance capable of weighing to 0.5 mg
6.2 Fixed-volume and adjustable-volume pipettes to be calibrated
6.3 Pipette tips
6.4 Deionized water
6.5 Pipette calibration binder containing calibration forms
6.6 Labels
6.7 Beakers
6.8 Calibrated thermometer

Section 7.0 Procedures

7.1 *Gravimetric Calibration* - For gravimetric calibration, room temperature water, an analytical balance capable of weighing to 0.5 mg, and a calibrated thermometer are required. An example of a completed pipette calibration form is attached at the end of this SOP.

7.1.1 Fill in the header information on the pipette calibration form. If using an adjustable-volume pipette, set the delivery volume as desired.

7.1.2 Pipet at least five replicate aliquots of the water into a tared container, and record the weight of each aliquot on the calibration form. Other relevant information on the form such as water temperature, analyst, calibration date, etc., will also be recorded.

7.1.3 Calculate and record the mean weight of the replicates.

7.1.4 Calculate and record the absolute recovery (accuracy) of the pipette delivery.

7.1.5 Calculate and record the coefficient of variation (precision) of the replicates.

7.1.6 The absolute recovery must be 95%–105%, and the coefficient of variation must be less than 2.0% unless the manufacturer specifies less stringent acceptance criteria.

7.1.7 Place a label with the date and the next calibration due date on the pipette.

7.1.8 Place the calibration form in the appropriate binder.

7.2 Maintenance

7.2.1 Pipettes that fail calibration must be cleaned, adjusted, and re-calibrated. Refer to the manufacturer's literature for detailed cleaning and adjustment directions.

EXAMPLE OF A COMPLETED PIPETTE CALIBRATION FORM

Date: 1/20/06   Analyst:
Pipette: Gilson P1000   ID #: K14282C
Thermometer ID #: C11997   Calibration Due
Water Temp.: 24°C   Density (g/mL): 0.9973
Weight Set: C10816   Calibration Due: 12/20/05

Balance Calibration	
Balance Used: AE240	ID #: X51781
Weight Used (mg/g)	Balance Reading (g)

Replicate	Setting 500 µL — Wt. Water Delivered (g)	Setting ____ µL — Wt. Water Delivered (g)
1	0.5016	
2	0.5032	
3	0.5029	
4	0.5029	
5	0.5036	
6	0.5026	
7	0.5031	
8	0.5039	
9	0.5027	
10	0.5032	
Mean	0.5030	
Absolute Recovery %	$\dfrac{Mean}{(WaterDensity)(DeliveredVol, mL)} \times 100$ $\dfrac{0.5030}{(0.9973)(0.500)} \times 100 = 100.87\%$	
CV	$\dfrac{St. Deviation}{Mean} \times 100$ $\dfrac{6.2191 \times 10^{-4}}{0.5030} \times 100 = 0.12\%$	
Acceptance	PASS	

*See CRC Handbook of Chemistry and Physics for water density values at different temperature.

Pipette Calibration Due: 7/20/06

7.2.2   Contact the laboratory coordinator if non-routine maintenance is required.

7.2.3   A pipette calibration binder contains all the forms will be kept in the laboratory.

Section 8.0 Records

Records of all calibration activities will be recorded in the pipette calibration forms, and the forms are placed in a three-ring binder in the laboratory. Calibrated pipettes will be labeled with dates and initials.

REFERENCES

1. Ahmed, F. E., Protein-based methods: Eludication of the principles, In *Testing of genetically modified organisms in food*, Ahmed, F. E., Ed., Haworth Press, Binghamton, NY, pp. 117–146, 2004.

2. Thomzig, A., Cardone, F., Kruger, D., Pocchiari, M., Brown, P., and Beekes, M., Pathological prion protein in muscles of hamsters and mice infected with rodent-adapted BSE or vCJD, *J. Gen. Virol.*, 87, 251, 2006.

3. Van Emon, J. M. and Lopez-Avila, V., Immunochemical methods for environmental Analysi, *Anal. Chem.*, 64, 79A, 1992.

4. Gee, S. J., Hammock, B. D., and Van Emon. J. M., A user's guide to environmental immunochemical analysis, EPA/540/R-94/509, March 1994.

5. Hage, D. S., Review: Survey of recent advances in analytical applications of immunoaffinity chromatography, *J. Chromatogr. B*, 715, 3, 1998.

6. Marco, M. P., Gee, S. J., and Hammock, B. D., Immunochemical techniques for environmental analysis. Antibody production and immunoassay development, *TrAC Trends Anal. Chem.*, 14, 415, 1995.

7. Weller, M. G., Immunochromatography techniques-a critical review, *Fresenius J. Anal. Chem.*, 366, 636, 2000.

8. Van Emon, J. M., Immunochemical application in environmental science, *J. AOAC Int.*, 84 (1), 125, 2001.

9. Lee, N. A. and Kennedy, I. R., Environmental monitoring of pesticides by immunoanalytical techniques: Validation, current status, and future perspectives, *J. AOAC Int.*, 84, 1393, 2001.

10. Blake, R. C. II, Pavlov, A. R., Khosraviani, M., Ensley, H. E., Kiefer, G. E., Yu, H., Li, X., and Blake, D. A., Novel monoclonal antibodies with specificity for chelated uranium (VI): Isolation and binding properties, *Bioconjug. Chem.*, 15, 1125, 2004.

11. Van Emon, J. M., Seiber, J. N., and Hammock, B. D., Application of an enzyme-linked immunosorbent assay to determine paraquat residues in milk, beef, and potatoes, *Bull. Environ. Contam. Toxicol.*, 39, 490, 1987.

12. Van Emon, J. M., Hammock, B., and Seiber, J. N., Enzyme-linked immunosorbent assay for paraquat and its application to exposure analysis, *Anal. Chem.*, 58, 1866, 1986.

13. Johnson, J. C. and Van Emon, J. M., Quantitative enzyme-linked immunosorbent assay for determination of polychlorinated biphenyls in environmental soil and sediment samples, *Anal. Chem.*, 68 (1), 162, 1996.

14. Dankwardt, A. and Hock, B., Enzyme immunoassays for analysis of pesticides in water, *Food Technol. Biotechnol.*, 35, 165, 1997.

15. Shackelford, D. D., Young, D. L., Mihaliak, C. A., Shurdut, B. A., and Itak, J. A., Practical immunochemical method for determination of 3,5,6-trichloro-2-pyridinol in human urine: Applications and considerations for exposure assessment, *J. Agric. Food Chem.*, 47, 177, 1999.

16. Nichkova, M., Galve, R., and Marco, M. P., Biological monitoring of 2,4,5-trichlorophenol: Evaluation of an enzyme-linked immunosorbent assay for the analysis of water, urine and serum samples, *Chem. Res. Toxicol.*, 15 (11), 1371, 2002.

17. Nichkova, M., Park, E., Koivunen, M. E., Kamita, S. G., Gee, S. J., Chuang, J. C., Van Emon, J. M., and Hammock., B. D., Immunochemical determination of dioxins in sediment and serum samples, *Talanta*, 63, 1213, 2004.

18. Biagini, R. E., Murphy, D. M., Sammons, D. L., Smith, J. P., Striley, C. A. F., and MacKenzie, B. A., Development of multiplexedd fluorescence microbead immunosorbent assays (FMIAs) for pesticide biomonitoring, *Bull. Environ. Contam. Toxicol.*, 68, 470, 2002.
19. Guomin, S., Huang, H., Stoutamire, D. W., Gee, S. J., Leng, G., and Hammock, B. D., A sensitive class specific immunoassay for the detection of pyrethroid metabolites in human urine, *Chem. Res. Toxicol*, 17, 218, 2004.
20. Van Emon, J. M., Reed, A. W., Yike, I., and Vesper, S. J., ELISA measurement of stachylysin in serum to quantify human exposures to the indoor mold Stachybotrys chartarum, *J. Occup. Environ. Med.*, 56 (6), 582, 2003.
21. Brecht, A., Piehler, J., Lang, G., and Gauglitz, G., A direct optical immunosensor for atrazine detection, *Anal. Chim. Acta.*, 311, 289, 1995.
22. Dzantiev, B. B. and Zherdev, A. V., Electrochemical immunosensors for determination of the pesticide 2,4-dichlorophenoxyacetic acid and 2,4,5-trichlorophenoxyacetic acids, *Biosens. Bioelectron.*, 11, 179, 1996.
23. Skladal, P. and Kalab, T. A., Multichannel immunochemical sensor for determination for 2,4-dichlorophenoxyacetic acid, *Anal. Chim. Acta.*, 316, 73, 1995.
24. Mulchandani, A., Chen, W., Mulchandani, P., Wang, J., and Rogers, P. K., Biosensors for direct determination of organophosphate pesticides, *Biosens. Bioelectronics*, 16 (45), 225, 2002.
25. Mauriz, E., Calle, A., Lechuga, L. M., Quintana, J., Montoya, A., and Manclús, J. J., Real-time detection of chlorpyrifos at part per trillion levels in ground, surface and drinking water samples by a portable surface plasmon resonance immunosensor, *Anal. Chim. Acta.*, 561, 40, 2006.
26. Centi, S., Laschi, S., Franek, M., and Mascini, M., A disposable immunomagnetic electrochemical sensor based on functionalised magnetic beads and carbon-based screen-printed electrodes (SPCEs) for the detection of polychlorinated biphenyls (PCBs), *Anal. Chim. Acta.*, 53B, 205, 2005.
27. Chuang, J. C., Van Emon, J. M., Reed, A. W., and Junod, N., Comparison of immunoassay and gas chromatography-mass spectrometry methods for measuring 3,5,6-trichloro-2-pyridinol in multiple sample media, *Anal. Chim. Acta.*, 517, 177, 2004.
28. Gámiz-Gracia, L., García-Campana, A. M., Soto-Chinchilla, J. J., Huertass-Perez, J. F., and Gonzáles-Casado, A., Analysis of pesticides by chemiluminescence detection in liquid phase, *Trends Anal. Chem.*, 24 (11), 927, 2005.
29. Kramer, P., Franke, A., Zherdev, A. V., Yazynina, E. V., and Dzantiev, B. B., Comparison of two express immunotechniques with polyelectrolyte carriers, ELISA and FIIAA, for the analysis of atrazine, *Talanta*, 65 (2), 324, 2005.
30. Van Emon, J. M., Gerlach, C. L., and Bowman, K., Bioseparation and bioanalytical techniques in environmental monitoring, *J. Chromatogr. B*, 715, 211, 1998.
31. Li, J. S., Li, X. W., and Hu, H. B., Immunoaffinity column cleanup procedure for analysis of ivermectin in swine liver, *J. Chromatogr. B.*, 696, 166, 1997.
32. Wu, Z., Junsuo, L., Zhu, L., Luo, H., and Xu, S., Multiresidue analysis of avermectins in swine liver by immunoaffinity extraction and liquid chromatographymass spectrometry, *J. Chromatogr. B*, 755, 361, 2001.
33. Delaunay, N., Pichon, V., and Hennion, M., Immunoaffinity solid-phase extraction for the trace analysis of low molecular mass analytes in complex sample matrices, *J. Chromatogr. B*, 745, 15, 2000.
34. Concejero, M. A., Galve, R., Herradon, B., Gonzalez, M., and de Frutos, M., Feasibility of high-performance immunochromatography as an isolation method for PCBs and other dioxin-like compounds, *Anal. Chem.*, 73, 3119, 2001.
35. Carrasco, P. B., Escola, R., Marco, M. P., and Bayona, J. M., Development and application of immunoaffinity chromatography for the determination of the triazinic biocides in seawater, *J. Chromatogr. A*, 909, 61, 2001.
36. Dalluge, J., Hankemaier, T., Vreuls, R., and Brinkman, U., On-line coupling of immunoaffinity-based solid-phase extraction and gas chromatography for the determination of s-triazines in aqueous samples, *J. Chromatogr. A*, 830, 377, 1999.

37. Ferrer, I. M., Hennion, M. C., and Barcelo, D., Immunosorbents coupled on-line with liquid chromatrography/atmospheric pressure chemical ionization/mass spectrometry for the part per trillion level determination of pesticides in sediments and natural waters using low preconcentration volumes, *Anal. Chem.*, 69, 4508, 1997.
38. Thomas, D., Beck-Westermeyer, M., and Hage, D., Determination of atrazine in water using tandem high-performance immunoaffinity chromatography and reversed phase liquid chromatography, *Anal. Chem.*, 66, 3823, 1994.
39. Thomas, D. H., Lopez-Avila, V., Betowski, L. D., and Van Emon, J., Determination of carbendazim in water by high-performance immunoaffinity chromatography on-line with high-performance liquid chromatography with diode-array or mass spectrometric detection, *J. Chrom. A*, 724, 207, 1996.
40. Huwe, J. K., Shelver, W. L., Stanker, L., Patterson, D. G. Jr., and Turner, E. W., On the isolation of polychlorinated dibenzo-p-dioxins and furans from serum samples using immunoaffinity chromatography prior to high-resolution gas chromatography-mass spectrometry, *J. Chromatogr. B. Biomed. Sci. Appl.*, 757, 285, 2001.
41. Thomas, D. H., Rakestraw, D. J., Schoeniger, J. S., Lopez-Avila, V., and Van Emon, J., Selective trace enrichment by immunoaffinity capillary electrochromatograpy on-line with capillary zone electrophoresis-laser-induced fluorescence, *Electrophoresis*, 20 (1), 57, 1999.
42. Van Emon, J. M., Seiber, J. N., and Hammock, B. D., In *Immunoassay Techniques for Pesticide Analysis in Analytical Methods for Pesticides and Plant Growth Regulators, Advanced Analytical Techniques*, Sherma, J., Ed., Vol. XVII, Academic Press, Inc., San Diego, CA, pp. 217–263, 1989.
43. Shan, G., Lipton, C., Gee, S. J., and Hammock, B. D., Immunoassay, biosensors and other nonchromatographic methods, In *Handbook of Residue Analytical Methods for Agrochemicals*, Lee, P. W., Ed., Wiley, Chinchester, pp. 623–679, 2002.
44. Hammock, B. D. and Mumma, R. O., In *Potential of Immunochemical Technology for Pesticide Analysis in Pesticide Analytical Methodology*, Harvey, J. J. and Zweig, G., Eds., American Chemical Society, Washington, DC, pp. 321–352, 1980.
45. Landsteiner, L., *The Specificity of Serological Reactions*, Pulb., New York, Dover Publications, Inc., pp. 1104, 1962.
46. Kohler, G. and Milstein, C., Continuous cultures of fused cells screening antibodies of predefined specificities, *Nature*, 349, 495, 1975.
47. Kramer, K., Synthesis of a group-selective antibody library against haptens, *J. Immunol. Methods*, 266, 211, 2002.
48. Lipman, N. S., Jackson, L. R., Trudel, L. J., and Weis-Garcia, F., Monoclonal versus polyclonal antibodies: Distinguishing characteristics, applications and information resources, *ILAR J.*, 46, 258, 2005.
49. Deryabina, M. A., Yakovleva, Y. N., Popova, V. A., and Eremin, S. A., Determination of the herbicide acetochlor by fluorescence polarization immunoassay, *J. Anal. Chem.*, 60 (1), 80, 2005.
50. Botchkareva, A. E., Eremin, S. A., Montoya, A., Manclús, J. J., Mickova, B., Rauch, P., Fini, F., and Girotti, S., Development of chemiluminescent ELISAs to DDT and its metabolites in food and environmental samples, *J. Immunol. Methods*, 283, 45, 2003.
51. Jiang, T., Halsall, H. B., Heineman, W. R., Giersc, T., and Hock, B., Capillary enzyme immunoassay with electrochemical detection for the determination of atrazine in water, *J. Agric. Food Chem.*, 43, 1098, 1995.
52. Hampl, J., Hall, M., Mufti, N. A., Yao, Y. M., MacQueen, D. B., Wright, W. H., and Cooper, D. E., Upconverting phosphor reporters in immunochromatographic assays, *Anal. Biochem.*, 288 (2), 176, 2001.
53. Voller, A., Bidwell, D. E., and Bartlett, A., Microplate enzyme immunoassays for the immunodiagnosis of virus infections, In *Manual of Clinical Immunology*, Rose, N. and Friedman, H., Eds., American Society for Microbiology, Washington, DC, pp. 506–512, 1976.
54. Gee, S. J., Hammock, B. D., and Van Emon, J. M., *Environmental Immunochemical Analysis for Detection of Pesticides and Other Chemicals, a User's Guide*, Noyes Publications, New Jersey, NY, 1996.

55. http://www.epa.gov/sw-846/main.htm, EPA SW846 Test Methods.
56. Chuang, J. C., Miller, L. S., Davis, D. B., Peven, C. S., Johnson, J. C., and Van Emon, J. M., Analysis of soil and dust samples for polychlorinated biphenyls by enzyme-linked immunosorbent assay (ELISA), *Anal. Chim. Acta.*, 376, 67, 1998.
57. Chuang, J. C., Pollard, M. A., Chou, Y.-L., and Menton, R. G., Evaluation of enzyme-linked immunosorbent assay for the determination of polycyclic aromatic hydrocarbons in house dust and residential soil, *Sci. Total Environ.*, 224, 189, 1998.
58. Chuang, J. C., Van Emon, J. M., Chou, Y.-L., Junod, N., Finegold, J. K., and Wilson, N. K., Comparison of immunoassay and gas chromatography-mass spectrometry for measurement of polycyclic aromatic hydrocarbons in contaminated soil, *Anal. Chim. Acta.*, 486, 31, 2003.
59. Kolosova, A. Y., Park, J., Eremin, S. A., Park, S., Kang, S., Shim, W., Lee, H., Lee, Y., and Chung, D., Comparative study of three immunoassays based on monoclonal antibodies for detection of the pesticide parathion-methyl in real samples, *Anal. Chim. Acta.*, 511 (2), 323, 2004.
60. Chuang, J. C., Van Emon, J. M., Durnford, J., and Thomas, K., Development and evaluation of an enzyme-linked immunosorbent assay (ELISA) method for the measurement of 2,4- dichlorophenoxyacetic acid in human urine, *Talanta*, 67, 658, 2004.
61. Altstein, M., Bronshtein, A., Glattstein, B., Zeichner, A., Tamiri, T., and Almog, J., Immunochemical approaches for purification and detection of TNT traces by antibodies entrapped in a sol-gel matrix, *Anal. Chem.*, 73, 2461, 2001.
62. Rejeb, S. B., Cleroux, C., Lawrence, J. F., Geay, P., Wu, S., and Stavinski, S., Development and characterization of immunoaffinity columns for the selective extraction of a new developmental pesticide: thifluzamide, from peanuts, *Anal. Chim. Acta.*, 432 (2), 193, 2001.
63. Khan, A. A., Akhtar, S., and Husain, Q., Simultaneous purification and immobilization of mushroom tyrosinase on an immunoaffinity support, *Process. Biochem.*, 40 (7), 2379, 2005.
64. Ghildyah, R. and Kariofillis, M., Determination of triasulfuron in soil affinity chromatography as a soil extract cleanup procedure, *J. Biochem. Bioph. Methods*, 30, 207, 1995.
65. Shelver, W. L., Shan, G., Gee, S. J., Stanker, L. H., and Hammock, B. D., Comparison of immunoaffinity column recovery patterns of polychlorinated dibenzo-p-dioxins/polychlorinated dibenzofurans on columns generated with different monoclonal antibody clones and polyclonal antibodies, *Anal. Chim. Acta.*, 457, 199, 2002.
66. Lei, Y., Mulchandani, P., Chen, W., and Mulchandani, A., Direct determination of p-nitrophenyl substituent organophosphorous nerve agents using recombinant pseudomonas putida IS444-modified clark oxygen electrode, *J. Agri. Food Chem.*, 53 (3), 524, 2005.
67. Yhou, Y., Yu, M. B., Shiu, E., and Levon, K., Potentiometric sensing of chemical warfare agents: Surface imprinted polymer integrated with an indium tin oxide electrode, *Anal. Chem.*, 76 (10), 2689, 2004.
68. Dill, K., Montgomery, D. M., Oleinikov, A. V., Ghindilis, A. L., and Schwarzkopf, K. R., Immunoassays based on electrochemical detection using microelectrode arrays, *Biosens. Bioelectron.*, 20, 736, 2004.
69. Theodore, M. L., Jackman, J., and Bethea, W. L., Counterproliferation with advanced microarray technology, *Johns Hopkins APL Technical Digest*, 25 (1), 38, 2004.
70. Petrenko, V. A. and Vodyanoy, V. Y., Phage display for detection of biological threat agents, *J. Microbio. Methods*, 53 (2), 253, 2003.
71. Anderson, L. I., Hardenborg, E., Sandberg-Stall, M., Moller, K., Henriksson, J., Bramsby-Sjostrom, I., Olsson, L. E., and Abdel-Rehim, M., Development of a molecularly imprinted polymer based solid-phase extraction of local anaesthetics from human plasma, *Anal. Chim. Acta.*, 526, 147, 2004.
72. Ansell, R. J., Molecularly imprinted polymers in pseudoimminoassay, *J. Chromatography B*, 804 (1), 151, 2004.
73. Tombelli, S., Minunni, M., and Mascini, M., Analytical applications of aptamers, *Biosens. Bioelectron.*, 20, 224, 2004.
74. Bekheit, H. K. M., Lucas, A. D., Gee, S. J., Harrison, R. O., and Hammock, B. D., Development of an enzyme-linked immunosorbent assay for the ß-exotoxin of bacillus thuringiensis, *J. Agric. Food Chem.*, 41, 1530, 1993.

75. Harlow, E. and Lane, D., *Using antibodies. A Laboratory Manual*, Cold Spring Harbor Laboratory Press, Cold Spring Harbor, NY, 1999.
76. Rodbard, D., Mathematics and statistics of ligand assays: An illustration guide, In *Ligand Assay, Analysis of International Developments on Isotopic and Non-Isotopic Immunoassay*, Langan, J. and Clapp, J. J., Eds., Masson, New York, pp. 45–99, 1981.
77. *Molecular Devices SOFTmax® PRO User Manual*, Sunnyvale, CA, Molecular Devices,1998.
78. Crowther, J. R., *The ELISA Guide Book, Methods in Molecular Biology, 149*, Humana Press, Totowa, NJ, 2001.
79. Morgan, M. K., Sheldon, L. S., Croghan, C. W., Jones, P. A., Robertson, G., Chuang, J. C., Wilson, N. K., and Lyu, C., Exposures of preschool children to chlorpyrifos and its degradation product 3,5,6-trichloro-2-pyridinol in their everyday environments, *J. Expos. Anal. Environ. Epidemiol.*, 15, 297, 2005.
80. Lebowitz, M. D., O'Rourke, M. K., Gordon, S., Moschandreas, D. J., Buckley, T., and Nishioka, M., Population-based exposure measurements in Arizona: A phase I field study in support of the National Human Exposure Assessment Survey, *J. Expos. Anal. Environ. Epidemiol.*, 5, 297, 1999.
81. Pang, Y., MacIntosh, D. L., Camann, D. E., and Ryan, P. B., Analysis of aggregate exposure to chlorpyrifos in the NHEXAS-Maryland investigation, *Environ. Health Perspect.*, 110 (3), 235, 2002.
82. Franek, M., Kolar, M., Granatova, M., and Nevorankova, Z., Monoclonal ELISA for 2,4-dichlorophenoxyacetic acid: Characterization of antibodies and assay optimization, *J. Agric. Food Chem.*, 42, 1369, 1994.
83. Thomas, K., Chapa, G., Croghan, C., Jones, P., Dosemeci, M., Coble, J., Alavanja, M., Hoppin, J., Sandler D., Assessing exposure classification in the agricultural health study, International Society of Exposure Analysis 2004, Philadelphia, PA, October, 17–21, 2004.
84. Van den Berg, M., Birnbaum, L., Bosveld, B. T. C., Brunstrom, B., Cook, P., Feeley, M., Giesy, J. P., et al., Toxic equivalency factors (TEFs) for PCBs, *Environ. Health Perspect.*, 106, 775, 1998.
85. Method 1613, Revision B, Tetra- through Octa-Chlorinated Dioxins and Furans by Isotope Dilution HRGC/HRMS, EPA 821-B-94-005Q, 1994.
86. Okochi, M., Ohta, H., Taguchi, T., Ohta, H., and Matsunaga, T., Construction of an electrochemical probe for on chip type flow immunoassay, *Electrochemical Acta.*, 51 (5), 952, 2005.
87. Lodes, M. J., Suciu, D., Elliott, M., Stover, A. G., Ross, M., Caraballo, M., Dix, K., et al., Use of semiconductor-based oligonucleotide microarrays for influenza a virus subtype identification and sequencing, *J. Clinical Microbiol.*, 44, 1209, 2006.
88. Rodriguez-Mozaz, S., Lopez de Alda, M. J., and Barcelo, D., Fast and simultaneous monitoring of organic pollutants in a drinking water treatment plant by a multi-analyte biosensor followed by LC-MS validation, *Talanta*, 69 (2), 384, 2006.
89. Glass, T. R., Saiki, H., Joh, T., Taemi, Y., Ohmura, N., and Lackie, S. J., Evaluation of a compact bench top immunoassay analyzer for automatic and near continuous monitoring of a sample for environment contaminants, *Biosens. Bioelectron.*, 20 (2), 397, 2004.
90. Tschmelak, J., Proll, G., Riedt, J., Kaiser, J., Kraemmer, P., Barzaga, R., Wilkinson, J., et al., Automated water analyzer computer supported system (AWACSS) Part I: Project objectives, basic technology, immunoassay development software design and networking, *Biosens. Bioelectron.*, 20 (8), 1499, 2005.
91. Yalow, R. S. and Berson, S. A., Immunoassay of endogenous plasma insulin in man, *J. Clin. Invest.*, 39, 1157, 1960.
92. Engvall, E. and Perlmann, P., Enzyme-linked immunosorbent assay (ELISA) quantitative assay of immunoglobulin, *G. Immunochem.*, 8, 871, 1971.
93. Vunakis, H. V. and Langone, J. J., Eds., *Method in Enzymology Immunochemical Techniques, Part A*, Vol. 70, Academic Press, Inc., New York, 1980.
94. Van Emon, J. M. and Mumma, R. O., Eds., *Immunochemical Methods for Environmental Analysis ACS symposium series 442*, ACS Press, Washington, DC, 1990.
95. Chan, W. C. W. and Nie, S. M., Quantum dot bioconjugates for ultrasensitive nonisotopic detection, *Science*, 281, 2016, 1998.

2 Directed Evolution of Ligand-Binding Proteins

K. Kramer, H. Geue, and B. Hock

CONTENTS

2.1 INTRODUCTION

Protein-based receptor molecules that are capable of selectively binding a particular ligand or group of related ligands are extensively applied in analysis, diagnostics, therapy, nanotechnology, and a multitude of new interdisciplinary areas currently emerging in biotechnology. The selective binding of the target molecule is based upon molecular recognition, making use of the noncovalent interaction of amino-acid residues at the interface of the receptor binding area and the corresponding sites of the ligand. There are several binding forces driving the specific association of receptor and ligand such as ionic, van-der-Waals, and hydrophobic interactions as well as hydrogen and Ca^{2+} bridges [1]. The affinity is dependent on the additive effect of these interactions exerted by a particular pairing of receptor and ligand. Because these multiple forces are exclusively effective on short molecular distances, they require a complementary topography at the interface between receptor and ligand.

Two fundamentally different types of binding proteins can be distinguished with respect to their implementation into analytical devices applied in the majority of environmental monitoring. For the first type of binders, the analytical information is essentially hinged with their natural molecular structure. This implies that any alteration of the wild-type protein affecting the binding properties will lead to a loss of valuable information. As a consequence, genetic engineering is confined, e.g., to heterologous expression or production of fusion proteins in order to install additional functions to the receptor. However, each kind of manipulation must be carefully controlled with respect to potential changes in the binding properties, ideally by comparison with the wild-type protein in the natural cellular microenvironment.

Classical examples for these types of binding proteins are hormone receptors such as G-protein-, ion-channel-, or enzyme-coupled surface receptors or intracellular receptors triggering signal transduction pathways. These receptors can deliver valuable information on the biological impact of binding ligands because they occupy sensitive key sites in the complex network of extra- and intracellular signaling pathways [2,3]. Interestingly, several of these receptors are characterized by a high degree of permissivity for ligand binding, e.g., they are capable of binding structurally diverse ligands [4].

In contrast, the second type of receptors applied in environmental analysis is characterized by a very low degree of permissivity, resulting in selective binding of structurally defined ligands. Taking this to an extreme, monoselective proteins are found in this group, exclusively binding a single ligand at a reasonable level of affinity. A second feature of these proteins is that they are amenable to molecular modifications, which may have an effect on the affinity and selectivity for a particular ligand. These modifications can be performed by altering the gene encoding the protein receptor that is accomplished by subjecting the corresponding DNA strand to mutational processes utilizing appropriate tools of genetic engineering.

Because the interaction between receptor and ligand depends on multiple noncovalent interactions, successful optimization of molecular recognition necessitates a precise and individual sequence modification with respect to a distinct ligand. This gives rise to the questions of how the type and location of suitable sequence alterations can be reliably identified in an efficient manner and how the alterations are then implemented into the receptor molecule.

Two basically different strategies evolved for the improvement of receptors. The first one, termed as rational design, is a computational approach. It relies on structural analysis in order to identify amino acid positions at the receptor protein that are contributing to the ligand binding. Computer models can be delineated from sequence information or from crystal structure [5]. However, models exclusively based on sequence information are not necessarily representing the real conformation of the ligand-binding site, whereas precise conformational data obtained by crystal structures yet require significant experimental efforts. The template model forms the basis for the calculation of improved in silico-variants suggesting favorable amino acid substitutions. The proposed variants are then produced by site-directed mutagenesis and tested in order to verify the improvement in ligand binding.

There are still serious problems encountered with the pure rational design strategy. For instance, it still remains problematic to predict the functional impact of simultaneous mutations at multiple sites of the protein. This problem results in the inability to predetermine the effect of complex matrix-based mutational solutions (i.e., the interplay of widely dispersed mutations). In addition, depending on the applied algorithm, rational design may exclude alternative solutions for achieving improvements, solutions that are more likely to be found by directed evolution as outlined below. As a consequence, rational design is frequently complemented by evolutionary techniques in the laboratory [6]. Principles of rational design will not be discussed here; there are numerous reviews addressing this subject [7,8].

The second strategy for designing receptors in vitro is based on the evolutionary concept of variation and selection. This approach is not dependent on a precise knowledge of the receptor conformation. Similarly, as in the case of the in vivo molecular evolution of protein families, the receptor gene is diversified in an initial step by mutational procedures. The mutated-gene repertoire is subsequently expressed and screened or selected for improved variants. It is increasingly acknowledged that particular randomization processes are hinged with a corresponding outcome in an evolutionary experiment in vitro [9]. For example, directed evolution often discovers beneficial mutations at sites that are not considered in the first instance. Rather than directly modifying the active sites of enzymes or the binding region of antibodies (Abs), mutations often occur far from these areas and affect catalysis or binding through subtle, long-range interactions [10]. As with enzymes, directed evolution experiments have yielded novel mutations throughout the frame of the antibody (Ab) molecule [5,11].

The rationale behind the concept evolved as modern evolutionary theory in the first half of the twentieth century when population biologists started to describe the natural selection of gene populations. They established mathematical models for the evolution of genes under the influence of recursive rounds of mutation, recombination, and selection [12]. Applying principles of natural selection to the evolution of genes in the laboratory, fundamental evolutionary theorems predict that the fitness of a gene will evolve most rapidly in a population of high genetic variability that is exposed to selection pressure [13]. The methodology of choice in a directed evolution experiment is therefore to construct a library of gene variants and screen or select improved variants on the level of the encoded proteins.

Essential steps of the strategy are depicted in Figure 2.1. The success of molecular evolution in the laboratory is crucially dependent on the choice of an appropriate protein scaffold for the

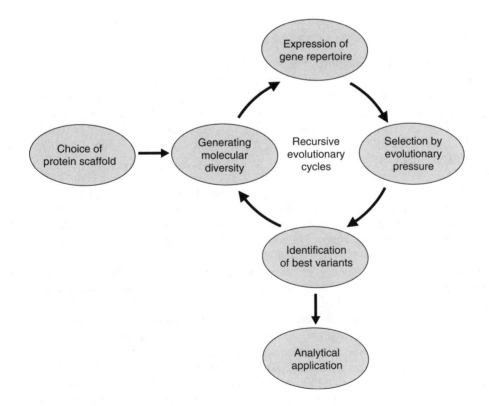

FIGURE 2.1 Principle of directed evolution.

considered analytical task. Structural aspects have to be taken into account like the potential number of residues involved in binding that are directly affecting the interaction of receptor and ligand. However, further properties such as the functional folding of the receptor and stability are of equal importance. The next step involves the generation of a repertoire of variants, providing a high level of functional diversity. The polypeptide repertoire is subsequently expressed in appropriate systems. Some of the most successful systems are those where the genetic information is linked with the functional properties of the encoded protein. Then, the diverse repertoire is exposed to a selection pressure; for example, a certain level of affinity to the ligand is required for the protein variant in order to be selected. This step corresponds to the survival of individuals in a population of organisms in natural selection events. Independent from the particular experimental method, it is definitely important that the experimental procedure applied for exerting the selection pressure in the laboratory exactly operates at molecular mechanisms that are modulating the considered parameter (e.g., selectivity, affinity). Moreover, further parameters potentially interfering with the selection goal have to be excluded or reduced to a minimum. Interfering parameters that are not taken into account may lead to the inefficient selection of otherwise appropriate protein repertoires. Following selection, the best variants are identified by characterizing the ligand-binding properties. If the selected variants do not equal the required analytical properties, one or several variants are used as templates for subsequent molecular diversification in order to generate a new protein repertoire that can be selected again. The evolutionary approach might necessitate several rounds of variation and selection until the required properties are achieved. In addition, at each stage, fundamental alterations of the experimental protocol or the underlying strategy may be required in order to obtain protein receptors meeting the desired quality in ligand-binding in a perfect manner. Subsequently, an introduction is given on principles of the individual steps in directed molecular evolution, which are highlighted in Figure 2.1.

2.2   CHOICE OF PROTEIN SCAFFOLD

The choice of a particular protein scaffold constitutes the initial step in directed evolution. Polypeptides are generally considered to be scaffolds in the context of structural modification if they provide an intrinsic conformational stability to tolerate molecular transformations at their bioactive sites, resulting in a modulation rather than a loss of characteristic functions (e.g., ligand binding). The scaffold concept emerged from developments in Ab engineering and is now extended to other polypeptide structures that are amenable to directed evolution (cf. below; recommended review: Nygren and Skerra [14]).

Depending on the particular nature of the protein scaffold, essential features of the final analytical performance and applicability in environmental monitoring are ultimately determined. For instance, choosing an enzyme scaffold will dictate the potential of the underlying measuring principle. The catalytic center of an enzyme can be subjected to molecular evolution in order to optimize the rate of substrate turnover or substrate selectivity. Examples for enzyme optimization by genetic engineering have been reviewed [15–18].

Because of the high efficiency of the natural immune response, the first artificial affinity reagents were based on Abs. Ab molecules are members of the immunoglobulin superfamily. They consist of two identical heavy chains and two identical light chains (Figure 2.2). Two antigen-binding F_{ab} fragments are assembled each by the complete light chain combined with the corresponding part of the first two N-terminal heavy-chain domains. Two F_{ab} fragments are linked via flexible hinge regions with the crystallizable fragment F_c that harbors the remaining constant heavy-chain domains C_H2 and C_H3. The globular structure of Ab domains is due to the characteristic immunoglobulin fold [20]. Anti-parallel ß-strands form a typical double layer in each domain that is stabilized by hydrophobic interactions and conserved intra-domain disulfide bonds. The antigen-binding site is localized at the N-terminal moiety of the variable regions. Because of

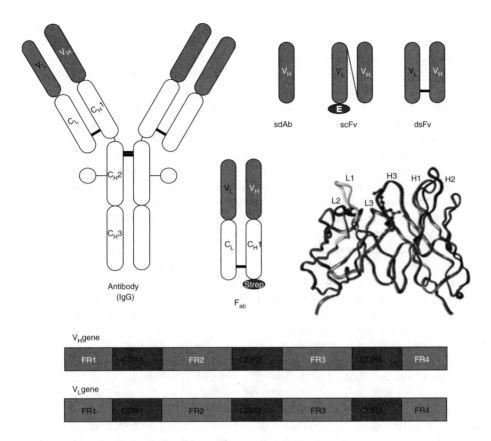

FIGURE 2.2 Domain structure of native IgG antibody molecule (top left) and genetically engineered sdAb, scFv, disulfide-stabilized Fv (dsFv), and F_{ab} fragments. Some of the fragments are genetically fused to affinity tags, e.g., Strep tag (Strep), E tag (E). The middle right shows a ribbon model of the V_H and V_L domain with pesticide analyte complexed in the Ab binding site (L1-3: V_L CDR1-3; H1-3: V_H CDR1-3). The bottom shows segmentation of V_H and V_L encoding genes into FR FR1-4 and the hypervariable CDRs CDR1-3. The ribbon model was kindly provided by S. Hörsch. (From Hörsch, S., Diploma thesis, Institute of Technical Biochemistry, University of Stuttgart, Stuttgart, Germany, 1998.)

the frequency of sequence variation between different Ab molecules, sections with hypervariable or complementarity determining regions (CDRs) and conserved or framework regions (FRs) can be distinguished for these domains [21,22]. Each of the variable (V) regions contains three CDR loops that are embedded into four sections of the FR (cf. Figure 2.2). In addition to their enhanced sequence variability, most of the CDR loops are characterized by significant differences in length [23].

The binding of the antigen is crucially determined by the three CDRs of each V domain. The composition and conformation of the six CDRs (three from the light- and three from the heavy-chain V domain) determine the topography of the antigen-binding site and, hence, the recognition of the ligand. Additional interactions by residues of the framework region are reported for discrete ligands [24].

The majority of genetically engineered Ab-based affinity reagents were fragments such as F_{ab} fragments [25] and single-chain F_v (scFv) fragments (cf. Figure 2.2) that are stabilized by a short peptide-linker [26,27]. Alternatively, chain dissociation of F_v fragments is prevented by engineered cysteine residues forming interdomain disulphide bridges [28] and helix-stabilized Ab fragments [29]. In addition, libraries based on single V_H [30] and single V_L domains [31] have been described (cf. Figure 2.2).

Currently, the potential of alternative Ab molecules is exploited. For instance, single-domain Ab (sdAb) is naturally occurring in camelids and sharks. The large CDR loop of camelids that is stabilized by an intraloop disulfide bond is considered a critical component in providing high affinity for the corresponding sdAb [32]. Similar, a single-domain heavy-chain library based on Ab of the isotype novel antigen receptor (IgNAR) was prepared from nurse sharks (*Ginglymostoma cirratum*). sdAb binders of high stability and with high affinity to hen egg-white lysozyme have been selected from this library [33].

In addition to Ab domains, so-called Ab mimics have been reported to be amenable to molecular optimization. The tenth fibronectin type III domain ($_{10}$Fn3), a monomeric member of the immunoglobulin superfamily, was used as a scaffold for library synthesis by randomizing three exposed loops [34]. $_{10}$Fn3 variants were obtained by directed evolution that bound tumor necrosis factor α with dissociation constants as low as 20 pM. The extracellular domain of human cytotoxic T-lymphocyte associated antigen, CTLA4, represents another Ab mimic. Libraries based on the CTLA4 scaffold were created by introducing diversity within the CDR3-like loop of CTLA-4 [35].

In addition to immunoglobulins and immunoglobulin-like polypeptides, some specific affinity reagents have been selected from small, globular protein scaffolds not related to Ab. For instance, the α-helical Z-domain of protein A has been used as a scaffold for library synthesis. The engineered proteins are designated as *affibodies*. Initially, low affinities of isolated sequence variants [36] to the target molecule *Taq* DNA polymerase were improved in subsequent directed-evolution experiments to the nanomolar range [37].

Similarly, the bilin-binding protein (BBP), a lipocalin from the butterfly *Pieris brassicae*, was used as a scaffold [38]. The protein has a conserved β-barrel core formed by eight antiparallel β-strands. Short loops are connecting the individual strands. The four loops on one end of the barrel confer binding to the natural ligand bilin. Sixteen residues of the four binding loops were randomized for initial library synthesis. Affinities in the nanomolar range for digoxigenin of the lead isolate were further improved by selective mutation of particular binding loop regions. The highest affinity of these anticalins were reported to be as low as 800 pM for digoxin and 600 pM for digitoxin [39].

Another type of protein scaffold is the ankyrin repeat (AR). AR proteins are composed of several 33 amino-acid repeats stacked in a row. Each repeat comprises a β-turn followed by two antiparallel α-helices and a C-terminal loop reaching the β-turn of the adjacent loop [40]. Synthetic libraries were developed by randomizing amino-acid positions at the β-turn and the short hinge connecting the two α-helices. Two or three AR modules were flanked at their N and C termini with capping AR of defined sequence [41]. The AR libraries yielded high-affinity binding variants with K_D in the nanomolar range against various protein targets.

A recent report described the generation of a library based on a stable variant of green fluorescent protein (GFP) [42]. These fluorobodies are very attractive candidates for analytical applications because they combine the fluorescence of the wild-type protein with the binding characteristics of Ab. However, they still suffer from moderate affinity level.

These and even examples for smaller biomolecules, e.g., zinc fingers [43,44], knottins [45], and Kunitz domains [46], show that natural proteins are promising candidates for library construction if the mode of molecular recognition is well understood. Each of these scaffolds are characterized by features such as molecular size, stability, solubility, type of natural ligands, ligand-contacting area, and maximum number of amino-acid residues that tolerate sequence randomization. Therefore, it is essential to define the most appropriate polypeptide architecture for the considered analytical task before starting with the evolutionary optimization.

2.3 CREATING MOLECULAR DIVERSITY

Two fundamentally different approaches can be distinguished for the generation of molecular diversity: first, the cloning of natural repertoires, and second, the generation of synthetic repertoires by in vitro methods. Whereas the first method is based on naturally occurring genes, the second strategy takes advantage of introducing alterations that do not necessarily exist in nature into a protein scaffold.

2.3.1 CLONING NATURAL DIVERSITY

A popular example for the application of natural diversity is the cloning of Ab gene repertoires that are isolated from donor organisms. The corresponding repertoire can be unbiased, meaning that the immune system of the donor organism was not challenged by an antigen. These naïve repertoires theoretically harbor Ab for any target structure, however, at a moderate affinity level for the majority of binding molecules. In contrast, biased Ab repertoires can be cloned from immune sources. The latter strategy benefits from in vivo mechanisms of the immune system because Ab variable genes encoding the antigen binding domains (BDs) are modified during the secondary immune response in the microenvironment of lymphoid germinal centers by somatic hypermutation. Appropriate variants are subsequently selected from this pool of mutant immunoglobulins upon their improved affinity to the antigen [47]. Therefore, immunizing an organism with a specific antigen serves as an in vivo preselection of potent Ab genes. Ab derived from immune sources is characterized by increased levels of affinity and selectivity toward a particular molecular target that was applied for the immunization process. The probability to contain selective, high-affinity Ab species is therefore enhanced, whereas the diversity to other proteins is concomitantly reduced.

2.3.2 CREATING SYNTHETIC REPERTOIRES

In contrast to repertoires of natural diversity, synthetic repertoires are either produced entirely synthetically or by isolating a gene of interest from natural sources and subjecting it subsequently to sequence diversification in vitro. Frequently applied in vitro methods include random and directed nucleotide alterations or recombination techniques [48]. An inherent feature of the latter strategy is the exchange of gene fragments rather than single nucleotides.

2.3.2.1 Random Techniques

The introduction of point mutations by error-prone polymerase chain reaction (epPCR) is a straight-forward method of altering gene sequences. Nucleotides throughout the entire gene are randomly exchanged (cf. Figure 2.3). The frequency for introducing false nucleotides in the amplicons is relatively low if standard polymerase chain reaction (PCR) conditions are applied. Therefore, the intrinsic error rate of *Taq* polymerase is raised in epPCR protocols, e.g., by substituting the cofactor magnesium by manganese, unbalancing nucleotide compositions or employing mutagenic polymerases [49,50]. Specific combinations of these experimental parameters result in mutation rates of up to one out of five bases [51].

Although the point mutations are theoretically distributed in a random manner, epPCR is prone to various bias effects. (1) Most polymerases favor specific transitions (e.g., *Taq* polymerase A/T or T/A) [52]. This bias can be compensated by simultaneous application of polymerases with different mutational biases or polymerases with uniform mutational spectra (e.g., Mutazyme®, Stratagene, Inc.). (2) Another kind of amplification bias occurs because DNA molecules become over-represented in the final library if they are already copied in early epPCR cycles. Several parallel epPCRs can be performed in order to minimize this bias [48]. (3) A third source of bias evolves from the degenerated nature of the genetic code. A single-point mutation does not necessarily result in a different amino acid encoding. Usually, either two- or three-point mutations are required at a single codon site in order to access the entire panel of amino-acid permutations. The random insertions/deletion (RID) method offers an appropriate means to compensate this codon bias. Here, nucleotides are deleted at random positions, and subsequently, a defined mixture of oligonucleotides is inserted that provides the twenty amino acids [53].

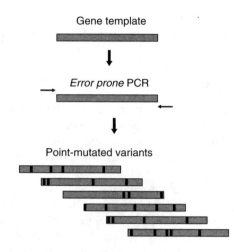

FIGURE 2.3 Randomization of a gene template by epPCR. PCR primers are depicted as horizontal arrows. Randomized point mutations caused by amplification errors are indicated by vertical black lines in the resulting gene variants.

2.3.2.2 Oligonucleotide-Directed Techniques

In contrast to random methods, oligonucleotide-directed procedures insert synthetic sequences into the gene of interest at specific sites. This is basically accomplished by introducing degenerate oligonucleotide primers at defined positions of the target gene by PCR (cf. Figure 2.4). If a mixture of all four nucleotides is used for primer synthesis, again the problem of codon bias is encountered as described above. Moreover, there exists a risk of introducing all three stop codons that results in the expression of truncated protein fragments. A mixture of NNT and NNC trinucleotides can be used in order to prevent the incorporation of translational stops. However, this also reduces the number of encoded amino acids to fifteen. Alternatively, the nucleotide base adenine can be excluded for the randomization at the third codon position for primer synthesis. By this strategy, all twenty amino acids are represented, whereas only a single-stop codon is included in the oligonucleotide primers [48]. The trinucleotide phosphoramidites technique solves the problem of codon bias and insertion of stop codons in the most stringent manner to date. Instead of synthesizing nucleotide by nucleotide, the primers are produced by entire codons that results in the definitive elimination of translation stops within the gene [54].

The incorporation of oligonucleotides into the target gene is commonly performed by various PCR methods such as strand overlap extension (SOE) [55] or megaprimer-based PCR [56]. However, directed methods generally encounter the amplification bias problem because primer sequences with higher similarity to the template DNA are more efficiently incorporated than oligonucleotides that differ substantially. In addition, primers that are mutated near the 3′-end are incorporated less efficiently than primers altered at the 5′-end. Again, this effect can be compensated by performing parallel amplifications in several different PCR preparations combined with a minimum number of PCR cycles.

For the transformation of more than a single site, multiple rounds with different primers have to be performed. Several more complex procedures have been developed for this purpose. One of them is based on the production of several overlapping gene segments with mutagenic primers by PCR and reconstruction of the entire gene by an overlap extension reaction. An amplification bias can be avoided if the number of gene fragments is not higher than four. In this case, facile strand extension can be performed instead of PCR amplification with external primers. In contrast, a high

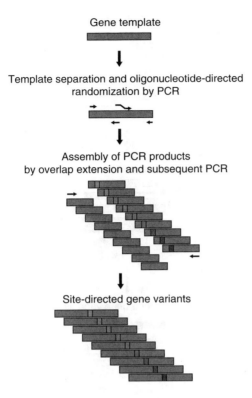

Gene template

↓

Template separation and oligonucleotide-directed randomization by PCR

↓

Assembly of PCR products by overlap extension and subsequent PCR

↓

Site-directed gene variants

FIGURE 2.4 Oligonucleotide-directed randomization of gene template by overlap extension. PCR primers are depicted as horizontal arrows. Two of the primers are partially complementary and overlap in the area upstream of the randomized section. The longer one of the two primers contains the randomized sequence that is indicated by the inclination of the arrow. After assembly and final amplification, full-length gene variants are obtained that are mutated at a defined gene section (indicated by color differences).

number of gene segments again requires amplification by PCR because the yield of entire gene products drops with an increasing number of gene segments. In this case, the inevitable PCR will lead to an over-representation of early amplified genes.

2.3.2.3 Recombination Techniques

Recombination methods deal with already existing gene fragments of DNA libraries. The fragments are exchanged between different parental genes and recombined to yield filial genes encoding new proteins with potentially modified characteristics. DNA shuffling is a prevalent procedure for homologous recombination [57]. Library genes are digested with DNAse (cf. Figure 2.5). The resulting mixture of DNA fragments is consecutively melted, annealed, and extended until an adequate amount of full-length DNA can be amplified by PCR. In a similar approach, staggered extension process (StEP) [58,59], the DNA fragments are added stepwise to the end of a growing strand. Following partial elongation, the fragments are dissociated by melting, annealed to a different template DNA, and subsequently amplified by PCR. The process is repeated until the full-length gene is completed. Random chimeragenesis by transient templates (RACHITT) is a technology that produces higher diversity than StEP or DNA shuffling but requires several additional steps [60]. All but one parental gene is fragmented, and the single-stranded fragments are reassembled using the opposite strand of the unfragmented gene as template. Mismatched sections are removed. The corresponding fragments are then extended and ligated afterward to obtain the

Template gene pool

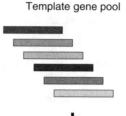

Random fragmentation by DNAse digest
and combinatorial assembly

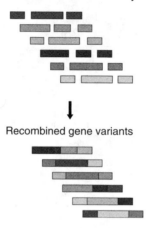

Recombined gene variants

FIGURE 2.5 Random recombination of gene pool. Enzymatic fragmentation of genes and subsequent combinatorial assembly results in recombined gene variants derived from different parental genes.

full-length gene. Following destruction of the template strand, the chimeras are completed to double-stranded DNA.

All of these recombinational methods rely on existing parental gene diversity and are therefore limited. Alternative techniques designated as assembly of designed oligonucleotides [61] or synthetic shuffling [62] generate recombined full-length genes entirely from synthetic DNA. Therefore, they are not dependent on template DNA structures and enable the synthesis of genes de novo.

In contrast to homologous recombination methods where almost identical DNA sequences are required at the crossover points of gene segments, nonhomologous recombination does not depend on sequence similarity in the parental genes. The incremental truncation for the creation of hybrid enzymes (ITCHY) technology uses exonucleases to truncate two different DNA templates from opposite sites. The 5'-truncated gene is religated with the 3'-end digested gene at random positions. Although this method effectively recombines the remaining fragments, only a single recombination results from two template sequences [63].

The cre/lox system is an attractive in vivo recombination technology. This system is based on a site-specific DNA recombinase that is encoded on the cyclization recombination (cre) gene, and the corresponding 34-base-pair-long target sequences known as locus of X-over P1 (loxP). The recombinase excises DNA segments that are located between two lox sites. DNA fragments from homologous lox sites can be recombined at high efficiency. In contrast, recombination events are very rare at heterologous sites because the excised DNA is subsequently degraded. The presence or introduction of lox sites in the encoding gene sequence is required for this technology. In addition, if the host organism (e.g., *E. coli*) for protein expression does not contain its own cre gene, it may be introduced via a phage vector [64].

2.4 SELECTION STRATEGIES

Genes encoding highly diverse sequence repertoires of ligand-binding proteins can be selected either by physical or genetic techniques. The most frequently applied physical selection methods comprise phage display [65,66], cellular display on the surface of bacteria or yeast [67,68], ribosome display [69], and mRNA display [70]. These procedures are all characterized by the physical linkage of proteins and their encoding genes that can be reamplified for further processing.

2.4.1 PHAGE DISPLAY

Phage display is the most commonly applied method for the selection of ligand-binding proteins. Ab [71,72], DNA-binding proteins [43], hormones [73], and many other proteins have been selected by phage display.

The genome of a filamentous phage consists of a single-stranded DNA molecule encapsulated in a coat of various proteins. The DNA encoding the protein of interest is fused to one of these phage-coat protein genes and displayed upon expression on the phage surface as fusion with the coat protein (cf. Figure 2.6) [65]. The most commonly used coat proteins are pVIII and pIII. The latter is essential for host infection and is located at the phage tip, whereas pVIII composes the major part of the phage coat. Depending on the cloning strategy and coat protein, recombinant phages display a single copy or multiple copies of the recombinant protein. For instance, in a phagemid vector, a single protein copy is ideally displayed on the surface of recombinant phage particles if the gene of interest is fused to gene III. In contrast, a fusion to pIII results in the presentation of multiple copies. The number of displayed copies has a direct impact on the affinity level of subsequently selected proteins. Monovalent display enables the selection for intrinsic affinity, whereas in multivalent display systems, the selection is based on functional affinity that is evoked by multiple interactions of displayed proteins per phage particle with the selection surface. In the latter case, low-affinity clones are co-selected [66].

For the selection of the most favorable binding proteins, the ligand is prevalently immobilized on a solid phase, e.g., at the surface of a polystyrene tube, microtiter plate well or pin and incubated with the phage library. Those phages that do not bind to the ligand are removed by washing, whereas recombinant phages that recognize the ligand are subsequently eluted [74]. Employing biotin-labeled ligands is another selection strategy. Ligand-binding phages can be separated via the biotin moiety using streptavidin-coated magnetic beads [75]. The selectively infective phage (SIP) strategy is a further selection method [76,77]. The receptor protein is fused to C-terminal domains of the pIII coat protein. The recombinant phages are lacking any wild type pIII with the N-terminal N1 domain that is necessary for infection of *E. coli*. Infectivity is exclusively restored upon the specific interaction of the displayed receptor protein and the ligands that are present in the selection vessel as ligand-N1 fusion construct. Therefore, affinity selection is combined with the capability for reinfection.

2.4.2 CELL SURFACE DISPLAY

In cell surface display, the protein library is fused to cellular membrane proteins of bacteria, yeast, or mammalian cells [78]. The membrane protein, usually a lipoprotein, is anchored in the cell membrane and presents the desired protein on the cell surface (cf. Figure 2.6). For instance, a common anchor protein utilized in yeast display is the cell surface receptor a-agglutinin [68]. Protein selection is performed by fluorescence-activated cell sorting (FACS). The ligand is labeled for this purpose with fluorescent markers [79]. State-of-the-art flow cytometers can analyze and sort 50,000 cells per second, providing a rapid and high performance sampling of receptor libraries [78].

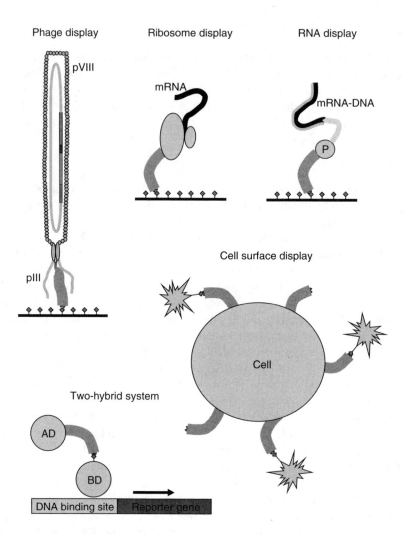

FIGURE 2.6 Selection strategies based on phage, ribosome, RNA, and cell display as well as on the yeast two hybrid system. Selection is based on the specific interaction between ligand (rhomb) and receptor. The ligand is either immobilized on a solid phase (phage, ribosome, and mRNA display) or conjugated with a DNA-BD (in yeast two hybrid system) or a fluorescent marker (cell surface display), pIII and pVIII, minor and major phage suface protein; AD, activation domain; BD, DNA-binding domain; P, puromycin.

2.4.3 RIBOSOME DISPLAY

Ribosome display is an in vitro display system [69,80,81]. In vitro display libraries comprise up to 10^{14} different proteins. In contrast to phage or surface display on bacteria and yeast, the in vitro approach does not depend on a transformation step for creating a selectable library of protein variants. Therefore, the size is not limited by the transformation efficiency of DNA into a host organism.

A DNA library encoding the desired protein is fused to a C-terminal tether and transcribed into mRNA in vitro. The mRNA and the nascent polypeptide are not released from the ribosome because of the lack of a stop codon. The receptor protein is noncovalently linked to the mRNA by the ribosome (cf. Figure 2.6). This ternary complex is further stabilized by high magnesium concentrations and low temperature. The complexes, connecting phenotype and genotype, can be selected on immobilized ligands in a manner similar to the one described above for phage display.

Following elution and dissociation of the ternary complex, the mRNA is released and can be used for subsequent cycles of selection or expression in host organisms [82].

2.4.4   RNA DISPLAY

The noncovalent complexes of ribosome display are relatively unstable and may dissolve during stringent selection steps. RNA display is considered a major technical improvement in this respect [83]. Here, the encoded protein is covalently linked to the mRNA (cf. Figure 2.6) and is therefore less susceptible to dissociation than the ternary ribosome complex. The DNA library is transcribed to mRNA that is covalently conjugated by enzymatic ligation or photocrosslinking to the antibiotic puromycin via a short DNA spacer. The translation of mRNA in vitro is interrupted at the DNA spacer. The puromycin subsequently enters the ribosome, and by mimicking the aminoacyl end of tRNA, it is covalently linked to the nascent protein. The mRNA-coupled protein can be separated from the ribosome and purified. Then, a RNA-DNA hybrid is produced by reverse transcription. The resulting constructs are selected against immobilized ligands (cf. Figure 2.6) and can be utilized after a final PCR amplification either for subsequent rounds of selection or for analyses [70,84]. This method has been successfully applied, for instance, to convert a DNA chip into a protein chip by hybridization of mRNA-protein to the DNA probes [85].

In ribosome display and in RNA display techniques, the protein-encoding genes are usually cloned after selection into bacterial expression systems in order to produce the corresponding protein. However, this confers an inherent problem on in vitro techniques in general because not all proteins selected by in vitro display are well expressed in host organisms [74].

2.4.5   GENETIC SELECTION

All physical selection methods require significant quantities of the ligand to perform selection and screening, whereas genetic methods rely on the in situ synthesis of, and subsequent interaction between, binding ligand and target to confer a selectable phenotype. The receptor and a peptide- or protein-ligand are synthesized in vivo and interact in the host cell [74].

Various yeast hybrid systems have been designed that utilize the transcriptional activation mechanism for selection [83,86]. The DNA-BD of the transcription activator is fused to the N-terminus of the receptor protein, whereas the activation domain (AD) of the transcription activator is fused to the potential ligand (cf. Figure 2.6). The BD binds to a promoter, but transcription is exclusively activated if the BD is connected to the AD by the interaction of receptor and ligand. If no receptor-ligand binding occurs, the reporter gene transcription is not triggered. The concept facilitates the distinction between cells harboring potent and weak binders.

A similar concept is pursued by the protein complementation assay [87,88]. In this case, the gene coding for an essential enzyme (e.g., dihydrofolate reductase or β-lactamase) is deliberately cleaved into two fragments. Each of the resulting domains is fused to either the receptor or the ligand. The resulting fusion proteins are coexpressed, and the enzyme restores functionality if the domains get into close contact via the receptor-ligand interaction.

2.5   IMPACT OF RANDOMIZATION TECHNIQUE ON DIRECTED EVOLUTION

It is obvious that the optimization of proteins for bioanalytical applications is challenging if the degree of conformational variability is considered. For instance, the variable domain of an antibody consists of approximately 110 amino acid residues that cover a theoretical diversity of 20^{110} sequence variants on the amino acid level. The number of variants is further increased on the gene level because there are up to six different codons for particular amino acids. Stringent optimization procedures are guided by both computational and experimental methods.

Numerous predictive models were established for the *in silico* analysis of the restricted diversity that can be technically achieved in the laboratory. These include the assessment of the completeness and functional diversity of libraries generated by epPCR, oligonucleotide-directed mutagenesis, and in vitro recombination techniques [89]. Computational pre-screens are aiding directed evolution, e.g., by identifying protein regions that are most likely to yield beneficial mutations [90] or by predicting functional subunits of polypeptides that are able to be recombined with minimal disruption of the overall three-dimensional protein architecture [91]. This work has essentially improved the understanding of fundamental mechanisms that underlie directed evolution experiments.

From a theoretical point of view, directed evolution can be visualized as a random walk through a sequence space (Figure 2.7). The sequence space consists of all possible sequences of a given number of residues (i.e., polypeptides encoded in a gene library). The sequences are arranged so

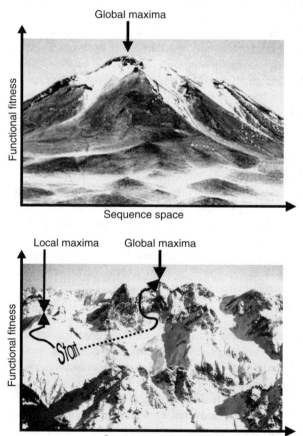

FIGURE 2.7 Sequence space of a smooth (top) and rugged (bottom) fitness landscape. Each point in the two-dimensional projection of sequence space corresponds to a defined amino acid sequence. Functional fitness adds a third dimension to the fitness landscape. The smooth landscape (top) contains a single global peak of functional fitness. From any starting point, a stringent combination of mutagenesis and selection is likely to access the global optimum (arrow). In contrast, rugged landscapes (bottom) are characterized by multiple maxima that are separated from local maxima by clefts of low fitness. Starting a random walk at any site of one of these local optima will therefore converge in a sub-optimal increase in functional fitness (permanent line). In order to approach the global maxima from a starting point at any local maxima, the crossing of the cleft in between requires an initial mutational jump (dotted line).

that the smallest unit of distance separates mutants differing by only one amino acid residue. Related sequences, therefore, are proximal to one another in the sequence space. Each amino acid combination (i.e., any single point in the sequence space) has an associated fitness (e.g., affinity for a particular ligand). This third dimension produces a characteristic fitness landscape. Efficient experimental approaches are characterized by stringent navigation through a given landscape towards increasing fitness values [92].

Conclusions derived so far from experimental examples as well as from computational models seem to manifest some parameters affecting the exhaustive exploration of the sequence space. The topology of the fitness landscape is probably the most crucial component of in vitro protein optimization [93]. If the shape is very smooth with a single paramount optimum (Figure 2.7, top), then it is very likely to access the global optimum by accepting those mutations that contribute to an increase in fitness. The situation is completely different if the landscape is rugged with a multitude of local maxima (Figure 2.7, bottom). In this case, the probability for accessing the global optimum (i.e., the ultimate level of functional optimization for a given sequence space) is significantly reduced if all randomly generated uphill steps are accepted. Thus, the optimization process becomes technically more challenging or even impossible [94]. In order to attain an initial level at the global maxima from a starting point at any local maxima, the cleft, separating individual fitness peaks, has to be crossed. However, similar as for evolution in vivo only mutations will be accepted in a directed evolution experiment, which are either neutral or provide an increase in fitness. These mutations are positively selected or "survive." In contrast, variants with decreased fitness are rejected or "die," simply because they are not considered to be useful. Individual members of the rejected fraction, however, may include valuable sequence variants that have the potential to bridge the gap between any local and the global maximum [95].

The question of how to achieve the sequence transformation from a local to the global maximum and from lower levels uphill to the peak of the global maximum is essentially addressed by the employed randomization technique. epPCR, oligonucleotide-directed mutagenesis, and in vitro recombination techniques differ substantially in this respect. epPCR produces a library of mutants from a parent sequence through random point mutations along the entire gene. This randomization technique enables a very restricted movement in a given sequence space. This is mainly due to the low incidence of nucleotide substitutions that result in a sparse sampling of a less defined region in the landscape. It is obvious that the probability is comparably small for any single random mutation to improve functional fitness. Furthermore, the probability of improvement rapidly decreases when multiple simultaneous mutations are made. Finally, amino acid substitutions are severely restricted by the negligible probability that two or three mutations occur in a single codon and the significant biases of epPCR (cf. above) [90,96].

These effects can be overcome by intense mutagenesis of a limited number of positions by directed randomization methods [97]. This mutagenesis corresponds to site-directed mutagenesis on the experimental level. The procedure consists of installing all twenty amino acids at defined points of interest and searching the resultant library for improved mutants. In landscape terms, a very small region will be intensively covered by a point-mutated library. In comparison, cassette-mutagenesis installs a random peptide sequence in a specific region of the polypeptide. Here, the sampling becomes less thorough with the increasing length of the randomized section that is a consequence of technical limitations in library size. However, the strategy benefits from sampling a larger region of the sequence space. The challenge in these directed methods is to identify the residues where such experiments are likely to be successful because beneficial mutations can appear far from sites that would be heuristically predicted (e.g., catalytic sites of enzymes) [98,99].

The situation is again completely different for recombinational methods. Natural proteins are composed by a limited number of structural subunits such as helices, hairpins, or other folds that have evolved over billions of years in the context of the surrounding sequence [100]. This may explain why beneficial mutations frequently occur at residues sharing the fewest interactions with the protein backbone [96]. These so-called uncoupled positions can be identified by structural analysis or by sequence comparison utilizing comprehensive databases. Appropriate databases are either general protein databases such as Swiss-Prot or specialized databases focusing on particular protein families like the immunoglobulin superfamily, e.g., IMGT [101] or Ab [102].

Once the location of uncoupled positions is identified, pools of gene segments can be shuffled, for instance, by mixing repertoires of functional core folds or independent protein domains. Taking Abs as an example, the directed shuffling of functional V_H and V_L domains (cf. Figure 2.2) can be considered as a recombination at uncoupled positions. This mutation strategy is most likely to produce the highest degree of functional recombination events derived from different parental V_H-V_L pairings. In contrast, random gene shuffling will occur at random sites (Figure 2.5). This, in turn, results in recombinations with higher probability at coupled than at uncoupled positions. Coupled positions within V_H- and V_L repertoires are predominantly located within the FRs that are flanking individual CDRs (cf. Figure 2.2). Therefore, the probability of obtaining recombined V_H-V_L gene pairings of retained or even improved functionality (e.g., ligand binding) is drastically reduced compared to directed shuffling of functional domains. Concomitantly, the extent of recombinations undergoing mutational death is significantly enhanced in randomly shuffled antibody libraries.

Finally, the efficient recombination by DNA shuffling in vitro as compared to the sexual recombination in vivo may be explained by the fact that shuffling accesses not only natural pairwise recombinations but also poolwise recombinations of multiple parental molecules. The benefit of poolwise recombinations is obvious: an enormous combinatorial potential is realized, enabling access to a vast area of the fitness landscape [6].

2.6 GENETIC ENGINEERING OF ANTIBODIES IN ENVIRONMENTAL ANALYSIS

Sensitivity and selectivity of bioanalytical assays in environmental monitoring essentially depend on the properties of the biorecognition elements to be used for analyte binding. Biological structures are usually derived from subcellular components. These include enzymes, Abs, hormone receptors, DNA, membrane components, and even organelles or entire cells. Immunochemical assay systems employing Abs as binding proteins are effective tools for the analysis of a wide variety of analytes ranging from low molecular weight xenobiotics (e.g., pesticides, xenoestrogens) to complex proteins (e.g., structures of pathogenic microorganisms). Many applications of Ab have been reported in the area of biosensor development and application (for reviews, see Cooper [103], Rogers [104], Sharpe [105], Paitan et al. [106], Rodriguez-Mozaz et al. [107]).

The proper functioning of sensors essentially depends on the immobilization of the respective ligands, Ab or coating conjugates, on the sensing surface, their correct orientation and homogeneity. Recombinant approaches for the synthesis of immunoreagents can accomplish the demand for homogeneous preparations in virtually unlimited amounts. In addition, genetic engineering enables the modification of available structures, e.g., alteration of binding properties, attachment of anchor groups, or improvement of stability.

A growing number of research groups utilizes recombinant Ab technology for environmental applications. The prevalently reported strategy for Ab production consists of the direct cloning and functional expression of Ab-encoding genes derived from hybridoma cell lines. The reason for this is obvious: (1) Groups engaged in Ab production frequently have access to hybridoma cell lines. (2) Hybridoma cells are secreting monoclonal Ab of defined analytical characteristics. Thus, the success of a cloning experiment can be easily validated by comparing the analytical properties of parental monoclonal Ab and their recombinant derivatives.

Recombinant Ab synthesis in the environmental area were described for the first time at the beginning of the 1990s [108]. This pioneering attempt in environmental analysis still encountered a multitude of technical challenges at that time. Thereafter, recombinant Ab fragments were synthesized from hybridoma cells against a panel of relevant xenobiotics like diuron [109], paraquat

[110], atrazine [111], cyclohexanedione [112], parathion [113], dioxine [114], picloram [115,116], mecoprop [117], chlorpyrifos [118], coplanar polychlorinated biphenyls [119], and others.

Like the parental monoclonal Ab, these recombinant Ab fragments were appropriate for the quantitative detection of xenobiotics. They were either similar to the parental monoclonal Ab [114,118], or they showed altered analytical characteristics [110,111]. The latter can be mainly attributed to the intrinsic feature of hybridoma cells to contain Ab genes from both the Ab-secreting B cell as well as the myeloma fusion partner that is necessary for immortalization. Altered binding characteristics are predominantly caused by recombining a V gene of a B cell with a myeloma V gene, by unintentionally introducing mutations during the cloning steps, and by misfolding of the eukaryotic protein in heterologous expression systems, just to mention a few reasons.

2.6.1 ANTIBODY LIBRARIES IN ENVIRONMENTAL ANALYSIS

In contrast to the increasing use of hybridoma cells as a V-gene source, libraries comprising an entire immune repertoire are yet very rarely reported in the environmental area. However, just at the level of repertoire cloning, the potential of recombinant techniques can be utilized in full consequence by simultaneously eliminating artifacts evoked by the presence of myeloma genes. In initial attempts, a naïve human Ab library was applied in order to isolate Ab fragments against diuron [120]. Because the isolated Ab fragments reacted very weakly with the free analyte, appropriate fragments for the application in trace amount analysis were not obtained. Thereafter, an Ab library was generated by cloning splenocyte genes from immunized rabbits [121]. The best clone obtained from this library showed an IC_{50} of ca 50 µg/l for the detection of atrazine. Finally, a library obtained from immunized sheep was described. This library contained remarkably sensitive clones for the detection of atrazine at the ppt-level [122]. In contrast to the medical area that is more or less bound to the use of human Ab repertoires, it is obvious that environmental applications can essentially benefit from individual features of a broader panel of immune sources.

2.6.2 ANTIBODY FRAGMENTS SELECTED FROM NATURAL B CELL REPERTOIRES

A specific problem encountered with low molecular weight target molecules like many xenobiotics is that the corresponding Ab libraries frequently suffer from high background levels of irrelevant Ab genes. This is a consequence of the immunization where the small nonimmunogenic target molecule (immunologically designated as *hapten*) is coupled to a large immunogenic carrier protein. In order to address this inherent shortcoming of conventional libraries, functional hapten-selective Ab genes were enriched by immunomagnetic separation prior to cloning [123]. The method takes advantage of membrane-bound receptor molecules on the surface of B cells, i.e., transmembrane receptor complexes that have identical ligand-binding characteristics as the secreted Ab. These surface receptors can be tagged by target molecules covalently linked to paramagnetic particles. Therefore, the target-specific cells were removed from bulk cultures by magnetic force. Ab genes from the magnet-bound B cell fraction were subsequently cloned into a phage display vector.

Because the B cells were derived from a group of mice immunized with different *s*-triazine derivatives, the resulting library contained a range of Abs against defined members of the *s*-triazine family. This was shown by subsequent isolation of Ab variants by means of phage display technology. The clones were selective for *s*-triazines containing a tert-butyl group, i.e., terbutryn and terbuthylazine as well as those *s*-triazines bearing an isopropylamino residue, i.e., atrazine and propazine [123]. The reaction kinetics are detailed in Table 2.1 for three clones that were selective for *s*-triazine herbicides containing a tert-butyl group. The binding constants are in the range of affinity-matured Ab that is obtained during the secondary immune response.

TABLE 2.1
Association Rate k_a, Dissociation Rate k_d, and Equilibrium Dissociation Constant K_D of the Three Terbuthylazine-Selective Clones BUT -4, -8, and -56 Obtained from the s-Triazine-Specific Ab Library

Clone	k_a [M^{-1}s^{-1}]	k_d [s^{-1}]	K_D [M]
BUT-4	8.49×10^3	2.87×10^{-4}	3.38×10^{-8}
BUT-8	3.51×10^3	2.49×10^{-4}	7.09×10^{-8}
BUT-56	4.09×10^3	3.28×10^{-4}	8.01×10^{-8}

The values for k_a, k_d, and K_D were measured utilizing the BIAcore 2000™ system.

2.6.3 Antibody Fragments Selected from Fully Synthetic Library

In contrast to natural immune repertoires, semisynthetic or fully synthetic Ab repertoires can be utilized in order to obtain specific binding molecules. A decisive advantage of fully synthetic libraries is that processing and potential modifications of the target molecule during the immune response in vivo is definitely avoided. This team employed the fully synthetic human combinatorial Ab library HuCAL® (MorphoSys, Inc., Martinsried, Germany). The library is based on consensus sequences for each of those seven variable heavy-chain (V$_H$) and variable light-chain (V$_L$) germline families that are most frequently used in human immune response. Diversity is created by replacing the V$_H$ and V$_L$ complementarity-determining regions CDR3 of the master genes by CDR3 library cassettes, generated from mixed trinucleotides [54], and biased toward natural human Ab CDR3 sequences [124].

The HuCAL® library was selected by means of phage display for clones expressing Ab fragments for the detection of the herbicide glyphosate (N-(phosphonomethyl)glycine). Glyphosate has gained increasing importance in the context of genetically modified herbicide-tolerant crops and is meanwhile the best-selling herbicide worldwide [125]. However, the analysis by conventional chemical methods is difficult because appropriate chromatographic methods necessitate laborious pre-column derivatization. On the other hand, there are only a few descriptions of immunochemical approaches in literature that offer a means for facile glyphosate analysis avoiding the derivatization-step [126–128]. The IC$_{50}$ of the best clone that was isolated from the synthetic library was determined to be 5.8 µg/l in a glyphosate ELISA (cf. Figure 2.8; Kramer unpublished). However, the maximum signal reduction in the assay was approximately 60%, still retaining a significant level of background noise.

The HuCAL® library was further used to obtain Ab fragments for the diagnostics of food-borne pathogens such as *Listeria monocytogenes*, *Escherichia coli*, *Campylobacter spp.*, *Bacillus cereus*, *Staphylococcus aureus*, and *Salmonella spp.* Pathogen-specific clones were enriched by employing peptides conjugated with carrier proteins. The peptide sequences were delineated from pathogen-associated invasion factors, e.g., bacteria toxins. Selective Ab fragments were identified for each of the six pathogens because any of them was covered by up to four different peptides (cf. Table 2.2) [129]. The diagnostic properties of the Ab fragments were evaluated with culture supernatants obtained from pathogenic bacteria strains. These supernatants contained the native bacteria toxins. In the case of *L. monocytogenes*, *B. cereus*, and *S. aureus*, the recombinant Ab fragments bound selectively to the native protein in a noncompetitive ELISA format (Table 2.2).

2.6.4 Optimizing Antibodies by Directed Evolution

The alteration of the affinity profile of existing Ab by genetic engineering is considered a powerful instrument for circumventing classical Ab production schemes that require extended immunization

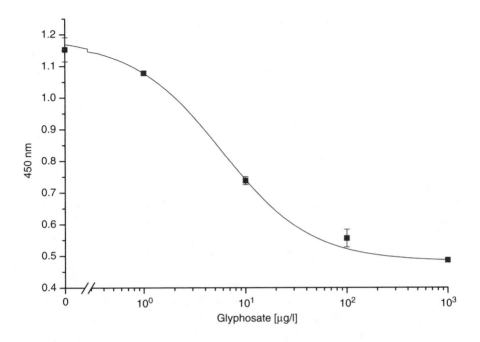

FIGURE 2.8 ELISA calibration curve based on the HuCAL®-derived clone Gly 12 for the quantitative detection of glyphosate. Sepharose beads are employed as solid phase for the ELISA.

periods [11,130,131]. Although this point is frequently raised in environmental research, it has seldom resulted in experimental success. However, it should be noted that remarkable positive results concerning this aspect are rarely documented for recombinant Ab fragments in general. In the environmental area, the optimization of Ab fragments in vitro was addressed by both directed evolution [123] and rational design [132–134]. In the latter case, a triple mutant F_{ab} fragment was generated with an affinity five-fold greater toward an atrazine derivative as compared to the wild-type [134].

The principles of directed evolution in vitro are very similar to the mechanisms of somatic mutation in vivo. The Ab repertoire of the primary immune response in vivo is predominantly the result of recombining germ line genes of the V, D, J clusters (i. e., the genes constituting elements of

TABLE 2.2
Binding of HuCAL® Ab Fragments Selected with Pathogen-Related Peptides to Peptide and Native Toxin

Microorganism	Target Antigen	Peptide Binding	Antigen Binding
Bacillus cereus	Hemolysin lytic component	+	+
Campylobacter spp.	Cytolethal distending toxin	+	−
Escherichia coli	EHEC hemolysin	+	−
Listeria monocytogenes	Invasion-associated protein p60	+	+
Salmonella spp.	Invasion protein D	+	−
Staphylococcus aureus	Enterotoxin type B	+	+

+, indicates binding; −, indicates no binding. Native antigen was derived from culture supernatants of the corresponding microorganism.

the finally expressed V_H and V_L genes). The primary response yields Ab with generally low affinities. Thereafter, Ab-variable genes of the primary repertoire are modified during the secondary immune response in the microenvironment of lymphoid germinal centers by somatic hypermutation. Physiologically appropriate variants are subsequently selected from this pool of mutant immunoglobulins upon their improved affinity to the antigen [47].

In vitro, Ab genes are subjected to iterative evolutionary cycles of mutation and selection until they meet the requirements for the designated application. The gene repertoire of the group-selective Ab library presented above was employed for subsequent optimization of individual Ab molecules by directed evolution [123]. It was expected that this library would substantially facilitate the engineering of desired Ab specificities and affinities to any member of the triazine group without the need of new immunizations.

The Ab clone IPR-7 was used as a template for the optimization process. This clone was initially selected from the group-selective library and provided the highest affinity to s-triazines bearing an isopropylamino residue, i.e., atrazine and propazine (Table 2.3) [135]. Chain shuffling was applied as a directed, recombinatorial approach for improving the affinity of this clone. At the outset, the light chain of the template Ab IPR-7 was shuffled with the heavy-chain repertoire of the group-specific library and subsequently selected by phage display. Then, the heavy chain of the best binder (IPR-26) was shuffled against the library light chain repertoire. The kinetic data of the Ab variants are detailed in Table 2.3 [123]. The equilibrium dissociation constants of the Ab variants approach the typical K_D level of affinity matured Ab in vivo [136]. The optimized variant IPR-83 showed a seventeen-fold increase in affinity as compared to the template Ab IPR-7.

Interestingly, sequence analysis of the shuffled clones revealed a bias of amino acid substitutions from the template IPR-7 to the optimized variant IPR-83 in the 5'-moiety of the V genes, including the first two CDRs and the flanking frame regions [123]. This is in contrast to a series of Ab-optimization experiments that is primarily targeting the variation of the CDR3 regions at the 3'-moiety of the Ab variable domain [11,66,132]. The V_H CDR3 region is generally considered to constitute the key determinant for the antigen selectivity [137]. However, the distribution of sequence alterations obtained in the presented Ab optimization is consistent with proposed models for mutational mechanisms during the secondary immune response in vivo [138–139]. In addition, experimental data obtained from in vivo immune repertoires confirm that individual sites at the V genes are prone to hypermutation. These mutational hot-spots for affinity maturation are strategically located at the CDR1 and CDR2 rather than in the CDR3 loop [140,141]. Therefore, the applied in vitro optimization strategy resulted in a distribution of sequence alterations; this strategy is fitting very well into the current knowledge of natural affinity maturation.

The success of the strategy can be partly attributed to the applied chain-shuffling procedure. Shuffling of functional V_L and V_H domains is very likely to generate a repertoire of new functional

---

TABLE 2.3
Association Rate Constant k_a, Dissociation Rate Constant k_d, and Equilibrium Dissociation Constant K_D for Atrazine-Selective Clones IPR-7 (Template Ab), IPR-26 (VH-Shuffled Ab), and IPR-83 (V_L-Shuffled Ab)

Clone	k_a $(M^{-1}s^{-1})$	k_d (s^{-1})	$K_D = k_d/k_a$ (M)
IPR-7	1.38×10^5	1.75×10^{-3}	1.27×10^{-8}
IPR-26	2.10×10^5	1.93×10^{-3}	9.20×10^{-9}
IPR-83	6.73×10^5	5.02×10^{-4}	7.46×10^{-10}

The values for k_a, k_d, and K_D were measured utilizing the BIAcore 2000™ system.

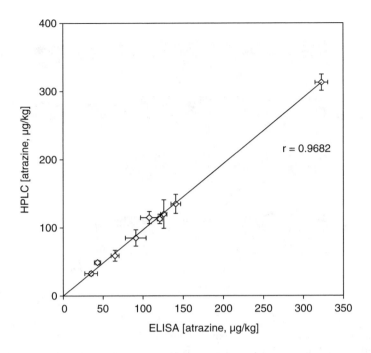

FIGURE 2.9 Atrazine analyses of soil samples by ELISA employing the mutant antibody IPR-83. The samples were collected from the top layer (0–5 cm) of corn fields in Southern Germany in May/June, subsequent to seasonal application and immediately processed. Validation was performed by HPLC. HPLC data were kindly provided by Dr. J. Lepschy, Bayerische Landesanstalt für Bodenkultur und Pflanzenbau, Freising, Germany. (From Kramer, K., *Environ. Sci. Technol.*, 36, 4892, 2002.)

heterodimers. The concept corresponds to splicing at uncoupled sites flanking the globular domains rather than at coupled sites that are located intradomain (cf. above).

The applicability of the optimized Ab variant IPR-83 was tested by measurement of environmental samples. IPR-83 was applied to determine atrazine contaminations of soil samples collected in Southern Germany. Although atrazine has been banned in Germany by the European Community in 1991, environmental contaminations have been observed during the last years as a result of illegal applications. The corresponding threshold for atrazine is 100 μg/kg soil. The immunochemical analysis was complemented by HPLC analyses as reference methods for validation (cf. Figure 2.9). The ELISA measurements were consistent with the HPLC data within the experimental error [135]. Therefore, the engineered scF$_v$ mutants proved to be suitable for the application in environmental analyses under real sample conditions.

2.6.5 GENETIC ENGINEERING BEYOND AFFINITY AND SELECTIVITY

In addition to affinity and selectivity, the stability of Ab fragments is a crucial aspect in the analytical practice. Appropriate Ab fragments have to retain entire functionality upon prolonged storage periods and/or at elevated temperatures as well as under unfavorable assay conditions, e.g., extrinsic factors like organic solvents. Particularly, the latter aspect is a priority criterion for analyzing samples that contain significant amounts of organic solvents such as methanol. Organic solvents are frequently added in order to extract contaminants from samples for the subsequent analysis.

As an example, a selection of Ab fragments from the HuCAL® library was investigated by incubation at varying concentrations of methanol, and their functionality was subsequently investigated by ELISA. A content of 40% (v/v) methanol in the sample did not affect the binding

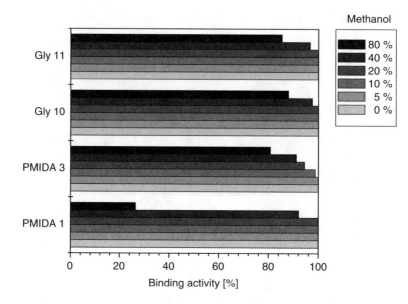

FIGURE 2.10 Methanol sensitivity of pesticide-selective Ab clones isolated from HuCAL® library. Binding activity was evaluated by measuring the maximum and minimum signal in ELISA with sample buffer containing 0%–80% (v/v) methanol according to this formula: $(A_{max} - A_{min}$ at x% MetOH)/$(A_{max} - A_{min}$ at 0% MetOH)×100%.

potential of the majority of the glyphosate-selective Ab fragments in a significant manner (Figure 2.10) [129]. Even a content of 80% (v/v) methanol showed a relevant signal reduction in ELISA for only one of the four Ab clones. This is noteworthy because in similar investigations based on conventionally produced polyclonal and monoclonal Abs, this team observed that the majority of Abs lose their binding capability already at much lower concentrations in the range between 5% and 15% (v/v) organic solvent [129].

The thermal stability of the Ab fragments can be used as an indicator for their long-term storage properties and molecular integrity at unfavorable temperature conditions. For this purpose, Ab fragments were incubated for 24 h at various temperatures, and their functionality was again tested by ELISA. The corresponding results for the pathogen-selective Ab are presented in Figure 2.11 [129]. Almost all Ab fragments were stable up to temperatures of 37°C and 50°C, respectively. Most of the Abs showed reduced or even completely lost ligand binding at higher temperatures. These observations are consistent with analogous experiments performed with conventional monoclonal Abs derived from hybridoma cultures in this laboratory [142]. However, one out of the six fragments depicted in Figure 2.11 retained full functionality even after incubation at 80°C, indicating an exceptional stability with respect to the considered protein family.

Sequence alignment with human germline genes revealed that the immunoglobulin-variable heavy-chain genes of the clones providing the highest stabilities could be predominantly assigned to the V_H3 gene family, indicating that the stability of Ab fragments may be an inherent structural feature of particular germline families. These results are well in line with investigations of the biophysical properties of Ab fragments in the group of Plückthun employing the HuCAL® library [143]. Ab fragments containing the variable domain combinations H3κ3 and H5κ3 showed superior stability. Combination with λ light chains also exhibited high levels of stability, depending on the particular amino acid sequence of the CDR-L3.

In the example shown above, stable clones have been directly isolated from the synthetic library. However, stability can also be integrated as a parameter in a directed evolution experiment.

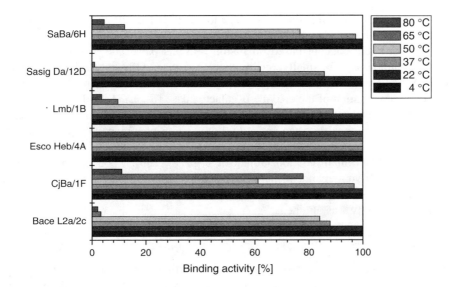

FIGURE 2.11 Thermal stability of peptide-selective Ab clones isolated from HuCAL® library. Ab fragments were incubated for 24 h at temperatures ranging from 4°C to 80°C. Binding activity was subsequently evaluated by measuring the maximum and minimum signal in ELISA according to this formula: $(A_{max} - A_{min}$ temperature treated Ab)/$(A_{max} - A_{min}$ of untreated control)$\times 100\%$.

This is accomplished by including organic solvents or exerting thermal pressure during affinity selection of a library repertoire.

2.6.6 ANTIBODY FORMATS AND FUSION PROTEINS IN ENVIRONMENTAL ANALYSIS

Ab fragments prepared for environmental analysis are predominantly produced either in the scF$_v$ [112,113] or F$_{ab}$ format [108,120]. An uncommon Ab format is also reported where scF$_v$ fragments are extended by an unpaired human C$_L$ domain [144]. The unpaired C$_L$ domain functions as an affinity tag for the purification of the recombinant proteins as well as for the dimerization of monovalent scF$_v$ fragments.

Vectors used in Ab selection are generally not the most appropriate ones for subsequent steps of genetic engineering. As a result, some groups have created modular systems where selected variable Ab genes can be excised from selection vectors and re-cloned into downstream vectors that have other useful properties. A generic concept for the optional transformation from the scF$_v$ into the F$_{ab}$ format by simultaneously maintaining the Ab binding properties was described in the environmental area [145]. The transformation into F$_{ab}$ fragments is recommended if the Ab was obtained from a scF$_v$ library. Therefore, the undesired tendency of scF$_v$ fragments for dimerization is avoided. Generally, the merits of dimerization or multimerization of recombinant Ab fragments, e.g., by leucine zipper extensions for environmental analysis, are ambivalent. Whereas this option provides an efficient means for the enhancement of functional affinity for high molecular-weight antigens in medical applications, no beneficial effects with respect to the affinity are expected for small target molecules in environmental analysis.

In contrast, Ab fragments that are fused with marker proteins offer the advantage of a reduced number of assay steps in analytical test formats. The genetically engineered fusion of the Ab-binding function with marker enzymes was already reported for environmental analysis [146]. The corresponding vector contains an alkaline phosphatase and generic restriction sites for the convenient one-step cloning of scF$_v$ fragments obtained from Ab libraries. Similar, a gene encoding green fluorescent protein was inserted into a vector harboring a picloram-specific

Ab fragment [147]. The resulting fluobody elegantly avoids the enzyme-substrate reaction for colorimetric detection that is required in a conventional ELISA.

Finally, a remarkable concept based on fusion proteins paves the way for noncompetitive homogeneous immunoassays for small target molecules instead of the standard competitive heterogeneous ELISA. The assay utilizes the antigen-dependent reassociation of Ab variable domains and concomitant complementation of the enzyme β-galactosidase (β-gal) [148]. In a proof of principle approach, the reassociation of two fusion proteins was monitored by restoring the enzymatic activity. The first protein consists of an anti 4-hydroxy-3-nitrophenylacetyl V_H fragment fused to an N-terminal deletion mutant of β-gal ($V_H\Delta\alpha$), and the second protein correspond to the V_L fragment fused to a C-terminal deletion mutant of β-gal ($V_L\Delta\omega$). Upon mixing of the reagents with the sample, an analyte-dependent increase in reassociation and, thus, in enzymatic activity, was observed. Compared to the corresponding heterogeneous open-sandwich ELISA, a 1000-fold improvement in sensitivity was attained.

2.6.7   HOST ORGANISMS

Besides the crucial aspects of directed improvement of binding properties, the flexibility in Ab fragment design, and the genetic combination of Ab with marker proteins, genetic Ab engineering offers the choice between various host organisms. The decision for a particular expression host, however, depends on the benefits and drawbacks of the individual systems employed. For analytical applications, Ab fragments are definitely sufficient that comprise the entire antigen binding site. Even bacteria are suited to optimize the yield of functionally expressed Ab fragments by a careful choice of the expression vector and the optimization of the expression conditions in order to meet the requirements for applications in industrial scale [145]. In an alternative expression system, a F_{ab}-fragment encoding gene was used for the functional expression of recombinant herbicide Ab in yeast cells. In this approach, the Ab fragments do not need any further processing by affinity purification for the majority of applications in environmental analysis [149]. The Ab fragments are directly translocated into the culture medium by a leader sequence inserted into the expression vector. Thereafter, they can be directly harvested from the culture supernatant by simple concentration procedures because the culture supernatant contains very low amounts of contaminating yeast proteins. Finally, the functional expression of pesticide-selective Ab fragments in transgenic plants of *Nicotiana tabacum* cv. Samsun NN and cell suspension cultures was reported [150]. Although "Molecular Farming" offers advantages with respect to the economic mass production of recombinant proteins, ecological deliberations may hamper the full potential of this approach (e.g., outdoor cultivation).

2.7 OUTLOOK

Acceptance of recombinant protein technology in general will be essentially driven by the facile access to genetic Ab engineering, either as kit systems including training on the basic technique or as a service of specialized biotechnology plants. Meanwhile, there are commercial Ab libraries existing (e.g., HuCAL® [124], Tomlinson I and J [151]) that are accessible for environmental analysis. Some of these libraries are claimed to be universal, suggesting that the libraries provide Ab, which are covering a huge panel of various specificities. However, even at their best, these libraries represent nothing more than basic immune repertoires for the isolation of primary binders. Their subsequent optimization by means of directed evolution strategies and/or rational design probably remains an integral part in the synthesis of appropriate Ab for the majority of immunochemical applications in environmental analysis. Therefore, one of the vital goals for the future is the simple and cheap access to evolutionary technologies for tailored binders with predefined properties such as selectivity, affinity, stability, and more. Massive parallel processing combined with high-throughput strategies as well as high-content screening will have a beneficial impact on this new era of receptor production.

ACKNOWLEDGMENTS

The authors would like to thank MorphoSys for providing the HuCAL® library. Further thanks are addressed to Mrs. H. Geltl and Mrs. A. Hubauer for their excellent technical assistance. Financial support was obtained from the EC (DG XII Environment and Climate 1994-8, project number ENV4-CT96-0333, Envirosense project) and the Bavarian Government (Project number BFS 306/98).

REFERENCES

1. Van Oss, C. J., Antibody-antigen intermolecular forces, In *Encyclopedia of Immunology*, Roitt, I. M. and Daves, P. J., Eds., Vol. 1, Academic Press, London, p. 97, 1992.
2. Bauer, E. R., Bitsch, N., Brunn, H., Sauerwein, H., and Meyer, H. H., Development of an immuno-immobilized androgen receptor assay (IRA) and its application for the characterization of the receptor binding affinity of different pesticides, *Chemosphere*, 46, 1107, 2002.
3. Vollenbroeker, B., Fobker, M., Specht, B., Bartetzko, N., Erren, M., Spener, F., and Hohage, H., Receptor assay based on surface plasmon resonance for the assessment of the complex formation activity of cyclosporin A and its metabolites, *Int. J. Clin. Pharmacol. Ther.*, 41, 248, 2003.
4. Seifert, M., Luminescent enzyme-linked receptor assay for estrogenic compounds, *Anal. Bioanal. Chem.*, 378, 684, 2004.
5. Chen, Y., Wiesmann, C., Fuh, G., Li, B., Christinger, H. W., McKay, P., de Vos, A. M., and Lowman, H. B., Selection and analysis of an optimized anti-VEGF antibody: Crystal structure of an affinity-matured Fab in complex with antigen, *J. Mol. Biol.*, 293, 865, 1999.
6. Tobin, M. B., Gustafsson, C., and Huisman, G. W., Directed evolution: The 'rational' basis for 'irrational' design, *Curr. Opin. Struct. Biol.*, 10, 421, 2000.
7. Marshall, S. A. et al., Rational design and engineering of therapeutic proteins, *Drug Discov. Today*, 8, 212, 2003.
8. Eijsink, V. G. et al., Rational engineering of enzyme stability, *J. Biotechnol.*, 30, 105, 2004.
9. Rowe, L. A. et al., A comparison of directed evolution approaches using the β-glucuronidase model system, *J. Mol. Biol.*, 332, 851, 2003.
10. Spiller, B. et al., A structural view of evolutionary divergence, *Proc. Natl. Acad. Sci. USA*, 96, 12305, 1999.
11. Boder, E. T., Midelfort, K. S., and Wittrup, K. D., Directed evolution of antibody fragments with monovalent femtomolar antigen-binding affinity, *Proc. Natl. Acad. Sci. USA*, 97, 10701, 2000.
12. Fisher, R. A., *The Genetical Theory of Natural Selection*, 2nd ed., Dover Publications, Inc., New York, 1958.
13. Kurtzman, A. L. et al., Advances in directed protein evolution by recursive genetic recombination: Applications to therapeutic proteins, *Curr. Opin. Biotechnol.*, 12, 361, 2001.
14. Nygren, P. A. and Skerra, A., Binding proteins from alternative scaffolds, *J. Immunol. Methods*, 290, 3, 2004.
15. Brakmann, S., Discovery of superior enzymes by directed molecular evolution, *Chem. Bio. Chem.*, 2, 865, 2001.
16. Arnold, F. H. and Georgion, G., *Directed Enzyme Evolution: Screening and Selection Methods*, Humana Press, Totowa, NJ, 2003.
17. Jestin, J. L. and Kaminski, P. A., Directed enzyme evolution and selections for catalysis based on product formation, *J. Biotechnol.*, 113, 85, 2004.
18. Williams, G. J., Nelson, A. S., and Berry, A., Directed evolution of enzymes for biocatalysis and the life sciences, *Cell. Mol. Life Sci.*, 61, 3034, 2004.
19. Hörseh, S., Diploma thesis, Institute of Technical Biochemistry, University of Stuttgart, Stuttgart, Germany, 1998.
20. Poljak, R. J. et al., Three-dimensional structure of the Fab fragment of a human immunoglobulin at a 2.8-Å resolution, *Proc. Natl. Acad. Sci. USA*, 70, 3305, 1973.

21. Wu, T. T. and Kabat, E. A., An analysis of the sequences of the variable regions of Bence-Jones proteins and myeloma light chains and their implication for antibody complementarity, *J. Exp. Med.*, 132, 211, 1970.

22. Kabat, E. A., et al., *Sequences of Proteins of Immunological Interest*, 5th Ed., U. S. Department of Health and Human Services, Public Health Service, National Institutes of Health (NIH Publication No 91–3242), Bethesda, 1991.

23. Padlan, E. A., The anatomy of the antibody molecule, *Mol. Immunol.*, 31, 169, 1994.

24. Tulip, W. R. et al., Refined crystal structure of the influenza virus N9 neuraminidase-NC41 Fab complex, *J. Mol. Biol.*, 227, 122, 1992.

25. Hoogenboom, H. R. et al., Multi-subunit proteins on the surface of filamentous phage: Methodologies for displaying antibody (Fab) heavy and light chains, *Nucleic Acids Res.*, 19, 4133, 1991.

26. Marks, J. D. et al., By-passing immunization. Human antibodies from V-gene libraries displayed on phage, *J. Mol. Biol.*, 222, 581, 1991.

27. Barbas, C. F. D. et al., Semisynthetic combinatorial antibody libraries: A chemical solution to the diversity problem, *Proc. Natl. Acad. Sci.*, 89, 4457, 1992.

28. Glockshuber, R. et al., A comparison of strategies to stabilize immunoglobin Fv fragments, *Biochemistry*, 29, 1362, 1990.

29. Arndt, K. M., Müller, K. M., and Plückthun, A., Helix-stabilized Fv (hsFv) antibody fragments: Substituting the constant domains of a Fab fragment for a heterodimeric coiled-coil domain, *J. Mol. Biol.*, 312, 221, 2001.

30. Reiter, Y. et al., An antibody single-domain phage display library of a native heavy chain variable region: Isolation of functional single-domain VH molecules with a unique interface, *J. Mol. Biol.*, 290, 685, 1999.

31. Van den Beucken, T. et al., Building novel binding ligands to B7.1 and B7. 2 based on human antibody single variable light chain domains, *J. Mol. Biol.*, 310, 591, 2001.

32. Desmyter, A. et al., Antigen specificity and high affinity binding provided by one single loop of a camel single-domain antibody, *J. Biol. Chem.*, 276, 26285, 2001.

33. Dooley, H., Flajnik, M. F., and Porter, A. J., Selection and characterization of naturally occurring single-domain (IgNAR) antibody fragments from immunized sharks by phage display, *Mol. Immunol.*, 40, 25, 2003.

34. Xu, L. et al., Directed evolution of high-affinity antibody mimics using mRNA display, *Chem. Biol.*, 9, 933, 2002.

35. Hufton, S. E. et al., Development and application of cytotoxic T-lymphocyte-associated antigen 4 as a protein scaffold for the generation of novel binding ligands, *FEBS Lett.*, 475, 225, 2000.

36. Nord, K. et al., Binding proteins selected from combinatorial libraries of an a-helical bacterial receptor domain, *Nat. Biotechnol.*, 15, 772, 1997.

37. Gunneriusson, E. et al., Affinity maturation of a Taq DNA polymerase specific affibody by helix shuffling, *Protein Eng.*, 12, 873, 1999.

38. Beste, G. et al., Small antibody-like proteins with prescribed ligand specificities derived from the lipocalin fold, *Proc. Nat. Acad. Sci.*, 96, 1898, 1999.

39. Schlehuber, S. and Skerra, A., Lipocalins in drug discovery: From natural ligand-binding proteins to 'anticalins.', *Drug Discov. Today*, 10, 23, 2005.

40. Sedgwick, S. G. and Smerdon, S. J., The ankyrin repeat: A diversity of interactions on a common structural framework, *Trends Biochem. Sci.*, 24, 311, 1999.

41. Binz, H. K. et al., High-affinity binders selected from designed ankyrin repeat protein libraries, *Nat. Biotechnol.*, 22, 575, 2004.

42. Zeytun, A. et al., Retraction: Fluorobodies combine GFP fluorescence with the binding characteristics of antibodies, *Nat. Biotechnol.*, 22, 601, 2004.

43. Choo, Y. and Klug, A., Selection of DNA binding sites for zinc fingers using rationally randomized DNA reveals coded interactions, *Proc. Natl. Acad. Sci. USA*, 91 1994.

44. Jamieson, A. C., Kim, S. H., and Wells, J. A., In vitro selection of zinc fingers with altered DNA-binding specificity, *Biochemistry*, 33, 5689, 1994.

45. Smith, G. P. et al., Small binding proteins selected from a combinatorial repertoire of knottins displayed on phage, *J. Mol. Biol.*, 277, 317, 1998.

46. Roberts, B. L. et al., Directed evolution of a protein: Selection of potent neutrophil elastase inhibitors displayed on M13 fusion phage, *Proc. Natl. Acad. Sci. USA*, 89, 2429, 1992.
47. Rajewsky, K., Clonal selection and learning in the antibody system, *Nature*, 381, 751, 1996.
48. Neylon, C., Chemical and biochemical strategies for the randomization of protein encoding DNA sequences: Library construction methods for directed evolution, *Nucleic Acids Res.*, 32, 1448, 2004.
49. Xu, H. et al., Random mutagenesis libraries: Optimization and simplification by PCR, *Biotechniques*, 27, 1102, 1999.
50. Cirino, P. C., Mayer, K. M., and Umeno, D., Generating mutant libraries using error-prone PCR, *Methods Mol. Biol.*, 231, 3, 2003.
51. Zaccolo, M. and Gherardi, E., The effect of high-frequency random mutagenesis on in vitro protein evolution: A study on TEM-1 β-lactamase, *J. Mol. Biol.*, 285, 775, 1999.
52. Cline, J. and Hogrefe, H. H., Randomize gene sequences with new PCR mutagenesis kit, *Strategies*, 13, 157, 2002.
53. Murakami, H., Hohsaka, T., and Sisido, M., Random insertion and deletion mutagenesis, *Methods Mol. Biol.*, 231, 53, 2003.
54. Virnekäs, B. et al., Trinucleotide phosphoramidites: Ideal reagents for the synthesis of mixed oligonucleotides for random mutagenesis, *Nucleic Acids Res.*, 22, 5600, 1994.
55. Ho, S. N. et al., Site-directed mutagenesis by overlap extension using the polymerase chain reaction, *Gene*, 77, 51, 1989.
56. Miyazaki, K., Creating random mutagenesis libraries by megaprimer PCR of whole plasmid (MEGAWHOP), *Methods Mol. Biol.*, 231, 23, 2003.
57. Stemmer, W. P. C., DNA shuffling by random fragmentation and reassembly: In vitro recombination for molecular evolution, *Proc. Natl. Acad. Sci. USA*, 91, 10747, 1993.
58. Zhao, H. et al., Molecular evolution by staggered extension process (StEP) in vitro recombination, *Nat. Biotechnol.*, 16, 258, 1998.
59. Aguinaldo, A. M. and Arnold, F. H., Staggered extension process (StEP) in vitro recombination, *Methods Mol. Biol.*, 231, 105, 2003.
60. Coco, W. M., RACHITT: Gene family shuffling by random chimeragenesis on transient templates, *Methods Mol Biol.*, 231, 111, 2003.
61. Zha, D., Eipper, A., and Reetz, M. T., Assembly of designed oligonucleotides as an efficient method for gene recombination: A new tool in directed evolution, *Chembiochem.*, 4, 34, 2003.
62. Ness, J. E. et al., Synthetic shuffling expands functional protein diversity by allowing amino acids to recombine independently, *Nat. Biotechnol.*, 20, 1251, 2002.
63. Lutz, S., Ostermeier, M., and Benkovic, S. J., Rapid generation of incremental truncation libraries for protein engineering using alphaphosphothioate nucleotides, *Nucleic Acids Res.*, 29, E16, 2001.
64. Sblattero, D. and Bradbury, A., Exploiting recombination in single bacteria to make large phage antibody libraries, *Nat. Biotechnol.*, 18, 75, 2000.
65. Smith, G. P., Filamentous fusion phage: Novel expression vectors that display cloned antigens on the virion surface, *Science*, 228, 1315, 1985.
66. Smith, G. P. and Scott, J. K., Libraries of peptides and proteins displayed on filamentous phage, *Methods Enzymol.*, 217, 228, 1993.
67. Georgiou, G. et al., Display of heterologous proteins on the surface of microorganisms: From the screening of combinatorial libraries to live recombinant vaccines, *Nat. Biotechnol.*, 15, 29, 1997.
68. Boder, E. T. and Wittrup, K. D., Yeast surface display for screening combinatorial polypeptide libraries, *Nat. Biotechnol.*, 15, 553, 1997.
69. Hanes, J. and Plückthun, A., In vitro selection and evolution of functional proteins by using ribosome display, *Proc. Natl. Acad. Sci. USA*, 94, 4937, 1997.
70. Roberts, R. W. and Szostak, J. W., RNA-peptide fusions for the in vitro selection of peptides and proteins, *Proc. Natl. Acad. Sci. USA*, 94, 12297, 1997.
71. Vaughan, T. J. et al., Human antibodies with sub-nanomolar affinities isolated from a large non-immunised phage display library, *Nat. Biotechnol.*, 14, 309, 1996.
72. McCafferty, J. et al., Phage antibodies: Filamentous phage displaying antibody variable domains, *Nature*, 348, 552, 1990.
73. Lowman, H. B. and Wells, J. A., Affinity maturation of human growth hormone by monovalent phage display, *J. Mol. Biol.*, 234, 564, 1993.

74. Bradbury, A. et al., Antibodies in proteomics I: Generating antibodies, *Trends Biotechnol.*, 21, 275, 2003.

75. Hawkins, R. E. et al., Selection of phage antibodies by binding affinity: Mimicking affinity maturation, *J. Mol. Biol.*, 226, 889, 1992.

76. Dueñas, M. and Borrebaeck, C. A., Clonal selection and amplification of phage displayed antibodies by linking antigen recognition and phage replication, *Biotechnology*, 12, 999, 1994.

77. Jung, S. et al., Selectively infective phage (SIP) technology: Scope and limitations, *J. Immunol. Methods*, 231, 93, 1999.

78. Wittrup, K. D., Protein engineering by cell-surface display, *Curr. Opin. Biotechnol.*, 12, 395, 2001.

79. Daugherty, P. S., Iverson, B. L., and Georgiou, G., Flow cytometric screening of cell-based libraries, *J. Immunol. Methods*, 243, 211, 2000.

80. Mattheakis, L. C., Bhatt, R. R., and Dower, W. J., An in vitro polysome display system for identifying ligands from very large peptide libraries, *Proc. Natl. Acad. Sci. USA*, 91, 9022, 1994.

81. He, M. and Taussig, M. J., Antibody-ribosome-mRNA (ARM) complexes as efficient selection particles for in vitro display and evolution of antibody combining sites, *Nucleic Acids Res.*, 25, 5132, 1997.

82. Amstutz, P. et al., In vitro display technologies: Novel developments and applications, *Curr. Opin. Chem. Biol.*, 12, 400, 2001.

83. Lin, H. and Cornish, V. W., Screening and selection methods for large-scale analysis of protein function, *Angew Chem. Int. Ed.*, 41, 4402, 2002.

84. Nemoto, N. et al., In vitro virus: Bonding of mRNA bearing puromycin at the $3'$-terminal end to the C-terminal end of its encoded protein on the ribosome in vitro, *FEBS Lett.*, 414, 405, 1997.

85. Weng, S. et al., Generating addressable protein microarrays with PROfusione covalent mRNA-protein fusion technology, *Proteomics*, 2, 48, 2002.

86. Fields, S. and Song, O., A novel genetic system to detect protein-protein interactions, *Nature*, 340, 245, 1989.

87. Pelletier, J. N. et al., An in vivo library-versus-library selection of optimized protein-protein interactions, *Nat. Biotechnol.*, 17 (683), 99, 1999.

88. Mössner, E., Koch, H., and Plückthun, A., Fast selection of antibodies without antigen purification: Adaptation of the protein fragment complementation assay to select antigen-antibody pairs, *J. Mol. Biol.*, 308, 115, 2001.

89. Patrick, W. M., Firth, A. E., and Blackburn, J. M., User-friendly algorithms for estimating completeness and diversity in randomized protein-encoding libraries, *Protein Eng.*, 16, 451, 2003.

90. Voigt, C. A. et al., Computational method to reduce the search space for directed protein evolution, *Proc. Natl. Acad. Sci. USA*, 98, 3778, 2001.

91. Voigt, C. A. et al., Protein building block preserved by recombination, *Nat. Struct. Biol.*, 9, 553, 2002.

92. Eigen, M., The origin of genetic information: Viruses as models, *Gene*, 135, 37, 1993.

93. Kauffman, S. and Levin, S., Towards a general theory of adaptive walks on rugged landscapes, *J. Theor. Biol.*, 128, 11, 1987.

94. Macken, C. A. and Perelson, A. S., Protein evolution on rugged landscapes, *Proc. Natl. Acad. Sci. USA*, 86, 6191, 1989.

95. Bolon, D. N., Voigt, C. A., and Mayo, S. L., De novo design of biocatalysts, *Curr. Opin. Chem. Biol.*, 6, 125, 2002.

96. Voigt, C. A., Kauffman, S., and Wang, Z. G., Rational evolutionary design: The theory of in vitro protein evolution, *Adv. Protein. Chem.*, 55, 79, 2001.

97. Skandalis, A., Encell, L. P., and Loeb, L. A., Creating novel enzymes by applied molecular evolution, *Chem. Biol.*, 4, 889, 1997.

98. Moore, J. C. and Arnold, F. H., Directed evolution of a para-nitrobenzyl esterase for aqueousorganic solvents, *Nat. Biotechnol.*, 14, 458, 1996.

99. Miyazaki, K. et al., Directed evolution of temperature adaptation in a psychrophilic enzyme, *J. Mol. Biol.*, 297, 1015, 2000.

100. Söding, J. and Lupas, A. N., More than the sum of their parts: On the evolution of proteins from peptides, *Bioessays*, 25, 837, 2003.

101. Lefranc, M. P. et al., IMGT, the international ImMunoGeneTics information system, *Nucleic Acids Res.*, 33, D593, 2005.
102. Honegger, A. and Plückthun, A., Yet another numbering scheme for immunoglobulin variable domains: An automatic modeling and analysis tool, *J. Mol. Biol.*, 309, 657, 2001.
103. Cooper, M. A., Optical biosensors in drug discovery, *Nat. Rev. Drug Discov.*, 1, 515, 2002.
104. Rogers, K. R., Principles of affinity-based biosensors, *Mol. Biotechnol.*, 14, 109, 2000.
105. Sharpe, M., It's a bug's life: Biosensors for environmental monitoring, *J. Environ. Monit.*, 5, 109N, 2003.
106. Paitan, Y. et al., On-line and in situ biosensors for monitoring environmental pollution, *Biotechnol. Adv.*, 22, 27, 2003.
107. Rodriguez-Mozaz, S. et al., Biosensors for environmental monitoring of endocrine disruptors: A review article, *Anal. Bioanal. Chem.*, 378, 588, 2004.
108. Ward, V. K. et al., Cloning, sequencing and expression of the Fab fragment of a monoclonal antibody to the herbicide atrazine, *Protein Eng.*, 6, 981, 1993.
109. Bell, C. W. et al., Recombinant antibodies to diuron. A model for the phenylurea combining site, In *Immunoanalysis of Agrochemicals: Emerging Yechnology*, Nelson, J. O., Karu, A. E., and Wong, R. B., Eds., ACS Symposium Series 586, Washington, 1995.
110. Graham, B. M., Porter, A. J., and Harris, W. J., Cloning, expression and characterisation of a single-chain antibody fragment to the herbicide paraquat, *J. Chem. Technol. Biotechnol.*, 63, 279, 1995.
111. Byrne, F. R. et al., Cloning, expression and characterization of a single-chain antibody specific for the herbicide atrazine, *Food Agric. Immunol.*, 8, 19, 1996.
112. Webb, S. R., Lee, H., and Hall, J. C., Cloning and expression of *Escherichia coli* of an anti-cyclohexanedione single-chain variable antibody fragment and comparison to the parent monoclonal antibody, *J. Agric. Food Chem.*, 45, 535, 1997.
113. Garrett, S. D. et al., Production of a recombinant anti-parathion antibody (scFv); stability in methanolic food extracts and comparison to an anti-parathion monoclonal antibody, *J. Agric. Food Chem.*, 45, 4183, 1997.
114. Lee, N., Holtzapple, C. K., and Stanker, L. H., Cloning, expression, and characterization of recombinant Fab antibodies against dioxin, *J. Agric. Food Chem.*, 46, 3381, 1998.
115. Yau, K. Y. F. et al., Bacterial expression and characterization of a picloram-specific recombinant Fab for residue analysis, *J. Agric. Food Chem.*, 46, 4457, 1998.
116. Tout, N. L. et al., Synthesis of ligand-specific phage-display scFv against the herbicide picloram by direct cloning from hyperimmunized mouse, *J. Agric. Food Chem.*, 49, 3628, 2001.
117. Strachan, G. et al., Reduced toxicity of expression, in *Escherichia coli*, of antipollutant antibody fragments and their use as sensitive diagnostic molecules. J Appl Microbiol. 87:410, 1999.
118. Alcocer, M. J. C. et al., Functional scFv antibody sequences against the organophosphorus pesticide chlorpyrifos, *J Agric Food Chem.*, 48, 335, 2000.
119. Chiu, Y. W. et al., Derivation and properties of recombinant Fab Ab to coplanar polychlorinated biphenyls, *J. Agric. Food Chem.*, 48, 2614, 2000.
120. Karu, A. E. et al., Recombinant antibodies to small analytes and prospects for deriving them from synthetic combinatorial libraries, *Food Agric. Immunol.*, 6, 277, 1994.
121. Li, Y. et al., Selection of rabbit single-chain Fv fragments against the herbicide atrazine using a new phage display system, *Food Agric. Immunol.*, 11, 5, 1999.
122. Charlton, K., Harris, W. J., and Porter, A. J., The isolation of super-sensitive anti-hapten antibodies from combinatorial antibody libraries derived from sheep, *Biosens. Bioelectron.*, 16, 639, 2001.
123. Kramer, K., Synthesis of a group-selective antibody library against haptens, *J. Immunol. Methods.*, 266, 211, 2002.
124. Knappik, A. et al., Fully synthetic human combinatorial antibody libraries (HuCAL) based on modular consensus frameworks and CDRs randomized with trinucleotides, *J. Mol. Biol.*, 296, 57, 2000.
125. Baylis, A. D., Why glyphosate is a global herbicide: Strengths, weaknesses and prospects, *Pest Manag. Sci.*, 56, 299, 2000.
126. Rubio, F. et al., Comparison of a direct ELISA and an HPLC method for glyphosate determinations in water, *J. Agric. Food Chem.*, 51, 691, 2003.

127. Lee, E. A. et al., Linker-assisted immunoassay and liquid chromatography/mass spectrometry for the analysis of glyphosate, *Anal. Chem.*, 74, 4937, 2002.
128. Clegg, B. S., Stephenson, G. R., and Hall, J. C., Development of an enzyme-linked immunosorbent assay for the detection of glyphosate, *J. Agric. Food Chem.*, 47, 5031, 1999.
129. Kramer, K., Unpublished data. 2003.
130. Daugherty, P. S. et al., Quantitative analysis of the effect of the mutation frequency on the affinity maturation of single chain Fv antibodies, *Proc. Natl. Acad. Sci. USA*, 97, 2029, 2000.
131. Hanes, J. et al., Picomolar affinity antibodies from a fully synthetic naive library selected and evolved by ribosome display, *Nat. Biotechnol.*, 18, 1287, 2000.
132. Wyatt, G. M. et al., Alteration of the binding characteristics of a recombinant scFv anti-parathion antibody: 1. Mutagenesis targeted at the VH CDR3 domain, *Food Agric. Immunol.*, 11, 207, 1999.
133. Chambers, S. J. et al., Alteration of the binding characteristics of a recombinant scFv anti-parathion antibody: 2. Computer modeling of hapten docking and correlation with ELISA binding, *Food Agric. Immunol.*, 11, 219, 1999.
134. Kusharyoto, W. et al., Mapping of a hapten-binding site: Molecular modeling and site-directed mutagenesis study of an anti-atrazine antibody, *Protein Eng.*, 15, 233, 2002.
135. Kramer, K., Evolutionary affinity and selectivity optimization of a pesticide-selective antibody utilizing a hapten-selective immunoglobulin repertoire, *Environ. Sci. Technol.*, 36, 4892, 2002.
136. Winter, G. et al., Making antibodies by phage display technology, *Annu. Rev. Immunol.*, 12, 433, 1994.
137. Xu, J. L. and Davis, M. M., Diversity in the CDR3 region of VH is sufficient for most antibody specificities, *Immunity*, 13, 37, 2000.
138. Reynaud, C. A. et al., Introduction: What mechanism(s) drive hypermutation? *Semin. Immunol.*, 8, 125, 1996.
139. Steele, E. J., Rothenfluh, H. S., and Blanden, R. V., Mechanism of antigen-driven somatic hypermutation of rearranged immunoglobulin V(D)J genes in the mouse, *Immunol. Cell Biol.*, 75, 82, 1997.
140. Jolly, C. J. et al., The targeting of somatic hypermutation, *Semin. Immunol.*, 8, 159, 1996.
141. Green, N. S., Lin, M. M., and Scharff, M. D., Somatic hypermutation of antibody genes: A hot spot warms up, *Bioessays*, 20, 227, 1998.
142. Hock, B. et al., Stabilisation of immunoassays and receptor assays, *J. Mol. Catal. B: Enzym.*, 7, 115, 1999.
143. Ewert, S. et al., Biophysical properties of human antibody variable domains, *J. Mol. Biol.*, 325, 531, 2003.
144. Grant, S. D., Porter, A. J., and Harris, W. J., Comparative sensitivity of immunoassays for haptens using monomeric and dimeric antibody fragments, *J. Agric. Food Chem.*, 47, 340, 1999.
145. Kramer, K. et al., A generic strategy for subcloning antibody variable regions from pCANTAB 5 E into pASK85 permits the economic production of F_{ab} fragments and leads to improved recombinant protein stability, *Biosen Bioelectron.*, 17, 305, 2002.
146. Rau, D., Kramer, K., and Hock, B., Single-chain Fv antibody-alkaline phosphatase fusion proteins produced by one-step cloning as rapid detection tools for ELISA, *J. Immunoassay Immunochem.*, 23, 129, 2002.
147. Kim, I. S. et al., Green fluorescent protein-labeled recombinant fluobody for detecting the picloram herbicide, *Biosci. Biotechnol. Biochem.*, 66, 1148, 2002.
148. Yokozeki, T. et al., A homogeneous noncompetitive immunoassay for the detection of small haptens, *Anal. Chem.*, 74, 2500, 2002.
149. Lange, S., Schmitt, J., and Schmid, R. D., High-level expression of the recombinant, atrazine-specific Fab fragment K411B by the methylotrophic yeast Pichia pastoris, *J. Immunol. Meth.*, 255, 103, 2001.
150. Longstaff, M. et al., Expression and characterisation of single-chain antibody fragments produced in transgenic plants against the organic herbicides atrazine and paraquat, *Biochim. Biophys. Acta.*, 1381, 147, 1998.
151. de Wildt, R. M., Antibody arrays for high-throughput screening of antibody-antigen interactions, *Nat. Biotechnol.*, 18, 989, 2000.

3 In Vitro Monoclonal Antibody Production: Academic Scale

Frances Weis-Garcia

CONTENTS

3.1 INTRODUCTION

The ability to produce homogenous antibody preparations was first documented 30 years ago by Kohler and Milstein when they generated of the first hybridomas [1] via the somatic cell fusion of antibody secreting B cells with an immortalized myeloma cell line. Since this groundbreaking discovery, scientists have harnessed the specific antigen binding capacity of monoclonal antibodies (MAbs) in a multitude of in vitro and in vivo applications. MAbs are key reagents in ELISAs (enzyme linked immunoabsorbent assays), western blots, immuopreciptations, immunostaining of cells and tissues, immunoaffinity purification as well as neutralization, activation, or depletion of cells in vitro and in vivo [2–5]. Additionally, MAbs are utilized in the clinical setting where they facilitate radioimmunoimaging procedures and also are the cornerstone of such immunotherapeutics as the breast cancer drug Herceptin® [6–13].

All of these applications have created an enormous demand for MAbs at a reasonable cost. Some procedures require less than a milligram, whereas others consume many milligrams and even a few grams. Low concentration MAb stocks (μg/ml) have always been relatively easy to obtain because hybridomas naturally secrete significant amounts MAb into the culture media, about 10–50 µg/ml [2]. For many years, the only efficient means of generating highly concentrated MAb stocks (mg/ml) or large quantities (few mg to a gram) was via ascites [2,14]. This process involves inducing an inflammatory response in the peritoneal cavity of mice with Pristane, preparing it for the hybridoma cells that are injected 10–14 days later. MAb rich ascitic fluid accumulates in the peritoneal cavity and is drained (tapped) before the animal experiences difficulty walking. This procedure is considered painful and distressing to the mouse [15,16]. In vitro MAb production systems became a viable alternative to ascites by providing the capacity of generating milligrams of functional MAb per ml of culture media. This result prompted some European countries and the United States to ban or restrict the use of mice for ascites productions, respectively [17,18].

Currently, many academic core facilities use commercially available bioreactors to meet their goal of producing up to a gram of MAb at concentrations that rival ascites (1–10 mg MAb/ml [2]) with relative ease and at a reasonable cost. The following section summarizes in vitro MAb production approaches that can be brought into laboratories competent in tissue culture techniques with minimal effort. A detailed protocol for culturing hybridomas in dialysis-based, dual-chamber bioreactors is provided. A comparison of the productivity of and MAb concentrations obtained from the various systems is shown.

3.2 IN VITRO MAB PRODUCTION SYSTEMS

Academic core facilities and research laboratories have many techniques available for culturing hybridomas and producing MAbs. The best system for an individual group strikes a balance between the quantity (micrograms, milligrams, or grams) and quality (purity, concentration) of the MAb required for a particular line of experiments, along with the cost incurred by and technical skill level required for the chosen method. This section provides a springboard from which an academic or other laboratory can begin to match its needs to some of the in vitro MAb production options currently available. Table 3.1 outlines four MAb production scales or levels, facilitating

TABLE 3.1
MAb Production Scales

	Level 1	Level 2	Level 3	Level 4
MAb Amount	1–5 mg	5–20 mg	20–1,000 mg	100 to > 1,000 mg
mg MAb /ml	0.01–0.10	0.01–0.40	0.20–3.5	0.50–6.00
MAb Purity	Negligible	Negligible	20%–80%	20%–80%
Culture Vessels	Standard tissue culture vessels	Gas permeable bags	Dialysis-based, dual chamber bioreactors	Dialysis-based, hollow fiber bioreactors
	(Wells, flasks, roller bottles and spinner flasks)	(VetraCell™ and Wave Bioreactor®)	(miniPERM® and CELLine™)	(FiberCell™)
Technical Skill	Low	Low to moderate	Moderate	Moderate
Media Adaptation	Not necessary	Not necessary	Recommended	Recommended
Principle Cost (other than standard tissue culture equipment)	FBS	Bioreactor, FBS (if used), rocker (if needed)	Bioreactor, FBS (if used), roller (if needed)	Bioreactor, FBS (if used), pumping device
Labor	Negligible	Negligible	< 2 hours / week / hybridoma	< 3 hours / week / hybridoma

a comparison of the various methodologies. These are loose groupings rather than rigid categories that can easily blend into each other depending upon how well a hybridoma produces and which culture conditions are employed, i.e., the use of serum free media, supplements, the amount of fetal bovine serum (FBS), and the culture vessel chosen.

3.2.1 STANDARD TISSUE CULTURE VESSELS

The easiest and cheapest way to produce a functional MAb stock simply is to allow the hybridoma to grow to saturation in standard media such as RPMI-1640 or DMEM-high glucose supplemented with 5%–10% FBS. Several references exist outlining protocols using tissue culture flasks, roller bottles, and spinner flasks [2,19–21]. As with all tissue culture techniques, it is best to start with a viable ($>90\%$), mycoplasma free, actively growing hybridoma culture. Then, grow the hybridoma until the media have been exhausted, at which point many of the cells may have died. Media containing phenol red becomes yellow (acidic) soon after the cells have reached saturation, usually about 1.2 to 2×10^6 cells/ml. Centrifuge the cells and particulate debris away from the exhausted/over-conditioned media at $900 \times g$, and collect the MAb-enriched supernatant that is ready for use and/or storage. Roller bottles and spinner flasks provide larger culture volume options. The shearing force of these approaches can be an issue for some hybridomas, either preventing the cells from growing at all or significantly decreasing the viability of the culture. In these cases, commercially available supplements such as CellProtect(Greiner Bio-One) can minimize this negative effect.

MAb concentrations in the exhausted/over-conditioned media vary greatly between individual hybridomas, easily ranging from 0.01 to 0.10 µg/ml [2,22]. MAbs produced in this basic manner are more than concentrated enough for many immunological techniques where the required working MAb concentration is very low. The bovine serum albumin (BSA) and immunoglobulins (Ig) from the FBS added to the initial culture media as well as the cellular debris are irrelevant contaminants for many applications because they will not interfere with MAb-antigen binding and are washed away during the protocol. Western blots, immunoprecipitations, and fluorescence-activated cell scanning or sorting (FACS) need very little MAb, a working concentration around ~1 µg/ml [3,4] is usually sufficient. Therefore, exhausted/over-conditioned media provide researchers a relatively effortless means of generating microgram to milligram quantities of impure MAb.

When small amounts of pure MAb are required, such as MAbs conjugated to fluorophores for FACS, the MAb can be affinity purified with a Protein G coupled resin, (Pierce Biotechnology, GE Healthcare, etc.). The main caveat to purifying MAbs from media supplemented with FBS is the contaminating bovine Ig will co-purify. When necessary, this can be eliminated by using serum replacement supplements (i.e., Nutridoma-CS from Roche Applied Science) or removing the bovine Ig from the FBS with a Protein G coupled resin before addition to culture media. Hybridoma serum-free media formulas are another option that are readily available from all the major media suppliers (i.e., Invitrogen, Hyclone, Sigma, and Beckon Dickenson). MAb yields from serum-free cultures may be slightly diminished, but this is not as much of an issue when used in conjunction with one of the bioreactor systems described in the following sections.

3.2.2 GAS PERMEABLE BAGS

The VectraCell™ (BioVecta) and Wave Bioreactor® (Wave Biotech) are gas permeable bags in which hybridoma cells can be grown to saturation, in batch culture, inside a standard CO_2 incubator. The VectraCell™ was previously sold as the iMAB, for which a study has been published [23]. The Wave Bioreactor® can also function as a continuous perfusion culture. Under this format, cells are grown inside the gas permeable bag, conditioned media is removed and fresh culture media is added back at set times for the duration of the production (Figure 3.1). Unlike the VectraCell™, the Wave Bioreactor® is rocked to maximize oxygen and carbon dioxide exchange. VectraCell™

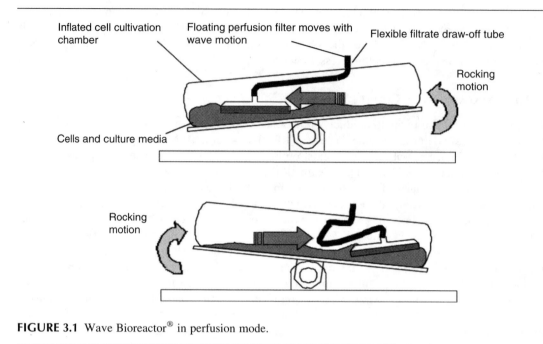

FIGURE 3.1 Wave Bioreactor® in perfusion mode.

product literature cites that hybridoma cells cultured with Hybridoma-SFM (Invitrogen) in its gas permeable bags yield 0.01–0.4 mg MAb/ml. Similar concentrations (0.14–0.26 mg MAb/ml) are also cited for the Wave Bioreactor® [24]. The major advantage to gas permeable bags over standard tissue culture in flasks or roller bottles is reduced labor without decreasing the volume of MAb enriched culture media. Gas permeable bags as large as 3 liters are available, allowing bulk production with much less labor than traditional culture methods.

3.2.3 DIALYSIS-BASED BASIC BIOREACTORS

The concept of segregating hybridoma cells away from the bulk of the culture media with a dialysis membrane has been evaluated for over 20 years [25–36]. Today, many academic core facilities have found that commercially available dialysis-based bioreactors meet their goal of producing up to a gram of MAb at concentrations that rival ascites (1–10 mg MAb/ml [2]) with relative ease and at a reasonable cost.

The IVSS classic miniPERM® (VWR International) and Integra CELLine™ (Fisher Scientific) are dialysis-based, dual chamber bioreactors [33–38]. The two chambers are separated by a dialysis membrane that has a 12.5 or 10 KDa molecular weight cut off (MWCO), respectively (Figure 3.2 and Figure 3.3). The smaller compartment (cell chamber) confines the hybridoma and MAb it secretes in a minimum volume of culture media, allowing both to become very concentrated. The miniPERM® production module is a 35 ml culture chamber, and the cell compartment of the CELLine™ can hold 10–15 ml (CL350) or 15–40 ml (CL1000), depending upon the unit. The larger chamber (nutrient module) houses the bulk of the media upon which the hybridoma feeds and into which its small metabolic by-products are able to dilute away from the cells. The miniPERM® can hold up to 550 ml. The CELLine™ CL350 and CL1000 have maximum volumes of 350 and 1000 ml, respectively. The bottom of each cell compartment has a gas permeable support that permits oxygen and carbon dioxide exchange with the atmosphere of the CO_2 incubator. Gases exchange also occurs with media in the nutrient compartment through a vented cap on

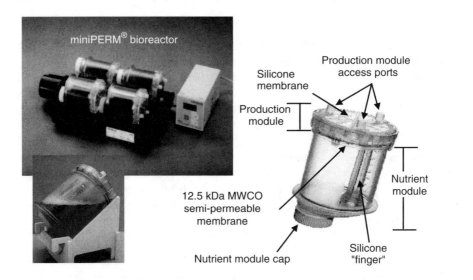

FIGURE 3.2 miniPERM® Bioreactor system.

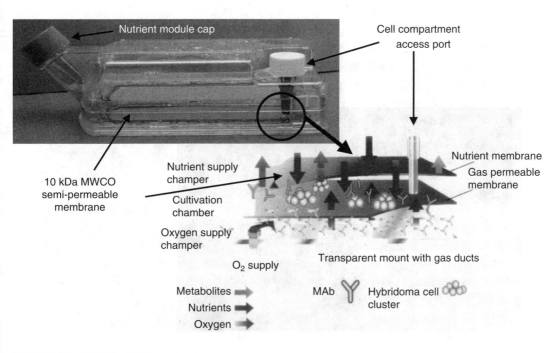

FIGURE 3.3 CELLine™ Bioreactor system.

both systems. In addition, miniPERM® has a silicone finger that protrudes into the nutrient module to maximize gas exchange.

When the media from the cell compartment is harvested it will contain 0.20–3.5 mg MAb/ml and occasionally as high as 5 mg MAb/ml [34,37] (and data within this chapter). After the initial harvest, the miniPERM® and CELLine™ vessels can produce an average of 40–70 mg of MAb per week per bioreactor (data in this chapter), respectively. The MAb in the harvested media or bioreactor supernatant is rather pure, ranging from 20% for poor producers to 80% for high secreting hybridomas (data not shown). Although there are subtle differences in the overall design of the two systems, the main one is motion. The modified roller bottle style of the miniPERM® requires a simple turning devise inside a standard CO_2 incubator. The CELLine™ provides a stationary culture environment similar to regular culture flasks. They can be stacked on top of each other, minimizing space requirements. Barring contamination, membrane rupture, or an out growth of non-producing cells, both systems can continuously produce MAb over several weeks, if not months, providing research groups with a steady source of concentrated MAb or allowing hundreds of milligrams of MAb to accumulate over time with minimal labor.

3.2.4   DIALYSIS-BASED HOLLOW FIBER BIOREACTORS

Hollow fiber bioreactors take dialysis-based MAb production to the next level [39–42]. This design involves culturing hybridoma cells with in a cylindrical cartridge through which many porous, hollow fibers pass (Figure 3.4). Media pumps through the densely packed fiber bundles, providing nutrients and removing cellular metabolites, much like the capillaries of the circulatory system. The concept of using these artificial capillaries to culture various kinds of cells in vitro at high densities was published [43] before the first hybridomas were even generated [1]. Hybridoma cells and MAbs remain concentrated within the extracapillary space (ECS) because, as with the dual chamber bioreactors, the pore size is only large enough for small proteins, cellular metabolites, and ions to pass. Barring contamination, these systems can last 6–12 months.

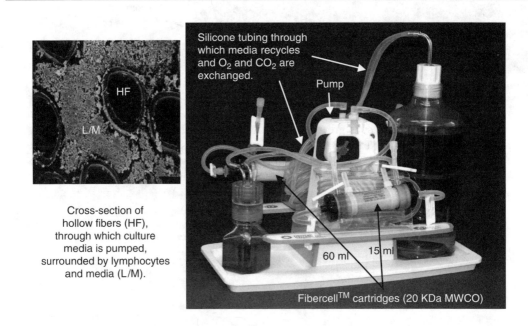

FIGURE 3.4 Fibercell™ Bioreactor system.

The FiberCell™ (FiberCell Systems, Bellco Biotechnology), previously marketed as the CellMAX™, is probably the best-suited hollow fiber system on the market for academic environments or small start-up biotechnology companies [42,44]. It is reasonably priced (approximately $425 for a 12 ml cartridge), the harvests contain highly concentrated MAb, and the technical skill level required to bring it into a group is moderate, assuming tissue culture is an established technique within the group. A few reusable bottle top fittings and a pumping station that fits inside standard CO_2 incubators are all that is needed to get started (Figure 3.4b). The MAb production protocol for this system has been refined over the years and is available from FiberCell Systems with a set of video segments demonstrating how to use the unit. Because hollow fiber bioreactors can cultivate more than 10^8 cells/ml, it is not surprising that MAb concentrations of 0.5–5.0 mg MAb/ml are achieved with this system on a regular basis [42]. This result translates into 18–180 mg MAb/week when the 12 ml cartridge is harvested every 2 days and 100–1000 mg MAb/week from the 70 ml cartridge when similarly cultured. Like the other dialysis-based systems, the FiberCell™ also yields a low endotoxin product. FiberCell™ cartridges and CELLine™ flasks are reusable for culturing the same hybridoma cell line later, but care must be taken to thoroughly flush the cartridge or flask before storage.

In contrast to the miniPERM® and CELLine™ systems, the molecular weight cut off of the membrane in the FiberCell™ averages 20 KDa. Therefore, 30 KDa proteins can cross the fiber, however, not as efficiently as a 10 KDa one. The pores in the miniPERM® and CELLine™ membranes are much smaller, restricting molecules larger than 12.5 and 10 KDa, respectively, from crossing the membrane with 95% efficiency. Consequently, the hollow fibers of the FiberCell™ allow transforming growth factor-β (TGF-β) to dialyze away from the hybridoma into the nutrient media, lowering its concentration around the cells. Hybridomas are generally growth inhibited by this multifunctional cytokine [42,45]. Consequently, hybridomas that secrete TGF-β and have an intact TGF-β signaling pathway should grow better in FiberCell™ cartridges than the dual-chamber models.

3.2.5 ADDITIONAL PRACTICALITY OF MAB-ENRICHED BIOREACTOR SUPERNATANTS

Because the miniPERM®, CELLine™, and FiberCell™ systems can all yield hundreds of milligrams of very concentrated, rather pure MAb, purification is potentially much easier. Many experimental applications require purified MAbs because the MAb either needs to be biochemically manipulated (i.e., conjugated or proteolytic digested) or the other components of the spent media (i.e., growth factors produced by the hybridoma, cellular debris, or serum proteins) interfere with the experiment that utilizes the MAb (i.e., in vitro cell studies). Starting with concentrated material always simplifies downstream purification and MAb supernatants from these three bioreactors provide this edge. If a harvest is greater than 1 mg MAb/ml, the MAb can often be purified by sodium sulfate precipitation (Table 3.2). More than 60% of mouse and rat IgG antibodies tested by the Sloan Kettering Institute-Monoclonal Antibody Core Facility (SKI-MACF) at Memorial Sloan Kettering Cancer Center can be purified to >95% purity with this very inexpensive and low-tech method (see the legend of Table 3.2 for the protocol). When it works, the average yield is around 45%. When sodium sulfate precipitation fails to purify the MAb, the harvested bioreactor supernatants can be purified by Protein G chromatography because the time necessary to load the MAb onto the Protein G resin decreases as the MAb concentration increases. Because the time the MAb remains on the column tends to be inversely proportional to the yield, shorter load times should also maximize recovery.

MAb harvested as bioreactor supernatants can be used for in vivo animal studies as well. For example, scientists can manipulate the immune system of a mouse with MAbs (i.e., depleting CD4 positive T cells with clone GK1.5) or test a MAb for pre-clinical therapeutic efficacy by injecting the MAb bioreactor supernatants into animals without any further processing. These types of experiments require large quantities (0.1–2 mg MAb/mouse) of concentrated

TABLE 3.2
MAb Sodium Sulfate Precipitation Efficiency from Bioreactor Supernatant

Species	Isotype	Average		# MAbs		
		Yield	Purity	Precipitated to >95% purity	Precipitated	Tested
Mouse	IgG1	45%	90%	8	12	17
	IgG2a	39%	85%	2	6	8
	IgG2b	45%	91%	1	3	6
Rat	IgG1	53%	95%	0	1	2
	IgG2a	50%	93%	2	11	15
	IgG2b	45%	91%	2	7	10
	Overall	46%	91%	63% that precipitated were >95% pure	69% of MAbs tested precipitated	
Hamster	Ig	39%	91%	2	8	9
Mouse	IgM	43%	95%	1	1	2
Rat	IgM	50%	91%	0	1	2

The results presented in this table are from sodium sulfate precipitations performed by the Sloan Kettering Institute Monoclonal Antibody Core Facility at Memorial Sloan-Kettering Cancer Center. Bioreactor supernatants produced as described in the final section of this chapter were pooled to no less than 1 mg MAb/ml and centrifuged at $13,000 \times g$ for 15 min. Two sequential precipitations were performed at 37°C, for 30 min., under constant slow stirring. The first precipitation occurred with 18%–20% (w/v) sodium sulfate and the second used 16%–18% (w/v). Each precipitation was centrifuged at $13,000 \times g$ for 15 min. at room temperature, and the pellet was resuspended in ultra pure water to either one third the original volume for the first pellet and for the second pellet, one fifth of the original volume. The MAb was finally dialyzed into phosphate buffered saline. This protocol is a revised version of one obtained from Dr. Peter Cresswell for purification of IgGs and IgMs from ascitic fluid.

MAb (>1mg MAb/ml) that have very little, if any, endotoxins. Bioreactor supernatants meet all these objectives. MAbs produced with the miniPERM® and CELLine™ bioreactors at the SKI-MACF contain less than 1 endotoxin unit/ml. The FiberCell™ is another good option when low endotoxin MAbs are the ultimate goal.

3.3 MINIPERM® VERSUS CELLINE™

The Integra CELLine™ and IVSS classic miniPERM®, available through Fisher and VWR, respectively, are basic dialysis-based bioreactors (Figure 3.2 and Figure 3.3) whose simple designs cultivate hundreds of millions of cells producing milligrams of MAb in each harvest. (For an additional overview of these systems, refer to the Dialysis-based Basic Bioreactor section of this chapter, the manufacturer's web sites, and refs [33–38].)

The classic miniPERM® comes with a 12.5 KDa MWCO, a 35 ml cell module, and a nutrient module that holds up to 550 ml. The CELLine™ can be purchased in two sizes, the CL 350 and CL 1000. The capacities are 10–15 ml or 15–40 ml of cells, and the cells feed off a maximum of 350 ml or 1 L of media, respectively. Dialysis membrane rupture is an issue for both systems and care should be taken not to over stretch the membrane.

There is a significant difference in the cost of these two systems. Because the miniPERM® is basically a modified roller bottle, it requires an initial investment in a turning devise to rotate the bioreactor within a CO_2 incubator. The list price for an assembled miniPERM® is approximately $460 each, when purchased as a kit containing four complete units. Buying cases of nutrient or production modules separately and assembling them on site can save some money. The CELLine™ CL 1000 units cost about \$175 per bioreactor when purchased as a case of three. CELLine™ bioreactors have an added monetary advantage because they can be recycled, as can the FiberCell™, by simply rinsing the cell chamber and storing it in phosphate-buffered saline. Because the rate of contamination in recycled CELLine™ flasks is higher, it is wise not to re-inoculate units carried in parallel with cells from a recycled bioreactor.

3.3.1   MAB PRODUCTION PROTOCOL FOR THE CELLINE™ AND MINIPERM® BIOREACTORS

The SKI-MACF has produced MAbs from over 160 hybridomas using the protocol outlined in Section 3.3.1.1. All hybridomas are the products of a mouse myeloma fusion with mouse, rat, or hamster B cells. Most hybridomas easily adapt to the production media and produce concentrated MAb when cultured in the bioreactors as described below. Less than 1% of all attempted adaptations either fail to grow or simply do not produce enough MAb (>0.2 mg MAb/ml) on the first attempt. Re-derivation of a stable homogeneous cell population by sub-cloning usually resolves poor producers as well as some that fail to grow. Additional support for the protocol comes from the fact that every one of the MAbs produced with this method has been functional in the assays for which the MAb was selected, whether they be in vitro or in vivo techniques.

Dr. Howard Petrie initially developed the in vitro MAb production protocol for the mini-PERM® when he began the SKI-MACF in 1994. What is outlined here reflects subsequent refinements to the basic dual-chamber bioreactor protocol. Minor system specific differences for each bioreactor unit are noted within the protocol. Similar production methods are published [33–38]. Integra Biosciences, the CELLine™ manufacturer, also provides a detailed culture protocol.

3.3.1.1   Production Media

This method utilizes the serum-free media Hybridoma-SFM (Invitrogen) to minimize the fetal bovine serum costs, as well as, the protein content it adds, specifically bovine Igs. Other serum-free or animal-free formulas can be used such as BD Cell™ MAb, but no other media currently tested seems to work as well as Hybridoma-SFM. Although most hybridomas adapt into Hybridoma-SFM alone, this protocol supplements it with 0.5% FBS, providing a smooth transition into the bioreactor for 99% of hybridomas without having to pamper the culture or resort to sub-cloning. To minimize the amount of bovine Ig from the FBS contaminating the MAb harvests, FBS lots pre-selected for ultra low levels of bovine Ig are preferred. They are available from most commercial vendors upon requested (Invitrogen, Hyclone, Sigma, etc.) and generally are not significantly more expensive. They may cost 25%–40% more money, but when the media only contain 0.5% FBS, the impact on the total cost of the media is minimal. It is a good laboratory practice to screen FBS stocks before purchasing many vendors supply trial lots for testing. With that said, the SKI-MACF has not noticed significant differences in ultra-low bovine Ig lots of FBS it has tested. Antibiotics can be added, but it is not recommended because if a baseline bacterial contamination is propagated for a harvest or two, before it takes over the culture vessel, those harvests will contain significant amounts of endotoxin from the bacteria.

3.3.1.2   Media Adaptation for MAb Production

To make the adaptation process easier, passage an actively growing, mycoplasma-free hybridoma culture daily (usually a 1:1 ratio) into Hybridoma-SFM plus 1% FBS, referred here as Standard Growth Media. For the rare hybridoma that has a harder time adjusting, carrying it as a small culture in a 24-well plate where the cells tend to stay settled on the plastic is usually enough pampering to permit later scale up into flasks. During adaptation into the Standard Growth Media, a sub-population of cells that do not produce MAb may occasionally overgrow the culture. It is wise to confirm that the culture is still producing MAb prior to scaling up for freezing and bioreactor inoculation. Sub-cloning of the parental stock after adaptation to 1% FBS usually resolves any problem of lost or low expression by finding a stable producer or eliminating the non-productive subpopulation, respectively.

3.3.1.3   Bioreactor Inoculation

On Friday (Day 0), prepare and inoculate each bioreactor type accordingly:

1. Assemble the miniPERM® in a sterile hood, making sure that all clips are properly set; otherwise, media will leak out of the nutrient module. See package insert for further instructions.
2. When adding or withdrawing media and cells into or from the cell compartments use a:
    a. 60 cc syringe for the miniPERM®,
    b. 25 ml pipette for the CELLine™.
3. Pre-wet the membrane by adding 25 ml of Production Media (Hybridoma-SFM plus 0.5% FBS) to the nutrient compartment and 15 ml to the cell compartment.
    a. If the dialysis membrane of the miniPERM® is ruptured, the first 25 ml added to the nutrient module will leak into the cell module. It is best to see if this happens before adding the cells.
4. Inoculate the cell compartment with the hybridoma resuspended in Production Media accordingly:
    a. CL 1000: $35–65 \times 10^6$ viable cells in a final volume of 15 ml ($2–5 \times 10^6$ viable cells/ml)
       * If the membrane is pre-wet at least one hour before inoculation, 30 ml of cells at the same density noted above can be added from the start.
       * Ensure that all large and as many small bubbles as possible are removed from the cell compartment because bubbles decrease the surface area through which molecules on either side of the membrane can be exchanged.
    b. miniPERM®: $50–100 \times 10^6$ viable cells in a final volume of 30 ml cells
       * Anti-shearing supplements are not necessary when using Hybridoma-SFM. Productions can be started with fewer cells. If the cell number is significantly less, the first scheduled harvest will have lower than normal MAb concentrations and may not be worth harvesting.
5. Fill the nutrient compartment with production media
    a. miniPERM® with 350 ml
    b. CELLine™ with 650 ml.
6. Incubate until Monday (Day 3) at 37°C and 7% CO_2
    a. The miniPERM® needs to rotate, 5 rpm is sufficient. The cells simply need to remain suspended.

3.3.1.4   Bioreactor Harvest

Harvest the MAb rich media and passage of the hybridoma cells (re-inoculation) every Monday (Day 3, 10, 17, etc.) and Thursday (Day 7, 14, 21, etc.) accordingly:

1. For the CELLine™ only, open the nutrient cap slightly to allow air to flow in and out.
2. Resuspend the cells by removing and replacing the cells a few times.
3. Transfer all the cells to a sterile 50 ml conical tube.
4. Perform a cell count and determine viability.

5. Centrifuge the harvested cells and media at $300 \times g$ for 3–5 min.
6. Transfer the supernatant to a new sterile 50 ml conical tube. Centrifuge the supernatant at $900 \times g$ to remove large particulate matter and aliquot the harvested bioreactor supernatant into sterile tube(s) for analysis, use, and/or storage.
7. Resuspend the cell pellet in fresh Production Media and re-inoculate each bioreactor unit with 30 ml of cells accordingly:
   a. CL 1000: 12×10^6 or 6×10^6 viable cells/ml for Monday or Thursday re-inoculations, respectively.
   b. miniPERM®: 33% or 25% of the cells for Monday or Thursday re-inoculations, respectively.
   c. If the harvest does not contain $>7 \times 10^6$ viable cells/ml for the miniPERM® or 12×10^6 viable cells/ml from the CELLine™, consider putting all the media and cells back into the cell compartment until the next scheduled harvest. If this occurs after the first harvest, the cells are not fairing well. If an alternative hybridoma is not an option, increasing the serum or adding serum replacement supplements such as Nutridoma-CS (Roche Applied Sciences), Hybridoma Fusion and Cloning Supplement (Roche Applied Sciences) or Hybridoma Cloning Factor (BioVeris) can promote cell survival in the bioreactor.
8. Incubate, as in step 6 of Section 3.3.1.3, until the following day when the nutrient compartment media is changed.

3.3.1.5 Nutrient Compartment Media Exchange

Aspirate all the media from the nutrient compartment and replace with fresh Production Media every Tuesday (Day 4, 11, 18, etc.) and Friday (Day 8, 15, 22, etc.).

- miniPERM®: always receives 350–550 ml, depending upon how low the pH has gone since the last feeding.
- CELLine™: receives 650 ml when there are three days between harvest and 1 L whenever there are four days between harvests.

Do not use a Pasteur pipette because it can rupture the membrane directly or, if it breaks inside, the glass chips can theoretically cause small holes.

Some hybridomas consume more glucose and change the pH faster than this feeding timetable provides. Therefore, more frequent media changes may be necessary if the color is more yellow than orange or if the glucose goes below 200 on a standard blood glucometer (available at any pharmacy). With this twice-a-week exchange, most hybridomas should have more than enough nutrients to last 3–4 days. Regular testing of glucose levels is usually not necessary.

3.3.2 TIPS FOR LARGER SCALE MAB PRODUCTIONS

When more than a gram of MAb is required, the FiberCell™ is a great option. If a few hundred milligrams to a gram is needed, multiple miniPERM® or CELLine™ units can be carried in parallel, or the systems can be maintained for months until the target amount is reached. Each research group will need to decide which is more important to them, saving the cost of the bioreactor(s) and spending more time and labor maintaining the continuous production over months or saving time and some labor by simultaneously culturing more bioreactors with the same hybridoma. When multiple units are concurrently producing the same MAb, it is best to scale up early because some hybridomas do lose expression over time. Remember, if recycled CELLine™ bioreactors are used for multiple bioreactor productions, it is best not to use the cells harvested from the re-used units for the re-inoculations because they have a higher chance of carrying contamination that has not yet been noticed. This is especially true if antibiotics are employed.

3.3.3 MAb Production Comparison

The following data come from MAb productions performed by the SKI-MACF using the protocol described in the previous section. Each harvest was evaluated by SDS-polyacrlymide electrophoresis (10% pre-cast gel from BioRAD) under reducing conditions. The gels were stained with Commassie Blue R-250, de-stained, digitally scanned, and analyzed by densitometry using BioRAD's Quantity One or its predecessor, Molecular Analyst. Purified mouse polyclonal IgG from Sigma (5, 10, and 15 µg per lane) was used as the standard to determine MAb concentrations and purity.

Data in Figure 3.5 and Figure 3.6 comes from at least 85 different hybridomas (mouse/mouse, rat/mouse and hamster/mouse) cultured in the miniPERM® and CELLine™ bioreactors, giving rise to over 160 independent productions for each system. Most of these productions were not performed side by side. Data for the fraction that were performed side by side is consistent with the numbers presented here. Figure 3.5 and Figure 3.6 demonstrate how well each system produces MAb under the culture conditions outlined in the previous section. Complementary graphs between each figure have identical x- and y-axes to facilitate the comparison. The average value (black) and standard deviations (grey) for each harvest (day 3, 6, 10, 13, 17, etc.) are plotted for the following variables monitored during a MAb production in the SKI-MACF for quality control purposes:

- MAb concentration of the bioreactor supernatant (top, center)
- Mg MAb/day/bioreactor (top, right)
- viable cells/ml harvested (bottom, left)
- cellular viability (bottom, center)
- total number of cells (bottom, right).

The milligrams of MAb being produced /day/bioreactor is a measure of productivity. It is a useful tool to estimate when a particular production may reach the target amount and be terminated prior to actually having quantitated the MAb in the final harvest.

One significant difference between MAb productions in these two bioreactors is the concentration of viable cells at the time of harvest (bottom, left). Using this production protocol, the CELLine™ can sustain more viable cells per ml ($\sim 22.5 \times 10^6$/ml) than the miniPERM® ($\sim 15 \times 10^6$/ml). This is not due to the cells being more viable overall (bottom center graphs) because, on average, the cells plateau around 70% viability by the second scheduled harvest (day 6) in both bioreactors. Rather, a higher total number of cells is achieved in the CELLine™ (bottom, right). Not surprisingly, maintaining more viable cells per ml correlates well with a higher MAb concentration in CELLine™ than miniPERM® harvests, approximately 1.5 mg MAb/ml verses 1.0 mg MAb/ml, respectively. Because the inoculation and harvest volumes for both units were identical (30 ml), this translates into the CELLine™ being a more productive system (average about 12 mg/day/bioreactor) than the miniPERM® (average around 7 mg/day/bioreactor) when this protocol is used. M.P. Bruce et al. [36] also found more concentrated harvests from the CELLine™ when they measured functional MAb, but they came to the opposite conclusion with respect to productivity because, in their analysis, the miniPERM® yielded more MAb in the end. This discrepancy is likely because they compared the CL 350 with classic miniPERM® that have significantly different cell compartment volumes, 15 and 35 ml, respectively. Although more concentrated MAb was harvested from a CELLine™ production, twice the volume was harvested from the miniPERM® and inturn more total MAb. The data in this chapter compare the CL 1000 and classic miniPERM® using identical harvest volumes. This is most likely the source of the differing conclusions.

The overall higher MAb concentrations and yields from the CELLine™ hold true when looking at individual clones as well. Table 3.3 provides a comparison of 30 hybridomas. 30% of these

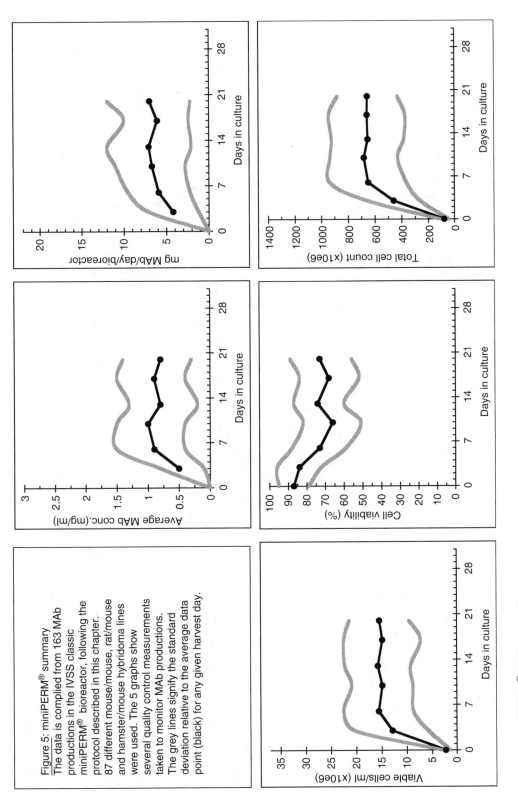

Figure 5: miniPERM® summary
The data is complied from 163 MAb productions in the IVSS classic miniPERM® bioreactor, following the protocol described in this chapter. 87 different mouse/mouse, rat/mouse and hamster/mouse hybridoma lines were used. The 5 graphs show several quality control measurements taken to monitor MAb productions. The grey lines signify the standard deviation relative to the average data point (black) for any given harvest day.

FIGURE 3.5 miniPERM® MAb production summary.

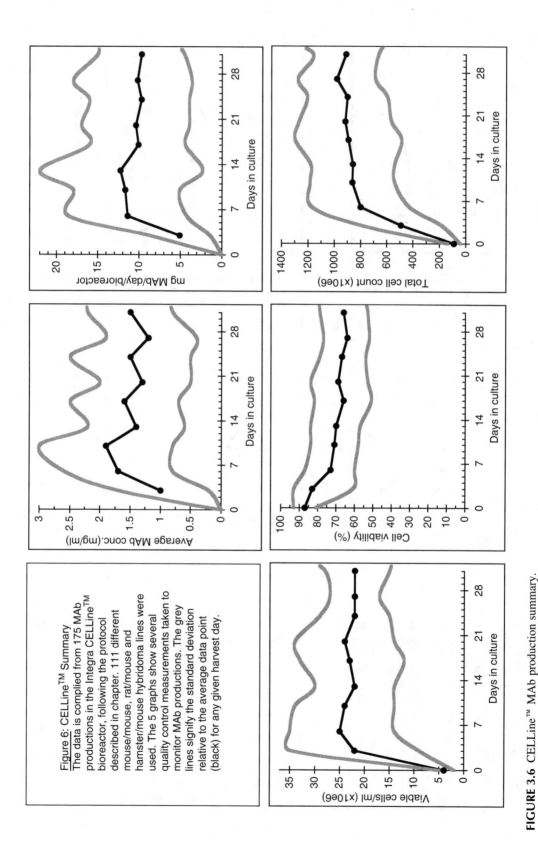

Figure 6: CELLine™ Summary
The data is compiled from 175 MAb productions in the Integra CELLine™ bioreactor, following the protocol described in chapter. 111 different mouse/mouse, rat/mouse and hamster/mouse hybridoma lines were used. The 5 graphs show several quality control measurements taken to monitor MAb productions. The grey lines signify the standard deviation relative to the average data point (black) for any given harvest day.

FIGURE 3.6 CELLine™ MAb production summary.

TABLE 3.3
CELLine™ versus miniPERM®: MAb Concentration and Production Efficiency

Hybridoma Clone	Isotype	Preferred Bioreactor	mg MAb/ml		mg MAb/day/bioreactor	
			miniPERM	CELLine	miniPERM	CELLine
9E10	Mouse IgG 1	–	0.6	0.7	4.2	4.6
DA6.147	Mouse IgG 1	C	0.7	1.7	5.7	10.5
Private 1	Mouse IgG 1	C	0.5	1.4	4.2	6.3
Private 2	Mouse IgG 1	C	1.1	3.0	9.1	21.8
Private 3	Mouse IgG 1	C	0.9	1.9	6.1	12.8
Private 4	Mouse IgG 1	–	0.9	1.0	6.7	7.4
104–2	Mouse IgG 2a	C	0.8	1.8	5.5	13.4
A–20	Mouse IgG 2a	C	1.1	2.2	6.1	14.5
W6/32	Mouse IgG 2a	C	0.6	1.7	4.6	9.7
Y3P (cl.10)	Mouse IgG 2a	C	0.3	0.7	2.1	4.6
PK136 (cl.30)	Mouse IgG 2b	–	1.1	1.0	8.1	7.6
11B11	Rat IgG 1	C	1.1	2.6	9.1	15.2
PC61.5.3	Rat IgG 1	C	0.6	1.0	4.1	6.2
1C10	Rat IgG 2a	C	1.2	3.3	8.6	24.3
1D3	Rat IgG 2a	C	0.2	0.5	1.8	5.5
53.6–72	Rat IgG 2a	C	1.7	3.1	11.6	17.8
FGK45	Rat IgG 2a	–	1.0	1.5	8.1	8.3
KT3	Rat IgG 2a	–	1.3	1.9	10.1	13.9
Private 5	Rat IgG 2a	–	2.0	2.1	13.4	13.6
2.43	Rat IgG 2b	M	1.4	1.2	10.0	7.5
2.4G2	Rat IgG 2b	C	1.1	2.4	8.4	17.1
GK1.5	Rat IgG 2b	–	0.9	1.3	6.7	8.1
Private 6	Rat IgG 2b	–	1.0	1.2	8.6	10.1
Private 7	Rat IgG 2b	C	1.0	2.0	8.4	18.6
Rb6–8C5	Rat IgG 2b	C	2.9	3.9	22.3	36.4
Ter 119	Rat IgG 2b	C	0.9	3.3	6.6	26.4
RA3–3A1/6.1	Rat IgM	M	1.4	0.6	9.2	3.9
Private 8	Hamster Ig	–	0.3	0.5	2.5	3.4
H57–597	Hamster Ig	C	0.9	2.6	6.1	19.3
MR-1	Hamster Ig	–	1.3	1.2	10.1	9.6
	Average		1.0	1.8	7.6	12.6

Comparison of 30 different hybridomas produced as described in this chapter. The data presented are the average MAb concentration or mg MAb/day/bioreactor for the specific hybridoma cultured either in the miniPERM™ or the CELLine™ bioreactors. A dash in column 3 denotes productions where there is no difference between the two systems (30%). An "M" in the third column signifies highlights productions where the cells produced more MAb when cultured in the miniPERM™ (7%). A "C" in the third column denotes hybridomas that secrete more MAb in the CELLine™ (63%). "*Private*" is used in the place of some hybridoma names because the Sloan Kettering Institute, Monoclonal Antibody Core Facility at Memorial Sloan Kettering Cancer Center produced these clones at the exclusive request of certain researchers and does not have permission to disclose the clone name.

hybridomas produced MAb equally well in either bioreactor (noted by a dash). Only 7% of these 30 hybridomas produced more MAb in miniPERM® (M) and 63% (C) were more productive in the CELLine™. For the clones that preferred the CELLine™, the MAb concentration and productivity were, on average, double what was harvested from the miniPERM®, specifically 2.1 versus 1.0 mg MAb/ml and 15.2 versus 7.3 mg/day/bioreactor, respectively. In addition, all 162 MAbs produced by the SKI-MACF are functional in the in vitro or in vivo application for which they were selected.

3.4 CONCLUDING REMARKS

There are more in vitro MAb production options available beyond what has been outlined here such as stirred-tank, airlift, fixed-bed, fibrous-bed, acoustic wave, and more advanced hollow fiber systems [22,37]. This chapter has concentrated on those techniques that most laboratories already performing basic tissue culture techniques could bring into their group with minimal effort. They include producing MAb as exhausted/over-conditioned media from hybridomas grown in flasks, roller bottles, or spinner flasks as well as producing MAb as bioreactor supernatants using the commercially available miniPERM®, CELLine™, and FiberCell™ systems. A comparison of the basic dual-chamber dialysis bioreactors indicates the CELLine ™ is a more cost effective and productive system than the miniPERM® when following the culture protocol described here. The FiberCell™ cartridges yield more concentrated MAb faster with a little more cost and not much more effort and technical skill. Therefore, this more advanced dialysis-based system is a good option for those groups who consistently need tens of milligrams more than a gram of MAb. No matter how much MAb an academic core facility or research laboratory needs to produce, in vitro MAb production methods have developed well enough that ascites should rarely, if ever, be employed.

REFERENCES

 1. Kohler, G. and Milstein, C., Continuous culture of fused cells secreting antibodies with predefined specificity, *Nature*, 256, 495–497, 1975.
 2. Harlow, E., and Lane, D., Antibodies. In *A Laboratory Manual*, Cold Spring Harbor Press, Cold Spring Harbor, NY, pp. 271–275, 1988
 3. Harlow, E., and Lane, D., Using antibodies. In *A Laboratory Manual*, Cold Spring Harbor Press, Cold Spring Harbor, NY, pp. 271–273, 1999.
 4. Givan, A. L., Flow Cytometery. In *First Principles*, Wiley-Liss, New York, pp. 1–273, 2001.
 5. Lipman, N. S., Jackson, L. R., Trudel, L. J., and Weis-Garcia, F., Monoclonal versus polyclonal antibodies: Distinguishing characteristics, applications and information resources, *ILAR J.*, 46, 258–268, 2005.
 6. Britz-Cunningham, S. H. and Adelstein, S. J., Molecular targeting with radionuclides: State of the science, *J. Nucl. Med.*, 44, 1945–1961, 2003.
 7. Francis, R. J. and Begent, R. H. J., Monoclonal antibody targeting therapy: An overview, in *Targeted Therapy for Cancer*, Syrigos, K. N. and Harrington, K. J., Eds., Oxford University Press, New York, pp. 30–46, 2003.
 8. Kipriyanov, S. M., Generation of antibody molecules through antibody engineering, in *Methods in Molecular Biology, Recombinant Antibodies for Cancer Therapy, Methods and Protocols*, Welschof, M. and Krauss, J., Eds., Humana Press, Totowa, NJ, pp. 3–25, 2003.
 9. Waldman, T. A., Immunotherapy: Past, present and future, *Nature Med.*, 9, 269–277, 2003.
 10. Borjesson, P. K. E., Postemab, E. J., deBreea, R., Roos, J. R., Leemansa, C. R., Kairemod, K. J. A., and van Dongen, J. A. M. S., Radioimmunodetection and radioimmunotherapy of head and neck cancer, *Oral Oncol.*, 40, 761–772, 2004.
 11. Casadevall, A., Dadachova, E., and Pirofski, L. A., Passive antibody therapy for infectious diseases, *Nature Rev. Microbiol.*, 2, 265–703, 2004.
 12. Chester, K., Pedley, B., Tolner, B., Violet, J., Mayer, V., Sharma, S., Boxer, G., Green, A., Nagl, A., and Begent, R., Engineering antibodies for clinical applications in cancer, *Tumor Biol.*, 25, 91–98, 2004.
 13. Pelegrin, M., Gros, L., Dreja, H., and Piechaczyk, M., Monoclonal antibody-based genetic immunotherapy, *Curr. Gene Ther.*, 4, 347–356, 2004.
 14. Jackson, L. R., Trudel, L. J., Fox, J. G., and Lipman, N. S., Monoclonal antibody production in murine ascites: II. Production characteristics, *Lab. Anim. Sci.*, 49, 81–86, 1999.
 15. Jackson, L. R., Trudel, L. J., Fox, J. G., and Lipman, N. S., Monoclonal antibody production in murine ascites: I. Clinical and pathologic features, *Lab. Anim. Sci.*, 49, 70–80, 1999.

16. Peterson, N. C., Behavioral, clinical and physiologic analysis of mice used for ascites monoclonal antibody production, *Comp. Med.*, 50, 516, 2000.

17. Falkenberg, F. W., Monoclonal antibody production: Problems and solutions 74th Forum in Immunology, *Res. Immunol.*, 149, 542–547, 1998.

18. Institute for Laboratory Animal Research, National Research Council, *Monoclonal antibody production A Report of the Committee on Methods of Producing Monoclonal Antibodies*, National Academy Press, Washington DC, 1999.

19. Freshney, I. R., Cell culture of animal cells. In *A Manual of Basic Technique*, 3rd ed., Wiley, New York, 1994 pp 369–377

20. Voigt, A. and Zintl, F., Hybridoma cell growth and anti-neuroblastoma monoclonal antibody production in spinner flasks using a protein-free medium with microcarriers, *J. Biotechnol.*, 68, 213–226, 1999.

21. Yokoyama, W. M., Monoclonal antibody supernatant and ascites fluid production, in *Current Protocols in Immunology*, Coligan, J. E., Kruisbeek, A. M., Margulies, D. H., Shevach, E. M., and Strober, W., Eds., Wiley, New York, 2000. Unit 2.6.

22. Yang, S. T., Luo, J., and Chen, C., A fibrous-bed bioreactor for continuous production of monoclonal antibody by hybridoma, *Adv. Biochem. Engin./Biotechnol.*, 87, 61–96, 2004.

23. Lipski, L. A., Witzleb, M. P., Reddington, G. M., and Reddington, J. J., Evaluation of small to moderate scale in vitro monoclonal antibody production via the use of the i-MAb gas-permeable bag system, *Res. Immunol.*, 149, 547–552, 1998.

24. Ohashi, R., and Hamel, J.F., Perfusion culture in the wave reactor for the production of a monoclonal antibody. Waterside Conference, 2003, http://www.wavebiotech.com/pdfs/literature/mit_waterside_2003.pdf.

25. Sjogren-Jansson, E. and Jeansson, S., Large-scale production of monoclonal antibodies in dialysis tubing, *J. Immunol. Methods*, 84, 359–364, 1985.

26. Kasehagen, C., Linz, F., Kretzmer, G., Scheper, T., and Schugerl, K., Metabolism of hybridoma cells and antibody secretion at high cell densities in dialysis tubing, *Enzyme. Microb. Technol.*, 13, 873–881, 1991.

27. Sjogren-Jansson, E., Ohlin, M., Borrebaeck, C. A., and Jeansson, S., Production of human monoclonal antibodies in dialysis tubing, *Hybridoma*, 10, 411–419, 1991.

28. Witt, S., Ziegler, B., Blumentritt, C., Schlosser, M., and Ziegler, M., Production of monoclonal antibodies in serum-free medium in dialysis tubing, *Allerg Immunol. (Leipz)*, 37, 67–74, 1991.

29. Pannell, R. and Milstein, C., An oscillating bubble chamber for laboratory scale production of monoclonal antibodies as an alternative to ascitic tumours, *J. Immunol. Methods*, 146, 43–48, 1992.

30. Falkenberg, F. W., Hengelage, T., Krane, M., Bartels, I., Albrecht, A., Holtmeier, N., and Wuthrich, M., A simple and inexpensive high density dialysis tubing cell culture system for the in vitro production of monoclonal antibodies in high concentration, *J. Immunol. Methods*, 165, 193–206, 1993.

31. Falkenberg, F. W., Weichert, H., Krane, M., Bartels, I., Palme, M., Nagels, H. O., and Fiebig, H., In vitro production of monoclonal antibodies in high concentration in a new and easy to handle modular minifermenter, *J. Immunol. Methods*, 179, 13–29, 1995.

32. Jaspert, R., Geske, T., Teichmann, A., Kassner, Y. M., Kretzschmar, K., and L'age-Stehr, J., Laboratory scale production of monoclonal antibodies in a tumbling chamber, *J. Immunol. Methods*, 178, 77–87, 1995.

33. Nagel, A., Koch, S., Valley, U., Emmrich, F., and Marx, U., Membrane-based cell culture systems—an alternative to in vivo production of monoclonal antibodies, *Dev. Biol. Stand.*, 101, 57–64, 1999.

34. Trebak, M., Chong, J. M., Herlyn, D., and Speicher, D. W., Efficient laboratory-scale production of monoclonal antibodies using membrane-based high-density cell culture technology, *J. Immunol. Methods*, 230, 59–70, 1999.

35. Scott, L. E., Aggett, H., and Glencross, D. K., Manufacture of pure monoclonal antibodies by heterogeneous culture without downstream purification, *BioTechniques*, 31, 666–668, 2001.

36. Bruce, M. P., Boyd, V., Duch, C., and White, J. R., Dialysis-based bioreactor systems for the production of monoclonal antibodies-alternatives to ascites production in mice, *J. Immunol. Methods*, 264, 59–68, 2002.

37. Dewar, V., Voet, P., Denamur, F., and Smal, J., Industrial implementation of in vitro production of monoclonal antibodies, *ILAR J.*, 46, 307–313, 2005.

38. Entrican, G., and Young, G., Growing hybridomas, In *Methods in Molecular Biology: Immunological Protocols*, 3rd Ed., Vol. 295, Humana Press, Inc., Totowa, NJ, pp. 55–70, 2005.
39. Schonherr, O. T., van Gelder, P. T., van Hees, P. J., van Os, A. M., and Roelofs, H. W., A hollow fiber dialysis system for the in vitro production of monoclonal antibodies replacing in vivo production in mice, *Dev. Biol. Stand.*, 66, 211–220, 1987.
40. Dhainaut, F., Bihoreau, N., Meterreau, J. L., Lirochon, J., Vincentelli, R., and Mignot, G., Continuous production of large amounts of monoclonal immunoglobulins in hollow fibers using protein-free medium, *Cytotechnology*, 10, 33–41, 1992.
41. Jackson, L. R., Trudel, L. J., Fox, J. G., and Lipman, N. S., Evaluation of hollow fiber bioreactors as an alternative to murine ascites production for small scale monoclonal antibody production, *J. Immunol. Methods*, 189, 217–231, 1996.
42. Cadwell, J. J. S., New developments in hollow fiber cell culture, *Amer. Biotech. Lab.*, July 14–17, 2000.
43. Knazek, R. A., Gullino, P. M., Kohler, P. O., and Dedrick, R. L., Cell culture on artificial capillaries: An approach to tissue growth in vitro, *Science*, 178 (56), 65–66, 1972.
44. Jackson, L. R., Trudel, L. J., and Lipman, N. S., Small scale monoclonal antibody production in vitro: Methods and resources, *Lab. Animal*, 28, 38–50, 1999.
45. Richards, S. M., Garman, R. D., Keyes, L., Kavanagh, B., and McPherson, J. M., Prolactin is an antagonist of TGF-b activity and promotes proliferation of murine B cell hybridomas, *Cell Immunol.*, 184, 85–91, 1998.

4 Antibodies to Heavy Metals: Isolation, Characterization, and Incorporation into Microplate-Based Assays and Immunosensors

Diane A. Blake, Robert C. Blake II, Elizabeth R. Abboud, Xia Li, Haini Yu, Alison M. Kriegel, Mehraban Khosraviani, and Ibrahim A. Darwish

CONTENTS

4.1   INTRODUCTION

A heavy metal can be defined as any metallic chemical element with a relatively high density. At certain sites, heavy metals, including mercury (Hg), cadmium (Cd), lead (Pb), uranium (U), and copper (Cu) have been released into the soil or groundwater at levels that are toxic to plant and animal life. Unlike carbon-based environmental contaminants that will eventually be degraded and removed, metals that are deposited in the environment can persist for very long periods of time. When bound to soils and sediments, metals are relatively non-toxic except to bottom-feeding aquatic life [1]. Unfortunately, any number of ill-defined episodes, including changes in weather patterns and hydrology, changes in soil or water pH, or release of organics into the environment can mobilize metals and greatly increase their toxicity. Sites contaminated with heavy metals require long-term stewardship that includes regular monitoring for heavy metal mobilization.

Of the analytical techniques [2–6] currently used to measure metal ions, inductively coupled plasma atomic emission spectroscopy (ICPAES) is the most commonly employed [7,8]. This technique routinely affords good sensitivity; however, the analysis is expensive, and large sample volume is often required to achieve maximum sensitivity. In addition, the time required for analysis can be prolonged because the samples must be analyzed sequentially and may sit in a queue for long periods of time. Unless coupled with mass spectrometry, high performance liquid chromatography, or capillary electrophoresis [9,10], ICPAES cannot provide information on metal ion speciation. Analyses conducted using such coupled instruments are particularly expensive and often less sensitive than those conducted using each individual instrument.

Antibody-based techniques are alternative approaches for metal ion analyses. These approaches are attractive to local and governmental agencies because they have significant advantages over the traditional analytical methods. Immunoassays are remarkably quick, reasonably portable to the analysis site, and simple to perform. Because most of the technology of an immunoassay is built into the antibody that is at the core of this detection system, sample pretreatment is usually minimal, and the instrumentation designed to transduce the antibody binding event to a measurable signal can be relatively small and inexpensive. The immunoassay and instrument can also be formatted for high throughput analysis. In addition, studies have shown that the use of immunoassays during remediation processes can reduce analysis costs by 50% or more [11]. Although most of the commercial immunoassays are directed toward complex organic chemicals, peptides, and proteins [12–17], this technique is theoretically applicable to any analyte, including a metal ion, if a suitable antibody can be generated.

Over the past ten years, this laboratory has generated over 20 monoclonal antibodies directed toward different chelated metal ions by immunizing mice with the corresponding metal–chelate complexes. This article reviews the techniques used to generate and characterize these antibodies and describe how they may be formulated into field portable assays designed to provide inexpensive, portable, near real-time analyses of heavy metals in environmental samples.

4.2 PREPARATION OF THE IMMUNOGEN

4.2.1 CHOOSING THE CORRECT CHELATOR

Heavy metals do not elicit immune responses; however, this laboratory and others discovered that when a heavy metal was added to a chelator covalently conjugated to a protein, the resulting metal–chelate complex comprised an epitope that was immunogenic and could elicit an immune response [18–24]. In general, the tighter the complex between a metal and a chelator, the greater the likelihood that the complex will survive in vivo to stimulate an immune response. This laboratory has used published critical stability constants of metal–chelate interactions to guide the choice of chelator for a particular metal ion [25]. It should be noted, however, that the published constants have been extrapolated to extremes of pH where the chelator is completely ionized (the conjugate base); at physiological pH and ionic strength, the actual stability constant can be lower than the published value by several orders of magnitude. The chelators used in these studies, (e.g., EDTA, DTPA, cyclohexyl-DTPA, 2,9-dicarboxyl-1,10-phenanthroline) have relatively open structures (Figure 4.1) that facilitate diffusion-controlled binding of the metal to the chelator. This rapid formation of the metal–chelate complex is required in a fast antibody-based assay. Crown ether derivatives like DOTA (1,4,7,10-tetraazacyclodecane-1,4,7,10-tetraacetic acid) often require incubation of the metal and chelate for prolonged periods of time (more than 24 h) and/or at high temperatures for metal–chelate complex formation. Such requirements have limited the enthusiasm for crown ether-based chelators in immunoassays designed for near real-time analysis. Because metal–chelate complexes by themselves are too small to stimulate the immune system, high molecular weight immunogens were constructed using bifunctional chelators as shown in Figure 4.1. Although the bifunctional chelators shown in Panels a–c were originally synthesized as a means to target radionuclides to cancer cells [26–28], our team discovered that they could also be used to generate antibodies to metal–chelate complexes. In the team's experience, the ease of developing a monoclonal antibody with specificity for a single metal (or pair of metals) was highly dependent upon the bifunctional chelator used to generate the immunogen. Immunogens prepared from bifunctional EDTA (Figure 4.1, Panel a) were relatively efficient in eliciting antibodies with specificity to Hg^{2+}, Cd^{2+}, and Cu^{2+} but failed to produce antibodies with specificity to Pb^{2+}. It hypothesized that the Pb^{2+} was released from the chelate during the immune processing of the antigen, and when bifunctional chelators with higher affinity for Pb^{2+} were chosen (Figure 4.1, Panels b and c), the team was able to generate antibodies that recognized Pb^{2+}–chelate complexes [23,29,30]. Despite several attempts, antibodies with specificity for chelated UO_2^{2+} using immunogens prepared from the bifunctional chelators shown in Panels a–c of Figure 4.1 could not be generated. A study of the binding affinities of several chelators for UO_2^{2+} finally led the team to the phenanthroline derivative shown in Panel d. This chelator that bound to UO_2^{2+} approximately 1000

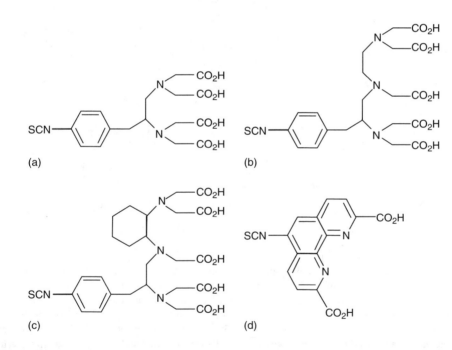

FIGURE 4.1 Bifunctional chelators used in the generation of antibodies to metal-chelate complexes. (a) Bifunctional derivative of EDTA (1-(4-isothiocyanatobenzyl)ethylenediamine-*N,N,N′,N′*-tetraacetic acid), available commercially from Dojindo Molecular Technologies, Inc., Gaithersburg, MD, was used to generate antibodies with specificity for chelated Cd^{2+}, Hg^{2+}, and Cu^{2+}. (b) Bifunctional derivative of DTPA, (2-(4-isothiocyanatobenzyl)-diethylenetriamine-*N-N-N′-N′-N″*-pentaacetic acid), available commercially from Macrocyclics, Dallas, TX, was used to generate antibodies with specificity for chelated Pb^{2+} and Co^{2+}. (c) Bifunctional derivative of cyclohexyl-DTPA (2-*p*-isothiocyanatobenzyl-*trans*-cyclohexyldiethylenetriamine *N-N-N′-N′-N″*-pentaacetic acid, CHX-A-DTPA) was synthesized as described in Brechbiel and Gansow. This compound was used to generate antibodies to chelated Pb^{2+}. (d) Bifunctional derivative of 2,9-dicarboxyl-1,10-phenanthroline (5-isothiocyanato-1,10-phenanthroline-2,9-dicarboxycylic acid), was synthesized as described in Blake et al. and used to generate antibodies to chelated UO_2^{2+}. (From Brechbiel, M. W. and Gansow, O. A., *J. Chem. Soc. Perkin Trans.*, 1, 1992; and Blake II, R. C., Pavlov, A. R., Khosraviani, M., Ensley, H. E., Kiefer, G. E., Yu, H., Li, X., and Blake, D. A., *Bioconjug. Chem.*, 15, 2004.)

times more tightly than did EDTA also turned out to be very efficient in eliciting antibodies with high affinity and specificity for chelated UO_2^{2+} [24].

4.2.2 Synthesis of Protein Conjugates

Once the optimal chelator had been chosen, the immunogen was prepared as shown in Figure 4.2. The metal-free bifunctional chelator was first mixed with a 1.1-fold molar excess of the metal of interest, then with the carrier protein. Although our team has always followed strict procedures for working under metal-free conditions during the preparation of its immunogens [31], loading the chelator with metal before synthesis and purification of the conjugate ensured that the majority of the chelate molecules coupled to the carrier protein contained the metal of interest, not some undesirable trace metal scavenged by the chelate during synthesis. The isothiocyanate on the bifunctional chelate reacted with unprotonated amino groups on the carrier protein to form a stable thioureido group [32]; because the protonated amino groups are unreactive, the reaction rate increased with increasing pH. Care was taken to choose a buffer that did not contain primary amino groups (for example, Tris) as they would interfere with the coupling reaction. In the team's

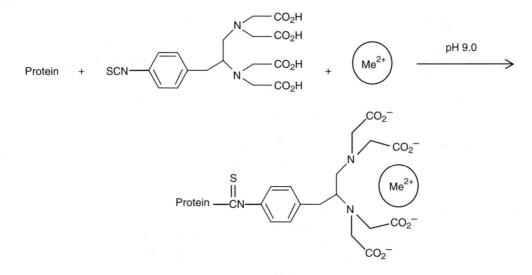

FIGURE 4.2 Preparation of metal–chelate–protein conjugates for use as immunogens and immunoassay reagents. For conjugation using bifunctional EDTA- and DTPA-based reagents, a 1.1-fold molar excess of a single metal ion (Cd^{2+}, Pb^{2+}, Cu^{2+}, Hg^{2+}) was added to the chelator before reaction with the protein. Conjugation reactions normally contained 10–20 mg/mL of protein and 1.5–3.0 mM of the bifunctional chelate, buffered to pH 9.0–9.5 with HEPES or borate (50–100 mM). After reaction for 18 h at room temperature, the metal–chelate–protein conjugate was separated from low molecular weight reaction products by ultrafiltration or size exclusion chromatography, and the extent of substitution on the epsilon amino groups of lysyl residues was determined as described in Cavot and Tainturier. Immunogens and immunoassay reagents were prepared using keyhole limpet hemocyanin (KLH) and bovine serum albumin (BSA), respectively, as the protein carrier. The addition of UO_2^{2+} to the bifunctional phenanthroline derivative (Panel d in Figure 4.1) at the mM concentrations required for the conjugation reaction caused precipitation. These protein conjugates were therefore synthesized and stored as metal-free derivatives, and UO_2^{2+} (1.1-fold molar excess) was added after dilution of the protein conjugates to working concentrations (0.5–50 µg/mL). (From Blake, D. A., Chakrabarti, P., Khosraviani, M., Hatcher, F. M., Westhoff, C. M., Goebel, P., Wylie, D. E., and Blake II., R. C., *J. Biol. Chem.*, 271, 1996; Blake II, R. C., Delehanty, J. B., Khosraviani, M., Yu, H., Jones, R. M., and Blake, D. A. *Biochemistry*, 42, 2003; Khosraviani, M., Blake II, R. C., Pavlov, A. R., Lorbach, S. C., Yu, H., Delehanty, J. B., Brechbiel, M. W., and Blake, D. A. *Bioconjug. Chem.*, 11, 2000; Chakrabarti, P., Hatcher, F. M., Blake II, R. C., Ladd, P. A., and Blake, D. A. *Anal. Biochem.*, 217, 1994; and Cayot, P. and Tainturier, G. *Anal. Biochem.*, 249, 1997.)

experience, both borate and HEPES buffers have performed well in the synthesis of protein conjugates [24,30,33].

The standard practice in this laboratory has been to synthesize and characterize the conjugate that will serve as the immunogen (metal–chelate–keyhole limpet hemocyanin) and the carrier protein to be employed for screening assays (metal–chelate–bovine serum albumin) at the same time. Keyhole limpet hemocyanin has been the carrier protein of choice for immunization; this highly immunogenic protein stimulates T cell-dependent responses and is often used as a carrier for low molecular weight antigens [34,35]. Bovine serum albumin was chosen as the carrier protein for screening assays because it is structurally dissimilar to keyhole limpet hemocyanin, readily soluble, and available in a highly purified, metal-free form. For some assay formats, the team also synthesized metal–chelate-horseradish peroxidase conjugates [36,37]; in those experiments, the pH of the reaction mixture was lowered to 8.5 to preserve enzyme activity. After an overnight reaction, the conjugates were purified to remove low molecular weight reaction products, and the extent of free lysine substitution was characterized using the trinitrobenzenesulfonic acid method [38]. Carrier proteins with between 15% and 50% of derivatized lysyl residues normally provided adequate performance as immunogens and in assays.

4.3 ANIMAL IMMUNIZATION, CELL FUSION, AND HYBRIDOMA SCREENING

Purified immunogens (20–50 µg/animal/injection) were emulsified in adjuvant and intraperitoneally injected into mice at approximately two-week intervals as shown in Figure 4.3. Both Ribi™ and TiterMax™ adjuvants were well-tolerated by the animals and evoked similar immune responses. Immunization of six animals has permitted this group to evaluate the polyclonal antibody response and to choose the animal with the best response for subsequent preparation of hybridomas. Once the animal had been sacrificed, splenocytes were prepared and fused with myeloma cells (X63-Ag 8.653 or SP2/Ag 14 lines). The reagents required for cell fusion and subsequent cell expansion were supplied in the ClonaCell-HY™ kit available from StemCell Technologies (Vancouver, BC). Individual hybridoma colonies were selected from the semisolid growth medium and cultured in microwell plates. Antibody activity was assessed directly from the culture supernatants of the microwell plates.

Supernatant screening was the most critical step in the identification of monoclonal antibodies useful for environmental analysis. The team's screening protocol was designed to rapidly and efficiently identify antibodies with the following properties

1. Little or no reactivity with the metal-free chelator. Most metals in environmental samples exist as complexes with other compounds (inorganic anions, humics, etc.). In environmental immunoassays, relatively high concentrations of metal-free chelator were used to pull the metal ion of interest away from the natural complexants in the environmental sample and into a form recognized by its antibodies (the metal–chelate complex). During the monoclonal antibody screening procedures, only those clones that synthesized antibodies with low affinity for the metal-free chelator were expanded and further characterized. During more than ten years of study on the binding properties of these monoclonals, our team has discovered that antibodies that bound very tightly to the metal-chelate complex (with sub-nanomolar equilibrium dissociation constants, K_d's) were also more likely to bind with relatively high affinity to the metal-free chelator [24]. Antibodies that bound to the metal-loaded chelate at least fifty times more tightly than to the metal-free chelator could usually be formulated into useful environmental assays.
2. Strong reactivity with the metal–chelate complex of interest. A second characteristic of the ideal antibody was a tight and specific binding to the metal–chelate complex in the

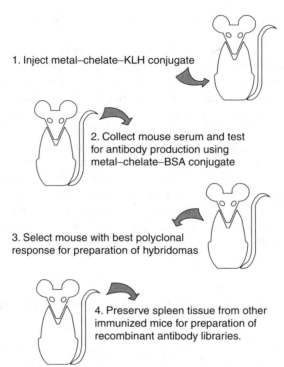

1. Inject metal–chelate–KLH conjugate

2. Collect mouse serum and test for antibody production using metal–chelate–BSA conjugate

3. Select mouse with best polyclonal response for preparation of hybridomas

4. Preserve spleen tissue from other immunized mice for preparation of recombinant antibody libraries.

FIGURE 4.3 Immunization protocol and generation of antibodies. (1) Metal–chelate–KLH conjugates were emulsified in RIBI or TiterMax adjuvant according to the manufacturers' protocols, and 50 μg aliquots of emulsified conjugate were intraperitoneally injected at approximately 2-week intervals. (2) After four–five injections, a small amount of blood (0.05–0.1 mL) was collected from the tail of each animal, and the polyclonal antisera were assessed for the ability to bind to the immobilized metal–chelate–BSA conjugate by indirect ELISA and by the competitive two-step ELISA method shown in Figure 4.4a, below. (3) The animal that displayed the best polyclonal response was sacrificed, and its splenocytes were used in the preparation of hybridomas. (4) Additional immunized animals were reserved until a suitable hybridoma had been generated; these animals were then sacrificed and their spleens preserved for use in preparation of recombinant antibody libraries.

     immunogen. The binding affinity of the antibody to the complex was directly related to the ultimate sensitivity of the final assay. Antibodies that bound to the metal–chelate complex with K_d values of less than or equal to 10^{-8} usually proved useful in immunoassay formats.

3. Little or no binding with metals most likely to contaminate the environmental sample. The specificity of the final assay was also directly related to antibody affinity for other chelated metal ions. In these monoclonal antibody screening protocols, this group looked for at least a ten-fold difference between the affinity of the metal–chelate of interest and the affinity of other metal–ion complexes likely to be present in the sample matrix. Monoclonal antibodies that did not exhibit this ideal behavior could still be formulated into useful assays through the use of masking reagents that specifically bound to interfering metals and removed them from the binding equilibrium [21]. Different distance matrices were also used to compare the molecular shapes of metal–chelate complexes in an attempt to predict which chelate complexes might provide the best metal ion specificity when used as immunogens [39]. This approach has been limited, however, by the scarcity of available crystal structures and the inability of current molecular modeling programs to accurately represent the electron orbitals of heavy metal atoms.

4. Insensitivity to changes in ionic strength and pH. The effects of sample matrix in immunoassay and immunosensor performance has been a challenging problem in the development of environmental immunoassays because it is difficult to predict what types of interfering substances will be encountered in a given sample matrix. In this laboratory, antibody binding at a variety of pHs and ionic strengths have been routinely assessed in an attempt to minimize the effect of the sample matrix and to maximize precision and reproducibility [22,40].

In order to identify those antibodies that exhibit these desirable characteristics, a streamlined screening assay has been developed that has permitted the rapid analysis of the large number of supernatants that are generated during the production and screening of hybridoma clones. Each culture supernatant was diluted with an equal volume of one of the four following reagent mixtures: HBS buffer (10 mM HEPES, pH 7.2, 140 mM NaCl, 10 mM KCl); HBS buffer containing metal-free chelator (10 mM for EDTA or DTPA, 2 μM for DCP); HBS buffer containing the metal-free chelator plus 5–50 ppm of the metal ion used to prepare the immunogen; and HBS buffer containing the metal-free chelate plus 5–50 ppm of a mixture of other metal ions (usually other metals of environmental interest, e.g., Pb^{2+}, Ni^{2+}, Zn^{2+}, Cd^{2+}, and/or Cu^{2+}). HEPES buffer was used in these experiments because of its negligible metal binding capacity [41]. These mixtures were subsequently incubated for an hour in microwells coated with the metal–chelate–BSA conjugate. After a wash step to remove unbound primary antibody, an enzyme-labeled secondary antibody was added, and color was subsequently developed.

Hybridoma supernatants that contained antibodies with desirable characteristics demonstrated patterns of color formation as shown in Figure 4.4a. They bound tightly to the metal–chelate–BSA conjugate coated in the microwell (buffer), and the addition of metal-free chelator (buffer + chelator) either had no effect on the binding or actually stimulated binding to the immobilized conjugate. A decrease in absorbance in the well that contained both chelate and the metal ion that was used in the synthesis of the immunogen (buffer + chelator + metal) indicated that soluble metal–chelate complex was capable of inhibiting binding to the immobilized metal–chelate–BSA conjugate. The final well (buffer + chelator + mixture) was included to assess the metal ion specificity of the culture supernatant. On occasion, the pattern of color formation shown in Figure 4.4b was also detected, indicating that antibodies to metal–chelate complexes other than those used for immunization had been generated by the animal. Hybridomas whose supernatants demonstrated either of the patterns shown in Figure 4.4 were expanded and recloned by limiting dilution. These hybridomas were subsequently cultured either as an ascites or in a CellLine apparatus to generate large quantities of purified antibody for subsequent binding studies and assay development.

4.4 CHARACTERIZATION OF THE BINDING PROPERTIES OF ANTIBODIES TO METAL–CHELATE COMPLEXES

Detailed studies of the binding properties of purified antibodies generated in this laboratory have greatly enhanced the ability to formulate these antibodies into workable assays. The equilibrium dissociation constants of many of the antibodies have been determined for both metal-free and metal-loaded chelators (see Table 4.1 and references therein). These studies have defined the apparent specificity of the antibodies for both the metal ion and the chelator and, in addition, provided an estimate of the likely sensitivity of the eventual immunoassay. It is evident and generally acknowledged that the sensitivity of an immunoassay is directly related to the affinity of the antibody for its antigen: the higher the affinity, the more sensitive the assay. Bimolecular association and unimolecular dissociation rate constants have also been determined for many of the higher affinity antibody–antigen interactions. The values of the individual dissociation rates for

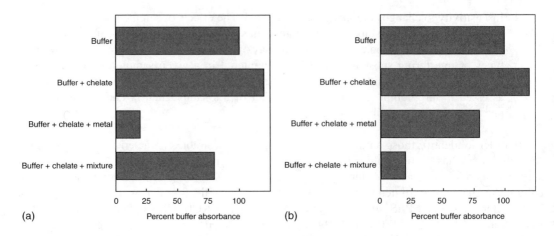

FIGURE 4.4 Representative data from screens of hybridoma supernatants. Panel (a). Pattern of color formation that indicates that a hybridoma is making a monoclonal antibody specific for the soluble form of the metal–chelate complex used in the immunogen. Panel (b). Pattern of color formation that indicates that a hybridoma is making a monoclonal antibody that may have specificity for another soluble metal–chelate complex. In these experiments, hybridoma supernatants were mixed with an equal volume of one of the following: buffer. HBS (10 mM HEPES, pH 7.2, 140 mM NaCl, 10 mM KCl); buffer + chelator. HBS buffer containing metal-free chelate (10 mM for EDTA or DTPA, 2 μM for DCP); buffer + chelator + metal. HBS buffer containing the metal-free chelate plus 5–50 ppm of the metal ion used to prepare the immunogen; buffer + chelator + mixture. HBS buffer containing the metal-free chelate plus 5–50 ppm of a mixture of other metal ions (usually other metals of environmental interest, e.g., Pb^{2+}, Ni^{2+}, Zn^{2+}, Cd^{2+}, and/or Cu^{2+}). These mixtures were subsequently incubated for an hour in microwells coated with metal–chelate–BSA conjugate. After a wash step to remove unbound primary antibody, an enzyme-labeled secondary antibody was added, and color was developed (see Panel a of Figure 4.5, below for more details). Data were expressed as percent of buffer control. For hybridoma screening, these analyses were normally performed as singlets.

these higher affinity interactions control the time required to reach equilibrium; in some rapid sensor formats, the time required to reach equilibrium can be rate-limiting for the assay [42]. Equilibrium and kinetic binding studies for antibodies to metal–chelate complexes have been performed on the KinExA 3000™ instrument using standard assay protocols developed in our group's laboratories [20,29,43,44].

Table 4.1 provides a partial list of the equilibrium dissociation constants determined for fourteen different antibodies generated by the laboratory. For most of the antibodies in the table, equilibrium dissociation constants have been determined for a wide variety of different combinations of individual chelates and metals ions; the reader is directed to the cited references for more comprehensive descriptions of the binding affinities of these antibodies.

4.5 FORMULATION OF ENVIRONMENTAL ASSAYS

All of the environmental immunoassays for metals developed in the laboratory are based upon the same general principal. Metal-free chelator was added to an environmental sample in quantities sufficient to complex the metal of interest. When the antibody used for the analysis had been raised using a chelator that bound relatively non-specifically to a variety of metals (e.g., EDTA or DTPA), the concentration of chelator needed to be high enough to complex all the divalent metal ions in the sample. If the antibody had been raised using a chelate that bound relatively specifically to a subset of metals in the sample (e.g., the interaction of DCP with UO_2^{2+}), then significantly lower

TABLE 4.1
Antibodies and Antibody Fragments that Recognize Metal–Chelate Complexes

Clone Designation	Metal–Chelates Recognized with Highest Affinity[a]	K_d (M)		Reference
		Metal-Loaded Chelator	Metal-Free Chelator	
2A81G5	Cd(II)-EDTA	2.1×10^{-8}	$> 10^{-1}$	[20]
	Hg(II)-EDTA	2.6×10^{-8}		[20]
A4	Hg(II)-EDTA	3.6×10^{-9}	$> 10^{-1}$	[39]
	Cd(II)-EDTA	1.5×10^{-8}		
E5	Cd(II)-EDTA	1.6×10^{-9}		[39]
	Hg(II)-EDTA	3.6×10^{-9}		
rE5Fab	AmBz-EDTA-Cd(II)	1.45×10^{-11}	$> 10^{-1}$	[23]
scFv-A10	Cd(II)-EDTA- Protein conjugate	1.6×10^{-7}	$> 10^{-3}$	[56]
	Cd(II)-EDTA	1.3×10^{-3}	$> 10^{-1}$	
2C12	Pb(II)-CHXDTPA	8.4×10^{-9}	2.1×10^{-7}	[30]
	Pb(II)-DTPA	1.0×10^{-5}	2.1×10^{-3}	
	Ca(II)-DTPA	2.8×10^{-6}	2.1×10^{-3}	
5B2	AmBz-EDTA-Pb(II)	9.5×10^{-10}	8.0×10^{-6}	[29]
	AmBz-EDTA-Ca(II)	8.9×10^{-6}		
	Pb(II)-DTPA	3.9×10^{-7}	1.9×10^{-6}	
	Ca(II)-DTPA	1.1×10^{-6}		
r5B2Fab	Pb(II)-DTPA	2.6×10^{-6}	nd	[23]
	Ca(II)-DTPA	nd		
15B4	Co(II)-DTPA	5.2×10^{-8}	$> 10^{-3}$	[57]
	Ni(II)-DTPA	2.7×10^{-7}		
	Zn(II)-DTPA	2.5×10^{-7}		
8A11	U(VI)-DCP	5.5×10^{-9}	3.7×10^{-6}	[24]
10A3	U(VI)-DCP	2.4×10^{-9}	2.8×10^{-6}	[24]
12F6	U(VI)-DCP	9.1×10^{-10}	1.5×10^{-7}	[24]
1A4	Cu(II)-EDTA	2.2×10^{-10}	6.9×10^{-9}	This publication
4B33	Cu(II)-EDTA	2.2×10^{-9}	1.25×10^{-8}	This publication

[a] The following abbreviations were used in this table: EDTA, ethylenediamine-N,N,N',N'-tetraacetic acid; CHXDTPA, *trans*-cyclohexyldiethylenetriamine N-N-N'-N'-N''-pentaacetic acid; DTPA, diethylenetriamine-N-N-N'-N'-N''-pentaacetic acid: AmBzEDTA, (1-(4-aminobenzyl)ethylenediamine-N,N,N',N'-tetraacetic acid); DCP, 2,9-dicarboxyl-1,10 phenanthroline; rFab, recombinant Fab fragment; scFv, single chain variable fragment.

concentrations of chelate were required. Antibody was added, and the binding of antibody to the metal–chelate complex was allowed to proceed to equilibrium.

This binding event could be transformed into a quantifiable signal in a number of ways. Three of the assay formats used in this laboratory are illustrated in Figure 4.5. The first format, shown in Panel a, is very similar to the assay used to screen hybridomas. First, the metal–chelate–BSA conjugate was coated onto microwell plates. Next, the antibody and soluble metal–chelate derived from the environmental sample were added to the plate, and the immobilized and soluble metal–chelate complex competed for antibody binding sites. In the third step, each micro-well was washed to remove unbound primary antibody. An enzyme-labeled anti-species secondary antibody was subsequently added in step four, and after a second wash to remove unbound secondary antibody, enzyme substrate was added to generate a colored signal. This format has been used in the development of assays for both cadmium and uranium [21,22,40].

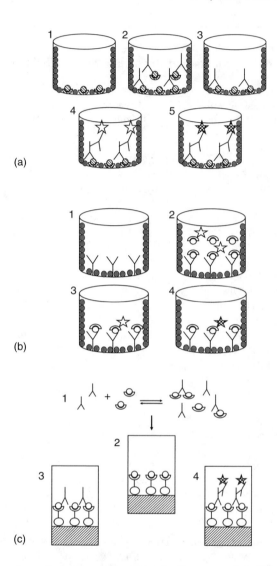

FIGURE 4.5 Assay formats for analysis of metals using antibodies to metal-chelate complexes. Panel (a), two-step competitive ELISA. (1) The wells of a microtitre plate were coated with a metal-chelate-BSA conjugate and nonspecific binding sites were blocked with BSA. (2) The sample (or standard) was mixed with chelator and a known concentration of primary antibody, and the mixture was added to the microwell where the immobilized and the soluble antigens compete for primary antibody binding. (3) Unbound antibody was washed away. (4) Bound primary antibody was detected with an HRP-conjugated anti-species secondary antibody. (5) Reaction color is developed by addition of HRP substrate. Panel (b), one-step competitive ELISA. (1) The wells of a microtitre plate were coated with primary antibody to the metal–chelate complex and non-specific binding sites were blocked with BSA. (2) The sample (containing metal–chelate complex) was spiked with a known concentration of HRP-conjugated metal–chelate complex, and this mixture was allowed to bind to the immobilized primary antibody. Enzyme-linked and unconjugated metal–chelate complexes competed for binding to the immobilized antibody. (3) Excess unbound antigen is washed away. (4) Reaction color is developed by addition of HRP substrate. Panel (c), Sensor Assay. (1) Sample (or standard) containing metal-chelate complex was mixed with known concentrations of primary antibody, and the mixture was allowed to come to equilibrium. (2) The equilibrated antibody–antigen solution was then passed rapidly over the metal–chelate complex immobilized on the surface of beads packed into a capillary bed. (3) Only those antibodies without metal-chelate complex in their binding sites were available to bind to the immobilized antigen; unbound antibody-antigen complex was washed away. (4) Bound primary antibodies captured in the capillary bed were detected by using a fluorescently-labeled anti-species secondary antibody.

Although the format described above could accurately assess metal ion concentrations in environmental samples, the total time for the analysis was 3–4 h because each of the two antibody incubation steps required 60 min. In an effort to streamline the analysis, a one-step assay format [36,37] was developed as shown in Figure 4.5, Panel b. In this format, the purified metal–chelate specific antibody was coated onto the microwell. A metal–chelate–horseradish peroxidase conjugate and the soluble metal–chelate complex were then allowed to compete for antibody binding sites. After a wash step to remove the unbound metal–chelate–horseradish peroxidase conjugate, enzyme substrate was added to generate a colored signal. This format required only one 60-minute incubation step, and the total assay time was reduced from 3 to 4 h to approximately 90 min. This method as originally published (36, 37) required relatively large quantities of purified anti-metal antibody to coat the microwell plate (12–24 µg/96-well plate). In an effort to reduce antibody consumption, the original one-step assay has been modified by first coating the microwell with 2 µg/ml of purified anti-mouse antibody and then adding a 5–10-fold lower concentration of the purified anti-metal–chelate antibody. Control experiments showed that this modification provided data identical to the published method while reducing consumption of the purified anti-chelate antibody by 80%–90% [45].

Representative data obtained using either the two-step or one-step microwell format are shown in Figure 4.6, Panel a. The data obtained in the microwell assays conformed to the log-linear concentration dependence characteristic of such competitive immunoassays. These data were fit to the following equation

$$y = a0 - \frac{(a1 * x)}{(a2 + x)} \tag{4.1}$$

where y was the observed absorbance, x was the metal ion concentration, $a0$ was the absorbance in the absence of metal ion, $a1$ was the difference between the absorbance in the absence of metal ion and at a saturating concentration of metal ion, and $a2$ was the IC_{50}, the metal ion concentration that produced a 50% inhibition of signal.

In an effort to further streamline these analyses for field applications, a sensor format was developed [46–48] as shown in Panel c of Figure 4.5. This sensor consisted of a capillary

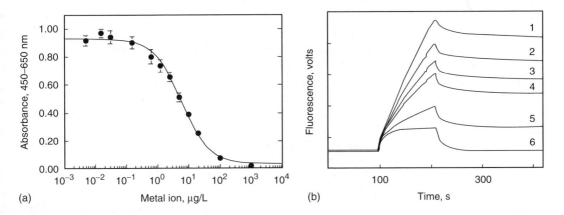

FIGURE 4.6 Representative data from different assay formats. Panel (a). Representative data developed using either the 2-step or 1-step microwell assay formats described in Figure 4.4. As the concentration of metal-chelate complex in the sample increases (x-axis), the relative signal decreases (y-axis). Panel (b). Representative data from the sensor assay format. Each trace is derived from the equilibrium binding between the primary antibody and a single antigen concentration. As the concentration of antigen increases, the fluorescence decreases, so *1* represents a mixture with the lowest concentration of antigen, and *6* represents one with the highest antigen concentration.

flow/observation cell fitted with a microporous screen through which various solutions were drawn under negative pressure. Uniform particles larger than the average pore size of the screen were deposited above the screen to form a microbead column. The antibody was mixed with the soluble metal–chelate complex derived from the environmental sample, and the assay components were allowed to come to equilibrium (1–2 min). This mixture was then passed over the microbead column that contained an immobilized version of the metal-chelate complex. Only those antibody molecules with unoccupied binding sites were available to bind to the immobilized ligand on the surface of the beads; antibodies whose binding sites were already occupied with ligand were not. After a buffer wash to remove unbound primary antibodies from the microbeads, a fluorescently labeled anti-species antibody was then introduced, and excess unbound secondary antibody was removed by a buffer wash. An light-emitting diode (LED) was used to excite the fluorophore bound to the microbeads, and a photodiode measured the amount of fluorescence emanating from the microbead column. The amount of antibody bound to the microbeads was inversely proportional to the amount of metal–chelate complex in the sample because the binding of the antibody to the complex reduced the free antibody concentration in a dose-dependent fashion.

Representative data from the sensor format is shown in Figure 4.6, Panel b. The instrument began monitoring fluorescence starting from the introduction of the antibody-antigen mixture. The instrument response from 0 to 95 s corresponded to the background signal generated while the unlabeled equilibrium mixture was exposed to and washed out of the packed microbead column. The time period from 95 to 215 s corresponded to the exposure of the microbeads to a fluorescently labeled anti-mouse antibody, and the time from 215 s to 400 s corresponded to a buffer wash that removed excess unbound secondary antibody from the microbeads. When the equilibrium mixture contained enough metal-chelate complex to saturate all the enzyme-binding sites (trace 6 in Figure 4.6, Panel b), the instrument response resembled a square wave corresponding to the fluorescence of the secondary antibody during its transient passage through the microbeads in the observation cell. The signal failed to completely return to background levels; this indicated that a small (0.5%) amount of the secondary antibody remained bound nonspecifically to the microbead column. When no metal-chelate complex was added to the antibody mixture applied to the column (trace 1 of Figure 4.6, Panel b), all of the antibody binding sites were available for interaction with the immobilized complex on the microbeads. The instrument response from 95 to 215 s reflected the sum of two contributions: the fluorescence of unbound secondary antibody in the interstitial regions among the beads and that of the labeled secondary antibody that had bound to the primary antibody captured by the immobilized metal–chelate complex. Binding of the secondary antibody was an ongoing process that produced a positive slope in this portion of the curve. When the excess unbound secondary antibody was washed from the beads (instrument response from 215 to 420 s), the signal that remained was the sum of the labeled secondary antibody bound to the primary antibody plus the small amount due to non-specific binding. Instrument traces 2–5 in Figure 4.6, Panel b represent concentrations of soluble metal–chelate complex intermediate between those of zero and saturation.

Theoretically, information about the concentration of metal-chelate complex in an environmental sample could be derived either from the positive slope in the 95 to 215 s interval of the trace or from the average value of the plateau in the interval from 215 to 420 s. The slope parameter have been used to quantify metal–chelate concentrations in alpha and beta prototypes of a field-portable, hand-held sensor and the average plateau values to quantify metal–chelate concentrations using the KinExA 3000™ and the autonomous in-line sensor (see Figure 4.7 and [48,49]).

Four versions of this sensor are presently available in this laboratory as shown in Figure 4.7. Panel a is a photograph of the KinExA 3000™, a research instrument commercially available from Sapidyne Instruments Inc. (Boise, ID) [43]. The user assembles various antibody-antigen equilibrium mixtures and prepares the antigen-coated microbeads. Subsequent steps in the analysis, including preparation of the microbead column, introduction of the antibody solutions and buffer washes, are controlled by a computer interfaced to the instrument. A second version of this sensor,

FIGURE 4.7 Photographs of four sensors available for immunoanalyses. Panel (a). The KinExA 3000™, a research instrument available commercially from Sapidyne Instruments Inc. (Boise, ID). Panel (b). The beta prototype of a handheld sensor designed for field-portable analyses. An updated version of this sensor is presently under development. Panel (c). The beta prototype of an in-line sensor designed to autonomously monitor industrial processes or environmental remediation, also available commercially from Sapidyne Instruments Inc. Panel (d). Alpha prototype of a sensor designed to operate in underwater buoys or autonomous underwater vehicles.

shown in Figure 4.7, Panel b, is a handheld device for field analysis [48]. The user prepares a mixture of the environmental sample, chelator, and antibody and loads it into the syringe located on the right side of the instrument. The sensor, controlled by a Palm™ device, pushes the sample through a prepacked microbead column located inside the light-tight disposable cassette unit at the top of the syringe. The instrument monitors the accumulation of fluorescently labeled antibody on the microbead column and converts arbitrary fluorescence units to the concentration of metal in the environmental sample; this concentration calculation is based upon calibrators previously run on the instrument. A third version of this sensor, shown in Panel c of Figure 4.7, is the beta prototype of an in-line sensor designed to autonomously monitor industrial processes or environmental remediation. This instrument has the ability to autonomously run a standard curve and prepare environmental samples for analysis [47]. The user prepares concentrated stock solutions and antigen-coated microbeads and loads them into a refrigerated compartment of the instrument. The instrument then autonomously prepares antibody-antigen mixtures from these stocks or from a combination of stocks and environmental samples via a programmable computer interface. The final sensor in development in this laboratory is shown in Figure 4.7, Panel d. This sensor was designed to operate in underwater buoys or autonomous underwater vehicles. It contains five prepacked microbead columns and can be programmed via the attached Palm™ device to analyze a sample for either five different types of analytes or to perform multiple assays of a single analyte over a prearranged period of time. A smaller version of this sensor is also under development for miniaturized autonomous underwater vehicles.

4.6 ANALYSIS OF ENVIRONMENTAL AND SERUM SAMPLES

These antibody-based assays for heavy metals have thus far seen limited use in the field, primarily because the antibodies are not yet available commercially and have not been formulated into easy-to-use kits. 2A81G5 was the first anti-metal monoclonal antibody isolated in this laboratory, and most of this group's analytical applications have utilized this antibody. 2A81G5 has been used in both a two-step and one-step ELISA format to measure cadmium in environmental water samples spiked with cadmium [22,36]. Representative data from these experiments are shown in Figure 4.8. Panel a shows a comparison of results obtained when environmental water samples were spiked with 0.36–5.37 μM cadmium and assayed either by the two-step ELISA or by graphite furnace atomic absorption spectroscopy (AAS). In general, the results from the immunoassay correlated well with the values obtained from AAS, and the immunoassay correctly identified minimally, moderately, and heavily contaminated water samples. There was some positive bias in the immunoassay, as indicated by the nonzero intercept of the graph in Panel a; however, such a positive bias may be acceptable in an assay designed as a field-portable screening tool.

The one-step ELISA that required approximately one hour to perform [36] also showed better sensitivity for cadmium than did the two-step assay (compare the x-axes in Panels a and b of Figure 4.8). In spike and recovery experiments using the one-step procedure, the immunoassay was actually superior to graphite furnace AAS in accurately determining the amount of cadmium that had been spiked into each environmental water sample (Figure 4.8, Panel b). A variation of the one-step ELISA has also been developed for serum [37]; the serum ELISA has been used to demonstrate significant differences in the levels of serum cadmium in patients with pancreatic cancer as compared with age-matched controls [45].

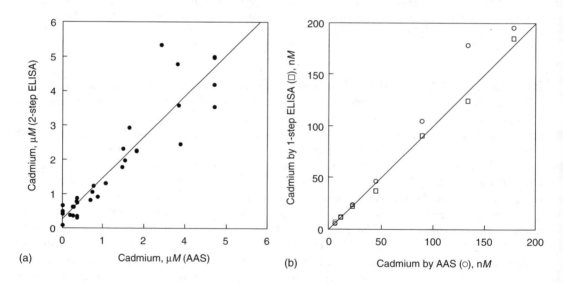

(a)   (b)

FIGURE 4.8 Comparison of ELISA and atomic absorption results in the analysis of spiked environmental water samples for cadmium. Panel (a). Cadmium was added to water samples collected on different dates from Bayou Trepagnier and analyzed for cadmium using either the two-step ELISA or the graphite furnace AAS. The cadmium concentrations determined by two-step ELISA were plotted on the y-axis, and the cadmium as determined by AAS was plotted on the x-axis. Linear regression analysis of the data generated a line with a slope of 0.951, an intercept of 0.2 μM, and a correlation coefficient of 0.931. Panel (b). Cadmium analysis using the one-step ELISA. Cadmium was spiked into water taken from Bayou Trepanier at concentrations from 5.6 nM to 178 nM, and the samples were subsequently analyzed by one-step ELISA or graphite furnace AAS. The solid line shows the theoretical line obtained if all samples exhibited a 100% recovery of cadmium; ($\bigcirc$) cadmium values determined by graphite furnace AAS; ($\square$) cadmium values determined by the one-step ELISA. Bayou Trepagnier was chosen to test the cadmium immunoassays because its water chemistry is typical of polluted bayous found in southern Louisiana. (From Khosraviani, M., Pavlov, A. R., Flowers, G. C. and Blake, D. A. *Environ. Sci. Technol.*, 32, 1998.; and Darioish, I. A. and Blake, D. A., *Anal Chem.*, 73, 2001.)

4.6.1 FUTURE RESEARCH DIRECTIONS

As investigators who have employed antibodies in the development of several new analytical methodologies, our group has discovered that the performance of any new ELISA kit or immunosensor is largely dependent upon the antibodies incorporated into the assay. An important aim of our present research is to develop new technologies in antibody engineering that will enhance antibody function in the ongoing immunosensor program. Recombinant antibodies have potential advantages over monoclonal antibodies produced by standard hybridoma technology. The cloned genes represent a stable, recoverable source for antibody production. In addition, the recombinant format offers opportunities for protein engineering that could enhance antibody performance and for mechanistic studies that could lead to a better understanding of the nature of the antibody–antigen interaction. The nucleotide and deduced amino-acid sequences have been determined for the light and heavy chain variable regions of most of the metal–chelate-specific monoclonals generated by this laboratory. These sequences have been expressed in a variety of recombinant forms in order to determine if any of the recombinant forms might provide an improved reagent for immunoassays and immunosensors.

Monoclonal antibodies to metal–chelate complexes have been expressed in all of the structural variations shown in Figure 4.9. The heavy and light chain variable regions of two antibodies, 5B2 and E5, were expressed as single chain fragments (scFv, d in Figure 4.9) using both bacterial and mammalian expression vectors. Although scFv protein was expressed in both systems, neither antibody retained binding activity toward metal-chelate complexes when expressed as an scFv. When the light and heavy chain variable regions of E5 were then expressed in the scAb format (c in Figure 4.9), this recombinant fragment regained between 15% and 20% of its binding activity. It was hypothesized that the binding site was compromised in these single chain formats and that the more correct alignment of the light and heavy chains possible with a Fab or F(ab)₂ format would increase the stability of the protein–ligand interaction and would increase the affinity of the recombinant antibodies for metal–chelate complexes.

Indeed, when the 5B2 and E5 antibodies were expressed as recombinant F(ab)₂ fragments (shown in Figure 4.8b), they regained their binding activities. This recombinant format was

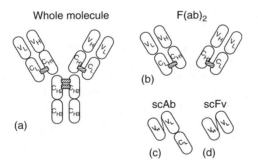

FIGURE 4.9 Antibodies and antibody fragments used in immunoassays. (a) Intact IgG molecule that contains two light chains (_L_) and two heavy chains (_H_). The hypervariable regions of the light and heavy chain are designated with a V; the constant regions are designated with an H. (b) F(ab)₂ fragment that consists of the entire light chain and the variable and first constant region of the heavy chain. This fragment can be prepared either by proteolysis of an intact IgG molecule or via recombinant technology. (c) Single-chain antibody fragment (scAb) usually produced via recombinant technology in bacterial cells. This fragment contains the variable region of the heavy chain and the variable and constant region of the light chain connected by a linker peptide. (d) scFv variable also recombinantly produced that consists of variable regions of the light and heavy chain connected by a flexible linker peptide disulfide bonds are shown as cross-hatched bands. (Adapted from Roitt, I., Brostoff, J. and Male, D. *Immunology,* 5th ed., Mosby International, London, 1998.)

generated by ligating the complete light chain and the variable and first constant region of the heavy chain into a commercially available dicistronic mammalian expression vector [23]. These techniques have been used to express recombinant $F(ab)_2$ fragments for three different monoclonal antibodies: 5B2, E5, and 12F6 (Table 4.1). In all three cases, the recombinant proteins had binding properties identical to the $F(ab)_2$ prepared by proteolysis of the native monoclonal antibody [23]. Molecular models of the heavy and light chain variable domains were constructed according to the canonical structures method detailed by Morea et al. [50]. The participation of specific residues in antigen recognition could then be assessed using site-directed mutagenesis.

In a separate series of experiments, a recombinant phage-displayed antibody library has also been prepared using RNA isolated from the spleens of sheep and rabbits immunized with specific metal–chelate complexes. Phage-display libraries produced from an immunized source are inclined to include variable genes specific for the immunized antigen(s), many that are already affinity-matured. An antibody fragment specific for the UO_2^{2+} – DCP complex has been isolated from this combined phage display library. Whereas the binding affinity of this antibody fragment for UO_2^{2+} – DCP was not as high as that of monoclonal antibodies, the beauty of antibody phage display technology is that it allows for the potential manipulation and maturation of the antibody's binding affinity that may drastically improve and ultimately surpass that of monoclonal antibodies.

A number of techniques can be employed to both produce and select antibodies with improved specificities and affinities. Error-prone polymerase chain reaction followed by selection with phage-display increased the affinity of an scFv against lysozyme five-fold [51]. Schier et al. [52,53] used chain shuffling to increase the binding affinity of an scFv for tumor antigen c-*erb*B-2 by six-fold from 16 nM to 2.5 n*M*. Chain shuffling involves introducing a variety of light chains to a heavy chain specific for a particular antigen. Once the antibodies with the greatest binding affinity are chosen, a portion of the heavy chain is replaced. As the CDRH3 loop is most likely to confer specificity for an antigen, this portion of the variable heavy chain is retained [54]. Site-directed mutagenesis can also be employed to improve antibody affinity [54], but as shown by error-prone PCR affinity maturation, some changes in the antibody sequence that improved binding affinity were not anticipated and centered mainly on framework residues and those residues in the CDRs that were not in direct contact with the antigen [55].

4.7 CONCLUSIONS

Antibody-based assays for metal ions are a promising new technology that will permit the real-time, on-site assessment of heavy metal contamination. In the research programe discribed, the isolation and characterization of antibodies directed toward metal–chelate complexes was performed in parallel with the development of autonomous and miniaturized field-portable sensors that incorporate these antibodies as the recognition element. The availability of such sensors will significantly decrease the cost of site monitoring and greatly improve risk assessment efforts.

ACKNOWLEDGMENTS

This research was supported by the Office of Science (BER) U.S. Department of Energy, Grant No. DE-FG02-98ER62704 (to D.A.B.) by grants from the Office of Naval Research (N00014-06-1-0307) and U.S. Geological Survey (05HQAQ 0109) to the Tulane/Xavier Center for Bioenvironmental Research and by Subcontract No. 93603-001-04 3F from the Los Alamos National Laboratory via the Office of Nuclear Nonproliferation, U.S. Department of Energy. Stipend support for E.R. Abboud was provided NOAA via the Coastal Restoration and Enhancement through the Science and Technology (CREST) Program and by a GRO Predoctoral Fellowship (MA91657401) from the U.S. Environmental Protection Agency.

REFERENCES

1. Devi, M. and Fingerman, M., Inhibition of acetylcholinesterase activity in the central nervous system of the red swamp crayfish, *Procambarus clarkii*, by mercury, cadmium, and lead, *Bull. Environ. Contam. Toxicol.*, 55, 746–750, 1995.

2. MacCarthy, P., Klusman, R. W., Cowling, S. W., and Rice, J. A., Water analysis, *Anal. Chem.*, 67, 525R–582R, 1995.

3. Santelli, R. E., Gallego, M., and Valcarcel, M., Preconcentration and atomic absorption determination of copper traces in waters by online adsorption-elution on an activated carbon minicolumn, *Talanta*, 41, 817–823, 1994.

4. Corsi, M., Cristoforetti, G., Hidalgo, M., Legnaioli, S., Palleschi, V., Salvetti, A., Tognoni, E., and Vallebona, C., Application of laser-induced breakdown spectroscopy technique to hair tissue mineral analysis, *Appl. Opt.*, 42, 6133–6137, 2003.

5. Basta, N. T., Ryan, J. A., and Chaney, R. L., Trace element chemistry in residual-treated soil: key concepts and metal bioavailability, *J. Environ. Qual.*, 34, 49–63, 2005.

6. Shaw, M. J. and Haddad, P. R., The determination of trace metal pollutants in environmental matrices using ion chromatography, *Environ. Int.*, 30, 403–431, 2004.

7. Meyer, G. A., ICP; Still a panacea for trace metal analysis, *Anal. Chem.*, 59, 1345–1354, 1987.

8. Komaromy-Hiller, G., Flame, flameless, and plasma spectroscopy, *Anal. Chem.*, 71, 338R–342R, 1999.

9. Alvarex-Llamas, G., De laCampa, M. R. F., and Sanz-Medel, A., ICP–MS for specific detection in capillary electrophoresis, *Trends Anal. Chem.*, 24, 28–36, 2005.

10. Cornelis, R., Caruso, J., Crews, H., and Heumann, K., *Handbook of Elemental Speciation Techniques and Methodology*, Wiley, Chichester, U.K., 2003.

11. Szurdoki, F., Jaeger, L., Harris, A., Kido, H., Wengatz, I., Goodrow, M. H., Szekacs, A. *et al.*, Rapid assays for environmental and biological monitoring, *J. Environ. Sci. Health, Part B: Pestic. Food Contam., Agric. Wastes*, 31, 451–458, 1996.

12. Van Emon, J. M., Gerlach, C. L., and Bowman, K., Bioseparation and bioanalytical techniques in environmental monitoring, *J. Chromatogr. B, Biomed. Sci. & Appl.*, 715, 211–228, 1998.

13. Bushway, R. J., Perkins, L. B., Fukal, L., Harrison, R. O., and Ferguson, B. S., Comparison of enzyme-linked immunosorbent assay and high-performance liquid chromatography for the analysis of atrazine in water from Czechoslovakia, *Arch. Environ. Contam. Toxicol.*, 21, 365–370, 1991.

14. Abad, A., Moreno, M. J., Pelegri, R., Martinez, M. I., Saez, A., Gamon, M., and Montoya, A., Determination of carbaryl, carbofuran and methiocarb in cucumbers and strawberries by monoclonal enzyme immunoassays and high-performance liquid chromatography with fluorescence detection. An analytical comparison, *J. Chromatogr. A*, 833, 3–12, 1999.

15. Baek, N. H., Evaluation of immunoassay tests in screening soil contaminated with polychlorinated biphenyls, *Bull. Environ. Contam. Toxicol.*, 51, 844–851, 1993.

16. Biagini, R. E., Tolos, W., Sanderson, W. T., Henningsen, G. M., and MacKenzie, B., Urinary biomonitoring for alachlor exposure in commercial pesticide applicators by immunoassay, *Bull. Environ. Contam. Toxicol.*, 54, 245–250, 1995.

17. Giraudi, G. and Baggiani, C., Immunochemical methods for environmental monitoring, *Nucl. Med. Biol.*, 21, 557–572, 1994.

18. Reardan, D. T., Meares, C. F., Goodwin, D. A., McTigue, M., David, G. S., Stone, M. R., Leung, J. P., Bartholomew, R. M., and Frincke, J. M., Antibodies against metal chelates, *Nature*, 316, 265–268, 1985.

19. Blake, D. A., Chakrabarti, P., Hatcher, F. M., and Blake, R. C. II, Enzyme immunoassay to determine heavy metals, *Proceedings of the 17th Annual EPA Conference on Pollutants in the Environment*, U.S. EPA Office of Water, Washington, DC, pp. 293–316, 1994.

20. Blake, D. A., Chakrabarti, P., Khosraviani, M., Hatcher, F. M., Westhoff, C. M., Goebel, P., Wylie, D. E., and Blake, R. C. II, Metal binding properties of a monoclonal antibody directed toward metal–chelate complexes, *J. Biol. Chem.*, 271, 27677–27685, 1996.

21. Blake, D. A., Pavlov, A. R., Khosraviani, M., and Flowers, G. C., Immunoassay for cadmium in ambient water samples, In *Current Protocols in Field Analytical Chemistry*, Lopez-Avila, V., Ed., Wiley, New York, pp. 1.1–1.10, 1998.

22. Khosraviani, M., Pavlov, A. R., Flowers, G. C., and Blake, D. A., Detection of heavy metals by immunoassay: Optimization and validation of a rapid, portable assay for ionic cadmium, *Environ. Sci. Technol.*, 32, 137–142, 1998.
23. Delehanty, J. B., Jones, R. M., Bishop, T. C., and Blake, D. A., Identification of important residues in metal-chelate recognition by monoclonal antibodies, *Biochemistry*, 42, 14173–14183, 2003.
24. Blake, R. C. II, Pavlov, A. R., Khosraviani, M., Ensley, H. E., Kiefer, G. E., Yu, H., Li, X., and Blake, D. A., Novel monoclonal antibodies with specificity for chelated uranium(VI): Isolation and binding properties, *Bioconjug. Chem.*, 15, 1125–1136, 2004.
25. Martell, A. E.,. Smith, R. M., NIST critically selected stability constants of metal complexes version 5.0, In *NIST Standard Reference Data*, National Institutes of Standards and Technology, Gaithersburg, MD, 1998.
26. DeRiemer, L. H., Meares, C. F., Goodwin, D. A., and Diamanti, C. I., BLEDTA: Tumor localization by a bleomycin analogue containing a metal-chelating group, *J. Med. Chem.*, 22, 1019–1023, 1979.
27. Brechbiel, M. W., Beitzel, P. M., and Gansow, O. A., Purification of *p*-nitrobenzyl C-functionalized diethylenetriamine pentaacetic acids for clinical applications using anion-exchange chromatography, *J. Chromatogr. A.*, 771, 63–69, 1997.
28. Brechbiel, M. W. and Gansow, O. A., Synthesis of C-functionalized *trans*-cyclohexyldiethylenetria-minepenta-acetic acids for labelling of monoclonal antibodies with the bismuth-212 α-particle emitter, *J. Chem. Soc. Perkin Trans.*, 1, 1173–1178, 1992.
29. Blake, R. C. II, Delehanty, J. B., Khosraviani, M., Yu, H., Jones, R. M., and Blake, D. A., Allosteric binding properties of a monoclonal antibody and its Fab fragment, *Biochemistry*, 42, 497–598, 2003.
30. Khosraviani, M., Blake, R. C. II, Pavlov, A. R., Lorbach, S. C., Yu, H., Delehanty, J. B., Brechbiel, M. W., and Blake, D. A., Binding properties of a monoclonal antibody directed toward lead-chelate complexes, *Bioconjug. Chem.*, 11, 267–277, 2000.
31. Thiers, R. E., Contamination in trace element analysis and its control, *Methods Biochem. Anal.*, 5, 273–335, 1957.
32. Wong, S. S., *Chemistry of Protein Conjugation and Cross-Linking*, CRC Press, Boca Raton, 1993.
33. Chakrabarti, P., Hatcher, F. M., Blake, R. C. II, Ladd, P. A., and Blake, D. A., Enzyme immunoassay to determine heavy metals using antibodies to specific metal-EDTA complexes: Optimization and validation of an immunoassay for soluble indium, *Anal Biochem.*, 217, 70–75, 1994.
34. Harris, J. R. and Markl, J., Keyhole limpet hemocyanin (KLH): A biomedical review, *Micron*, 30, 597–623, 1999.
35. Harris, J. R. and Markl, J., Keyhole limpet hemocyanin: Molecular structure of a potent marine immunoactivator. A review, *Eur. Urol.*, 37 (Suppl. 3), 24–33, 2000.
36. Darwish, I. A. and Blake, D. A., One-step competitive immunoassay for cadmium ions: Development and validation for environmental water samples, *Anal. Chem.*, 73, 1889–1895, 2001.
37. Darwish, I. A. and Blake, D. A., Development and validation of a one-step immunoassay for determination of cadmium in human serum, *Anal Chem.*, 74, 52–58, 2002.
38. Cayot, P. and Tainturier, G., The quantification of protein amino groups by the trinitrobenzenesulfonic acid method: a reexamination, *Anal Biochem.*, 249, 184–200, 1997.
39. Jones, R. M., Yu, H., Delehanty, J. B., and Blake, D. A., Monoclonal antibodies that recognize minimal differences in the three-dimensional structures of metal-chelate complexes, *Bioconjug. Chem.*, 13, 408–415, 2002.
40. Blake, D. A., Pavlov, A. R., Yu, H., Khosraviani, M., Ensley, H. E., and Blake, R. C. II, Antibodies and antibody-based assays for hexavalent uranium, *Anal. Chim. Acta.*, 444, 11–23, 2001.
41. Good, N. E., Winget, G. D., Winter, W., Connolly, T. N., Izawa, S., and Singh, R. M. M., Hydrogen ion buffers for biological research, *Biochemistry*, 5, 467–477, 1966.
42. Braden, B. C., Goldman, E. R., Mariuzza, R. A., and Poljak, R. J., Anatomy of an antibody molecule: Structure, kinetics, thermodynamics and mutational studies of the antilysozyme antibody D1.3, *Immunol. Rev.*, 163, 45–57, 1998.
43. Blake, R. C. I., Pavlov, A. R., and Blake, D. A., Automated kinetic exclusion assays to quantify protein binding interactions in homogenous solution, *Anal Biochem.*, 272, 123–134, 1999.
44. Blake, R. C. II and Blake, D. A., Kinetic exclusion assay to study high-affinity binding interactions in homogeneous solutions, In *Methods in Molecular Biology: Antibody Engineering-Methods and Protocols*, Lo, B. K. C., Ed., Humana Press, Towata, NJ, pp. 417–430, 2003.

45. Kriegel, A. M., Soliman, A. G., El-Ghawalby, N., Ezzat, F., Soultan, A., Abdel-Wahab, M., Fathy, O., et al., Serum cadmium levels in pancreatic cancer patients from the East Nile Delta region of *Egypt Environmental Health Perspectives*, 2005, Forthcoming.

46. Blake, D. A., Jones, R. M., Blake, R. C. II, Pavlov, A. R., Darwish, I. A., and Yu, H., Antibody-based sensors for heavy metal ions, *Biosens. Bioelectron.*, 16, 799–809, 2001.

47. Yu, H., Jones, R. M.,Blake, D. A., An immunosensor for autonomous in-line detection of heavy metals: Validation for hexavalent uranium. *Int. J. Env. Anal. Chem.*, 2005, Forthcoming.

48. Blake, D. A., Yu, H., and Blake, R. C. II, Development of rapid, portable immunoassays for heavy metals in acid mine drainage, In *Biohydrometallurgy and the Environment-IBS 2001*, Amils, R., Ed., Elsevier, Amsterdam, pp. 533–540, 2002.

49. Glass, T. R., Saiki, H., Blake, D. A., Blake, R. C. II, Lackie, S. J., and Ohmura, N., Use of excess solid-phase capacity in immunoassays: advantages for semicontinuous, near-real-time measurements and for analysis of matrix effects, *Anal. Chem.*, 76, 767–772, 2004.

50. Morea, V., Tramontano, A., Rustici, M., Chothia, C., and Lesk, A. M., Conformations of the third hypervariable region in the VH domain of immunoglobulins, *J. Mol. Biol.*, 275, 269–294, 1998.

51. Johnson, K. S. and Hawkins, R. E., Affinity maturation of antibodies using phage display, In *Antibody Engineering: a Practical Approach*, Chiswell, D. J., Ed., IRL Press, New York, pp. 41–58, 1996.

52. Schier, R. and Marks, J. D., Efficient in vitro affinity maturation of phage antibodies using BIAcore guided selections, *Hum. Antibodies Hybridomas*, 7, 97–105, 1996.

53. Schier, R., Bye, J., Apell, G., McCall, A., Adams, G. P., Malmqvist, M., Weiner, L. M., and Marks, J. D., Isolation of high-affinity monomeric human anti-c-erbB-2 scFv using affinity-driven selection, *J. Mol. Biol.*, 255, 28–43, 1996.

54. Adams, G. P. and Schier, R., Generating improved single-chain Fv molecules for tumor targeting, *J. Immunol. Methods*, 231, 249–260, 1999.

55. Hawkins, R. E., Russell, S. J., Baier, M., and Winter, G., The contribution of contact and non-contact residues of antibody in the affinity of binding to antigen. The interaction of mutant D1.3 antibodies with lysozyme, *J. Mol. Biol.*, 234, 958–964, 1993.

56. Kriegel, A. M., Blake, D.A., Forthcoming. Antibody fragments with specificity for metal-chelate complexes from human semi-synthetic phage display libraries, In *Recent Research developments in bioconjugate chemistry*, ed. A. Gayathri. Trivandrum, India: Transworld Research Network, 2005.

57. Blake, D. A., Blake, R. C. II, Khosraviani, M., and Pavlov, A. R., Metal ion immunoassays, *Anal. Chim. Acta.*, 376, 13–19, 1998.

5 Molecular Imprinting for Small Molecules

Zoe Cobb and Lars I. Andersson

CONTENTS

5.1 INTRODUCTION

Recent years have seen an extensive interest in the field of molecular imprinting [1] where, in particular, there has been much research activity into analytical separation applications. These applications include the use of molecularly imprinted polymers (MIPs) as antibody mimics in pseudo-immunoassay, affinity sorbents in solid-phase extraction (SPE), highly selective stationary

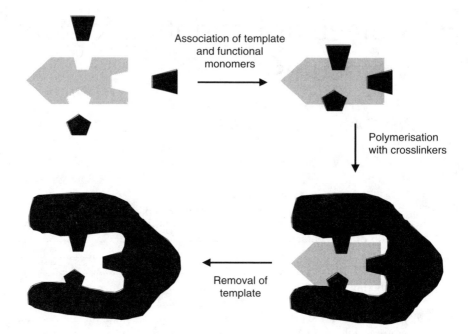

FIGURE 5.1 The formation of molecular imprints occurs when functional and cross-linking monomers are allowed to polymerise in the presence of a template. The functional monomers are selected so that they interact and form complexes with the template in the pre-polymerization solution. The arrangement of the functional monomers around the template is then retained in position by polymerization with an excess of cross-linker that forms the bulk of the resultant insoluble polymer. Removal of the template leaves behind imprinted sites with a steric (size and shape) and chemical (complementary chemical functionality) memory for the template. These imprints can selectively rebind the original template from a mixture.

phases in liquid chromatography and capillary electrochromatography, and selective barriers in chemical sensing. In many cases, MIPs have been shown to give highly competitive, if not improved, results over traditional materials. The main attraction of the technique is its apparent simplicity (Figure 5.1). In theory, the synthesis of a polymer in the presence of a template molecule and subsequent removal of the template creates a robust material with memory sites that have the ability to selectively rebind the original template from a mixture. In principle, MIPs can be made with selectivity for essentially any of a diverse range of analyte species such as drug enantiomers, pesticides, hormones, toxins, short peptides, and nucleic acids. Whereas for biomacromolecules, antibody technology will in the foreseeable future remain the obvious alternative, molecular imprinting may offer a viable alternative for small molecules. In these instances, antibody preparation requires conjugation of the hapten to a carrier protein that often changes the structural properties of the antigen exposed to the immune system. Therefore, the antibodies obtained may be directed against a structure subtly different to the intended one. In some instances, MIP selectivity profiles are better than those reported of monoclonal antibodies. MIPs are inherently more robust than antibodies, and they can be employed for separation of the analyte in matrices ranging from pure organic solvents to biological fluids.

5.2 PREPARATION OF IMPRINTED POLYMERS

5.2.1 FORMATION OF MOLECULAR IMPRINTS

Two principally different approaches, covalent and non-covalent imprinting, can be used to achieve complexation between the monomers and the template. Covalent imprinting [2,3] utilizes reversible covalent bonds and requires a synthesis step to prepare the imprint species consisting of functional monomers covalently bound to the template. After polymerization, these bonds then must be cleaved, usually by hydrolysis, to produce the imprinted sites. Non-covalent imprinting [4,5] is a self-assembly approach that employs interactions such as hydrogen bonds, ionic bonds, and hydrophobic interactions between complementary chemical functionalities of the template and functional monomers. It is generally accepted that the non-covalent approach is experimentally easier and more flexible with respect to chemical structures amenable to imprinting. Also, the functional monomers can be selected from a wider range of commercially available chemicals. Other molecular imprinting approaches include stoichiometric non-covalent imprinting [3,6] that employs multiple hydrogen bonds between template and monomer and thereby gives rise to very stable template-monomer complexes. Although associations are very strong in aprotic solvents, the template can be dissociated from the imprint by addition of water or alcohols. The lack of excess monomers results in the location of all monomer functional groups within the imprint sites. Semi-covalent imprinting employs covalent binding in order to synthesize imprinted polymers, and the subsequent recognition mechanism employs non-covalent binding of the analyte to the imprinted site [7,8]. This result is achieved by cleaving off the template, often simultaneously removing a sacrificial spacer between template and monomer, to vacate the imprinted site leaving behind a functional group able to non-covalently interact with the template molecule in the same position. Molecular imprinting based on metal ion co-ordination [9,10] involves the formation of ternary template-metal ion-monomer complexes. The interactions are strong under conditions used for polymerization that again minimizes formation of non-specific binding sites and offers a high degree of versatility for binding of biologically relevant template functionalities.

Non-covalent imprinting will be the main focus of discussion in this chapter as it has been used for the majority of applications discussed in depth later. Although appearing relatively simple, this method of synthesis encompasses a range of factors that can affect the morphology, binding capacity, affinity, and surface properties of the resulting polymer. There are, however, a number of rules of thumb, discussed later, that, when applied, generally result in imprinted sites within a polymer matrix [11,12].

5.2.2 TEMPLATE

Whereas a large number of low-molecular weight compounds are amenable to non-covalent imprinting, there are certain requirements of a template molecule. It should be both soluble in the monomer solution and also be chemically inert under the required polymerization conditions. Does the template contain any double bonds or any other chemical functionality that may interfere with free radical polymerization? Is the template stable to UV or heat used to activate the initiator? It is necessary to consider these questions; and if the answers are satisfactory, then molecular imprinting is theoretically possible.

5.2.3 FUNCTIONAL MONOMERS

Functional monomers carry chemical functionalities that engage in complementary interactions with the template in solution and during polymerization, e.g., a hydrogen bond donor binds with a hydrogen bond acceptor. These interactions remain in the resultant imprints in correct spatial positions, and together with the complementary shape of the imprint, form the basis for the molecular recognition observed (Figure 5.1). Typically, the functional monomer is chosen from any of the vinyl, methacrylate, acrylate, methacrylamide, or acrylamide monomer classes (Table 5.1). It is possible to use more than one type of monomer as long as the types readily copolymerize with each other during the free radical polymerization [11]. The most common monomer is methacrylic acid (MAA) that is both a hydrogen bond donor and a hydrogen bond acceptor and can form hydrogen bonds with a variety of template functionalities. Hence, MAA is

TABLE 5.1
Representative Monomers Used for Molecular Imprinting of Small Molecules

Monomer	Properties
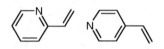 Methacylic acid (MAA)	Acidic monomer with pKa = 4.6. Participates in ionic interactions and hydrogen bonding. Generally applicable.
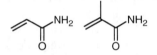 Trifluoromethacrylic acid (TFMAA)	More acidic monomer than MAA with pKa = 2.1. Participates in ionic interactions and hydrogen bonding.
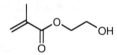 2- and 4-vinylpyridine (VPy)	Basic monomers with pKa = 5.0 and 5.4, respectively. Participates in ionic interactions and hydrogen bonding.
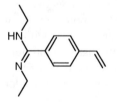 Acrylamide and methacrylamide	Neutral monomers. Participates in strong hydrogen bonding.
	Neutral monomer. Participates in hydrogen bonding. Renders MIP surface hydrophilic.
2-Hydroxyethylmethacrylate (HEMA)	
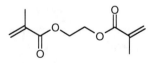 N,N´-diethyl(4-vinyl-phenyl)amidine	Strongly basic monomer with pKa = 11.9. Forms strong ionic bonds with acidic groups, such as carboxylic acids and phosphonic acids.
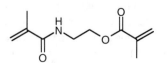 Ethylene glycol dimethacrylate (EGDMA)	The most commonly used cross-linking monomer.
N,O-bismethacryloyl ethanolamine (NOBE)	Cross-linking monomer, which contains acrylamide functionality. Participates in hydrogen bonding and can be used alone without additional functional monomer.

(continued)

TABLE 5.1 *(Continued)*

Monomer	Properties
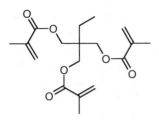	Trifunctional cross-linker. Useful for chromatographic applications.

Trimethylolpropane trimethacrylate (TRIM)

applicable to imprinting of a diverse range of template structures. A comparison of the binding properties of dansyl-L-phenylalanine-MIPs prepared by copolymerization of MAA and/or 2-vinyl-pyridine (2VPy) with EGDMA showed that the 2VPy-MIP gave increased enantiomeric selectivity over the MAA-MIP [13]. The combination of 2VPy and MAA in the same MIP gave an even greater increase in enantiomer selectivity. The monomer 2-trifluoromethacrylic acid (TFMAA) is more acidic than MAA, and as a result, is expected to form stronger interactions with basic templates [14]. However, a MIP prepared using both MAA and TFMAA showed an unexpectedly higher affinity for the prometryn template than a polymer containing solely MAA or TFMAA [15]. Acrylamide is a neutral monomer that provides strong hydrogen bonding interactions and has, for some templates, been shown to give improved enantiomer recognition and load capacities over MAA based MIPs [16]. Several studies have used 2-hydroxyethylmethacrylate (HEMA) [17] and an interesting alternative is monomers with an urea functionality [18]. Basic monomers proven to form selective imprints of acidic templates include the amidine-based monomer N,N'-diethyl(4-vinyl-phenyl)amidine [19], 4-vinylpyridine, and 1-vinylimidazole [20].

The functional monomer to template ratio influences selectivity, binding capacity, and affinity of the resulting polymer and should be optimized for each system. Typically, (non-optimized) ratios of 4:1 are employed [or rather two monomers per chemical functionality (amine, amide, carboxylic acid, hydroxyl, carbamate, etc.) of the template molecule]. In equilibrium binding studies, very high ratios are acceptable [21,22]. Morphine-MIPs prepared with an MAA to template ratio of 50:1 have been employed in radio-ligand binding studies without significant loss of binding capacity or selectivity, and a ratio of 150:1 still gave measurable binding curves with reduced capacity [21]. This extends the applicability of non-covalent imprinting to templates that are expensive, not available in large quantities, or are poorly soluble. For chromatographic separations, the situation is more complex. Although imprinting efficiency increases with increased ratio, non-specific binding also increases with more functional monomer residues being randomly distributed on the polymer surface. At high monomer-to-template ratios, non-specific binding dominates leading to long retention times and a complete loss of selectivity. An optimal ratio that balances capacity, selectivity, and non-specific binding must be experimentally found.

5.2.4 CROSS-LINKING MONOMERS

The cross-linker performs three main functions within the MIP: it helps to permanently fix the functional monomers that interact with the template in place within the imprinted site (Figure 5.1); it gives mechanical stability to the polymer; and it helps to control the morphology of the polymer matrix. Ethylene glycol dimethacrylate (EGDMA) is the most commonly used cross-linker [2–5].

It is cheap, and it produces MIPs with excellent mechanical, thermal, and chemical stability [23] and often with good selectivity. There are several alternatives that have been successfully used such as the trifunctional cross-linkers trimethylolpropane trimethacrylate (TRIM) and pentaerythritol triacrylate that were reported to produce MIPs with good chromatographic enantiomer selectivity, higher load capacity, and better resolution than the corresponding EGDMA MIPs [24]. However, literature also provides evidence of the reverse where nicotine-imprinted MAA/EGDMA polymers had a better selectivity than MAA/TRIM polymers [25]. Another popular cross-linker is divinyl-benzene (DVB), and a metsulfuron-methyl imprinted TFMAA/DVB polymer has been reported to give lower non-specific binding than the TFMAA/EGDMA MIP [26]. The hybrid cross-linker N,O-bismethacryloyl ethanolamine (NOBE) that contains an acrylamide functionality eliminates the need for additional functional monomers, and the utilization of NOBE alone is reported to provide polymers with enantiomer selectivity better than traditional EGDMA/MAA systems [27,28]. However, despite these discoveries, EGDMA still tends to be the cross-linker of choice in non-covalent imprinting.

The ratio of cross-linker to functional monomer is another important factor that affects polymer morphology and, to some extent, imprint selectivity [2,29]. A ratio of approximately 5:1 or alter-natively in excess of 80 mol% cross-linker is considered to be the optimum and most commonly used composition. Recently, factorial design approaches to finding the optimal ratio predicted an optimum at 55:10:1 (cross-linker to functional monomer to template) for imprinting of sulfametha-zine in MAA/EGDMA polymers [30] although this may be system specific rather than a general optimum condition for any given template.

5.2.5 SOLVENTS

In polymer synthesis, template, functional monomers, and cross-linking monomers are all dissolved in a solvent otherwise known as a porogen. The amount and type of solvent has a large influence on physical properties and morphology, including pore volume, pore size, and surface area observed for the resulting MIP [5]. For instance, surface area varied from 3.5 to 256 m^2/g when the solvent was chloroform and acetonitrile, respectively, for polymers imprinted with the same template using the same composition of MAA and EGDMA and prepared under otherwise identical conditions [31]. Additionally, the solvent has a large influence on the strength of the binding interactions that stabilize the complexes between monomers and template present in the pre-polymerization mixture [5]. For monomers that interact via hydrogen bonding and electrostatic interactions with the template, maximal efficiency of imprint formation occurs when the polymerization is performed using an aprotic, apolar solvent. Depending on solubility of the imprint species, typical solvents employed are toluene, chloroform, dichloromethane, and acetonitrile. This ensures maximal strength of the polar non-covalent interactions employed that are strongly dependent on the polarity of the solvent. If instead hydrophobic interactions are being exploited, then a solvent such as water should be used to encourage the interaction between the template and functional monomers. This is a less popular approach, and there are only a few successful literature reports including a 2,4-dichlorophenoxyacetic acid-MIP based on 4VPy and EGDMA that was prepared in a mixture of methanol and water (4:1; v/v) [32].

5.2.6 POLYMERIZATION AND WORK UP

Once the imprinting solution is prepared, the initiator is added. Then, in order to prevent inhibition of the free radical polymerization, any dissolved oxygen is removed either by degassing with an inert gas or by freeze thaw treatment. The initiator added is commonly 2,2'-azo-bisisobutyronitrile (AIBN) or 2,2'-azo-bis(2,4-dimethylvaleronitrile) (ABDV). Polymerization is started by thermo-chemical decomposition of the initiator into free radicals at 60°C or 40°C, respectively, or photochemically by UV irradiation at 5°C or even down to −30°C or below where low temperature seems to be beneficial for imprint formation [33]. Finally, the resultant polymer must be worked up. Polymer development involves extensive solvent extraction to remove the template followed by crushing and sieving the robust polymer to produce small particles (<25 μm or 25–50 μm in size). Unfortunately, the wash process rarely removes 100% of the template, and often, traces remain embedded within the polymer matrix [34]. This may later lead to template bleed during use, so it is sometimes beneficial to use a dummy template approach where a structural analogue of the analyte is used as the template [35,36].

As a direct result of the large number of factors that require optimization in molecular imprinting, recently, there has been a trend toward the use of combinatorial chemistry techniques [37,38]. A large number of small-scale polymers are prepared, often in HPLC vials or in a 96-well plate format, and evaluated. In this way, the optimal conditions are identified with high speed and synthesis of the polymer is scaled up for more detailed evaluation. Stringent experimental design and multivariate analysis increase precision of the optimization and allow better understanding of the interdependence of variables [39]. NMR is an excellent technique to study stoichiometry and structure of pre-polymerization complexes [40–42], including the influence of cross-linking monomer template interaction and functional monomer self-association [43]. The data generated provide detailed information about the system and valuable input into polymer design. Computational chemistry allows the development of a virtual library of functional monomers and screening against a template using molecular modeling software [44].

5.2.7 ALTERNATIVE POLYMERIZATION TECHNIQUES

There are five commonly used polymer synthesis methods for molecular imprinting: suspension, precipitation, multi-step swelling, core-shell emulsion, and bulk polymerization [45]. Bulk polymerization is performed as described in the previous text, and the resultant polymer is produced in the form of a monolith that must be crushed and sieved in order to produce particles of the required size. This process is time consuming, and it results in a relatively low yield of irregularly sized and shaped particles with many fines (<10 μm). Therefore, alternative, less time-consuming methods that produce MIPs with similar, or better, binding properties than bulk polymerization are urgently required.

1. Multi-step swelling polymerization involves the initial swelling of polystyrene seed particles in an activating solvent followed by a second step swelling in an emulsion of initiator, toluene, and polyvinyl alcohol in water, a third step swelling in a dispersion of monomers and template in water, and finally, polymerization [46]. The procedure produces uniformly sized particles that could selectively separate the enantiomers of the template in aqueous-rich media [46].

2. Suspension polymerization of monomers and template dissolved in small droplets of an organic solvent in an inert liquid fluorocarbon as the continuous phase is initiated after achieving an emulsion of the imprinting mixture in the presence of a perfluoropolymeric surfactant [47]. Adaptation of the procedure to work with different porogenic solvents and polymerization conditions resulted in a controlled average bead size between 5 and 50 μm with enantiomer separation abilities and binding capacities similar to those of bulk polymers [48].

3. Precipitation polymerization is based on the precipitation of the growing polymeric chains, out of the solvent, in the form of particles as they grow more and more insoluble in a continuous organic medium [49]. It is a technique initially used for the production of monodisperse copolymer particles [49], and only recently has it been brought to the attention of molecular imprinters [50]. When compared to conventional bulk polymerization, the resulting monodisperse particles of 0.2 μm had higher binding site densities and allowed faster ligand transfer rates to the binding sites [50].

4. Core-shell emulsion polymerization uses submicron cores produced by batch emulsion polymerization in a second-stage emulsion polymerization of an imprinting mixture to create an imprinted layer (shell) over the seed core particle [51]. The particles produced by this method are often < 100 nm.

A comparative study performed on these five techniques evaluated the selective binding to each of the imprinted systems in organic and aqueous solvents [52]. Under aqueous conditions, the selective rebinding of radioactive ligands varied as follows: two step swelling ≈ suspension ≈ bulk > core shell > precipitation. Rebinding from a toluene solution was somewhat different with precipitation > suspension > bulk > core shell > two step swelling. Although free radical polymerization is the major method for the preparation of MIPs, other approaches include condensation polymerization to form polyurethane based MIPs [53], cross-linking of pre-formed polymers [54,55], sol-gel technology [56,57], and imprinting in liquid crystalline polysiloxane materials [58].

It is also possible to prepare imprinted films [59], and there are at least three main approaches to the formation of MIP films: (1) Sandwiching where imprinting solution is trapped between a non-adhesive surface and a treated surface. An example is the polymerization of a 2 μm MIP layer between the surface of a gold electrode and a quartz cover disk [60]; (2) Addition of the imprinting solution to the sensor surface and spin coating to remove excess solution before polymerization under nitrogen gas using UV radiation [61]; and (3) Preparation of composite films by filling the pores of another film or membrane, for instance, polymerization of the imprinting solution on a glass filter [62].

5.3 PSEUDOIMMUNOASSAYS

5.3.1 INTRODUCTION TO PSEUDOIMMUNOASSAY

The inherent selectivity of biological antibodies and their high affinity for the antigen are frequently exploited in order to perform quantitative analysis using various forms of immunoassay [63]. A characteristic for many immunoassays is their ability to detect minute amounts of analyte in a complex biological matrix such as plasma, serum, or urine without prior extensive sample pre-treatment. Although immunoassays are extremely sensitive, the antibodies used, like many other biological macromolecules, are prone to being unstable, expensive, and sometimes very difficult to generate. The use of MIPs as an alternative to antibodies in pseudo-immunoassays has been widely researched over recent years. MIPs are stable, robust, relatively cheap, and easy to produce, and they can, in principle, be used in aqueous buffers over a wide pH range as well as in organic solvents. A disadvantage is that, at present, molecular imprinting is generally limited to small molecules.

The first example of a MIP-based pseudoimmunoassay is a radio-labeled competitive assay reported by Vlatakis et al. [64]. Imprinted polymers prepared against theophylline and diazepam demonstrated strong binding of the original template and cross-reactivity profiles similar to those of antibodies. The assay, however, required extraction of the analyte from plasma into an organic solvent where the actual assay was performed. This was typical of the early assays as the selective binding to the imprints was particularly strong in organic solvents. Selective binding has since been extended to aqueous samples, including plasma samples.

5.3.2 MIP CHARACTERISATION

One particular use of pseudoimmunoassay is for the characterization of imprinted polymers. In the literature, MIP binding selectivity is most commonly characterized using LC and retention analysis although the use of pseudoimmunoassay can give more information in a shorter time period

[42,52,65–67]. Initial characterization examines binding capacity and binding selectivity of the imprinted polymer as well as the relative contribution of selective and non-specific binding modes with the imprinted sites and polymer surface, respectively. Using pseudoimmunoassays, binding capacity can be characterized in polymer titration experiments, where a constant amount of probe (typically, the radio-labeled form of the template) is added to increasing concentrations of polymer. Following incubation in a suitable solvent (an organic solvent or an aqueous buffer) until equilibrium, the polymer particles are separated, and the unbound probe in the supernatant is quantified. The binding capacity of the polymer can be defined as the PC_{50}-value that is the polymer concentration where 50% of the probe is bound (Figure 5.2). This is an empirical value that, in principle, is related to the number of binding sites with a certain or higher affinity. The concentration of probe determines which range of affinity of the sites is probed.

The selectivity of the imprints is characterized using competitive binding experiments that evaluate the MIP's ability to bind molecules structurally related or unrelated to the template. Increasing concentrations of ligand compete with probe for binding to a limited and constant number of imprinted sites, and any binding of the competing ligand results in displacement of the probe into the supernatant (Figure 5.3). The selectivity of the MIP toward different structural analogues can be expressed as a series of IC_{50}-values that are the respective ligand concentration where 50% of the probe is displaced from the polymer.

In some instances, non-specific binding with polymer surface is present. In water, non-specific adsorption is due to hydrophobic interactions with the hydrophobic polymer surface and in apolar organic solvents because of hydrogen bonding and electrostatic interactions with polar functionalities distributed at the polymer surface. Whereas the binding to a MIP is the sum of both selective and non-specific binding modes, the binding of the probe to a non-imprinted reference polymer prepared in the absence of a template is equivalent to the non-specific binding to the MIP surface. Therefore, imprint binding can be calculated as the total binding to the MIP minus any non-specific binding observed for the reference polymer.

Non-covalent molecularly imprinted polymers exhibit a polyclonal distribution of binding sites from high numbers of low affinity sites to low numbers of high affinity sites. Depending on the conditions used for its characterization, differing binding capacities and selectivity profiles may be observed for the same MIP [42]. Notably, chromatography uses relatively high concentrations of ligand, and the MIP appears less selective than when it is characterized using radio-ligand binding

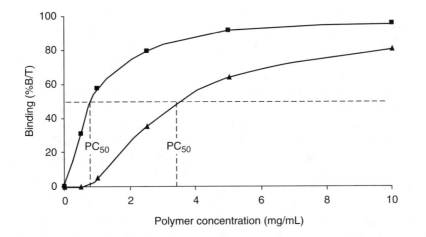

FIGURE 5.2 The PC_{50}-value is the polymer concentration where 50% of the added probe is bound. A MIP with higher affinity and binding capacity for the probe will give rise to a lower PC_{50}-value as a lower concentration of polymer is needed to bind 50% of the probe.

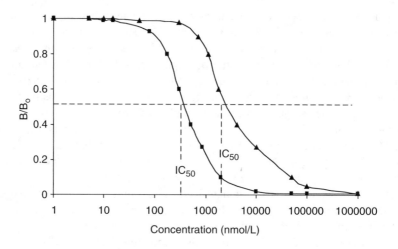

FIGURE 5.3 The IC_{50}-value is the ligand concentration where 50% of the probe is displaced from the polymer. A lower IC_{50}-value indicates that the polymer is more selective to that ligand as 50% displacement of the probe occurs at a lower concentration.

studies that are performed over a much larger concentration regime. All of the above parameters are empirical and directly determined from experimental data, and they require no assumption with respect to any binding site affinity distribution model. As long as the conditions are consistent between experiments, PC_{50}- and IC_{50}-values can be used to compare imprinting recipes as well as to compare re-binding efficiency between varying incubation conditions. For a more thorough discussion on models for characterization of the binding properties of MIPs, see [68] and original work primarily by the groups of Guiochon [69,70] and Shimizu [71,72].

5.3.3 METHOD DEVELOPMENT

5.3.3.1 Application to Organic Solvent-Based Samples

Most MIP syntheses are organic solvent-based. Subsequent characterization studies on imprint rebinding often are conducted using organic solvents as the incubation medium. Also, for assay of samples, the use of organic solvents may be beneficial such as following extraction of solid samples and following clean up of liquid samples by solvent extraction or in situations where the analyte is very poorly soluble in aqueous buffers. A common perception is that optimal binding is achieved in the porogen used for polymer synthesis. The use of matched solvents has been shown to result in superior imprint rebinding selectivity [5,73]. In practice it is, however, often more fruitful to optimize binding strength as lower limits of detection and higher binding capacities may be achieved. For imprint molecular recognition based on polar interactions such as hydrogen bonding and electrostatic interactions (such as is the case for the frequently studied MAA-type MIPs), binding strength correlates with polarity of the solvent used. A less polar solvent gives a higher binding strength and a higher apparent binding capacity when measured as PC_{50}-values. Binding strength increases (PC_{50}-value decreases) when the solvent is changed from acetonitrile to dichloromethane to toluene to heptane, either as a pure solvent or as mixtures. Also non-specific adsorption to randomly distributed monomer residues increases with decreased polarity of the solvent, and small amounts of a polar and protic modifier can be added to overcome this effect. Such modifiers typically include, in order of increasing ability to reduce non-specific adsorption, ethanol, methanol and acetic acid. These modifiers also show increasing ability to interfere with the selective imprint-ligand binding, and their concentration should be optimized for each system under study. Additions of small amounts of acetic acid in acetonitrile or toluene have been used to reduce the non-specific binding while maintaining the selective imprint-binding of morphine [74], yohimbine [75], and (S)-propranolol [65] to their respective MIPs. Optimal binding of corticosteroids to cortisol- and corticosterone-MIPs was achieved in mixtures of tetrahydrofuran and *n*-heptane with a small amount of acetic acid [76], and toluene-acetonitrile was demonstrated to be the optimal binding solvent for atrazine [77].

5.3.3.2 Application to Aqueous Samples

With regards to application of pseudoimmunoassay to aqueous samples, an early breakthrough was made with the example of a morphine-MIP that bound morphine with high affinity and a selectivity profile comparable to those of antibodies [74]. Because of the hydrophobic nature of many MIPs, problems with non-specific adsorption to the polymer surface are often encountered. Also, water may strongly interfere with the hydrogen bonding interactions between imprint and ligand. These two effects of water have been reported to result in a loss in binding [75] or a reduction in selectivity [76]. If the non-specific binding is too strong, the majority of binding events will occur through adsorption to the polymer surface and the selectivity associated with the imprinted sites may remain obscured. Reduction of the hydrophobic interactions can be achieved through the use of organic modifiers such as methanol, ethanol, and acetonitrile [65,66]. A second means of reducing non-specific adsorption is the use of detergents where non-charged detergents such as Tween 20, Triton X-100, and Brij 35 can reduce efficiently non-specific binding while leaving selective imprint-analyte interactions intact [78]. Buffer pH also may influence the extent of non-specific adsorption, where an increase in pH may lead to an increase in seen for adsorption of basic compounds to MAA-type polymers [65,66,74] and a decrease with increased pH for adsorption of acidic compounds to VPy type MIPs [79]. A detailed investigation into binding of (S)-propranolol under aqueous conditions studied the effects of pH, ethanol concentration, buffer concentration, and ionic strength, and it found that optimal selective binding was achieved in sodium citrate (25 mM, pH 6) containing 2% ethanol [65]. A study into the combined use of a surfactant and an organic solvent as additives found optimal selective binding of bupivacaine to a bupivacaine-MIP in citrate buffer (50 mM, pH 5) containing 5% ethanol and 0.05% Tween 20 [66]. An assay system for 2,4-D based on a VPy type MIP used phosphate buffer (20 mM, pH 7) containing 0.1% triton X-100 [32]. Whereas these additives reduce non-specific binding to the polymer surface, they also may affect selective imprint-ligand binding, and it is necessary to optimize additive levels to achieve minimal degrees of non-specific binding while maintaining high levels of selective binding.

The selectivity of the imprint-analyte interactions is tuned by the solvent composition of the incubation medium. In apolar solvents such as toluene and chloroform, the imprints show relatively better recognition of the parts of the ligand molecule that can participate in hydrogen bonding, whereas in water, the imprints demonstrate better recognition of hydrophobic parts of the molecule. A study on MAA type S-propranolol-MIPs showed that the MIPs had excellent enantioselectivity in toluene, and under aqueous conditions, high substrate selectivity in the presence of structurally similar compounds was observed [65]. For this system, enantioselectivity requires recognition of the relative positions of several hydrogen bonding groups around the chiral carbon, whereas substrate-selectivity requires recognition of a hydrophobic naphthyl ring on propranolol. For analogous reasons, a bupivacaine-MIP showed greater selectivity for bupivacaine over structural analogues in aqueous buffers than in toluene [66]. In contrast, binding studies on a morphine-MIP performed in toluene gave better selectivity against the analogues codeine and normorphine than that reported for monoclonal antibodies, and weaker and less selective binding was observed in water [74]. Significantly higher cross-reactivity was observed for rebinding to an atrazine-MIP under aqueous conditions than under organic solvent conditions [77]. This shows that the conditions for an assay may be chosen so as to obtain a selectivity pattern more suitable for a particular analysis problem.

5.3.4 ENZYME-LABELED, FLUORESCENT AND OTHER PROBES

Whereas most generic studies into imprint molecular recognition have used a radio-labeled form of the ligand, a wider acceptance of pseudoimmunoassay requires access to more popular detection systems such as those based on fluorescent, electroactive, or enzyme-labeled probes. The first enzyme labeled assay was reported by Surugiu et al. who presented an assay for the herbicide 2,4-D [79]. The reporter probe was 2,4-D conjugated to tobacco peroxidase and used with either colorimetric or chemiluminscence detection. Selective binding of the enzyme conjugate with the 2,4-D-MIP was achieved, resulting in a competitive assay for 2,4-D. Further work by the same group demonstrated the possibility of transferring this method to 96- or 384-well microtitre plate formats where poly(vinylalcohol) was used as a glue to coat MIP microspheres in the wells [80]. Again, it was a competitive assay, and the bound fraction of 2,4-D tobacco peroxidase conjugate was quantified. Following the addition of a chemiluminescent substrate, luminol, light emission was measured by a CCD camera. Because of irregularities in the manually prepared polymer coatings, the sensitivity and reproducibility of the assay decreased when 384-well plates were used over 96-well plates. A flow injection system also was developed by covalently attaching the MIP to the inner wall of glass capillaries and regeneration of the MIP capillary after each measurement allowed consecutive measurements of a large number of samples [81].

Additional enzyme labeled assays include microtitre plates coated by the polymerization of 3-aminophenylboronic acid in the presence of epinephrine that resulted in the grafting of a thin MIP layer to the polystyrene surface of the wells [82]. Using an epinephrine-horseradish peroxidase (HRP) conjugate, an enzyme-linked assay for competitive binding studies was developed. Over a range of pH-values, the epinephrine-conjugate bound stronger to the MIP than to a non-imprinted reference polymer. Photografting has been used to coat polymer beads with a biotin-imprinted layer that showed superior affinity for a biotin-HRP conjugate over HRP [83] and HRP-labeled microcystin-LR was used to evaluate the affinity and cross-reactivity of microcystin-MIPs [84]. These methods were not as sensitive as some antibody based enzyme-linked assays, but the results demonstrate a great potential for the use of MIPs in enzyme-linked pseudoimmunoassay. A potential problem is the large size and bulkiness of the enzyme that may impede diffusion of the probe into the fine pores of the MIP and dramatically reduce binding with the imprinted sites. An example is the observed inability of an atrazine-MIP to recognize an enzyme labeled form of the template, and instead, a smaller fluoresceine-labeled triazine derivative was used as the probe [85]. Increased accessibility can be produced by the use of particles with imprinted sites located on the surface of the polymer such as MIP microspheres [79] and surface imprinted beads [83]. Also, HRP has been used as a high-molecular weight refractive label for surface plasmon resonance (SPR) detection of domoic acid using a thin MIP film coated on an SPR chip [86].

Two distinctly different approaches may be applied when developing a fluorescence assay: the use of a fluorescence-labeled derivative of the analyte; and the use of a structural analogue of the analyte that is fluorescent and selectively binds to the imprinted sites. A chloramphenicol-methyl red conjugate [87] and a dansylated analogue of chloramphenicol [88] were demonstrated to selectively bind to chloramphenicol-MIPs and used in competitive flow-injection assays (FIA) for determination of chloramphenicol. A fluorescein labeled 2,4-D derivative bound solely in a non-specific manner to the 2,4-D-MIP, and a non-related fluorescent probe, based on coumarin and with some structural features in common with the analyte, was found to selectively bind to the imprints and could be used for the development of a competitive assay for 2,4-D [89].

Electroactive structural analogues have been used as probes for the development of competitive assays for 2,4-D. One study investigated two electrochemically active compounds:

2, 4-dichlorophenol and homogentisic acid, of which the latter bound weakly, but selectively, with the imprints and could be used as a competitive probe on disposable screen-printed electrodes [90]. Another study identified a close structural analogue, 2-chloro-4-hydroxyphenoxyacetic acid that was used with a MIP prepared against the probe rather than the analyte on screen-printed electrodes for detection of 2,4-D [91].

Covalent incorporation of a scintillation monomer into the MIP allowed development of a homogenous radioactive assay that used the MIP itself as the detection system and did not require separation of bound and unbound ligand [92]. Imprinted microspheres were prepared where the selective binding of radio-labeled S-propranolol was translated into a radioluminescent signal that could be measured in both aqueous buffers and organic solvents [93]. Conventional radioactive assay could be simplified by the use of magnetic beads that were separated from the incubation solution by application of a magnetic field [94].

5.3.5 ANALYTICAL APPLICATIONS

Although great progress has been made in the development of assay systems, few studies have dealt with analysis of real samples. In early studies, pseudoimmunoassay of biosamples was performed after a prior liquid-liquid extraction of the analyte. Theophylline could be determined in patient serum samples of clinically significant concentrations and with good correlation with an established immunoassay method [64]. Diazepam in serum [64] and cyclosporin A in whole blood [95] could be similarly assayed, and preliminary data indicated that chloramphenicol could be detected in bovine serum [87]. Once the efficient use of pseudoimmunoassay under aqueous conditions was demonstrated, the first assay performed directly on samples without a prior clean up step for determination of S-propranolol in human plasma with a high degree of accuracy and precision was soon developed [96]. Environmental assays include 2,4-D in tap water samples [80].

5.4 SOLID PHASE EXTRACTION

5.4.1 INTRODUCTION TO SPE

In order to perform trace analysis of analytes contained in complex biological or environmental matrices, a sample pre-treatment step such as liquid-liquid extraction, protein precipitation, or solid phase extraction is often required. A more efficient sample clean-up that removes the majority of the matrix components simplifies the downstream analytical separation or immunoassay and facilitates accurate and sensitive quantification. Solid-phase extraction (SPE) involves passing the sample through a packed bed of extraction material that ideally binds the analyte quantitatively while most matrix components are not retained. Separation on most commercial SPE materials is based on physicochemical retention on the functionalized surface (C2, C8, C18, ion exchange, etc.), and the SPE column may not only retain the analyte but also other matrix components. Immunosorbents [97,98] and MIPs are more selective materials that rely on affinity interactions and potentially offer a more efficient extraction and cleaner extracts. Characteristic for both types of affinity materials are their high selectivity and affinity where selectivity can be pre-determined for a particular analyte by the choice of antigen used for antibody generation or the choice of template used for MIP preparation. Molecular imprint-based solid-phase extraction (MISPE) is a relatively young technique with the first reported study being that of Sellergren where a pentamidine-MIP was used for online sample enrichment of a spiked urine sample [99]. Since then, MIPs have been extensively investigated for use as an alternative to traditional SPE materials and, to date, have been successfully used for on-line and off-line extractions of both environmental and biological samples (Table 5.2).

TABLE 5.2
Selected Examples of the Applications of MISPE to Environmental, Bioanalytical and Food Control Analysis

Analyte	Type of Sample	Treatment Prior to MISPE	Analytical Separation and Detection	Concentration Range	References
Off-line mode extraction of aqueous samples					
Amobarbital	Human urine	Dilution with water	LC-UV diode array	0.2 µg/mL	[120]
Benzo[a]pyrene	Tap and lake water	Addition of 20% acetonitrile	LC-fluorescence detection	1–2 ng/mL	[121]
	Instant coffee	Dissolve in 50% acetonitrile		1–100 ng/g	
β-agonists	Calf urine	Dilution with water, enzymatic hydrolysis	LC-ESI-ion trap MS-MS	0.25–1 ng/mL	[122,123]
	Bovine meat	Digestion, hydrolysis, extraction into ethyl acetate, re-dissolution in water-methanol (4:1; v/v), de-fatting with heptane	LC-APCI-ion trap MS-MS	0.5–5 ng/g	[124]
Bisphenol A	River water	None	LC-fluorescence detection	3.3–120 ng/mL	[125]
Bupivacaine and ropivacaine	Human plasma	Dilution with citrate buffer pH 5	GC-NPD	3.9–1000 nmol/L	[67,105]
Tap water and river water		Oxidising UV-photolysis in the presence of H_2O_2	Flame atomic absorption spectroscopy	0.21–30 µg/L	[126]
Chlorophenoxy-acetic acids	River water	Acidification to pH 4	CZE-UV	2–10 µg/L	[127]
Chlorotriazine pesticides	Ground water	None	LC-UV diode array	~20 µg/L	[128]
	Sediment samples	Soxhlet extraction with methanol			
Clenbuterol	Calf urine	Dilution with ammonium acetate pH 6.7	LC-UV diode array	0.5–100 ng/mL	[129]
Compound A	Dog plasma	Protein precipitation with acetonitrile, dilution with 1 M chloroacetic acid in water-methanol (9:1; v/v)	LC-fluorescence detection	4–400 ng/mL	[119]
Copper ions	Sea water and certified reference seawater	None	Flame atomic absorption spectroscopy	0.4–25 ng/mL	[130]
Diphenylphosphate	Urine	Dilution with 50 mM citrate buffer pH 3.0 containing 10% acetonitrile	LC-ESI-ion trap MS	130–260 ng/mL	[131]
Hydroxycoumarin	Urine	None	CZE-UV diode array	10–50 µg/mL	[132]
M47070	Human plasma	Protein precipitation with acetonitrile	LC-fluorescence detection	2–60 ng/mL	[133,134]
Naphthalene sulfonates	River water	None	LC-UV	5–100 µg/L	[135]
Naproxen	Human urine	Acidification to pH 3	LC-UV	9–110 µg/L	[136]

Analyte	Matrix	Sample treatment	Detection method	Range/LOD	Ref.
Phenytoin	Plasma	None	LC-UV	2.5–40 µg/mL	[137]
Propranolol	Dog plasma, rat bile and human urine	None	LC-UV	2.5 µg/mL	[106]
Quercetin	Merlot red wine	None	LC-UV	8.8 mg/L	[138]
Sulfonylureas	Surface water, rainwater and drinking water	Treatment with EDTA	LC-UV diode array	50 ng/L	[139]
	Soil	Extraction with NaHCO$_3$ pH 7.8 and treatment with EDTA		50 µg/kg	
Triazines	Tap water, mineral water, diluted industrial effluent water	None	LC-UV diode array	1 µg/L	[140]
Off-line mode extraction after prior extraction into organic solvent					
Atrazine	Beef liver homogenate	Extraction with chloroform	LC-UV / ELISA	0.005–0.5 ppm / 0.005–0.5 ppm	[141]
Chlorotriazine pesticides	Tap and groundwater	Extraction on PS-DVB SPE disks and re-dissolution in toluene	MEKC-UV diode array	0.1–0.5 µg/L	[142]
	Soil	Extraction in acetone and re-dissolution in toluene		100 µg/L	
	Corn	Extraction in acetonitrile and re-dissolution in toluene		100 µg/L	
Clenbuterol	Liver samples	Matrix solid-phase dispersion on C18-sorbent and elution with acetonitrile-1% acetic acid	LC-electrochemical detection	5 ng/g	[102]
Fenuron	Wheat, barley, potato and carrot	Extraction with acetonitrile, re-dissolution in toluene	LC-UV diode array	100 ng/g	[143]
Monosulfuron	Soil	Extraction with 0.2 M ammonia in water-methanol (1:1; v/v) and re-dissolution in acetonitrile	LC-UV	0.1–2.5 µg/g	[144]
Nerve agent degradation products	Human serum	Extraction in acetonitrile	CE-UV	0.2–10 µg/g	[145]
Ochratoxin A	Red wine	Extraction on C18 cartridges, elution with methanol	LC-fluorescence detection	0.033–1 ng/mL	[146]

(continued)

TABLE 5.2 *(Continued)*

Analyte	Type of Sample	Treatment Prior to MISPE	Analytical Separation and Detection	Concentration Range	References
5-OH-PhIP	Human urine	Dilution with sodium acetate pH 5.5, hydrolysis with β-glucuronidase/arylsulfatase, extraction on Chromabond C$_{18e}$ column, elution with methanol and re-dissolution in 0.1% formic acid in 80% methanol	LC-ESI-ion trap MS^3		[147]
Triazines	Peas, potatoes and corn	Extraction with acetonitrile, re-dissolution in toluene, extraction on non-imprinted polymer	LC-UV diode array	20 ng/g	[148]
	Grape fruit juice	Extraction on SDB polymeric sorbent, re-dissolution in DCM-1% methanol	LC-UV diode array	10 µg/L	[149]
	Soil	Microwave assisted solvent extraction into DCM-methanol (9:1; v/v), re-dissolution in DCM-1% methanol	LC-UV diode array	20 ng/g	[149]
On-line mode extraction					
Alprenolol	Rat plasma	None	LC-fluorescence	12.5–250 ng/mL	[150]
Bisphenol A	River water	None	LC-fluorescence detection	25–1000 ng/L	[151]
	Lake water	None	LC-electrochemical detection	20 ng/L	[152]
Caffeine	Human urine, instant coffee and beverage	Dilution with water	LC-UV	0.18–1.8 µg/mL	[143]
Chlorotriazine pesticides	River water and aquacheck samples	Extraction on C18 RAM-SPE and elution with acetonitrile	LC-APCI-MS	0.1–2 ng/mL	[113]
Ibuprofen and naproxen	Rat plasma	None	LC-UV	0.2–50 µg/mL	[110]
4-Nitrophenol and 4-chlorophenol compounds	River water	Acidification to pH 2.5	LC-UV	1–100 µg/L	[116,154, 155]
Propranolol	Human serum	None	LC-UV	0.5–100 µg/mL	[117]
Triazine herbicides	Water containing humic acid	Extraction on C18 SPE column and elution with acetonitrile	LC-UV	0.5 ng/mL	[156]
	Apple extract	Extraction in methanol, re-dissolution in buffer, extraction on C18 SPE column and elution with acetonitrile		20 ng/mL	

Analyte	Matrix	Sample treatment	Detection	Range	Ref.
	Urine	Extraction on C18 SPE column and elution with acetonitrile		20 ng/mL	
Verapamil and gallopamil	Plasma, urine and cell culture medium	Extraction on RAM column, elution with acetonitrile	LC-ESI ion trap MS	25–500 ng/mL	[115]
Extraction with direct detection					
4-Aminopyridine	Human serum	Extraction in chloroform	UV	2.5–100 μg/mL	[157]
Cephalexin	Human plasma and serum	Extraction on C18 SPE column, elution with methanol, dilution with chloroform	LC-UV	1–20 μg/mL	[158]
			ESI-MS	0.25–25 μg/mL	[159]
Chloramphenicol	Skimmed milk and full cream milk	Deproteinisation with 15% trichloroacetic acid in water	Square wave voltammetry	9.7–485 μg/L	[160]
Metformin	Human plasma	Protein precipitation with acetonitrile- phosphate buffer pH 7 (9:1; v/v)	UV	0.1–10 μg/mL	[161]
Nicotine	Tobacco	Extraction in methanol -0.1N NaOH (1:1) and dilution with methanol	UV	1.8–1000 μg/mL	[162]
Pirimicarb	Tap water, spring water, river water and sea water	None	Differential pulse voltammetry	71.5 μg/L	[163]
Theophylline	Human serum	Extraction in chloroform	UV	0.25–1000 μg/mL	[118,164]

5.4.2   METHOD DEVELOPMENT

5.4.2.1   Solvent Systems and Solvent Switch

MISPE method development has used either aqueous samples after appropriate pre-treatment steps such as pH-adjustment or protein precipitation of plasma samples (Table 5.2) or in particular in many early MISPE studies application of samples dissolved in a non-polar organic solvent. To achieve this, the analyte first must be extracted into an organic solvent that may be done by solvent extraction of solid samples (soil, tissue, vegetables), liquid-liquid extraction (serum, plasma), or SPE using a hydrophobic column (various types of water samples, plasma) (Table 5.2). In most instances, highly selective binding can be achieved in apolar organic solvents, sometimes in the presence of a polar additive such as acetic acid, formic acid, ethanol, or methanol [43,100–102], to suppress non-specific binding.

In order to avoid a time-consuming extraction of analyte into organic solvents prior to MISPE, considerable work has been done to investigate the possibilities of direct application of aqueous samples to the MIP material. Upon extraction of aqueous samples, quantitative retention of the analyte is achieved in many instances through a combination of imprint binding and non-specific hydrophobic adsorption to the polymer surface. Then, a selective wash step is required to improve MIP selectivity and to remove all other adsorbed sample components. The strength and selectivity of the imprint-analyte binding is tuned by the solvent properties of the surrounding medium (aqueous vs. non-aqueous, buffer pH, additives, etc.). Also non-specific retention varies with the conditions used and under aqueous conditions, non-specific physicochemical retention is mainly due to hydrophobic interactions. The selective imprint-analyte binding that is mainly due to hydrogen bonding and electrostatic interactions is strong in apolar solvents where non-specific hydrophobic adsorption is weak. Therefore, a solvent switch [67,103], e.g., to acetonitrile or dichloromethane, changes the retention conditions to the normal-phase mode that leads to re-distribution of the analyte to the imprinted sites and washing off of non-related structures (Figure 5.4 and Figure 5.5). For large sample volumes, e.g., in environmental analysis, the strong hydrophobic adsorption can be employed for capturing the analyte from the aqueous sample passed through the column. A subsequent solvent switch assures a selective MISPE method ([104] and references in Table 5.2). The decision to

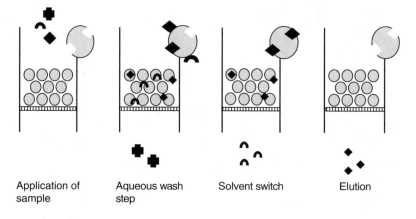

| Application of sample | Aqueous wash step | Solvent switch | Elution |

FIGURE 5.4 A typical protocol for extraction of aqueous samples on a MIP column. After activation of the column, the aqueous sample is applied. This is followed by an aqueous wash step to remove any hydrophilic matrix components. A solvent switch using an organic wash step then removes any hydrophobic matrix components non-specifically bound to the polymer surface and encourages selective binding of the analyte to the imprinted sites. Finally, the analyte is eluted using either acid or base in solvent or water.

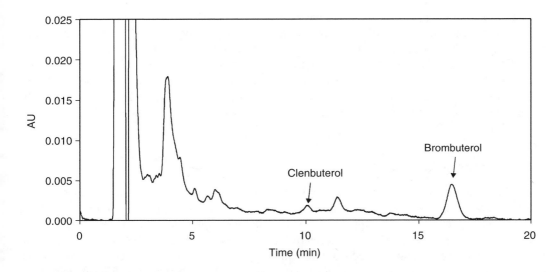

FIGURE 5.5 Chromatogram from the analysis of an extract of 5 mL calf urine spiked with clenbuterol at 0.5 ng/mL. The sample was applied to a brombuterol-MIP, and the column was washed with 1 mL of water, 1 mL of acetonitrile-acetic acid (98:2; v/v), 1 mL of 0.5 M ammonium acetate pH 5, and 1 mL of acetonitrile-water (7:3; v/v). The analyte was eluted using methanol-trifluoroacetic acid (99:1; v/v) and analyzed on a BetaBasic C18 column with UV detection. (Reprinted from Blomgren, A., Berggren, C., Holmberg, A., Larsson, F., Sellergren, B., and Ensing, K., *J. Chromatogr. A*, 975, 2002. With permission.)

use organic or aqueous conditions for a MISPE method relies on the type of sample and analyte, i.e., whether it is necessary or optimal to use pre-treatment steps using organic solvents and retention and selectivity on the MIP under different conditions.

Because of the often strong imprint-analyte affinity, difficulties in effecting quantitative elution of the analyte have been encountered in some instances [35,105–107]. This is most pronounced for extraction of basic analytes containing amino functionalities on MAA-type MIPs where typically eluents consisting of (sometimes high percentages of) acetic acid, formic acid, TFA, or TEA in acetonitrile or methanol are used. For neutral compounds, weak acids, and weak bases, complete elution may occur simply by treatment with polar solvents, mixtures of polar solvents and water, or acidic water.

On occasion, the clean up from this approach is not sufficient as in the determination of triazines in corn samples using solvent extraction followed by MISPE. As a result, a two-step MISPE approach was employed where a non-imprinted polymer was used to retain some interfering matrix compounds [100]. The analytes were eluted from the non-imprinted polymer and transferred for further clean up on the imprinted polymer. A possibility to speed up SPE in the off-line mode is the parallel extraction of samples using a 96-well MISPE format as was reported by Chassaing et al. [119]. The high throughput 96-well format resulted in improved reproducibility, and the clean MISPE extracts allowed the method to be validated at a better sensitivity than a conventional SPE assay.

5.4.2.2 Non-Specific Adsorption

Because of the hydrophobic nature of many polymers, extraction of aqueous samples often results in moderately and highly lipophilic compounds being non-specifically adsorbed to the MIP surface. This hydrophobic driven adsorption can be reduced while largely maintaining the integrity of the binding of analyte to the imprinted sites by the addition of an organic modifier such as ethanol, methanol, or acetonitrile [65,66] or a detergent such as Tween 20, Triton X-100, or Brij 35 [78] to the buffer. Also, buffer pH may influence the extent of non-specific adsorption where increased pH results in increased adsorption of basic compounds to MAA type polymers [65,66], and decreased pH results in increased adsorption of acidic compounds to VPy type polymers [32]. An example is an MAA-based bupivacaine-MIP for which optimal selective binding was achieved at pH 5 in the presence of 5% ethanol and 0.5% Tween 20 [66]. Non-specific adsorption also may be reduced by the use of small columns, thereby reducing the polymer surface area available for lipophilic adsorption. As most MISPE applications are for trace analysis, binding capacity does not seem to be a problem.

5.4.2.3 Template Bleeding

One of the main technical problems with the use of MIPs in SPE is leaching, or bleeding, of template molecules during elution of the extracted analyte. It is not unusual to observe template peaks when extracting blank samples as a result of the incomplete removal of the template after polymerization. Whereas often, template recovery of $\geq 99\%$ is obtained, this still leaves a significant amount of the template embedded within the polymer matrix. This leftover amount can leach during an extraction process and give artificially high levels of analyte, resulting in inaccurate quantification when performing trace analysis. A study into the influence of various post-polymerization treatments on bleeding compared extraction techniques such as thermal annealing, microwave-assisted extraction, Soxhlet extraction, and supercritical fluid desorption [34]. Although microwave-assisted extraction using trifluoroacetic acid or formic acid was found to be the most efficient extraction technique, polymer degradation and loss of selectivity also were observed. After solvent extraction, the lowest levels of bleeding were achieved when the polymer was washed with pure organic acids such as TFA or acetic acid. None of these methods, however, eradicated the problem of bleeding; they only reduced the bleed to acceptable levels.

In order to circumvent problems with template bleeding, a structural analogue of the analyte or a dummy template can be used as the imprint molecule (Table 5.3). The alternative template should possess common structural features with the analyte and give rise to imprints with the ability to selectively bind the target analyte(s) (Figure 5.6). Following extraction and elution, the analyte and leaking template then are separated via a chromatographic technique for quantification. An early example is the use of a close structural analogue of sameridine to form a MIP for the extraction of sameridine where the MIP was found to bind analyte, internal standard, and template with similar strength [35]. Dibutylmelamine was used as a dummy template [36] for the preparation of a MIP for class selective extraction of triazine herbicides from environmental samples [108]. A per-deuterated analogue of bisphenol A was used for the preparation of a bisphenol A-selective MIP [109]. Other examples are seen in Table 5.3. A possible limitation of this approach is the availability of two structural analogues for use as template and internal standard. Control over template bleeding is particularly important in a trace analysis, and each method development must include a confirmation that template bleeding does not interfere with the assay giving rise to poor accuracy and precision. The risk is most severe for off-line extraction protocols using fresh material for each extraction, whereas for on-line protocols, the MIP column is constantly washed by the continuous flow, and bleeding can be reduced to below detection levels.

5.4.2.4 Aqueous Compatible MIPs

Recently, there has been an increased effort to produce more aqueous compatible MIPs. The addition of the hydrophilic monomers glycerol monomethacrylate and glycerol dimethacrylate at the final stage of polymerization can render the polymer surface more hydrophilic [110]. The resultant MIPs were successfully applied to the extraction of rat plasma samples. Serum albumin was quantitatively recovered in the breakthrough volume demonstrating their hydrophilic properties with low non-specific hydrophobic adsorption. Another technique uses the replacement of a

TABLE 5.3
Selected Examples of Structural Analogues Used for Molecular Imprinting

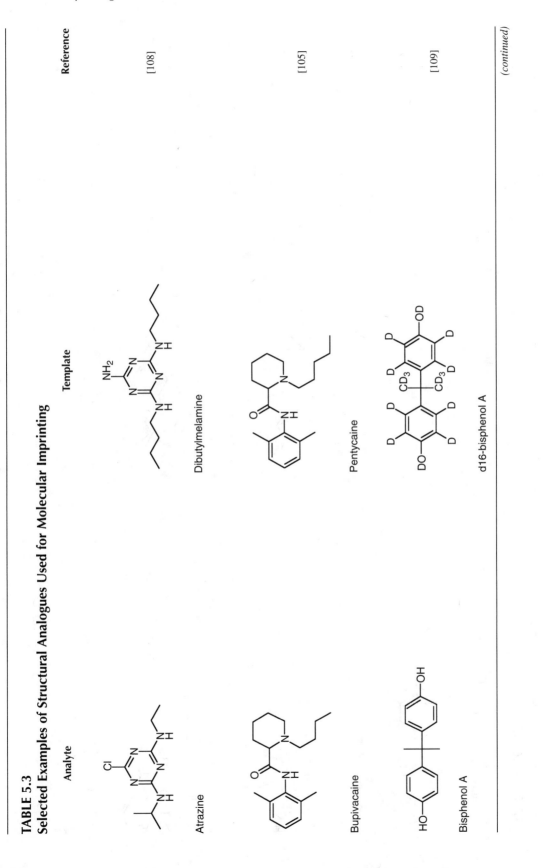

Analyte	Template	Reference
Atrazine	Dibutylmelamine	[108]
Bupivacaine	Pentycaine	[105]
Bisphenol A	d16-bisphenol A	[109]

(continued)

TABLE 5.3 *(Continued)*

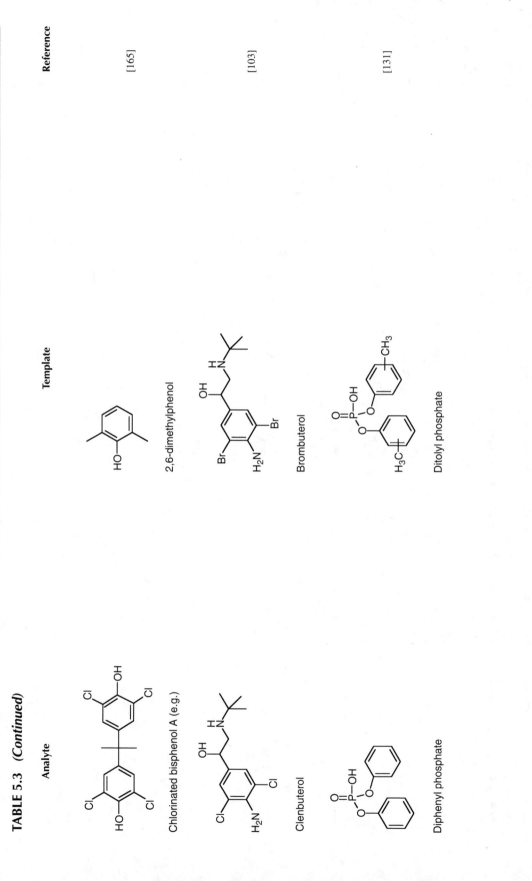

Analyte	Template	Reference
Chlorinated bisphenol A (e.g.)	2,6-dimethylphenol	[165]
Clenbuterol	Brombuterol	[103]
Diphenyl phosphate	Ditolyl phosphate	[131]

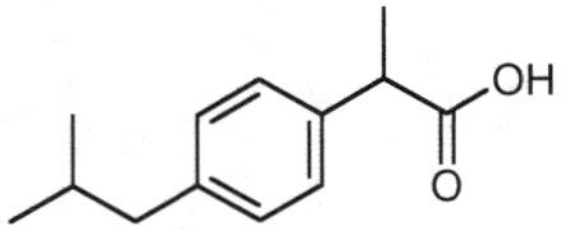

[110]

S-Naproxen

[120]

Amobarbital

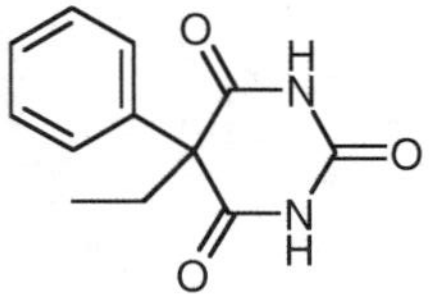

[35]

Analogue

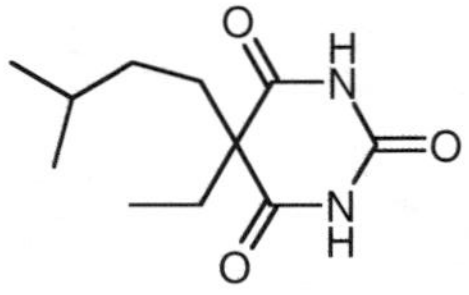

[166]

Hyoscyamine

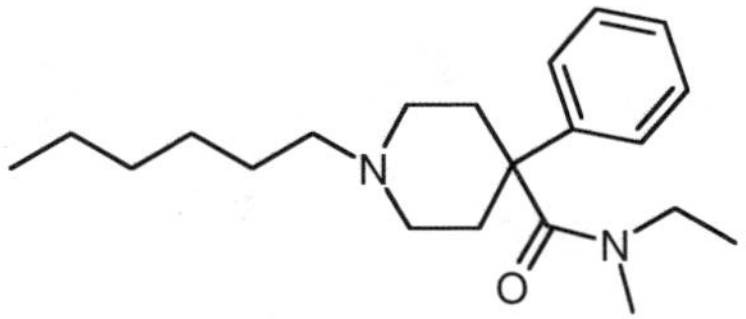

S-Ibuprofen

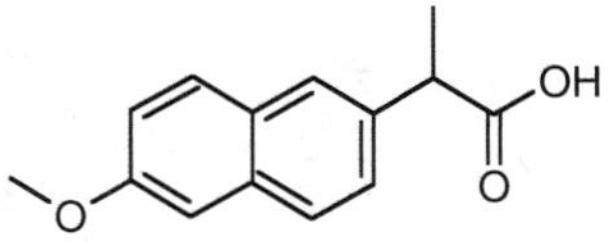

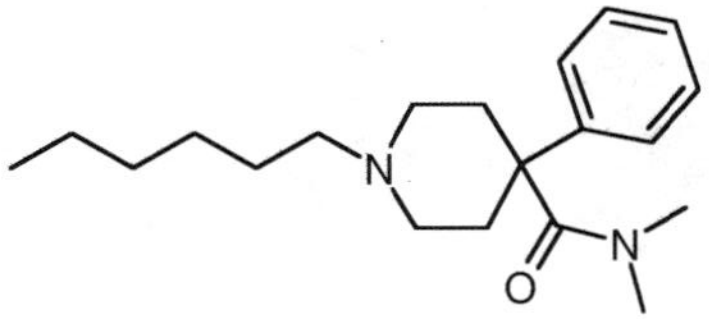

Phenobarbital

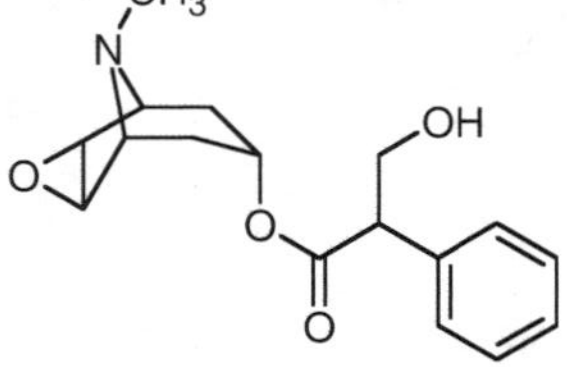

Sameridine

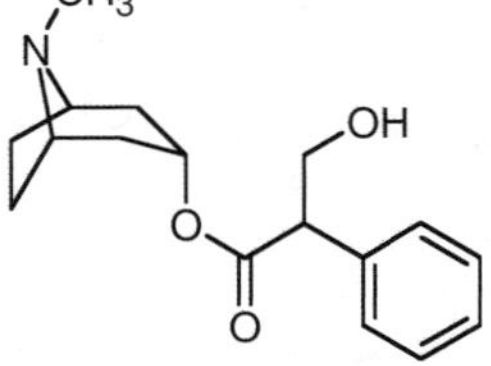

Scopolamine

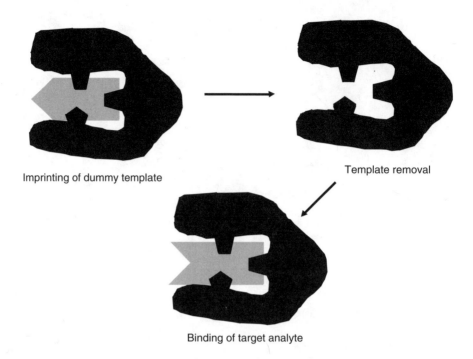

Imprinting of dummy template

Template removal

Binding of target analyte

FIGURE 5.6 Imprinting of a dummy template produces imprints that recognize structural analogues when the same binding interactions are formed as for the original template. Variations in non-binding parts of the analogue are allowed.

portion of the cross-linking monomer EGDMA with 2-hydroxyethylmethacrylate (HEMA). Incorporation of HEMA in the polymer matrix created a hydrophilic bupivacaine-MIP that, in pure phosphate buffer pH 7.4, showed strong retention of the template while non-specific binding was reduced compared with a standard MAA-EGDMA MIP [111]. Successful extraction of plasma samples was demonstrated. Other aqueous compatible MIPs include MIP membranes with low non-specific binding synthesized as a MIP layer grafted onto a commercial low binding membrane (hydrophilised PVDF) at optimum cross-linker ratios [112].

5.4.2.5 On-Line Extraction Techniques

Although the majority of the work published on MISPE has been performed using off-line SPE, various on-line extraction protocols offer possibilities to automate and to perform the entire sample preparation and analysis in one operation and in a short period of time. The sample can be introduced either directly on the MIP column or via a prior trapping column that captures the analyte and transfers it to a solvent where MIP-analyte binding is selective (Table 5.2). In a multidimensional extraction of chlorotriazines [113], the sample was loaded onto a restricted access material (RAM) column that retained low molecular weight compounds on its hydrophobic inner surface while high molecular weigh compounds were directed to waste. The retained small compounds were then transferred in an organic solvent from the RAM column onto the MIP column where only the triazines were retained. The triazines were next transferred to the analytical separation system in acidic water containing an organic solvent modifier. An advantage of this approach is that selective binding can be achieved in the load solvent without the need for additional wash steps. This approach has been applied to the extraction of tramadol from human plasma [114] and to varying degrees validated for extraction of verapamil from plasma and urine [115] and triazines from river water [113]. Direct on-line injection of samples onto the MIP column is also possible and

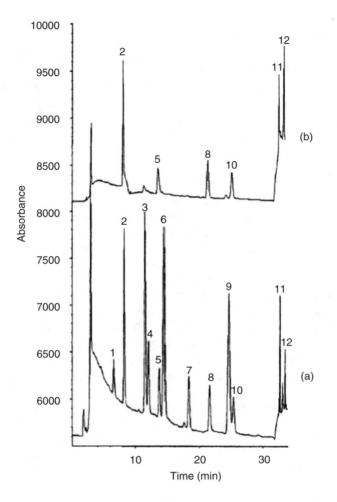

FIGURE 5.7 Chromatogram obtained by on-line MISPE with a 4-chlorophenol-MIP of 10 mL Ebro river water (pH 2.5) spiked at 10 μg/L with 12 phenolic compounds. (a) With washing using 4 mL water (pH 2.5). (b) After an additional wash step using 0.1 mL of dichloromethane only phenols with a 4-chloro group were retained. The analytes were desorbed using 1% acetic acid in acetonitrile and analysed on a C18 column with UV detection. (Reprinted from Caro, E., Marcé, R. M., Cormack, P. A. G., Sherrington, D. C., and Borrull, F., *J. Chromatogr. A*, 995, 2003. With permission.)

has been demonstrated for extraction of 4-chlorophenols and 4-nitrophenol (Figure 5.7) [116]. The use of optimal loading conditions and a solvent switch wash, however, is required to enhance the selectivity of binding to the polymers. A highly interesting achievement is the use of a MIP with a hydrophilic external surface as a RAM-MIP for direct injection of plasma samples in a coupled-column system [110]. The versatility of an automated and on-line MIP solid-phase micro extraction (SPME) system was demonstrated by the determination of propranolol in serum [117]. Advantages of on-line MISPE are drastic reduction or elimination of the leaching problem and the ability to process and pre-concentrate large sample volumes.

A simple technique that relies solely on the highly efficient sample clean up achievable on MIPs uses on-line extraction with direct detection by UV, mass spectrometry, fluorescence, or voltammetry (Table 5.2). Extraction on a MIP micro-column with pulsed elution and direct in-line UV-detection was first realized by Mullet and Lai [118]. The technique involves the application of a sample using a mobile phase that encourages selective binding of the analyte to the imprinted sites and the elimination of non-specific adsorption of interfering components by differential pulsed elution using successive aliquots of solvent with increasing polarity and protic nature. Pulsed elution of analyte in a very narrow band allows sensitive UV detection online. Theophylline was extracted from serum into chloroform, and an aliquot of the organic layer was injected onto the MIP micro-column in a mobile phase of chloroform [118]. Interfering drugs were eliminated by an intermediate wash with a pulse of acetonitrile and a pulse of methanol quantitatively desorbed the bound theophylline.

5.5 CONCLUSIONS

Two main attractions of molecular imprinting are its apparent simplicity of polymer synthesis, in the presence of a template molecule where subsequent removal of the template produces a material with memory sites, and the highly selective rebinding of the original template from a mixture. The chemistries available for imprinting are expanding with novel monomers being added to the toolbox that provide greater ability to synthesize selective MIPs against an increasing number of target structures. A current effort is the exploration of different polymerization techniques to produce MIPs in any desired format, from microbeads of highly defined shape and narrow size range, to thin polymer membranes, and to MIPs with hydrophilic external surface for better compatibility with biological matrices. Also, the future will see more MIPs with built-in detection systems where binding gives rise to fluorescent, scintillation, or other type of signal. Increased use of structural analogue and dummy template imprinting circumvents problems with template leakage and produces MIPs with class-selective recognition. As seen by the variety of pseudoimmunoassay and solid-phase extraction applications discussed in this review, MIPs can be employed for separation of the analyte in matrices ranging from pure organic solvents to complex environmental and biological samples. Several MISPE studies have made direct comparisons with conventional SPE materials and demonstrated superior cleanup using the MIP material. This allows the use of MIPs for pre-concentration of very large sample volumes and simplification of the downstream analytical separation system as well as for single analyte analysis with direct quantification of the MISPE eluate by UV, fluorescence, or mass spectrometry detection.

REFERENCES

1. Sellergren, B., Ed, *Man-Made Mimics of Antibodies and their Applications in Analytical Chemistry*, Elsevier, Amsterdam, 2001.
2. Wulff, G., Molecular imprinting in cross-linked materials with the aid of molecular templates—a way towards artificial antibodies, *Angew. Chem. Int. Ed. Engl.*, 34, 1812, 1995.
3. Wulff, G. and Biffis, A., Molecular imprinting with covalent or stoichiometric non-covalent interactions, In *Molecularly Imprinted Polymers. Man-Made Mimics of Antibodies and their Applications in Analytical Chemistry*, Sellergren, B., Ed., Elsevier, Amsterdam, pp. 71–111, 2001.
4. Mosbach, K. and Ramström, O., The emerging technique of molecular imprinting and its future impact on biotechnology, *Bio/Technology*, 14, 163–170, 1996.
5. Sellergren, B., The non-covalent approach to molecular imprinting. In *Molecularly Imprinted Polymers. Man-Made Mimics of Antibodies and their Applications in Analytical Chemistry*, Sellergren, B., Ed., Elsevier, Amsterdam, pp. 113–184, 2001.
6. Wulff, G. and Knorr, K., Stoichiometric non-covalent interaction in molecular imprinting, *Bioseparation*, 10, 257–276, 2002.
7. Whitcombe, M. J., Rodriguez, M. E., Villar, P., and Vulfson, E. N., A new method for the introduction of recognition site functionality into polymers prepared by molecular imprinting: synthesis and characterization of polymeric receptors for cholesterol, *J. Am. Chem. Soc.*, 117, 7105–7111, 1995.
8. Klein, J. U., Whitcombe, M. J., Mulholland, F., and Vulfson, E. N., Template-mediated synthesis of a polymeric receptor specific to amino acid sequences, *Angew. Chem. Int. Ed. Engl.*, 38, 2057–2060, 1999.

9. Striegler, S., Designing selective sites in templated polymers utilizing coordinative bonds, *J. Chromatogr. B.*, 804, 183–195, 2004.

10. Chen, G. H., Guan, Z. B., Chen, C. T., Fu, L. T., Sundaresan, V., and Arnold, F. H., A glucose-sensing polymer, *Nat. Biotechnol.*, 15, 354–357, 1997.

11. Cormack, P. A. G. and Zurutuza Elorza, A., Molecularly imprinted polymers: synthesis and characterisation, *J. Chromatogr. B.*, 804, 173–182, 2004.

12. Sellergren, B. and Hall, A. J., Fundamental aspects on the synthesis and characterization of imprinted network polymers, In *Molecularly Imprinted Polymers. Man-Made Mimics of Antibodies and their Applications in Analytical Chemistry*, Sellergren, B., Ed., Elsevier, Amsterdam, pp. 21–57, 2001.

13. Ramstrom, O., Andersson, L. I., and Mosbach, K., Recognition sites incorporating both pyridinyl and carboxy functionalities prepared by molecular imprinting, *J. Org. Chem.*, 58, 7562–7564, 1993.

14. Matsui, J., Doblhoff-Dier, O., and Takeuchi, T., 2-(Trifluoromethyl)acrylic acid: a novel functional monomer in non-covalent molecular imprinting, *Anal. Chim. Acta*, 343, 1–4, 1997.

15. Matsui, J., Miyoshi, Y., and Takeuchi, T., Fluoro-functionalised molecularly imprinted polymers selective for herbicides, *Chem. Lett.*, 1007–1008, 1995.

16. Yu, C. and Mosbach, K., Molecular imprinting utilizing an amide functional group for hydrogen bonding leading to highly efficient polymers, *J. Org. Chem.*, 62, 4057–4064, 1997.

17. Sreenivasan, K., Effect of the type of monomers of molecularly imprinted polymers on the interaction with steroids, *J. Appl. Pol. Sci.*, 68, 1863–1866, 1998.

18. Hall, A. J., Achilli, L., Manesiotis, P., Quaglia, M., De Lorenzi, E., and Sellergren, B., A substructure approach toward polymeric receptors targeting dihydrofolate reductase inhibitors 2. Molecularly imprinted polymers against Z-L-glutamic acid showing affinity for larger molecules, *J. Org. Chem.*, 68, 9132–9135, 2003.

19. Wulff, G. and Schönfeld, R., Polymerizable amidines—Adhesion mediators and binding sites for molecular imprinting, *Adv. Mater.*, 10, 957–959, 1998.

20. Kempe, M., Fischer, L., and Mosbach, K., Chiral separation using molecularly imprinted hetero-aromatic polymers, *J. Mol. Recogn.*, 6, 25–29, 1993.

21. Mayes, A. G. and Lowe, C. R., Optimization of molecularly imprinted polymer for radioligand binding assays, In *Drug-Development Assay Approaches Including Molecular Imprinting and Bio-markers*, Reid, E., Hill, M., and Wilson, I. D., Eds., The Royal Society of Chemistry, Cambridge, U.K., pp. 28–36, 1998.

22. Yilmaz, E., Mosbach, K., and Haupt, K., Influence of functional and cross-linking monomers and the amount of template on the performance of molecularly imprinted polymers in binding assays, *Anal. Commun.*, 36, 167–170, 1999.

23. Svenson, J. and Nicholls, I. A., On the thermal and chemical stability of molecularly imprinted polymers, *Anal. Chim. Acta*, 435, 19–24, 2001.

24. Kempe, M., Antibody-mimicking polymers as chiral stationary phases in HPLC, *Anal. Chem.*, 68, 1948–1953, 1996.

25. Zander, Å., Findlay, P., Renner, T., Sellergren, B., and Swietlow, A., Analysis of nicotine and its oxidation products in nicotine chewing gum by a molecularly imprinted solid phase extraction, *Anal. Chem.*, 70, 3304–3314, 1998.

26. Zhu, Q.-Z., Haupt, K., Knopp, D., and Niessner, R., Molecularly imprinted polymer for metsulfuron-methyl and its binding characteristic for sulfonylurea herbicides, *Anal. Chim. Acta*, 468, 217–227, 2002.

27. Sibrian-Vasquez, M. and Spivak, D. A., Enhanced enantioselectivity of molecularly imprinted polymers formulated with novel cross-linking monomers, *Macromolecules*, 36, 5105–5113, 2003.

28. Sibrian-Vasquez, M. and Spivak, D. A., Molecular imprinting made easy, *J. Am. Chem. Soc.*, 126, 7827–7833, 2004.

29. Sellergren, B., Molecular imprinting by noncovalent interactions. Enantioselectivity and binding capacity of polymers prepared under conditions favouring the formation of template, *Makromol. Chem.*, 190, 2703–2711, 1989.

30. Davies, M. P., de Biasi, V., and Perrett, D., Approaches to the rational design of molecularly imprinted polymers, *Anal. Chim. Acta*, 504, 7–14, 2004.

31. Sellergren, B. and Shea, K. J., Influence of polymer morphology on the ability of imprinted network polymers to resolve enantiomers, *J. Chromatogr.*, 635, 31–49, 1993.

32. Haupt, K., Dzgoev, A., and Mosbach, K., Assay system for the herbicide 2,4-Dichlorophenoxy acetic acid using a molecularly imprinted polymer as an artificial recognition element, *Anal. Chem.*, 70, 628–631, 1998.

33. O'Shannessy, D. J., Ekberg, B., and Mosbach, K., Molecular imprinting of amino-acid derivatives at low-temperature (0°C) using photolytic homolysis of azobisnitriles, *Anal. Biochem.*, 177, 144–149, 1989.

34. Ellwanger, A., Berggren, C., Bayoudh, S., Crecenzi, C., Karlsson, L., Owens, P. K., Ensing, K., Sherrington, D., and Sellergren, B., Evaluation of methods aimed at complete removal of template from molecularly imprinted polymers, *Analyst*, 126, 784–792, 2001.

35. Andersson, L. I., Paprica, A., and Arvidsson, T., A highly selective solid-phase extraction sorbent for preconcentration of sameridine made by molecular imprinting, *Chromatographia*, 46, 57–62, 1997.

36. Matsui, J., Fujiwara, K., and Takeuchi, T., Atrazine-selective polymers prepared by molecular imprinting of trialkylmelamines as dummy template species of atrazine, *Anal. Chem.*, 72, 1810–1813, 2000.

37. Takeuchi, T., Fukuma, D., and Matsui, J., Combinatorial molecular imprinting: an approach to synthetic polymer receptors, *Anal. Chem.*, 71, 285–290, 1999.

38. Lanza, F., Sellergren, B., Method for synthesis and screening of large groups of molecularly imprinted polymers, *Anal. Chem.*, 71:2092-2096.

39. Navarro-Villoslada, F., San Vicente, B., and Moreno-Bondi, M. C., Application of a multivariate analysis to the screening of molecularly imprinted polymers for bisphenol A, *Anal. Chim. Acta.*, 504, 149–162, 2004.

40. Sellergren, B., Lepistö, M., and Mosbach, K., Highly enantioselective and substrate-selective polymers obtained by molecular imprinting utilizing noncovalent interactions—NMR and chromatographic studies on the nature of recognition, *J. Am. Chem. Soc.*, 110, 5853–5860, 1988.

41. Takeuchi, T., Dobashi, A., and Kimura, K., Molecular imprinting of biotin derivatives and its application to competitive binding assay using nonisotopic labeled ligands, *Anal. Chem.*, 72, 2418–2422, 2000.

42. Karlsson, J. G., Karlsson, B., Andersson, L. I., and Nicholls, I. A., The roles of template complexation and ligand binding conditions on recognition in bupivacaine molecularly imprinted polymers, *Analyst*, 129, 456–462, 2004.

43. Ansell, R. J. and Kuah, K. L., Imprinted polymers for chiral resolution of (+/−)-ephedrine: understanding the pre-polymerization equilibrium and the action of different mobile phase modifiers, *Analyst*, 130, 179–187, 2005.

44. Piletsky, S. A., Karim, K., Piletska, E. V., Day, C. J., Freebairn, K. W., Legge, C., and Turner, A. P. F., Recognition of ephedrine enantiomers by molecularly imprinted polymers designed using a computational approach, *Analyst*, 126, 1826–1830, 2001.

45. Pérez-Moral, N. and Mayes, A. G., Novel MIP formats, *Bioseparation*, 10, 287–299, 2002.

46. Haginaka, J., Takehira, H., Hosoya, K., and Tanaka, N., Molecularly imprinted uniform-sized polymer-based stationary phase for naproxen—Comparison of molecular recognition ability of the molecularly imprinted polymers prepared by thermal and redox polymerization techniques, *J. Chromatogr. A*, 816, 113–121, 1998.

47. Mayes, A. G. and Mosbach, K., Molecularly imprinted polymer beads: Suspension polymerization using a liquid perfluorocarbon as the dispersing phase, *Anal. Chem.*, 68, 3769–3774, 1996.

48. Ansell, R. J. and Mosbach, K., Molecularly imprinted polymers by suspension polymerization in perfluorocarbon liquids, with emphasis on the influence of the porogenic solvent, *J. Chromatogr. A.*, 787, 55–66, 1997.

49. Ober, C. K. and Lok, K. P., Formation of large monodisperse copolymer particles by dispersion polymerization, *Macromol.*, 20, 268–273, 1987.

50. Ye, L., Cormack, P. A. G., and Mosbach, K., Molecularly imprinted monodisperse microspheres for competitive radioassay, *Anal. Commun.*, 36, 35–38, 1999.

51. Pérez, N., Whitcombe, M. J., and Vulfson, E. N., Molecularly imprinted nanoparticles prepared by core-shell emulsion polymerization, *J. Appl. Polym. Sci.*, 77, 1851–1859, 2000.

52. Pérez-Moral, N. and Mayes, A. G., Comparative study of imprinted polymer particles prepared by different polymerization methods, *Anal. Chim. Acta.*, 504, 15–21, 2004.

53. Dickert, F. L., Achatz, P., and Halikias, K., Double molecular imprinting—a new sensor concept for improving selectivity in the detection of polycyclic aromatic hydrocarbons (PAHs) in water, *Fresenius J. Anal. Chem.*, 371, 11–15, 2001.

54. Wizeman, W. J. and Kofinas, P., Molecularly imprinted polymer hydrogels displaying isomerically resolved glucose binding, *Biomaterials*, 22, 1485–1491, 2001.

55. Matsui, J., Tamaki, K., and Sugimoto, N., Molecular imprinting in alcohols: Utility of a pre-polymer based strategy for synthesizing stereoselective artificial receptor polymers in hydrophilic media, *Anal. Chim. Acta*, 466, 11–15, 2002.

56. Marx, S. and Liron, Z., Molecular imprinting in thin films of organic-inorganic hybrid sol-gel and acrylic polymers, *Chem. Mater.*, 13, 3624–3630, 2001.

57. Fernández-González, A., Laíño, R. B., Diaz-García, M. E., Guardia, L., and Vialem, A., Assessment of molecularly imprinted sol-gel materials for selective room temperature phosphorescence recognition of nafcillin, *J. Chromatogr. B*, 804, 247–254, 2004.

58. Marty, J. D., Mauzac, M., Fournier, C., Rico-Lattes, I., and Lattes, A., Liquid crystal polysiloxane networks as materials for molecular imprinting technology: Memory of the mesomorphic organization, *Liq. Cryst.*, 29, 529–536, 2002.

59. Ulbricht, M., Membrane separations using molecularly imprinted polymers, *J Chromatogr. B*, 804, 113–125, 2004.

60. Haupt, K., Noworyta, K., and Kutner, W., Imprinted polymer-based enantioselective acoustic sensor using a quartz crystal microbalance, *Anal. Commun.*, 36, 391–393, 1999.

61. Blanco-Lopez, M. C., Lobo-Castanon, M. J., Miranda-Ordieres, A. J., and Tunon-Blanco, P., Voltammetric sensor for vanillylmandelic acid based on molecularly imprinted polymer-modified electrodes, *Biosens. Bioelectron.*, 18, 353–362, 2003.

62. Piletsky, S. A., Piletskaya, E. V., Elgersma, A. V., Yano, K., Karube, I., Parhometz, Y. P., and El'skaya, A. V., Atrazine sensing by molecularly imprinted membranes, *Biosens. Bioelectron.*, 10, 959–964, 1995.

63. Price, C. P. and Newman, D. J., Eds., *Principles and Practice of Immunoassay* 2nd ed., Macmillan Reference Ltd, London, 1997.

64. Vlatakis, G., Andersson, L. I., Müller, R., and Mosbach, K., Drug assay using antibody mimics made by molecular imprinting, *Nature*, 361, 645–647, 1993.

65. Andersson, L. I., Application of molecular imprinting to the development of aqueous buffer and organic solvent based radioligand binding assays for (S)-propranolol, *Anal. Chem.*, 68, 111–117, 1996.

66. Karlsson, J. G., Andersson, L. I., and Nicholls, I. A., Probing the molecular basis for ligand-selective recognition in molecularly imprinted polymers selective for local anaesthetic bupivacaine, *Anal. Chim. Acta*, 435, 57–64, 2001.

67. Andersson, L. I., Hardenborg, E., Sandberg-Ställ, M., Möller, K., Henriksson, J., Bramsby-Sjöström, I., Olsson, L. I., and Abdel-Rehim, M., Development of a molecularly imprinted polymer based solid-phase extraction of local anaesthetics from human plasma, *Anal. Chim. Acta*, 526, 147–154, 2004.

68. Umpleby, R. J., Baxter, S. C., Rampey, A. M., Rushton, G. T., Chen, Y. Z., and Shimizu, K. D., Characterization of the heterogeneous binding site affinity distributions in molecularly imprinted polymers, *J. Chromatogr. B*, 804, 141–149, 2004.

69. Sajonz, P., Kele, M., Zhong, G. M., Sellergren, B., and Guiochon, G., Study of the thermodynamics and mass transfer kinetics of two enantiomers on a polymeric imprinted stationary phase, *J. Chromatogr. A*, 810, 1–17, 1998.

70. Kim, H. J. and Guiochon, G., Comparison of the thermodynamic properties of particulate and monolithic columns of molecularly imprinted copolymers, *Anal. Chem.*, 77, 93–102, 2005.

71. Umpleby, R. J., Bode, M., and Shimizu, K. D., Measurement of the continuous distribution of binding sites in molecularly imprinted polymers, *Analyst*, 125, 1261–1265, 2000.

72. Rampey, A. M., Umpleby, R. J., Rushton, G. T., Iseman, J. C., Shah, R. N., and Shimizu, K. D., Characterization of the imprint effect and the influence of imprinting conditions on affinity, capacity, and heterogeneity in molecularly imprinted polymers using the Freundlich isotherm-affinity distribution analysis, *Anal. Chem.*, 76, 1123–1133, 2004.

73. Spivak, D., Gilmore, M. A., and Shea, K. J., Evaluation of binding and origins of specificity of 9-ethyladenine imprinted polymers, *J. Am. Chem. Soc.*, 119, 4388–4393, 1997.

74. Andersson, L. I., Müller, R., Vlatakis, G., and Mosbach, K., Mimics of the binding sites of opioid receptors obtained by molecular imprinting of enkephalin and morphine, *Proc. Natl. Acad. Sci. U.S.A.*, 92, 4788–4792, 1995.

75. Berglund, J., Nicholls, I. A., Lindbladh, C., and Mosbach, K., Recognition in molecularly imprinted polymer α_2-adrenoreceptor mimics, *Bioorg. Med. Chem. Lett.*, 6, 2237–2242, 1996.

76. Ramström, O., Ye, L., and Mosbach, K., Artificial antibodies to corticosteroids prepared by molecular imprinting, *Chem. Biol.*, 3, 471–477, 1996.

77. Siemann, M., Andersson, L. I., and Mosbach, K., Selective recognition of the herbicide atrazine by noncovalent molecularly imprinted polymers, *J. Agric. Food Chem.*, 44, 141–145, 1996.

78. Andersson, L. I., Abdel-Rehim, M., Nicklasson, L., Schweitz, L., and Nilsson, S., Towards molecular-imprint based SPE of local anaesthetics, *Chromatographia*, 55, S65–S69, 2002.

79. Surugiu, I., Ye, L., Yilmaz, E., Dzgoev, A., Danielsson, B., Mosbach, K., and Haupt, K., An enzyme-linked molecularly imprinted sorbent assay, *Analyst*, 125, 13–16, 2000.

80. Surugiu, I., Danielsson, B., Ye, L., Mosbach, K., and Haupt, K., Chemiluminscence imaging ELISA using an imprinted polymer as the recognition element instead of an antibody, *Anal. Chem.*, 73, 487–491, 2001.

81. Surugiu, I., Svitel, J., Ye, L., Haupt, K., and Danielsson, B., Development of a flow injection capillary chemiluminscent ELISA using an imprinted polymer instead of an antibody, *Anal. Chem.*, 73, 4388–4392, 2001.

82. Piletsky, S. A., Piletska, E. V., Chen, B., Karim, K., Weston, D., Barrett, G., Lowe, P., and Turner, A. P. F., Chemical grafting of molecularly imprinted homopolymers to the surface of microplates. Application of artificial adrenergic receptor in enzyme-linked assay for ß-agonists determination, *Anal. Chem.*, 72, 4381–4385, 2000.

83. Piletska, E., Piletsky, S., Karim, K., Terpetschnig, E., and Turner, A., Biotin-specific synthetic receptors prepared using molecular imprinting, *Anal. Chim. Acta.*, 504, 179–183, 2004.

84. Chianella, I., Lotierzo, M., Piletsky, S. A., Tothill, I. E., Chen, B., Karim, K., and Turner, A. P. F., Rational design of a polymer specific for microcystin-LR using a computational approach, *Anal. Chem.*, 74, 1288–1293, 2002.

85. Piletsky, S. A., Piletska, E. V., Bossi, A., Karim, K., Lowe, P., and Turner, A. P. F., Substitution of antibodies and receptors with molecularly imprinted polymers in enzyme-linked and fluorescent assays, *Biosens. Bioelectron.*, 16, 701–707, 2001.

86. Lotierzo, M., Henry, O. Y. F., Piletsky, S., Tothill, I., Cullen, D., Kania, M., Hock, B., and Turner, A. P. F., Surface plasmon resonance sensor for domoic acid based on grafted imprinted polymer, *Biosens. Bioelectron.*, 20, 145–152, 2004.

87. Levi, R., McNiven, S., Piletsky, S. A., Cheong, S. H., Yano, K., and Karube, I., Optical detection of chloramphenicol using molecularly imprinted polymers, *Anal. Chem.*, 69, 2017–2021, 1997.

88. Suárez-Rodríguez, J. L. and Díaz-García, M. L., Fluorescent competitive flow-through assay for chloramphenicol using molecularly imprinted polymers, *Biosens. Bioelectron.*, 16, 955–961, 2001.

89. Haupt, K., Mayes, A. G., and Mosbach, K., Herbicide assay using an imprinted polymer-based system analogous to competitive fluoroimmunoassays, *Anal. Chem.*, 70, 3936–3939, 1998.

90. Kröger, S., Turner, A. P. F., Mosbach, K., and Haupt, K., Imprinted polymer-based sensor system for herbicides using differential-pulse voltammetry on screen-printed electrodes, *Anal. Chem.*, 71, 3698–3702, 1999.

91. Schöllhorn, B., Maurice, C., Flohic, G., and Limoges, B., Competitive assay of 2, 4-dichlorophenoxy-acetic acid using a polymer imprinted with an electrochemically active tracer closely related to the analyte, *Analyst*, 125, 665–667, 2000.

92. Ye, L. and Mosbach, K., Polymers recognizing biomolecules based on a combination of molecular imprinting and proximity scintillation: A new sensor concept, *J. Am. Chem. Soc.*, 123, 2901–2902, 2001.

93. Ye, L., Surugiu, I., and Haupt, K., Scintillation proximity assay using molecularly imprinted microspheres, *Anal. Chem.*, 74, 959–964, 2002.

94. Ansell, R. J. and Mosbach, K., Magnetic molecularly imprinted polymer beads for drug radioligand binding assay, *Analyst*, 123, 1611–1616, 1998.

95. Senholdt, M., Siemann, M., Mosbach, K., and Andersson, L. I., Determination of cyclosporin A and metabolites total concentration using a molecularly imprinted polymer based radioligand binding assay, *Anal. Lett.*, 30, 1809–1821, 1997.

96. Bengtsson, H., Roos, U., and Andersson, L. I., Molecular imprint based radioassay for direct determination of S-propanolol in human plasma, *Anal. Commun.*, 34, 233–235, 1997.

97. Hennion, M. C. and Pichon, V., Immuno-based sample preparation for trace analysis, *J. Chromatogr. A*, 1000, 29–52, 2003.

98. Stevenson, D., Immuno-affinity solid-phase extraction, *J. Chromatogr. B*, 745, 39–48, 2000.

99. Sellergren, B., Direct drug determination by selective sample enrichment on an imprinted polymer, *Anal. Chem.*, 66, 1578–1582, 1994.

100. Cacho, C., Turiel, E., Martín-Esteban, A., Pérez-Conde, C., and Cámara, C., Clean-up of triazines in vegetable extracts by molecularly- imprinted solid-phase extraction using a propazine-imprinted polymer, *Anal. Bioanal. Chem.*, 376, 491–496, 2003.

101. Chapuis, F., Pichon, V., Lanza, F., Sellergren, B., and Hennion, M. C., Retention mechanism of analytes in the solid-phase extraction process using molecularly imprinted polymers—Application to the extraction of triazines from complex matrices, *J. Chromatogr. B*, 804, 93–101, 2004.

102. Crescenzi, C., Bayoudh, S., Cormack, P. A. G., Klein, T., and Ensing, K., Determination of clenbuterol in bovine liver by combining matrix solid phase dispersion and molecularly imprinted solid phase extraction followed by liquid chromatography/electrospray ion trap multiple stage mass spectrometry, *Anal. Chem.*, 73, 2171–2177, 2001.

103. Blomgren, A., Berggren, C., Holmberg, A., Larsson, F., Sellergren, B., and Ensing, K., Extraction of clenbuterol from calf urine using a molecularly imprinted polymer followed by quantitation by high-performance liquid chromatography with UV detection, *J. Chromatogr. A*, 975, 157–164, 2002.

104. Matsui, J., Okada, M., Tsuruoka, M., and Takeuchi, T., Solid-phase extraction of a triazine herbicide using a molecularly imprinted synthetic receptor, *Anal. Commun.*, 34, 85–87, 1997.

105. Andersson, L. I., Efficient sample pre-concentration of bupivacaine from human plasma by solid-phase extraction on molecularly imprinted polymers, *Analyst*, 125, 1515–1517, 2000.

106. Martin, P., Wilson, I. D., Morgan, D. E., Jones, G. R., and Jones, K., Evaluation of a molecular-imprinted polymer for use in the solid phase extraction of propranolol from biological fluids, *Anal. Commun.*, 34, 45–47, 1997.

107. Olsen, J., Martin, P., Wilson, I. D., and Jones, G. R., Methodology for assessing the properties of molecular imprinted polymers for solid phase extraction, *Analyst*, 124, 467–471, 1999.

108. Matsui, J., Fujiwara, K., Ugata, S., and Takeuchi, T., Solid-phase extraction with a dibutylmelamine-imprinted polymer as triazine herbicide-selective sorbent, *J. Chromatogr. A*, 889, 25–31, 2000.

109. Sambe, H., Hoshina, K., Hosoya, K., and Haginaka, J., Direct injection analysis of bisphenol A in serum by combination of isotope imprinting with liquid chromatography-mass spectrometry, *Analyst*, 130, 38–40, 2005.

110. Haginaka, J. and Sanbe, H., Uniform-sized molecularly imprinted polymers for 2-arylpropionic acid derivatives selectively modified with hydrophilic external layer and their applications to direct serum injection analysis, *Anal. Chem.*, 72, 5206–5210, 2000.

111. Dirion, B., Cobb, Z., Schillinger, E., Andersson, L. I., and Sellergren, B., Water-compatible molecularly imprinted polymers obtained via high-throughput synthesis and experimental design, *J. Am. Chem. Soc.*, 125, 15101–15109, 2003.

112. Sergeyeva, T. A., Matuschewski, H., Piletsky, S. A., Bendig, J., Schedler, U., and Ulbricht, M., Molecularly imprinted polymer membranes for substance-selective solid-phase extraction from water by surface photo-grafting polymerization, *J. Chromatogr. A*, 907, 89–99, 2001.

113. Koeber, R., Fleischer, C., Lanza, F., Boos, K. S., Sellergren, B., and Barcelo, D., Evaluation of a multidimensional solid-phase extraction platform for highly selective on-line cleanup and high-throughput LC-MS analysis of triazines in river water samples using molecularly imprinted polymers, *Anal. Chem.*, 73, 2437–2444, 2001.

114. Boos, K. S. and Fleischer, C. T., Multidimensional on-line solid-phase extraction (SPE) using restricted access materials (RAM) in combination with molecular imprinted polymers (MIP), *Fresenius J. Anal. Chem.*, 371, 16–20, 2001.

115. Mullett, W. M., Walles, M., Levsen, K., Borlak, J., and Pawliszyn, J., Multidimensional on-line sample preparation of verapamil and its metabolites by a molecularly imprinted polymer coupled to liquid chromatography-mass spectrometry, *J. Chromatogr. B*, 801, 297–306, 2004.

116. Caro, E., Marcé, R. M., Cormack, P. A. G., Sherrington, D. C., and Borrull, F., On-line solid-phase extraction with molecularly imprinted polymers to selectively extract substituted 4-chlorophenols and 4-nitrophenol from water, *J. Chromatogr. A*, 995, 233–238, 2003.

117. Mullett, W. M., Martin, P., and Pawliszyn, J., In-tube molecularly imprinted polymer solid-phase microextraction for the selective determination of propranolol, *Anal. Chem.*, 73, 2383–2389, 2001.

118. Mullett, W. M. and Lai, E. P. C., Determination of theophylline in serum by molecularly imprinted solid-phase extraction with pulsed elution, *Anal. Chem.*, 70, 3636–3641, 1998.

119. Chassaing, C., Stokes, J., Venn, R. F., Lanza, F., Sellergren, B., Holmberg, A., and Berggren, C., Molecularly imprinted polymers for the determination of a pharmaceutical development compound in plasma using 96-well MISPE technology, *J. Chromatogr. B*, 804, 71–81, 2004.

120. Hu, S. G., Wang, S. W., and He, X. W., An amobarbital molecularly imprinted microsphere for selective solid-phase extraction of phenobarbital from human urine and medicines and their determination by high-performance liquid chromatography, *Analyst*, 128, 1485–1489, 2003.

121. Lai, J. P., Niessner, R., and Knopp, D., Benzo[a]pyrene imprinted polymers: Synthesis characterization and SPE application in water and coffee samples, *Anal. Chim. Acta*, 522, 137–144, 2004.

122. Widstrand, C., Larsson, F., Fiori, M., Civitareale, C., Mirante, S., and Brambilla, G., Evaluation of MISPE for the multi-residue extraction of beta-agonists from calves urine, *J. Chromatogr. B*, 804, 85–91, 2004.

123. Fiori, M., Civitareale, C., Mirante, S., Magaro, E., and Brambilla, G., Evaluation of two different clean-up steps to minimise ion suppression phenomena in ion trap liquid chromatography-tandem mass spectrometry for the multi-residue analysis of beta agonists in calves urine, *Anal. Chim. Acta*, 529, 207–210, 2005.

124. Kootstra, P. R., Kuijpers, C. J. P. F., Wubs, K. L., van Doorn, D., Sterk, S. S., van Ginkel, L. A., and Stephany, R. W., The analysis of beta-agonists in bovine muscle using molecular imprinted polymers with ion trap LCMS screening, *Anal. Chim. Acta*, 529, 75–81, 2005.

125. San Vicente, B., Villoslada, F. N., and Moreno-Bondi, M. C., Continuous solid-phase extraction and preconcentration of bisphenol A in aqueous samples using molecularly imprinted columns, *Anal. Bioanal. Chem.*, 380, 115–122, 2004.

126. Liu, Y. W., Chang, X. J., Wang, S., Guo, Y., Din, B. J., and Meng, S. M., Solid-phase extraction and preconcentration of cadmium(II) in aqueous solution with Cd(II)-imprinted resin (poly-Cd(II)-DAAB-VP) packed columns, *Anal. Chim. Acta*, 519, 173–179, 2004.

127. Baggiani, C., Giovannoli, C., Anfossi, L., and Tozzi, C., Molecularly imprinted solid-phase extraction sorbent for the clean-up of chlorinated phenoxyacids from aqueous samples, *J. Chromatogr. A*, 938, 35–44, 2001.

128. Ferrer, I., Lanza, F., Tolokan, A., Horvath, V., Sellergren, B., Horvai, G., and Barcelo, D., Selective trace enrichment of chlorotriazine pesticides from natural waters and sediment samples using terbuthylazine molecularly imprinted polymers, *Anal. Chem.*, 72, 3934–3941, 2000.

129. Blomgren, A., Berggren, C., Holmberg, A., Larsson, F., Sellergren, B., and Ensing, K., Extraction of clenbuterol from calf urine using a molecularly imprinted polymer followed by quantitation by high-performance liquid chromatography with UV detection, *J. Chromatogr. A*, 975, 157–164, 2002.

130. Say, R., Birlik, E., Ersöz, A., Yilmaz, F., Gedikbey, T., and Denizli, A., Preconcentration of copper on ion-selective imprinted polymer microbeads, *Anal. Chim. Acta*, 480, 251–258, 2003.

131. Möller, K., Crescenzi, C., and Nilsson, U., Determination of a flame retardant hydrolysis product in human urine by SPE and LC-MS. Comparison of molecularly imprinted solid-phase extraction with a mixed-mode anion exchanger, *Anal. Bioanal. Chem.*, 378, 197–204, 2004.

132. Walshe, M., Howarth, J., Kelly, M. T., O'Kennedy, R., and Smyth, M. R., The preparation of a molecular imprinted polymer to 7-hydroxycoumarin and its use as a solid-phase extraction material, *J. Pharm. Biomed. Anal.*, 16, 319–325, 1997.

133. Martin, P. D., Jones, G. R., Stringer, F., and Wilson, I. D., Comparison of normal and reversed-phase solid phase extraction methods for extraction of beta-blockers from plasma using molecularly imprinted polymers, *Analyst*, 128, 345–350, 2003.

134. Martin, P. D., Jones, G. R., Stringer, F., and Wilson, I. D., Comparison of extraction of a [beta]-blocker from plasma onto a molecularly imprinted polymer with liquid-liquid extraction and solid phase extraction methods, *J. Pharm. Biomed. Anal.*, 35, 1231–1239, 2004.

135. Caro, E., Marcé, R. M., Cormack, P. A. G., Sherrington, D. C., and Borrull, F., Molecularly imprinted solid-phase extraction of naphthalene sulfonates from water, *J. Chromatogr. A*, 1047, 175–180, 2004.

136. Caro, E., Marcé, R. M., Cormack, P. A. G., Sherrington, D. C., and Borrull, F., A new molecularly imprinted polymer for the selective extraction of naproxen from urine samples by solid-phase extraction, *J. Chromatogr. B*, 813, 137–143, 2004.

137. Bereczki, A., Tolokan, A., Horvai, G., Horvath, V., Lanza, F., Hall, A. J., and Sellergren, B., Determination of phenytoin in plasma by molecularly imprinted solid-phase extraction, *J. Chromatogr. A*, 930, 31–38, 2001.

138. Molinelli, A., Weiss, R., and Mizaikoff, B., Advanced solid phase extraction using molecularly imprinted polymers for the determination of quercetin in red wine, *J. Agric. Food Chem.*, 50, 1804–1808, 2002.

139. Zhu, Q. Z., DeGelmann, P., Niessner, R., and Knopp, D., Selective trace analysis of sulfonylurea herbicides in water and soil samples based on solid-phase extraction using a molecularly imprinted polymer, *Environ. Sci. Technol.*, 36, 5411–5420, 2002.

140. Chapuis, F., Pichon, V., Lanza, F., Sellergren, S., and Hennion, M.-C., Optimization of the class-selective extraction of triazines from aqueous samples using a molecularly imprinted polymer by a comprehensive approach of the retention mechanism, *J. Chromatogr. A*, 999, 23–33, 2003.

141. Muldoon, M. T. and Stanker, L. H., Molecularly imprinted solid phase extraction of atrazine from beef liver extracts, *Anal. Chem.*, 69, 803–808, 1997.

142. Turiel, E., Martín-Esteban, A., Fernández, P., Pérez-Conde, C., and Cámara, C., Molecular recognition in a propazine-imprinted polymer and its application to the determination of triazines in environmental samples, *Anal. Chem.*, 73, 5133–5141, 2001.

143. Tamayo, F. G., Casillas, J. L., and Martin-Esteban, A., Highly selective fenuron-imprinted polymer with a homogeneous binding site distribution prepared by precipitation polymerization and its application to the clean-up of fenuron in plant samples, *Anal. Chim. Acta*, 482, 165–173, 2003.

144. Dong, X. C., Wang, N., Wang, S. L., Zhang, X. W., and Fan, Z. J., Synthesis and application of molecularly imprinted polymer on selective solid-phase extraction for the determination of mono-sulfuron residue in soil, *J. Chromatogr. A*, 1057, 13–19, 2004.

145. Meng, Z. H. and Qin, L., Determination of degradation products of nerve agents in human serum by solid phase extraction using molecularly imprinted polymers, *Anal. Chim. Acta*, 435, 121–127, 2001.

146. Maier, N. M., Buttinger, G., Welhartizki, S., Gavioli, E., and Lindner, W., Molecularly imprinted polymer-assisted sample clean-up of ochratoxin A from red wine: Merits and limitations, *J. Chromatogr. B*, 804, 103–111, 2004.

147. Frandsen, H., Frederiksen, H., and Alexander, J., 2-Amino-1-methyl-6-(5-hydroxy-)phenylimidazo[4 5-b]pyridine (5- OH-PhIP) a biomarker for the genotoxic dose of the heterocyclic amine 2-amino-1-methyl-6-phenylimidazo[4 5- b]pyridine (PhIP), *Food Chem. Toxicol.*, 40, 1125–1130, 2002.

148. Cacho, C., Turiel, E., Martín-Esteban, A., Pérez-Conde, C., and Cámara, C., Clean-up of triazines in vegetable extracts by molecularly-imprinted solid-phase extraction using a propazine-imprinted polymer, *Anal. Bioanal. Chem.*, 376, 491–496, 2003.

149. Chapuis, F., Pichon, V., Lanza, F., Sellergren, B., and Hennion, M. C., Retention mechanism of analytes in the solid-phase extraction process using molecularly imprinted polymers—Application to the extraction of triazines from complex matrices, *J. Chromatogr. B*, 804, 93–101, 2004.

150. Sanbe, H. and Haginaka, J., Restricted access media-molecularly imprinted polymer for propranolol and its application to direct injection analysis of beta-blockers in biological fluids, *Analyst*, 128, 593–597, 2003.

151. Sanbe, H., Hosoya, K., and Haginaka, J., Preparation of uniformly sized molecularly imprinted polymers for phenolic compounds and their application to the assay of bisphenol A in river water, *Anal. Sci.*, 19, 715–719, 2003.

152. Watabe, Y., Kondo, T., Morita, M., Tanaka, N., Haginaka, J., and Hosoya, K., Determination of bisphenol A in environmental water at ultra- low level by high-performance liquid chromatography with an effective on-line pretreatment device, *J. Chromatogr A*, 1032, 45–49, 2004.

153. Theodoridis, G., Zacharis, C. K., Tzanavaras, P. D., Themelis, D. G., and Economou, A., Automated sample preparation based on the sequential injection principle—Solid-phase extraction on a molecularly imprinted polymer coupled on-line to high-performance liquid chromatography, *J. Chromatogr. A*, 1030, 69–76, 2004.

154. Masqué, N., Marcé, R. M., Borrull, F., Cormack, P. A. G., and Sherrington, D. C., Synthesis and evaluation of a molecularly imprinted polymer for selective on-line solid-phase extraction of 4-nitrophenol from environmental water, *Anal. Chem.*, 72, 4122–4126, 2000.

155. Caro, E., Masqué, N., Marcé, R. M., Borrull, F., Cormack, P. A. G., and Sherrington, D. C., Noncovalent and semi-covalent molecularly imprinted polymers for selective on-line solid-phase extraction of 4-nitrophenol from water samples, *J. Chromatogr. A*, 963, 169–178, 2002.

156. Bjarnason, B., Chimuka, L., and Ramström, O., On-line solid-phase extraction of triazine herbicides using a molecularly imprinted polymer for selective sample enrichment, *Anal. Chem.*, 71, 2152–2156, 1999.

157. Mullett, W. M., Dirie, M. F., Lai, E. P. C., Guo, H. S., and He, X. W., A 2-aminopyridine molecularly imprinted polymer surrogate micro-column for selective solid phase extraction and determination of 4-aminopyridine, *Anal. Chim. Acta*, 414, 123–131, 2000.

158. Lai, E. P. C. and Wu, S. G., Molecularly imprinted solid phase extraction for rapid screening of cephalexin in human plasma and serum, *Anal. Chim. Acta*, 481, 165–174, 2003.

159. Wu, S. G., Lai, E. P. C., and Mayer, P. M., Molecularly imprinted solid phase extraction-pulsed elution-mass spectrometry for determination of cephalexin and [alpha]-aminocephalosporin antibiotics in human serum, *J. Pharm. Biomed.*, 36, 483–490, 2004.

160. Mena, M. L., Agüí, L., Martinez-Ruiz, P., Yáñez-Sedeño, P., Reviejo, A. J., and Pingarrón, J. M., Molecularly imprinted polymers for on-line clean up and preconcentration of chloramphenicol prior to its voltammetric determination, *Anal. Bioanal. Chem.*, 376, 18–25, 2003.

161. Feng, S. Y., Lai, E. P. C., Dabek-Zlotorzynska, E., and Sadeghi, S., Molecularly imprinted solid-phase extraction for the screening of antihyperglycemic biguanides, *J. Chromatogr. A*, 1027, 155–160, 2004.

162. Mullett, W. M., Lai, E. P. C., and Sellergren, B., Determination of nicotine in tobacco by molecularly imprinted solid phase extraction with differential pulsed elution, *Anal. Commun.*, 36, 217–220, 1999.

163. Mena, M. L., Martinez-Ruiz, P., Reviejo, A. J., and Pingarrón, J. M., Molecularly imprinted polymers for on-line preconcentration by solid phase extraction of pirimicarb in water samples, *Anal. Chim. Acta*, 451, 297–304, 2002.

164. Mullett, W. M. and Lai, E. P. C., Rapid determination of theophylline in serum by selective extraction using a heated molecularly imprinted polymer micro-column with differential pulsed elution, *J. Pharm. Biomed. Anal.*, 21, 835–843, 1999.

165. Kubo, T., Hosoya, K., Watabe, Y., Ikegami, T., Tanaka, N., Sano, T., and Kaya, K., Polymer-based adsorption medium prepared using a fragment imprinting technique for homologues of chlorinated bisphenol A produced in the environment, *J. Chromatogr. A*, 1029, 37–41, 2004.

166. Theodoridis, G., Kantifes, A., Manesiotis, P., Raikos, N., and Tsoukali-Papadopoulou, H., Preparation of a molecularly imprinted polymer for the solid- phase extraction of scopolamine with hyoscyamine as a dummy template molecule, *J. Chromatogr. A*, 987, 103–109, 2003.

6 Aptamer-Based Bioanalytical Methods

Sara Tombelli, Maria Minunni, and Marco Mascini

CONTENTS

6.1 INTRODUCTION

Aptamers are artificial nucleic acid ligands that can be generated against amino acids, drugs, proteins and other molecules. They are isolated from complex libraries of synthetic nucleic acids by an iterative process of adsorption, recovery and amplification called systematic evolution of ligands by exponential enrichment (SELEX) [1].

Several reviews on aptamers have appeared in the literature in the last 10 years, after the first publications on their selection [2,3] and their initial employment in the therapeutic field [4,5]. In one of the first reviews [6], aptamers were presented as a recent entry in the field of selective binding and molecular recognition and the authors forecasted the expansion of their application beyond clinical diagnostics and therapeutic monitoring to a broader area of analytical chemistry. Later, the potential application of aptamers as alternative ligands in diagnostic assays [7], biosensors [8–10] and other analytical techniques [11,12] have been presented.

In a very recent review, several critical aspects of aptamers have been documented [13] with a particular focus on the problems related to the selection process and to the generation of universal rules for the analytical use of already-selected aptamers.

The present review will follow this line, examining in detail the possible important aspects when using aptamers in bioanalytical methods. The selection process (SELEX) and the molecules for which aptamers have been selected will also be considered. Moreover, several important aspects that should be considered when developing bioanalytical methods based on aptamers will be discussed, with particular emphasis on the immobilization process and on the assay protocol.

6.2 APTAMER SELECTION

The SELEX process [2,3] is described in Figure 6.1. It involves iterative cycles of selection and amplification (usually 12–18 cycles) starting from a library of oligonucleotides with different sequences. A candidate mixture of nucleic acids of differing sequences is generated by a standard DNA-oligonucleotide synthesizer, including a randomized region (usually 30–40 nucleotides long) flanked by fixed regions (each member of the library contains the same sequences at the same location). The fixed sequences are the polymerase chain reaction (PCR) primers' binding sites and, for RNA aptamers selection, the 5′ constant region contains the T7 promoter sequence, onto which T7 RNA polymerase binds for the in vitro transcription of the library into RNA. Selections are frequently carried out with RNA pools due to the known ability of RNAs to fold into complex structures that can be a source of diversity of RNA function, but single-stranded DNA pools can also yield aptamers. Single-stranded DNAs are also known to fold in vitro into structures containing stem loops, internal loops, etc., even if the structures are less stable than the corresponding RNA structures.

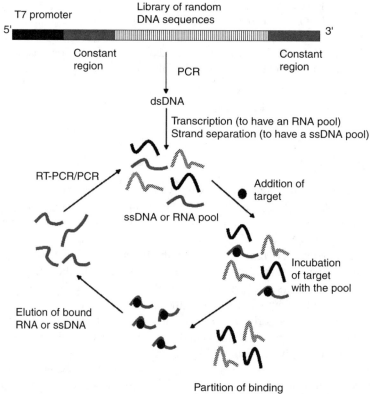

FIGURE 6.1 Scheme of the SELEX process for the in vitro selection of nucleic acid aptamers. A library of DNA oligonucleotides containing a portion of randomized sequence is synthesized. The library is converted into dsDNA by PCR using the 5′ and 3′ constant regions for primers annealing, and then converted into ssDNA by strands separation or into RNA by in vitro transcription using the T7 RNA polymerase. The target analyte is mixed with the nucleic acid pool and, after incubation, the nonspecific or low-affinity binding nucleic acid molecules are removed by washing steps. The captured RNA/DNA molecules are eluted, recovered, and amplified by RT-PCR/PCR to obtain a new, enriched DNA library. The whole cycle is repeated until obtaining a specific population of RNA/DNA that is finally isolated and characterized.

The library of nucleic acids is then mixed with the target molecule to search for the small fraction of individual sequences of the pool binding to the target (usually only 5% of the pool has this ability at this stage). Usually, this incubation step is performed in solution for proteins or, in the case of small molecules, with the target immobilized on a solid support to generate an affinity matrix (i.e., a column). After mixing, the complex obtained from the interaction pool-ligand is isolated by nitrocellulose filtering [3], immunoprecipitation [14], or separation on polyacrylamide gels [15]. When a small molecular target is immobilized on a solid support, the separation can be performed by simple washing steps. In this way, the nucleic acid molecules with the highest affinity for the target are separated from the nonbinding species. The oligonucleotides that are selected during partitioning are amplified (PCR for DNA aptamers and RT-PCR for RNA aptamers) to create a new mixture enriched in those nucleic acid molecules having a relatively high affinity for the target. After several iterations of the selection process (12–18 cycles), performed under increasingly stringent conditions, the required enrichment in the pool of the high-affinity sequences at the expense of the low-affinity binders is reached, eventually resulting in one or more potential candidates with the best performance (highest affinity). The enriched library (10–30 individual sequences) is finally cloned and sequenced. Individual sequences are investigated for their ability (affinity constants values) to bind to the target.

There are a number of issues that need to be considered for a good selection process, among which the complexity of the library, the chemistry of the nucleotides, and the design of the constant regions being the most important as they determine the final product of the selection process [16,17]. Considering the complexity, or the molecular diversity, of the library, the starting pool must be large enough to generate a high probability of selecting an active aptamer. If the randomized oligonucleotides are N in length and generated from y different nucleotides, then the complexity is given by the formula y^N. Usually, for practical reasons regarding the DNA synthesis chemistry, the maximum number of sequences that can be screened is 10^{13}–10^{15}. Diversity can also be introduced during the cycles of selection, and especially during the PCR amplification step, due to the possible introduction of mutations into the selected oligonucleotides.

The second important factor is the chemistry of the nucleotides to be used [16], which can strongly influence the result of the selection process. The nature of the nucleotides defines the range of possible three-dimensional structures that an aptamer can fold into, as well as the stability for degradation. The high number of secondary (and tertiary) structural motifs with different shapes that aptamers can assume are based on the ability of nucleotide bases to interact with each other through canonical Watson–Crick, as well as unusual base pairing [18]. Moreover, regarding the selection of RNA aptamers, ribonucleotides $2'$-fluoro- or $2'$-amino-modified ribonucleotides can be used instead of naturally occurring ribonucleotides because they can be incorporated by T7 RNA polymerase. The SELEX process has been conducted with a normal RNA pool and with a pool containing $2'$-amino-pyrimidines; due to these different libraries, aptamers with different sequences and structural motifs have been selected [19].

Because the nucleases that are most abundant in biological fluids are the pyrimidine-specific nucleases, introduction of specific modifications at the $2'$-position of pyrimidine nucleotides ($2'$-amino and $2'$-fluoro functional groups) protects an RNA oligonucleotide from degradation, increasing the half-life up to 15 h [20]. Because $2'$-amino and $2'$-fluoro CTP and UTP can be incorporated into in vitro transcribed RNA, these modifications can be introduced into the combinatorial library. In this way, it has been possible to select aptamers with enhanced stability in biological fluids from libraries containing modified pyrimidines with $2'$-amino and $2'$-fluoro [21].

A second option for RNA aptamer stabilization is represented by a post-selection modification and exploits enantiomeric aptamers known as *spiegelmers* (from the German word for mirror). This process initially involves the creation of a chemical mirror image of the target, followed by the selection of aptamers to this mirror image and the creation of a chemical mirror image of the SELEX-selected aptamer. The spiegelmer will bind the target but will not be susceptible to normal enzymatic cleavage [19,22].

The third important parameter for a correct selection process is the constant region (primers) design. Aptamers consist of a random DNA or RNA sequence that is the core region, flanked by two constant regions: $5'$ and $3'$ flanking sequences. These function as sites for primer hybridization for Klenow extension, cDNA synthesis, polymerase chain reaction amplification, and T7 RNA polymerase transcription, all of which are essential for the SELEX protocol. The design of the constant regions should ensure a correct amplification of the desired fragments, i.e., primers should anneal strongly to the template and should not form secondary structures or dimers [17]. These are the normal properties for primer design in common PCR protocols, but their importance is greater for a SELEX process where many selection and amplification rounds are performed [16].

One of the major drawbacks of aptamers is the SELEX process itself in terms of time required for a complete selection and factors influencing the result of the selection [13]. Usually, a selection process can take around 15 cycles of selection and amplification and each round can be accomplished in 2–3 days. This means that a typical SELEX experiment may be completed in approximately 3 months, which is at least faster than the time required to generate a cell line to produce a specific monoclonal antibody and purify it [7]. Several research groups have worked for the development of a faster selection process. An automated SELEX process has been presented [23–26] in which a multiple selection is performed on microtiter plates with a robotic workstation accommodating 8 selections in parallel and completing about 12 rounds in 2 days. The system is also capable of performing the selection of aptamers for 8 different targets in a run.

An automated platform has also been reported [27,28] for the generation of photoaptamers with $5'$-iodo- or $5'$-bromo-substituted bases in a 96-well format. With this new photochemical SELEX (PhotoSELEX) method, modified ssDNA aptamers capable of photo-crosslinking the target molecule (human basic fibroblast growth factor (bFGF)) have been identified. The method is based on the incorporation of a modified nucleotide activated by absorption of light, in place of a native base in either RNA- or in ssDNA-randomized oligonucleotide libraries. The aptamers selected with this method have the ability to form a photoinduced covalent bond with the target molecule and have, for this reason, greater sensitivity and specificity than those aptamers selected through conventional selection methodologies.

In addition, capillary electrophoresis (CE) was coupled to the SELEX process to reduce the number of selection cycles. Nonequilibrium capillary electrophoresis of equilibrium mixtures (NECEEM) has been recently presented as an effective method to reduce the number of cycles required for the selection [29]. In a previous paper, a different research group [30] has also demonstrated that the use of CE to separate the binding sequences for the nonactive aptamer can shorten the SELEX process to only four rounds, thereby reducing the time required for the whole selection process to only 2–4 days.

Recently, advances in the in vitro selection methods have provided signalling aptamers applicable for biosensors and allosteric ribozymes regulated by both small ligands and proteins [31].

6.3 APTAMERS AND ANTIBODIES

For several decades, antibodies have been the most-used biomolecules for target recognition in bioanalytical methods. Although antibodies and aptamers can both bind the relative target with high affinity and specificity, aptamers can offer advantages over antibodies that make them very promising for this application [9,12]. The main advantage is that aptamers overcome the use of animals or cell lines for the production of the molecules. Antibodies against molecules that are not immunogenic are difficult to generate. On the contrary, aptamers are isolated by in vitro methods that are independent of animals: an in vitro combinatorial library can be generated against any target. In addition, generation of antibodies in vivo means that the animal immune system selects the sites on the target protein to which the antibodies bind. The in vivo parameters restrict the identification of antibodies that can recognize targets only under physiological conditions, limiting the extension to which the antibodies can be functionalized and applied. Moreover, the aptamer selection process can be manipulated to obtain aptamers that bind a specific region of the target and with specific binding properties under different binding conditions. After selection, aptamers are produced by chemical synthesis and highly purified by eliminating the batch-to-batch variation found when using antibodies. In addition, by chemical synthesis, modifications in the aptamer can be introduced that enhance the stability, affinity, and specificity of the molecules. Often, the kinetic parameters of the aptamer-target complex can be changed for higher affinity or specificity. Another advantage over antibodies can be seen in the higher temperature stability of aptamers; in fact, antibodies are large proteins that are sensitive to temperature and can undergo irreversible denaturation. On the contrary, aptamers are very stable; they can recover their native active conformation after denaturation.

The primary limitation on the use of aptamers (mainly RNA aptamers) in bioanalytical methods has been their nuclease sensitivity, which is very critical for their use in ex vivo and in vivo applications [22]. However, it has been shown that the stability of such molecules can be improved by chemical modification of the ribose ring at the $2'$-position [32].

Several authors have reported a direct comparison between aptamers and antibodies that are specified for the same molecule when used as biorecognition elements in biosensors or biochips. An anti-IgE DNA aptamer was compared with the monoclonal antibody for the same molecule (IgE) in a quartz crystal biosensor [33]. The two receptors were immobilized onto the gold surface of the sensor following a similar procedure and sensitivity, specificity, and stability of the two biosensors were compared. They both reached a detection limit of 0.5 nM for IgE, but the aptamer-based one showed a 10-fold extended linear range (Figure 6.2a), probably due to the better immobilization of the small aptamer in a dense and well-oriented layer relative to the bigger antibody molecule. Moreover, regeneration and recycling of the aptamer-based biosensor was possible in contrast with the antibody layer. When examining the possible regeneration of the sensor, in the case of the antibody, the receptor layer was probably damaged or irreversibly unfolded after regeneration and all the subsequent injections of IgE resulted in reduced signals. In contrast, the anti-IgE aptamer tolerated repeated cycles of regeneration and binding with little loss of sensitivity (Figure 6.2b). A similar comparison has been performed on a piezoelectric biosensor using an RNA aptamer specific for HIV-1 Tat protein and the corresponding monoclonal antibody [34]. In this case, an extended linear range was obtained with the antibody, but the aptamer-based sensor exhibited better sensitivity. The reproducibility of the two sensors was comparable and a good specificity was observed with both of the receptors.

To determine whether aptamers exhibit a specific affinity that rivals antibodies, the individual rupture force of IgE-anti IgE monoclonal antibody was compared to that of IgE-anti IgE aptamer by atomic force microscopy (AFM) [35]. IgE was immobilized on the AFM tip and the modified tip was used to measure the interactions with different substrates modified with the aptamer or the antibody under the same experimental conditions. The average rupture forces of IgE-aptamer and IgE-antibody were calculated and compared. Comparing the results obtained from the same AFM tip revealed that the rupture forces for the aptamer were always larger than those of the IgE-antibody interaction. These results revealed the high affinity of the aptamer for the protein that can even surpass that of the antibody for the same molecule.

Besides antibodies, the selection process itself, with the amplification step, gives some advantages to aptamers with respect to other nonnatural receptors, such as oligopeptides, that cannot be amplified during their selection procedure.

6.4 APTAMER IMMOBILIZATION SCHEMES

The procedure to fix the aptamer to the biosensor/bioanalytical device surface is of paramount importance to obtain an ordered layer that is able to exploit the flexibility of the bioreceptor as much

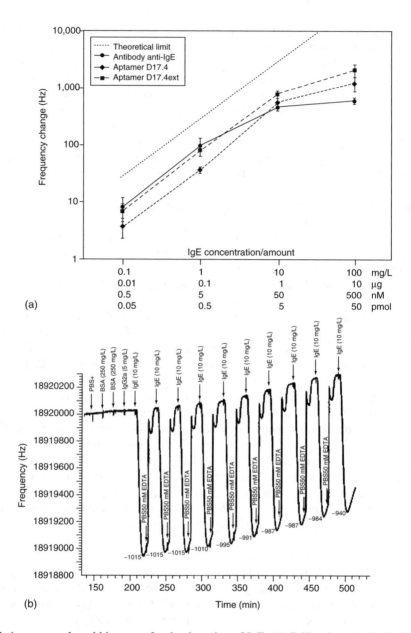

FIGURE 6.2 An aptamer-based biosensor for the detection of IgE. (a) Calibration plot for IgE; comparison between two different IgE-binding aptamers and the anti-IgE antibody. (b) Sensorgram for different binding/regeneration cycles between the aptamer and IgE (regeneration with EDTA 50 mM). (From Liss, M., Petersen, B., Wolf H., and Prohaska, E., *Analytical Chemistry*, 74, 4488, 2002.)

as possible without altering its structure and its affinity for the target molecule. Aptamers can bind to their targets with high affinity and they can discriminate between closely related targets [31]. This is due to the adaptive recognition: aptamers, unstructured in solution, fold upon associating with their molecular targets into molecular architectures in which the ligand becomes an intrinsic part of the nucleic acid structure. The immobilization of the aptamer on a solid support must avoid any steric hindrance or constraint that could prevent the folding of the aptamer in the correct conformation.

Usually, the aptamer molecule (RNA and DNA) is immobilized via a modifying molecule present at the $3'$ or the $5'$ end: the most frequent modifications are a thiol [36–39], or an amino group [35,40] or a biotin molecule [41–43]. A beacon aptamer-based biosensor has been developed by Bang et al. [44] to detect thrombin via an electrochemical method by using the amino-group modification. Beacon aptamers are single-stranded oligonucleotides containing a short stem and a loop region. The short stem structure, possessing a complementary sequence at both ends of the beacon aptamer, can be used as a transducing element because upon aptamer-target-molecule binding, the broken stem structure leads to changes in the vibrational energy of the fluorescent tagging or intercalating molecules at the stem structure. Gold electrodes were modified with a beacon aptamer fixed at the $5'$ end via a linker containing a primary aliphatic amine. The authors demonstrated the high specificity of the sensor, the possibility of regeneration, and a detection limit for the protein of 11 nM. The interaction between streptavidin and a $5'$ or $3'$ biotinylated aptamer is one of the most widely used immobilization protocols in the development of different biosensors and/or other bioanalytical techniques.

An aptamer-based biosensor array for multiplex analysis of proteins has been developed using $5'$-biotinylated aptamers specific for several proteins with relevance to cancer (inosine mono-phosphate dehydrogenase (IMPDH), vascular endothelial growth factor (VEGF) and basic fibroblast growth factor (bFGF)) [45]. Fluorescence polarization anisotropy was used for solid-phase measurements of target–protein binding. The dissociation constant, k_d, for the target–sensor interaction ($k_d = 15$ nM) was similar to that from solution-phase experiments ($k_d = 26$ nM), suggesting that the immobilization of the aptamer did not affect its functionality. The same immobilization technique was adopted in a recent work on aptamer-based arrays published by Ellington and coworkers [46]. The authors report the adaptation of a chip-based microsphere array (the electronic taste chip) to aptamer receptors. The system was composed of a flow cell connected to a fast-performance liquid chromatograph pump and a fluorescence microscope for observation. The flow cells contained silicon chips with multiple wells in which beads modified with the sensor elements were deposited. Commercially available streptavidin agarose beads were modified with $5'$-biotinylated aptamers; RNA anti-ricin aptamers were used to demonstrate the possibility of quantifying the labeled protein. A sandwich assay format was also optimized using anti-ricin antibodies, to directly detect the unlabelled protein. In the assay, the aptamer was bioti-nylated during transcription by introducing biotinylated dinucleotides into the reaction mixture. After immobilization, the aptamer was put in contact with the solution containing fluorescently labeled ricin. The fluorescence intensities of the captured proteins were used to construct a cali-bration plot for ricin and a detection limit of 8 μg/mL was calculated. The estimated k_d of the aptamer–protein complex was 1.24 μM, higher than the previously reported k_d (7.4 nM) [47]. In the sandwich assay, the anti-ricin aptamer acted as a capture reagent and unlabeled ricin bound to the aptamer could interact with a fluorophore-labelled antibody that served as a reporter. A detection limit of 320 ng/mL was achieved in this format with high selectivity, as demonstrated by the absence of a fluorescence signal from negative controls lacking ricin. The sensor was reusable after denaturation and removal of the bound protein with a washing solution containing 7 M urea.

In 2002, Liss et al. [33], in developing a quartz crystal biosensor designed to detect IgE, compared the immobilization of a $5'$- or a $3'$-biotinylated aptamer with a protocol based on the use of an amino-modified IgE-specific aptamer. The biosensors modified by coupling $5'$- or $3'$-amino-modified aptamers to 3,3$'$-dithiodipropionic acid di(N-succinimidyl ester) (DSP) activated surfaces, were specific for IgE, but they were not as sensitive as the antibody-based sensors or the biotinylated aptamer-based sensor. On the contrary, both the $5'$- and the $3'$-biotinylated aptamers showed better specific and sensitive target binding. Moreover, a dissociation constant of 3.6 nM was found for the aptamer that was extended both at the $5'$ and the $3'$ ends by adding a GCGC sequence with respect to the original aptamer ($k_d = 10$ nM).

The $5'$-modified aptamer-coated biosensor was also demonstrated to specifically detect IgE in complex protein samples, such as low-fat milk, meat extract, and brain–heart bouillon.

TABLE 6.1

Variants of the Thrombin-Binding Aptamers Used in the Development of an ELAA (Enzyme-Linked Aptamer Assay)

Thrombin Aptamer Variants	Aptamer Sequence
5′-biotinylated with spacer	5′-biot-6C-GGTTGGTGTGGTTGGT
3′-biotinylated with spacer	GGTTGGTGTGGTTGGT-6C-biot-3′
5′-biotinylated without spacer	5′-biot-GGTTGGTGTGGTTGGT
3′-biotinylated without spacer	GGTTGGTGTGGTTGGT-Biot-3′

Source: From Baldrich, E., Restrepo, A., and O'Sullivan, C. K., *Analytical Chemistry*, 76, 7053, 2004.

A detailed study on the effect of the aptamer immobilization on an enzyme-linked aptamer assay (ELAA) was performed by Baldrich et al. (2004) [48] by using four variants (Table 6.1) of a biotinylated thrombin aptamer. An additional 6C-linker was included or not in 3′- and 5′-biotinylated aptamers to study the effect of this spacer arm on the flexibility of the aptamer. Immobilization through the 3′ end improved the thrombin binding relative to immobilization via the 5′ end. The authors suggested that this could be due to the interference of the 5′ modification with the correct folding of the aptamer and to the improved stability of the 3′-modified aptamer. Addition of a linker correlated with improved target binding in both cases, probably due to the decrease of steric hindrance resulting in a better folding and target recognition. The same results were obtained when performing a competitive assay using the same four variants. The highest sensitivity (5.2×10^{-5} AU/nM) and the best detection limit (2 nM) were obtained when using the 3′-biotinylated aptamer with the spacer arm.

Some recent works report the use of this immobilization protocol for the development of biochips (arrays) and stationary phases. In particular, a lysozyme-specific aptamer with biotin at the 5′ end was immobilized on streptavidin-coated magnetic beads and used with an electrochemical approach (chronopotentiometric stripping measurements) of the captured target molecule with a detection limit of 7 nM [49]. RNA aptamers specific for the same target molecule, lysozyme, were also spotted on streptavidin-coated microarrays slides. In this case, they were directly transcribed with biotinylated-guanoside 5′-monophosphate to produce 5′-biotinylated molecules. A detection limit of 70 fM was obtained in this case and the microarray retained the specificity for the target protein of a 10,000-fold excess of T-4 cell lysate protein [50].

A biotinylated anti-L-arginine D-RNA aptamer was also used as target-specific chiral stationary phase and employed in an HPLC system [51]. Usually, when dealing with biotinylated aptamers, a thermal treatment of the molecule is applied before immobilization to unfold the aptamer strand and make the biotin label available for the interaction with streptavidin; this can also be used to unfold existing 3D structures that can interfere with the target binding. The thermal treatment usually consists of the incubation of the aptamer at high temperature, followed by rapid cooling in ice to block the molecule in its unfolded structure [52]. A thermal treatment (90°C for 1 min, in ice for 10 min) was applied to the RNA aptamer specific for HIV-1 Tat protein before its immobilization on piezoelectric quartz crystals [34]. A lower experimental detection limit (0.25 ppm) was obtained with respect to the biosensor with the nontreated aptamer and the reproducibility was improved, without altering the specificity of the system. A similar treatment (95°C for 3 min, chilling in ice) was applied to a biotinylated anti-IgE DNA aptamer for the development of another piezoelectric biosensor [33]. No comparison to the same biosensor with a nontreated aptamer was reported.

Other authors reported a slightly different treatment involving heating the aptamer at 70°C or 85°C and then slowly cooling it to room temperature to allow the molecule to fold into its active conformation. This treatment (70°C) was used with biotinylated RNA aptamers specific for lysozyme before their immobilization onto streptavidin-coated microarray slides [50] or onto micromachined chips for an electronic tongue sensor array [46]. Brumbt et al. [51] treated an anti-arginine RNA aptamer at 85°C before its immobilization for the development of a chiral stationary phase. A very interesting study has been conducted and reported on this subject by Baldrich et al. [48], who examined the effect of aptamer denaturation with a thrombin-specific DNA aptamer to be immobilized on streptavidin-coated plates for a direct enzyme-linked aptamer assay. The assay was conducted with plates modified with the aptamer with and without a thermal treatment before its immobilization (94°C for 10 min, rapid cooling in ice) and the results were compared in terms of target binding. According to this comparison, aptamer denaturation did not improve the efficiency of the assay, suggesting that the aptamer adopts the same folding before and after the denaturation and that the presence of thrombin promotes a certain folding. They suggested that this denaturation treatment may be necessary in other cases when, unlike the thrombin aptamer, the molecule is known to adopt very stable conformations that compete with a less stable structure needed for target binding.

An alternative to the immobilization method based on the biotin/streptavidin interaction can be found in the modification of the aptamer by a thiol group at one of the two ends. Savran et al. [53] employed a thiol-modified aptamer specific for *Thermus aquaticus* (Taq) DNA polymerase, immobilized on the surface of a cantilever for a micromechanical detection of the protein. By using interferometry, the differential bending between this cantilever and a reference cantilever modified with a nonspecific sequence was determined. The system was characterized in terms of sensitivity performing binding experiments at seven different Taq concentrations; a k_d of 15 pM was calculated. Recently, a $5'$-thiol-modified aptamer specific for human IgE mixed with cysteamine to create a hybrid modified layer has been immobilized on an array of electrodes and used in an electrochemical impedance spectroscopy method [54]. The same electrodes were examined by atomic force microscopy (AFM) after the immobilization of the aptamer and the binding of the target. AFM images demonstrated that IgE could specifically bind to the aptamer and stand above the monolayer surface. All these research works, centered on aptamer-based bioanalytical methods created by using several different immobilization protocols, suggested that the procedure to fix the aptamer to the solid surface should be studied and optimized for each different aptamer. This demonstrates that the properties of the aptamer have a strong influence on the sensor/assay performance and should be taken into account each time a new aptamer is used for an analytical application [48].

6.5 APTAMER-BASED ASSAY PROTOCOLS

6.5.1 APTAMER-BASED DETECTION OF PROTEINS

The advancement in the use of protein-specific aptamers, together with the identification of important medical marker proteins, could facilitate the development of new bioanalytical methods to be applied in proteomics and the detection or treatment of diseases.

A great amount of RNA or DNA aptamers have been selected currently against specific proteins, some of which are reported in Table 6.2 [4,5,24,43,55–69]. In the following sections, several protocols that have been optimized for the aptamer-based detection of proteins are reported.

6.5.1.1 ELONA—Enzyme Linked Oligonucleotide Assay

The use of ELONA was patented by NeXstar Pharmaceuticals (now Gilead Sciences) and applied to the detection of human VEGF, a protein secreted by tumors that generates the growth of new blood vessels [60]. The system followed one of the possible formats that can be exploited in ELONA or mixed ELISA/ELONA assays [8] (Figure 6.3a). The format consists of the use of antibodies as capture molecules and labeled aptamers against the antibody–analyte immunocomplex as reporter molecules. Anti-VEGF monoclonal antibody was immobilized onto microtiter

TABLE 6.2
Selected Protein-Binding Aptamers

Target Protein	Type of Aptamer	Reference
Thrombin	DNA	4,55
IgE	DNA	5
NF-Kb	RNA	56,57
Lysozyme	DNA	24
HIV-1 Tat protein	RNA	58,59
Vesicular endothelial growth factor (VEGF)	RNA	60
CD4 antigen	RNA	61
HIV-1 gag protein	RNA	62
L-selectin	DNA	63
Iron regulatory protein (IRP)	RNA	64
HIV-1 rev protein	RNA	65
SelB E. coli protein	RNA	66
Platelet-derived growth factor (PDGF)	DNA	67
Colicin E3	RNA	68
Thyroid transcription factor (TTF1)	DNA	43
Acetylcholine receptor	RNA	69

plates and then reacted with samples containing different amounts of VEGF. Fluoresceinated anti-VEGF RNA aptamers were bound to the captured VEGF and then reacted with anti-fluorescein antibodies conjugated with alkaline phosphatase for a chemiluminescent signal transduction. Standard solutions of VEGF in the concentration range of 31–8000 pg/mL were detected and the assay was able to detect VEGF in human serum with results similar to a standard antibody-based ELISA.

A similar mixed assay has been presented for the detection of thrombin using a hybrid immunobead aptamer test [70]. Also in this case, the aptamer has the function of detection reagent. Immunomagnetic beads were coated with anti-thrombin antibodies and reacted with an incubated mixture of thrombin to be detected and the anti-thrombin 5′-biotinylated DNA aptamer. Europium-labeled streptavidin was finally added to the beads and the reaction was read by time-resolved

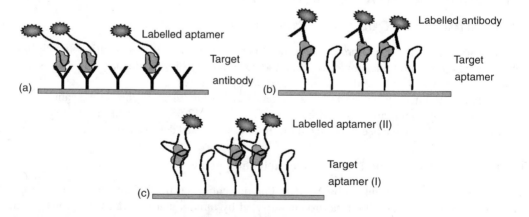

FIGURE 6.3 Different formats for enzyme linked oligonucleotide assays (ELONA). (a) The antibody is immobilized onto a solid support and employed as capture molecule; the labeled aptamer is the reporter molecule. (b) The aptamer is the capture reagent and the labeled antibody is the reporter molecule. (c) The aptamer (two different or the same) is employed as both capture and, after its labeling, as reporter molecule.

fluorescence. The immunofluorometric aptamer assay showed sensitivity for thrombin within the physiological parameters. Very important parameters were illustrated when developing such an assay. The authors tested different traditional 96-well assay formats using the same aptamer and the same detection principle, but a good sensitivity could be observed only in a very specific format with the formation of the aptamer–thrombin complex in solution. This could be related to possible general limitations in the use of aptamers in classical 96-well format tests, but it could also be due to more specific problems connected to the thrombin aptamer or to aptamers with similar structural motifs. These findings demonstrate that some aptamers could be more susceptible to assay conditions than the corresponding antibody or other aptamers; the results also confirmed that the protocol employed when using aptamers in such assays should be carefully studied and that optimal operating conditions could be different from one aptamer to another [13,48,70].

In this regard, a very interesting systematic evaluation has been reported on the parameters affecting the performance of the ELONA (here called ELAA—enzyme linked aptamer assay) tests [48]. Different assay formats using a thrombin-specific aptamer were examined together with the effect of several parameters such as pH, incubation time, temperature, and aptamer denaturation (discussed in Section 6.4). The mixed antibody–aptamer sandwich assay was tested in the direct, the indirect, and the inverse format. In the indirect sandwich format, the thrombin-binding aptamer was immobilized onto streptavidin-coated microtiter plates and the captured thrombin was recognized by a sheep anti-thrombin antibody and detected by an anti-sheep antibody labeled with horseradish peroxidase (HRP). In the inverse format, the antibody was immobilized and the captured thrombin was revealed by the HRP-labeled aptamer. Good results were obtained with both the direct and indirect assays with a detection limit for thrombin below 1 nM, but a very high LOD was obtained with the inverse format. An LOD of 3.5 nM could be reached only with a preincubation step of the protein with the aptamer before being added to the immobilized antibody. A competition assay was also performed that was based on the immobilization of the aptamer with binding of labeled and nonlabeled thrombin. A higher LOD (2–9 nM) was observed respect to the other assays.

For all these assay formats [48] the best performances were obtained when the aptamer was immobilized through its $3'$ end (see discussion on this subject in Section 4 of this review). HEPES 10 mM, pH 8.0 was found to be the optimal working buffer and no influence was observed by the absence or presence of K^+ in the solution. A decreased binding efficiency was instead observed following addition of Na^+ or Mg^{2+} to the KCl-containing buffer.

6.5.1.2 Aptasensors

The application of aptamers as biocomponents in biosensors offers a multitude of advantages over classical affinity sensing methods based on antibodies. The advantages include the possibility of easily regenerating the function of immobilized aptamers, their homogeneous preparation, and the possibility of using different detection methods due to easy labeling [7–9,12]. The advantages of aptamers as recognition element for biosensing have been presented in conjunction with colorimetric methods [71], acoustic waves [33,34,72,73] and surface plasmon resonance [74,75] transduction. These applications have been extensively reviewed [9,10,12], but very little information has been reported about aptamer-based sensors with electrochemical transduction, which will be examined in the following section.

6.5.1.3 Electrochemical Aptamer-Based Protein Detection

Electrochemical transduction is a relatively new method employed in aptamer-based bioanalytical procedures. Several advantages of this method have been presented, especially those related to proteome analysis and to the possibility of fabrication of integrated systems such as microarrays [76]. One of the first aptamer-based electrochemical sensors has been reported by Ikebukuro et al. [76] in a sandwich assay format for the detection of thrombin (Figure 6.3c). The possibility of detecting this protein in a sandwich test is related to the existence of two thrombin aptamers that bind the protein in different positions [4,55]. In this assay, one of the aptamers (15-mer, thiol-modified at the 5' end) was immobilized onto gold electrodes and the other aptamer (29-mer) was used for the detection after its labeling with pyrroquinoline quinine glucose dehydrogenase ((PQQ)GDH). The electrochemical signal was generated upon the addition of glucose. A linear relationship between the current and thrombin concentration was observed in the range 40–100 nM and the lower detection limit was 10 nM. The same target protein (thrombin) has been detected using the specific aptamer and other electrochemical methods, such as the measurement of the charge consumption from the electrode covered by a DNA thrombin-specific aptamer to an electro-chemical indicator (methylene blue, MB), bound to the protein [77]. The MB electrochemical marker has also been used with a beacon-type aptamer biosensor for thrombin detection [44]. In this case, the 5'-amino-modified beacon aptamer was fixed onto the surface of gold electrodes and MB was intercalated into the beacon sequence. The binding of thrombin causes a conformational change of the aptamer and the release of the intercalated MB with a decrease in the current intensity. A detection limit for thrombin of 10 nM was achieved by both of the proposed methods.

Impedance spectroscopy represents another electrochemical technique that has been used combined with the thrombin-binding aptamer [78]. The 5'-thiol-modified aptamer was immobilized onto thin-film gold electrodes and the formation of the aptamer–protein complex was detected via monitoring the changes in interfacial electron transfer resistance of electrochemical impedance spectroscopy. A lower detection limit for thrombin was achieved with this technique, 0.1 nM, with a detection range spanning three orders of magnitude (pM–µM).

Another label-free electrochemical method has been recently presented by Kawde et al. [49]. The detection of lysozyme has been achieved by combining the protein-specific aptamer immobi-lized onto magnetic particles in combination with the electrochemical measurement of the captured protein (oxidation of tyrosine and tryptophan residues) (Figure 6.4). This reagentless label-free detection cannot be accomplished with traditional immunoassays due to the presence of the elec-troactive residues both in the target protein and in the antibody. This method has been presented as an alternative technique for the development of protein biochips.

6.5.2 APTAMER-BASED DETECTION OF SMALL MOLECULES

Aptamers that recognize small molecules are increasingly applied as tools in bioanalytical methods [79]. The plethora of aptamers that have been selected to complex molecules of low molecular weight leads to the possible use of these aptamers not only in diagnostic assays, but also with a wider range of applications, such as environmental analytical chemistry. Some of the small molecules for which aptamers have been selected are reported in Table 6.3 [71,80–101].

Small-molecule-binding aptamers have been used in affinity chromatography as a stationary phase for retaining and separating the target molecule from complex samples. An anti-adenosine DNA aptamer was evaluated as a stationary phase and adenosine was successfully monitored in microdialysis samples [81]. Up to 6 µL of 1.2 µM adenosine could be injected onto a 7 cm-long column (I.D. 150 µm) without loss of adenosine and a detection limit of 30 nM was achieved. The robustness of the stationary phase (at least 200 injections) demonstrated that such columns could be used for chemical monitoring and for high-throughput analysis.

Many other works are related to bioanalytical aptamer-based methods with a colorimetric, fluorescence, or luminescence transduction. Stojanovic and Landry [71] used an aptamer specific for cocaine to develop a colorimetric probe for this molecule. A pool of 35 different dyes was first screened to examine their binding to the aptamer in the presence or absence of cocaine. One of these dyes, diethylthiotricarbocyanine iodide, was chosen to construct the colorimetric sensor. After incubation of the dye with the aptamer, cocaine addition caused a displacement of the dye with an attenuation of absorbance proportional to the concentration of the added cocaine (Figure 6.5). No change in the visible spectra of the aptamer-dye complex was observed upon addition of cocaine

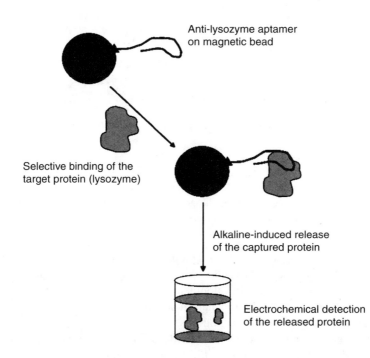

Anti-lysozyme aptamer
on magnetic bead

Selective binding of the
target protein (lysozyme)

Alkaline-induced release
of the captured protein

Electrochemical detection
of the released protein

FIGURE 6.4 Electrochemical aptamer-based detection of lysozyme. The protein specific aptamer is immobilized onto magnetic particles serving as capture reagent. After alkaline-induced release of the captured lysozyme, the electrochemical measurement is accomplished by detecting the oxidation of tyrosine and tryptophan residues of the released protein. (From Kawde, A., Rodriguez, M.C., Lee, T.M.H., and Wang, J., *Electrochemistry Communications*, 7, 537, 2005.)

TABLE 6.3
Selected Small-Molecule-Binding Aptamers

Target Molecule	Type of Aptamer	Reference
Adenosine	DNA	80,81
Cocaine	DNA	71,82
Aromatic amines (methylenedianiline)	RNA	83
Chloroaromatics (4-chloroaniline, 2,4,6-trichloroaniline, pentachlorophenol)	DNA	84
ATP	DNA	85,86
Theophyllin	RNA	87,88
Arginine	RNA	89,90
Neomycin	RNA	91,92
Flavin mononucleotide FMN	RNA	93,94
Streptomycin	RNA	95
Tetracycline	RNA	96
Biotin	RNA	97
Organic dyes	DNA	98
Moenomycin A	RNA (modified)	99
S-adenosylhomocysteine	RNA	100
Dopamine	RNA	101

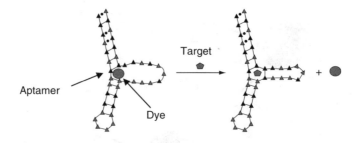

FIGURE 6.5 Scheme of an aptamer-based assay for the detection of cocaine. After incubation of a specific dye with a cocaine-binding aptamer, cocaine addition causes a displacement of the dye with an attenuation of absorbance proportional to the concentration of the added cocaine. (From Stojanovic, M. N. and Landry, D. W., *Journal of the American Chemical Society*, 124, 9678, 2002.)

metabolites such as benzoyl ecgonine, indicating a high selectivity of the colorimetric probe. The demonstrated ability of these receptors to undergo visible changes upon recognition of their ligands in this simple assay could result in the development of very simple and inexpensive colorimetric analytical methods, including spot tests for small molecules.

Another small-molecule detection method based on aptamers [86] has been reported as an extension of a previously reported method for the detection of proteins. The method was based on aptamers and a DNA molecular light-switching complex, $[Ru(phen)_2(dppz)]^{2+}$, and applied to the detection of ATP as a model target. $[Ru(phen)_2(dppz)]^{2+}$ has no luminescence in aqueous solution but exhibits strong luminescence when intercalating into the nonaqueous pocket of DNA duplex. The complex can intercalate into the folded aptamer and luminesce. The binding of the target molecule to the aptamer changes its conformation, leading to a change in luminescence. The selectivity of the method has been tested by comparing the luminescence changes due to the binding of ATP and other ATP analogues such as UTP, CTP, and GTP. The three analogues failed to cause a significant change in luminescence at all tested concentrations. Titration of ATP was also carried out and a detection limit of 20 nM was achieved, which resulted in comparable or better results with respect to other ATP detection assays [85].

The development of aptamer-based arrays would represent a great advance, especially in diagnostics and in metabolomic studies. A prototype RNA array was presented to detect, in parallel, different analytes in complex mixtures and provide a chemical fingerprint of the sample [94]. Seven RNA switches were used to construct an array for the detection of Co^{2+}, $3',5'$-cyclic guanosine monophosphate (cGMP), $3',5'$-cyclic cytosine monophosphate (cCMP), $3',5'$-cyclic adenosine monophosphate (cAMP), flavin mononucleotide (FMN), and theophylline. The resulting array was able to detect the different analytes in complex mixtures and also to characterize different *E. coli* strains by detecting naturally produced cAMP in bacterial culture media.

6.6 CONCLUSIONS

Aptamers have been presented as a valid alternative for biorecognition elements in bioanalytical methods. The main aspects related to their selection through the SELEX protocol have been evidenced, focusing mainly on those parameters, such as the complexity of the library and the chemistry of the nucleotides, that can highly influence the result of the selection. Aptamers can be considered to have several advantages with respect to antibodies, such as overcoming of the use of animals for their selection and production. Moreover, when comparing aptamers and other biomimetic molecules, such as oligopeptides, the possibility of amplification during the selection process represents an important benefit.

In this review, several applications based on aptamers have been reported, focusing on the parameters that need to be optimized when developing such assays (i.e. immobilization protocols, etc.). Different bioanalytical methods based on aptamers have been considered both for the detection of proteins or small molecules. From the examination of the different protocols employed in such assays, one important point must be emphasized: the nature, conformation, and sequence of each aptamer should be carefully considered because optimal working conditions can remarkably vary from one aptamer to another.

If all these important findings, together with the shortening of the time required for the selection process, are taken into consideration, aptamers can actually represent the alternative for the development of bioanalytical methods with the possibility of producing new multianalyte aptamer-based arrays.

REFERENCES

1. James, W., Aptamers. In *Encyclopaedia of Analytical Chemistry*, Mayers, R. A., Ed., Chichester, John Wiley & Sons Ltd, p. 4848, 2000.
2. Ellington, A. D. and Szostak, J. W., In vitro selection of RNA molecules that bind specific ligands, *Nature*, 346, 818–852, 1990.
3. Tuerk, C. and Gold, L., Systematic evolution of ligands by exponential enrichment: RNA ligands to bacteriophage T4 DNA polymerase, *Science*, 249, 505–510, 1990.
4. Bock, L. C., Griffin, L. C., Latham, J. A., Vermaas, E. H., and Toole, J. J., Selection of single-stranded DNA molecules that bind and inhibit human thrombin, *Nature*, 355, 564–566, 1992.
5. Wiegand T. W., Williams, P. B., Dreskin, S. C., Jouvin, M. H., Kinet, J. P., and Tasset, D., High affinity oligonucleotide ligands to human IgE inhibit binding to FCe receptor I, *Journal of Immunology*, 157, 221–230, 1996.
6. McGown, L. B., Joseph, M. J., Pitner, J. B., Vonk, G. P., and Linn, C. P., The nucleic acid ligand. A new tool for molecular recognition, *Analytical Chemistry*, 663A–668A, 1995.
7. Jayasena, S., Aptamers: an emerging class of molecules that rival antibodies in diagnostics, *Clinical Chemistry*, 45, 1628–1650, 1999.
8. O'Sullivan, C. K., Aptasensors—the future of biosensing? *Analytical and Bioanalytical Chemistry*, 372, 44–48, 2002.
9. Luzi, E., Minunni, M., Tombelli, S., and Mascini, M., New trends in affinity sensing: aptamers for ligand binding, *Trends in Analytical Chemistry*, 22, 810–818, 2003.
10. Tombelli, S., Minunni, M., and Mascini, M., Analytical applications of aptamers, *Biosensors and Bioelectronics*, 20, 2424–2434, 2005.
11. Clark, S. L. and Remcho, V. T., Aptamers as analytical reagents, *Electrophoresis*, 23, 1335–1340, 2002.
12. You, K. M., Lee, S. H., Im, A., and Lee, S. B., Aptamers as functional nucleic acids: in vitro selection and biotechnological applications, *Biotechnology and Bioprocess Engineering*, 8, 64–75, 2003.
13. Mukhopadhyay, R., Aptamers are ready for the spotlight, *Analytical Chemistry*, 115A–118A, 2005.
14. Tsai, D. E., Harper, D. S., and Keene, J. D., U1-snRNP-A protein selects a ten nucleotide consensus sequence from a degenerate RNA pool presented in various structural contexts, *Nucleic Acids Research*, 19, 4931–4936, 1991.
15. Blackwell, T. K. and Weintraub, H., Differences and similarities in DNA-binding preferences of MyoD and E2A protein complexes revealed by binding site selection, *Science*, 250, 1104–1110, 1990.
16. Sampson, T., Aptamers and SELEX: the technology, *World Patent Information*, 25, 123–129, 2003.
17. Marshall, K. A. and Ellington, A. D., In vitro selection of RNA aptamers, *Methods in Enzymology*, 318, 193–214, 2000.
18. James, W., Nucleic acid and polypeptide aptamers: a powerful approach to ligand discovery, *Current Opinion in Pharmacology*, 1, 540–546, 2001.
19. Jiang, L. and Patel, J. D., Saccharide-RNA recognition in an aminoglycoside antibiotic-RNA aptamer complex, *Chemical Biology*, 4, 35–50, 1997.

20. Heidenreich, O. and Eckstein, F., Hammerhead ribozyme-mediated cleavage of the long terminal repeat RNA of human immunodeficiency virus type 1, *Journal of Biological Chemistry*, 267, 1904–1909, 1992.

21. Kusser, W., Chemically modified nucleic acid aptamers for in vitro selections: evolving evolution, *Journal of Biotechnology*, 74, 27–38, 2000.

22. Famulok, M., Mayer, G., and Blind, M., Nucleic acid aptamers—from selection in vitro to application in vivo, *Accounts of Chemical Research*, 33, 591–599, 2000.

23. Cox, J. C., Rudolph, P., and Ellington, A. D., Automated RNA selection, *Biotechnology Progress*, 14, 845–850, 1998.

24. Cox, J. C. and Ellington, A. D., Automated selection of anti-protein aptamers, *Bioorganic and Medicinal Chemistry*, 9, 2525–2531, 2001.

25. Cox, J. C., Hayhurst, A., Hesselberth, J., Davidson, E. A., Sooter, L. J., Bayer, T. S. et al., Automated selection of aptamers against protein targets translated in vitro: from gene to aptamer, *Nucleic Acids Research*, 30, e108, 2002a.

26. Cox, J. C., Rajendran, M., Riedel, T., Davidson, E. A., Sooter, L. J., Bayer, T. S., Schmitz-Brown, M. et al., Automated acquisition of aptamer sequences, *Combinatorial Chemistry and High Throughput Screening*, 5, 289–299, 2002b.

27. Brody, E. N. and Gold, L., Aptamers as therapeutic and diagnostic agents, *Journal of Biotechnology*, 74, 5–13, 2000.

28. Golden, M. C., Collins, B. D., Willis, M. C., and Koch, T. H., Diagnostic potential of PhotoSELEX-evolved ssDNA aptamers, *Journal of Biotechnology*, 81, 167–178, 2000.

29. Berezovski, M., Drabovich, A., Krylova, S. M., Musheev, M., Okhonin, V., Petrov, A., and Krylov, S. N., Nonequilibrium capillary electrophoresis of equilibrium mixtures: a universal tool for development of aptamers, *Journal of the American Chemical Society*, 127, 3165–3171, 2005.

30. Mendonsa, S. D. and Bowser, M. T., In vitro evolution of functional DNA using capillary electrophoresis, *Journal of the American Chemical Society*, 126, 20–22, 2004.

31. Rajendran, M. and Ellington, A. D., Selecting nucleic acids for biosensor applications, *Combinatorial Chemistry and High Throughput Screening*, 5, 263–270, 2002.

32. Pieken, W., Olsen, D. B., Benseler, F., Aurup, H. H., and Eckstein, F., Kinetic characterization of ribonuclease-resistant $2'$-modified hammerhead ribozymes. *Science*, 253, 314–317, 1991.

33. Liss, M., Petersen, B., Wolf, H., and Prohaska, E., An aptamer-based quartz crystal protein biosensor, *Analytical Chemistry*, 74, 4488–4495, 2002.

34. Minunni, M., Tombelli, S., Gullotto, A., Luzi, E., and Mascini, M., Development of biosensors with aptamers as bio-recognition element: the case of HIV-1 Tat protein, *Biosensors and Bioelectronics*, 20, 1149–1156, 2004.

35. Jiang, Y., Zhu, C., Ling, L., Wan, L., Fang, X., and Bai, C., Specific aptamer–protein interaction studied by atomic force microscopy, *Analytical Chemistry*, 75, 2112–2116, 2003.

36. Dick, L. W. and McGown, L. B., Aptamer-enhanced laser desorption/ionization for affinity mass spectrometry, *Analytical Chemistry*, 76, 3037–3041, 2004.

37. Stadtherr, K., Wolf, H., and Lindner, P., An aptamer-based protein biochip, *Analytical Chemistry*, 77, 3437–3443, 2005.

38. Connor, A. C. and McGown, L. B., Aptamer stationary phase for protein capture in affinity capillary chromatography, *Journal of Chromatography A*, 1111, 115–119, 2006.

39. Baldrich, E., Acero, J. L., Reekmans, G., Laureyn, W., and O'Sullivan, C. K., Displacement ezyme linked aptamer assay, *Analytical Chemistry*, 77, 4777–4784, 2005.

40. Farokhzad, O. C., Khademhosseini, A., Jon, S., Hermmann, A., Cheng, J., Chin, C., Kielyuk, A. et al., Microfluidic system for studying the interaction of nanoparticles and micropartcles with cells, *Analytical Chemistry*, 77, 5453–5459, 2005.

41. Romig, T. S., Bell, C., and Drolet, D. W., Aptamer affinity chromatography: combinatorial chemistry applied to protein purification, *Journal of Chromatography B*, 731, 275–284, 1999.

42. Michaud, M., Jourdan, E., Villet, A., Ravel, A., Grosset, C., and Peyrin, E., A DNA aptamer as a new target-specific chiral selector for HPLC, *Journal of the American Chemical Society*, 125, 8672–8679, 2003.

43. Murphy, M. B., Fuller, S. T., Richardson, P. M., and Doyle, S. A., An improved method for the in vitro evolution of aptamers and applications in protein detection and purification, *Nucleic Acids Research*, 31, e110, 2003.
44. Bang, G. S., Cho, S., and Kim, B., A novel electrochemical detection method for aptamer biosensors, *Biosensors and Bioelectronics*, 21, 863–870, 2005.
45. McCauley, T. G., Hamaguchi, N., and Stanton, M., Aptamer-based biosensor arrays detection and quantification of biological macromolecules, *Analytical Biochemistry*, 319, 244–250, 2003.
46. Kirby, R., Cho, E. J., Gehrke, B., Bayer, T., Park, Y. S., Neikirk, D. P., McDevitt, J. T. et al., Aptamer-based sensor arrays for the detection and quantitation of proteins, *Analytical Chemistry*, 76, 4066–4075, 2004.
47. Hesselberth, J. R., Miller, D., Robertus, J., and Ellington, A. D., In vitro selection of RNA molecules that inhibit the activity of Ricin A-chain, *Journal of Biological Chemistry*, 275, 4937–4942, 2000.
48. Baldrich, E., Restrepo, A., and O'Sullivan, C. K., Aptasensor development: elucidation of critical parameters for optimal apamer performance, *Analytical Chemistry*, 76, 7053–7063, 2004.
49. Kawde, A., Rodriguez, M. C., Lee, T. M. H., and Wang, J., Label-free bioelectronic detection of aptamer–protein interactions, *Electrochemistry Communications*, 7, 537, 2005.
50. Collett, J. R., Cho, E. J., Lee, J. F., Levy, M., Hood, A. J., Wan, C., and Ellington, A. D., Functional RNA microarrays for high-throughput screening of antiprotein aptamers, *Analytical Biochemistry*, 338, 113–123, 2005.
51. Brumbt, A., Ravelet, C., Groset, C., Ravel, A., Villet, A., and Peyrin, E., Chiral stationary phase based on a biostable L-RNA aptamer, *Analytical Chemistry*, 77, 1993–1998, 2005.
52. Ducongè, F., Di Primo, C., and Toulmè, J.-J., Is a closing GA pair a rule for stable loop–loop RNA complexes? *Journal of Biological Chemistry*, 275, 21287–21294, 2000.
53. Savran, C. A., Knudsen, S. M., Ellington, A. D., and Manalis, S. R., Micromechanical detection of proteins using aptamer-based receptor molecules, *Analytical Chemistry*, 76, 3194–3198, 2004.
54. Xu, D., Xu, D., Yu, X., Liu, Z., He, W., and Ma, Z., Label-free electrochemical detection for aptamer-based array electrodes, *Analytical Chemistry*, 77, 5107–5113, 2005.
55. Tasset, D. M., Kubik, M. F., and Stiner, W., Oligonucleotide inhibitors of human thrombin that bind distinct epitopes, *Journal of Molecular Biology*, 272, 688–698, 1997.
56. Lebruska, L. L. and Mather, L. L., Selection and characterization of an RNA decoy for transcription factor NF-kB, *Biochemistry*, 38, 3168–3174, 1999.
57. Yang, X., Li, X., Prow, T. W., Reece, L. M., Bassett, S. E., Luxon, B. A., Herzog, N. K. et al., Immunofluorescence assay and flow-cytometry selection of bead-bound aptamers, *Nucleic Acids Research*, 31, e54, 2003.
58. Yamamoto, R., Katahira, M., Nishikawa, S., Baba, T., Taira, K., and Kumar, P. K. R., A novel RNA motif that binds efficiently and specifically to the Tat protein of HIV and inhibits the trans-activation by Tat of transcription in vitro and in vivo, *Genes to Cells*, 5, 371–388, 2000.
59. Yamamoto, R. and Kumar, P. K. R., Molecular beacon aptamer fluoresces in the presence of tat protein of HIV-1, *Genes to Cells.*, 5, 389–396, 2000.
60. Drolet, D. W., Moon-Mcdermott, L., and Romig, T. S., An enzyme-linked oligonucleotide assay, *Nature Biotechnology*, 14, 1021–1025, 1996.
61. Kraus, E., James, W., and Barclay, A. N., Cutting edge: novel RNA ligands able to bind CD4 antigen and inhibit CD4$^+$T lymphocyte function, *Journal of Immunology*, 160, 5209–5212, 1998.
62. Lochrie, M. A., Waugh, S., Pratt, D. G., Clever, J., Parslow, T. G., and Polisky, B., In vitro selection of RNAs that bind to the human immunodeficiency virus type-1 gag polyprotein, *Nucleic Acids Research*, 25, 2902–2910, 1997.
63. Hicke, B. J., Watson, S. R., Koenig, A., Lynott, C. K., Bargatze, R. F., and Chang, Y. F., DNA aptamers block L-selectin function in vivo, *Journal of Clinical Investigation*, 98, 2688–2692, 1996.
64. Lisdat, F., Utepbergenov, D., Haseloff, R. F., Blasig, I. E., Stocklein, W., Scheller, F. W., and Brigelius-Flohe, R., An optical method for the detection of oxidative stress using protein–RNA interaction, *Analytical Chemistry*, 73, 957–962, 2001.
65. Xu, W. and Ellington, A. D., Anti-peptide aptamers recognize amino acid sequence and bind a protein epitope, *Proceedings of the National Academy of Science USA*, 94, 7475–7480, 1997.

66. Klug, S. J., Huttenhofer, A., Kromayer, M., and Famulok, M., In vitro and in vivo characterization of novel mRNA motifs that bind special elongation factor SelB, *Proceedings of the National Academy of Science USA*, 94, 6676–6681, 1997.

67. Fang, X., Cao, Z., Beck, T., and Tan, W., Molecular aptamer for real-time oncoprotein PDGF monitoring by fluorescence anisotropy, *Analytical Chemistry*, 73, 5752–5757, 2001.

68. Hirao, I., Harada, Y., Nojima, T., Osawa, Y., Masaki, H., and Yokoyama, S., In vitro selection of RNA aptamers that bind to colicin E3 and structurally resemble the decoding site of 16S ribosomal RNA, *Biochemistery*, 43, 3214–3421, 2004.

69. Ulrich, H., Ippolito, J. E., Paga'n, O., Eterovic, V. A., Hann, R. M., Shi, H., Lis, J. T. et al., In vitro selection of RNA molecules that displace cocaine from the membrane-bound nicotinic acetylcholine receptor, *Proceedings of the National Academy of Science USA*, 95, 14051, 1998.

70. Rye, P. D. and Nustad, K., Immunomagnetic DNA aptamer assay, *Biotechnology Techniques*, 30, 290–295, 2001.

71. Stojanovic, M. N. and Landry, D. W., Aptamer-based colorimetric probe for cocaine, *Journal of the American Chemical Society*, 124, 9678–9679, 2002.

72. Groneowld, T. M. A., Glass, S., Quandt, E., and Famulok, M., Monitoring complex formation in the blood-coagulation cascade using aptamer-coated SAW sensors, *Biosensors and Bioelectronics*, 20, 2044–2048, 2005.

73. Schlensog, M. D., Gronewold, T. M. A., Tewes, M., Famulok, M., and Quandt, E., A love-wave biosensor using nucleic acids as ligands, *Sensors and Actuators B*, 101, 308–315, 2004.

74. Tombelli, S., Minunni, M., Luzi, E., and Mascini, M., Aptamer-based biosensors for the detection of HIV-1 Tat protein, *Bioelectrochemistry*, 67, 135–141, 2005.

75. Kawakami, J., Hirofumi, I., Yukie, Y., and Sugimoto, N., In vitro selection of aptamers that act with Zn^{2+}, *Journal of Inorganic Biochemistry*, 82, 197–206, 2000.

76. Ikebukuro, K., Kiyohara, C., and Sode, K., Novel electrochemical sensor system for protein using the aptamers in sandwich manner, *Biosensors and Bioelectronics*, 20, 2168–2172, 2005.

77. Hianik, T., Ostatna, V., Zajacova, Z., Stoikova, E., and Evtugyn, G., Detection of aptamer–protein interactions using QCM and electrochemical indicators methods, *Bioorganic and Medicinal Chemistry Letters*, 15, 291–295, 2005.

78. Cai, H., Lee, T. M. H., and Hsing, I. M., Label-free protein recognition using an aptamer-based impedance measurement assay, *Sensors and Actuators B*, 114, 433, 2006.

79. Famulok, M., Oligonucleotide aptamers that recognize small molecules, *Current Opinion in Structural Biology*, 9, 324–329, 1999.

80. Deng, Q., Watson, C. J., and Kennedy, R. T., Aptamer affinity chromatography for rapid assay of adenosine in microdialysis samples collected in vivo, *Journal of Chromatography A*, 1005, 123–130, 2003.

81. Deng, Q., German, I., Buchanan, D. D., and Kennedy, R. T., Selective retention and separation of adenosine and its analogs by affinity LC using an aptamer stationary phase, *Analytical Chemistry*, 73, 5415–5421, 2001.

82. Stojanovic, M. N., de Prada, P., and Landry, D. W., *Journal of the American Chemical Society*, 123, 4928–4931, 2001.

83. Brockstedt, U., Uzarowska, A., Montpetit, A., Pfau, W., and Labuda, D., In vitro evolution of NA aptamers recognizing carcinogenic aromatic amines, *Biochemical and Biophysical Research Communications*, 313, 1004–1007, 2004.

84. Bruno, J. G., In vitro selection of DA to chloroaromatics using magnetic microbead-based affinity separation and fluorescence detection, *Biochemical and Biophysical Research Communications*, 234, 117–120, 1997.

85. Jhaveri, S. D., Kirby, R., Conrad, R., Maglott, E. J., Bowser, M., Kennedy, R. T., and Glick, G., Designed signaling aptamers that transduce molecular recognition to changes in fluorescence intensity, *Journal of the American Chemical Society*, 122, 2469–2473, 2000.

86. Wang, J., Jiang, Y., Zhou, C., Fang, X., and Aptamer-based, A. T. P., Aptamer-based ATP assay using a luminescent light switching complex, *Analytical Chemistry*, 77, 3542–3546, 2005.

87. Jenison, R. D., Gill, S. C., Pardi, A., and Polisky, B., High-resolution molecular discrimination by RNA, *Science*, 263, 1425–1429, 1994.

88. Frauendorf, C. and Jaschke, A., Detection of small organic analytes by fluorescing molecular switches, *Bioorganic and Medicinal Chemistry*, 9, 2521–2524, 2001.
89. Geiger, A., Burgstaller, P., Von der Eltz, H., Roeder, A., and Famulok, M., RNA aptamers that bind L-ariginine with sub-micromolar dissociation constants and high enantioselectivity, *Nucleic Acids Research*, 24, 1029–1036, 1996.
90. Harada, K. and Frankel, A. D., Identification of two novel arginine binding DNAs, *EMBO Journal*, 14, 5798–5811, 1995.
91. Wallis, M. G., Von Ahsen, U., Schroeder, R., and Famulok, M., A novel RNA motif for neomycin recognition, *Chemical Biology*, 2, 543–552, 1995.
92. Famulok, M. and Huttenhofer, A., In vitro selection analysis of neomycin binding RNAs wit a mutagenized pool of variants of the 16S rRNA decoding region, *Biochemistry*, 35, 4265–4270, 1996.
93. Burgstaller, P. and Famulok, M., Isolation of RNA aptamers for biological cofactors by in vitro selection, *Angewandte Chemie International Edition England*, 33, 1084–1087, 1994.
94. Seetharaman, S., Zivarts, M., Sudarsan, N., Breaker, R. R., and Immobilized, R. N. A., Immobilized RNA switches for the analysis of complex chemical and biological mixtures, *Nature Biotechnology*, 19, 336–341, 2001.
95. Bachler, M., Schroeder, R., and von Ahsen, U., StreptoTag: a novel method for the isolation of RNA-binding proteins, *RNA*, 5, 1509–1516, 1999.
96. Berens, C., Thain, A., and Schroeder, R., A tretracycline-binding RNA aptamer, *Bioorganic and Medicinal Chemistry*, 9, 2549–2556, 2001.
97. Wilson, C., Nix, J., and Szostak, J. W., Functional requirements for specific ligand recognition by a biotin-binding RNA pseudoknot, *Biochemistry*, 37, 14410–14419, 1998.
98. Ellington, A. D. and Szostak, J. W., Selection in vitro of single-stranded DNA molecules that fold into specific ligan-binding structures, *Nature*, 355, 850–854, 1992.
99. Schurer, H., Stembera, K., Knoll, D., Mayer, G., Blind, M., Forster, H. H., Famulok, M. et al., Aptamers that bind to the antibiotic moenomycin A, *Bioorganic and Medicinal Chemistry*, 9, 2557–2563, 2001.
100. Gebhardt, K., Shokraei, A., Babaie, E., and Lindqvist, B. H., RNA aptamers to S-adenosylhomo-cysteine: kinetic properties, divalent cation dependency and comparison with anti-S-adenosylhomocysteine antibody, *Biochemistry*, 39, 7255–7265, 2000.
101. Mannironi, C., Di Nardo, A., Fruscoloni, P., and Tocchini-Valentini, G. P., In vitro selection of dopamine RNA ligands, *Biochemistry*, 36, 9726–9734, 1997.

7 Surface Imprinting: Integration of Recognition and Transduction

Yanxiu Zhou, Bin Yu, and Kalle Levon

CONTENTS

7.1   INTRODUCTION

In recent years, molecular imprinting technology [1–16] has achieved great success in the recognition and detection of chemically and biologically important substances from inorganic [17–19] and organic compounds [20–24], toxins [25–27], proteins [28–33], and viruses [28,34–35] to microorganisms [36–39]. This technology offers a cheap, robust, and versatile platform, thus proven to be the best alternative recognition elements for biological receptors in sensor construction. Especially when there is not a biological recognition component available for the target molecule, the imprinted polymer is the only solution. Molecularly imprinted polymers (MIPs) are prepared by polymerization of functional monomers in the presence of a molecular template. The elution of the template results in cavities that are complementary in size, shape, and chemical functionality with the template. These complementary cavities allow rebinding of target molecules with a high specificity, sometimes comparable to that of antibodies [40]. MIPs are currently attracting growing attention in sensor development because of the successful utilization of both the enthalpic and the entropic contributions of the binding process [15].

Although a large variety of molecular imprinted polymer-based sensors are now under active investigation [41–60], this technique suffers from certain inherent limitations that may stem from the fact that the transduction mechanism is still separated from the binding, whereas in natural processes, both recognition and transduction are integrated, e.g., in the function of an ion channel in a natural membrane. Besides, most of the imprints in a bulk polymer are not accessible even with the use of a porogen. It has been observed that only 15% of the cavities formed during imprinting could be reoccupied upon rebinding of template molecules [8]. It also takes a longer time for the target to be captured by those imprints deeply buried inside the polymer. This would result in a final sensor with a slow response. Therefore, the mimicking of nature by combining the recognition and transduction by imprinting the templates directly on the surface of a transducer (surface imprinting) might address aforementioned obstacles and fulfill the principle of a sensor. Surface imprinting technique could produce the imprints at or close to the surface of the imprinted materials. When a surface imprinting was carried out on the surface of a transducer, such an integrated sensor construction would more closely resemble a natural sensing process.

In this chapter, the discussion will only focus on this group's recent achievements in surface imprinting and sensors based on surface imprinted polymers (SIP). The development on of MIP technology can be found in recent reviews [41,50,52–87].

7.2 HISTORY OF SIPS

Sagiv [88] first reported surface imprinting by self-assembled monolayers of alkylsiloxanes to form molecular recognition sites. Mixed monolayers formed onto polar surfaces from anhydrous organic solutions. One of the components, e.g., a fatty acid, is physically adsorbed, whereas the other component (the silane) is covalently bonded to solid polar substrates [88–90]. The covalent poly-siloxane monolayer-surface binding with intralayer cross-linking was shown to result in an unusual mechanical, chemical, and electrical stability [91]. Furthermore, the silane monolayers were shown to be perfectly stable under conditions that may cause a major deterioration of the structure of fatty acid films [92].

The fact that the mixed monolayers contained both physisorbed and chemisorbed components made the removal of the physisorbed components possible. It was proposed that the resulted skeleton monolayers have holes of molecular dimensions that may be used for free adsorption sites. For exploring this opportunity, the mixed monolayers were formed in the presence of surfac-tant dyes that were only physically adsorbed on the substrate surface in contrast to the monolayers that were covalently linked. The dye molecules were washed away, leaving holes in the polymer-ized silane network [90]. The initial size and distribution of holes in the skeletonized monolayer were preserved because of the covalent bonding of the octadecyltrichlorosilane (OTS) molecules to the surface of the substrate [93]. This experiment can be considered as the first successful surface molecular imprinting experiment.

Tabushi et al. have applied this concept to imprint alkanes in the aforementioned mixed monolayers [94,95]. In their studies, they used n-hexadecane as the template and implanted the templates with the OTS molecules on a SnO_2 electrode surface. The formed monolayer was able to sense molecules with alkyl tails, indicating the presence of the space induced by the templates. This application confirmed the recognition components, but it did not take advantage of the molecular imprinting technology. The detection of the affinity binding by guest recognition was exemplified by Mosbach et al. using ellipsometry for the binding measurements [96]. Kallury [97] followed

Tabushi's work of surface imprinting in two-dimensional octadecylsilyl (ODS) films against hexadecane to study the interaction between resulting cavities and analytes, e.g., steric acid octadecylamine by reflectance Fourier transform infrared spectroscopy, ellipsometry, and x-ray photoelectron spectroscopy. The results provided the evidence of the active surface-containing cavities left by hexadecane bound with the hydrophobic tails incorporated in the film and the polar head-groups-up oriented away from the surface. Similar studies have been applied to prepare recognition sites for porphyrin [98], amino acids, nucleotides, cholesterol [99], and 1,2-epoxydodecane [100].

Formation of organic monolayers such as adsorption of alkanethiols on gold has been used to develop a sensor containing two-dimensional molecular recognition sites composed of multiple chemical species. A photochemically imprinted sensor for 6-[(4-carboxymethyl)phenoxy]-5-12-naphthacene quinine, trans-quinine, has been reported by self-assembled monolayer (SAM) [101]. This method was based on the self-assembly of trans-quinone photoisomerizable monolayer with 1-tetradecanethiol onto a gold surface. Photoisomerization of the trans-quinone monolayer produced the ana-quinone. Thereafter, the ana-quinone could be photochemically removed, and it could generate molecular recognition sites in a monolayer array for the trans-quinone. Mirsky [102] has reported a comparatively simple, yet effective, way of using a spreader bar approach to generate molecular recognition sites for barbituric acid by co-adsorption of thiobarbituric acid and dodecanethiol onto a gold substrate. Without removal of the incorporated thiobarbituric acid, the artificial chemosensor generated a high selectivity for barbituric acid, a molecule of similar shape to the spreader bar. The template cannot be moved without destroying the SAM layer. Recently, a SAM MIP sensor for L-serine [103] has been constructed by self-assembling of L-cysteine and L-serine onto the surface of gold microelectrodes. After removal of the template by 10 mM HCl, the imprinted SAM films presented moderate enantioselectivity and sensitivity towards L- and D-serine.

Conducting polymers could be very promising materials for molecular imprinting as conducting polymeric films can be grown adherently onto the surface of electrodes with any shape and thickness. Moreover, the resulting polymer could be a three-dimensional network of intrinsically conducting macromolecular wires to communicate between polymer and transducer when there is a reaction happening within the polymer backbone [104]. The conducting polymer possesses a memory effect for the anions of the electrolyte used in the preparation process [105,106]. An electrochemically mediated imprinting method to imprint nitrate into polypyrrole was described by Hutchins et al. [107]. This approach was based on the polymerization of pyrrole in the presence of $NaNO_3$ with pores complementary to the size of the target analyte ions. The resulted sensor exhibited a high selectivity for nitrate over other anions. However, the resulting nitrate template was not washed out from the polymer, and whether the polymerized film retained its host cavity was not investigated. Polypyrrole has been employed to imprint charged and neutral species such as caffeine and amino acids [108–111]. These templates were removed from polypyrrole by dedoping via overoxidization [108,109] or phosphate buffer (pH 7.0) [112]. The over-oxidized film lost its conductivity, but it displayed good ionic conductivity and high recognition ability. Poly(o-phenylenediamine) (PPD) has been employed to imprint neutral templates of glucose and sorbitol [113,114]. Yu [22] described a novel way to preserve the imprinted cavity shape by self-assembling of 2-mercaptobenzimidazole (2-MBI) onto the surface of gold substrate before electropolymerizing and employing electropolymerized molecularly imprinted polymer as a conducting macromolecular wire to produce a sensor device. The observation that the sensor could only retain 50% of the original response after 10 days storage suggested that even a covalently bound conducting polymer at the surface of a substrate could not provide long-term stability to the sensor. More recently, Liao [115] used a chiral histidine as the template to introduce enantiorecognition sites into the electropolymerized polyacrylamide membranes on Au or Au-coated quartz crystal electrodes. The electrodes can selectively rebind their corresponding template molecules but produced different frequency output when exposed to same amount of histidine entiomers.

The observations of Sagiv and coworkers could only create specific geometrical features to molecules with ODS film that could be physically adsorbed on polar solid surface, e.g., surfactant. This feature limits further application of this method. The molecularly imprinted SAM and electrosynthesized MIPs seem to fulfill the criterion of surface imprinting, but none of them could be a generally applicable approach to target molecules with various properties. Therefore, it is extremely important to establish a method that could be applicable to various molecular structures.

7.3 SURFACE IMPRINTED ODS SENSORS

7.3.1 General Procedure for Sensor Construction

As shown in Scheme 7.1, an indium tin oxide (ITO)-coated glass electrode is used as a transducer for electroanalytical detection. It is pretreated according to a literature method [88] to introduce hydroxyl groups on the surface. Templates and a long chain silane molecule, OTS, are co-adsorbed on the polar surface of ITO glass plates from the chloroform/carbon tetrachloride solution at room temperature ($20°C \pm 1°C$) for a period of time as shown in Scheme 7.1. OTS formed a thin film by the formation of the covalent silicon-oxygen bond in the presence of template molecules. Recognition cavities were created on the surface of ITO-coated glass plates upon removal of the embedded template molecules with chloroform. The host structure can be designed at the molecular level by carefully selecting the reaction conditions and parameters. Potentiometry was employed to investigate the binding between cavities and target analytes. A potentiometer is simple, inexpensive, portable, and easy to operate, making it promising in field detection.

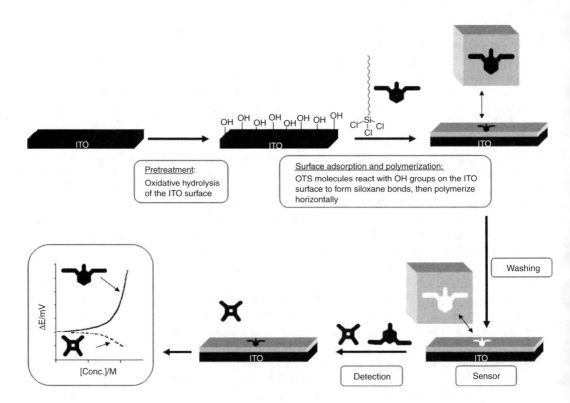

SCHEME 7.1 Schematic of the formation of molecular recognition sites on the surface of ITO substrate by surface imprinting using an ODS matrix and their use in sensing with potentiometry.

7.3.2 CHEMICAL SENSOR BASED ON SIPS

7.3.2.1 Chiral Electrochemical Sensor for the Enantiorecognition of Amino Acids

It is a great challenge to design electrochemical sensors to discriminate between enantiomers, especially for amino acids, the essential substances in protein metabolism, as well as in pharmaceutical and food products. Current detection sensor systems without a separation process require enzymes. The high specificity of enzymes guaranteed a high reliability of these sensors. However, because of the presence of the biological components, such biosensors suffered with long-term stability and irreversible deactivation at high temperatures or under harsh chemical environments. Enantioselective recognition sites were generated on the ITO surface with one chiral isomer of N-carbobenzoxy-aspartic acid (N-CBZ-Asp) (Scheme 7.2) as the template for the surface imprinting technique. Such a sensor could recognize the isomer that had been used as the template from the racemic mixture [116]. The coefficients of enantiomeric selectivity of the surface imprinted sensors for the counter isomers were 0.004–0.009. The sensors translated the enantio-selective recognition event into a potential change. Similar chiral sensors were also prepared for glutamic acid and underivatized aspartic acid. An excellent detection approach was achieved despite the fact that the amino acids all had a very poor solubility in the solvent used (CCl_4/$CHCl_3$).

7.3.2.2 Detection of Pesticides

Various pesticides such as pyrethroids are of environmental concern because of their acute toxicity to many aquatic species and relatively long persistence in aquatic systems. Traditionally, analytical methods for their detection require time-consuming sample preparation and sophisticated instruments such as high-performance liquid chromatography (HPLC), capillary electrophoresis, and gas chromatography [117–119]. The pyrethroid insecticides permethrin, cypermethrin, and deltamethrin are structurally similar compounds (Scheme 7.2). They are neutral, uncharged molecules,

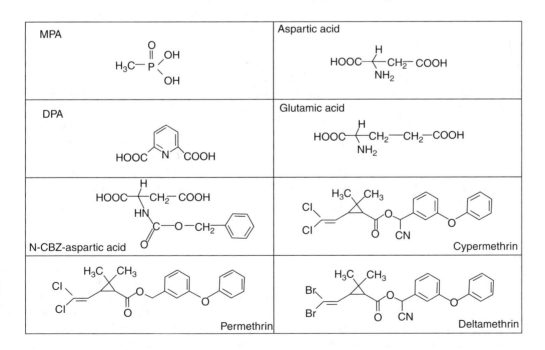

SCHEME 7.2 Chemical structures of MPA, DPA, aspartic acid, glutamic acid, N-CBZ-Asp, permethrin, cypermethrin, and deltamethrin.

and they are not electroactive. Thus, pyrethroids are not supposed to be directly detected by any electrochemical technique. However, by integration of surface imprinting and transduction, potentiometric molecular sensors for these pyrethroids could be constructed (see Figure 7.3b and Figure 7.4).

7.3.2.3 Detection of Nerve Agents

Chemical warfare agents such as Sarin (isopropyl methylphosphonofluoridate), Soman (pinacolyl methylphosphonofluoridate), and VX (O-ethyl-S-2-diisopropylaminoethyl methylphospho-nothioate) are among the most toxic of all known substances [120]. These and other toxic nerve agents eventually degrade in the environment to methylphosphonic acid (MPA). Thus, small portable devices that can quickly detect this compound are desirable. Using MPA as the template, a selective, sensitive sensor for these nerve agents was developed based on a SIP approach [121].

7.3.3 BIOSENSOR BASED ON SIPs

7.3.3.1 *Bacillus* Spores Detection

A fast and cheap alarm system is extremely desirable for detection of biological warfare agents such as *Bacillus anthracis* spores [122,123]. SIP technology perfectly matches this requirement. It is well known that dipicolinic acid (2,6-pyridinedicarboxylic acid; DPA) is a major constituent of bacterial endospores, comprising 5%–14% of their dry weight after extraction [124]. Thus, the biomarker DPA was chosen as the template to produce a SIP sensor for DPA. The stable SIP layer was advantageous when the sensor was used to analyze real-world samples [125].

7.3.4 PROPERTIES OF ODS SURFACE IMPRINTING

OTS has been used to octadecylsilylate materials having surfaces with hydroxyl groups such as silica in high-performance liquid chromatography. The silanized layer provides hydrophobicity to those stationary phase materials to prevent the interaction with polar analytes [126–129]. However, ionic molecules could be inserted within it because the ionic molecules were in a low polarity medium [116]. Furthermore, the templates are not necessary to be surfactants in SIPs as required in Sagiv's observation [88]. The imprinted molecules can be any molecules, including small molecules, for example, methylphosphoric acid (MPA) [121], aspartic acid (Asp) [116], glutamic acid (Glu) [116], dipicolinic acid (DPA) [125], to N-CBZ-Asp [116], permethrin, cypermethrin, or deltamethrin as shown in Scheme 7.2. However, these templates cannot be adsorbed on the surface of the ITO film. Therefore, the only possibility is that the templates were entrapped within the ODS film when OTS was polymerized and chemically bound to the ITO surface. Thus, the interaction between OTS and the template molecule might be only hydrophobic interaction because the OTS does not have any functional group to interact with those templates before or during the imprinting process (see Mechanism below). In this new approach, a cross-linking agent is not required, nor is a solvent porogen.

7.3.5 SOLUBILITY OF TEMPLATES

Pyrethroid insecticides such as permethrin, deltamethrin, and cypermethrin could be dissolved in the OTS-CHCl$_3$/CCl$_4$ templating solution, whereas the solubility of those acidic template molecules in the above imprinting solvent was low, solubility $\sim 10^{-6}$ M. However, the good solubility of the template in the OTS-CHCl$_3$/CCl$_4$ templating solution does not increase the amount of imprints of pyrethroids on the surface of the ITO film as the optimal concentration of those acidic molecules [116,121,125] falls in the same concentration ranges as permethrin (24.6–30.8 mM) and arachidic acid [88]. Thus, the ratio of the template to OTS on the ITO surface during the coadsorption/polymerization process is independent of the solubility of the template molecules in the templating solution. In other words, the SIP-method is suitable to both soluble and insoluble analyte molecules.

7.3.6 TEMPLATE REMOVAL

Removal of the templates completely from the polymer is very important in producing of binding sites with high affinity and selectivity. Conventional MIPs were prepared by bulk polymerization followed by mechanical grinding into small particles. The template was often trapped deep within the polymer matrix and was difficult to wash away. Producing smaller and thinner MIPs ensures the imprinted sites reside at the surface of the polymer or transducer, and enables better removal of the original template and faster diffusion and complexation of the target molecule with the imprinted polymer sites. Moreover, it will secure efficient adsorption of guest molecules when the molecular imprinted sites are at the surface. The relatively weak hydrophobic interaction between templates and OTS and the thin layer of ODS make it even easier to quantitavely remove templates from the ODS film by rinsing with chloroform. x-ray Photoelectron spectroscopy (XPS) has been used to monitor the quantity of the chloroform rinsing, and the results prove that the molecular template can be completely removed [116].

7.3.7 PERFORMANCE OF SURFACE IMPRINTED ODS SENSOR

7.3.7.1 Selectivity

The most effective way to study sensor selectivity is through developing a chiral sensor since enantiomers are made up of the same atoms connected by the same sequence of bonds but are different three-dimensional structures. If chiral sensors could be fabricated, it could be easier to develop sensors for other analytes. Figure 7.1 shows the results of enantiorecognition studies of the N-CBZ-L-Asp and N-CBZ-D-Asp sensors to the corresponding amino acids. N-CBZ-L-Asp was selectively bound to the N-CBZ-L-Asp sensor (Figure 7.1a, curve ($\bigcirc$)) where the N-CBZ-L-Asp

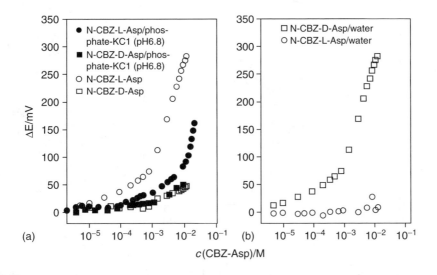

FIGURE 7.1 Enantioselective potentiometric responses of (A) N-CBZ-L-Asp sensor and (B) N-CBZ-D-Asp sensor sensors for N-CBZ-L-Asp ($\bigcirc$,$\bullet$), N-CBZ-D-Asp ($\square$, $\blacksquare$) in water or in 0.1 M phosphate-0.1 M KCl (pH 6.8). [OTS] = 8.0×10^{-1} mM. [N-CBZ-Asp]/(CHCl$_3$/CCl$_4$) = 3.7×10^{-2} M. The immersion time of the ITO plates in the OTS and N-CBZ-Asp dispersion is 3 min.

was used as the template molecule during the surface imprinting process. Almost no binding was seen for N-CBZ-D-Asp (Figure 7.1a, curve (□)). A similar phenomenon was observed for the N-CBZ-D-Asp sensor (Figure 7.1b). When the isomer, N-CBZ-D-Asp, was used as the template molecule, it gave a film the recognition ability for the D-isomer only. The hosts with the ODS film approximated the shapes of template molecules.

Only the cavities at the molecular scale or the monoclonic molecularly imprinted hole, exhibited enantioselectivity [8,9]. The minimum percentage of monoclonal cavities in relation to all the cavities was estimated to be 90% based on the ratio of the potential of the two enantiomers to the N-CBZ-D-Asp sensor (Figure 7.1b). This indicated that 10% of the surface is not ideal or is a nonspecific binding site. Otherwise, the N-CBZ-D-Asp sensor could not give any potentiometric response to N-CBZ-L-Asp.

This simple molecularly imprinted thin film offers a greater level of control over material structure compared to three-dimensional materials because of the high orientation of the film structure, and it can be used to detect molecular details in a more precise manner. With ODS films as the matrix, this group prepared recognition sites using 2,6-pyridinedicarboxylic acid (dipicolinic acid, DPA) as the template molecule. The resulting film had an enhanced affinity for DPA, and it excluded the binding of other pyridinedicarboxylic acids (Figure 7.2) where the only difference is the position of the two carboxyl groups on the pyridine ring. The DPA sensor is capable of producing different potentiometric response behaviors toward different structural features. For instance, 2,6-pyridinedicarboxylic acid (DPA) has a very close structure

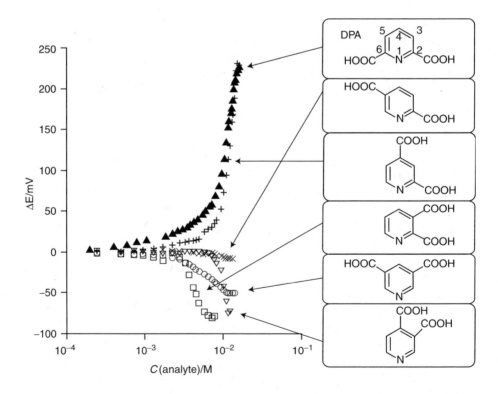

FIGURE 7.2 Potentiometric responses of DPA ODS/ITO sensor to DPA (▲), 2,5-pyridinedicarboxylic acid (×), 2,4-pyridinedicarboxylic acid (+), 2,3-pyridinedicarboxylic acid (□); 3,5-pyridinedicarboxylic acid (○) and 3,4-pyridinedicarboxylic acid (▽) in 0.15 M NaCl–PBS buffer (pH 7.2). [OTS] $= 8.0 \times 10^{-1}$ mM. [DPA]/(CHCl$_3$/CCl$_4$) $= 3.3 \times 10^{-2}$ M. The immersion time of the ITO plates in the OTS and DPA dispersion is 3 min.

to 2,4-pyridinedicarboxylic acid. The DPA sensor generated almost the same amount of potential output but with a different slope (Figure 7.2, curve (▲,†)). These results show that immobilizing recognition sites directly on the surface of the ITO transducer by SIPs enable a sensor to detect the fine structural differences of analytes.

This integrated sensor system not only recognized the difference between enantiomers or structurally similar molecules, but it also has the chemical architecture that rejected similar structures and smaller molecules. The N-CBZ-L-Asp sensor (Figure 7.3a curve (○)) has no affinity for aspartic acid (Asp) (Figure 7.3a curve (♦)) that does not have the sterically bulky CBZ- group. The lack of a significant potential difference for Asp indicates the ODS film shows a distinguishable affinity for N-CBZ-L-Asp, and the host structure is accurate. When the electrode was imprinted with neutral pesticide deltamethrin with the same approach, the potential difference was similar to the previous experiments in the case of the template molecules, and the potential alteration for cypermethrin solution did not lead to any false positive response either as shown in Figure 7.3b.

This SIP technique has an amazing capability of creating a robust receptor polysiloxane film with very high specificity and affinity for small molecules of interest. The degradation product of many nerve agents is MPA, a small molecule. The only difference between MPA and phosphoric acid (H_3PO_4) is that one of the OH groups in H_3PO_4 is replaced by a CH_3 in MPA. When using the same protocol to imprint ODS films with MPA, the selective binding for MPA was confirmed using potentiometry. Figure 7.3(c) shows a distinguishable affinity for MPA (Figure 7.3(c) curve (○)) and a lack of significant potential response for phosphoric acid (Figure 7.3(c) curve (□)), a stronger acid than MPA. Moreover, comparison of the response with (sensor) and without template molecule (control) during molecular imprinting showed that the selective binding affinity of sensors for the imprinted molecules exists [116,121,125]. Therefore, even when there are not chemical functionalities present in the imprinted host matrix as common for MIPs [1–16], the imprinted ODS films are able to discriminate target molecules from isomers, small molecules, or structurally similar compounds. In other words, chemical functionalization of a monomer to form specific and definable interactions with the print molecule is not a critical step to make highly selective recognition sites within the imprinted polymer.

7.3.7.2 Sensitivity

The sensitivity of the current generation of MIP sensors is generally 100- to 1000-fold lower than other types of biosensors because these arrangements do not reach the criterion of a biosensor—with the recognition component and transducer in close proximity [10]. In order to tackle this problem, this group investigated the potentiometric response behavior of a bare ITO and a SIP sensor on an ITO to a simple proton from hydrochloric acid. Both an imprinted N-CBZ-L-Asp sensor on ITO (with ODS film) and a bare ITO-coated electrode (without ODS film, transducer) responded to protons with discontinuous slopes at different pH ranges, and the overall slope was more than 59 mV dec^{-1} [116], demonstrating that the polymer film did not decrease the sensitivity of the sensor. In other words, the sensitivity of the MIP-based sensor comes from the transducer, not the MIP itself.

The heterogeneity of the ITO surface accounts for the discontinuous slopes in the response, and indications for heterogeneous surface composition, including differences in nano-scales have been recently found [116,130–135]. The lower slope at low analyte concentration range on the ITO could furnish a low limit of detection (LOD) for analytes. Figure 7.3b curve (×) is a calibration curve of a deltamethrin sensor to deltamethrin. The LOD was about 1.0×10^{-14} M. The conventional dynamic range of analyte concentrations are 10^{-7} to 1 M in potentiometry [136]. The high sensitive nanoscale transducer ITO and surface imprinting are required for attaining wider dynamic ranges and superior detection limits.

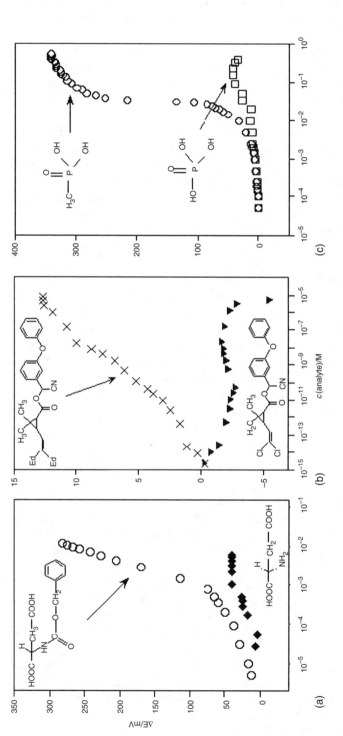

FIGURE 7.3 Potentiometric responses of (A) N-CBZ-L-Asp-cavity sensor for N-CBZ-L-Asp (○) and Asp (◆) in water, other conditions as shown in Figure 7.1; (B) deltamethrin sensor to deltamethrin (×) and cypermethrin (▼) in 0.1 M phosphate-buffered saline (PBS pH 7.6), [OTS] = 8.0×10⁻¹ mM, [deltamethrin]/[CHCl₃-CCl₄] = 3.21×10⁻² M. Adsorption time: 4 min; (C) MPA sensor to MPA (○) and phosphoric acid (□) in 0.15 M NaCl-0.1 M PBS buffer (pH 7.2). [OTS] = 8.0×10⁻¹ mM, [MPA] (CHCl₃/CCl₄) = 2.5×10⁻² M. The immersion time of the ITO plates in the OTS and MPA dispersion is 3 min.

7.3.7.3 Stability and Response Time

Significant effort has been expanded in MIP development for many years in order to circumvent some of the major drawbacks of biological recognition elements such as a lack of storage and operational stability. This group's SIP sensors listed in Table 7.1 retained 90% of their original response after several hundred measurement. After being stored for two years, the DPA sensor yielded only approximately -2.5% relative potentiometric difference for 0.0104 M DPA.

As previously mentioned, bulk imprinted polymers used in sensors as porous materials [137,138] have disadvantages in terms of slow mass transfer and kinetics. The sites of molecular imprinting located at the surface of an ITO transducer could address these obstacles. The kinetics of substrate rebinding and recognition by surface imprinted ODS films depend on the interactions involved. The weaker hydrophobic interactions between ODS film and target molecule, the electrostatic interactions between molecule and oxide on the surface of ITO film, no chemical functionality involved in the formed cavity for recognition, two-dimensional molecular recognition ODS film and the most important, one dimension of the cavity is the transducer ITO, all of those factors enable fast rebinding kinetics. The response time is within minutes for SIPs-based sensors (Table 7.1) indicated by fast response of the templated ODS films toward target molecules.

7.3.7.4 Methodology and the Functionality of Transducer

Employing the interaction between artificial cavity host and guest molecule as a basis for detection is clearly a demand for new analytical technologies. The challenge is to identify methods for assaying the molecular recognition as current analytical methods may not satisfy the need of tracing complex formation during the detection process. Figure 7.1(A) curves (•,■) are the calibration curves for the enantiomers of N-CBZ-Asp dissolved in 100 mM phosphate buffer-0.1 M KCl (pH 6.8). The results illustrated that the N-CBZ-L-Asp sensor was not able to differentiate them very well. While using pure water instead of supporting electrolyte and buffer solution as the medium, the sensor produces a significant potential difference between these enantiomers (Figure 7.1a curves (○,□)). Therefore, the conventional principle of using buffer and supporting electrolytes for controlling pH of the solution and the migration of the ionic species was not suitable for self-electrolyte analytes, e.g., amino acids. Thus, a new method, supporting electrolyte and buffer-free potentiometry was proposed for electrochemical enantiodiscrimination of chiral amino acid SIP sensors.

As described above, the transducer plays a paramount role in the sensitivity of a sensor. Searching for the right transducer for analyte molecules is equally important as investigating the other components of the sensor such as the recognition element [139]. Actually, nanoscale transducer ITOs not only provide high sensitivity and low LOD compared to SIPs-based sensors, but also have a high resolution for analytes that regular transducers do not [140]. When the surface was

TABLE 7.1
Surface Imprinted ODS Sensors' Properties

Sensor	Response Time (s)	Reproducibility RSD (%)	Lifetime Measurement (Times)	Signal Remained (%)
N-CBZ-Asp	160 s/0.20 mM	2.60	200	92
MPA	50 s/15 mM	2.36	210	92
DPA	40 s/15.7 mM	2.79	550	90
Permethrin	40 s/1 ppb	2.42	170	105.5

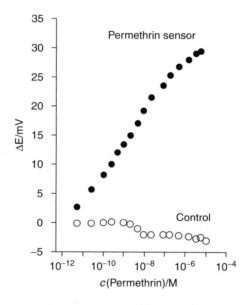

FIGURE 7.4 Potentiometric response of permethrin on the permethrin ODS/ITO electrodes ($\bullet$) with and ($\bigcirc$) without the template. [OTS] $= 8.0 \times 10^{-1}$ mM, [permethrin]/(CHCl$_3$/CCl$_4$) $= 2.82 \times 10^{-2}$ M. The immersion time of the ITO plates in the OTS and permethrin dispersion is 3 min.

imprinted with the insecticide, permethrin with an ODS film, it produced a significant potential alteration with permethrin (Figure 7.4 curve ($\bullet$)). Without the neutral template, the control only gave a - 3.1 mV potential difference (Figure 7.4 curve ($\bigcirc$)). The same phenomenon was observed in Figure 7.3b for the deltamethrin sensor. With an ITO transducer, neutral molecules could be detected by potentiometry, demonstrating that the nanoscale transducer plays an important role in sensor performance.

7.3.7.5 Mechanism

There are two distinct molecular imprinting approaches, covalent [141–143] and non-covalent imprinting [7,144–146]. Generally, the methodology of molecular imprinting is based on the principle of using functional monomers to form a complex with the template through covalent or non-covalent interactions. Subsequent removal of the template creates the polymer's binding sites with the precise spatial arrangement of functional groups to ensure highly selective recognition of the target molecule. Shape, size, and functional complementarity are the two basic principles for molecular imprinting that have been proposed [10,147]. However, for this SIP-based sensor system, no complex is formed between the OTS film and templates as the OTS film interacts only with the hydroxyl groups on the surface of ITO. Figure 7.5c is the infrared spectrum of the reaction product of ODS and N-CBZ-L-Asp in CHCl$_3$/CCl$_4$ at 20°C (using the same conditions the surface imprinted sensors were fabricated) [116]. The two sharp peaks at 2850 and 2920 cm^{-1} are the CH$_2$ symmetric and asymmetric stretching vibrations [148]. The 1702 cm^{-1} band attributed to the carbonyl groups of N-CBZ-L-Asp (Figure 7.5, curve (b)) did not change after reacting with ODS (Figure 7.5, curve (a)), demonstrating no adduct formed between them (Figure 7.5, curve (c)). Only at an elevated temperature could the reaction between the N-CBZ-L-Asp template and OTS be observed (Figure 7.5, curve (d)). The peak of the carbonyl group shifts from 1702–1735 cm^{-1}, indicating that the carboxylic group of the N-CBZ-L-Asp covalently bound to Si–Cl in the poly-siloxane film. The experiments with the construction of sensors were carried out at room temperature [116] or even lower [121,123]; therefore, there is minimal chance to form a product

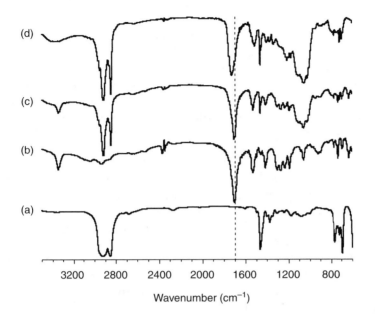

FIGURE 7.5 FT-IR spectra of (a) ODS, (b) L-CBZ-Asp, (c) reaction product from the mixture of ODS and CBZ-L-Asp in $CHCl_3/CCl_4$ (2:3 v/v) at 20°C (the same solution used to prepare the sensors [91]), and (d) Reaction product from the mixture of ODS and CBZ-L-Asp in $CHCl_3/CCl_4$ (2:3 v/v) at 40°C.

between the molecular template and OTS. The same phenomenon was observed in the case of arachidic acid [88].

Several research groups have demonstrated that the number of stable complexes formed between monomer and template in the pre-polymerization solution are quite low [8,149–150]. The binding sites formed in this case could be attributed to the organization of ODS and templates during the adsorption process. The only interaction between them is a hydrophobic interaction [116,121,123,151]. Actually, hydrophobic interactions are strong in water and should make the biggest contribution to subsequent binding [152–154]. Thus, hydrophobicity can boost the selectivity of the sensor without functional groups involved. Therefore, the interaction between the imprint molecule and ODS film can be divided into two parts: the hydrophobic interaction with the ODS layer and the electrostatic binding with the surface oxides. The hydrophobic interaction with the polymer formation around the cavities left by the template provides selectivity to the sensor; only those molecules with the same geometrical properties as the template molecules could enter the ODS layer. In other words, the molecular recognition is attributed to binding sites that complement molecules in size and shape functionality, not chemical functional complementarity. The electrostatic interaction with the surface oxides furnishes chemical interaction energy between the target molecule and the modified surface, and the ITO transducer converts the binding into an easy readout.

7.4 CONCLUSION AND OUTLOOK

This group's results demonstrate that in spite of the simplicity of ODS films, surface imprinting is straightforward and generally applicable to various molecular structures. The selectivity and affinity acquired from imprinting the specific recognition sites on the surface of the transducer are on par with natural binding entities such as antibodies. The inherent processing flexibility enables ODS films to be readily fabricated, whereas the mild conditions enable template molecules to be readily introduced within the ODS matrix without any danger of thermal or chemical decomposition.

The great advantage of this surface imprinting approach to form specific recognition sites on the surface is its versatility, generality, and simplicity. Perhaps, most significant is that the imprinting of templates with low solubility in the polymerization solvent becomes possible. The robust recognition elements (imprints in a ODS film) as well as a simple instrument (potentiometer) makes this integrated device a great sensor for field use to detect chemical warfare and/or biological warfare agents. This method is applicable to both planar substrates and curved surfaces, and it should also be available for variety of transducer surfaces such as an optical fiber. Moreover, because an ODS film is very thin and can be produced cleanly and reproducibly, it can be used for the development of optical and acoustic sensors. Last, this group's results highlight the importance of considering not only chemical and physical interactions for recognition, but also methodology and transducer type when designing a chemical sensor. It is anticipated that this work will lead to promising developments in the next generation of biomimetic sensors.

So far, surface imprinting has achieved more success for small molecular targets. There is certainly a challenge when the templates become larger. Generally speaking, the larger the template, the less rigid it becomes. More investigation has to be done to prove the imprints are really complementary toward the template when the size of the template becomes big such as for bacteria or viruses [28,34–39]. In addition, the size and shape for live biological species vary with the environment, and the complete removal of bacteria or viruses is not an easy task at the current developmental stage, adding an additional challenge toward the application of surface imprinting and molecular imprinting techniques.

ACKNOWLEDGMENTS

Financial support from the Defense Advanced Research Projects Agency (DARPA, grant #0660076225) and the Environmental Protection Agency (EPA, grant # QT-RT-03-001078) is gratefully acknowledged.

REFERENCES

1. Wulff, G. and Sarhan, A., Use of polymers with enzyme-analogous structures for the resolution of racemates, *Angew. Chem. Int. Ed.*, 11, 341–344, 1972.
2. Wulff, G. and Sarhan, A., Method of preparing polymers analogous to enzymes. German Patent application (Offenlegungsschrift) DE-A 2242796, 1974, *Chem. Abstr.*, 83, P 60300W, 1974.
3. Wulff, G., Sarhan, A., and Zarbrocki, K., Enzyme-analogue built polymers and their use for the resolution of racemates, *Tetrahedron. Lett.*, 14 (44), 4329–4332, 1973.
4. Takagishi, T. and Klotz, I. M., Macromolecule-small interactions; introduction of additional binding sites in polyethyleneimine by disulfide cross-linkages, *Biopolymers*, 11, 483–491, 1972.
5. Sellergen, B., Imprinted polymers with memory for small molecules, proteins, or crystals, *Angew. Chem. Int. Ed.*, 39 (6), 1031–1039, 2000.
6. Andersson, L., Sellergren, B., and Mosbach, K., Imprinting of amino acid derivatives in macroporous polymers, *Tetrahedron Lett.*, 25 (45), 5211–5214, 1984.
7. Sellergen, B., Lepisto, M., and Mosbach, K., Highly enantioselective and substrate-selective polymers obtained by molecular imprinting utilizing noncovalent interactions. NMR and chromatographic studies on the nature of recognition, *J. Am. Chem. Soc.*, 110 (17), 5853–5860, 1988.
8. Wulff, G., Molecular imprinting in cross-linked materials with the aid of molecular templates—A way towards artificial antibodies, *Angew. Chem. Int. Ed.*, 34 (17), 1812–1832, 1995.
9. Mosbach, K., Molecular imprinting, *Trends Biochem. Sci.*, 19 (1), 9–14, 1994.
10. Kriz, D., Ramström, O., and Mosbach, K., Molecular imprinting—new possibilities for sensor technology, *Anal. Chem.*, 69 (11), 345A–349A, 1997.
11. Norrlöw, O., Glad, M., and Mosbach, K., Acrylic polymer preparations containing recognition sites obtained by imprinting with substrates, *J. Chromatogr.*, 299, 29–41, 1984.
12. Yan, M., *Molecularly Imprinted Materials: Science and Technology*, CRC, Boca Raton, FL, 2004.

13. Komiyama, M., Takeuchi, T., Mukawa, T., and Asanuma, H., In *Molecular Imprinting: From Fundamentals to Applications*, Komiyama, M., Takeuchi, T., Mukawa, T., and Asanuma, H., Eds., Wiley-VCH, Weinheim, 2003.

14. *Templated Organic Synthesis*, Diederich, F. and Stang, P. J., Eds., Wiley-VCH, Weiheim, 2000.

15. Wulff, G. and Biffis, A., In *Molecularly Imprinted Polymers, Man-Made Mimics of Antibodies and their Application in Analytical chemistry*, Sellergren, B., Ed., Elsevier, New York, 2001.

16. Haupt, K., In *Molecular and Ionic Recognition with Imprinted Polymers*, Bartsch, R. A. and Maeda, M., Eds., American Chemical Society, Washington, DC, 1998.

17. Kanazawa, R., Mori, K., Tokuyama, H., and Sakohara, S., Preparation of thermosensitive microgel adsorbent for quick adsorption of heavy metal ions by a temperature change, *J. Chem. Eng. Jpn*, 37 (6), 804–807, 2004.

18. Murray, G. M., *Molecularly Imprinted Polymer Solution Anion Sensors*, US Patent 6749, 811, 2004.

19. Gladis, J. M. and Rao, T. P., Effect of porogen type on the synthesis of uranium ion imprinted polymer materials for the preconcentration/separation of traces of uranium, *Microchimica Acta.*, 146 (3–4), 251–258, 2004.

20. Matsuguchi, M. and Uno, T., Molecular imprinting strategy for solvent molecules and its application for QCM-based VOC vapor sensing, *Sens. Actuators, B.*, 113 (1), 94–99, 2006.

21. Liu, F., Liu, X., Ng, S.-C., and Chan, H. S.-O., Enantioselective molecular imprinting polymer coated QCM for the recognition of -tryptophan, *Sens. Actuators B.*, 113 (1), 234–240, 2006.

22. Gong, J. L., Gong, F. C., Kuang, Y., Zeng, G. M., Shen, G. L., and Yu, R. Q., Capacitive chemical sensor for fenvalerate assay based on electropolymerized molecularly imprinted polymer as the sensitive layer, *Anal. Bioanal. Chem.*, 379, 302–307, 2004.

23. Zhou, H. J., Zhang, Z. J., He, D. Y., Hu, Y. F., Huang, Y., and Chen, D. L., Flow chemiluminescence sensor for determination of clenbuterol based on molecularly imprinted polymer, *Anal. Chim. Acta.*, 523, 237–242, 2004.

24. Yin, F., Capacitive sensors using electropolymerized o-phenylenediamine film doped with ion-pair complex as selective elements for the determination of pentoxyverine, *Talanta*, 63 (3), 641–646, 2004.

25. Barasc, M., Ogier, J., Durant, Y., and Claverie, J., Preparation of molecularly surface imprinted polymeric nanoparticles for the direct detection of saxitoxin in water by quartz crystal microbalance, *Polym. Preprints, American Chemical Society, Division of Polymer Chemistry*, 46 (2), 1146–1147, 2005.

26. Kubo, T., Hosoya, K., Watabe, Y., Tanaka, N., Takagi, H., Sano, T., and Kaya, K., Interval immobilization technique for recognition toward a highly hydrophilic cyanobacterium toxin, *J. Chromatogr. B.*, 806 (2), 229–235, 2004.

27. Kubo, T., Hosoya, K., Watabe, Y., Tanaka, N., Sano, T., and Kaya, K., Toxicity recognition of hepatotoxin, homologues of microcystin with artificial trapping devices, *J. Environ. Sci. Health, Part A: Toxic/Hazard. Subst. Environ. Eng.*, 39, 2597–2614, 2004.

28. Tai, D.-F., Lin, C.-Y., Wu, T.-Z., and Chen, L.-K., Recognition of dengue virus protein using epitope-mediated molecularly imprinted film, *Anal. Chem.*, 77 (16), 5140–5143, 2005.

29. Shi, H., Tsai, W. B., Garrison, M. D., Ferrari, S., and Ratner, B. D., Template-imprinted nanostructured surfaces for protein recognition, *Nature*, 398, 593–597, 1999.

30. Lin, T. Y., Hu, C. H., and Chou, T. C., Determination of albumin concentration by MIP–QCM sensor, *Biosen. Bioelectron.*, 20, 75–81, 2004.

31. Guo, T. Y., Xia, Y. Q., Hao, G. J., Song, M. D., and Zhang, B. H., Adsorptive separation of hemoglobin by molecularly imprinted chitosan beads, *Biomaterials*, 25, 5905–5912, 2004.

32. Rachkov, A. and Minoura, N., Towards molecularly imprinted polymers selective to peptides and proteins. The epitope approach, *Biochmica et Biophysica Acta*, 1544, 255–266, 2001.

33. Hart, B. R. and Shea, K. J., Synthetic peptide receptors: molecularly imprinted polymers for the recognition of peptides using peptide-metal interactions, *J. Am. Chem. Soc.*, 123, 2072–2073, 2001.

34. Hayden, O., Bindeus, R., Haderspock, C., Mann, K.-J., Wirl, B., and Dickert, F. L., Mass-sensitive detection of cells, viruses and enzymes with artificial receptors, *Sens. Actuators B.*, 91, 316–319, 2003.

35. Nassif, N., Bouvet, O., Rager, M. N., Roux, C., Coradin, T., and Livage, J., Living bacteria in silica gels, *Nature Mater.*, 1, 42–44, 2002.
36. Osten, D. E., Jang, J., and Park, J. K., Surface plasmon resonance coupled with molecularly imprinted polymers for detecting microcystin-LR, Abstracts of Papers, 230th ACS National Meeting, Washington, DC, United States, Aug. 28-Sept. 1, 2005, ENVR-198.
37. Dickert, F. L., Hayden, O., Bindeus, R., Mann, K. J., Blaas, D., and Waigmann, E., Bioimprinted QCM sensors for virus detection-screening of plant sap, *Anal. Bioanal. Chem.*, 378, 1929–1934, 2004.
38. Hayden, O. and Dickert, F. L., Selective microorganism detection with cell surface imprinted polymers, *Adv. Mater.*, 13, 1480–1483, 2001.
39. Das, K., Penelle, J., Rotello, V. M., and Nusslein, K., Specific recognition of bacteria by surface-templated polymer films, *Langmuir*, 19, 6226–6229, 2003.
40. Vlatakis, G., Andersson, L. I., Muller, R., and Mosbach, K., Drug assay using antibody mimics made by molecular imprinting, *Nature*, 361, 645–647, 1993.
41. Mahony, J. O., Nolan, K., Smyth, M. R., and Mizaikoff, B., Molecularly imprinted polymers-potential and challenges in analytical chemistry, *Anal. Chim. Acta.*, 534, 31–39, 2005.
42. Haupt, K. and Mosbach, K., Molecularly imprinted polymers in chemical and biological sensing, *Biochem. Soc. Trans.*, 27 (2), 344–350, 1998.
43. Vidyasankar, S. and Arnold, F. H., Molecular imprinting: selective materials for separations, sensors and catalysis, *Curr. Opin. Biotechnol.*, 6, 218–224, 1995.
44. Yano, K. and Karube, I., Molecularly imprinted polymers for biosensor applications, *Trends Anal. Chem.*, 18 (3), 199–204, 1999.
45. Piletsky, S. A., Panasyuk, T., Piletskaya, E. V., Nicholls, I. A., and Ulbricht, M., Receptor and transport properties of imprinted polymer membranes-a review, *J. Membr. Sci.*, 157, 263–278, 1999.
46. Haupt, K. and Mosbach, K., Molecularly imprinted polymers and their use in biomimetic sensors, *Chem. Rev.*, 100, 2495–2504, 2000.
47. Zimmerman, S. G., Wendland, M. S., Rakow, N. A., Zharov, I., and Suslick, K. S., Synthetic hosts by monomolecular imprinting inside dendrimers, *Nature*, 418, 399–403, 2002.
48. Piletsky, S. A. and Turner, A. P. F., Electrochemical sensors based on molecularly imprinted polymers, *Electroanalysis*, 14 (5), 317–323, 2002.
49. Haupt, K., Moleculy imprinted polymers: the next generation, *Anal. Chem.*, 75, 376A–383A, 2003.
50. Zimmerman, S. C. and Lemcoff, N. G., Synthetic hosts via molecular imprinting–are universal synthetic antibodies realistically possible?, *Chem. Commun.*, 5 (14), 5–14, 2004.
51. Haupt, K., Imprinted polymers-tailor-made mimics of antibodies and receptors, *Chem. Commun.*, 2, 171–178, 2003.
52. Kindschy, L. M. and Alocilja, E. C., A review of molecularly imprinted polymers for biosensor development for food and agricultural applications, *Trans. ASAE.*, 47, 1375–1382, 2004.
53. Adhikari, B. and Majumdar, S., Polymers in sensor applications, *Prog. Poly. Sci.*, 29 (7), 699–766, 2004.
54. Monk, D. J. and Walt, D. R., Optical fiber-based biosensors, *Anal. Bioanal. Chem.*, 379, 931–945, 2004.
55. Trojanowicz, M. and Wcislo, M., Electrochemical and piezoelectric enantioselective sensors and biosensors, *Anal. Lett.*, 38, 523–547, 2005.
56. Henry, O. Y. F., Cullen, D. C., and Piletsky, S. A., Optical interrogation of molecularly imprinted polymers and development of MIP sensors: a review, *Anal. Bioanal. Chem.*, 382, 947–956, 2005.
57. Nakamura, H. and Karube, I., Current research activity in biosensors, *Anal. Bioanal. Chem.*, 377, 446–468, 2003.
58. Ye, L. and Haupt, K., Molecularly imprinted polymers as antibody and receptor mimics for assays, sensors and drug discovery, *Anal. Bioanal. Chem.*, 378, 1887–1897, 2004.
59. Pap, T., Horvath, V., and Horvai, G., Molecularly imprinted polymers for analytical chemistry, *Chem. Anal.*, 50, 129–137, 2005.
60. Sadecka, J. and Polonsky, J., Molecularly imprinted polymers in analytical chemistry, *Chem. Listy*, 99, 222–230, 2005.
61. Shinkai, S. and Takeuchi, M., Molecular design of synthetic receptors with dynamic, imprinting, and allosteric functions, *Bull. Chem. Soc. Jpn*, 78, 40–51, 2005.

62. Marty, J.-D. and Mauzac, M., Molecular imprinting: state of the art and perspectives, *Adv. Polym. Sci.*, 172, 1–35, 2005.

63. Thiesen, P. H. and Niemeyer, B., Customised adsorbent materials in the application spectra of Bio-, medicine- and environmental technology, *Chem. Ing. Tech.*, 77, 373–383, 2005.

64. Ramos, L., Ramos, J. J., and Brinkman, U. A. T., Miniaturization in sample treatment for environmental analysis, *Anal. Bioanal. Chem.*, 381, 119–140, 2005.

65. Chapuis, F., Pichon, V., and Hennion, M. C., Molecularly imprinted polymers: developments and applications of new selective solid-phase extraction materials, *LC-GC Europe*, 17 (7), 408–417, 2004.

66. Dalko, P. I. and Moisan, L., In the golden age of organocatalysis, *Angew. Chem. Int. Ed.*, 43, 5138–5175, 2004.

67. Dmitrienko, S. G., Irkha, V. V., Kuznetsova, A. Y., and Zolotov, Y. A., Use of molecular imprinted polymers for the separation and preconcentration of organic compounds, *J. Anal. Chem.*, 59, 808–817, 2004.

68. Liu, H. Y., Row, K. H., and Yan, G. L., Monolithic molecularly imprinted columns for chromatographic separation, *Chromatographia*, 61, 429–432, 2005.

69. Turiel, E. and Martin-Esteban, A., Molecular imprinting technology in capillary electrochromatography, *J. Sep. Sci.*, 28, 719–728, 2005.

70. Guihen, E. and Glennon, J. D., Recent highlights in stationary phase design for open-tubular capillary electrochromatography, *J. Chromatogr. A.*, 1044, 67–81, 2000.

71. Hilder, E. F. F. and Frechet, J. M., Development and application of polymeric monolithic stationary phases for capillary electrochromatography, *J. Chromatogr. A.*, 1044, 3–22, 2004.

72. Palmer, C. P. and McCarney, J. P., Developments in the use of soluble ionic polymers as pseudo-stationary phases for electrokinetic chromatography and stationary phases for electrochromatography, *J. Chromatogr. A.*, 1044, 159–176, 2004.

73. Quaglia, M., Sellergren, B., and De Lorenzi, E., Approaches to imprinted stationary phases for affinity capillary electrochromatography, *J. Chromtaogr. A.*, 1044, 53–66, 2004.

74. Hu, S. G., Li, L., and He, X. W., Molecularly imprinted polymers: a new kind of sorbent with high selectivity in solid phase extraction, *Prog. Chem.*, 17, 531–543, 2005.

75. Haginaka, J., Molecularly imprinted polymers for solid-phase extraction, *Anal. Bioanal. Chem.*, 379 (3), 332–334, 2004.

76. Kashyap, N., Kumar, N., and Kumar, M., Hydrogels for pharmaceutical and biomedical applications, *Crit. Rev. Therapeutic Drug Carrier Sys.*, 22, 107–149, 2005.

77. Hilt, J. Z. and Byrne, M. E., Configurational biomimesis in drug delivery: molecular imprinting of biologically significant molecules, *Adv. Drug Delivery Rev.*, 56, 1599–1620, 2004.

78. Kandimalla, V. B. and Ju, H. X., New horizons with a multi dimensional tool for applications in analytical chemistry—Aptamer, *Anal. Lett.*, 37, 2215–2233, 2004.

79. Lanza, F. and Sellergren, B., Molecularly imprinted polymers via high-throughput and combinatorial techniques, *Macromol. Rapid Commun.*, 25, 59–68, 2004.

80. Lisichkin, G. V., Novotortsev, R. Y., and Bernadyuk, S. Z., Chemically modified oxide surfaces capable of molecular recognition, *Colloid J.*, 66, 387–399, 2004.

81. Liu, Y., Lantz, A. W., and Armstrong, D. W., High efficiency liquid and super-/subcritical fluid-based enantiomeric separations: an overview, *J. Liq. Chromatogr. Relat. Technol.*, 27 (7-9), 1121–1178, 2004.

82. Mastrorilli, P. and Nobile, C. F., Supported catalysts from polymerizable transition metal complexes, *Coord. Chem. Rev.*, 248 (3–4), 377–395, 2004.

83. Walcarius, A., Mandler, D., Cox, J. A., Collinson, M., and Lev, O., Exciting new directions in the intersection of functionalized sol-gel materials with electrochemistry, *J. Mater. Chem.*, 15, 3663–3689, 2005.

84. Lu, Y. K. and Yan, X. P., Preparation and application of molecularly imprinted sol-gel materials, *Chin. J. Anal. Chem.*, 33, 254–260, 2005.

85. Diaz-Garcia, M. E. and Laino, R. B., Molecular imprinting in sol-gel materials: Recent developments and applications, *Microchim. Acta*, 149, 19–36, 2005.

86. Ruckert, B. and Kolb, U., Distribution of molecularly imprinted polymer layers on macroporous silica gel particles by STEM and EDX, *Micron*, 36 (3), 247–260, 2005.

87. Tada, M. and Iwasawa, Y., Chemical design and in situ characterization of active surfaces for selective catalysis, *Annu. Rev. Mater. Res.*, 35, 397–426, 2005.
88. Sagiv, J., Organized monolayers by adsorption I. Formation and structure of oleophobic mixed monolayers on solid surface, *J. Am. Chem. Soc.*, 102, 92–98, 1980.
89. Sagiv, J., Organized monolayers by adsorption II. Molecular orientation in mixed dye monolayers built on anisotropic polymeric surface, *J. Isr. J. Chem.*, 18, 339–345, 1979.
90. Sagiv, J., Organized monolayers by adsorption III. Irreversible adsorption and memory effects in skeletonized silane monolayers, *Isr. J. Chem.*, 18, 346–353, 1979.
91. Neizer, L. and Sagiv, J., A new approach to construction of artificial monolayer assemblies, *J. Am. Chem. Soc.*, 105, 674–676, 1983.
92. Gun, J., Iscovici, R., and Sagiv, J., On the formation and structure of self-assembling monolayers. II. A comparative study of Langmuir-Blodgett and adsorbed films using ellipsometry and IR reflection-absorption spectroscopy, *J. Colloid Interface Sci.*, 101, 201–213, 1984.
93. Polymeropoulos, E. E. and Sagiv, J., Electrical conduction through adsorbed monolayer, *J. Chem. Phys.*, 69, 1836–1848, 1978.
94. Tabushi, I., Kurihara, K., Naka, K., Yamamura, K., and Hatakeyama, H., Supramolecular sensor based on SnO_2 electrode modified with octadecylsilyl monolayer having molecular binding sites, *Tetrahedron Lett.*, 28 (37), 4299–4302, 1987.
95. Yamamura, K., Hatakeyama, H., Naka, K., Tabushi, I., and Kurihara, K., Guest selective molecular recognition by an octadecylsilyl monolayer covalently bound on a SnO_2 electrode, *Chem. Commun.*, 79–81, 1988.
96. Andersson, L. I., Mandenius, C. F., and Mosbach, K., Studies on guest selective molecular recognition on an octadecyl silylated silicon surface using ellipsometry, *Tetrahedron Lett.*, 29, 5437–5440, 1988.
97. Kallury, K. M. R., Thompson, M., Tripp, C. P., and Hair, M. L., Interaction of silicon surfaces silanized with octadecylchlorosilanes with octadecanoic acid and octadecanamine studied by ellipsometry, x-ray photoelectron spectroscopy, and reflectance Fourier transform infrared spectroscopy, *Langmuir*, 8, 947–954, 1992.
98. Kim, J. H., Cotton, T. M., and Uphaus, R. A., Electrochemical and Raman characterization of molecular recognition sites in assembled monolayers, *J. Phys. Chem.*, 92, 5575–5578, 1988.
99. Starodub, N. F., Piletsky, S. A., Lavryk, N. V., and El'skaya, A. V., Template sensors for low weight organic molecules based on SiO_2 surface, *Sens. Actuators B.*, 13–14, 708–710, 1993.
100. Binnes, R., Gedanken, A., and Margel, S., Self-assembled monolayer coatings as a new tool for the resolution of racemates, *Tetrahedron Lett.*, 35, 1285–1288, 1994.
101. Lahav, M., Katz, E., Doron, A., Patolsky, F., and Willner, I., Photochemical imprint of molecular recognition sites in monolayers assembled on Au electrodes, *J. Am. hem. Soc.*, 121, 862–863, 1999.
102. Mirsky, V. M., Hirsch, T., Piletsky, S. A., and Wolfbeis, O. S., A spreader-bar approach to molecular architecture: formation of stable artificial chemoreceptors, *Angew. Chem. Int. Ed.*, 38, 1108–1110, 1999.
103. Huan, S., Shen, G., and Yu, R., Enantioselective recognition of amino acid by differential pulse voltammetry in molecularly imprinted monolayers assembled on Au electrode, *Electroanalysis*, 16 (12), 1019–1023, 2004.
104. Zhou, Y., Yu., B, and Levon, K., The role of cysteine residues in electrochemistry of cytochrome c at a polyaniline modified electrode, *Syn. Met.*, 142, 137–141, 2004.
105. Dong, S., Sun, Z., and Lu, Z., Chloride chemical sensor based on an organic conducting polypyrrole polymer, *Analyst*, 113 (10), 1525–1528, 1988.
106. Wang, J., Chen, S., and Lin, M. S., Use of different electropolymerization conditions for controlling the size-exclusion selectivity at polyaniline, polypyrrole and polyphenol films, *J. Electroanal. Chem.*, 24 (1-2), 231–242, 1989.
107. Hutchins, R. S. and Bachas, L. G., Nitrate-selective electrode developed by electrochemically mediated imprinting/doping of polypyrrole, *Anal. Chem.*, 67, 1654–1660, 1995.
108. Deore, B., Chen, Z., and Nagaoka, T., Potential-induced enantioselective uptake of amino acid into molecularly imprinted overoxidized polypyrrole, *Anal. Chem.*, 72, 3989–3994, 2000.
109. Deore, B., Chen, Z., and Nagaoka, T., Overoxidized polypyrrole with dopant complementary cavities as a new molecularly imprinted polymer matrix, *Anal. Sci.*, 15, 827–828, 1999.

110. Spurlock, L. D., Jaramillo, A., Praserthdam, A., Lewis, J., and Brajter-Toth, A., Selectivity and sensitivity of ultrathin-purine-templated overoxidized polypyrrole film electrodes, *Anal. Chim. Acta.*, 336, 37–46, 1996.

111. Shiigi, H., Okamura, K., Kijima, D., Hironaka, A., Deore, B., Sree, U., and Nagaoka, T., Fabrication process and characterization of a novel structural isomer sensor, *Electrochem. Solid-State Lett.*, 6, H1–H3, 2003.

112. Ramanavicine, A., Finkelsteinas, A., and Ramanavicius, A., Molecularly imprinted polypyrrole for sensor design, *Mater. Sci.*, 10, 18–23, 2004.

113. Malitesta, C., Losito, I., and Zambonin, P. G., Molecularly imprinted electrosynthesized polymers: new materials for biomimetic sensors, *Anal. Chem.*, 71, 1366–3994, 1999.

114. Feng, L., Liu, Y., Tan, Y., and Hu, J., Biosensor for the determination of sorbitol based on molecularly imprinted electrosynthesized polymers, *Biosen. Bioelectron.*, 19, 1513–1519, 2004.

115. Liao, H., Zhang, Z., Nie, L., and Yao, S., Electrosynthesis of imprinted polyacrylamide membranes for the stereospecific L-histidine sensor and its characterization by AC impedance spectroscopy and piezoelectric quartz crystal technique, *J. Biochem. Biophys. Methods*, 59, 75–87, 2004.

116. Zhou, Y., Yu, B., and Levon, K., Potentiometric sensing of chiral amino acids, *Chem. Mater.*, 15 (14), 2774–2779, 2003.

117. Karcher, A. and El Rassi, Z., Capillary electrophoresis of pesticides: V. Analysis of pyrethroid insecticides via their hydrolysis products labeled with a fluorescing and UV absorbing tag for laser-induced fluorescence and UV detection, *Electrophoresis*, 18, 1173–1179, 1997.

118. Beltran, J., Peruga, A., Pitarch, E., Lopez, F. J., and Hernandez, F., Application of solid-phase microextraction for the determination of pyrethroid residues in vegetable samples by GC-MS, *Anal. Bioanal. Chem.*, 376, 502–511, 2003.

119. Bast, G. E., Taeschner, D., and Kampffmeyer, H. G., Permethrin absorption not detected in single-pass perfused rabbit ear, and absorption with oxidation of 3-phenoxybenzyl alcohol, *Arch. Toxicol.*, 71, 179–186, 1997.

120. Martin, T., and Lobert, S., Chemical warfare toxicity of nerve agents. *Critical Care Nurse*, 23, 15–20, 2003.

121. Zhou, Y., Yu, B., Shiu, E., and Levon, K., Potentiometric sensing of chemical warfare agents: Surface imprinted polymer integrated with an indium tin oxide electrode, *Anal. Chem.*, 76 (10), 2689–2693, 2004.

122. Zhou, Y., Yu, B., and Levon, K., *Biosensor and Method of Making Same*, WO Patent 2005059507 A2, 2005.

123. Zhou, Y., Yu, B., and Levon, K., *Bacterial Biosensors*, US Patent Application, *WO Patent* 2005067425 A2, 2005.

124. Murrell, W. G., In *The Bacterial Spores*, Gould, G. W. and Hurst, A., Eds., Academic Press, London, pp. 215–273, 1969.

125. Zhou, Y., Yu, B., and Levon, K., Potentiometric sensor for dipicolinic acid, *Biosen. Bioelectronics.*, 20, 1851–1855, 2005.

126. Unger, K. K., *Porous Silica*, Elsevier, Amsterdam, 1989.

127. Nawrocki, J. and Buszewski, B., Influence of silica surface chemistry and structure on the properties, structure and coverage of alkyl-bonded phases for high-performance liquid chromatography, *J. Chromatogr. A.*, 449, 1–24, 1988.

128. Dorsey, J. G. and Dill, K. A., The molecular mechanism of retention in reversed-phase liquid chromatography, *Chem. Rev.*, 89 (2), 331–346, 1989.

129. *Chemically Modified Surfaces*, Leyden, D. E., Ed., Gordon and Breach, New York, 1986 (Vol. 1).

130. Kulkarni, A. K., Schulz, K. H., Lim, T. -S., and Khan, M., Electrical, optical and structural characteristics of indium-tin-oxide thin films deposited on glass and polymer substrates, *Thin Solid Films*, 308–309, 1–7. 1997.

131. Li, G. J. and Kawi, S., Synthesis, characterization and sensing application of novel semiconductor oxides, *Talanta*, 45 (4), 759–766, 1998.

132. Xu, C., Tamaki, J., Miura, N., and Yamazoe, N., Grain size effects on gas sensitivity of porous SnO_2-based elements, *Sens. Actuators B.*, 3, 147–155, 1991.

133. Cui, Y., Wei, Q., Park, H., and Lieber, C. M., Nanowire nanosensors for highly sensitive and selective detection of biological and chemical species, *Science*, 293, 1289–1292, 2001.

134. Weimar, U. and Gőpel, W. A. C., Measurements on tin oxide sensors to improve selectivities and sensitivities, *Sens. Actuators B.*, 26 (1–3), 13–18, 1995.

135. *Electrochemistry of Nanomaterials*, Hodes, G., Ed.; Wiley-VCH Verlag GmbH, Weinheim, p. VI, 2001.

136. Monk, P. M. S., *Fundamentals of Electroanalytical Chemistry*, John Wiley and Sons, New York, 2001.

137. Kriz, D., Kempe, M., and Mosbach, K., Introduction of molecularly imprinted polymers as recognition elements in conductometric chemical sensors, *Sens. Actuators B.*, 33 (1–3), 178–181, 1996.

138. Kriz, D., Ramstrom, O., Svensson, A., and Mosbach, K., Introducing biomimetic sensors based on molecularly imprinted polymers as recognition elements, *Anal. Chem.*, 67 (13), 2142–2144, 1995.

139. Janata, J. and Bezegh, A., Chemical sensors, *Anal. Chem.*, 60 (12), 62R–74R, 1988.

140. Zhou, Y., YU, B., Levon, K., and Nagaoka, T., Enantioselective recognition of aspartic acid by chiral ligand exchange potentiometry, *Electroanalysis*, 16 (11), 955–960, 2004.

141. Wulff, G., Molecular recognition in polymers prepared by imprinting with templates, In *Polymeric Reagents and Catalysts*, Ford, W. T., Ed., Vol. 308, American Chemical Society, ACS Symposium Series, Washington, DC, pp. 186–230, 1986.

142. Wulff, G., The role of binding-site interactions in the molecular imprinting of polymers, *Trends Biotechnol.*, 11 (3), 85–87, 1993.

143. Shea, K. J. and Sasaki, D. Y. J., On the control of microenvironment shape of functionalized network polymers prepared by template polymerization, *J. Am. Chem. Soc.*, 111 (9), 3442–3444, 1989.

144. Ramstrom, O., Andersson, L. I., and Mosbach, K., Recognition sites incorporating both pyridinyl and carboxy functionalities prepared by molecular imprinting, *J. Org. Chem.*, 58, 7562–7564, 1993.

145. Mosbach, K. and Ramström, O., The emerging technique of molecular imprinting and its future impact on biotechnology, *BioTechnol*, 14, 163–170, 1996.

146. Sellergren, B., Enantiomer separations using designed imprinted chiral phases, In *A Practical Approach to Chiral Separations by Liquid Chromatography*, Subramanian, G., Ed., John Wiley & Sons: Weinheim, Weinheim, pp. 151–184, 1994.

147. Mosbach, K., Toward the next generation of molecular imprinting with emphasis on the formation, by direct molding, of compounds with biological activity (biomimetics), *Anal. Chim. Acta*, 435 (1), 3–8, 2001.

148. Rangnekar, V. M. and Oldham, P. B., Fourier transform infrared study of a C_{18} modified quartz surface, *Spectrosc. Lett.*, 22 (8), 993–1005, 1989.

149. Andersson, H. S. and Nicholls, I. A., Spectroscopic evaluation of molecular imprinting polymerisation systems, *Bioorg. Chem.*, 25 (3), 203–211, 1997.

150. Nicholls, I. A., Adbo, K., Andersson, H. S., Andersson, P. O., Ankarloo, J., Hedin-Dahlström, J., Jokela, P. et al., Can we rationally design molecularly imprinted polymers?, *Anal. Chim. Acta.*, 435 (1), 9–18, 2001.

151. Katz, A. and Davis, M. E., Investigations into the mechanisms of molecular recognition with imprinted polymers, *Macromolecules*, 32 (12), 4113–4121, 1999.

152. Haupt, K., Dzgoev, A., and Mosbach, K., Assay system for the herbicide 2,4-dichlorophenoxyacetic acid using a molecularly imprinted polymer as an artificial recognition element, *Anal. Chem.*, 70 (3), 628–631, 1998.

153. Piletsky, S. A., Andersson, H. S., and Nicholls, I. A., Combined hydrophobic and electrostatic interaction-based recognition in molecularly imprinted polymers, *Macromolecules*, 32 (3), 633–636, 1999.

154. Striegler, S., Carbohydrate recognition in cross-linked sugar-templated poly(acrylates), *Macromolecules*, 36 (4), 1310–1317, 2003.

8 Phages as Biospecific Probes

Valery A. Petrenko and Jennifer R. Brigati

CONTENTS

8.1 INTRODUCTION

Many pathogens are currently detected only upon clinical manifestations of disease, followed by culture tests and biochemical reactions.[1–5] A challenge of biological toxins may be revealed by a bioassay, in which toxin-specific antibodies protect mice against the toxin present in the sample, as reviewed in Arnon et al.[6] These microbiological, biochemical, and animal tests dominate in all clinical and bio-monitoring laboratories as gold standards because they are sensitive, specific, and accurate.[7] However, new criteria for biological monitoring—fast, sensitive, accurate, and inexpensive—depreciate the value of the traditional procedures and call for new concepts and methods. Phage display is one of the novel concepts permitting development of diagnostic and detection probes that may meet the demanding criteria for biological monitoring.[8–10] This chapter is based on the course *Combinatorial Biochemistry and Phage Display* that has been taught by the authors during the 2003–2005 school years for graduate students at Auburn University. It contains an updated analysis of the literature,[8,10,25] protocols, and methods that are routinely used in this laboratory for development of diagnostic and detection probes.

8.2 DIAGNOSTIC AND DETECTION PROBES

The choice of the detection probes for various analytical platforms is dictated by their specificity, selectivity, performance, operational robustness, storage and environmental stability. In most platforms, the probes are polyclonal or monoclonal antibodies.[11,12] However, polyclonal antibodies (pAbs) obtained from immunized animals recognize all antigens to which the animal has been exposed in the past. Furthermore, the pAbs often lack appropriate affinity and specificity against the desired native target because of the limitations of immunization procedures where bacteria, viruses, or toxins must be inactivated before immunizing an animal and where spore germination processes occur in vivo. These limitations make it difficult to standardize and certify pAbs for routine applications. Ideally, pAbs used for detection of a biological agent should be affinity purified on columns with the immobilized agent or its isolated antigens, but this dramatically increases costs.[13] In reality, pAb are fractionated to obtain a pure fraction of IgG, but this procedure does not assure the selectivity and specificity of the preparation. Monoclonal antibodies (mAb) are more selective, but their application may be hindered by high production costs and inherent sensitivity to unfavorable environmental conditions.[14,15] Furthermore, because of the limited number of clones generated by hybridoma techniques, it may be difficult to obtain a mAb with the desired specificity, for example, binding of a characteristic epitope marker of a biological agent.[16] Both pAb and mAb can be directed against epitopes that are shared by different microorganisms that may cause cross-reactions and ambiguous detection results.[17–21] The shortcomings of antibodies have created a need for a new type of probe—substitute antibodies—antigen binding molecules with engineered immunoglobulin or non-immunoglobulin scaffolds that might be selected and evolutionary improved in vitro, bypassing the biased immune system[22–24] as reviewed in Smith and Petrenko.[8]

   It was recently shown that substitute antibodies can be recruited from huge multibillion clone libraries of peptides or mutagenized functional domains of various proteins (including single-chain Fv and Fab fragments of antibodies) displayed on the surface of bacteriophages. This powerful molecular evolution technique was named phage display (reviewed in Smith and Petrenko; Petrenko and Smith; Barbas III et al., and Hoogenboom et al.)[8,25–27] In contrast to the hybridoma technique that relies on the screening of a limited number of clones, display technologies operate with billions of potential binders and allow genetic manipulations and molecular evolution to control specificity and selectivity of the probes.[28,29] The phage technique can also be used to enhance probe affinity[30–34] (reviewed in Marvin and Lowman)[35] and stability.[36,37]

8.3 PHAGE DISPLAY TECHNIQUE

Here, a brief outline of phage display technology that has been sufficient for this group's purposes will be given. The reader can find more detailed information on this subject in recent reviews.[8,25] Phage display technology is most fully developed in the Ff class of filamentous phage that includes three wild-type strains: f1;[38] M13;[39] and fd.[40] The Ff phages are flexible, thread-like particles approximately 1 μm long and 6 nm in diameter (reviewed in Rodi et al.).[41] The bulk of their protective tubular capsid (outer shell) consists of 2700 identical subunits of the 50-residue major coat protein pVIII arranged in a helical array with a fivefold rotational axis and a coincident twofold screw axis with a pitch of 3.2 nm. The pVIII protein constitutes 97% of protein and 87% of total virion mass.[42] Each pVIII subunit is largely α-helical and rod-shaped; its axis lies at a shallow angle to the long axis of the virion.[40] About half of its 50 amino acids are exposed to the solvent, the other half being buried in the capsid (for review of phage structure, see Rodi et al.; Marvin et al.)[41,43] At the leading tip of the particle—the end that emerges first from the cell during phage assembly—the outer tube is capped with five copies each of minor coat proteins pVII and pIX (encoded by genes *VII* and *IX*); five copies each of minor coat proteins pIII and pVI (encoded by genes *III* and *VI*) cap the trailing end. The capsid encloses a single-stranded DNA (ssDNA)—the viral or plus strand whose length is 6407–6408 nucleotides in wild-type strains, but is not constrained by the geometry of the helical capsid. Longer or shorter plus strands—including recombinant genomes with foreign DNA inserts—can be accommodated in a capsid whose length matches the length of the enclosed DNA by including proportionally fewer or more pVIII subunits.

In 1985, recombinant DNA technology was applied to phage coat protein pIII to fashion a new type of molecular chimera: the fusion phage that underlies today's phage display technology.[44] To create the phage chimeras, a foreign coding DNA is spliced in-frame into a phage coat protein gene so that the guest peptide encoded by the DNA sequence is fused to a coat protein and thereby displayed on the exposed surface of the virion. Soon after the first demonstration of the phage display, this novel technique was applied to the major coat protein pVIII of the Ff phage originating a new class of landscape phage[45,46] (reviewed in Petrenko and Smith).[25] A phage display library is an ensemble of up to 10 billion such phage clones, each harboring a different foreign DNA and, therefore, displaying a different guest peptide on the virion surface. The foreign coding sequence can be derived from a natural source, or it can be designed and synthesized chemically. For instance, phage libraries displaying billions of random peptides can be readily constructed by splicing degenerate synthetic oligonucleotides into the coat protein gene (reviewed in Fellouse and Pal).[47]

Surface exposure of guest peptides underlies affinity selection, a defining aspect of phage display technology (reviewed in Dennis).[48] A target binding molecule that is generically referred to here as the selector is immobilized on a solid support of some sort (e.g., on a magnetic bead or on the polystyrene surface of an enzyme-linked immunosorbent assay (ELISA) well) and exposed to a phage display library. Phage particles whose displayed peptides bind the selector are captured on the support, and they can remain there while all other phages are is washed away. The captured phage, generally a minuscule fraction of the initial phage population, can then be eluted from the support without destroying phage infectivity, and it can be propagated or cloned by infecting fresh bacterial host cells. A single round of affinity selection is able to enrich for selector-binding clones by many orders of magnitude; a few rounds suffice to survey a library with billions or even trillions of initial clones for exceedingly rare guest peptides with particularly high affinities for the selector. After several rounds of affinity selection, individual phage clones are propagated, and their ability to bind the selector is confirmed.

Phage antibodies are a special type of phage display construct where the displayed peptide is an antibody molecule, or more exactly, a domain of the antibody molecule that includes the site that binds antigen (see Figure 8.1 and legend for explanation). A phage-antibody library includes billions of clones, displaying billions of antibodies with different antigen specificities (reviewed in

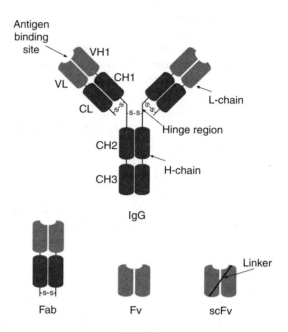

FIGURE 8.1 Display formats of antibodies. In a native immunoglobulin G (IgG) molecule, two identical heavy (H) and two identical light (L) chains are covalently linked by disulfide–S–S–bonds. In the H and L chains, a number of independently folded domains of about 110 amino acids each provide a modular structure to the IgG molecule. Antigen binding domains Fv ($\sim$ 25 kDa) can be stabilized by introducing disulfide bonds. Single-chain Fv fragments (scFv) are contiguous polypeptides consisting of the heavy- (V_H) and light- (V_L) chain variable domains linked by a 15- to 20-amino-acid flexible linker. They inherit the specificity of the parental antibody but are smaller than intact IgG (30 kDa as opposed to 150 kDa). Fab fragments ($\sim$ 50 kDa) are cleaved from native IgG by papain and consist of the light chain coupled with a fragment of the heavy chain. Phage display libraries may be assembled from the variable (V) regions expressed by B-lymphocytes from an individual with a particular immune response (immune libraries), or a nonimmunized individual (native libraries). In synthetic libraries, randomized antigen-binding regions CD3 are spliced in vitro to the cloned V genes. (Adapted from Petrenko, V. A. and Sorokulova, I. B., *J. Microbiol. Methods*, 58 (2), 147–168, 2004.)

Fellouse and Sidhu; Berry and Popkov; Dobson et al.; Hoogenboom).[49–52] A single phage-antibody library can serve as the source of cloned antibodies against an unlimited array of antigens. After selection of phage binders, the antibody gene(s) can be transferred to a high-level expression system in order to produce the antibody in usable quantities in a recombinant DNA host. Selected antibodies can be fused to reporter molecules[53–57] or subjected to molecular evolution procedures to improve their affinity, specificity,[29,58] stability,[37] and avidity.[59]

Because selection is based solely on affinity, many toxic and biological threat agents that could not be used to immunize animals without their prior inactivation can nevertheless serve as native antigens in this artificial immune system. Furthermore, phage display allows selection of antibodies recognizing unique epitopes on biological agents that may be missed in hybridoma screening.[16] Another advantage of phage display over the hybridoma technique is that the quantity of antigens required for selection of phage antibodies may be amazingly small (1–10 ng),[62] and the properties of selected probes can be further improved by affinity maturation and molecular evolution.[31,60,61] Thus, for many purposes, this system may well come to replace natural immunity in animals.[62]

General principles and numerous applications of phage display technology are summarized in recent reviews[8,25] and manuals.[26,63,64] Here, attention will be paid to the application of phage display for development of diagnostic and detection probes.

8.4 NEW BIORECOGNITION ENTITIES THAT HAVE EMERGED FROM PHAGE DISPLAY

8.4.1 PEPTIDES

Peptides are short chains of amino acids connected sequentially by peptide bonds. Usually, chains of 2–50 amino acids are classified as peptides, whereas longer chains are classified as proteins although the boundary between these two classes of polymers is blurry. Because peptides can be chemically synthesized in large scale as pure standard products or can be fused to carrier proteins and expressed at high levels, they are considered potentially valuable as diagnostic probes.[65] Commonly, primarily selected peptides demonstrate modest affinity to target receptors, but their performance may be enhanced by maturation through iterative mutagenesis and selection.[66] High affinity ligands are more likely to be selected from disulfide constrained peptide libraries rather than from linear peptide libraries[67] (reviewed in Smith and Petrenko; Petrenko and Smith).[8,25] Diagnostic peptide probe technology is still in the infant stages of development, and its prospects may be evaluated after accumulation of a sufficient pool of data. Nevertheless, the promise of this technology may be confirmed by several examples of its successful use, as listed in Table 8.1 and outlined below.

Detection of viral particles by phage-borne peptide probes can be exemplified by the work of Gough, Cockburn, and Whitelam[68] The authors selected several proline-rich peptides that specifically bind to the coat protein of cucumber mosaic virus (CuMV) from a pVIII-fused random 9-mer library. Some of the selected peptides were produced in *Escherichia coli* cells as fusions to the N-terminus of thioredoxin. Both the phage-displayed and thioredoxin-fusion versions of the peptides could detect purified virus and virus present in crude leaf extracts from infected plants. The fused peptide probes detected as little as 5 ng of CuMV in dot blots and did not bind to other plant viruses. Because peptides isolated by phage display can be fused to carrier proteins and expressed at high levels, this approach may provide a cheap, adaptable, and convenient source of diagnostic proteins.

Peptide probes were derived against spores of *Bacillus* species in non-biased selection experiments (without depletion of libraries or blocking undesirable binding sites) from 7- and 12-mer pIII-fused libraries.[69–71] Fluorescence-activated cell sorting (FACS) demonstrated that the probes bind spores (but not vegetative cells) in a highly selective manner: they bind only to the target spores and not to spores of phylogenetically similar species. For example, the peptide probe ATYPLPIRGGGC coupled with a highly fluorescent protein phycoerythrin (PE) was shown in FACS to bind well to spores of multiple strains of *Bacillus anthracis* while binding poorly or not at all to spores of phylogenetically similar species. Thus, allowing one to distinguish between spores of *B. anthracis* and the other *Bacillus* species.

Detection of bacterial cells of *Mycobacterium paratuberculosis* with phage-derived peptides in artificially contaminated milk was demonstrated recently by Stratmann and colleagues.[72] Peptides were selected from a commercial 12-mer pIII-fused M13 phage display library in a nonbiased selection with whole bacterial cells, and these peptides were immobilized to magnetic beads through biotin–streptavidin linkage. The beads were used for detection of *M. paratuberculosis* by capture PCR with primers specific for the insertion element IS*Mav*2. The method competes well with previously invented immunomagnetic PCR,[73] and solves the problem of standardization inherent for polyclonal antibodies.

A selected phage itself can be used as a probe in a detection device without chemical synthesis of the displayed peptide or its fusion to a carrier protein. For example, phage can be used in an automated fluorescence-based sensing assay[74] or FACS[71] by conjugating some of the thousands of exposed reactive amino-groups on the phage surface with Cys5, Alexa, or other fluorescent labels. In this format, phage can successfully compete with antibody-derived probes. Thus, a phage-probe binding staphylococcal enterotoxin B (SEB) that was selected from a pIII-fused random 12-mer

TABLE 8.1
Diagnostic and Therapeutic Probes against Biological Threats Agents Developed Using Display Technologies

Probe//Library	Target	Binding Test	Sensitivity
Peptides			
Toxins			
Peptide//pIII 12-mer library[76]	Botulinum neurotoxins	n.d.	n.d.
Phage//pIII 12-mer library[139]	Staphylococcal enterotoxin B (SEB),	Fluorimmunoassay on the plates	1.4 ng/well
Peptide//*Escherichia coli* FliTrx 12-mer library (Invitrogen)[176]	Ricin	Surface plasmon resonante (SPR), BiaCore 3000 (Pharmacia)	Kd 1 μM
Viruses			
Phage and peptide fused with thioredoxin//pVIII 9-mer library[68]	Cucumber mosaic virus (CuMV)	Dot blot analysis	5 ng of CaMV
Phage-fused peptide//pIII 8-mer library[142]	HBsAg	Phage-linked immune-absorbent assay[141]	~1ng HBsAg
Phage-fused peptide//pIII 7-mer peptide library[177]	HBsAg	Enzyme linked immunosorbent assay (ELISA)	Kd 2.9±0.9 nM
Peptides//pIII 10-mer peptide library[179]	White spot syndrome virus (WSSV)	ELISA	K_{aff} = 5.08–8.54 nM
Bacteria			
Peptide, phage//pIII 12-mer library[143]	*Mycobacterium avium* subsp. *paratuberculosis*	Paramagnetic beads capture PCR	10–100 cell/mL
Spores			
Peptides, phage//pIII 7- and 12-mer libraries[65,140,144]	*Bacillus (anthracis, subtilis, cereus, globigii)*	Fluorescence-activated cell sorting (FACS) with PE-coupled peptides or Alexa-labeled phage; fluorescence microscopy	10^7 spores
Antibodies			
Toxins			
scAb, expressed in *E. coli*//immune library[145–147]	Botulinum toxins	SPR, flow cytometry, ELISA, immuno-chromatographic assay	Kd = 2nM
Purified scFv//nonimmune libraries[148]	Botulinum toxins	SPR	Kd = 26–72 nM
Viruses			
Fab//human immune Fab-library[160–162]	Ebola virus	n.d.	n.d.
Fab//human immune Fab-library[153]	Epstein-Barr virus	n.d.	n.d.

(continued)

TABLE 8.1 *(Continued)*

Probe//Library	Target	Binding Test	Sensitivity
scFv//Human Synthetic VH+VL Library [164,165]	Grapevine virus B	DAS-ELISA	n.d.
Fab//human immune Fab-library[166]	Hepatitis B virus	n.d.	n.d.
scFv//human immune scFv-library[154,155]	Hepatitis C virus (HCV)	ELISA	n.d.
Fab//human immune Fab-library[157, 158]	Herpes simplex virus	n.d.	n.d.
Fab//human immune Fab-library[167]	human cytomegalovirus HCMV	n.d.	n.d.
scFv//human nonimmune scFv-library, affinity maturation[33]	Human cytomegalovirus (HCMV)	Surface plasmon resonance sensor	Ka $= 4.3 \times 10^7 \, M^{-1}$
Fab//human immune Fab-library[149,150, 145]	Human immuno-deficiency virus (HIV) type 1	Surface plasmon resonance, virus neutralization test	Kd $= 7.7 \times 10^{-10}$ M; IC50 $= 10^{-9}$–10^{-10} M
Fab//human immune Fab-library[156]	Measles virus	Radioimmunoprecipitation assays	n.d.
scFv//human mAb-derived scFv-library[163]	Rabies virus	n.d.	n.d.
Fab//human Fab-library[151,152]	Respiratory syncytial virus	n.d.	n.d.
Fab//simian immune Fab-library[159]	Simian immuno-deficiency virus (SIV)	n.d.	n.d.
Fab//human immune Fab-library[97]	Vaccinia virus	ELISA	n.d.
scFv//constructed from mAb[100]	Venezuelan equine encephalitis virus (VEE)	Enzyme assay (IFA) with a light addressable potentiometric sensor (LAPS)	30 ng/mL
scFv//mouse immune scFv library[178]	White spot syndrome virus (WSSV)	ELISA	$K_{aff} = 2.02 \pm 0.42 \times 10^9 \, M^{-1}$.
Bacteria			
scFv//mice immune scFv-library[83]	*Brucella melitensis*	ELISA	n.d.
scFv//large human nonimmune scFv-library[171]	*Chlamydia trachomatis*	ELISA, dot-blot, immunoblot	n.d.
scFv//library from hybridoma clones specific for the LPS and flagella antigen[170]	*E. coli* O157:H7	ELISA	n.d.
scFv//rabbits immunized with rec. intimin and EspA[99]	*E. coli* O157:H7	ELISA, Western blotting, and dot blots	n.d.
scFv//"Griffin 1" human synthetic library[82]	*Listeria monocytogenes*	Bioelectrochemical Sensor	500 cell/mL

(continued)

TABLE 8.1 *(Continued)*

Probe//Library	Target	Binding Test	Sensitivity
scFv//semisynthetic scFv-library[168]	*Moraxella catarrhalis*	Immunoblot, ELISA	n.d.
scFv//mutagenized murine mAb-derived scFv-library[60,169]	*Salmonella* serogroup B O-polysaccharide	EIA, SPR	$Ka = 4 \times 10^7 \ M^{-1}$
scFv//semisynthetic scFv library[91]	*Streptococcus suis*	n.d.	n.d.
Spores			
Phage-fused scFv//Naïve human scFv[172]	Genus *Bacillus*	Fluorescence microscopy, ELISA, competition ELISA	10^7 cfu/mL (in inhibition test)
Landscape Phage			
Bacteria			
Phage//f8/8[10,127,128]	*Salmonella typhimurium*	ELISA, FACS, QCM biosensor, fluorescence and electron microscopy	100 cells/mL
Spores			
Phage//f8/8[126,173]	*B. anthracis*	ELISA, coprecipitation, magnetostrictive biosensor	n.d.
Phage//f8/8[132]	*Phytophthora capsic zoospores*	Coprecipitation	n.d.

Note: n.d., not defined; all antibody display libraries in the table are pIII-fused

library and control anti-SEB antibody showed comparable sensitivity of detection of SEB (1.4 ng/ well) in a plate-based fluorimmunoassay.[74] However, in SEB coated fibers, the antibodies generated a much higher and more specific signal than phage. Fusion of a peptide to the pIII minor coat protein located on the tip of the phage capsid is probably not optimal for obtaining phage probes because this expression format does not take advantage of the avidity effect gained when the binding peptides are multivalently displayed on the major coat protein pVIII[23] (reviewed in Smith and Petrenko; Petrenko and Smith; Mammene et al.)[8,25,75] Binding peptides were selected from pVIII fusion peptide libraries against botulinum neurotoxins[76] and many other molecular targets (reviewed in Petrenko et al.),[46] but their potential in detection of the target molecules was not evaluated.

8.4.2 ANTIBODIES

The display of diverse antibody domains on the filamentous phage and the selection of phage binding to a target antigen provide a powerful means of isolating an antibody with a predetermined specificity (reviewed in Fellouse and Sidhu; Berry and Popkov; Dobson et al.; Hoogenboom).[49–52] Effective display formats for antibodies are immunoglobulin variable fragments (Fv) with a disulfide linked VH–VL pair, single-chain Fv with VH and VL domains connected through a peptide linker, and *Fabs*—light chains connected by disulfide bonds with VH1–CH1 part of a heavy chain as illustrated in Figure 8.1. There are several examples where phage antibodies selected against biological threat agents have been used beneficially in various detection platforms (reviewed in Iqbal, Mayo, Bruno, Bronk, Batt, and Chambers).[77]

Emanuel et al. constructed an immune phage antibody library using mRNA from mice immunized with botulinum toxin.[16,22] The phage antibodies selected from the library were expressed individually in *E. coli* cells and isolated in a highly purified form by metal chelate affinity chromatography at levels of 0.5–1.6 mg/L LB broth. Recombinant antibodies revealed a high affinity to the neurotoxin (0.9–7 nM) and demonstrated higher performance than monoclonal antibodies in a variety of assay formats including surface plasmon resonance, flow cytometry, ELISA, and hand-held immunochromatographic assay. Another group used similar methods to isolate anti-botulinum toxin scFvs with affinities of 26–72 nM from a very large nonimmune human phage antibody library.[78]

In the example above, mRNA for the antibody phage library was isolated from animals immunized with a toxoid of botulinum neurotoxin and therefore contained species demonstrating high affinity for the intrinsic antigen. In an alternative approach, phage antibodies were selected from a universal synthetic antibody library.[49,79] Antibodies from this universal library typically have moderate affinities in the 10^{-6}–10^{-7} M^{-1} range. However, because the library consists of 5×10^8 potentially different heavy chains paired to a single VL chain, it lends itself to combinatorial mutagenesis of the light chain. New sub-libraries are rich reservoirs for selecting antibodies with higher affinity for the target antigen. This process of in vitro evolution[10] mimics affinity maturation of antibodies in natural immune systems and allows adjustment of affinities of the primarily selected antibodies to the required level. For example, Pini with co-authors successfully used affinity maturation for development of antibodies against cytomegalovirus (HCMV) with an affinity of 4.3×10^7 M^{-1}.[33] Earlier, Barbas III, and colleagues used molecular evolution techniques to enhance the affinity of an antibody for gp120 protein of human immunodeficiency virus (HIV) 6–8 times compared with its parental antibody.[30]

As a paradigm for production of human antibodies against spores of *B. anthracis*, Zhou and co-authors investigated selection of human antibodies against native *Bacillus* spores from a naïve human antibody-phage library.[80] The authors discovered human anti-spore antibodies in the direct non-biased selection procedure by bio-panning of phage binding suspended spores of *Bacillus subtilis*. The selected antibodies were used for obtaining highly specific fluorescent probes that allowed identification of *Bacillus* spores by fluorescence microscopy with high resolution and sensitivity.

A large, nonimmune scFv library was successfully used for selection of probes against surface components on elementary bodies (EB) of *Chlamydia trachomatis*.[81] ScFv antibodies selected on the surface of purified EBs enabled the detection and identification of a variety of EB-associated antigens, some of which were chlamydial in origin whereas others were host-cell antigens. ScFv antibody specificity and selectivity toward different serotypes and species of *Chlamydia* were assayed by ELISA, immunoblot, and immunofluorescence testing.

In an attempt to use phage-displayed antibodies as probes for immunoelectrochemical detection of bacterial cells, Benhar and co-authors selected anti-*Listeria monocytogenes* scFv from the "Griffin 1" human synthetic library.[82] Both purified scFv and phage-derived antibodies were studied as probes in an enzyme linked immuno-filtration assay (ELIFA) where complexes of bacteria and labeled probe are separated from excess probe by filtration followed by electrochemical measurement of horseradish peroxidase (HRP) activity. Whereas the work demonstrates the feasibility of using phage-derived antibody probes in microbial detectors, the electrochemical platform itself is not suitable in its current format for continuous monitoring and requires essential improvements.

Phage antibodies were obtained against *Brucella abortus*, a Gram-negative facultative intracellular pathogen responsible for zoonoses worldwide that is considered a potential offensive biological warfare weapon.[83] Detection of this bacterium by immunoassay methods is hampered by the prevalence of the O:9 epitope on its surface because this epitope is also found on another pathogen, *Yersinia*.[84] As a result, many existing anti-*Brucellae* sera exhibit cross-reactivity with the *Yersinia*. Phage-borne scFv against *B. abortus* were selected by panning an immune library against radiated bacterial cells absorbed onto plastic tubes at room temperature overnight—conditions favoring selection of strong non-polysaccharide targeting binders. Indeed, the selected antibody allows distinguishing between *Brucella* and *Yersinia*, in contrast to a control monoclonal antibody demonstrating the reaction with both strains.

One important advantage that phage display technology offers over immunization and hybridoma techniques is that it allows the use of subtractive selection to obtain phage probes against differentially expressed structures on the surface of different cell types.[85–89] Boel and colleagues[87] first used subtractive selection to obtain phage antibodies against differentially expressed structures on phenotypically dissimilar strains of prokaryotic cells; complement-resistant and complement-sensitive strains of the gram-negative diplococcus *Moraxella* (*Branhamella*) *catarrhalis*. The authors used competitive panning where the target strain absorbed to MaxiSorp tubes was treated with the semisynthetic phage library preincubated with the competing strain.[90] It was shown by Western blotting analysis and inhibition ELISA that all phage antibodies were directed against the outer membrane protein of *M. catarrhalis* that earlier was found in multiple isolates of complement-resistant, but not complement-sensitive, strains. In another example, De Greeff and co-authors applied a subtractive selection procedure to isolate antibody fragments that can differentiate pathogenic and nonpathogenic strains of *Streptococcus suis*.[91] The library was subtracted by its treatment with the nonpathogenic strain and then applied to the pathogenic strain in the biopanning procedure. The selected phage antibody was found to recognize extra cellular factor (EF) that is differentially expressed by pathogenic and nonpathogenic strains. These results demonstrate a very high selectivity of the procedure.

Phage display has also been used to isolate recombinant antibodies against hepatitis C,[92,93] herpes simplex virus,[94] human cytomegalovirus,[95] rabies virus,[96] vaccinia virus,[97] Ebola virus,[98] and other viruses (Table 8.1). Although these anti-viral antibodies were developed primarily for prophylaxis and treatment of viral diseases, they may be considered also as prospective diagnostic probes.

Antibodies selected from phage display libraries can be genetically fused to reporter proteins such as fluorescent proteins,[53,55] alkaline phosphatase,[54,56,99] or streptavidin.[57] These bifunctional proteins possessing both antigen-binding capacity and marker activity can be obtained from transformed bacteria and used for one-step immunodetection of biological agents. The single chain antibodies (scFv) can also be genetically biotinylated and than immobilized to a streptavidin-linked matrix.[100]

Despite its promise, however, phage antibody technology is not without difficulties. The last step, expressing the selected antibody genes to make usable quantities of antibody, has proven particularly troublesome because it differs idiosyncratically from one antibody to another.[22,83,101–103] Besides, recombinant antibodies are sensitive to many stresses, so their engineered modification may be required to enhance their stability in non-physiological conditions[36,37,104–107] (reviewed in Worn and Pluckthun).[34] These concerns have led researchers to consider various non-immunoglobulin scaffolds for artificial antibodies such as minibodies, affibodies, etc.[108–116] (references in Legendre and Fastrez[115] reviewed in Hoess).[117]

8.5 LANDSCAPE PHAGE AS SUBSTITUTE ANTIBODIES IN IMMUNOASSAYS

Foreign peptides were displayed on pVIII soon after pIII display was pioneered[45,118,119] (reviewed in Petrenko and Smith).[25] The pVIII-fusion phages display the guest peptide on every pVIII subunit, increasing the virion's total mass by up to 20% (Figure 8.2), yet they retain the ability to infect *E. coli* and form phage progeny. Such particles were eventually given the name *landscape* phage to emphasize the dramatic change in surface architecture caused by arraying thousands

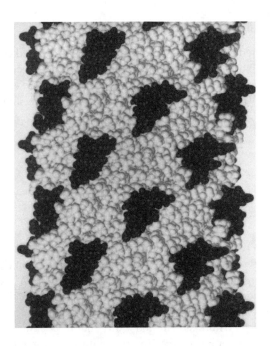

FIGURE 8.2 Structure of the landscape phage. Foreign peptides are pictured with dark atoms; their overall arrangement corresponds to the phage model.[43] (Adapted from Petrenko, V. A., Smith, G. P., Gong, X., and Quinn, T., *Protein Eng.*, 9 (9), 797–801, 1996.)

of copies of the guest peptide in a dense, repeating pattern around the tubular capsid.[46] The phage body serves as an interacting scaffold and helps to maintain a particular conformation of the guest peptides, creating a defined organic surface structure (landscape) that varies from one phage clone to the next. A landscape library is a huge population of such phages, encompassing billions of clones with different surface structures and biophysical properties.[46,115,120] Phages can be selected from landscape phage display libraries with affinities for a wide range of simple targets such as β-galactosidase, streptavidin, and neutravidin[23,46] as well as for more complex targets such as prostate cancer cells,[115,121,122] malignant glial cells,[85,123,124] or serum antibodies from patients with Lyme disease.[125] Landscape phages have been shown to substitute for antibodies in different analytical tests against various antigens and receptors,[23,85,120,121] including live bacterial cells,[110,126,127] and serve as gene-delivery vehicles,[124] detection probes in biosensors,[9,23,128] and affinity matrices for isolation of the targeted proteins.[123] In this laboratory, phages with affinities for spores[126] and vegetative bacterial cells[10,127] have been selected from landscape phage display libraries. In the following pages, the selection of these probes will be reviewed and the protocols used for their identification and characterization will be detailed.

8.6 DEVELOPMENT OF LANDSCAPE PHAGE PROBES AGAINST INFECTIOUS AGENTS

There are a variety of assays available for the detection of biological threats, but few have been adapted for continuous monitoring of the environment. Immunoassay and biosensor-based detection systems are the best prospects for continuous monitoring systems, but they require specific, selective, and robust diagnostic probes by which the pathogen can be detected. In this subsection, the development of a new class of phage-derived probes—landscape phage—that may fit well to the strong requirements of environmental monitoring is described.

8.6.1 LANDSCAPE LIBRARIES

Phage probes may be selected from the landscape library f8/8 described earlier[46] or the library f8/9 recently developed in this laboratory (Kouzmitcherva et al. unpublished). These large (10^9-clone) landscape libraries were constructed by splicing the underlined degenerate coding sequence as shown into the beginning of the pVIII coat-protein gene, replacing wild-type codons 2–4 (in f8/8 library), or 2–5 (in f8–9 library):

```
DNA      GCAGNKNNKNNKNNKNNKNNKNNKNNKCCCGCAAAAGCGGCCTTTGAC...

pVIII    A X X X X X X X X X P A K A A F D...

         1 a b c d e f g h i 6 7 8 9 10 11 12...
```

where each N represents an equal mixture of all four nucleotides (A, G, C, and T) and represents a K-mixture of G and T. As a result of this modification, every pVIII subunit in a phage is five amino acids longer than the wild-type and displays a random sequence of eight (f8/8) or nine (f8/9) amino acids: X_9 above. In any single clone, the random peptide is the same in every particle, but every clone displays a unique random peptide. The peptides are arranged regularly around the outside of the virion, occupying a substantial fraction of the surface (Figure 8.2). In another library, amino acids in positions 12–19 of pVIII are substituted with random peptides.[120]

To avoid possible contamination of the libraries with the vector phage that may hinder the selection procedure for construction of the new generation libraries, this group designed the vector f8–6[25] that

- Have four unique sites of restriction in p8-coding area ...PstI...BamHI...NheI...MluI... and one or two TAG-stop codons at the beginning of the gene *VIII*
- Grow in amber suppressor E and D where pVIII gains Gln or Ser in the N-terminus AEQQDPAKAA...; AESSDPAKAA...
- Do not grow in sup (-) strains
- Provide selection of recombinant phages with fused random foreign peptides that grow in sup (-) strain
- Show very low reversion to WT type in sup (-) strain $\ll 10^{-7}\%$

In using these vectors many different landscape phage libraries can be obtained by splicing of degenerate (randomized) DNA fragments into the gene *VIII* as outlined in Petrenko et al.; Fellouse and Pal; and Petrenko et al.[46,47,120] Amplified libraries can be obtained from the inventor's laboratory.

8.6.2 SELECTION OF TARGET-SPECIFIC PHAGE PROBES

Target-specific phage probes can be identified from phage display libraries with panning and coprecipitation protocols that will be described in detail later in this text. Regardless of the protocol used to select target-specific phage probes, certain basic methods are necessary for manipulation of the phage display library. These basic methods include preparation of starved host cells, phage amplification, and titering of phage through physical (UV absorption) and/or biological (colony-forming or plaque-forming assay) means.

Phages are viruses and cannot be grown without a host bacterial cell. Each phage has its own unique host-range, so it is important to make sure an appropriate host cell is used for the phage display library. For the fd-tet derived phages used in this lab, *E. coli* K91BluKan is used as a host cell. This bacterium (chromosome *thi* and sex Hfr-C) was derived from strain K38, first by its curing of λ prophage[174] (resulting in K91 strain), and then by introducing the mini-kan hopper element in the lacY gene using the vector λNK1105, resulting in a LacY⁻ and Kan^R phenotype; it has *lacZΔM15* mutation and probably is *lacI^Q* (super-repressor; George P. Smith, personal communication).[175] The strain has stable Hfr phenotype and does not require selective maintenance. It is excellent for propagating filamentous phage; gives large plaques with wt phage, small, but visible, plaques with infective members of fd-tet family; and requires (only) thiamine in minimal medium. Phage infection of this host cell is initiated by an interaction between the phage minor coat protein pIII and the host cell F-pilus. Growth and titering of phage is performed with specially prepared cultures of the host cell that have been concentrated, starved, and gently handled to ensure that pili are not removed by shear forces. Cells are prepared in this manner to achieve the highest infection efficiency possible. Protocol 8.7.1 at the end of this chapter describes a procedure for preparation of starved cells. This protocol and many of the basic phage manipulation protocols described at the end of this chapter are modifications of those published by Barbas.[129]

Amplification of phage, whether one is working with a sub-library or individual phage clones, is achieved by the growth of phage in the host cell. Protocol 8.7.2 at the end of this chapter describes large-scale amplification and purification of individual phage clones. The procedure can be easily scaled down for smaller scale amplification, and the polyethylene-glycol-based purification procedure can also be used for purification of the phage amplified after each round of selection during a selection procedure.

In order to efficiently carry out selection and characterization protocols, it is necessary to know the concentration of stock phage solutions. The concentration of filamentous phages can be determined spectrophotometrically (physical titer) using the general formula[129]

$$\text{virions (vir)/mL} = (A_{269} \times 6 \times 10^{16})/\text{number of nucleotides in the phage genome}$$

For the recombinant phage from the library f8/8 (9198 nucleotides), the specific formula is

$$\text{absorbance unit (AU)}_{269} = 6.5 \times 10^{12} \text{ vir/mL}$$

The concentration of infective phage particles (biological titer) of a phage solution can be determined by infection of host cells, plating, and counting of plaques or colonies. To titer phages from the f8/8 library, starved K91BlueKan cells are mixed with phage and then spread on a tetracycline-containing agar plate. The recombinant phages carry the gene necessary for tetra-cycline resistance so only those cells infected by phage form colonies on the plate.[129] This procedure is described in detail in Protocol 8.7.3. Other phages may be titered based on their ability to form visible areas of clearing (plaques) in a lawn of bacteria (see Philippa O'Brienor others for protocol).[130] The biological titer of recombinant phage is often lower then the physical titer.

A variety of protocols for the selection of target-specific phage probes can be used. In this laboratory, landscape phage clones that bind to *B. anthracis* spores were selected through a panning procedure where the phage library was incubated with immobilized *B. anthracis* spores, non-bound phages were washed away, and bound phages were eluted with mild acid.[126] To derive probes against vegetative bacterial cells of *Salmonella typhimurium*, three differing selection procedures were used, including a panning procedure similar to that used for selection of spore-specific probes, a high-stringency panning procedure, and a coprecipitation procedure where complexes of bacterial cells and phage were formed in solution and separated by centrifugation.[127] In each of these selection procedures multiple rounds of selection are performed where the phage clones that bound to the target in the initial selection procedure are amplified to create a sub-library, and the sub-library is used as input for the next round of selection. The numbers of infective phage particles present in the library (or sub-library in the second and subsequent rounds of selection) washes and eluate of each round of selection are monitored because an increase in phage recovery is an indicator that there has been an increase in the representation of phage clones in the sub-library that are capable of binding to the target.

After 2–5 rounds of selection, randomly picked clones are isolated, and a segment of genomic DNA encoding the displayed peptide is sequenced. This can be achieved by PCR-amplifying the appropriate DNA fragment using phage from colonies (or plaques) as the template (there is no need to amplify phage or purify DNA.)[127] If selection has occurred, it is expected that a limited number of unique phage clones will be isolated and that some of these clones will carry peptides with common motifs.

8.6.3 SELECTION ANALYSIS

During this group's efforts to identify selective *B. anthracis*-binding landscape phages, it monitored the evolution of the landscape phage display library f8/8. The changes that occur in a phage display library during selection were analyzed by sequencing a population of phage clones following each round of selection and analyzing the gathered data with programs found on RELIC: the Receptor Ligands Contacts bioinformatics server for combinatorial peptide analysis.[131] The programs found on RELIC are designed to search for common motifs within a set of sequences, estimate the diversity of groups of peptides (overall and at each position of the peptide), calculate the prevalence of specific amino acids in a group of peptides, and calculate the information content of individual peptides within a group.

This group studied the selection of target-specific landscape phage clones in depth, examining changes in the library occurring during each round of selection following different types of library depletion and target-bound phage elution. By sequencing a portion of the library after each round of selection, the group was able to monitor changes in diversity and prevalence of amino acids, occurrence of rare peptides in the library, and development of common motifs throughout the selection process. It also examined the influence of propagation on the appearance of dominant motifs in a selection procedure.

The diversity of the library was found to decrease with each round of selection while the number of isolated clones with high information content (low probability of being selected at random) tended to increase following each round of selection. These two factors may be useful in determining the success of a selection procedure as a whole with this type of phage display library. However, neither the information content of a displayed peptide nor the prevalence of a phage clone displaying a given peptide in the library could predict the target binding strength of a phage clone. This does not necessarily mean that there is no correlation between the affinity of selected peptides and information content, but the interaction between a probe and a complex target is dependant upon both affinity of the probe for its receptor and the abundance of the receptor on the target, thus, the only way to identify the best target binder in a group of selected clones is to perform target binding assays with each clone.

8.6.4 CHARACTERIZATION OF SELECTED PHAGE PROBES

Phage-capture assays, coprecipitation assays, FACS, fluorescent microscopy, and ELISAs can be used to evaluate the specificity and selectivity with which selected phage bind to their target. The characterization and use of phage probes in ELISAs will be discussed in a separate section because of its potential real-world applications. Here, the other assays used for characterization of phage probes will be discussed.

Although the procedures used for selection of target-specific phage probes are designed to identify phages that bind specifically to their target, the specificity of phage must be confirmed prior their use in a diagnostic assay. Here, *specificity* is defined as the ability of the recombinant phage to interact with its selector because of the presence of a specific peptide sequence displayed on the surface of the phage and *selectivity* as the ability of a recombinant phage clone to preferentially interact with the selector in comparison to other potential targets. The phage-capture assay, a procedure very similar to a panning selection procedure, is used to determine relative binding of phage clones to immobilized target. In this procedure, target spores or bacteria are adsorbed to plastic; a solution of potentially target-specific phage (many copies of one phage clone) is incubated with the adsorbed target; non-bound phage are washed away; bound phage are eluted; and the eluted phage are quantified through biological titering.[85,126] Vector phage (phage not carrying any peptide fused to their coat proteins) and unrelated phage probes (phage probes selected against other targets that carry peptides with different amino acid sequences than those being tested) are used as controls for non-specific binding. If the isolated phage probes are selective for the target, their recovery will be much greater (> 10-fold) than the recovery of the control phages. Some optimization of this procedure may be necessary to reduce background binding (varying the number of washes and the concentration of blocking agents such as Tween 20 and BSA), but if after optimization the recovery of selected phage clones is still equal to or lower than the recovery of controls, the selection procedure may have failed to isolate target-specific probes.

To determine the selectivity of phage probes for the selector bacteria or spores versus other bacteria or spores, a coprecipitation assay is often used. This test has been used in numerous published studies for analysis of phage binding to zoospores, bacterial spores, and vegetative bacteria.[69,126,127,132] In the coprecipitation assay, selected phages are mixed with spores/bacteria of various species in solution, and spore- or bacteria-bound phages are collected by a low speed centrifugation prior to enumeration. The lowest centrifugation speed that will pellet the spores/bacteria is used to reduce the possibility of phage clumps precipitating on their own. In addition to vector phage and unrelated phage controls, a phage-only (no target) control is often used in these experiments to be sure that phage clumping and precipitation is not occurring. Typically, several spores/bacteria that are closely related to the target organism as well as several more distantly related spores/bacteria are selected for testing. Refer to Protocol 8.7.4 for an example of a basic coprecipitation experiment.

The specificity and selectivity of phage probes can also be evaluated using FACS and fluorescence microscopy.[127] Phage can be fluorescently labeled with an Alexa Fluor (Molecular Probes) with a density of ~300 dye molecules per phage. For the FACS assay, fluorescently labeled phage are incubated with spores/bacteria at room temperature for 1 h and then centrifuged as in the coprecipitation test. The pellet is then washed and analyzed by FACS and fluorescence microscopy. The complex of phage and bacteria or spores can also be visualized directly by fluorescence microscopy.

Basic selection protocols do not necessarily encourage selectivity, so it is not unusual to select phage probes that bind to common bacterial or spore antigens. If phage probes selected in non-biased selection procedures have undesired cross-reactivity with non-target organisms, a subtractive (biased) procedure can be used to eliminate phages that bind to common antigens from the library. In subtractive phage display, phage clones that bind to common antigens are removed by pre-incubation of the library with non-target antigens before panning on the target or co-incubation of the library with target and non-target antigens. Subtractive procedures have been used to identify phage-borne probes that can be used to differentiate between different types of cells [85,133–135] and bacteria.[80,91] In this laboratory, unique *B. anthracis* spore-specific phage probes, featuring different peptide motifs other than those identified in the initial non-biased procedure were identified by mixing the phage library with spores of *cereus*, *thuringiensis*, and/or *B. subtilis*, precipitating spore-bound phage with centrifugation, and then using the depleted supernatant as input for a normal selection procedure.

8.6.5 USE OF PHAGE PROBES IN ELISA

Enzyme-linked immunosorbent assays are commonly used for detection of biological agents and agent-specific antibodies. Enzyme linked immunosorbent assays have been developed in many formats, broadly categorized as direct, for the detection of agents/antigens, and indirect, for the detection of agent/antigen-specific antibodies. Within these broad categories, there are multiple techniques, including sandwich assays where a target antigen is captured by one antibody and detected with a second (different or identical) antibody and direct competitive assays where antigen in the test sample competes with labeled antigen to bind capture antibodies.

In traditional direct ELISAs, a capture antibody binds the target antigen from solution, a detector antibody labeled with an enzyme binds the captured antigen (or a non-labeled antigen-specific antibody and a secondary enzyme-linked antibody specific for IgG can be used), and a chromogenic or chemiluminescent substrate is added. Phage probes can be used as substitutes for antibodies in direct ELISA. Initial demonstration of the ability of phage probes to capture their targets was done with simple molecules such as β-galactosidase, streptavidin, and neutravidin.[23] In these ELISAs, phage probes were adsorbed to the wells of a microtiter plate, and the target (β-galactosidase, streptavidin–alkaline phosphatase conjugate, or neutravidin–alkaline phosphatase conjugate) was captured from solution. Target presence was quantified by addition of the β-galactosidase substrate *o*-nitrophenyl-β-D-galactoside or the alkaline phosphatase substrate *p*-nitrophenyl phosphate.

A direct ELISA for the pathotyping of Newcastle disease virus was developed using a phage probe selected from a commercially available M13 pIII based phage display library.[136,137] In this ELISA, Newcastle disease virus from allantoic fluid was adsorbed to microtiter plates, and then the selected phage probe was added. Bound phage probe was detected with anti-M13 antibody conjugated to HRP.

Enzyme linked immunosorbent assays have been used to characterize landscape phage probes and have demonstrated the potential utility of phage probes as antibody substitutes for the specific detection of bacteria and spores.[126,127] In an ELISA designed to evaluate the specificity of phage probes selected against *B. anthracis* spores, wells of a microtiter plate were coated with phage, then incubated with biotinylated *B. anthracis* spores. Bound spores were detected with streptavidin conjugated alkaline phosphatase and *p*-nitrophenylphosphate. A similar spore-capture assay was also run with non-biotinylated *B. anthracis* spores with detection achieved by use of a *B. anthracis*-specific antibody. Phage probes were also used to detect spores attached to the wells of a 96-well plate; in this phage-capture assay, phage clones were added to wells coated with *B. anthracis* spores, and bound phages were detected with biotinylated anti-fd IgG, APSA, and *p*-nitrophenylphosphate. Similar phage-capture and bacteria-capture ELISAs were used to evaluate the specificity and selectivity of phage probes selected for binding to *S. typimurium*.[127] A basic protocol for a target-capture ELISA format is included at the end of this chapter (Protocol 8.7.6).

Phage probes that mimic antigens can be selected for use in indirect ELISA (detection of antibodies). Methods landscape phage probes with the ability to bind ConA-specific polyclonal antibodies were selected and shown to be selective through an inhibition ELISA.[46] Some of the selected phage probes carried identical epitopes as ConA, whereas others carried mimitopes that were not similar to the primary structure of ConA. Kouzmitcheva et al.[125] selected phage clones from an f88/15mer library (carrying random 15 amino acid peptides fused to ~150 out of the 4000 copies of pVIII on the surface of the phage) that bound specifically to serum antibodies from patients with Lyme Disease. An ELISA using four of the selected phage clones in different wells was able to distinguish between sera from ten Lyme Disease-positive and ten Lyme Disease-negative subjects.

Phage probes offer an alternative to antibodies for conferring selectivity to ELISAs for the detection of antibodies and antigens. The assays described above used additional target-specific or phage-specific antibodies to complete the immunoassay. It may also be possible to use phage probes in direct competitive and sandwich ELISAs, completely eliminating the need for antibodies, although this has not yet been demonstrated.

8.7   PROTOCOLS

Please note that protocols often indicate a range of concentrations, volumes, and centrifugation speeds. These steps will need to be optimized within this range for individual applications. The procedures will need further modification and optimization if one is working with a phage library that was not derived from fd-tet.

8.7.1   PREPARATION OF STARVED HOST CELLS

8.7.1.1   Materials

- Isolated colonies of *E. coli* K91BluKan on an NZY-kanamycin plate (11g Bacto-agar in 500 mL water, autoclaved, plus 500 mL $2\times$NZY broth and 100 µg/mL kanamycin)
- NZY broth (10 g NZ amine A, 5 g yeast extract, 5 g NaCl dissolved in 1 L water, pH 7.5, autoclaved)
- NZY broth containing 100 µg/ml Kanamycin
- 80 mM NaCl, autoclaved
- NAP buffer (80 mM NaCl, 50 mM $NH_4H_2PO_4$ pH 7.0 with NH_4OH), Filter sterilized

8.7.1.2   Methods

1. Grow K91BluKan cells in 2 mL NZY broth supplemented with 100 µg/mL Kanamycin in a 14 mL snap-cap tube overnight at 37°C with shaking (210 rpm).
2. Transfer 300 µL from the overnight culture to 20 mL NZY broth (no antibiotics) and grow in a 250 mL side-arm (12 mm) flask at 37°C with shaking (200 rpm) to mid log phase ($OD_{600} \sim 0.45$ as measured using the side arm cuvette). Incubate the culture with gentle shaking for an additional 5–8 min to allow any sheared F pili to regenerate. The final OD_{600} should be ~ 0.48 (in the side arm cuvette).
3. Pellet cells by centrifuging in sterile Oak Ridge tubes at 2200 rpm in a Sorvall SS34 rotor for 10 min at 4°C.
4. Pour off the supernatant and then very gently resuspend the cells in 20 mL 80 mM NaCl. DO NOT VORTEX. Transfer cells to a sterile 125-mL flask and incubate 45 min at 50 rpm, 37°C.
5. Centrifuge as in step 3.
6. Gently resuspend cells in 1 mL cold NAP buffer. The concentration of viable cells is $\sim 5 \times 10^9$/mL.
7. Store cells at 4°C. Cells are best if used within 24–48 h, but they may be used for less sensitive procedures (such as amplification of individual phage clones) for up to five days.

8.7.2   AMPLIFICATION OF INDIVIDUAL PHAGE CLONES

8.7.2.1   Materials

- NZY broth
- Tetracycline 20 mg/mL in 50% glycerol/water, v/v
- Starved K91 BluKan cells
- PEG/NaCl (100 g PEG 8000, 116.9 g NaCl, 475 mL water), autoclaved

8.7.2.2 Methods

8.7.2.2.1 Medium Scale Propagation and Purification of Phages

1. Inoculate 20 mL NZY containing 20 μg/mL tetracycline with a fresh single colony of phage-infected cells (see Protocol 8.7.3) in a 125-mL flask. Incubate the culture in a shaker-incubator at 200 rpm, 37°C for 16–24 h.
2. Pour the overnight culture into an Oak Ridge tube. Centrifuge the tube at 5,000 rpm in a Sorvall SS34 rotor for 10 min at 4°C.
3. Pour the supernatant into a fresh Oak Ridge tube. Centrifuge the tube at 8,000 rpm in a Sorvall SS34 rotor for 10 min at 4°C.
4. Pour the cleared supernatants into a fresh Oak Ridge tube. Add 3 mL PEG/NaCl and invert 100 times. Keep the tube on ice at 4°C for at least 4 h (or overnight).
5. Collect the precipitated phage by centrifuging the tube at 10,000 rpm in a Sorvall SS34 rotor for 15 min, RRR (Remove the supernatant and spin again briefly, then remove any remaining supernatant).
6. Add 1 mL TBS to dissolve the pellet.
7. Transfer the solution into a 1.5 mL microcentrifuge tube and centrifuge at maximum speed for 2 min to clear undissolved material.
8. Transfer the supernatant into a fresh labeled 1.5 mL microcentrifuge tube containing 150 μL PEG/NaCl. Invert the tube 100 times. Keep the tube on ice for at least 4 h (or at 4°C overnight).
9. Centrifuge the tube at maximum speed at 4°C for 10 min. Aspirate the supernatant, RRR.
10. Dissolve the pellet in 200 μL TBS (pH 7.5) by scraping and pumping with tip followed by vortexing.
11. Centrifuge at maximum speed for 1 min to clear undissolved material.
12. Transfer supernatant into a fresh 0.5 mL microcentrifuge tube. Store at 4°C.

8.7.2.2.2 Large Scale Propagation and Purification of Phages

1. Combine 1 μL phage with 10 μL starved K91 BlueKan cells and incubate 10 min at room temperature to allow infection to occur.
2. Add 190 μL NZY containing 0.2 μg/mL tetracycline to phage/cell mixture and incubate 40 min at 37°C to allow tetracycline resistance to develop.
3. Add this mixture to 1 L NZY + 20 μg/mL tetracycline in a 3 L flask. Incubate 24 h at 37°C with shaking at 200 rpm.
   NOTE: a well-separated colony of phage-infected bacterial cells obtained after phage titering (see Protocol 8.7.3 below) can be used to inoculate NZY/20 μg/mL tetracycline medium.
4. Pellet the cells by centrifugation in three 500 mL centrifuge bottles 10 min at 5,000 rpm in a Sorvall GS3 rotor. Transfer the supernatants (contains phage) to fresh bottles and centrifuge 10 min at 8,000 rpm in a Sorvall GS3 rotor. Transfer supernatant to fresh bottles.
5. Add 50 mL (15% volume) PEG/NaCl solution to each bottle and mix by inversion. Allow phage to precipitate overnight at 4°C or on ice for 4 h.
6. Pellet the phage by centrifuging bottles 40 min at 8,000 rpm in a Sorvall GS3 rotor. RRR (Remove the supernatant and spin again briefly, then remove any remaining supernatant).
7. Dissolve the phage pellet in 30 mL TBS (distributed among the bottles used for centrifugation).

8. Transfer the phage to a single tube and centrifuge 15 min at 10,000 rpm in a Sorvall S34 rotor to remove residual insoluble matter. Transfer to a clean tube.
9. Add 4.5 mL PEG/NaCl to the tube, invert it 100 times, then incubate on ice 4 h.
10. Centrifuge tube 15 min at 10,000 rpm in a Sorvall SS34 rotor. RRR.
11. Dissolve phage in 5 mL TBS. Centrifuge 10 min at 10,000 rpm in a Sorvall SS34 rotor to remove any precipitate and transfer to a fresh tube. Store phage at 4°C.

8.7.3 TITERING OF F8/8 PHAGE

8.7.3.1 Materials

- PBS (8 g/L NaCl, 0.2 g/L KCl, 1.44 g/L Na_2HPO_4, 0.24 g/L KH_2PO_4, pH 7.4)
- Starved *E. coli* K91BluKan cells
- NZY containing 0.2 µg/mL tetracycline
- NZY plates (11g Bacto-agar in 500 mL water, autoclaved, plus 500 mL 2×NZY broth) containing tetracycline (20 µg/mL)

8.7.3.2 Method

1. Create serial 10-fold dilutions of the phage to be titered in PBS.
2. For each dilution to be plated, mix 10 µL freshly prepared (<2 days old) starved cells and 10 µL diluted phage in a small sterile tube and incubate 15 min at room temperature to allow infection.
3. Add 180 µL of NZY containing 0.2 µg/mL tetracycline to each tube, then incubate 45 min at 37°C to allow tetracycline resistance to develop.
4. Plate the entire volume of each tube on NZY plates containing 20 µg/mL tetracycline and allow to dry before inverting the plates and incubating at 37°C overnight.
5. Count the number of colonies formed, and then multiply by the dilution factor ×100 to determine the phage titer expressed in colony-forming units per mL (cfu/mL).

8.7.4 SELECTION OF BACTERIA OR SPORE-BINDING PHAGE CLONES FROM A LANDSCAPE PHAGE DISPLAY LIBRARY

8.7.4.1 Materials

- Phage library f8/8, 4×10^{12} virions/mL ($\sim 10^9$ clones)
- Spores or vegetative bacterial cells
- Starved *E. coli* K91BluKan cells
- TBS (2.42 g/L Tris, 29.22 g/L NaCl, pH 7.5)
- TBS/1% BSA
- TBS/0.5% Tween 20
- Elution buffer (0.2 M glycine–HCl, pH 2.2, 1 mg/mL BSA, 0.1 mg/mL phenol red)
- DOC buffer (if desired) (2% w/v sodium deoxycholate, 10 mM Tris, 2 mM EDTA, pH 8.0)
- 1 M Tris–HCl, pH 8.9
- NZY broth
- Tetracycline (20 mg/mL)
- NZY plates

8.7.4.2 Method

1. Apply spores or concentrated freshly grown bacterial cells to a 35 mm Petri dish or several wells of a 96-well high-binding plate. Incubate the uncovered dish/plate at 37°C overnight or to dryness.
2. Block the dish/wells with 0.1%–1.0% BSA 1 h at room temperature.
3. Wash the dish/wells with TBS containing 0.5% Tween 20.
4. Add 10^{11} virions of phage library in 400 µL TBS containing 0.1% BSA and 0.5% Tween 20 to the dish (50 µL/well in the plate) and incubate 1 h at room temperature.
5. Wash the dish/wells 6–10 times with TBS containing 0.1%–0.5% Tween 20.
6. Elute surface-bound phage with elution buffer (400 µL for Petri dish, 100 µL/well in plate). Incubate 5–10 min at room temperature and transfer eluate to a single micro-centrifuge tube. If a deoxycholate fraction will be gathered, wash the wells or plate with a small amount of TBS and add this to the eluate. Centrifuge the eluate/wash briefly to remove any cells/spores that may have come loose during elution. Transfer supernatant to a fresh tube and neutralize with 1 M Tris (pH 9.1).
7. Concentrate eluate with a Centricon 100 kDa unit (Millipore Corp.)
8. If a deoxycholate fraction (lysis-internalized phage) is desired, add 250 µL (dish) or 50 µL (per well) DOC buffer to the dish/well and incubate 30 min at room temperature.
9. Add 100 µL starved K91BluKan cells to the concentrated, eluted phage. Add 1 mL starved K91BluKan cells to the phage in DOC buffer. Mix gently and incubate 10 min at room temperature to allow infection.
10. Transfer phage-infected cells to 20 mL NZY containing tetracycline (0.2 µg/mL) in a 125 mL flask and incubate 45 min at 37°C with shaking to allow tetracycline resistance to develop, then increase the tetracycline concentration in the flask to 20 µg/mL and continue shaking at 37°C overnight.
11. Isolate phage using a double PEG precipitation as described by Barbas et al. (see also Protocol 8.7.2 above).[129]

8.7.5 COPRECIPITATION ASSAY

8.7.5.1 Materials

- Selected phage clone(s)
- Spores or bacteria of interest
- TBS/20% Tween 20
- TBS/0.5% Tween 20
- Elution buffer
- TBS
- NZY broth containing 0.2 µg/mL tetracycline
- NZY plates containing 20 µg/mL tetracycline
- Starved *E. coli* K91BluKan cells

8.7.5.2 Methods

1. Dilute phages to a concentration of 10^6–10^9 cfu/mL in TBS. Heat diluted phage to 70°C for 10 min, then add 20% Tween 20 to a final concentration of 0.5% Tween 20. Centrifuge phage at 13,000 rpm (Microfuge 18 Centrifuge, Beckman Culture) for 15 min to precipitate any aggregated phage.

2. Combine phage in supernatant with bacterial cells from overnight cultures and/or spores in separate 1.5 mL tubes. Be sure to include phage-only controls and to also reserve a portion of the phage for titering. Total fluid volume in tubes should be at least 200 μL for optimum mixing.
3. Incubate all tubes for 1 h at room temperature on rotator.
4. Centrifuge all tubes for 10 min at 3500–5500 rpm (Microfuge 18 Centrifuge, Beckman Culture) to pellet bacteria/spores (use minimum speed that results in formation of a solid pellet).
5. Wash the pellet 5 times, very gently, with 200 μL TBS/0.5% Tween 20.
6. Resuspend the pellet in 200 μL elution buffer and incubate 10 min at room temperature with occasional vortexing. Centrifuge 10 min at 3500–5500 rpm (Microfuge 18 Centrifuge, Beckman Culture) to pellet spores/bacteria (phage should remain in supernatant). Transfer the supernatant to a fresh tube and neutralize with 38 μL 1 M Tris, pH 9.1.
7. Titer phage input and recovery as previously described (see Protocol 8.7.3 above).[129]

8.7.6 TARGET-CAPTURE ELISA

8.7.6.1 Materials

- Selected phage clone(s)
- Biotinylated spores of interest and streptavidin conjugated to alkaline phosphatase (APSA) and APSA diluent (0.05 M Tris–HCl, pH 7.5, 0.15 M NaCl, 0.1% Tween 20, 1 mg/mL BSA)

or

- Spores or bacteria of interest and target specific monoclonal or polyclonal antibodies and alkaline phosphatase conjugated secondary antibodies
- TBS/0.5% tween
- TBS/0.1% BSA
- p-Nitrophenyl phosphate tablets (5 mg substrate/tablet)
- p-Nitrophenyl phosphate diluent (5 mL 1 M diethanolamine buffer pH 9.8, 5 μL 1 M MgCl$_2$)

8.7.6.2 Method

1. Apply 3×10^{10} phage virions in 60 μL TBS to each well of a high-binding 96-well plate and allow phage to adsorb overnight at 4°C.
2. Wash the wells five times with TBS/0.5% Tween 20 using a plate washer to remove non-bound phage.
3. Apply 10^6–10^{10} spores/bacteria (biotinylated if possible) in 50 μL TBS/0.05% Tween 20 to each well. Incubate 2 h at room temperature with gently rocking.
4. Wash the wells again with TBS/0.5% Tween 20.
5. If biotinylated spores are not being used: add 45 μL target-specific monoclonal or polyclonal antibodies diluted with specific for the target diluted with TBS/0.05% Tween 20 to each well and incubate 1 h at room temperature with gentle rocking. Wash plate as previously described. Add 40 μL secondary antibody conjugated to alkaline phosphatase in TBS/0.5% Tween 20 to each well and incubate 1 h at room temperature with gentle rocking. Wash plate again.

6. If biotinylated spores are used: add 45 μL APSA (1 μg/mL in APSA diluent) to each well and incubate 1 h at room temperature with gentle rocking. Wash plate again.

7. Add 90 μL PNPP to each well and read plate on kinetic plate reader as previously described.[138]

REFERENCES

1. Hammerschmidt, S., Hacker, J., and Klenk, H. D., Threat of infection: Microbes of high pathogenic potential–strategies for detection, control and eradication, *Int. J. Med. Microbiol.*, 295 (3), 141–151, 2005.

2. Dennis, D. T., Inglesby, T. V., Henderson, D. A., Bartlett, J. G., Ascher, M. S., Eitzen, E., Fine, A. D., et al., Tularemia as a biological weapon: Medical and public health management, *JAMA*, 285 (21), 2763–2773, 2001.

3. Henderson, D. A., Inglesby, T. V., Bartlett, J. G., Ascher, M. S., Eitzen, E., Jahrling, P. B., Hauer, J., et al., Smallpox as a biological weapon: Medical and public health management, *JAMA*, 281 (22), 2127–2137, 1999.

4. Inglesby, T. V., Dennis, D. T., Henderson, D. A., Bartlett, J. G., Ascher, M. S., Eitzen, E., Fine, A. D., et al., Plague as a biological weapon: Medical and public health management, *JAMA*, 283 (17), 2281–2290, 2000.

5. Inglesby, T. V., Henderson, D. A., Bartlett, J. G., Ascher, M. S., Eitzen, E., Friedlander, A. M., Hauer, J., et al., Anthrax as a biological weapon: Medical and public health management, *JAMA*, 281 (18), 1735–1745, 1999.

6. Arnon, S. S., Schechter, R., Inglesby, T. V., Henderson, D. A., Bartlett, J. G., Ascher, M. S., Eitzen, E., et al., Botulinum toxin as a biological weapon: Medical and public health management, *JAMA*, 285 (8), 1059–1070, 2001.

7. Deisingh, A. K. and Thompson, M., Detection of infectious and toxigenic bacteria, *Analyst.*, 127 (5), 567–581, 2002.

8. Smith, G. P. and Petrenko, V. A., Phage display, *Chem. Rev.*, 97 (2), 391–410, 1997.

9. Petrenko, V. A. and Vodyanoy, V. J., Phage display for detection of biological threat agents, *J. Microbiol. Methods.*, 53 (2), 253–262, 2003.

10. Petrenko, V. A. and Sorokulova, I. B., Detection of biological threats. A challenge for directed molecular evolution, *J. Microbiol. Methods*, 58 (2), 147–168, 2004.

11. Luppa, P. B., Sokoll, L. J., and Chan, D. W., Immunosensors–principles and applications to clinical chemistry [Review], *Clin. Chim. Acta.*, 314 (1–2), 1–26, 2001.

12. Ziegler, C. and Gopel, W., Biosensor development, *Curr. Opin. Chem. Biol.*, 2 (5), 585–591, 1998.

13. Naimushin, A. N., Spinelli, C. B., Soelberg, S. D., Mann, T., Stevens, R. C., Chinowsky, T., Kauffman, P., Yee, S., and Furlong, C. E., Airborne analyte detection with an aircraft-adapted surface plasmon resonance sensor system, *Sens. Actuators B-Chem.*, 104 (2), 237–248, 2005.

14. Pancrazio, J. J., Whelan, J. P., Borkholder, D. A., Ma, W., and Stenger, D. A., Development and application of cell-based biosensors, *Ann. Biomed. Eng.*, 27 (6), 697–711, 1999.

15. Shone, C., Wilton-Smith, P., Appleton, N., Hambleton, P., Modi, N., Gatley, S., and Melling, J., Monoclonal antibody-based immunoassay for type A Clostridium botulinum toxin is comparable to the mouse bioassay, *Appl. Environ. Microbiol.*, 50 (1), 63–67, 1985.

16. Emanuel, P., O'Brien, T., Burans, J., DasGupta, B. R., Valdes, J. J., and Eldefrawi, M., Directing antigen specificity towards botulinum neurotoxin with combinatorial phage display libraries, *J. Immunol. Methods*, 193 (2), 189–197, 1996.

17. Vinogradov, E., Conlan, J. W., and Perry, M. B., Serological cross-reaction between the lipopolysaccharide O-polysaccharaide antigens of *E. coli* O157:H7 and strains of *Citrobacter freundii* and *Citrobacter sedlakii*, *FEMS Microbiol. Lett.*, 190 (1), 157–161, 2000.

18. Perry, M. B. and Bundle, D. R., Antigenic relationships of the lipopolysaccharides of Escherichia hermannii strains with those of *E. coli* O157:H7, *Brucella melitensis*, and *Brucella abortus*, *Infect. Immun.*, 58 (5), 1391–1395, 1990.

19. Lior, H. and Borczyk, A. A., False positive identifications of *Escherichia coli* O157, *Lancet*, 1 (8528), 333, 1987.
20. Bettelheim, K. A., Evangelidis, H., Pearce, J. L., Sowers, E., and Strockbine, N. A., Isolation of a *Citrobacter freundii* strain which carries the *Escherichia coli* O157 antigen, *J. Clin. Microbiol.*, 31 (3), 760–761, 1993.
21. Marsden, B. J., Bundle, D. R., and Perry, M. B., Serological and structural relationships between *Escherichia coli* O:98 and *Yersinia enterocolitica* O:11,23 and O:11,24 lipopolysaccharide O-antigens, *Biochem. Cell Biol.*, 72 (5–6), 163–168, 1994.
22. Emanuel, P. A., Dang, J., Gebhardt, J. S., Aldrich, J., Garber, E. A., Kulaga, H., Stopa, P., Valdes, J. J., and Dion-Schultz, A., Recombinant antibodies: A new reagent for biological agent detection, *Biosens. Bioelectron.*, 14 (10–11), 751–759, 2000.
23. Petrenko, V. A. and Smith, G. P., Phages from landscape libraries as substitute antibodies, *Protein Eng.*, 13 (8), 589–592, 2000.
24. Skerra, A., Engineered protein scaffolds for molecular recognition [Review], *J. Mol. Recognit.*, 13 (4), 167–187, 2000.
25. Petrenko, V. A. and Smith, G. P., Vectors and modes of display, In *Phage Display in Biotechnology and Drug Discovery*, Sidhu, S. S., Ed., Taylor & Francis, Boca Raton, FL, pp. 63–110, 2005.
26. Barbas, C. F. III, Barton, D. R., Scott, J. K., and Silverman, G. J., *Phage Display: A Laboratory Manual*, Cold Spring Harbor Laboratory Press, Cold Spring Harbor, New York, 2001.
27. Hoogenboom, H. R., de Bruine, A. P., Hufton, S. E., Hoet, R. M., Arends, J. W., and Roovers, R. C., Antibody phage display technology and its applications, *Immunotechnology*, 4 (1), 1–20, 1998.
28. Ditzel, H. J., Rescure of a broader range of antibody specificities using an epitope-masking strategy, In *Antibody Phage Display*, O'Brien, P. M., Ed., Humana Press, Totowa, NJ, pp. 179–186, 2002.
29. Short, M. K., Jeffrey, P. D., Demirjian, A., and Margolies, M. N., A single H: CDR3 residue in the anti-digoxin antibody 26-10 modulates specificity for C16-substituted digoxin analogs, *Protein Eng.*, 14 (4), 287–296, 2001.
30. Barbas, C. F., Hu, D., Dunlop, N., Sawyer, L., Cababa, D., Hendry, R. M., Nara, P. L., and Burton, D. R., In vitro evolution of a neutralizing human antibody to human immunodeficiency virus type 1 to enhance affinity and broaden strain cross-reactivity, *Proc. Natl Acad. Sci. U.S.A.*, 91 (9), 3809–3813, 1994.
31. Chowdhury, P. S., Targeting random mutations to hotspots in antibody variable domains for affinity improvement, In *Antibody Phage Display: Methods and Protocols*, O'Brien, P. M. and Aitken, R., Eds., Humana Press, Totowa, NJ, pp. 269–286, 2002.
32. Maynard, J. A., Chen, G., Georgiou, G., and Iverson, B. L., In vitro scanning-saturation mutagenesis, *Methods Mol. Biol.*, 182, 149–163, 2002.
33. Pini, A., Spreafico, A., Botti, R., Neri, D., and Neri, P., Hierarchical affinity maturation of a phage library derived antibody for the selective removal of cytomegalovirus from plasma, *J. Immunol. Methods*, 206 (1–2), 171–182, 1997.
34. Worn, A. and Pluckthun, A., Mutual stabilization of VL and VH in single-chain antibody fragments, investigated with mutants engineered for stability, *Biochemistry*, 37 (38), 13120–13127, 1998.
35. Marvin, J. S. and Lowman, H. B., Antibody humanization and affinity maturation using phage display, In *Phage Display in Biotechnology and Drug Discovery*, Sidhu, S. S., Ed., Taylor & Francis, Boca Raton, FL, pp. 493–528, 2005.
36. Jermutus, L., Honegger, A., Schwesinger, F., Hanes, J., and Pluckthun, A., Tailoring in vitro evolution for protein affinity or stability, *Proc. Natl Acad. Sci. U.S.A.*, 98 (1), 75–80, 2001.
37. Reiter, Y., Brinkmann, U., Jung, S. H., Lee, B., Kasprzyk, P. G., King, C. R., and Pastan, I., Improved binding and antitumor activity of a recombinant anti-erbB2 immunotoxin by disulfide stabilization of the Fv fragment, *J. Biol. Chem.*, 269 (28), 18327–18331, 1994.
38. Loeb, T., Isolation of bacteriophage specific for the F+ and Hfr mating types of *Escherichia coli* K12, *Science*, 131, 932–933, 1960.
39. Hofschneider, P. H., Untersuchungen uber "kleine" *E. coli* K12 Bacteriophagen M12, M13, und M20, *Z. Naturforschg.*, 18b, 203–205, 1963.
40. Marvin, D. A. and Hoffman-Berling, H., Physical and chemical properties of two new small bacterio-phages, *Nature (London)*, 197, 517–518, 1963.

41. Rodi, D. J., Mandava, S., and Makowski, L., Filamentous bacteriophage structure and biology, In *Phage Display in Biotechnology and Drug Discovery*, Sidhu, S. S., Ed., Taylor & Francis, Boca Raton, FL, pp. 1–61, 2005.

42. Berkowitz, S. A. and Day, L. A., Mass, length, composition and structure of the filamentous bacterial virus fd, *J. Mol. Biol.*, 102 (3), 531–547, 1976.

43. Marvin, D. A., Hale, R. D., Nave, C., and Citterich, M. H., Molecular models and structural comparisons of native and mutant class I filamentous bacteriophages Ff (fd, f1, M13), If1 and IKe, *J. Mol. Biol.*, 235 (1), 260–286, 1994.

44. Smith, G. P., Filamentous fusion phage: Novel expression vectors that display cloned antigens on the virion surface, *Science*, 228 (4705), 1315–1317, 1985.

45. Il'ichev, A. A., Minenkova, O. O., Tat'kov, S. I., Karpyshev, N. N., Eroshkin, A. M., Petrenko, V. A., and Sandakhchiev, L. S., Production of a viable variant of the M13 phage with a foreign peptide inserted into the basic coat protein, *Dokl Akad Nauk SSSR*, 307 (2), 481–483, 1989.

46. Petrenko, V. A., Smith, G. P., Gong, X., and Quinn, T., A library of organic landscapes on filamentous phage, *Protein Eng.*, 9 (9), 797–801, 1996.

47. Fellouse, F. A. and Pal, G., Methods for the construction of phage-displayed libraries, In *Phage Display in Biotechnology and Drug Discovery*, Sidhu, S. S., Ed., Taylor & Francis, Boca Raton, FL, pp. 111–142, 2005.

48. Dennis, M., Selection and screening strategies, In *Phage Display in Biotechnology and Drug Discovery*, Sidhu, S. S., Ed., Taylor & Francis, Boca Raton, FL, pp. 143–164, 2005.

49. Fellouse, F. A. and Sidhu, S. S., Synthetic antibody libraries, In *Phage Display in Biotechnology and Drug Discovery*, Sidhu, S. S., Ed., Taylor & Francis, Boca Raton, FL, pp. 709–740, 2005.

50. Berry, J. D. and Popkov, M., Antibody libraries from immunized repertoires, In *Phage Display in Biotechnology and Drug Discovery*, Sidhu, S. S., Ed., Taylor & Francis, Boca Raton, FL, pp. 529–658, 2005.

51. Dobson, C. L., Minter, R. R., and Hart-Shorrock, C. P., Naive antibody libraries from natural repertoires, In *Phage Display in Biotechnology and Drug Discovery*, Sidhu, S. S., Ed., Taylor & Francis, Boca Raton, FL, pp. 659–708, 2005.

52. Hoogenboom, H. R., Overview of antibody phage-display technology and its applications, In *Antibody Phage Display: Methods and Protocols*, O'Brien, P. M. and Aitken, R., Eds., Humana Press, Totowa, NJ, pp. 1–39, 2002.

53. Casey, J. L., Coley, A. M., Tilley, L. M., and Foley, M., Green fluorescent antibodies: Novel in vitro tools, *Protein Eng.*, 13 (6), 445–452, 2000.

54. Kerschbaumer, R. J., Hirschl, S., Kaufmann, A., Ibl, M., Koenig, R., and Himmler, G., Single-chain Fv fusion proteins suitable as coating and detecting reagents in a double antibody sandwich enzyme-linked immunosorbent assay, *Anal. Biochem.*, 249 (2), 219–227, 1997.

55. Morino, K., Katsumi, H., Akahori, Y., Iba, Y., Shinohara, M., Ukai, Y., Kohara, Y., and Kurosawa, Y., Antibody fusions with fluorescent proteins: A versatile reagent for profiling protein expression, *J. Immunol. Methods*, 257 (1–2), 175–184, 2001.

56. Muller, B. H., Chevrier, D., Boulain, J. C., and Guesdon, J. L., Recombinant single-chain Fv antibody fragment–alkaline phosphatase conjugate for one-step immunodetection in molecular hybridization, *J. Immunol. Methods*, 227 (1–2), 177–185, 1999.

57. Pearce, L. A., Oddie, G. W., Coia, G., Kortt, A. A., Hudson, P. J., and Lilley, G. G., Linear gene fusions of antibody fragments with streptavidin can be linked to biotin labelled secondary molecules to form bispecific reagents, *Biochem. Mol. Biol. Int.*, 42 (6), 1179–1188, 1997.

58. Chen, G., Dubrawsky, I., Mendez, P., Georgiou, G., and Iverson, B. L., In vitro scanning saturation mutagenesis of all the specificity determining residues in an antibody binding site, *Protein Eng.*, 12 (4), 349–356, 1999.

59. Kortt, A. A., Dolezal, O., Power, B. E., and Hudson, P. J., Dimeric and trimeric antibodies: High avidity scFvs for cancer targeting, *Biomol. Eng.*, 18 (3(Special Issue SI)), 95–108, 2001.

60. Deng, S. J., MacKenzie, C. R., Sadowska, J., Michniewicz, J., Young, N. M., Bundle, D. R., and Narang, S. A., Selection of antibody single-chain variable fragments with improved carbohydrate binding by phage display, *J. Biol. Chem.*, 269 (13), 9533–9538, 1994.

61. Worn, A. and Pluckthun, A., Stability engineering of antibody single-chain Fv fragments, *J. Mol. Biol.*, 305 (5), 989–1010, 2001.

62. Liu, B. and Marks, J. D., Applying phage antibodies to proteomics: Selecting single chain Fv antibodies to antigens blotted on nitrocellulose, *Anal. Biochem.*, 286 (1), 119–128, 2000.

63. Kay, B. K., Winter, J., and McCafferty, J., *Phage Display of Peptides and Proteins: A Laboratory Manual*, Academic Press, New York, 1996.

64. O'Brien, P. M. and Aitken, R., *Antibody Phage Display: Methods and Protocols*, 401 ed., Humana Press, Totowa, NJ, 2002.

65. Turnbough, C. L. Jr., Discovery of phage display peptide ligands for species-specific detection of *Bacillus* spores, *J. Microbiol. Methods*, 53 (2), 263–271, 2003.

66. Li, B., Tom, J. Y., Oare, D., Yen, R., Fairbrother, W. J., Wells, J. A., and Cunningham, B. C., Minimization of a polypeptide hormone, *Science*, 270 (5242), 1657–1660, 1995.

67. O'Neil, K. T., Hoess, R. H., Jackson, S. A., Ramachandran, N. S., Mousa, S. A., and DeGrado, W. F., Identification of novel peptide antagonists for GPIIb/IIIa from a conformationally constrained phage peptide library, *Proteins*, 14 (4), 509–515, 1992.

68. Gough, K. C., Cockburn, W., and Whitelam, G. C., Selection of phage-display peptides that bind to cucumber mosaic virus coat protein, *J. Virol. Methods*, 79 (2), 169–180, 1999.

69. Knurr, J., Benedek, O., Heslop, J., Vinson, R., Boydston, J., McAndrew, J., Kearney, J., and Turnbough, C. Jr., Peptide ligands that bind selectively to B. subtilis and closely related species, *Appl. Environ. Microbiol.*, 69, 6841–6847, 2003.

70. Steichen, C., Chen, P., Kearney, J. F., and Turnbough, C. L. Jr., Identification of the immunodominant protein and other proteins of the *Bacillus anthracis* exosporium, *J. Bacteriol.*, 185 (6), 1903–1910, 2003.

71. Turnbough, C. L. Jr., Discovery of phage display peptide ligands for species-specific detection of *Bacillus* spores, *J. Microbiol. Methods*, 53 (2), 263–271, 2003.

72. Stratmann, J., Strommenger, B., Stevenson, K., and Gerlach, G. F., Development of a peptide-mediated capture PCR for detection of *Mycobacterium avium* subsp. *paratuberculosis* in milk, *J. Clin. Microbiol.*, 40 (11), 4244–4250, 2002.

73. Grant, I. R., Pope, C. M., O'Riordan, L. M., Ball, H. J., and Rowe, M. T., Improved detection of *Mycobacterium avium* subsp. *paratuberculosis* in milk by immunomagnetic PCR, *Vet. Microbiol.*, 77 (3–4), 369–378, 2000.

74. Goldman, E. R., Pazirandeh, M. P., Mauro, J. M., King, K. D., Frey, J. C., and Anderson, G. P., Phage-displayed peptides as biosensor reagents, *J. Mol. Recognit.*, 13, 382–387, 2000.

75. Mammen, M., Choi, S. K., and Whitesides, G. M., Polyvalent interactions in biological systems—implications for design and use of multivalent ligands and inhibitors, *Angewandte Chemie. Int. Ed. Eng.*, 37 (20), 2755–2794, 1998.

76. Zdanovsky, A. G., Karassina, N. V., Simpson, D., and Zdanovskaia, M. V., Peptide phage display library as source for inhibitors of clostridial neurotoxins, *J. Protein Chem.*, 20 (1), 73–80, 2001.

77. Iqbal, S. S., Mayo, M. W., Bruno, J. G., Bronk, B. V., Batt, C. A., and Chambers, J. P., A review of molecular recognition technologies for detection of biological threat agents [Review], *Biosens. Bioelectron.*, 15 (11–12), 549–578, 2000.

78. Sheets, M. D., Amersdorfer, P., Finnern, R., Sargent, P., Lindquist, E., Schier, R., Hemingsen, G., et al., Efficient construction of a large nonimmune phage antibody library: The production of high-affinity human single-chain antibodies to protein antigens, *Proc. Natl Acad. Sci. U.S.A.*, 95 (11), 6157–6162, 1998.

79. Nissim, A., Hoogenboom, H. R., Tomlinson, I. M., Flynn, G., Midgley, C., Lane, D., and Winter, G., Antibody fragments from a 'single pot' phage display library as immunochemical reagents, *EMBO J.*, 13 (3), 692–698, 1994.

80. Zhou, B., Wirsching, P., and Janda, K. D., Human antibodies against spores of the genus *Bacillus*: A model study for detection of and protection against anthrax and the bioterrorist threat, *Proc. Natl Acad. Sci. U.S.A.*, 99 (8), 5241–5246, 2002.

81. Lindquist, E. A., Marks, J. D., Kleba, B. J., and Stephens, R. S., Phage-display antibody detection of *Chlamydia trachomatis*-associated antigens, *Microbiology*, 148 (Pt 2), 443–451, 2002.

82. Benhar, I., Eshkenazi, I., Neufeld, T., Opatowsky, J., Shaky, S., and Rishpon, J., Recombinant single chain antibodies in bioelectrochemical sensors, *Talanta*, 55 (5), 899–907, 2001.

83. Hayhurst, A., Happe, S., Mabry, R., Koch, Z., Iverson, B. L., and Georgiou, G., Isolation and expression of recombinant antibody fragments to the biological warfare pathogen *Brucella melitensis*, *J. Immunol. Methods*, 276 (1–2), 185–196, 2003.

84. Bundle, D. R., Gidney, M. A., Perry, M. B., Duncan, J. R., and Cherwonogrodzky, J. W., Serological confirmation of *Brucella abortus* and *Yersinia enterocolitica* O:9 O-antigens by monoclonal antibodies, *Infect. Immun.*, 46 (2), 389–393, 1984.

85. Samoylova, T. I., Petrenko, V. A., Morrison, N. E., Globa, L. P., Baker, H. J., and Cox, N. R., Phage probes for malignant glial cells, *Mol. Cancer Ther.*, 2 (11), 1129–1137, 2003.

86. Radosevic, K. and van Ewijk, W., Subtractive isolation of single-chain antibodies using tissue fragments, *Methods Mol. Biol.*, 178, 235–243, 2002.

87. Boel, E., Bootsma, H., de Kruif, J., Jansze, M., Klingman, K. L., van Dijk, H., and Logtenberg, T., Phage antibodies obtained by competitive selection on complement-resistant *Moraxella* (*Branhamella*) *catarrhalis* recognize the high-molecular-weight outer membrane protein, *Infect. Immun.*, 66 (1), 83–88, 1998.

88. Cai, X. H. and Garen, A., Antimelanoma antibodies from melanoma patients immunized with genetically-modified autologous tumor-cells—Selection of specific antibodies from single-chain Fv fusion phage libraries, *Proc. Natl Acad. Sci. U.S.A.*, 92 (14), 6537–6541, 1995.

89. Marks, J. D., Ouwehand, W. H., Bye, J. M., Finnern, R., Gorick, B. D., Voak, D., Thorpe, S. J., Hughes-Jones, N. C., and Winter, G., Human antibody fragments specific for human blood group antigens from a phage display library, *BioTechnology*, 11 (10), 1145–1149, 1993.

90. de Kruif, J., Boel, E., and Logtenberg, T., Selection and application of human single chain Fv antibody fragments from a semi-synthetic phage antibody display library with designed CDR3 regions, *J. Mol. Biol.*, 248 (1), 97–105, 1995.

91. de Greeff, A., van Alphen, L., and Smith, H. E., Selection of recombinant antibodies specific for pathogenic *Streptococcus suis* by subtractive phage display, *Infect. Immun.*, 68 (7), 3949–3955, 2000.

92. Chan, S. W., Bye, J. M., Jackson, P., and Allain, J. P., Human recombinant antibodies specific for hepatitis C virus core and envelope E2 peptides from an immune phage display library, *J. Gen. Virol.*, 77 (Pt 10), 2531–2539, 1996.

93. Plaisant, P., Burioni, R., Manzin, A., Solforosi, L., Candela, M., Gabrielli, A., Fadda, G., and Clementi, M., Human monoclonal recombinant Fabs specific for HCV antigens obtained by repertoire cloning in phage display combinatorial vectors, *Res. Virol.*, 148 (2), 165–169, 1997.

94. Sanna, P. P., Williamson, R. A., De Logu, A., Bloom, F. E., and Burton, D. R., Directed selection of recombinant human monoclonal antibodies to herpes simplex virus glycoproteins from phage display libraries, *Proc. Natl Acad. Sci. U.S.A.*, 92 (14), 6439–6443, 1995.

95. Williamson, R. A., Lazzarotto, T., Sanna, P. P., Bastidas, R. B., Dalla Casa, B., Campisi, G., Burioni, R., Landini, M. P., and Burton, D. R., Use of recombinant human antibody fragments for detection of cytomegalovirus antigenemia, *J. Clin. Microbiol.*, 35 (8), 2047–2050, 1997.

96. Muller, B. H., Lafay, F., Demangel, C., Perrin, P., Tordo, N., Flamand, A., Lafaye, P., and Guesdon, J. L., Phage-displayed and soluble mouse scFv fragments neutralize rabies virus, *J. Virol. Methods*, 67 (2), 221–233, 1997.

97. Schmaljohn, C., Cui, Y., Kerby, S., Pennock, D., and Spik, K., Production and characterization of human monoclonal antibody Fab fragments to vaccinia virus from a phage-display combinatorial library, *Virology*, 258 (1), 189–200, 1999.

98. Maruyama, T., Parren, P. W., Sanchez, A., Rensink, I., Rodriguez, L. L., Khan, A. S., Peters, C. J., and Burton, D. R., Recombinant human monoclonal antibodies to Ebola virus, *J. Infect. Dis.*, 179 (1), S235–S239, 1999.

99. Kuhne, S. A., Hawes, W. S., La Ragione, R. M., Woodward, M. J., Whitelam, G. C., and Gough, K. C., Isolation of recombinant antibodies against EspA and intimin of *Escherichia coli* O157:H7, *J. Clin. Microbiol.*, 42 (7), 2966–2976, 2004.

100. Hu, W. G., Thompson, H. G., Alvi, A. Z., Nagata, L. P., Suresh, M. R., and Fulton, R. E., Development of immunofiltration assay by light addressable potentiometric sensor with genetically biotinylated recombinant antibody for rapid identification of Venezuelan equine encephalitis virus, *J. Immunol. Methods*, 289 (1–2), 27–35, 2004.

101. Hayhurst, A. and Georgiou, G., High-throughput antibody isolation [Review], *Curr. Opin. Chem. Biol.*, 5 (6), 683–689, 2001.

102. Brichta, J., Hnilova, M., and Viskovic, T., Generation of hapten-specific recombinant antibodies: Antibody phage display technology: A review, *Vet. Med.*, 50 (6), 231–252, 2005.

103. Hayhurst, A., Improved expression characteristics of single-chain Fv fragments when fused down-stream of the *Escherichia coli* maltose-binding protein or upstream of a single immunoglobulin-constant domain, *Protein Expr., Purif.*, 18 (1), 1–10, 2000.

104. Strachan, G., Whyte, J. A., Molloy, P. M., Paton, G. I., and Porter, A. J. R., Development of robust, environmental, immunoassay formats for the quantification of pesticides in soil, *Environ. Sci. Technol.*, 34 (8), 1603–1608, 2000.

105. Brichta, J., Vesela, H., and Franek, M., Production of scFv recombinant fragments against 2,4-dichlorophenoxyacetic acid hapten using naive phage library, *Veterinarni Medicina*, 48 (9), 237–247, 2003.

106. Dooley, H., Grant, S. D., Harris, W. J., Porter, A. J., Shelton, S. A., Graham, B. M., and Strachan, G., Stabilization of antibody fragments in adverse environments immunomethods for detecting a broad range of polychlorinated biphenyls, *Biotechnol. Appl. Biochem.*, 28 (Part 1), 77–83, 1998.

107. Jung, S. and Pluckthun, A., Improving in vivo folding and stability of a single-chain Fv antibody fragment by loop grafting, *Protein Eng.*, 10 (8), 959–966, 1997.

108. McConnell, S. J. and Hoess, R. H., Tendamistat as a scaffold for conformationally constrained phage peptide libraries, *J. Mol. Biol.*, 250 (4), 460–470, 1995.

109. Nord, K., Nord, O., Uhlen, M., Kelley, B., Ljungqvist, C., and Nygren, P. A., Recombinant human factor VIII-specific affinity ligands selected from phage-displayed combinatorial libraries of protein A, *Eur. J. Biochem.*, 268 (15), 4269–4277, 2001.

110. Sollazzo, M., Venturini, S., Lorenzetti, S., Pinola, M., and Martin, F., Engineering minibody-like ligands by design and selection, *Chem. Immunol.*, 65, 1–17, 1997.

111. Bianchi, E., Folgori, A., Wallace, A., Nicotra, M., Acali, A., Phalipon, A., Barbato, G., et al., A conformationally homogeneous combinatorial peptide library, *J. Mol. Biol.*, 247 (2), 154–160, 1995.

112. Dennis, M. S., Herzka, A., and Lazarus, R. A., Potent and selective Kunitz domain inhibitors of plasma kallikrein designed by phage display, *J. Biol. Chem.*, 270 (43), 25411–25417, 1995.

113. Koide, A., Bailey, C. W., Huang, X., and Koide, S., The fibronectin type III domain as a scaffold for novel binding proteins, *J. Mol. Biol.*, 284 (4), 1141–1151, 1998.

114. Ku, J. and Schultz, P. G., Alternate protein frameworks for molecular recognition, *Proc. Natl Acad. Sci. U.S.A.*, 92 (14), 6552–6556, 1995.

115. Legendre, D. and Fastrez, J., Construction and exploitation in model experiments of functional selection of a landscape library expressed from a phagemid, *Gene*, 290, 203–215, 2002.

116. Martin, F., Toniatti, C., Salvati, A. L., Ciliberto, G., Cortese, R., and Sollazzo, M., Coupling protein design and in vitro selection strategies: Improving specificity and affinity of a designed beta-protein IL-6 antagonist, *J. Mol. Biol.*, 255 (1), 86–97, 1996.

117. Hoess, R. H., Protein design and phage display [Review], *Chem. Rev.*, 101 (10), 3205–3218, 2001.

118. Felici, F., Castagnoli, L., Musacchio, A., Jappelli, R., and Cesareni, G., Selection of antibody ligands from a large library of oligopeptides expressed on a multivalent exposition vector, *J. Mol. Biol.*, 222 (2), 301–310, 1991.

119. Greenwood, J., Willis, A. E., and Perham, R. N., Multiple display of foreign peptides on a filamentous bacteriophage. Peptides from *Plasmodium falciparum* circumsporozoite protein as antigens, *J. Mol. Biol*, 220 (4), 821–827, 1991.

120. Petrenko, V. A., Smith, G. P., Mazooji, M. M., and Quinn, T., Alpha-helically constrained phage display library, *Protein Eng.*, 15 (11), 943–950, 2002.

121. Romanov, V. I., Durand, D. B., and Petrenko, V. A., Phage display selection of peptides that affect prostate carcinoma cells attachment and invasion, *Prostate*, 47 (4), 239–251, 2001.

122. Romanov, V. I., Whyard, T., Adler, H. L., Waltzer, W. C., and Zucker, S., Prostate cancer cell adhesion to bone marrow endothelium: The role of prostate-specific antigen, *Cancer Res.*, 64 (6), 2083–2089, 2004.

123. Samoylova, T. I., Cox, N. R., Morrison, N. E., Globa, L. P., Romanov, V., Baker, H. J., and Petrenko, V. A., Phage matrix for isolation of glioma cell membrane proteins, *Biotechniques*, 37 (2), 254–260, 2004.

124. Mount, J. D., Samoylova, T. I., Morrison, N. E., Cox, N. R., Baker, H. J., and Petrenko, V. A., Cell targeted phagemid rescued by preselected landscape phage, *Gene*, 341, 59–65, 2004.

125. Kouzmitcheva, G. A., Petrenko, V. A., and Smith, G. P., Identifying diagnostic peptides for lyme disease through epitope discovery, *Clin. Diagn. Lab Immunol.*, 8 (1), 150–160, 2001.

126. Brigati, J., Williams, D. D., Sorokulova, I. B., Nanduri, V., Chen, I. H., Turnbough, C. L. Jr., and Petrenko, V. A., Diagnostic probes for *Bacillus anthracis* spores selected from a landscape phage library, *Clin. Chem.*, 50 (10), 1899–1906, 2004.

127. Sorokulova, I. B., Olsen, E. V., Chen, I. H., Fiebor, B., Barbaree, J. M., Vodyanoy, V. J., Chin, B. A., and Petrenko, V. A., Landscape phage probes for Salmonella typhimurium, *J. Microbiol. Methods*, 63, 55–72, 2005.

128. Olsen, E. V., Sorokulova, I. B., Petrenko, V. A., Chen, I. -H., Barbaree, J. M., and Vodyanoy, V. J., Affinity-selected filamentous bacteriophage as a probe for acoustic wave biodetectors of *Salmonella typhimurium*, *Biosens. Bioelectron*, 21, 1434–1442, 2006.

129. Barbas, C. F. III, Burton, D. R., Scott, J. K., and Silverman, G. J., *Phage Display: A Laboratory Manual*, Cold Spring Harbor Laboratory Press, Cold Spring Harbor, New York, 2001.

130. Philippa O'Brien, R. A., *Antibody Phage Display*, Humana Press, Totowa, NJ, 2001.

131. Mandava, S., Makowski, L., Devarapalli, S., Uzubell, J., and Rodi, D. J., RELIC–a bioinformatics server for combinatorial peptide analysis and identification of protein-ligand interaction sites, *Proteomics*, 4 (5), 1439–1460, 2004.

132. Bishop-Hurley, S. L., Mounter, S. A., Laskey, J., Morris, R. O., Elder, J., Roop, P., Rouse, C., Schmidt, F. J., and English, J. T., Phage-displayed peptides as developmental agonists for Phytophthora capsici zoospores, *Appl. Environ. Microbiol.*, 68 (7), 3315–3320, 2002.

133. Stausbol-Gron, B., Jensen, K. B., Jensen, K. H., Jensen, M. O., and Clark, B. F., De novo identification of cell-type specific antibody–antigen pairs by phage display subtraction. Isolation of a human single chain antibody fragment against human keratin 14, *Eur. J. Biochem.*, 268 (10), 3099–3107, 2001.

134. Van Ewijk, W., de Kruif, J., Germeraad, W. T., Berendes, P., Ropke, C., Platenburg, P. P., and Logtenberg, T., Subtractive isolation of phage-displayed single-chain antibodies to thymic stromal cells by using intact thymic fragments, *Proc. Natl Acad. Sci. U.S.A.*, 94 (8), 3903–3908, 1997.

135. Belizaire, A. K., Tchistiakova, L., St-Pierre, Y., and Alakhov, V., Identification of a murine ICAM-1-specific peptide by subtractive phage library selection on cells, *Biochem. Biophys. Res. Commun.*, 309 (3), 625–630, 2003.

136. Ramanujam, P., Tan, W. S., Nathan, S., and Yusoff, K., Pathotyping of Newcastle disease virus with a filamentous bacteriophage, *Biotechniques*, 36 (2), 296–300, 2004 see also p. 302

137. Ramanujam, P., Tan, W. S., Nathan, S., and Yusoff, K., Novel peptides that inhibit the propagation of Newcastle disease virus, *Arch. Virol.*, 147 (5), 981–993, 2002.

138. Yu, J. and Smith, G. P., Affinity maturation of phage-displayed peptide ligands, *Methods Enzymol*, 267, 3–27, 1996.

139. Goldman, E. R., Pazirandeh, M. P., Mauro, J. M., King, K. D., Frey, J. C., and Anderson, G. P., Phage-displayed peptides as biosensor reagents, *J. Mol. Recognit.*, 13 (6), 382–387, 2000.

140. Knurr, J., Benedek, O., Heslop, J., Vinson, R. B., Boydston, J. A., McAndrew, J., Kearney, J. F., and Turnbough, C. L., Peptide ligands that bind selectively to spores of *Bacillus subtilis* and closely related species, *Appl. Environ. Microbiol.*, 69 (11), 6841–6847, 2003.

141. Block, T., Miller, R., Korngold, R., and Jungkind, D., A phage-linked immunoadsorbant system for the detection of pathologically relevant antigens, *Biotechniques*, 7 (7), 756–761, 1989.

142. Lu, X., Weiss, P., and Block, T., A phage with high affinity for hepatitis B surface antigen for the detection of HBsAg, *J Virol. Methods*, 119 (1), 51–54, 2004.

143. Stratmann, J., Strommenger, B., Stevenson, K., and Gerlach, G. F., Development of a peptide-mediated capture PCR for detection of *Mycobacterium avium* subsp. *paratuberculosis* in milk, *J. Clin. Microbiol.*, 40 (11), 4244–4250, 2002.

144. Williams, D. D., Benedek, O., and Tournbough, C. L., Species-specific peptide ligands for the detection of *Bacillus anthracis* spores, *Appl. Environ. Microbiol.*, 69 (10), 6288–6293, 2003.

145. Zwick, M. B., Labrijn, A. F., Wang, M., Spenlehauer, C., Saphire, E. O., Binley, J. M., Moore, J. P., et al., Broadly neutralizing antibodies targeted to the membrane-proximal external region of human immunodeficiency virus type 1 glycoprotein gp41, *J. Virol.*, 75 (22), 10892–10905, 2001.

146. Emanuel, P. A., Dang, J., Gebhardt, J. S., Aldrich, J., Garber, E. A., Kulaga, H., Stopa, P., Valdes, J. J., and Dion-Schultz, A., Recombinant antibodies: A new reagent for biological agent detection, *Biosens. Bioelectron.*, 14 (10–11), 751–759, 2000.

147. Emanuel, P., O'Brien, T., Burans, J., DasGupta, B. R., Valdes, J. J., and Eldefrawi, M., Directing antigen specificity towards botulinum neurotoxin with combinatorial phage display libraries, *J. Immunol. Methods*, 193 (2), 189–197, 1996.

148. Sheets, M. D., Amersdorfer, P., Finnern, R., Sargent, P., Lindquist, E., Schier, R., Hemingsen, G., Wong, C., Gerhart, J. C., Marks, J. D., and Lindquist, E., Efficient construction of a large nonimmune phage antibody library: The production of high-affinity human single-chain antibodies to protein antigens, *Proc. Natl Acad. Sci. U.S.A.*, 95 (11), 6157–6162, 1998.

149. Barbas, C. F., Hu, D., Dunlop, N., Sawyer, L., Cababa, D., Hendry, R. M., Nara, P. L., and Burton, D. R., In vitro evolution of a neutralizing human antibody to human immunodeficiency virus type 1 to enhance affinity and broaden strain cross-reactivity, *Proc. Natl Acad. Sci. U.S.A.*, 91 (9), 3809–3813, 1994.

150. Burton, D. R., Barbas, C. F. III, Persson, M. A., Koenig, S., Chanock, R. M., and Lerner, R. A., A large array of human monoclonal antibodies to type 1 human immunodeficiency virus from combinatorial libraries of asymptomatic seropositive individuals, *Proc. Natl Acad. Sci. U.S.A.*, 88 (22), 10134–10137, 1991.

151. Barbas, C. F. III, Crowe, J. E. Jr., Cababa, D., Jones, T. M., Zebedee, S. L., Murphy, B. R., Chanock, R. M., and Burton, D. R., Human monoclonal Fab fragments derived from a combinatorial library bind to respiratory syncytial virus F glycoprotein and neutralize infectivity, *Proc. Natl Acad. Sci. U.S.A.*, 89 (21), 10164–10168, 1992.

152. Crowe, J. E., Firestone, C. Y., Crim, R., Beeler, J. A., Coelingh, K. L., Barbas, C. F., Burton, D. R., Chanock, R. M., and Murphy, B. R., Monoclonal antibody-resistant mutants selected with a respiratory syncytial virus-neutralizing human antibody fab fragment (Fab 19) define a unique epitope on the fusion (F) glycoprotein, *Virology*, 252 (2), 373–375, 1998.

153. Bugli, F., Bastidas, R., Burton, D. R., Williamson, R. A., Clementi, M., and Burioni, R., Molecular profile of a human monoclonal antibody Fab fragment specific for Epstein-Barr virus gp350/220 antigen, *Hum. Immunol.*, 62 (4), 362–367, 2001.

154. Plaisant, P., Burioni, R., Manzin, A., Solforosi, L., Candela, M., Gabrielli, A., Fadda, G., and Clementi, M., Human monoclonal recombinant Fabs specific for HCV antigens obtained by repertoire cloning in phage display combinatorial vectors, *Res. Virol.*, 148 (2), 165–169, 1997.

155. Chan, S. W., Bye, J. M., Jackson, P., and Allain, J. P., Human recombinant antibodies specific for hepatitis C virus core and envelope E2 peptides from an immune phage display library, *J. Gen. Virol.*, 77 (Pt 10), 2531–2539, 1996.

156. de Carvalho Nicacio, C., Williamson, R. A., Parren, P. W., Lundkvist, A., Burton, D. R., and Bjorling, E., Neutralizing human Fab fragments against measles virus recovered by phase display, *J.Virol.*, 76 (1), 251–258, 2002.

157. De Logu, A., Williamson, R. A., Rozenshteyn, R., Ramiro-Ibanez, F., Simpson, C. D., Burton, D. R., and Sanna, P. P., Characterization of a type-common human recombinant monoclonal antibody to herpes simplex virus with high therapeutic potential, *J. Clin. Microbiol.*, 36 (11), 3198–3204, 1998.

158. Sanna, P. P., Williamson, R. A., De Logu, A., Bloom, F. E., and Burton, D. R., Directed selection of recombinant human monoclonal antibodies to herpes simplex virus glycoproteins from phage display libraries, *Proc. Natl Acad. Sci. U.S.A.*, 92 (14), 6439–6443, 1995.

159. Glamann, J., Burton, D. R., Parren, P. W., Ditzel, H. J., Kent, K. A., Arnold, C., Montefiori, D., and Hirsch, V. M., Simian immunodeficiency virus (SIV) envelope-specific Fabs with high-level homologous neutralizing activity: Recovery from a long-term-nonprogressor SIV-infected macaque, *J. Virol.*, 72 (1), 585–592, 1998.

160. Maruyama, T., Rodriguez, L. L., Jahrling, P. B., Sanchez, A., Khan, A. S., Nichol, S. T., Peters, C. J., Parren, P. W., and Burton, D. R., Ebola virus can be effectively neutralized by antibody produced in natural human infection, *J. Virol.*, 73 (7), 6024–6030, 1999.

161. Maruyama, T., Parren, P. W., Sanchez, A., Rensink, I., Rodriguez, L. L., Khan, A. S., Peters, C. J., and Burton, D. R., Recombinant human monoclonal antibodies to Ebola virus, *J. Infect. Dis.*, 179 (1), S235–S239, 1999.

162. Meissner, F., Maruyama, T., Frentsch, M., Hessell, A. J., Rodriguez, L. L., Geisbert, T. W., Jahrling, P. B., Burton, D. R., and Parren, P. W., Detection of antibodies against the four subtypes of Ebola virus in sera from any species using a novel antibody-phage indicator assay, *Virology*, 300 (2), 236–243, 2002.

163. Muller, B. H., Lafay, F., Demangel, C., Perrin, P., Tordo, N., Flamand, A., Lafaye, P., and Guesdon, J. L., Phage-displayed and soluble mouse scFv fragments neutralize rabies virus, *J. Virol. Methods*, 67 (2), 221–233, 1997.

164. Griffiths, A. D., Williams, S. C., Hartley, O., Tomlinson, I. M., Waterhouse, P., Crosby, W. L., Kontermann, R. E., et al., Isolation of high affinity human antibodies directly from large synthetic repertoires, *EMBO J.*, 13 (14), 3245–3260, 1994.

165. Saldarelli, P., Keller, H., Dell'Orco, M., Schots, A., Elicio, V., and Minafra, A., Isolation of recombinant antibodies (scFvs) to grapevine virus B, *J. Virol. Methods*, 124 (1–2), 191–195, 2005.

166. Zebedee, S. L., Barbas, C. F. III, Hom, Y. L., Caothien, R. H., Graff, R., DeGraw, J., Pyati, J., et al., Human combinatorial antibody libraries to hepatitis B surface antigen, *Proc. Natl Acad. Sci. U.S.A.*, 89 (8), 3175–3179, 1992.

167. Williamson, R. A., Lazzarotto, T., Sanna, P. P., Bastidas, R. B., Dalla Casa, B., Campisi, G., Burioni, R., Landini, M. P., and Burton, D. R., Use of recombinant human antibody fragments for detection of cytomegalovirus antigenemia, *J. Clin. Microbiol.*, 35 (8), 2047–2050, 1997.

168. Boel, E., Bootsma, H., de Kruif, J., Jansze, M., Klingman, K. L., van Dijk, H., and Logtenberg, T., Phage antibodies obtained by competitive selection on complement-resistant *Moraxella* (*Branhamella*) *catarrhalis* recognize the high-molecular-weight outer membrane protein, *Infect. Immun.*, 66 (1), 83–88, 1998.

169. Deng, S. J., MacKenzie, C. R., Hirama, T., Brousseau, R., Lowary, T. L., Young, N. M., Bundle, D. R., and Narang, S. A., Basis for selection of improved carbohydrate-binding single-chain antibodies from synthetic gene libraries, *Proc. Natl Acad. Sci. U.S.A.*, 92 (11), 4992–4996, 1995.

170. Kanitpun, R., Wagner, G. G., and Waghela, S. D., Characterization of recombinant antibodies developed for capturing enterohemorrhagic *Escherichia coli* O157:H7, *Southeast Asian J. Trop. Med. Public Health*, 35 (4), 902–912, 2004.

171. Lindquist, E. A., Marks, J. D., Kleba, B. J., and Stephens, R. S., Phage-display antibody detection of *Chlamydia trachomatis*-associated antigens, *Microbiology*, 148 (art 2), 443–451, 2002.

172. Zhou, B., Wirsching, P., and Janda, K. D., Human antibodies against spores of the genus *Bacillus*: A model study for detection of and protection against anthrax and the bioterrorist threat, *Proc. Natl Acad. Sci. U.S.A.*, 99 (8), 5241–5246, 2002.

173. Wan, J. H., Li, Y. Q., Fiebor, B., Chen, I. H., Petrenko, V. A., and Chin, B. A., Detection of *Bacillus anthracis* spores by landscape phage-based magnetostrictive biosensors, *Abstr. Pap. Am. Chem. Soc.*, 229 (411-ANYL Part) U155–U155, 2005.

174. Lyons, L. B. and Zinder, N. D., The genetic map of the filamentous bacteriophage f1, *Virology*, 49, 45–60, 1972.

175. Smith, G. P., Phage-Display Vectors and Libraries Based on Filamentous Phage Strain fdtet, http://www.biosci.missouri.edu/smithgp/PhageDisplayWebsite/PhageDisplayWebsiteIndex.html (accessed September 8, 2006).

176. Khan, A. S., Thompson, R., Cao, C., and Valdes, J. J., Selection and characterization of peptide memitopes binding to ricin, *Biotechnol. Lett.*, 25, 1671–1675, 2003.

177. Tan, W. S., Tan, G. H., Yusoff, K., and Seow, H. F., A phage-displayed cyclic peptide that interacts tightly with the immunodominant region of hepatitis B surface antigen, *J. Clin. Virol.*, 34, 35–41, 2005.

178. Dai, H., Gao, H., Zhao, X., Dai, L., Zhang, X., Xiao, N., Zhao, R., and Hemmingsen, S. M., Construction and characterization of a novel recombinant single-chain variable fragment antibody against white spot syndrome virus from shrimp, *J. Immunol. Methods*, 279, 267–275, 2003.

179. Yi, G., Qian, J., Wang, Z., and Qi, Y., A phage-displayed peptide can inhibit infection by white spot syndrome virus of shrimp, *J. Gen. Virol.*, 84, 2545–2553, 2003.

9 Upconverting Phosphors for Detection and Identification Using Antibodies

David E. Cooper, Annalisa D'Andrea, Gregory W. Faris, Brent MacQueen, and William H. Wright

CONTENTS

9.1 INTRODUCTION

Upconverting phosphors are rare-earth-doped ceramic materials with the unique property of emitting visible light upon excitation with near-infrared light. This process is called upconversion and was first proposed in 1959 by Bloembergen [1] and experimentally demonstrated by Auzel in 1966 [2]. Following their discovery, these materials were studied for applications as light sources and as detection materials for near-infrared diode laser light. The first use of downconverting phosphors as labels in biological assays was in 1990 by Beverloo et al. [3] who first reported the advantages of using this class of materials. The use of upconverting phosphors as labels in biological assays was first considered by Zarling et al. in the early 1990s [4]. Since the mid-1990s, several groups have extensively studied the use of upconverting phosphors in a variety of assay formats and detection schemes. These include Tanke's group at Leiden University as well as researchers at OraSure Technologies (formerly STC) and SRI International. To date, upconverting phosphors have been used primarily in lateral flow assays to detect protein [5], bacterial [6], and nucleic acid [7] targets. However, they have also been used in microarray, microtiter plate, and bead-based assay formats.

Upconversion is a multiphoton process where two or more incident photons of low energy (infrared) produce a single photon of higher energy (visible). Rare-earth-emitting (e.g., Er, Ho, and Tm) and -absorbing (e.g., Yb, Er, and Sm) centers are coupled by substitution in the lattice of appropriate host crystals. Figure 9.1 shows a process where rare-earth ions (a sensitizer or absorber—typically ytterbium) absorbs infrared light. The absorbed energy is nonradiatively transferred to a second rare earth ion (activator or emitter) through two sequential steps. Two absorbed photons are required to produce each upconverting state. A photon of energy in the visible region of the spectrum is emitted when this second rare-earth ion spontaneously relaxes to its ground state.

Using different dopant-ion combinations makes possible a wide variety of phosphor compositions having the same absorption but different emission wavelengths, thereby yielding unique colors. Figure 9.2 shows the unique emission spectra of nine different upconverting phosphor compositions. The emission characteristics that can be a set of unique wavelengths for a given phosphor compound are determined by the nature of the absorber and emitter ions as well as that of

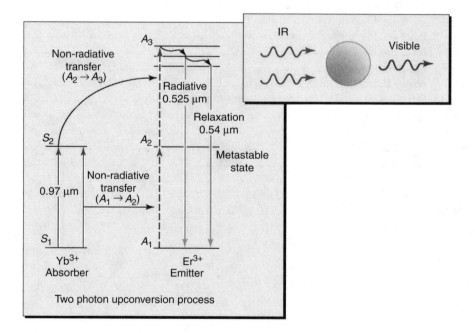

FIGURE 9.1 Two photon upconversion process in a YbEr phosphor.

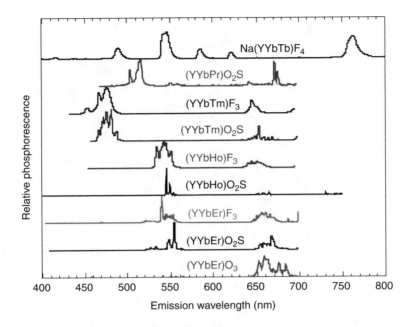

FIGURE 9.2 Emission spectra from nine different upconverting phosphor compositions.

the crystalline host and the ion densities. Some twenty, and perhaps more, unique phosphor compounds appear feasible. The narrow (typically 25–50 nm) linewidths of these emission spectra as well as the absorption linewidth, together with the wide separation between both, permit multiple analytes to be simultaneously detected in the same sample (i.e., multiplexed assays).

The emitted power varies nominally as the square of the excitation intensity as shown in Figure 9.3 for two different phosphor compounds. Saturation intensities on the order of 100 W/cm^2 are easily attainable with infrared diode lasers. These lasers are semiconductor devices produced by the same manufacturing technology currently used to fabricate the diode lasers used in compact disc players and for fiber-optic telecommunications. They are compact

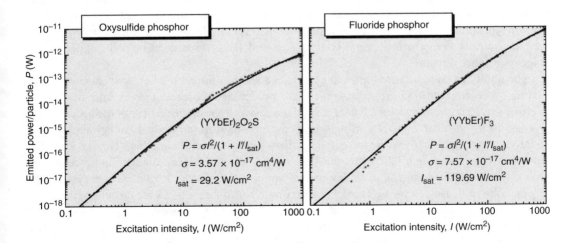

FIGURE 9.3 Power emission curves from oxysulfide and fluoride upconverting phosphor compositions.

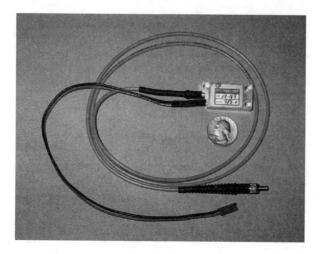

FIGURE 9.4 Compact 980 nm fiber-coupled diode laser excitation source.

(see Figure 9.4), high power (1 W), efficient electrical-to-optical converters (25%—4 W in yields 1 W out), have long lifetimes (greater than 10,000 h), can be fiber coupled, and are inexpensive. The noise-equivalent power specifications of readily available, reasonably priced, compact photo-multiplier detectors in the visible spectral region mean that single phosphor particles can be easily detected. This is an optical sensitivity threshold—the overall sensitivity of the assay where the phosphors are used will also depend on the assay efficiency, nonspecific binding (NSB), and other assay-related factors. The combination of compact laser diodes and photomultipliers enables the development of low cost and compact biodetection systems based on the upconverting phosphor technology.

Submicron microspheres of these materials are used as labels or reporters in sandwich immu-noassays (Figure 9.5). First, phosphor particles are coated with a biochemically inert substance such as silica. Antibodies (or other suitable ligands) are then conjugated to the silica surface at optimal densities using standard cross-linking chemistries. The resulting functionalized phosphor particles are used in a liquid sample to label captured target antigens with a unique phosphor color assigned for each different antigen. Target capture is achieved by immobilizing a set of different specific antibodies (again, one for each antigen) on a suitable substrate and exposing the substrate to the liquid sample. Illumination of the capture surface with near-infrared light results in upcon-version phosphorescence from captured, labeled targets. Because of numerous spectrally unique phosphor compositions, each type of captured target is labeled or color coded with a unique phosphor reporter.

The ability to do highly multiplexed assays on a single sample is a key advantage of upcon-verting over conventional reporters such as molecular fluorescent dyes and fluorescent microspheres. The second major advantage is that, because of the unique nature of the upconversion process itself, no other materials in nature upconvert—there is no optical background. Conse-quently, these materials can be detected in dirty environmental samples. This is a distinct advantage over fluorescent labels that must always be detected against the autofluorescent back-ground that is usually dominant in environmental samples. It is also an advantage over systems that rely on color or density changes. Third, because single phosphor particles are detected using diode laser excitation sources and conventional optical systems, small biosensors are possible that offer exquisite sensitivity. Finally, because their emission properties are a characteristic of the bulk ceramic material, the upconverting reporters are photochemically stable (i.e., they do not photo-bleach) and have long lifetimes (years). Hence, they are ideal reporters for use in the field for

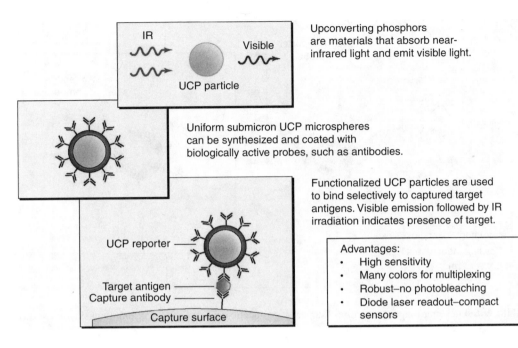

FIGURE 9.5 Overview of upconverting phosphor technology.

sensitive, real-time detection and identification of pathogens with minimum false alarms. Furthermore, their stability means that phosphor-based assays can be archived for subsequent analysis.

Phosphor particles are coated to obtain uniform surfaces with stable reproducible chemistry. A biologically active probe is then coupled (conjugated) to a coated phosphor particle to provide specificity to a particular antigen. To simultaneously detect more than one type of antigen (called multiplexing), different specific probes are conjugated to different unique color phosphors; as an example, the antibody for anthrax might be coupled to a blue emitting phosphor, and the antibody for plague might be coupled to a green emitting phosphor, and so on. The same laser wavelength (about 980 nm) when incident on the sample where these reagents are used would yield both green and blue spectral signals if both antigens are present. This multiplexing capability is the basis for high-throughput screening and is a critical step in decreasing the time to identify biowarfare or bioterrorism agents in the field. Both antibody and nucleic acid reagents have been produced (Figure 9.6).

A typical assay-based detection process (Figure 9.7) involves the use of capture surfaces prepared with capture sites for the target antigens. Upon introduction of the sample, target antigens are captured at these sites on a surface, followed by labeling with multiple colors of upconverting phosphors (with a unique color for each different antigen). The surface is then washed to remove any reporter-probe reagent not attached to the surface-captured antigens. Finally, a diode laser coupled with a photomultiplier or photodiode detector ascertains the presence of reporters through the detection of visible light emitted if a target antigen is present. Spectral discrimination, achieved through the use of a spectrometer or filters, determines which colors are present in the detected emitted visible light. Each unique color (may be a set of different wavelengths) that is measured means that a different antigen is present in the sample.

This review article primarily covers the immunoassay work performed at SRI International and focuses on biowarfare agent detection. The following sections review and discuss the optical properties of upconverting phosphors, the development of phosphor reagents and assays for biodetection, and some of the detector platforms used for assay readout.

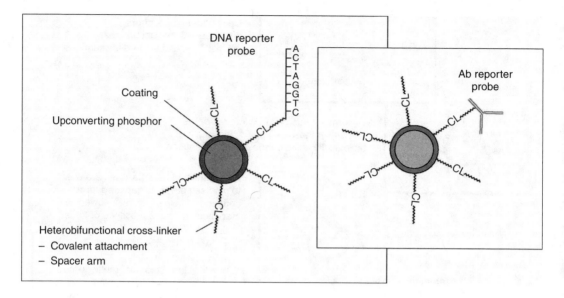

FIGURE 9.6 Phosphor reagent structure. Reagents for nucleic acid detection and immunoassay are shown.

9.2 SPECTROSCOPY AND DETECTION OF UPCONVERTING PHOSPHORS

9.2.1 Upconversion Process

One advantage of upconversion is the fact that the upconversion process is relatively uncommon, leading to very low background levels. Stokes Law states that fluorescence is emitted at longer wavelengths than the excitation light. For upconversion, also called anti-Stokes emission, the emitted light is at shorter wavelengths than the excitation light. That is, the emitted photons

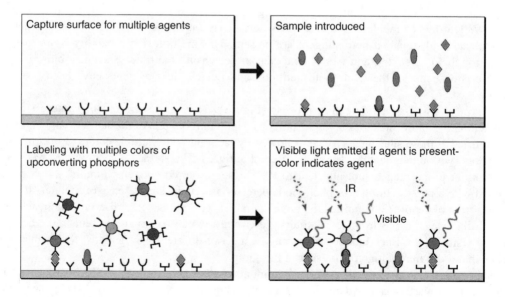

FIGURE 9.7 Schematic of the phosphor-based detection process. A sandwich assay format is used to capture and label specific biological agents.

have higher energy than the excitation photons. The extra energy can be provided through the conversion of two or more excitation photons for each emitted photon, such as occurs for upconverting phosphors, multi-photon fluorescence, coherent anti-Stokes Raman scattering (CARS), and harmonic generation. Of the materials that do not require prior energy storage, only upconverting phosphors provide good efficiency for moderate excitation intensities (less than 1 kW/cm^2). The primary reason for this is that the efficient lanthanide-based upconversion processes do not involve virtual states. That is, each input photon is absorbed into a real physical state of the lanthanide.

There is another type of upconverting phosphor based on electron traps that is used in cards for viewing infrared beams. The extra energy for upconversion comes from visible or other short-wavelength radiation that promotes electrons into the electron traps where they can subsequently emit visible radiation when stimulated by infrared light. Because continuous exposure to infrared light depletes the electron traps, these materials require frequent recharging with short wavelength light.

A number of processes can lead to upconversion in lanthanide materials [8–11]. An example of upconversion involving energy transfer is shown in Figure 9.8. This figure shows upconversion in materials containing high concentrations of ytterbium and erbium. The light is absorbed by the ytterbium (the sensitizer ion) and emitted by the erbium (the activator ion). Because the density of ytterbium and erbium ions is high, the ion–ion separation is small, and nonradiative energy transfer can occur much more rapidly than radiative transfer. When the phosphor is illuminated by an intense infrared light source near 980 nm, many excited state ytterbium ions are produced. A first energy transfer from ytterbium to erbium leads to excitation of the $^4I_{11/2}$ level in erbium. A subsequent ytterbium to erbium energy transfer leads to excitation of the $^4F_{7/2}$ level with a total energy of roughly twice that of the excitation photons.

The second excitation step in Figure 9.8 involves near-resonant nonradiative energy transfer. A number of additional processes can lead to the second excitation step including direct absorption of a second photon (excited state absorption or stepwise excitation), radiative energy transfer, nonresonant energy transfer or phonon-assisted energy transfer (wherein creation or destruction of the photon makes up the difference in energy of the activator and sensitizer ions), cooperative sensitization and luminescence (that are relatively weak processes), and cross relaxation nonradiative energy transfer (involved in the photon avalanche process) [12].

9.2.2 SPECTROSCOPIC PROPERTIES

Ytterbium can serve as a sensitizer for a number of activator ions including erbium, holmium, thulium, praseodymium, and terbium. Examples of emission from Yb^{3+}–Er^{3+}, Yb^{3+}–Ho^{3+},

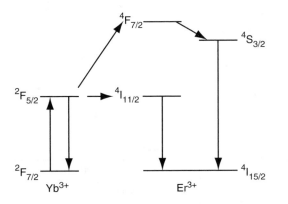

FIGURE 9.8 Energy diagram for upconversion with the Yb^{3+}–Er^{3+} pair.

Yb^{3+}–Pr^{3+}, and Yb^{3+}–Tm^{3+} are shown in Figure 9.9. The use of a common sensitizer ion allows excitation of different phosphor types using the same excitation wavelength, near 980 nm for ytterbium. The strongest emission from Yb^{3+}–Tm^{3+} occurs near 800 nm, although this is not shown in the figure. The 800 nm process requires two excitation photons for each emission photon, and the blue emission in Figure 9.9 requires three photons. Another three-photon process involves excitation of erbium near 1.5 μm with green emission where the erbium acts as both the sensitizer and activator.

Upconversion tends to be more efficient for low phonon energy materials because nonradiative loss processes are less favored. Exceptions to this apply when phonons are needed to compensate for the energy difference in nonresonant energy transfer. Very low phonon energies can be obtained with halides as anions, and fluorides are characterized by high efficiency. The heavier halides such as chlorides and bromides have even lower phonon energies. However, these are not widely used because they tend to be hydroscopic. For biological applications, oxysulfides are often used because they can readily be fabricated as submicron monodisperse particles, whereas fluorides tend to fuse into larger particles during activation of the phosphors.

The host material also influences the emission and excitation spectra of the upconverting phosphors. The more significant changes occur with different anions. The influence of the anions on emission from Yb^{3+}–Er^{3+} is shown in Figure 9.10. The shape of the emission profiles and the ratio of green to red emission dramatically change.

The excitation spectra of ytterbium-sensitized phosphors in an Y_2O_2S host are shown in Figure 9.11. Note that the excitation maxima have the same position for all activator ions. This is beneficial when different lanthanide pairs are used for multiplexing because all of the phosphors will have the same optimal excitation wavelength. The excitation spectra of the Yb^{3+}–Er^{3+} pair has additional spectral features as a result of the presence of absorption resonances in erbium that also occur near 980 nm, but this has little influence on the excitation peak. When using host materials with different anions, significant changes in the excitation spectra occur as shown in Figure 9.12.

The absolute emitted power per particle for 0.3-μm oxysulfide upconverting phosphors is shown in Figure 9.13. At low intensities, the emitted power increases with the square of the excitation intensity. That is, each curve has a slope of 2 when the logarithm of the emitted

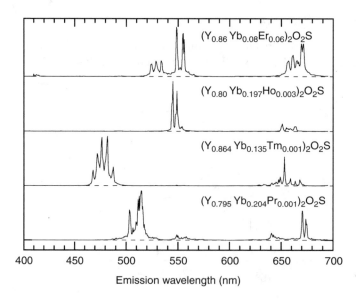

FIGURE 9.9 Emission spectra for erbium, holmium, thulium, and praseodymium in Y_2O_2S.

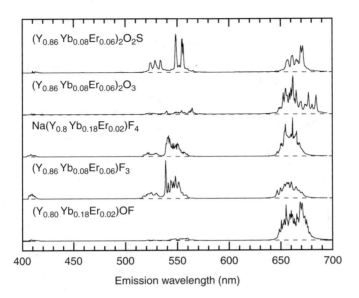

FIGURE 9.10 Emission spectra for Yb^{3+}–Er^{3+} pair in host materials with different anions.

power is plotted versus the logarithm of the excitation intensity. At higher intensities, the emitted power is linearly proportional to the excitation intensity (a slope of 1 in the log–log plot of Figure 9.13).

9.2.3　Time Response

One particularly unusual aspect of upconverting phosphors is that the temporal response depends on the excitation intensity. This behavior is shown in Figure 9.14 that shows the intensity dependence of the rise and fall times of the Yb^{3+}–Er^{3+} pair. Each point represents the $1/e$ rise or fall time when the excitation light is turned on or off in a discontinuous (step function) fashion. As the intensity is

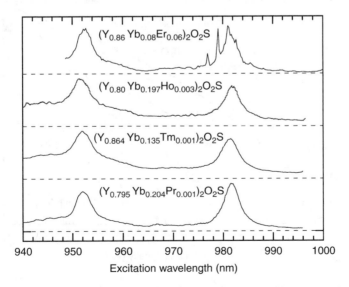

FIGURE 9.11 Excitation spectra for erbium, holmium, thulium, and praseodymium in Y_2O_2S.

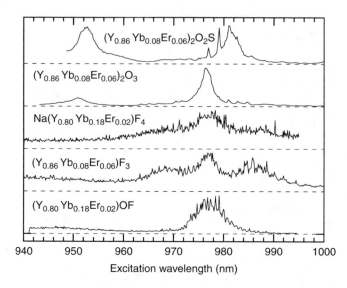

FIGURE 9.12 Excitation spectra for $Yb^{3+}-Er^{3+}$ pair in host materials with different anions.

increased, both the rise and fall times decrease. This behavior is very uncommon. The intensity-dependent fall time is particularly surprising as one would expect that once the excitation source is turned off, the emission decay would not depend on how the excited state was produced. The rise and fall times start to decrease for excitation intensities of a few W/cm^2, the same level where the intensity dependence of the emitted power dependence changes from quadratic to linear (see Figure 9.13). The reason for this behavior is understood from examination of the rate equations for upconversion and the dominating mechanism for loss of the energy from the excitation light. At low intensities, the primary loss is from decay of the singly excited states, and the time response is

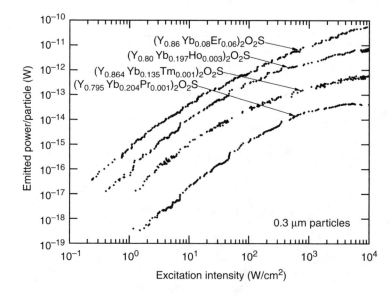

FIGURE 9.13 Absolute emitted power for different phosphor compositions as a function of excitation intensity.

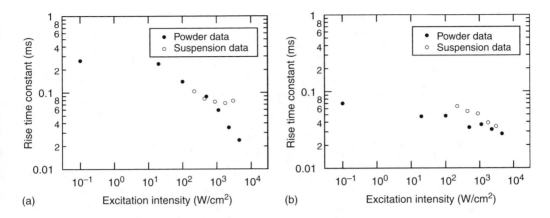

FIGURE 9.14 Variation in rise and decay times for green emission from $(Y_{0.86}Yb_{0.08}Er_{0.06})_2O_2S$.

determined by singly excited levels (the ytterbium $^4F_{5/2}$ level and erbium $^4I_{11/2}$ level in Figure 9.8). At high intensities, the number of doubly excited states becomes large enough that the primary energy loss occurs through decay of the doubly excited states (level $^4S_{3/2}$ populated through level $^4F_{7/2}$ of erbium). Because the decay rate of the doubly excited states is faster than the decay rate of singly excited states, the high-intensity decay rates are faster than the low-intensity decay rates. This same transition from singly excited states to doubly excited states also causes the change in the intensity dependence of the emitted power. When singly excited states dominate, the number of doubly excited states scales with the square of the number of singly excited states. Because the emitted power scales with the number of doubly excited states, the emitted power scales quadratically with the input intensity. In the high-intensity regime, the number of doubly excited states scales directly with the input intensity, and the output power scales linearly with the input intensity. The crossover for temporal response and emitted power both occur at the intensity for which the dynamics change from singly excited state dominance to doubly excited state dominance.

9.2.4 INSTRUMENTATION

The unusual properties of the upconverting phosphors have important implications on the instrumentation for their detection. Consider a light source focused to an area A. For conventional fluorophores with emission that linearly depends on the excitation intensity, focusing does not increase the total signal. Although the excitation rate for an illuminated fluorophore increases as $1/A$, the number of illuminated fluorophores decreases with A, leading to no net change in signal with focusing. The behavior for nonlinear fluorophores such as the upconverting phosphors is different. If the emission is proportional to the square of the intensity, the total signal increases as $1/A^2$ as the beam is focused more tightly. Although the number of illuminated phosphor particles decreases with A, the net signal increases as $1/A$ as the beam is focused. Thus, larger signals are obtained in tightly focused geometries. To obtain signal from a given area, much larger signals are generally obtained by scanning a focused beam over the area rather than using uniform illumination over the whole area. This generalization is no longer valid when the excitation beam intensity is high enough that the emission is linear (see Figure 9.13) or at high scan rates where the illumination times are short compared with the phosphor rise and fall times.

Upconverting phosphors have longer characteristic response times than other fluorophores such as fluorescent dyes. The decrease in the rise and fall times of upconverting phosphors with increasing intensity (Figure 9.14) makes higher intensity excitation advantageous for applications requiring faster time response such as flow cytometry [13,14].

Diode lasers are good light sources for excitation of upconverting phosphors as they provide moderate power ($\sim$ 1 W) from a compact package with good efficiency. The common wavelengths for exciting upconverting phosphors fortuitously happen to be popular wavelengths for optical communications (980 nm is used for pumping erbium fiber amplifiers and 1.55 μm is the location of the minimum-loss window for silica-based optical fibers). Near-infrared semiconductor laser systems can emit luminescence at shorter wavelengths in the visible spectral range. It may be necessary to filter out such luminescence to obtain the best performance. On the detection side, a notch filter may be used to block the near-infrared excitation light.

Although upconverting phosphors are excited in the near-infrared spectral region, detection is performed at shorter wavelengths. This is significant because detectors tend to have better performance in the visible spectral region and in the short near-infrared wavelengths out to approximately 1 μm. The photocathode materials used in photomultipliers and image intensifiers have better quantum efficiencies and lower background currents in this range. Similarly, silicon-based detectors have good performance in this region. With sensitive and low noise detectors and no upconverted background light in most situations, instrumentation can readily detect single upconverting phosphor particles. This means that assay performance is generally limited by factors such as NSB and probe affinity rather than the optical label and instrumentation.

9.3   REAGENT DEVELOPMENT

9.3.1   INTRODUCTION

An upconverting phosphor reagent is defined as a phosphor particle that has been coated for aqueous stability and functionalization followed by attachment of the biological probe. The process used in reagent preparation involves the following steps: (1) synthesizing the phosphor particle, (2) applying a coating to the phosphor surface, (3) attaching the biological probe (e.g., antibody), and (4) evaluating the finished reagent (Figure 9.15). Although the ceramic upconverting phosphor particles are considered to be very stable, passivation (silica coating) of the phosphor surface blocks exposure of the phosphor core to the aqueous solution and prevents possible lanthanide leaching. After passivation, the phosphors are coated with a silane having functional groups to which the antibody probes are attached by way of a crosslinker.

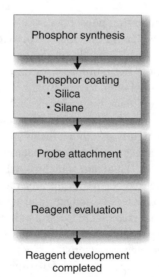

FIGURE 9.15 Reagent development flowchart.

Additionally, the functional groups, along with the crosslinker, provide tailorable dispersability of the phosphors in aqueous buffers. The crosslinker, the associated coupling conditions, and the blocking protocols are empirically determined by screening a number of phosphor reagents for best performance in an actual assay. Covalent attachment of the probe provides the strongest link to the phosphor particle. Alternatively, a phosphor reagent can be prepared by passively adsorbing the capture probe (i.e., antibody) onto the phosphor surface at any step after synthesis is complete.

This section describes the synthesis of submicron monodisperse phosphors, the sequential coating of phosphors with silica and an organoalkyoxysilane, and finally, the coupling of the probe to the phosphor with a crosslinker to complete the phosphor reagent. The stability of the finished phosphor reagent and its application to the assay device is discussed in Section 9.4.

9.3.2 PHOSPHOR SYNTHESIS

A goal in the development of the upconverting phosphor reagent was to obtain submicron, mono-disperse particles of uniform brightness. Initial efforts to prepare downconverting phosphors for use in an assay by ball milling larger crystals produced a polydisperse mixture with variable morphology [3]. It was also found that the luminescent output of the phosphors was diminished because of defects introduced to the crystalline structure of the phosphors during grinding [3,15]. Hence, this group was motivated to develop a synthesis process that would achieve its goal and also be scalable to the production of phosphors to support commercial development. Phosphors were made predominantly from yttrium oxysulfide doped with ytterbium and erbium; however, phosphors doped with holmium, thulium, or praseodymium in place of erbium were also used when more than one color was needed in an assay. Implementation of the process begins with the formation of spherical submicron phosphor particles by precipitation followed by heating to convert the material to an oxide. Conversion of the oxide to the oxysulfide was achieved by the addition of a flux and additional heating. The final step is the rearrangement of the crystal structure (particle activation) to achieve strong upconversion. Particle sizing was done after synthesis to characterize the phosphor size distribution.

9.3.2.1 Submicron Phosphor Particle Manufacturing

This group's method for preparing ytterbium–erbium oxysulfide phosphors [16] is summarized here; note that preparation of the other phosphors doped with different lanthanides follow a similar protocol. To produce the yttrium–ytterbium–erbium oxysulfide phosphors $(Y_{0.86}Yb_{0.08}Er_{0.06})_2O_2S$, the lanthanide nitrate salts were first stoichiometrically mixed together in aqueous solution. Phosphor particles were then formed by precipitation of the lanthanide ions from solution as a hydroxycarbonate using an excess of urea [17]. Next, the hydroxycarbonate phosphor precursor particles were converted to the oxide in a fluidized-bed reactor at 750°C in air. The oxide phosphors were converted to the oxysulfide form by resuspending the particles for 4 h using a gas mixture of H_2S, H_2O and N_2 at 850°C in the reactor. Finally, the oxysulfide phosphors were activated to increase their luminous efficiency by fluidizing the particles with argon for 30 min at 1450°C. The advantage of the fluidized bed is that, during the reaction, the particles are always moving with enough energy to minimize aggregation. The final result was the production of substantially mono-disperse, 0.4-μm-diameter phosphor particles (Figure 9.16). Oxysulfide phosphors with different emitter ions were prepared in a similar manner with substitution of the appropriate lanthanide ion for the emitter. Activated phosphors were stored desiccated under argon until later use.

Phosphor size is determined by the reaction conditions during the urea precipitation step. For some applications, smaller phosphors are desired because their increased diffusivity should improve assay kinetics. Preliminary studies using a combination of increased urea concentration and a decrease in the concentration of the metal ions during precipitation has shown that 200 nm phosphors can be synthesized [18] and that additional size reductions may be achievable using this approach.

FIGURE 9.16 Scanning electron micrograph of synthesized upconverting phosphors.

9.3.3 PHOSPHOR COATINGS

Synthesized phosphors without further surface treatment are normally hydrophobic. Left untreated, phosphors will agglomerate when suspended in aqueous buffer, a serious impediment for their use as an assay label. By coating the phosphor, the surface can be modified to increase hydrophilicity and tailor the dispersibility of the phosphors in common assay buffers. Additionally, functional groups can be introduced into the coating to allow both covalent and noncovalent attachment of the probe. A double coating process is used to prepare the phosphors for the functionalization step. A silica layer is first applied to passivate the phosphor followed by a silane layer to modify the surface for crosslinking.

9.3.3.1 Silica Coating

After investigating several approaches for applying the silica-based passivation layer, solution-based and chemical vapor deposition (CVD) techniques were selected because they provided the best coverage of the phosphor surface. The best results were obtained with a solution-based method to produce a coating of amorphous silica using established wet chemical methods [19]. The process involves the controlled hydrolysis condensation of a tetraethoxysilane in an alcohol suspension using a base catalyst to form the coating. The coating is then cured under mild heating to produce an amorphous silica layer nominally 20–30 nm thick (Figure 9.17) as measured by transmission electron microscopy (TEM) that completely covers the surface of the phosphor. Surface coverage was assessed by surface analysis methods, including Auger electron spectroscopy, energy dispersive x-ray (EDX) spectroscopy on a scanning electron microscope (SEM), x-ray fluorescence, and x-ray diffraction.

9.3.3.2 Silane Coating

The silica-coated phosphor is then treated with an organoalkoxysilane to generate the functional sites to be used in the covalent attachment of crosslinkers and, subsequently, antibodies. Silanes having amino, carboxyl, and sulfhydryl groups are typically used. The phosphor is coated with

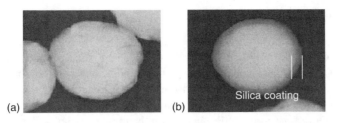

(a)   (b)

FIGURE 9.17 Transmission electron micrograph of synthesized upconverting phosphors (a) as prepared and (b) after silica coating. The thickness of the silica coating in (b) is 20–30 nm.

3-aminopropyltriethoxysilane using a protocol similar to published reports [20]. This reagent generates a surface with an average functional density of amines of 3.4 amines/nm^2 as detected using a colorometric analysis [21].

9.3.4 FUNCTIONALIZATION WITH CAPTURE PROBE

Probes can be attached to the phosphor either by adsorption or by covalent linkage. Whereas adsorption is the simplest approach, a covalent bond between the probe and the phosphor provides for increased stability and, in many cases, better control of the probe orientation. The field of bioconjugate chemistry is well developed [22], and conjugates with a variety of functional groups and lengths are widely available from multiple vendors. The choice of crosslinker and the method for attaching the phosphor to the antibody will depend on both the antibody and the target to be detected. In general, longer crosslinkers decrease steric hindrance between the antibody and the target [23], making it easier for the antibody to bind to large targets. Although it is possible to modify antibodies for linking (e.g., modification of an amine to a thiol or oxidation of the sugar on the Fc part of the antibody), this group has found that the best crosslinking strategies do not modify the antibody before it is attached to the crosslinker. In this case, the crosslinker should have an amine reactive group (i.e., carboxylic acid or aldehyde) to bind to the antibody.

9.3.4.1 Phosphor Reagent Production

The amine surface on the phosphor is converted to an aldehyde by reaction with terephthaldicarboxaldehyde (TPDCA), a rigid dialdehyde. At this point, antibodies are coupled to the surface of the particle via a spontaneous reaction between the aldehyde on the particle surface and an amine on the antibody followed by subsequent reduction of the Schiff base with sodium borohydride. To increase the antibody distance from the particle surface, a heterobifunctional polyethyleneglycol (PEG) crosslinker is used as an alternative to coupling the antibody directly to the TPDCA. The PEG crosslinker has an amine on one terminus that binds to the TPDCA and a carboxylic acid on the other terminus (PEG3400, Shearwater Polymers) [5]. The carboxylic acid is then converted to a succinimide ester using 1-ethyl-3-[3-dimethylaminopropyl]carbodiimide hydrochloride (EDC) and *N*-hydroxysulfosuccinimide [24]. The succinimide ester-modified linker then spontaneously reacts with an amine on the antibody. The antibody surface density on the phosphor was set to between 10,000 and 40,000 antibodies per square micrometer by diluting the antibody with bovine serum albumin (BSA). The upper limit for antibody density on the phosphor surface was estimated to be approximately 44,000 molecules/μm^2 from a calculation of the radius of gyration assuming a molecular weight of 150 kDa for an IgG antibody. After antibody conjugation, any remaining active sites on the phosphor are neutralized by the addition of an amine-containing protein such as BSA or a small hydrophilic amine-containing molecule such as tris(hydroxymethyl)aminoethane. Dextran has also been used as an alternative to PEG for crosslinking [5].

9.3.4.2 Phosphor Reagent Testing

The optical density of the phosphor is routinely measured during the production process to monitor for loss of material. Additionally, several rapid tests can be done to check the phosphor reagent before it used in an actual assay. The amount of antibody bound to the phosphor surface can be quantified using a fluorescently labeled antibody that is reactive to the species of the bound antibody and detects the complex by flow cytometry. By substituting a labeled target for the antibody, the number of functional antibody sites on the phosphor can be determined. Alternatively, a microtiter plate format can be used in place of the flow cytometer to check reagent functionality.

9.4 ASSAY DEVELOPMENT

9.4.1 INTRODUCTION

Assays using microscopic particles as labels such as upconverting phosphors can have greater sensitivity than molecular labels because each particle is the equivalent of hundreds to thousands of individually luminescent entities. Particulate labels are capable of providing signal amplification without the need for enzymes. Additionally, because upconversion produces a signal that is higher in energy than the light used to excite the label, assay detection is done against a background that is essentially zero. However, assay development using particulate labels, including upconverting phosphors, requires that attention be paid to several issues as discussed by Hall et al. [25] who identified three parameters regarding particulate labels that need to be considered during assay development. First is that low NSB is critical if the full advantage of signal amplification from particle labels is to be realized. Second, in accordance with mass-action law, the concentration of the labels used in the assay should be as high as possible. Third, they found that, contrary to conventional assay theory, there is an optimum antibody density for the surface of the label that may be less than the monolayer packing density. Although Hall et al. experimentally confirmed their findings with fluorescent microspheres, similar results have been viewed with upconverting phosphors.

9.4.2 ASSAY FORMATS

Although upconverting phosphors can, in principle, be used as labels in any type of immunoassay, they have predominantly been used in sandwich and antigen competition assays [26]. In general, sandwich assays require targets that have multiple epitopes for capture and labeling whereas competitive antigen assays, because only a single epitope is required for capture, favor small targets such as organic molecules. Sandwich assays, because they label bound target, are inherently more sensitive than competitive assays [27]. Using upconverting phosphors exploits the sensitivity advantage of sandwich assays because phosphors can be detected against a dark background. However, a multiplexed competitive assay for the detection of drugs of abuse using upconverting phosphors has been described [6]. This section focuses on sandwich upconverting phosphor assay development.

Upconverting phosphors have been used with several assay formats, including the lateral flow assay (LFA), microspheres (both polymer and magnetic), and microtiter plates. The LFA is a simple format that is inexpensive to produce and easy for a minimally trained person to use. They are the technological basis behind the home pregnancy test. Microspheres provide assay kinetics that approximate a solution-phase assay and can be separated easily from solution when used in heterogeneous assays. An additional advantage is that microspheres can be analyzed on a single-particle basis by flow cytometry. Microtiter plates represent a well-established assay format made popular by enzyme-linked immunosorbent assay (ELISA). Advantages include the availability of materials and protocols for assay development. A single upconverting phosphor reagent

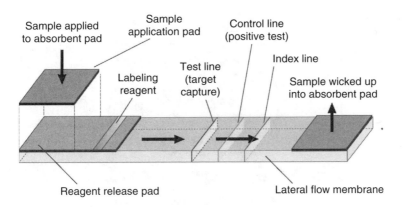

FIGURE 9.18 Schematic view of the lateral flow strip. The major components of the strip are (left to right) sample application pad, release pad, nitrocellulose membrane, and absorbent pad. Capture antibodies specific to the target and the antibodies on the upconverting phosphor particles are striped at the test and control lines, respectively. Phosphors comprising the index line are used as a fiduciary point when the wick is scanned by the reader. UCP reagent is deposited on the release pad and lyophilized before assembly onto the strip.

can be used with all of the above formats because they have in common a solid phase that is used to capture the target of interest.

Development of bead-based [14] and microtiter plate upconverting phosphor assays [5] have been previously described. This review primarily focuses on the LFA format. Figure 9.18 is a schematic of the structure of the lateral flow test strip this group designed and developed to detect a single pathogen target [5]. The test strip consists of a sample application pad, a reagent release pad (containing dried phosphor reagent), a nitrocellulose lateral flow membrane, an absorbent pad, and backing material. All materials for LFAs (except for the upconverting phosphor labeling reagents) are commercially available (e.g., Millipore, Scheicher & Schuell). The lateral flow membrane contains three distinct zones: a test line (*T*), a control line (*C*), and an index line (*I*). The test line contains antibody, specific to the target analyte of interest, adsorbed onto the nitrocellulose material. If target is present in the sample, it will be captured and labeled by upconverting phosphors at this position. The control line is designed to capture any phosphor particles that flow through the wick and serves as a test control (i.e., it indicates that the lateral flow test has worked properly). It is prepared by depositing antibodies on the nitrocellulose membrane that selectively binds to the antibodies crosslinked on the phosphor reporter particles. The index line contains a known amount of phosphor bound to the nitrocellulose membrane and serves as a detector index line (for positioning of the optical read head) and also for signal calibration.

Simultaneous detection of multiple targets is possible using upconverting phosphors in all three above-mentioned assay formats. Each target is identified by a unique phosphor color. Additional selectivity is achievable with the LFA and microsphere formats by capturing the target at a unique spatial location (LFA) or with a unique microsphere size. For example, up to nine targets could be identified in a microsphere assay format that combines three microsphere sizes with three phosphors.

9.4.3 Developing an Upconverting Phosphor-Based Assay

An overview of the assay development process is shown in Figure 9.19 using the LFA as an example. The sequence of steps begins with the deposition of the functional phosphor reagent to the release pad. Next, capture antibodies are striped on the selected nitrocellulose membrane and tested with the phosphor reagent. Once the phosphor reagent flows through the membrane with minimal NSB, target is added to the strip, and the assay is checked for sensitivity and specificity.

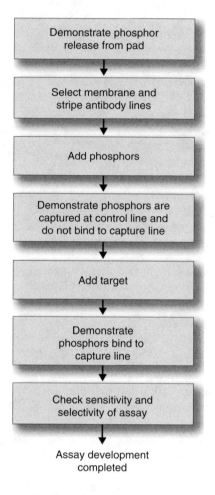

Demonstrate phosphor
release from pad

Select membrane and
stripe antibody lines

Add phosphors

Demonstrate phosphors are
captured at control line and
do not bind to capture line

Add target

Demonstrate
phosphors bind to
capture line

Check sensitivity and
selectivity of assay

Assay development
completed

FIGURE 9.19 Flowchart of the assay development process.

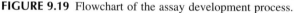

Additional assays are added to the strip in a sequential fashion to create a multiplexed test strip. Throughout the assay development process, feedback loops are followed in an iterative fashion to optimize the assay performance.

A well-performing immunoassay begins with the selection of antibodies. Antibody requirements include sensitivity and specificity for the target of interest. Because a solid-phase is used for both capture and detection of target, monoclonal or affinity-purified polyclonal antibodies optimized for use in ELISA are preferred. If possible, at least one of the antibodies should be a monoclonal that is preferentially attached to the phosphor. This orientation is preferred because crosslinking a monoclonal antibody to the phosphor should be more consistent (i.e., reproducible) compared with the linking of a polyclonal antibody because of the homogeneity of a monoclonal antibody preparation [26]. However, any antibody pairing must be empirically tested to determine the best orientation for use in any assay. Antibody selection is facilitated when there is a library of antibodies (especially monoclonals) from which to choose.

The phosphor reagent is deposited in a dry-down buffer (DDB) on the release pad. The buffer components must protect the phosphor reagent while dry and allow for good release when the reagent is rehydrated. Typical buffer compositions include large amounts of sugars such as sucrose (up to 20% w/v) that retard denaturation of antibodies [28] attached to the phosphors. This group has established a process to evaluate DDBs and release pad materials (such as glass fiber and

Accuflow-G). Release pads must effectively wick the sample without binding a significant amount of antigen or introducing interfering substances. As part of the assay development process, our group routinely (1) determines the amount of striped phosphor reagent per assay, (2) adjusts the DDB formulation and lyophilization process, and (3) optimizes blocking and sample treatment buffer as necessary to improve release of the dried reagent. Phosphor reagent was diluted in DDB composed of 10 mM borate buffer (pH 9.0) supplemented with 5% (w/v) BSA (Fraction V) and 20% (w/v) sucrose. The phosphor reagent was then striped onto Accuflow G release pads (Schleicher and Schuell) at 2 μL/s and dried for 1 h at 37°C to give a final phosphor weight of 200–400 ng per LFA strip, depending on the assay target.

The preparation of LFA strips is a multi-step process [5]. First is the selection and evaluation of the nitrocellulose membrane material. Membranes must have uniform flow and give sensitive detection with low variability on an intra-lot and lot-to-lot basis. Second is the striping of the membrane with the test and control line antibodies. Capture antibodies are deposited on the membrane in a buffer that facilitates adsorption of the protein on the nitrocellulose material. In some cases, the antibodies are striped in a buffer containing BSA, a protein that minimizes NSB when bound to the test line. Appropriate antibodies are diluted to 1–3 mg/mL in a 50 mM phosphate buffer (pH 7.5) along with 10 mg/mL BSA as a blocking and spacing agent. The antibody solutions are then dispensed onto a nitrocellulose card (40×200 mm HF135 material, Millipore) using an IVEK microstriper at a dispensing rate of 1 μL/cm. Additionally, an index line consisting of phosphors is striped 4 mm distal to the last antibody line. The index line is used by the biosensor to determine the location of the target capture and control line peaks from the scan data. The cards are then placed in an oven and dried 1 h at 37°C. Third is the selection of the absorbent pads at both ends of the strip. The application pad is selected based on its ability to absorb and regulate the flow of the applied test sample with minimal interference. The pad is soaked in running buffer (RB) containing 10 mM Hepes, 135 mM NaCl, 5 mM ethylenediaminetetraacetic acid (EDTA), 1% (w/v) BSA, 0.5% (v/v) Tween-20, and 0.2% (w/v) NaN_3. It is then lyophilized before it is attached to the other wick components. The RB components are released during rehydration to dynamically block the membrane strip during the assay. The absorbent pad at the end of the test pad is selected and sized according to the amount of fluid that is to flow through the strip. Finally, all of these components are laminated together and cut into strips before use.

Phosphor migration through the membrane is governed by the membrane pore size, membrane wetting agents (added during the membrane manufacturing process and also as part of the assay), quality of the phosphor reagent, the interaction of phosphors with the membrane, and the composition of the assay buffer. Flow agents such as sugars are used in the DDB to help the phosphors move through the membrane. Nitrocellulose membranes are sometimes pre-treated with 1% polyvinyl alcohol (PVA) as a blocking agent. The assay buffer contains additional blocking agents such as proteins and surfactants. Blocking agents such as BSA minimize binding of the phosphors to the test line in the absence of specific target by adsorbing to hydrophobic regions, crucial for reduction of background while retaining sensitivity. Additional blocking of NSB is provided by surfactants in the buffer (e.g., Tween-20).

In general, an empirical process is followed to determine the optimal assay conditions. This group's approach is to use a formal development process to quantitatively evaluate key variables in a pair-wise fashion over a realistic variable range. The development process is similar to the one previously discussed for phosphor reagent optimization. Table 9.1 lists several of the assay variables and a typical range over which optimization is performed.

A typical optimization experiment is performed by constructing a test matrix to compare two variables such as antibody concentration vs. crosslinker at several points (i.e., concentration of reagent, crosslinker type). Additional experimental matrices that would be generated include those investigating the pH of antibody coupling, type and concentration of blocking agent, type and concentration of quenching agent, and so forth. Initial optimization is performed over a wide range in the parameter space; subsequent experiments narrow down the parameter space to find

TABLE 9.1
Several Assay Optimization Variables and their Expected Range

Variable	Range
Membrane blocking agent (e.g., PVA)	0%–1% (w/v)
Blocking agent in assay buffer (e.g., BSA)	0.5%–5%
Surfactant (e.g., Tween-20)	0.01%–1%
Phosphor reagent on release pad	0.1–3 μg
Antibody striped on nitrocellulose	0.5–3 mg/ml

the optimum operating point. Optimization experiments are performed in the lateral-flow format; however, the phosphor reagent is initially applied to the release pad just before the experiment is started. Only later in the development process is the phosphor reagent lyophilized before use. The end result is the preparation of a phosphor reagent optimized for the detection of a specific target.

Solution stability of upconverting phosphor reagents depends on buffer composition [3]. Assay performance is impaired under conditions of reagent aggregation. This group has found that Hepes buffer, in comparison with phosphate buffers, minimizes phosphor agglomeration in solution [5]. The stability (or lifetime) of the phosphor reagent is determined, in large part, by the stability of the attached probe. In many cases, such as for the LFA, the reagent is applied to the assay device and dried before use for long-term stability. Excipients such as sucrose and BSA that are added to the reagent as it is dried down on the release pad, coat the phosphor reagent to prevent irreversible attachment of the phosphor to the release pad and aid in rehydration. Proper release of the phosphors is essential in preventing reagent pileup as the phosphors make the transition from the pad into the membrane. Phosphor pileup introduces variability in assay performance by restricting the flow of reagent and target past the test lines in the membrane strip. Pileup can be minimized by careful selection of the release pad, preparation of the phosphor reagent, and optimization of the dry-down and assay buffers. Additionally, LFA devices containing phosphor reagents are packaged in hermetically sealed bags containing desiccant and filled with inert gas to decrease reagent oxidation. Using these procedures, an assay shelf life of greater than six months when stored at room temperature and greater than a year when stored at 4°C has been achieved.

9.4.4 ASSAY ANALYSIS AND RESULTS

An important part of assay development is the development of algorithms to process the phosphor signal as the strip is scanned. The simplest approach is to integrate the area under the test line peak and to fit that data to a calibration curve to determine the concentration of target present. Whereas measurement of only the test line is sufficient for well-behaved assays, a more robust approach is to divide the area of the test line peak by the area under both the test and control peaks to normalize the signal to the amount of phosphor label that moved through the strip. Using the latter approach, the detection of target is treated as a hypothesis test between H_0, the hypothesis that the data represent only NSB, and H_1, the hypothesis that specific binding has occurred. In the case of the LFA for a single agent, it has been empirically found that under H_0, the test and control peaks, T and C, after baseline subtraction, are well represented by the linear model

$$T_k = a + bC_k + n_k \tag{9.1}$$

where k labels a data sample collected under the NSB condition, a and b are model coefficients, and n_k are model residuals assumed to be independent, and identically distributed as (approximately) normal with zero mean and variance σ^2. The model parameters a, b, and σ^2 are estimated from the

NSB data by standard linear regression analysis. Using the Neyman–Pearson detection criterion [29] to test for the presence of target binding at the test line leads to a test that rejects H_0 for

$$S = |T - (\hat{a} + \hat{b}C)| > \tau \tag{9.2}$$

where τ is a threshold chosen to provide a specified false detection probability P_{fa} as

$$P_{fa} = \frac{1}{\sqrt{2\pi\sigma^2}} \int_{\tau}^{\infty} \exp\left[-\frac{1}{2\sigma^2}(T - (\hat{a} + \hat{b}C)^2)\right] dT \tag{9.3}$$

This detection approach has worked well for data sets analyzed by us for a variety of bioagent assays.

Scan data for each lateral flow strip collected by the reader unit are analyzed by first identifying the test (T) and control (C) lines as well as a region between the two lines representing background phosphorescence. An average of five points at each line is taken and averaged. Equation 9.2 is then used to determine if the assay result exceeded the detection threshold τ. If so, a positive event is declared. Otherwise, the test result is considered negative. The test statistic, S, is computed in units of standard deviations from the NSB data. Thus, a test statistic of 2 indicates the result is 2σ above the null case.

An upconverting phosphor-based assay was developed for the detection of human chorionic gonadotropin (hCG) in an LFA format to demonstrate the potential of upconverting phosphors as a sensitive label. Commercially available antibodies having high affinity ($K_d \sim 10^{-11}$ M) and specificity to either the α or β chain of hCG were used in the assay. As shown in Figure 9.20, the group was able to detect hCG at concentrations as low as 10 pg/ml from a 100 μl sample. Also, the signal at the target line monotonically increased as the hCG concentration was increased from 10 pg/ml to 10 ng/ml. Detection sensitivities achieved with the phosphor labels were 10–100-fold greater in comparison to other labels such as colloidal gold [5].

This laboratory has also developed an upconverting phosphor-based lateral-flow-based assay for the detection of *Bacillus anthracis*. Figure 9.21 shows scans of test strips dosed with either 1×10^6 cfu/ml of *B. anthracis* spores or with no spores. Strips without target showed very little binding of phosphor reporter to the test line (the rightmost peak in Figure 9.21) but were captured at the control line, as expected, indicating that phosphors moved through the strip. When *B. anthracis* spores were present, strong binding of the phosphor reporter was seen at both target capture and control lines. This set of data demonstrates that micron-sized targets (i.e., *B. anthracis* spores) can flow through the nitrocellulose membrane and be detected at the test line with the phosphor labels.

Upconverting phosphors have been used to simultaneously detect two targets. In a previously reported experiment [5], two different phosphor colors were used to detect mouse IgG and ovalbumin in a LFA. Detection of ovalbumin was demonstrated in the presence of up to 1000 ng/ml mouse IgG. Table 9.2 shows the limit of detection achieved with the upconverting phosphor-based LFAs for a variety of target antigens. The data shown are for assays observed 15 min after the agent was applied to the test strip. For toxins and other molecular-sized targets (e.g., F1 antigen), the limit of detection is approximately 1 ng/ml. For bacterial agents (spores and cells), detection limits are between 10^4 and 10^6 cfu/ml.

This group has also developed flow cytometer immunoassays using upconverting phosphors to detect proteins (mouse IgG, ovalbumin), viruses (MS2 coliphage), bacterial cells (*Erwinia herbicola*), and spores (BG). Polystyrene or magnetic beads were used to capture the target of interest in a sandwich assay format. For detection, an antibody-functionalized phosphor was used or, indirectly, the target was first detected with a biotinylated antibody followed by neutravidin-coated phosphors. As an example, the group has demonstrated assay sensitivities of ~ 250 pg/ml

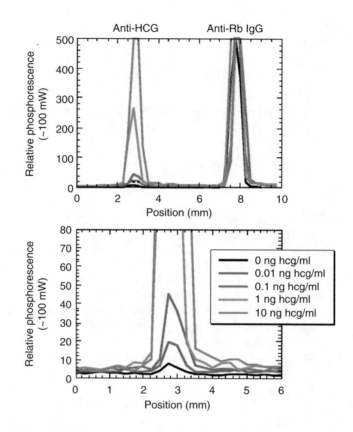

FIGURE 9.20 Detection of hCG by LFA. The upper graph shows the phosphorescence as a function of position along the assay strip. The anti-Rb IgG is the positive control line. The lower graph is an enlarged view of the anti hCG capture line and shows the dose response of the assay.

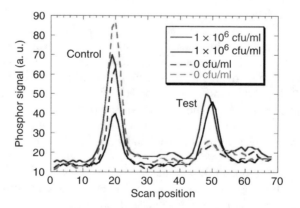

FIGURE 9.21 Detection of *B. anthracis* spores with upconverting phosphors in a lateral flow format. Lateral flow strips were inoculated with 10^6 cfu/ml of *B. anthracis* spores in Hepes buffer or with plain buffer (zero dose). Note that the spores are clearly distinguishable from the background at the test line.

TABLE 9.2
Upconverting Phosphor Sensitivities to Various Targets

Target Antigen	Limit of Detection[a]
Y. pestis (F1 antigen)	1 ng/ml
B. anthracis	6×10^5 cfu/ml
F. tularensis	2.5×10^5 cfu/ml
Vaccinia	10^6 pfu/ml
Ricin	1 ng/ml
hCG	10 pg/ml
BG spores	1×10^4 cfu/ml

[a] Upconverting phosphor labels in lateral flow assay format (15 min test).

for detection of mouse IgG [13,14]. In one experiment, mouse IgG was captured by magnetic beads coated with anti-mouse antibodies (Dynabeads M-450 sheep anti-mouse IgG, Dynal) at a concentration of 6.7×10^5 beads/ml. Detection of the captured mouse IgG was performed in two steps. First, a biotinylated anti-mouse antibody was attached to mouse IgG. Second, neutravidin coated phosphors at a concentration of 1.7×10^8/mL were used to detect the biotinylated antibody. Dose–response curves were generated by counting the capture beads that have phosphor signals that were above a background level defined by capture beads without phosphors present. A dot plot of green upconverted phosphorescence vs. side scatter was used to define the two regions consisting of unlabeled and labeled capture beads. The forward and/or side scatter histograms of the capture beads were used to gate the acquisition of 10,000 bead events during the analysis. From Figure 9.22, the dose response curve for the experiment, the detection sensitivity is estimated to be ∼250 pg/mL.

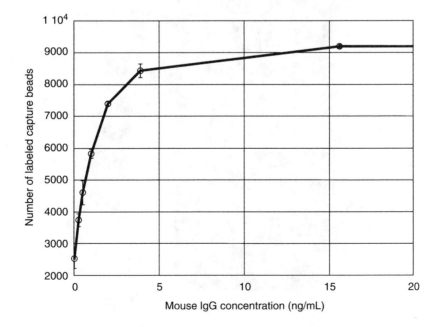

FIGURE 9.22 Detection of mouse IgG by flow cytometry using upconverting phosphor labels. The assay response is determined by counting the number of positively labeled capture beads as described in the text.

9.5 SENSOR PLATFORMS

This section reviews two instrument platforms developed at SRI International primarily for detection of biological warfare agents. These include a battery-operated handheld biosensor that reads LFAs and a compact flow cytometer that reads bead-based assays. Both instruments were designed for field operation for detection of environmental samples and are adaptable to other applications as well, including clinical diagnostics and drug testing.

9.5.1 HANDHELD SENSOR

The handheld sensor (Figure 9.23) is a small, battery-operated detector designed to read out upconverting phosphor lateral flow test strips described in the previous section. A schematic block diagram of the sensor is shown in Figure 9.24. The sensor is a three-channel device capable of simultaneously detecting the emissions from three different upconverting phosphors during a single scan of the lateral flow test strip. The sensor consists of a Palm PDA, a power supply board, a controller board, a laser driver board, and an optics module. The Palm PDA is used to control the overall operation of the sensor. It provides a graphical user interface for the operator and processes, displays, and archives the data obtained from each scan. The power supply board provides the power conversion and distribution for the sensor and operates from either a 3 V battery pack consisting of 10 standard 3 V camera batteries or from an external 3 V source. The controller board performs the necessary signal detection and analog processing of the detected light in each photomultiplier tube (PMT) channel. The board contains the requisite converters to digitize the analog signals and to control the PMT voltages. This board also controls the operation of the stepper motor used to scan the wick, and it provides the signal used for modulation of the laser and

FIGURE 9.23 Hand-held upconverting phosphor assay sensor. A Palm PDA provides the user interface to the instrument. A LFA strip in its protective black housing is shown to the right of the sensor.

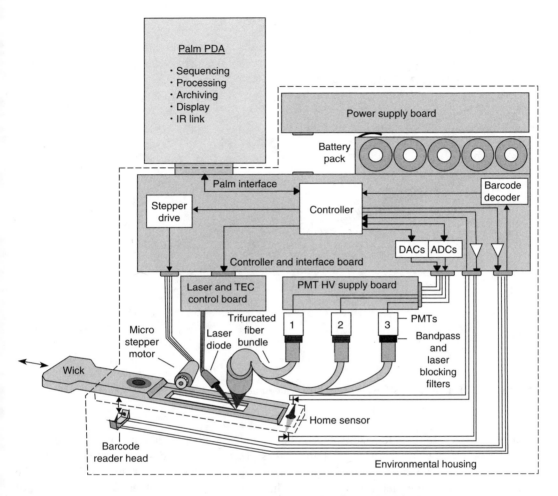

FIGURE 9.24 Hand-held sensor block diagram. Sensor operation is described in the text.

phase-sensitive detection of the photocurrents for each of the three optical detectors. The laser driver board provides the requisite current bias and modulation signal to the laser source and also regulates the diode laser temperature through a thermoelectric cooler (TEC) with which the laser is packaged. The optics module contains the laser diode and associated coupling optics, a trifurcated fiber optic bundle for collecting the emitted light from the lateral flow test strip, and three filtered PMTs. The laser source (Osram SPL-2F) is a 1.5 W CW 980 nm source provided in a TO-220 package with an integral TEC. An aspheric coupling lens is used to focus the excitation laser light on the lateral flow strip at an oblique angle ($\sim 45°$ to the normal). The fiber optic bundle consists of a large number of high numerical aperture optical fibers separated into three groups, each of which feeds into one of three PMTs (Hamamatsu R7400U). The filter stack placed in front of each PMT consists of a 980 nm blocking filter and a visible bandpass filter centered on the peak emission wavelength of a specific phosphor composition (510, 550, and 800 nm).

Figure 9.25 is a schematic illustration of the lateral flow wick assembly used with the sensor. The housing is made of injection molded black polypropylene designed to hold a lateral flow wick and associated application, reagent, and absorbing pads. The wicks are 4 mm wide by 25 mm long. One side of the wick housing has a corrugated section to provide for efficient coupling to the stepper motor. During operation of the sensor, the user inserts a wick into the sensor read port and activates the device through the Palm GUI. The sensor rapidly pulls the wick into the device to a home

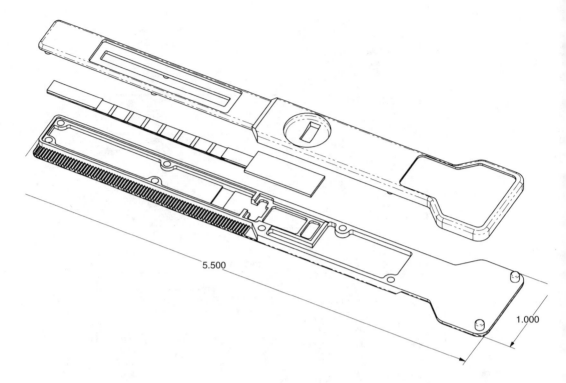

5.500

1.000

FIGURE 9.25 Hand-held sensor wick assembly. Scale dimensions are in inches.

position. The laser source is then activated, and the wick is slowly pushed out of the sensor housing while the optical signals from each of the PMTs are digitized and stored. After the scan is completed, the signals are processed using a maximum likelihood algorithm to decide if the target (or targets) being tested for are present. The operator can also look at the raw data scan of the wick to inspect the signal peaks at the capture, control, and index lines on the wick.

SRI produced a total of five handheld sensors and extensively tested them against various simulant targets. Sensor testing was conducted at SRI as well as at several outside laboratories and confirmed the limits of detection for the targets listed in Table 9.2.

9.5.2 FLOW CYTOMETER

A compact flow cytometer using upconverting phosphor technology was developed at SRI for pathogen detection in environmental samples. The instrument is a single-sample, five-parameter detector capable of detecting multiple pathogens simultaneously using a single, small laser diode excitation source to size latex (or magnetic) particles and excite multiple phosphors (3 colors) bound to antigens captured by the particles. The design for this instrument was based in part on an existing FacsCount™ flow cytometer manufactured by Becton–Dickinson (BD) Biosciences Division. Our group's design was also based on earlier work with a modified FacsCalibur™ benchtop cytometer that was used to support bead-based assay development with the phosphor reporters. The prototype instrument (Figure 9.26) consists of three separate interconnected components: the compact cytometer, a laptop computer (for data analysis and display), and bulk sheath fluid and waste containers.

The cytometer was designed and constructed using a modular approach and consists of a number of custom electronic, optical, and fluidic assemblies. While basing the compact cytometer

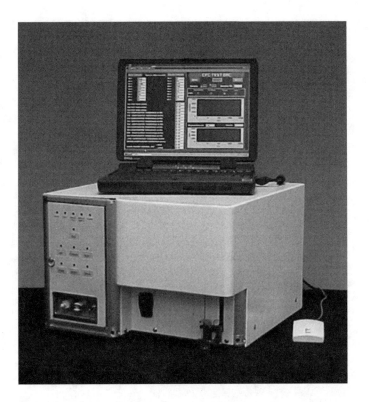

FIGURE 9.26 Compact flow cytometer. The laptop computer located on top is used to control the cytometer and analyze the acquired data.

on the BD FacsCount™ flow cytometer, the group preserved only the fluidic controls, the flowcell, and the optical plate (modified) from the original instrument. It then constructed a new frame, covers, electronics, control panel, and fluidics module. Custom control and GUI software was written to operate on a laptop PC. The finished cytometer weighs 36 lb., occupies a volume of 1.6 ft.^{3}, and requires less than 100 W of power for operation.

A schematic of the cytometer optics module is illustrated in Figure 9.27. A 980 nm diode laser, used as the excitation source, is focused to a 60×180 µm spot in the sample region of the flow cell, achieving power densities in excess of 1000 W/cm^2. This high power density is required because of the time dynamics of phosphor emission to achieve acceptable signal levels from upconverting phosphor particles during their short transit through the excitation beam. A custom optical assembly collects the light at 90° to the excitation beam and directs it to one of four detectors through a system of dichroic mirrors. Three filtered PMTs detect the light from three different phosphor emission bands (475, 550, and 660 nm) and a solid-state photodiode detects the elastically scattered excitation light. Light scattered in the forward direction by latex or magnetic particles is collected by a novel trifurcated fiber optic bundle (Figure 9.28). The bundle is sectored into concentric rings, and light from each ring is directed to a separate photodiode detector. This scheme permits measurement of the angular extent of the forward scattered light cone. Light detected by the central core of the fiber bundle is used as a laser power monitor and for system alignment. Measurement of both the forward and side scattered excitation light is used to determine the size of the latex (or magnetic) particles.

The compact upconverting phosphor flow cytometer was extensively tested and calibrated against standard capture beads, upconverting phosphors, and BG simulant assays. Using the

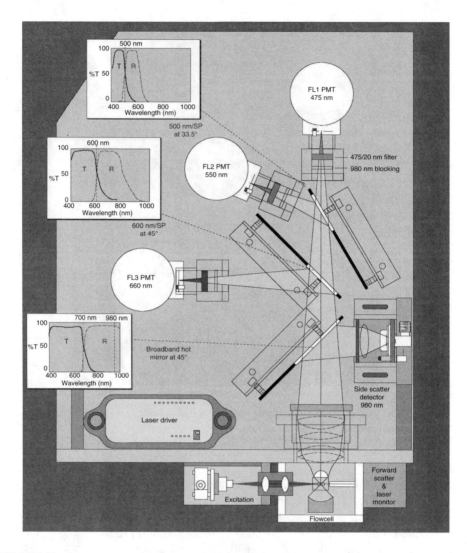

FIGURE 9.27 Flow cytometer optical schematic. See text for details.

combination of forward and side scatter, 2.8 and 4.5 μm diameter magnetic particles were easily distinguishable. This instrument could distinguish between three different upconverting phosphor compositions (erbium oxysulfide, erbium oxide, and thulium oxysulfide). A series of sensitivity and calibration runs with phosphor-labeled 4.5 μm magnetic beads demonstrated an instrumental detection limit of ~5 bound phosphors per particle.

9.6 CONCLUSIONS

Upconverting phosphor immunoassays have been developed for both lateral flow and bead-based formats, and high-sensitivity detection of a variety of biological targets in both formats has been demonstrated. Quantification of target concentration is easily accomplished using an optical reader (either hand-held sensor or flow cytometer) combined with standard curves. Additionally, upconversion of the emitted phosphor signal enables high-sensitivity detection in a variety of sample matrices with minimal optical background. However, upconverting phosphors, like other optical reporter systems, are antibody and format dependent and must be optimized for each application.

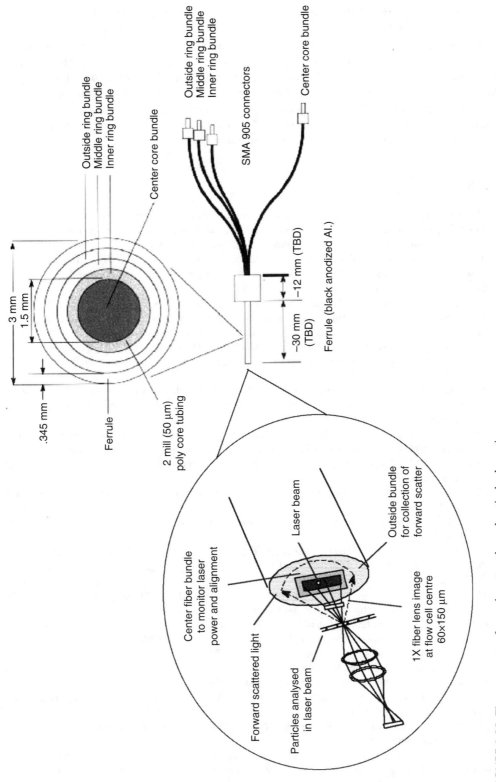

FIGURE 9.28 Flow cytometer forward scatter channel optical schematic.

ACKNOWLEDGMENTS

This work was sponsored by DARPA under Contracts MDA972-96-K-0004 and MDA972-97-C-0019 and the U.S. Army ERDEC under Contract DAAM01-95-C-0088. This work represents the contributions of many individual researchers. The authors are particularly indebted to our many colleagues at SRI and would like to especially acknowledge the contributions of Lisa Ahlberg, John Carrico, Michael Furniss, Mike Hall, Johannes Hampl, Jim Kane, Irina Kasakova, Paul Kojola, Nina Mufti, Gary Rundle, Angel Sanjurjo, Luke Schneider, Ken Shew, Tran Tran, Jan van der Laan, Russell Warren, Megan Yao, and David Zarling. The authors would also like to acknowledge many useful discussions with J. Michael Brinkley, Mildred Donlon, Hans Feindt, Keith Kardos, R. Sam Niedbala, Sal Salamone, Carleton Stewart, and Hans Tanke.

REFERENCES

1. Bloembergen, N., Solid state infrared quantum counters, *Phys. Rev. Lett.*, 2 (3), 84–85, 1959.
2. Auzel, F., Compteur quantique par transfert d'énergie entre deux ions de terres rares dans un tungstate mixte et dans un verre, *C.R. Acad. Sci. (Paris)*, 262, 1016–1019, 1966.
3. Beverloo, H. B., van Schadewijk, A., van Gelderen-Boele, S., and Tanke, H. J., Inorganic phosphors as new luminescent labels for immunocytochemistry and time-resolved microscopy, *Cytometry*, 11 (7), 784–792, 1990.
4. Zarling, D. A., Rossi, M. J., Peppers, N. A., Kane, J., Faris, G. W., Dyer, M. J., Ng, S. Y., and Schneider, L. V., Up-converting reporters for biological and other assays using laser excitation techniques. (U.S. Patent 5,674,698) Filed 30 March 1995, Issued 7 October 1997.
5. Hampl, J., Hall, M., Mufti, N. A., Yao, Y. M., MacQueen, D. B., Wright, W. H., and Cooper, D. E., Upconverting phosphor reporters in immunochromatographic assays, *Anal. Biochem.*, 288 (2), 176–187, 2001.
6. Niedbala, R. S., Feindt, H., Kardos, K., Vail, T., Burton, J., Bielska, B., Li, S., Milunic, D., Bourdelle, P., and Vallejo, R., Detection of analytes by immunoassay using up-converting phosphor technology, *Anal. Biochem.*, 293 (1), 22–30, 2001.
7. Corstjens, P., Zuiderwijk, M., Brink, A., Li, S., Feindt, H., Niedbala, R. S., and Tanke, H., Use of up-converting phosphor reporters in lateral-flow assays to detect specific nucleic acid sequences: a rapid, sensitive DNA test to identify human papillomavirus type 16 infection, *Clin. Chem.*, 47 (10), 1885–1893, 2001.
8. Auzel, F. E., Materials and devices using double-pumped phosphors with energy transfer, *Proc. IEEE*, 61, 758–786, 1973.
9. Auzel, F., Upconversion and anti-Stokes processes with f and d ions in solids, *Chem. Rev.*, 104, 139–173, 2004.
10. Wright, J. C., Up-conversion and excited state energy transfer in rare-earth doped materials, In *Radiationless Processes in Molecules and Condensed Phases*, Fong, F. K., Ed., Springer, Berlin, pp. 239–295, 1976.
11. Mita, Y., Infrared up-conversion phosphors, In *Phosphor Handbook*, Shionoya, S. and Yen, W. M., Eds.,, CRC Press, Boca Raton, FL, pp. 643–650, 1999.
12. Chivian, J. S., Case, W. E., and Eden, D. D., The photon avalanche—a new phenomenon in Pr(3+) based infrared quantum counters, *Appl. Phys. Lett.*, 35, 124–125, 1979.
13. Wright, W. H., Rundle, G. A., Mufti, N. A., Yao, Y. M., Carlisle, C. B., and Cooper, D.E., Flow cytometry with upconverting phosphor reporters. XIX Congress, International Society for Analytical Cytology, Colorado Springs, CO, (February 28-March 5), 1998.
14. Wright, W. H., Rundle, G. A., Mufti, N. A., Yao, Y. M., and Cooper, D. E., Flow cytometry with upconverting phosphor reporters, In *Optical Investigation of Cells In Vivo and In Vitro*, Farkis, D. L., Tromberg, B. J., and Leif, R. C., Eds., *Proceedings of SPIE 3260*, SRI International, Menlo Park, CA, pp. 245–254, 1998.
15. Beverloo, H. B., van Schadewijk, A., Bonnet, J., van der Geest, R., Runia, R., Verwoerd, N. P., Vrolijk, J., Ploem, J. S., and Tanke, H. J., Preparation and microscopic visualization of multicolor luminescent immunophosphors, *Cytometry*, 13 (6), 561–570, 1992.

16. Sanjurjo, A., Lau, K. -H., Lowe, D., Canizales, A., Jiang, N., Wong, V., Jiang, L. et al., Production of substantially monodisperse phosphor particles, (U.S. Patent 6,039,894) Filed 5 December 1997, Issued 21 March 2000.

17. Akinc, M. and Sordelet, D., Preparation of yttrium, lanthanum, cerium, and neodymium basic carbonate particles by homogeneous precipitation, *Adv. Ceramic Mat.*, 2, 232–238, 1987.

18. Li, S., Feindt, H., Giannaras, G., Scarpino, R., Salamone, S., and Niedbala, R. S., Preparation, characterization, and fabrication of uniform coated Y2O2S:RE3+ up-converting phosphor particles for biological detection applications, In *Nanoscale Optics and Applications*, Cao, G. and Kirk, W. P., Eds., *Proceedings of SPIE 4809*, SPIE, Bellingham, WA, pp. 100–109, 2002.

19. Ohmori, M. and Matijevic, E., Preparation and properties of uniform coated colloidal particles. VII. Silica on hematite, *J. Colloid. Interface Sci.*, 150, 594–598, 1992.

20. Brzoska, J. B., Azouz, I. B., and Rondelez, F., Silanization of solid substrates: a step toward reproducibility, *Langmuir*, 10, 4367–4373, 1994.

21. Moon, J. H., Kim, J. H., Kim, K., Kang, T., Kim, B., Kim, C., Hahn, J. H., and Park, J. W., Absolute surface density of the amine group of aminosilylated thin layers: Ultraviolet-visible spectroscopy, second harmonic generation and synchrotron-radiation photoelectron spectroscopy study, *Langmuir*, 13, 4305–4310, 1997.

22. Hermanson, G. T., *Bioconjugate Techniques*, Academic Press, San Diego, 1996.

23. Bieniarz, C., Husain, M., Barnes, G., King, C. A., and Welch, C. J., Extended length heterobifunctional coupling agents for protein conjugations, *Bioconjug. Chem.*, 7 (1), 88–95, 1996.

24. Grabarek, Z. and Gergely, J., Zero-length crosslinking procedure with the use of active esters, *Anal. Biochem.*, 185 (1), 131–135, 1990.

25. Hall, M., Kazakova, I., and Yao, Y. M., High sensitivity immunoassays using particulate fluorescent labels, *Anal. Biochem.*, 272 (2), 165–170, 1999.

26. Harlow, E. and Lane, D., *Antibodies: A Laboratory Manual*, Cold Spring Harbor Laboratory, Cold Spring Harbor, NY, 1988.

27. Ekins, R. P. and Chu, F. W., Multianalyte microspot immunoassay–microanalytical "compact disk" of the future, *Clin. Chem.*, 37 (11), 1955–1967, 1991.

28. Manning, M. C., Patel, K., and Borchardt, R. T., Stability of protein pharmaceuticals, *Pharm. Res.*, 6 (11), 903–918, 1989.

29. Middleton, D., *An Introduction to Statistical Communication Theory*, McGraw Hill, New York, 1960.

10 Mathematical Aspects of Immunoassays

James F. Brady

CONTENTS

10.1 INTRODUCTION

Immunoassays have long been recognized as inexpensive, rapid, and sensitive analytical methods. Yalow and Berson's [1] original work using radiochemical tracers sparked activity in developing assays for medically important compounds. Limitations imposed on the spread of the technology by dependence on radioactive materials were removed by Engvall and Perlmann's substitution of enzyme-based signal generators [2]. Indeed, the recent thirty year anniversary of the latter's work recognized the world wide application of the technology to the analysis of biomedical and environmental analytes of interest. As new analytical needs were met, researchers were often creative with the means by which the methods and their results were mathematically described. This has resulted in multiple approaches to modeling immunoassay data. One need only make a cursory review of the literature to realize that the calculations and manner of processing data are as varied as the researchers themselves. Although each approach may be valid for a given circumstance, this complexity makes it difficult for the novice to appreciate the pitfalls associated with some models. Moreover, varied criteria for evaluation of data can be misleading.

    The purpose of this chapter is to explore selected approaches of applying mathematical models to immunoassay data; determine the limits of detection and quantitation; minimize analytical error; and examine several examples to illustrate interpretation of analytical results. Some suggestions on how to address various analytical issues will be made based on experience in the author's laboratory.

10.2   INTERPRETATION OF STANDARD CURVES

As with other physical phenomena that operate within defined ranges, enzyme immunoassay dose response curves reflect a sigmoidal response when plotted on a log/linear graph (Figure 10.1). The sigmoid can be regarded as consisting of three parts: a linear dose–response region bounded by two tails of lesser slope. These tails have greatly reduced slope relative to the center portion of the curve. The low dose tail (closest to the vertical axis) represents the response produced by insufficient concentration of analyte to compete with the enzyme label or tracer. All doses of analyte in this range (approximately 0.01–0.10 μg/ml in Figure 10.1) fail to inhibit tracer binding, thus producing a similar response. The high dose tail shows a similar behavior but for the opposite reason: excessive analyte saturates the antibody binding sites and inhibits tracer binding except for a background signal.

The vertical response axes can be plotted in several linear systems. The raw absorbances are displayed on the left axis whereas these data (B, binding of a specified dose) are often converted to the percentage of the maximum signal the assay can generate, designated as B_0. The percentage is expressed as B/B_0, $\%B/B_0$ or simply *Percent Bound* (Figure 10.2). Because the decrease in signal can be considered as a decrease in binding of the enzymatic signal generator, the axis has also been defined as *Percent Inhibition*. The scale on the plot is constant regardless of the nomenclature. Note that the linear dose response region correlates to approximately 20%–80% of the maximal signal (Figure 10.2).

Several significant differences between immunoassay and chromatographic response curves are evident in Figure 10.1. The most noticeable difference is that the response is inversely proportional to the dose rather than the reverse. Moreover, because the scale of the dose axis is logarithmic, there is no zero. Hence, the curve cannot be forced through the origin. Finally, because the linear dose–response region is bordered by tails of different slope, an assumption of continuous linearity outside any selected part of the curve cannot be made. Consequently, whatever part of the response curve is utilized as the calibration curve restricts extrapolation of results outside of that range.

The immunochemist is, therefore, faced with describing the observed behavior with a mathematical expression in order to interpolate responses from unknowns into concentrations. A number of expressions are shown in Table 10.1. These expressions ran the gamut from the relatively simplistic direct reporting of the data in the first equation listed in Table 10.1 to an increasingly complex transformation of the data. Equation 10.1 through Equation 10.9 reflect the observation

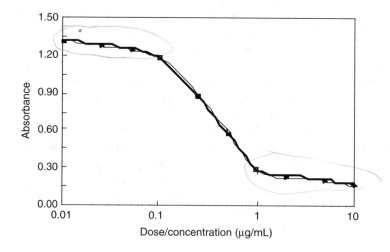

FIGURE 10.1 A representative immunoassay sigmoidal response curve based on absorbance data. An example of fitting Equation 10.9 to dose-response data.

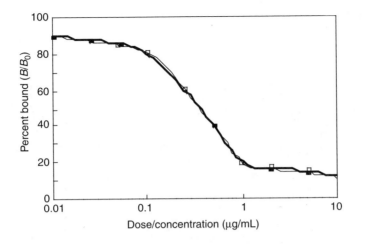

FIGURE 10.2 An immunoassay standard curve based on absorbance data. Where percentage bound equals the absorbance at a specific concentration (B) divided by the maximum signal the assay can generate (B_o). An example of fitting Equation 10.9 to dose-response data.

that immunoassays are log-linear assays, that is, the dose can be plotted on a logarithmic axis and the response plotted in a linear manner. Equation 10.1 through Equation 10.5 use a linear response axis with Equation 10.2 through Equation 10.5 converting the response variable to some form reflective of the maximum signal the assay can produce. Equation 10.6 through Equation 10.10 adopt a log or logit non-linear response axis. Log transforms can be used to compress a large response range but appear used for convenience as large response ranges are not typically encountered. Logit transforms, however, serve the specific purpose of linearizing the entire sigmoid response, effectively creating a functional relationship between dose and response over the entire dose range. The effect of this transform can be seen in Figure 10.3 where the labeling of the vertical axis inflates the slope of the sigmoidal tails (expansion of the response outside the 20%–80% range as shown in Figure 10.2) while compressing the central part of the curve. This action produces an apparent linear response over the full range of doses that does not correspond to the observed log/linear behavior. Whereas the precision of all calibration curves is greatest in the central portion of the curve and decreases as extreme values are reached [4], the loss of precision in the sigmoidal tails

TABLE 10.1
Mathematical Transformations Applied to Immunoassay Data [34]

Number	Equation	Reference
10.1	$y = m \log(x) + b$	[8–10]
10.2	$\% \, y = m \log(x) + b$	[11,12]
10.3	$B/B_0 = m \log(x) + b$	[13,14]
10.4	$\% \, B/B_0 = m \log(x) + b$	[15,16]
10.5	$\% \text{ Inhibition} = m \log(x) + b$	[17,18]
10.6	$\log(B/B_0) = m \log(x) + b$	[19]
10.7	$\text{logit}(y) = m \log(x) + b$	[20,21]
10.8	$\text{logit}(B/B_0) = m \log(x) + b$	[22,23]
10.9	$\text{logit}(\% \, B/B_0) = m \log(x) + b$	[24,25]
10.10	$y = (a - d)/(1 + (x/c)^b) + d$	[5]

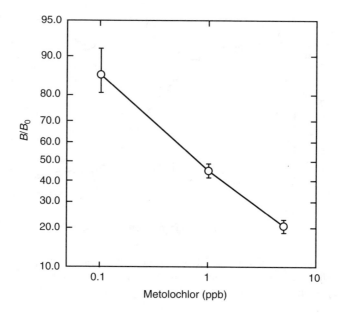

FIGURE 10.3 An immunoassay curve based on a log it transform-that linearlizes the entire sigmoidal curve.

in logit plots is especially apparent in the large error bars around the smallest standard whose mean correlates to approximately 85% B/B_0.

The final equation in Table 10.1 also utilizes the entire response range but does not impose linearity on the system (Figure 10.4). Instead, the four parameter log fit generates additional variables representing the minimum response [a], maximum response [d], inflection point in the central part of the curve [c], and slope at the inflection point [b] by iterative means [5].

In order to select a standard curve expression from among these choices, several criteria should be considered. Ease of use and transparency of data are paramount. Above all, keep it simple. Tijssen commented in 1985 that "easy comprehension" of data handling was his primary consideration, regarding mathematical manipulation of dose–response curves as a means of artificially compressing error and improving the data [3]. If the assay is designed to include only the dose–response region, there is no need to use an expression more complex than the function expressed in Equation 10.1. Regions other than the central portion of the dose–response region should not be

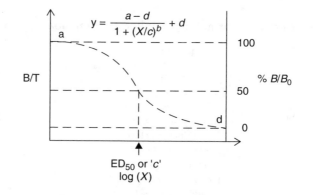

FIGURE 10.4 An immunoassay response curve based on the 4-parameter log fit where (a) is the minimum response (b) slope at the inflection point, (c) inflection point, and (d) is the maximum response.

used unless the precision of the assay in those areas has been demonstrated to be satisfactory for the goals of the analysis. Data derived from these ranges may also be difficult to reproduce. Thus reproducibility should be demonstrated empirically. Restricting the standard curve to the central portion of the response curve ensures that a true dose–response relationship exists, increases analytical precision, and enhances reproducibility. Another important consideration is the ability to verify that the mathematics are working properly. Analysts must produce their own evidence that the functions and other number-crunching aspects of the analysis perform as claimed. Function 10.1 can easily be replicated with a simple calculator or spreadsheet. Spreadsheets can be customized for individual studies and are a reliable method to produce data of known quality. Validation of Equation 10.2 through Equation 10.9 (Table 10.1) can also be accomplished in spreadsheet format. This process may be more involed for the 4-parameter log fit. This model cannot be transformed to a linear curve fit and nonlinear regression techniques are needed to fit the model, using a numerical iterative algorithm. This is typically done by using specialized statistical or mathematical software such as SAS. The 4-parameter curve fit is commonly used for immunoassay data generation. Several evaluation studies comparing immunoassay and gas chromatography/mass spectrometry methods have shown the data to be highly correlated. Thus, validating the 4-parameter curve fit for immunoassay data through comparative means.

10.3   REGIONS OF THE STANDARD CURVE

Once a standard curve has been developed and its function selected, the chemist must address the utility of the curve. Calibration curves can be characterized by their range of measurement, limit of detection (LOD), limit of quantitation (LOQ), and the potential region of non-quantifiable detections lying between these two parameters.

The range of measurement is the range of analyte concentrations bounded by the highest and lowest standards. Analytical results outside of this range are reported as less than the smallest standard or greater than the highest standard.

The sensitivity of immunoassays has been defined by the steepness of the calibration curve [3] or by a concentration or dose of analyte that can be statistically distinguished from a blank [6,7]. This discussion will focus on the latter because the former is not fixed, but changes, depending upon which part of the curve the measurement is taken. The concentration of analyte corresponding to the LOD has been determined by diverse means (Table 10.2). Several investigators have identified the LOD based on visual inspection of the curve [12,18,24]. Others selected an arbitrary percent of maximal signal as the LOD [11,14,15,19,26,27]. This practice can be traced to Midgely et al. [33], but ironically, this is not the preferred method of these authors. They suggested that an estimate of error in the overall assay be incorporated into the determination of the LOD. Two of the authors opted for the dose corresponding to the response calculated from subtracting a fixed number of

TABLE 10.2
Methods for Determining the Limit of Detection for Immunoassays [34]

	Method	References
1.	Visual inspection of curve	[12,18,24]
2.	80% bound	[11,26]
3.	90% bound	[14,15,27]
4.	95% bound	[19]
5.	Standard deviations from the mean of the blank	[20,28]
6.	Use $y_{min} = \overline{y_1} - RMSE(1.470)$	[29]
7.	To solve for smallest dose $t_{99}S_{LLMV}$	[30–32]

standard deviations of the measurement of the zero dose from the mean of those measurements [20,28]. In an attempt to satisfy the concern for a defined statistical separation and incorporate the suggestion of Midgley et al., Brady [34] modified an expression derived by Rodbard [35] for calculating the smallest response statistically distinguished from the response of the zero dose standard,

$$y_{min} = \bar{y}_1 - RMSE[1.470]$$

where y_{min} is the response of the smallest dose statistically different from the response of the blank, and $\bar{y}_1$ is the mean of the blank responses. The root mean squared error (RMSE) is an estimate of the standard deviation (or, equivalently, the standard error) of the "residuals" of the model fit. The residuals correspond to the differences between the actual response (e.g., absorbance) at a given dose and the response that is predicted by the fitted model (i.e., the standard curve regression function) at that dose. As the RMSE is reduced, the standard curve is considered a better fit to the observed measurements. When using the regression package in MicroSoft Excel 2003 [36], the RMSE is the standard error that is determined directly using the STEYX function and corresponds to the square root of the residual mean square (Residual *MS*). The value of y_{min} is inserted into the standard curve to solve for the corresponding dose. If the calculated dose is less than the smallest standard on the curve, the LOD defaults to that standard. Otherwise, the LOD assumes the calculated value. The LOD cannot assume a value less than the smallest standard because the calculated dose may lie outside of the dose–response region due to of the sigmoidal character of the response. Consequently, the calculated LOD assures the method developer that the smallest standard can be statistically distinguished from the response of the zero dose but does not indicate where the calculated value lies within the overall response of the immunoassay. A detailed discussion of this method can be found in Reference [4].

Finally, Glaser et al. [37] developed a method for determining the LOD based on the accuracy (standard deviation) of seven determinations of a sample of known concentration. This approach has been adopted by the U.S. Environmental Protection Agency (EPA) as a preferred technique [31], and it was applied to an atrazine immunoassay designed for regulatory use in drinking water [32].

The LOQ value is the "smallest concentration of analyte that can be measured with stated precision or accuracy" [7]. This concentration, therefore, serves as a "level above which quantitative results may be obtained with a specified degree of confidence" [6]. The LOQ is empirically determined by the analysis of samples fortified with a known concentration of analyte. These analyses must recover a pre-determined amount of the applied dose such as 70%–120% ± 20%. These fortification or procedural recovery samples are initially run to demonstrate proficiency with the method prior to sample analysis. They include all the steps the actual samples would be subjected to such as extraction from the matrix (e.g., solid phase or liquid–liquid extraction), evaporation, dilution, and solvent exchange. Once method proficiency has been established, recovery samples are always run concurrently with actual samples to demonstrate that each analytical set produces similar data quality. A control matrix (unfortified sample matrix that is used for the fortified samples) is analyzed along with the fortified matrix to detect any background concentration of analyte present in the control. Any background concentration of the analyte found in the control is subtracted from the amount measured in the fortified sample. Because analytical results are dependent on the skill of individual analysts, the LOQ is not an "intrinsic constant of the methodology" [6] and may vary between laboratories. Each laboratory and analyst must empirically demonstrate capability with a method by satisfactorily running these quality control samples with each analysis.

Analysts frequently include at least two fortified samples as part of the overall sample analysis. One fortified sample is prepared at the established LOQ and another at a higher concentration in the residue range anticipated in the actual samples. Thus, method proficiency can be demonstrated at

TABLE 10.3
Analytical Results of Procedural Recovery Samples in Human Urine

Fortification Level (ppb)	N	Mean Percent Recovery $\pm$ SD
1	15	103 ± 19
4	6	112 ± 19
5	10	102 ± 11
10	10	86 ± 16
20	4	92 ± 15

the method LOQ and at much higher concentrations. Brady et al. [38], for example, fortified control human urine at the LOQ (1.0 ppb) and at concentrations ranging up to 20 ppb to support analyses of the biomarker atrazine mercapturate in the urine of agricultural workers potentially exposed to atrazine (Table 10.3).

It is noteworthy to point out that the level of fortification used for determination of the LOQ must be higher than the LOD. If control samples are fortified with the concentration of the LOD, half of the sample results would be expected to be less than the LOD because data in the linear dose–response region are normally distributed. Thus, half of the results would be expected to be as non-detects (<LOD) and cannot be employed in calculations for any purpose, much less determination of the LOQ. This phenomenon is recognized to the extent that the Glaser method [37] adopted by the U.S. EPA [1993] specifies the fortification level used for determination of the LOD must be at least twice the concentration of the smallest analytical standard to ensure all results can be entered into the standard deviation calculation.

Consequently, the LOQ determination demarcates a region on the low end of the standard curve between the LOD and LOQ. Analytical results in this region are not supported by recovery samples and, hence, do not have the same quality of data as results that equal or exceed the recovery concentrations. Keith et al. [6] refer to the chromatographic equivalent of this part of the curve as the "region of less-certain quantitation." They recommend data from this part of the curve are reported as detections and given next to the LOD shown in parentheses. Analysts may choose other means, depending on the goals of the analysis, but should be aware of the softness of the data from this part of the curve. This region is of greater importance when data are entered into risk assessment calculations. Data collected in this region are assigned one-half the value of the LOQ for calculation purposes [39]. The reader is referred to Reference [39] for a comprehensive discussion of the topic.

Selection of an LOD or LOQ will impact the eventual data evaluation. The method developer usually designs a method to address a selected need, and the LOD and LOQ are set accordingly. If these parameters are established too low, some samples that are true negatives are likely to be scored as detections (in statistical terms, a type I error) [4]. On the other hand, if these parameters are too high, some samples actually containing residues are likely to be non-detections (type II error). These situations can also be described, respectively, as generating false positive and false negative results. Minimization of these errors can be achieved by prudent adjustment of the sensitivity of the assay to match the goals of the analysis.

10.4 SOURCES OF ERROR IN THE LABORATORY

Errors may be introduced at several points along the immunoassay analysis scheme. Common sources of potential errors and ways to minimize them are discussed here.

10.4.1 VALIDATION OF COMMERCIAL SOFTWARE

Manufacturers of microtiter plate readers have developed extensive software to govern the operation of their instruments. Despite their efforts, operation of commercial software is of particular concern because it produces analytical data within the confines of a virtual "black box". Analysts must be assured data reduction packages operate properly when employed in their laboratory. Manufacturers are aware of this need and include supporting documentation with the instrument. Analysts should not rely solely upon documentation generated outside of their laboratory. Analysts should perform their own quality control checks to validate the algorithms to generate data of known quality. Validation usually consists of replicating the calculations performed by the software by an alternative means such as a calculator or spreadsheet. Confirmatory analyses can also be used for method validation. Gerlach et al. [40] evaluated five commercial software systems and found erroneous data could easily be generated when comparison analyses were not employed.

This issue has been addressed in this laboratory in two ways for particular study applications. First, the plate reader is used solely as a data collection device. The absorbance files are stored and sent to a common drive for removal and processing at a later time. The files are stored as two types and in at least two locations to guard against loss of data. Using the plate reader in such a minimal fashion avoids having to validate each version of the manufacturer's software. Second, the data are processed in a spreadsheet developed within Syngenta Crop Protection, Inc. This sheet is customized to this laboratory's needs, written in Excel®, and contains three sheets, all of which contain the necessary quality control information to establish a paper trail for the analysis such as analyst name, notebook page, study number, and so forth. Having all the data in one document is a tremendous aid for the analyst when checking the results of an analysis, for an auditor to confirm the data, or for someone to re-construct an analysis years after it was performed. The first page indicates the locations of all solutions within the antibody-coated assay plate. Thus, absorbance measurements can be matched to individual samples, standards, or quality control samples. The measurements are displayed on the second page. The third page is in two portions with the calculated standard curve (using Excel® functions SLOPE, INTERCEPT, and RSQ to determine the slope, intercept, and coefficient of determination of the best fit line, respectively) displayed on the top half, and the remainder of the page devoted to calculation of analyte in the quality control samples and unknowns. Additional scratch pages are available in the event the data need to be converted to other units or presented in a particular manner required by the goals of the analysis. Each analysis is assigned a unique file name with the nomenclature of MM/DD/YY [Month/Day/Year] followed by a lower case letter, "a" being the first analysis of the day, "b" the second, etc. This nomenclature is not study specific but provides instantaneous recognition of when the analysis was performed. As with the absorbance data obtained from the plate reader, analytical set files are saved in locations that are backed up frequently to prevent loss of data. Validation of this sheet consisted of reproducing all calculations to generate the standard curve and its parameters, control and recovery values, and sample results and was accomplished with a hand-held calculator. Although time consuming, the validation process furnished evidence that the spreadsheet performed as required. Similar quality assurance/quality control measures can be adapted from the above procedures to fulfill the needs of individual projects.

10.4.2 PLATE READER REPRODUCIBILITY

Another aspect of plate reader performance is the ability to make consistent measurements across the plate. Manufacturers recognize the need for on-site verification of instrument performance and furnish specially designed tests or calibration plates, but these often utilize a limited number of wells that are measured at fixed, pre-determined wavelengths. The Molecular Devices $VERSA_{max}$ plate reader in this laboratory, for example, was equipped with a calibration plate not tested for the wavelength (450 nm) used in its assays. To meet the need for a calibration procedure for assays using 3,3',5,5'-tetramethyl-benzidene (TMB) substrate measured at 450 nm, a simple procedure was developed using readily available chemicals to quantitate the reproducibly of the reader across all wells of a microtiter plate. The procedure is as follows: Combine 1 mL of the TMB substrate, 0.1 mL of 10 mM $KMNO_4$, and swirl to mix; this will produce a strongly colored blue solution. Acidify this solution with 1.0 mL of 1.0 M HCl to convert the blue color to yellow. Dilute the yellowish product with 18 mL of H_2O. Add 200 μL of the dilute solution to each well of a new, uncoated plate with a multichannel pipette and measure the absorbance at 450 nm. Calculate the mean absorbance across all wells and determine the percent difference of each well from the mean. Calibration checks in this laboratory have shown the absorbances of all wells are typically within 1.5% of the mean.

10.4.3 Pipettor Calibration

Pipetting error is typically a major contributor to the overall measurement error in an immunoassay. Even when the assay is conducted robotically, data must be generated to know that the pipetting steps are delivering the prescribed volumes. This process involves the operation of the instrument and the technique of the analyst. Each pipettor should be calibrated gravimetrically by replicate measurements of water pipetted in a tared vial on an analytical balance. When the precision is satisfactory, which in this laboratory is to within 1% of the desired setting, the pipettor is deemed suitable for analytical work. A positive benefit incidentally derived from these procedures is that the technique of the analyst is honed by these calibrations. Consequently, each analyst must demonstrate a satisfactory pipetting capability prior to conducting laboratory work. Analysts new to immunochemistry are also introduced to bench chemistry only after their pipetting skill has been demonstrated. Pipettors should also undergo a professional calibration check on an annual basis. This check involves adjustments, cleaning, and replacement of worn tips or seals. All data generated during this process are stored as part of the laboratory maintenance records. (Appendix C in Chapter 1 presents a standard operating procedure for pipette calibration.)

10.4.4 Plate Layout

The physical process of setting up an analysis is often overlooked, but it is integral to obtaining valid data. An important goal of the analyst is to treat all ninety-six wells in the same fashion. Too often analytical standards, quality control samples, and unknowns are segregated from each other in

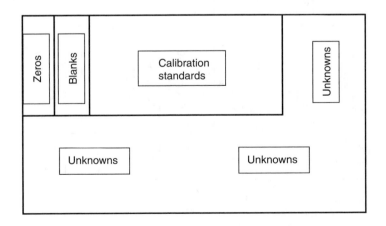

FIGURE 10.5 An undesirable design layout for a 96-well microtiter plate immunoassay that segregate, analytical standards, QC samples, and unknowns are actual samples.

Plate Layout Sheet

Set Name: 022406a	Study No: 6141-05
Analysis Date: 2/24/06	NB ref.: 7753/11
Analyst: JFB	

	1	2	3	4	5	6	7	8	9	10	11	12	
A	0 ppb		Sample 3		Sample 9		5.0		NU		NU		A
B	Control		Sample 4		Sample 10		NU		NU		NU		B
C	Control + 1.0 ng/mL		0.20		2.0		NU		NU		NU		C
D	0.05		Sample 5		Sample 11		NU		NU		NU		D
E	Control + 10 ng/mL		Sample 6		Sample 12		NU		NU		NU		E
F	Sample 1		Sample 7		Sample 13		NU		NU		NU		F
G	Sample 2		0.5		Sample 14		NU		NU		NU		G
H	0.10		Sample 8		Sample 15		NU		NU		NU		H
	1	2	3	4	5	6	7	8	9	10	11	12	

FIGURE 10.6 A suggested design lay-out for a 96-well microtiter plate immunoassay.

different parts of the microtiter plate (Figure 10.5) [41]. When an assay is performed under such conditions, the analyst works under the assumption that all wells are treated the same over the course of the assay but lacks evidence to support the assumption. Standards and samples are always mixed across the plate in this laboratory (Figure 10.6). Obtaining a satisfactory standard curve from standards spread across the plate reflects similar treatment of all wells. When the curve is run separate from the samples, there is no assurance that variable pipetting or well washing in different parts of the plate did not occur. It takes no additional effort to intersperse all solutions, and this practice is clearly beneficial to the quality of data.

10.5 INCUBATIONS

In the past, it was not unusual to report difficulties achieving reproducible data obtained from the exterior wells of a microtiter plate. Eventually, a hypothesis evolved that the solutions in these wells were of variable temperature relative to the solutions in interior wells. If so, non-uniform enzymatic catalysis could be responsible for producing varied colorimetric signals. The problem of poor data from the exterior wells can be resolved by conducting all incubations on a shaker (approximately 90 oscillations/min) encased in a cardboard box. The covered shaker was originally intended to shield the plate from drafts and to prevent spontaneous oxidation of substrate in the color development step. Use of the shaker was quickly extended to all incubations to make the incubation conditions constant and improve overall precision.

10.6 INTERPRETATION OF DATA

With the preceding discussions in mind, it can be assumed that immunoassay data have been collected. Now, the analyst must report the results and be prepared to defend or reject them. Several guidelines for the evaluation of immunoassay data have previously been discussed [34] and are summarized below.

The measurement step in the typical colorimetric enzyme-based immunoassay measures substrate converted to a chromogenic form that absorbs at a specific wavelength. In other words, the assay measures an indicator species and, in so doing, indirectly estimates the concentration of analyte. Thus, immunoassays do not directly measure analyte as is the case of chromatographic-based measurements. A consequence of indirect measurement is the inability or lack of evidence to identify the substance causing a decrease in signal output. The analytical standards contain, by definition, known concentrations of analyte. In contrast, samples have an unknown composition and may affect assay signal output in various ways. Presence of analyte or a compound of sufficient size, shape, and electrical charge to bind to the antibody at hand will specifically inhibit the binding of enzyme conjugate and decrease the signal generation. Inhibition of anti-atrazine antibodies is commonly caused by other structurally related triazines [32,34]. Another example comes from the eventual identification of an alachlor metabolite as the source of positive alachlor immunoassay results [42]. On the other hand, unknowns that decrease binding to antibodies or inhibit enzyme activity are regarded as non-specific inhibitors. Non-specific inhibition is conventionally thought of as prevention or interruption of antibody binding. Maintaining proper sample pH and minimizing the percentage of organic solvent in sample extracts are of primary concern as these can affect antibody performance. Experience with environmental substrates, however, has shown a variety of compounds that might be co-extracted with analyte that inhibit enzyme catalysis. This is not surprising for immunoassays utilizing horseradish peroxidase as the signal generator because the ubiquitous distribution of peroxidase isozymes in nature may have resulted in widespread pro-duction of peroxidase inhibitors. Investigators have noted cinnamic acid derivatives [43], conjugated linoleic acid [44], D-mannose [45], salicylic acid [46], ascorbic acid [47], and soybean extracts [48] as potential peroxidase inhibitors. Synthetic substances may also be responsible for enzyme inhibition. A recent investigation of non-specific inhibition by the constituents of raw and finished drinking water found some oxidizers added to finished drinking water decreased peroxidase-based signal output [49]. It was hypothesized these effects were due to a change in the oxidation state of the iron atom coordinated to the heme group in the active site of the enzyme. Unfortunately, time constraints did not permit confirmation of the cause.

Whatever the root cause of a positive immunoassay signal, the analyst is confronted with an interpretive problem when explaining positive results. This situation is not encountered when negative results are obtained. Negative results, as aptly described by Baker [50], are "unambiguous." Inhibition of any sort did not occur, and the sample can be regarded as not containing analyte, cross-reactive compounds, or so-called non-specific inhibitors.

A means of addressing the interpretive problem is to regard immunoassay results as analyte-equivalents. Thus, a signal equivalent to a concentration of analyte specified by the calibration curve can be regarded as responsible for the result unless confirmatory analyses indicate otherwise. In this manner, the analyst is treating the data in a conservative manner by assigning the maximum concentration of analyte that could be attributed to the test result. This concept would not apply, of course, to situations where the identity of the analyte is known such as in manufacturing or quality control environments.

10.7 MULTIANALYTE IMMUNOASSAYS

An interesting adjunct to this discussion is the development of multi-analyte immunoassays. In one scenario, an analyst is confronted with interpreting data resulting from mixtures of cross reactive analytes interacting in an assay with antibodies designed to react with a specific compound. Concentrations of unknowns cannot be directly determined from the standard curve because assay response cannot be attributed to a known sum of analytes. Most investigators have recommended modifying the standard curve expression based on previously measured cross reactivity behavior or apply sophisticated multivariate or neural net pattern recognition calculations [51–54]. Despite these approaches, they found up to 40% of all samples were not correctly determined and false negative results were observed. Significantly, only samples fortified with known amounts of analyte were used in these studies. Samples of initial unknown composition were not utilized. Until these techniques are improved to the extent that unknowns can be determined with accuracy similar to single analyte methods, such types of multianalyte methods will not provide practical solutions. Additional work is clearly warranted in this area. Alternate multianalyte methods such as microarrays (discussed in Chapter 18 and Chapter 19) or immunoassays employing multiple antibodies and signal resolution may be better alternatives.

10.8 CONCLUSION

This paper has examined some mathematical concepts and practical considerations related to immunoassays, including standard curves, instrument calibration, data processing subsequent to analysis, and interpretation of data. Some suggestions for minimizing sources of error have also been presented based on experience in the author's laboratory. Immunoassay is a powerful analytical tool, and it is hoped that the methods presented herein will assist analysts in deriving defensible, reproducible data.

REFERENCES

1. Yalow, R. S. and Berson, S. A., Immunoassay of endogenous plasma insulin in man, *J. Clin. Invest.*, 39, 1157–1175, 1960.
2. Engvall, E. and Perlmann, P., Enzyme-linked immunosorbent assay [ELISA]. Quantitative assay of immunoglobulin G, *Immunochemistry*, 8, 871–874, 1971.

3. Tijssen, P., *Practice and Theory of Enzyme Immunoassay*, Elsevier, Amsterdam, 5, 351, 391, 1985.
4. Lapin, L., *Statistics Meaning and Method*, Harcourt Brace Jovanovich Inc., New York, 290, 356, 1975.
5. Rodbard, D., Mathematics and statistics of ligand assays: An illustrated guide, In *Ligand Assay*, Langan, J. and Clapp, J. J., Eds., Masson Publishing, New York, pp. 45–101, 1981.
6. Keith, L. H., Crummett, W., Deegan, J. Jr., Libby, R. A., Taylor, J. K., and Wentler, G., Principles of environmental analysis, *Anal. Chem.*, 55, 2210–2218, 1983.
7. Rittenburg, J. and Dautlick, J., Quality standards for immunoassay kits, In *Immunoanalysis of Agrochemicals: Emerging Technologies*, Karu, A., Nelson, J., and Wong, R., Eds., American Chemical Society, Washington, DC, pp. 301–307, 1995.
8. Hunter, K. H. and Lenz, D. E., Detection and quantification of the organophosphate insecticide paraoxon by competitive inhibition enzyme immunoassay, *Life Sci.*, 30, 355–361, 1982.
9. Newsome, W. H., An enzyme-linked immunosorbent assay for metalaxyl in foods, *J. Agric. Food Chem.*, 33, 528–530, 1985.
10. Newsome, W. H., Development of an enzyme-linked immunosorbent assay for triadimefon in foods, *Bull. Environ. Contam. Toxicol.*, 36, 9–14, 1986.
11. Feng, P. C. C., Development of an enzyme-linked immunosorbent assay for alachlor and its application to the analysis of environmental water samples, *J. Agric. Food Chem.*, 38, 159–163, 1990.
12. Szurdoki, F., Bekheit, H. K. M., Marco, M.-P., Goodrow, M. H., and Hammock, B. D., Synthesis of haptens and conjugates for an enzyme immunoassay for analysis of the herbicide bromacil, *J. Agric. Food Chem.*, 40, 1459–1465, 1992.
13. Lawruk, T. S., Hottenstein, C. S., Herzog, D. P., and Rubio, F. M., Quantification of alachlor in water by a novel magnetic particle-based ELISA, *Bull. Environ. Contam. Toxicol.*, 48, 648–650, 1992.
14. Wigfield, Y. K. and Grant, R., Evaluation of an immunoassay kit for the detection of certain organochlorine [cyclodiene] pesticide residues in apple, tomato, and lettuce, *Bull. Environ. Contam.*, 49, 342–347, 1992.
15. Rinder, D. F. and Fleeker, J. R., A radioimmunoassay to screen for 2,4-dichlorophenoxyacetic acid and 2,4,5-trichlorophenoxyacetic acid in surface water, *Bull. Environ. Contam. Toxicol.*, 26, 375–380, 1981.
16. Schwalbe, M., Dorn, E., and Beyerman, K., Enzyme immunoassay and fluoroimmunoassay for herbicide diclofop-methyl, *J. Agric. Food Chem.*, 32, 734–741, 1984.
17. Van Emon, J., Hammock, B., and Seiber, J., Enzyme-linked immunosorbent assay for paraquat and its application to exposure analysis, *Anal. Chem.*, 58, 1866–1873, 1986.
18. Riggle, B. and Dunbar, B., Development of enzyme immunoassay for the detection of the herbicide norflurazon, *J. Agric. Food Chem.*, 38, 1922–1925, 1990.
19. Rubio, F. M., Itak, J. M., Scutellaro, A. M., Selisker, M. Y., and Herzog, D. P., Performance characteristics of a novel magnetic-particle-based enzyme-linked immunosorbent assay for the quantitative analysis of atrazine and related triazines in water samples, *Food Agric. Immunol.*, 3, 113–125, 1991.
20. Newsome, W. H. and Collins, P. G., Determination of imazamethabenz in cereal grain by enzyme-linked immunosorbent assay, *Bull. Environ. Contam. Toxicol.*, 47, 211–216, 1991.
21. Itak, J. A., Selisker, M. Y., Jourdan, S. W., Fleeker, J. R., and Herzogt, D. P., Determination of benomyl [as carbendazim] and carbendazim in water, soil, and fruit juice by a magnetic particle-based immunoassay, *J. Agric. Food Chem.*, 41, 2329–2332, 1993.
22. Lawruk, T. S., Lachman, C. E., Jourdan, S. W., Fleeker, J. R., Herzog, D. P., and Rubiot, F. M., Quantification of cyanazine in water and soil by a magnetic particle-based ELISA, *J. Agric. Food Chem.*, 41, 747–752, 1993.
23. Lawruk, T. S., Lachman, C. E., Jourdan, S. W., Fleeker, J. R., Herzog, D. P., and Rubiot, F. M., Determination of metolachlor in water and soil by a rapid magnetic particle-based ELISA, *J. Agric. Food Chem.*, 41, 1426–1431, 1993.
24. Ercegovich, C. D., Vallejo, R. P., Gettig, R. R., Woods, L., Bogus, E. R., and Mumma, R. O., Development of a radioimmunoassay for parathion, *J. Agric. Food Chem.*, 29, 559–563, 1981.
25. Huber, S. J., Improved solid-phase enzyme immunoassay systems in the ppt range for atrazin in fresh water, *Chemosphere*, 14, 1795–1803, 1985.
26. Marco, M. P., Gee, S. J., Cheng, H. M., Liang, Z. Y., and Hammock, B. D., Development of an enzyme-linked immunosorbent assay for carbaryl, *J. Agric. Food Chem.*, 41, 423–430, 1993.

27. Itak, J. A., Olson, E. G., Fleeker, J. R., and Herzog, D. P., Validation of a paramagnetic particle based ELISA for the quantitative determination of carbaryl in water, *Bull. Environ. Contam. Toxicol.*, 51 (2), 260–267, 1993.

28. Schlaeppi, J.-M., Fory, W., and Ramsteiner, K., Hydroxyatrazine and atrazine determination in soil and water by enzyme-linked immunosorbent assay using specific monoclonal antibodies, *J. Agric. Food Chem.*, 37, 1532–1538, 1989.

29. Brady, J. F., Validated immunoassay methods, In *Handbook of Residue Analytical Methods for Agrochemicals*, Lee, P. W., Ed., Vol. 2, Wiley, West Sussex, England, pp. 714–726, 2003.

30. Corley, J., Best practices in establishing detection and quantification limits for pesticide residues in foods, In *Handbook of Residue Analytical Methods for Agrochemicals*, Lee, P. W., Ed., Vol. 1, Wiley, West Sussex, England, pp. 59–74, 2003.

31. *Definition and procedure for the determination of the method detection limit, Rev. 1.11*, 40 CFR Part 136, Appendix B, U.S. Environmental Protection Agency, Office of the Fed. Reg., Nat. Archives Records Admin., Washington, DC (July 1, 1993).

32. Brady, J. F., Tierney, D. P., McFarland, J. E., and Cheung, M. W., Inter-laboratory validation study of an atrazine immunoassay, *J. Amer. Water Works Assoc.*, 93, 107–114, 2001.

33. Midgley, A. R., Niswender, G. D., and Rebar, R. W., Principles for the assessment of the reliability of radioimmunoassay methods (precision, accuracy, sensitivity, specificity), *Acta Endocrinol.*, (suppl. 142), 163–184, 1969.

34. Brady, J. F., Interpretation of immunoassay data, In *Immunoanalysis of Agrochemicals: Emerging Technologies*, Karu, A., Nelson, J., and Wong, R., Eds., American Chemical Society, Washington, DC, pp. 266–287, 1995.

35. Rodbard, D., Statistical estimation of the minimal detectable concentration ["sensitivity"] for radioligand assays, *Anal. Biochem.*, 90, 1–12, 1978.

36. Microsoft® Excel 2003, Microsoft Corporation, Redmond, Washington, 2003.

37. Glaser, J. A., Forest, D. L., McKee, G. D., Quave, S. A., and Budde, W. L., Trace analyses for wastewaters, *Environ. Sci. Technol.*, 15, 1426–1435, 1981.

38. Brady, J. F. et al., An immunochemical approach to estimating worker exposure to atrazine, In *Triazine Herbicides: Risk Assessment*, Ballantine, L. G., McFarland, J. E., and Hackett, D. S., Eds., American Chemical Society, Washington, DC, pp. 131–140, 1998.

39. *Assigning values to non-detected/non-quantified pesticide residues in human health food exposure assessments* 2000. U.S. Environmental Protection Agency, Office of Pesticide Programs, Washington, DC, March 23, 2000, Report No. PB 2000–105068.

40. Gerlach, R. W., White, R. J., Deming, S. N., Palasota, J. A., and Van Emon, J. M., An evaluation of five commercial immunoassay data analysis software systems, *Anal. Biochem.*, 212, 185–193, 1993.

41. Bunch, D. S., Rocke, D. M., and Harrison, R. O., Statistical design of ELISA protocols, *J. Immunol. Methods*, 132, 247–254, 1990.

42. Macomber, C., Bushway, R. J., Perkins, L. B., Baker, D. B., Fan, T. S., and Ferguson, B. S., Immunoassay screens for Alachlor in rural wells: false positives and an Alachlor soil metabolite, *J. Agric. Food Chem.*, 40 (8), 1450–1452, 1992.

43. Volpert, R., Osswald, W., and Elstner, E. F., Effects of cinnamic acid derivatives on indole acetic acid oxidation by peroxidase, *Phytochemistry*, 38, 19, 1995.

44. Cantwell, H., Devery, R., OShea, M., and Stanton, C., The effect of conjugated linoleic acid on the antioxidant enzyme defense system in rat hepatocytes, *Lipids*, 34, 833–839, 1999.

45. Soeiro, M. N., Paiva, M. M., Barbosa, H. S., Meirelles, M. N., and Araujo-Jorge, T. C., A cardio-myocyte mannose receptor system is involved in *Trypanosoma cruzi* invasion and is down-modulated after infection, *Cell Struct. Funct.*, 24, 149, 1999.

46. Ruffer, M., Steipe, B., and Zenk, M., Evidence against specific binding of salicylic acid to plant catalase, *FEBS Lett.*, 377, 175–180, 1995.

47. Takahama, U., Regulation of peroxidase-dependent oxidation of phenolics by ascorbic acid: different effects of ascorbic acid on the oxidation of coniferyl alcohol by the apoplastic soluble and cell wall-bound peroxidases from epicotyls of *vigna angularis*, *Plant Cell Physiol.*, 34, 809, 1993.

48. Sung, J. and Chiu, C., Lipid peroxidation and peroxide-scavenging enzymes of naturally aged soybean seed, *Plant Sci. Limerick*, 110, 45, 1990.

49. Brady, J. F., Determination of the performance characteristics of the modified Beacon Analytical Systems, Inc., atrazine tube kit for use in promulgated analytical method AG-625, "Atrazine in drinking water by immunoassay," Syngenta Crop Protection, Inc., Greensboro, N.C, 2004.

50. Baker, D., Bushway, R. J., Adams, S. A., and Macomber, C., Determination of the ethanesulfonate metabolite of alachlor in water by high-performance liquid chromatography, *Environ. Sci. Technol.*, 27, 562–564, 1993.

51. Jones, G., Wortberg, M., Kreissig, S. B., Bunch, D. S., Gee, S. J., Hammock, B. D., and Rocke, D. M., Extension of the four-parameter logistic model for ELISA to multianalyte analysis, *J. Immunol. Methods*, 177, 1–7, 1994.

52. Karu, A. E., Lin, T. H., Breiman, L., Muldoon, M. T., and Hsu, J., Use of multivariate statistical methods to identify immunochemical cross-reactants, *Food Agric. Immunol.*, 6, 371–384, 1994.

53. Robison-Cox, J. F., Multiple estimation of concentrations in immunoassay using logistic models, *J. Immunol. Methods*, 186, 79–88, 1995.

54. Fare, T. L., Itak, J. A., Lawruk, T. S., Rubio, F. M., and Herzog, D. P., Cross-reactivity analysis using a four-parameter model applied to environmental immunoassays, *Bull. Environ. Contam. Toxicol.*, 57, 367–374, 1996.

11 Immunochemical Techniques in Biological Monitoring

Raymond E. Biagini, Cynthia A. F. Striley, and John E. Snawder

CONTENTS

11.1 INTRODUCTION

Biological monitoring is the assessment of exposure to an agent through the measurement of biomarker(s) that result from contact with the agent(s). Examples are zinc protoporphyrin in blood, levels of which increase with lead exposure because lead inhibits the biosynthesis of heme; protein and DNA adducts of aromatic amines in blood that can both reflect the intensity of exposure and be correlated with the biologically effective dose; antibodies (Abs) produced against low-molecular-weight molecules—some chemicals, although not immunogenic in their own right because of small size and other limitations, may bind to constitutive polymers (such as host proteins) and become immunogenic, causing the production of specific antibody (Ab). Alternatively, such exposures may lead to the production of new antigenic determinants through non-adduct-forming reactions between the agent and selected protein-carrier molecules. Abs can be made to these modified proteins or to the parent hapten-conjugate. In both cases, the Abs may remain in the human system much longer than the toxicant that initiated their development.

Biomarkers have been trichotomized into biomarkers of exposure, effect, and susceptibility [1]. Biomarkers of exposure are defined as biomarkers that quantify the body burden of chemicals or metabolites of the initial xenobiotic. Biomarkers of effect detect functional change in the biological system under study. Biomarkers of susceptibility indicate the inter-individual variation based on host-susceptibility or genetic differences that can increase or decrease susceptibility. In some cases, a biomarker may simultaneously be a biomarker of exposure, effect, and susceptibility (such as specific IgE in allergies) [2].

When an individual is exposed to a chemical, he or she will receive an internal dose only if the chemical is absorbed into the body. Absorption can occur after dermal contact, inhalation, ingestion, or from a combination of those routes. The extent of absorption from an exposure and the rate of absorption depend on the properties of the chemical (especially its solubility in lipids and water) and the route(s) of exposure [3]. Once absorbed, a chemical is distributed and partitioned into various tissues as a result of tissue variations in pH, permeability, etc. Highly water-soluble chemicals may be distributed throughout the total body water, whereas more lipophilic substances may concentrate in the body fat or other lipid-rich tissues such as the brain [4]. The loss of chemical from the body can loosely be defined as elimination that depends on metabolism and excretion. Chemicals may be eliminated by numerous routes, including fecal, urinary, exhalation, perspiration, and lactation [5]. A chemical can be excreted from the body without metabolism, in which case, the parent compounds may be detectable in the urine, breath, fecal material, or other body fluid. In other cases, the chemical may be metabolized through oxidation, reduction, hydrolysis, or a combination of these processes, often followed by conjugation with an endogenous substrate [6]. Conjugation of a chemical or metabolite is a pathway for excretion. The more important conjugation reactions include glucuronidation, amino acid conjugation, acetylation, sulfate conjugation, and methylation [4]. Metabolism and excretion and the rates of metabolism and excretion can be affected by age, diet, general health status, race, and other factors. In general, the metabolic products will be more water soluble than the parent chemical [7–10]. Where metabolism yields more than one product, the relative amounts of each and the parent-metabolite ratios are affected by an individual's general health status, diet, genetic makeup, degree of hydration, time after exposure, and other factors.

The kidney is the major organ of excretion and is the primary route for water-soluble substances. These substances enter the urine by either glomerular filtration, tubular secretion, or sometimes both mechanisms. The rate of elimination is directly proportional to the serum chemical or metabolite concentration in first-order elimination [11]. Some xenobiotics such as ethanol have zero order elimination kinetics where, in general, the amount eliminated is independent of its concentration [12]. The relationship between the rate of elimination and serum concentration is linear, and the fraction of xenobiotic eliminated remains constant. The fraction eliminated per unit of time is the elimination rate constant (first order) given by the expression

$$C_t = C_0 e^{(-Kel t)}$$

where C_t is concentration at any time, $C_0 =$ the initial concentration, and Kel = the elimination rate constant. The half-life for elimination ($t_{1/2}$) is the time required for the amount of chemical (plasma concentration) in the body to decrease by half. It is expressed by

$$t_{1/2} = 0.693_{Kel}$$

The percentage of xenobiotic eliminated is (in a first order process) independent of the initial dose; therefore, after about five half-lives, over 95% of the xenobiotic will have been eliminated.

The most common matrices used for biological monitoring are exhaled air, blood, and urine, although saliva, tears, and breast milk can be used [13]. Monitoring exhaled air is limited to body burdens from volatile chemicals. Exhaled air monitoring is not suitable for chemicals inhaled as aerosols or for gases and vapors that decompose upon contact with body fluids or tissues or that are highly soluble in water such as ketones and alcohols [14]. Blood is the medium that transports chemicals and their metabolites in the body. Therefore, most biomarkers present in the body can be found in the blood during some period of time after exposure [14]. A chemical in the blood is in dynamic equilibrium with various parts of the body—the site of entry, tissues where the chemical is stored, and organs where it is metabolized or from which it is excreted. Thus, the concentration of a biomarker in the blood may differ between regions of the circulatory system. This would be the case during pulmonary uptake or elimination of a solvent that would cause differences in concentration between capillary blood (mainly arterial blood) and venous blood. Two advantages of blood monitoring are

1. The gross composition of blood is relatively constant between individuals. This eliminates the need to correct measured biomarker levels for individual differences.
2. Obtaining specimens is straightforward, and with proper care, can be accomplished with relatively little risk of contamination.

An important consideration in blood monitoring is that obtaining blood specimens requires an invasive procedure and should be performed only by trained persons.

Urine is more suitable for monitoring hydrophilic chemicals, metals, and metabolites than for monitoring chemicals poorly soluble in water. The concentration of the biomarker in urine is usually correlated to its mean plasma level during the period the urine dwells in the bladder [15]. In some instances, the urine concentration is affected by the amount of the biomarker stored in the kidneys. Examples are cadmium and chromium. The accuracy of the exposure estimate, using urine monitoring, depends upon the sampling strategy. The most influential factors are time of collection and urine output. Measurements from 24-h specimens are more representative than from spot samples and usually correlate better with intensity of exposure. However, collection, stabilization, and transportation of 24-h specimens in the field are difficult and often not feasible [4]. Determination of biomarkers in individual urine samples is confounded by urine dilution that can vary substantially with fluid intake and physical work load. In practice, this effect of urine dilution is reduced by adjusting the measured concentration of the biomarker to a normal value such as specific gravity [16,17]. This adjustment is made by multiplying the measured concentration of the biomarker by the ratio of $[(1.024\text{-}1)/(\text{sp.g.}\text{-}1)]$ where sp.g. is the specific gravity of the urine sample and 1.024 is the assumed normal specific gravity value. Creatinine concentration is the most frequently used adjustment. Creatinine is excreted by glomerular filtration at a relatively constant rate of 1.0–1.6 g/day. Urinary creatinine concentration can be determined by spectrometric or kinetic methods based on the Jaffé alkaline picrate reaction, enzymatic methods, and methods based on mass spectrometry and liquid chromatography [16]. The adjusted value is the quantity of the biomarker per unit quantity of creatinine. There are other considerations to be taken into account when adjusting urinalysis data for dilution. Adjustment to the creatinine level is not appropriate for compounds such as methanol that are excreted in the kidney primarily by tubular secretion. Because the mechanism of excretion of a biomarker can be altered if the urine is very concentrated or very dilute, measurements on samples, having creatinine concentrations outside the range 0.5–3 g/L or having specific gravities outside the range 1.010–1.030, are unreliable [16]. Adjustment for creatinine concentration, although correcting for dilution, introduces additional variation that must be considered when the data are evaluated. Among the factors affecting the rate of creatinine excretion are the muscularity of the subject, physical activity, urine flow, time of day, diet, pregnancy, and disease [16]. A biological monitoring analytical result is a determination of the level of the biomarker in the biological matrix from which the sample was taken at the time it was taken. Extrapolation from that datum to insight on the exposure of the individual requires knowledge of how the human body responds to the agent. Exposure can be estimated when a quantitative relationship between environmental level and biomarker level has been demonstrated. Health risk can be estimated when a quantitative relationship between a health effect and biomarker level has been demonstrated. Where knowledge of a biomarker is limited, one can only infer from its presence above the background level that exposure has occurred.

For a number of agents there exist published reference levels, termed *biological action levels* by the World Health Organization, that serve as guidelines for interpreting biological monitoring data. In the absence of published biomonitoring action levels, biomarker levels indicating occupational exposure have been inferred by comparison with the normal background levels of the biomarker. Biomonitoring action levels vary in their derivation, some being from correlations with exposure, others with health effects. These reference levels should be used only when one has full understanding of their derivation. When biomarker data are available for exposed and non-exposed populations that are otherwise similar, the upper limit of the range for the non-exposed population (two or three standard deviations) may serve as a reference level. Levels of biomarker significantly above that limit suggest exposure to the agent. For those biomarkers for which there is no measurable background level in non-exposed humans, this reference level is effectively the detection limit of the analytical method. In any case, levels of the biomarker above the reference level suggest there was exposure, but give no information on the potential health effect.

Biological monitoring data are subject to a number of sources of variability [4,18]. Rates at which an agent is taken up by the body, metabolized, and excreted vary from person to person and are affected by the person's age, sex, physical workload, medications [10], health, and diet [19]. The route of exposure can also affect uptake and metabolism. For example, absorption through the lungs is much faster than absorption through the skin. Thus, the appearance and elimination of a biomarker will be slower if the agent entered through the skin. If the biomarker is rapidly excreted, the optimum timing for collection of biological samples will be different for the two routes of entry. Differing individuals may use differing personal protective equipment and have differing personal work practices. Also, biomarkers can exist in both a free and a conjugated form, the relative proportions of which can vary substantially from person to person. For example, aniline is present in urine as both the free amine and as acetanilide, its acetyl derivative. Some individuals are genetically predisposed to excrete primarily free aniline; whereas others, primarily, acetanilide. There is also the possibility that concurrent exposure to several agents that compete for the same biotransformation sites in the body may occur. This may lead to altered metabolism and excretion that would change the relationship between exposure or health effect and the level of the biomarker [20]. In addition, concurrent exposure to several agents that are metabolized to the same biomarker will be additive. For example, trichloroacetic acid is a biomarker for trichloroethylene, 1,1,1-trichloroethane, and perchloroethylene.

Despite the source of the human specimens used for biological monitoring, acquisition and use of them are covered by federal guidelines (unless exempted) for the use of human specimens (45 CFR part 46- Protection of Human Subjects). These protections include Institutional Review Board (IRB) reviews of protocols involving human subjects with focused attention to informed consent. Depending on the type of biological monitoring measurements performed, analyses may also be impacted by The Clinical Laboratory Improvement Amendments of 1988 (CLIA 88). Key elements of a CLIA program include strict management of specimen collection, handling, storage, and transportation, thus ensuring sample integrity. Some commercial equipment and/or analytical kits used for biological monitoring are Food and Drug Administration (FDA) cleared/approved as in vitro diagnostic devices (IVD) through a process known as premarket notification (510(k) program) that is based on the Medical Device Amendments of 1976 [21]. Many enzyme linked immunosorbent assay (ELISA) assays are 510 K cleared as being essentially equivalent to previously cleared assays. In 1996, the FDA introduced a new IVD classification category called analyte-specific reagents (ASR) [22]. The FDA defines ASRs as "Ab, both polyclonal and monoclonal, specific receptor proteins, ligands, nucleic acid sequences, and similar reagents which, through specific binding or chemical reaction with substances in a specimen, are intended for use in a diagnostic application for identification and quantification of an individual chemical substance or ligand in biological specimens." In essence, the FDA recognized ASRs as the active ingredients of in-house tests, that, when used in combination with general purpose reagents (such as buffers or reactive materials without specific intended uses) and general purpose laboratory instruments, could be the basis for an assay developed and used by a single laboratory. In addition to those tests to which the regulatory oversight of the FDA applies, laboratories may develop and use in-house tests that are not regulated. Such tests may be useful as a tool in the diagnosis of disease; the responsibility for the validation of the test becomes that of the laboratory developing the test. However, there are no rules for validation of these tests. At a minimum, such validation should address evaluation of solid phase binding of Ag or Ab, primary and secondary incubation Ab times, the effect of interfering substances, and matrix effects.

Strict attention to specimen handling and collection is essential for quality data. The analytical laboratory should be consulted for sampling instructions. Analytical methods should provide specific directions on the collection, storage, and transportation of specimens to the laboratory. Adherence to these directions is of the utmost importance to ensure sample integrity. Timing of specimen collection should be appropriate. The method should include instructions for the timing of specimen collection, that is, whether specimens should be obtained during the work shift, at the end of the shift, or at some other time during the work-week. The longer the half-life of the xenobiotic, the less critical is the timing of the collection. The baseline of a biomarker should be evaluated when the toxicant accumulates in the body. The baseline should also be assessed if there is large inter-subject variability in the population. Care should be taken not to contaminate the specimen with either chemicals or bacteria. The proper preservative (for urine or blood samples) or anticoagulant (blood) should be used, when appropriate. Stability of the biomarker is maximized through proper storage and shipment of the specimen to the laboratory and proper storage by the laboratory. When dealing with human specimens, a biosafety program is essential. Pathogens, such as hepatitis B and human immunodeficiency virus (HIV), may be present in blood, saliva, semen, and other body fluids. Transmission can be by an accidental nick with a sharp object, exposure through open cuts, skin abrasions, and even dermatitis or acne and indirectly through contact with a contaminated environmental surface. Engineering controls that include mechanical or physical systems used to eliminate biological hazards must be available. These are items such as biosafety cabinets or self-sheathing needles. Employee work practices are essential to minimize exposure to pathogens. Good personal hygiene procedures and avoidance of needle recapping can lessen exposure to pathogens. Personal protective equipment such as gloves and masks should be used when necessary. Good housekeeping procedures that involve cleanup of the work area are essential to avoid contamination of the laboratory. Employees who have been identified as potential exposure candidates should be vaccinated for hepatitis B. Universal precautions that take into account the above measures should be practiced with every biological sample received. It is not possible to know if a particular sample may contain pathogens; therefore, each sample should be treated as if contaminated.

11.2 IMMUNOASSAYS

The classical chemical analysis paradigm used to identify and quantitate an analyte of interest includes isolation of the analyte, separation of the analyte from other potentially interfering substances, and quantitation by instrumental or other methods [23]. These classical methods have many shortcomings, including being highly labor intensive and requiring capital expenditures for expensive equipment, (i.e., gas chromatographs (GC), liquid chromatographs (LC), mass spectrometers (MS), or combinations of these instruments (GC–MS, LC-tandem-MS, etc.). In addition, recoveries during the separation and isolation phases of the paradigm may not be constant, and in some cases, may be associated with the level of analyte in the original sample, potentially yielding confounding systematic errors [24–26]. Despite these shortcomings, however, when adequately controlled, classical chemical biological monitoring has the capacity to quantitate the body burden of substances to the sub-ppb level.

Alternatives to classical chemical analyses are immunoassays. Immunoassays, especially enzyme immunoassays (EIAs) and ELISAs, are commonly used analytical techniques for clinical diagnostic measurements, drug screening, and measurements for evaluating exposure to environmental agents [7,8,23,27–42]. The first ELISA was described in 1971 [43]. Recently, immunoassays have been shown to be useful in evaluating exposure to bioterrorism agents [23,31] such as anthrax. Immunoassays are based on the formation and detection of immune complexes between antigens (Ags) and Ab. Ag are principally macromolecules (proteins, polysaccharides, nucleic acids) that can act as complete immunogens able to stimulate an immune response. Other substances that are too small to act as immunogens on their own (drugs, pesticides, etc.) have to be coupled to a macromolecular carrier molecule (usually a protein) to become immunogenic and elicit an immune response. These small molecules are called haptens. Many environmental agents (such as pesticides or pesticide metabolites) are haptens. The selection of the protein carrier used to form the hapten-protein conjugated immunogen is important (keyhole limpet hemocyanin, (KLH), a protein from the shelled keyhole limpet, is often used as a carrier protein as vertebrate exposure is unlikely) [44]. The number of haptens bound to the carrier, the chemistry of the conjugation reactions as well as other factors will all impact the final affinity and avidity of the resultant Abs. The purity of the hapten is also important as conjugation of closely related structures to the carrier may result in the formation of non-specific Ab. Spacer molecules are often used in preparation of haptens for conjugation [42] to attempt to increase the specificity of the Ab for the hapten portion of the conjugate. The ability of an Ab molecule to bind an Ag or a hapten is specifically controlled by structural and chemical interactions between the ligand and the Ab at the combining site [45]. The Ag–Ab interaction is reversible and does not involve formation of covalent bonds [45]. This interaction is controlled by the law of mass action

$$Ag + Ab = AgAb$$

and

$$K = \frac{AgAb}{[Ag][Ab]} \; mol^{-1}$$

where K, the affinity constant and AgAb, the Ag-Ab complex. High affinity constants, resulting from stronger Ag/Ab interactions, lead to lower limits of detection (LOD) in immunoassays.

The mammalian immune system has the capacity to produce five distinct classes of Ab (IgA, IgD, IgE, IgG, IgM). Immunoglobulins consist of two identical heavy chains (50–60 kDa) and two light chains (~ 25 kDa). Both the heavy and light chains have a variable region (V_H and V_L) whose sequence varies between Abs. The variable region is the portion where Ag binding occurs. The remainder of both chains is referred to as the constant region (C_H and C_L) because it has minimal variation in its amino acid sequence. This variation, however, distinguishes the two light chain subtypes (κ and λ) and the five heavy chain subisotypes (α, δ, γ, ε, and μ). Portions of the constant region are where the Ab binds to cells.

IgG (Figure 11.1) is the preponderant Ab class in most mammals, and as such, is the major Ab used in the development of EIAs. The Ab used in an ELISA can be polyclonal or monoclonal. Polyclonal Abs are usually prepared by injecting animals (usually rabbits) with Ag and adjuvant (a mixture that stimulates the immune response) and then collecting serum from the animals [6]. Polyclonal Abs may be further purified and isolated, yielding essentially monospecific polyclonal Abs [46]. Polyclonal Abs, as the name implies, are a mixture of immunoglobulins directed against specific epitopes present in an Ag (an epitope is the smallest fragment of an Ag to which an immune response can be directed; Ag can have numerous epitopes). The Ab response to each epitope is the result of clonal expansion of specific epitope directed B-lymphocytes.

Monoclonal Abs are produced by fusing tumor cells with cells that produce Ab (hybridoma). Hybridoma cells produce Ab to essentially one epitope, hence the name *monoclonal*. Monoclonal Abs provide a continuous and unlimited supply of a standardized reagent with defined specificity and assay characteristics [47].

Radioimmunoassays (RIA) use radiolabeled (e.g., iodine 125 (^{125}I)) reagents that detect the reaction between Ag and Ab. The presence of Ag-Ab reactions are measured using a gamma counter [48]. Most RIA have been replaced by ELISAs, sometimes referred to as EIA. In ELISAs, the solid support (usually a microtiter plate although other solid supports such as magnetic particles, microspheres, coated tubes, etc. have been used) binding of a reactant allows for separation of bound vs. unbound reactants by simple washing. The detector system in ELISAs is usually

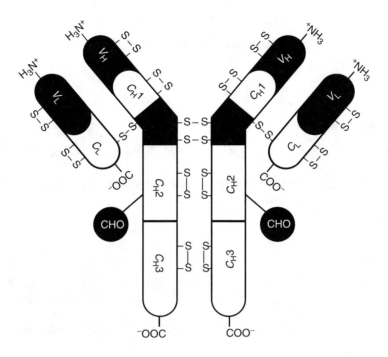

FIGURE 11.1 Structure of immunoglobulin G (IgG). IgG consists of two identical heavy chains (50–60 kDa) and light chains ($\sim$25 kDa). Both the heavy and light chains have variable regions (V_H and V_L) whose sequences vary between antibodies. The variable region is the portion where Ag binding occurs. The remainder of both chains is referred to as the constant region (C_H and C_L).

an enzyme (e.g., horseradish peroxidase, alkaline phosphatase) bound to a reactant (an Ab or analyte). Common chromogens (enzyme substrates) used in ELISAs include p-nitrophenyl phosphate and 2,2'-azino-di-(3-ethylbenz-thiazoline sulfonic acid, o-phenylenediamine, and tetramethylbenzidine).

ELISAs can be performed in many different formats (direct, indirect, capture, competitive, etc.). In the following descriptions, generic overviews of ELISA formats are given. Many differing variations of these generic formats have been utilized to detect numerous analytes, all of which would be too exhaustive for the present review. In a direct ELISA (Figure 11.2), the most basic ELISA format, an analyte (hapten, Ab, Ag) is attached to a solid support. Ab, specific for the analyte and containing a reporter system (usually an enzyme), is incubated with the captured analyte. After washing, a chromogen (enzyme substrate) is added and allowed to react, forming a colored product. In an indirect ELISA (Figure 11.2), analyte (hapten, Ab, Ag) is again attached to a solid support. A primary Ab, specific for the analyte, is incubated in the system and the excess removed by washing. A secondary labeled Ab, specific for the primary Ab, is added to the system and incubated. After washing, chromogen is added and the color measured in a spectrophotometer or other instrument. The amount of color produced is proportional to the amount of secondary Ab that was bound. ELISAs may also be designed in capture formats (Figure 11.3). In an Ag capture (sometimes called sandwich) ELISA, Ag is captured by Ag specific Ab that has been attached to the solid support. After washing, another labeled Ab, specific for another epitope on the Ag, is added. After incubation and washing, chromogen is added and the resultant color measured in a spectrophotometer. ELISAs may also be designed as Ab capture ELISAs that are performed in a similar fashion to Ag capture ELISAs, except that the analyte of interest is an Ab. Another format of ELISA is the competitive ELISA. In a competitive ELISA (Figure 11.3), the analyte (either Ab or

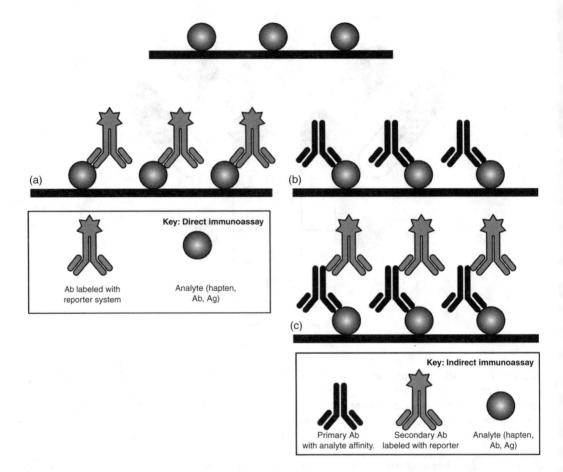

FIGURE 11.2 Direct and indirect immunoassay. In a direct assay (a), analyte (hapten, Ab, Ag) is bound to a solid support (i.e., bead or microplate). Reporter labeled Ab is introduced to the immobilized analyte, forming an Ag-Ab complex. After washing, the concentration of analyte is measured by radiometric, colorimetric, or fluorometric detection of the reporter system. In an indirect format (b, c), a primary Ab specific for the analyte is introduced to the solid support bound analyte. After washing, a secondary labeled Ab, specific for the primary Ab, is added to the system. The concentration of analyte is measured by radiometric, colorimetric, or fluorometric detection of the reporter system.

Ag) competes with labeled analyte for binding. With higher concentrations of analyte, less of the labeled analyte is bound, yielding a reduced signal. In a modification of this format (blocking ELISA), unlabeled analyte is added prior to the addition of labeled analyte. In most ELISAs, Ag/Ab is coated onto microwell plates by electrostatic attraction and van der Waals forces. Ag or Ab is diluted in coating buffers to assist in immobilizing them to the microplate. Commonly used coating solutions are sodium carbonate, Tris-HCl, and phosphate buffered saline. In order to minimize non-specific binding to the microtiter plates, solutions of proteins are used to block unbound sites. Commonly used blocking agents are bovine serum albumin, nonfat dry milk, casein, etc.

Body burdens from exposures to pesticides can be estimated from urinary analyses of pesticide parent/metabolites concentrations [9,49–53]. Pesticide applicators as well as others are often exposed to numerous unrelated pesticides, either sequentially or simultaneously. Classically, body burdens of pesticides are analyzed using chemical/instrumental analysis (CIM) or EIAs. Both of these technologies can usually be used to quantitate one analyte (or closely related groups of analytes in CIM) per assay. In addition, CIM assays usually need numerous cleanup and extraction steps before the sample can be introduced to the instrumentation. For example, the

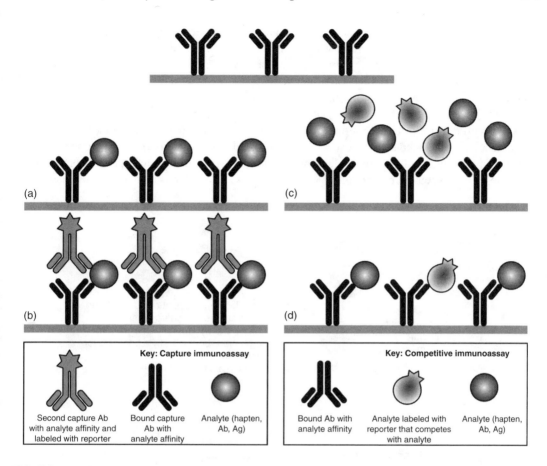

FIGURE 11.3 Capture and competitive immunoassay. In an Ag capture (sometimes called sandwich) assay, Ab, specific for the analyte, is bound to a solid support. Added analyte is bound by the first specific Ab (a). After washing, another labeled Ab, specific for another epitope on the Ag, is added (b). Concentration of analyte is measured by radiometric, colorimetric, or fluorometric detection of the reporter system. In a competitive assay, analyte and a reporter labeled analyte are allowed to compete for binding sites with the immobilized antibodies (c) and bind in relation to their relative concentrations (d). The concentration of analyte is measured by radiometric, colorimetric, or fluorometric detection of the reporter system. With higher concentrations of analyte, less of the labeled analyte is bound yielding reduced signal.

NIOSH Manual of Analytical Methods (NMAM) method for triazine herbicides and their metabolites in urine [54], using gas chromatography with a mass selective detector, has 39 steps from sample preparation to calculations. To perform multiple analyses for multiple unrelated pesticides by CIM could entail a prodigious amount of effort.

Alternatively, multiple analytes can be simultaneously measured using a multiplexed fluorescence covalent microbead immunoassay (FCMIA). In one example [23] of this method, three distinct spectrally addressable microspheres were coupled with three pesticide conjugates (glyphosate-ovalbumin, atrazine-bovine serum albumin, and metolachlor mercapturate-keyhole limpet hemocyanin) using 1-ethyl-3-(3 dimethyl-aminopropyl) carbodiimide hydrochloride (EDC) and N-hydroxysulfosuccinimide (sulfo-NHS). The primary Abs were anti-atrazine, anti-glyphosate, and anti-metolachlor mercapturate. To prepare standard curves, mixtures of atrazine, glyphosate, and metolachlor mercapturate were mixed with the conjugated microspheres and a mixture of primary Ab added. After a period of incubation, biotin labeled anti-rabbit IgG was added and allowed to incubate. After washing, streptavidin R-phycoerythrin was added and, after incubation and washing, the bead mixture was analyzed in a commercial instrument (see Figure 11.4). This type of assay is basically numerous competitive immunoassays being simultaneously performed using microspheres as solid supports. As the concentration of analyte increases, the reporter signal decreases. In this sytem, 5.6 μm polystyrene, divinyl benzene, and methacrylic acid, microspheres that have surface carboxylate functionalities, were used. Internally, the microspheres are dyed with red and infrared-emitting fluorochromes. The internal concentrations of each fluorochrome are proportioned such that spectrally addressable microsphere sets are obtained. Different Ags are covalently coupled to individual microsphere sets. When the microsphere sets are mixed, they can be analyzed with a standard bench-top flow cytometer or a commercially available

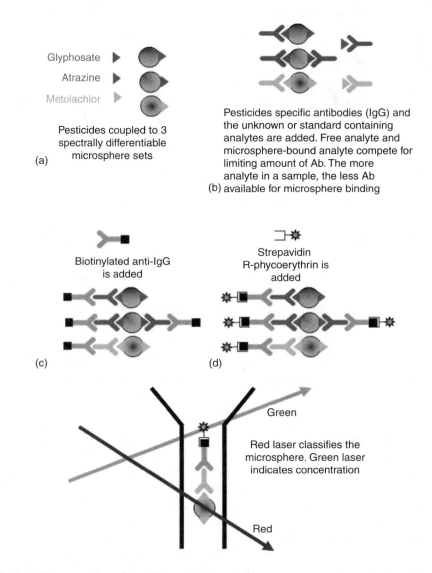

FIGURE 11.4 Diagram of a competitive multiplexed fluorescence covalent microbead immunoassay (FCMIA). (a) Pesticides coupled to spectrally differentiable microspheres. (b) Bound and free pesticide compete for limiting amount of Ab. (c) Bound Ab is reacted with biotinylated secondary Ab. (d) Reporter (streptavidin-phycoerythrin) is bound and the complex read in a flow cytometer. Red laser indicates bead bound while green laser indicates concentration.

dedicated instrument (Luminex, Inc., Austin, TX). The three major components of the system are a bench-top flow cytometer, microspheres, and computer hardware and software. The flow cytometer analyzes individual microspheres by size and fluorescence, distinguishing three fluorescent colors—green (530 nm), infrared (585 nm), and red (>650 nm)—simultaneously. Microsphere size, determined by 90-degree light scatter, is used to eliminate microsphere aggregates from the analysis. Infrared and red fluorescence are used for microsphere classification, and green fluorescence is used for analyte measurement [55].

The microspheres in a liquid suspension array technique can be conjugated with receptors [56], oligonucleotides [57], proteins [23,30], and Ab [58] such that studies of numerous biological interactions and assays can be performed [49,55,59–61]. In addition, the reporter signal from a FCMIA can be amplified by rolling circle DNA amplification. This is done by covalently attaching an oligonucleotide primer to the reporter Ab in the presence of circular DNA, DNA polymerase, and nucleotides. Amplification results in a long DNA molecule containing hundreds of copies of the circular DNA sequence that remain attached to the Ab. The amplified product is labeled in situ by hybridization with fluor-labeled oligonucleotides [62].

11.3 DATA ANALYSIS

Data can be analyzed from ELISA experiments where a set of known standard concentrations of analyte are measured and a relationship between the standards concentration and the ELISA system's response analyzed. This relationship, known as a standard curve, allows one to subsequently estimate the concentration of unknown samples. Many mathematical models have been used to construct ELISA standard curves, including logistic-log transforms [63], log–log transforms [64], four-parameter logistic–log curves, etc. [33]. The four-parameter logistic-log model (4-PL)

$$y = \frac{A - D}{1 + \left(\frac{x}{C}\right)^B}$$

where y is the response (optical density), x the analyte concentration, A and D the responses at zero and infinite dose, C the IC_{50} (the concentration giving 50% inhibition), and B a slope parameter [65] has been shown to be superior to log-log and other fits for immunoassays, even when R^2 values are high (>0.97). The 4-PL fit extends the range of the assay, thus providing a more precise measurement of analyte concentrations [66].

A practical method for assessing the quality of a standard curve fit is to calculate the concentrations of the standards after the regression has been completed [67,68]. This procedure is also known as standards recovery, and it is performed by calculating the concentration of each standard and then comparing it to the actual concentration using the formula

$$\frac{\text{Observed concentration from the } 4-\text{PL fit}}{\text{Expected concentration from analyte added}} \times 100$$

This method yields information about the relative error in the calculation of samples. It is most desirable to have each standard fall between 70 and 130% of the actual value, although more stringent ranges may be applied if greater accuracy is desired. The limitation of using back calculation as the sole method of evaluating goodness of fit is the existence of a bias toward the concentrations of the standards. More specifically, only the standard concentrations are used to assess the quality of the fit; the portions of the curve between each of the standard points are ignored [67]. Spiked recovery may also be used to assess the overall accuracy of an assay [69]. This method incorporates variables in assay preparation as well as the regression analysis. Samples are spiked with known concentrations of analyte and then analyzed to determine the closeness of the calculated value to the actual value. The chosen concentrations are usually between the concentrations of the standards, thus removing the bias inherent to the back calculation of standards method. The results are assessed in the same manner as the standards recovery, using the formula above. A spiked recovery value between 80 and 120% is considered acceptable. The disadvantage of this method is that it is affected by variables other than curve fitting. Errors in sample preparation or assay preparation (pipetting, adding reagents) may affect overall recovery. In addition, it is difficult to accurately spike low levels of analyte into samples because of the relative imprecision of pipets that deliver small volumes [70].

Estimates of least detectable doses (LDD) and minimum detectable concentrations (MDC) have also been calculated using numerous methods, including graphically from the intersection of the asymptote of a 4-PL regression's 95% confidence interval (CI) with the regression line [71] or as a multiple of standard deviations of the blank response [72,73]. In competitive assays where the response is expressed as $\%B/B_0$ data with $B =$ to the response of a standard and $B_0 =$ to the mean optical density measured for the blank, $90\%B/B_0$ is routinely used as the LDD [74].

Specificity is an important characteristic of any laboratory test, describing its ability to distinguish between true (or specific) and non-specific results. With immunoassay methods, interferences that affect specificity can be categorized into two major classes: (1) those that affect the binding event between the Ab and an Ag in a general way such as pH or ionic strength or (2) those substances that affect binding of Ag by competing for the specific binding site on the Ab. These specific interferences are often referred to as cross-reactants. In the analysis of pesticides and/or pesticide metabolites, it is often desirable to have high levels of cross-reactivity with related compounds and metabolites of the parent compounds so that broad screening can be performed. The specificity of an immunoassay may be characterized by adding increasing amounts of a potential cross-reacting substance to a sample and measuring the response in the immunoassay. The results of this experiment can be reported in several ways. One method of representing the comparative reactivity of these compounds is to determine the concentration of each compound required to displace the same amount of labeled Ag from Ab. For example, one commonly calculates cross-reactivity using the concentrations required to displace 50% of the label or 50% B/B_0. The concentration is called the ED_{50} (estimated dose at 50% B/B_0). A ratio of the resulting concentrations can be referred to as the "percent cross-reactivity at the ED_{50}." Cross-reactivity can also be calculated at other levels of displacement such as 20% (ED_{20}). Depending on the slope and shape of the response curve, the % cross-reactivity may be different at different displacement levels. Another method to report cross-reactivity may be to simply report the concentration of cross-reactant required to displace a given amount of labeled Ag. For example, one might report the concentration of cross-reactant required to displace 50% of the label (i.e., the ED_{50}). Again, different displacement levels can be used, but the absolute result and, possibly, the relative results will change. If one chooses the lowest level of displacement that can be reliably distinguished from zero displacement, the resulting concentrations could be represented as a LDD for each cross-reactant. Evaluation of cross-reactivity in poorly defined biological samples may be very complex [75].

Numerous extraneous factors can be present in a sample that may influence Ag-Ab binding, including pH, ionic strength, endogenous components such as enzymes, immunoglobulins, bile, and salts, and exogenous substances such as drugs, polymers, and detergents [76]. These factors contribute to matrix effects defined as follows: the influence of a sample property, other than analyte, on the measurement, and thereby on the measured values; and the physicochemical effect(s) of the matrix on the analytical method's ability to accurately measure an analyte [76].

To insure the integrity of immunoassay data, analytical quality control measures are an important component. Each analyst must take an independent responsibility for assuring that the analytical quality control system works. This can be accomplished by using known spiked samples that closely simulate the samples being analyzed with regard to concentration and interferences.

Because the analyst is most familiar with the methods being used and should know what range of recoveries to expect, problems with the system can be detected early. At a minimum, the following have to be evaluated [77]:

1. Blanks: Analyte free buffer or water.
2. Standards: Calibration curves should be prepared with triplicate points at each of at least five different concentration levels that bracket the concentration of actual samples to avoid extrapolation. Standards should be prepared in the same diluent as used for samples.
3. Blind Spiked Samples: Blind spiked samples are prepared by someone other than the analyst performing the measurement and are to provide an independent check on the accuracy and precision of the measurement.
4. Precision Analysis: Intra-assay and interassay coefficients of variation should be calculated, and their trends evaluated using quality control charting.

11.4 PESTICIDES

The United States Environmental Protection Agency estimates that 10,000–20,000 physician-diagnosed pesticide poisonings occur each year among the approximately 3,380,000 U.S. agricultural workers [78]. The Centers for Disease Control and Prevention, in their Second National Report on Human Exposure to Environmental Chemicals, measured urinary levels of pesticide parent/metabolites for organochlorines, organophosphates, carbamates, and herbicides as part of a report of biomonitoring exposure data for 116 environmental chemicals for the non-institutionalized, civilian U.S. population over the 2-year period 1999–2000. Results from that report showed results greater than the LODs for the majority of pesticide parent/metabolites measured [79]. Exposure to low doses of pesticide mixtures is thought to be related to chronic health effects in humans [80]. Human exposure to pesticides is multi-media and multi-route. Agricultural workers can be exposed to numerous pesticides for variable periods of time, at variable exposure levels, and by numerous routes (inhalation, dermal, ingestion). In addition, transfer exposures can occur from dermal or other contact with contaminated equipment and surfaces. Primary and transfer exposures are affected by weather conditions, type of applications, and work practices [6,9,10,30]. Estimates of pesticide exposures to equipment or clothing may be performed by analytical chemical analyses of elutions [81,82], whereas body burdens of pesticides are usually estimated by biological monitoring of urine samples [6,7,9]. EIAs have been used to measure pesticide concentrations in surface, rain, and groundwater for a variety of substances including alachlor, amitrole, atrazine, bentazon, bromacil, chlorodiamino-s-triazine, chlorsulfuron, clomazone, cyanazine, diethylatrazine, diclofop-methyl, 2,4-D, dichlorprop, diuron, hexazinone, hydroxyatrazine, imazamethabenz, imazaquin, isoproturon, maleic hydrazide, MCPB, metazachlor, methabenzthiazuron, metolachlor, molinate, monuron, norflurazon, paraquat, picloram, propazine, simazine, terbuthylazine, terbutryn, thiobencarb, triasulfuron, 2,4,5-T, trifluralin, fenitrothion, chlorpyrifos, heptachlor, methoprene, 1-naphthol, parathion, paraoxon, PCP, permethrin, pyrimiphos-methyl, quassin, 3,5,6-trichloro-2-pyridinol, benomyl, benzimidazole captan, chlorothalonil, fenpropimorph, iprodione, metalaxyl, myclobutanil, procymidone thiabendazole, triadimefon, and triazole [45]. The majority of these analyses were for parent compounds.

Urinary EIAs have been used to estimate body burdens of numerous pesticides [7,9,10,30,33,34,36,39,40,49–53,74,83–88]. Many commercial suppliers offer EIA kits for the measurement of pesticides in water and other matrices (e.g., EnviroLogix Inc., Portland, ME; Strategic Diagnostics Inc., Newark, DE; Abraxis LLC, Warminster, PA). EIA test kits are ideally suited when speed, simplicity, sensitivity, and low cost are important criteria. Immunoassays are appropriate when specific chemicals, or families of chemicals, are known or suspect, and the objective is to determine their presence, absence, or quantity contained within the sample. In some cases, commercial EIA kits, designed primarily to measure pesticide parent, can be used to screen for urinary metabolites. This is due to the apparent cross-reactivity of the Ab used in the commercial kit for the parent compound, also having affinity for the metabolite. For example, alachlor, MW 270, is too small to be immunogenic in its own right. To overcome this, most Ab for alachlor and other chloroacetanilide herbicides are raised against a derivatized chloroacetanilide that is coupled to a carrier macromolecule (usually a protein), forming a thioether linkage [89]. Polyclonal antisera to these alachlor-protein-thioethers would be expected to contain Ab to numerous antigenic determinants on the immunogen molecule, including the thioether region, probably with differing affinities and avidities for each antigenic determinant. Alachlor is metabolized to a mercapturate metabolite [90] that cross-reacts with some commercial Ab, actually showing an approximate 4x greater affinity than that shown for parent [7].

Commercial EIA test kits specifically designed for human pesticide exposure monitoring are also available. EnviroLogix Inc., Portland, ME offers test kits for the measurement of alachlor mercapturate, atrazine mercapturate, N,N-diethyl-m-toluamide (DEET), and metolachlor mercapturate in urine. These kits can be fast, accurate, and precise.

Conventional analytical techniques, including both classical instrumental methods and EIAs, although highly precise and accurate, are in reality, laboratory-based techniques [91]. Immunobiosensors (analytical devices with the potential for portability that combine the specificity of Ag–Ab interaction with a transducer that produces a signal proportional to the target analyte concentration) have been described [92]. The Ag-Ab complex is in close contact with a signal transducer (e.g., optical, amperometric, potentiometric, or acoustic) coupled to a data acquisition and processing system [93]. Immunobiosensors, because of their specificity, fast response times, low cost, portability, ease of use, and continuous real time signal, present distinct advantages over alternative methods of analyses [94].

The primary optical characteristics that are exploited in the development of immunobiosensors are fluorescence, chemiluminescence, and refractive index change. These optical effects can be measured by surface plasmon resonance (SPR) or evanescent wave effects [95,96]. For example, fluorescent fiber optic biosensors have optical fiber probes, each coated with an Ab specific for a particular analyte. Samples flow over the probes, followed by another Ab that has been attached to a fluorescent dye. If the fluorescent Ab binds to the captured agent, a fluorescent signal is generated at the surface of the probe (Figure 11.5).

Refractive index (the bending of light at the interface of two media) can also be exploited to measure Ag-Ab interactions. The way light interacts at an interface can be exploited to measure changes in surface conditions as occur when Ag binds to Ab on a surface and may be measured by SPR [97] (Figure 11.6). An SPR immunobiosensor is composed of an SPR transducer and a biological recognition element (e.g., Ab) that recognizes and is able to interact with the targeted analyte. The biomolecular recognition element is immobilized on the SPR transducer surface. When a liquid sample is brought in contact with the sensor surface, the interaction between the biomolecular recognition element and the analyte occurs, producing a change in the refractive index at the sensor surface. This in turn results in a change in the propagation constant of a surface plasmon excited at the sensor surface, and it is eventually measured by measuring a change in one of the characteristics of light interacting with the surface plasmon—resonant wavelength, resonant angle, intensity, phase, and polarization.

Piezoelectric immunobiosensors are based on a quartz crystal resonator (Figure 11.7), consisting of a disk with electrodes plated on it. Application of an external oscillating electric potential across the device induces an acoustic wave that propagates through the crystal. The frequency of the vibration can be determined by a frequency counter and is affected by changes in mass associated with Ag binding to the surface-immobilized Ab. This binding increases the mass of the crystal, decreasing the resonant frequency that can be measured [98].

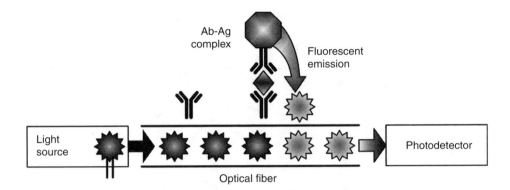

FIGURE 11.5 Fluorescence labeled biosensors. Fluorochrome molecules are used to label secondary antibodies that bind to the Ag in a sandwich format. The fluorochrome is excited by absorbing short-wavelength light, and then it emits light at a higher wavelength that can be detected by the biosensor transducer.

Light-addressable potentiometric sensors (LAPS) combine both electrochemical and electro-optical detection, measuring small pH differences (~ 0.01 pH units) on a semiconductor (Figure 11.8). The pH-sensing region of the instrument consists of a silicon layer wired into an electrical circuit. A LAPS measures an alternating photocurrent, generated when a light source such as a light emitting diode (LED), flashes rapidly. The current magnitude depends on the surface potential that, in turn, depends on the surface pH [99].

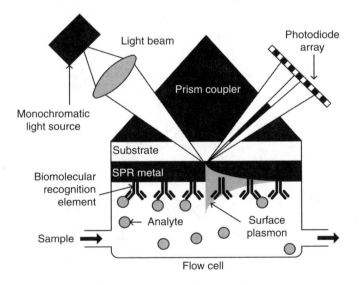

FIGURE 11.6 Surface Plasmon Resonance (SPR). A narrow-band convergent beam from a light-emitting diode is launched into a prism coupler and made incident onto a thin metal (gold) film. The angular component of light that fulfills the coupling condition excites a surface plasmon wave at the outer boundary of the metal film. The coupling produces a narrow dip in the angular spectrum of the reflected light; the precise angular position is determined using a computer-controlled position sensitive photodetector. When a solution containing analyte molecules is injected into the flow-cell, analyte molecules in the sample bind to the biomolecular recognition elements immobilized on the SPR sensor surface, producing a shift in the position of the dip in the angular spectrum of reflected light. The shift can be correlated with the concentration of analyte in the sample.

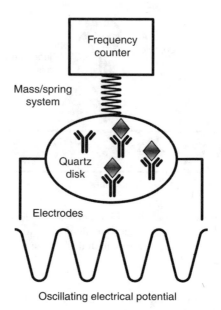

FIGURE 11.7 Piezoelectric biosensors. Piezoelectric (PZ) biosensors are based on a quartz crystal resonator, consisting of a disk with electrodes plated on it. The application of an external oscillating electric potential across the device induces an acoustic wave that propagates through the crystal. The frequency of the vibration can be determined by a frequency counter. The frequency is affected by changes in mass associated with Ag binding to the surface-immobilized Ab. Such binding increases the mass of the crystal, decreasing the resonant frequency of the crystal.

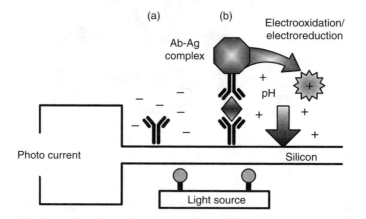

FIGURE 11.8 LAPS (Light addressable potentiometric sensor). A LAPS combines both electrochemical and electroptical detection, measuring small pH differences on a semiconductor. The pH-sensing region of the LAPS consists of a silicon layer wired into an electrical circuit. An alternating photocurrent is generated by the light source. The magnitude of the current depends on the surface potential that, in turn, depends on the surface pH. (a) No Ag bound to the primary Ab, the potential ($-$) is the potential generated by the light source. (b) Ag-labeled secondary Ab complex bound to the primary Ab catalyzes electrooxidation/electroreduction, thereby producing a pH change that affects the surface potential ($+$).

In conclusion, biological monitoring provides the basis for estimating an internal chemical dose by measuring parent compound and/or metabolite concentrations in selected body tissues and fluids. In contrast to biological monitoring that measures the compound or its metabolites in human tissue, biological effects monitoring (i.e., use of biomarkers) is used to detect evidence of chemical exposure by measuring a biochemical response, such as changes in enzyme activity [100]. In other words, chemical exposure is estimated based on an indicator property rather than through direct quantification of the chemical itself. This type of monitoring does not provide a direct measure of internal dose, but it can provide an indication of the potential for adverse effects. Dose cannot be estimated unless the correlation between exposure and biochemical response is well understood.

Biological monitoring has been classically performed by quantitative analyses for urinary-excreted or environmentally sampled chemicals by chemical/instrumental analysis (CIM) after extraction from urine or sample matrices (such as rinses of hands, patches on skin/clothing, sorbent materials, and filters). These procedures are costly, time consuming, labor intensive, and require the acquisition of high capital expenditure equipment and highly trained personnel, although they are usually highly specific. Alternatives to CIM are EIA where pesticides or their metabolites can be quantified in neat or diluted urine or water using Ab (usually polyclonal) directed against the pesticides or their metabolites. EIAs have been used to measure numerous types of analytes in both biological and environmental matrices over the last 35 years. EIAs have the benefit of being inexpensive, fast, and quantitative, and they can be simply performed on relatively inexpensive equipment. In many cases, EIAs have lower limits of quantitation than CIMs. Urinary pesticide/metabolite EIAs may have the disadvantages (in some cases) of not being specific or of suffering from matrix effects from urine, limiting their sensitivity by a factor of 10–100-fold. However, when speed and cost efficiency are evaluated, immunoassays significantly outperform conventional methods. The ability of immunoassays to be multiplexed (the measurement of numerous analytes simultaneously) is one factor that distinguishes EIAs from CIMs. Diversity in the chemical properties of mixtures has been shown to negatively impact recoveries when measuring multiple analytes by CIM [29]. The final outcome of these efforts are methods that are either sensitive and imprecise or precise and insensitive [24,101,102]. The future use of immunobiosensors, surface and liquid matrix arrays, and other cutting edge technologies (especially as these methods mature) for biological monitoring should allow for the almost instantaneous measurement of numerous analytes simultaneously with accuracy and precision.

ACKNOWLEDGMENTS

Mention of a product or company name does not constitute endorsement by NIOSH. This work was supported in part by an interagency agreement between NIOSH and NIEHS (Y1-ES-0001—Clinical Immunotoxicity). The content and conclusions of this chapter are those of the authors and do not necessarily represent the views of the National Institute for Occupational Safety and Health.

REFERENCES

1. Schulte, P. A., A conceptual and historical framework for molecular epidemiology, In *Molecular Epidemiology Principles and Practices*, Schulte, P. A. and Perera, F. P., Eds., Academic Press, San Diego, CA, pp. 3–44, 1993.
2. Biagini, R. E., Krieg, E. F., Pinkerton, L. E., and Hamilton, R. G., Receiver operating characteristics analyses of food and drug administration-cleared serological assays for natural rubber latex-specific immunoglobulin E antibody, *Clin. Diagn. Lab. Immunol.*, 8, 1145–1149, 2001.

3. Minh, T. B., Watanabe, M., Tanabe, S., Yamada, T., Hata, J., and Watanabe, S., Specific accumu-
    lation and elimination kinetics of tris(4-chlorophenyl)methane, tris(4-chlorophenyl)methanol, and
    other persistent organochlorines in humans from Japan, *Environ. Health Perspect.*, 109, 927–935,
    2001.
4. NIOSH, NIOSH manual of analytical methods (NMAM), DHHS (NIOSH), Cincinnati, OH, 1994.
5. Jaeger, A., Are kinetic investigations feasible in human poisoning? *Toxicol. Lett.*, 102–103,
    637–642, 1998.
6. Hines, C. J., Deddens, J. A, Striley, C. A., Biagini, R. E, Shoemaker, D. A., Brown, K. K., Mackenzie,
    B. A., and Hull, R. D., Biological monitoring for selected herbicide biomarkers in the urine of exposed
    custom applicators: application of mixed-effect models, *Ann. Occup. Hyg.*, 47, 503–517, 2003.
7. Biagini, R. E., Tolos, W., Sanderson, W. T., Henningsen, G. M., and MacKenzie, B., Urinary
    biomonitoring for alachlor exposure in commercial pesticide applicators by immunoassay, *Bull.
    Environ. Contam. Toxicol.*, 54, 245–250, 1995.
8. Biagini, R. E., Henningsen, G. M., MacKenzie, B., Sanderson, W. T., Robertson, S., and
    Baumgardner, E. S., Evaluation of acute immunotoxicity of alachlor in male F344/N rats, *Bull.
    Environ. Contam. Toxicol.*, 50, 266–273, 1993.
9. Sanderson, W. T., Biagini, R., Tolos, W., Henningsen, G., and MacKenzie, B., Biological moni-
    toring of commercial pesticide applicators for urine metabolites of the herbicide alachlor, *Am. Ind.
    Hyg. Assoc. J.*, 56, 883–889, 1995.
10. Sanderson, W. T., Ringenburg, V., and Biagini, R., Exposure of commercial pesticide applicators to
    the herbicide alachlor, *Am. Ind. Hyg. Assoc. J.*, 56, 890–897, 1995.
11. Albarellos, G. A., Kreil, V. E., and Landoni, M. F., Pharmacokinetics of ciprofloxacin after single
    intravenous and repeat oral administration to cats, *J. Vet. Pharmacol. Ther.*, 27, 155–162, 2004.
12. Norberg, A., Jones, A. W., Hahn, R. G., and Gabrielsson, J. L., Role of variability in explaining
    ethanol pharmacokinetics: research and forensic applications, *Clin. Pharmacokinet.*, 42, 1–31, 2003.
13. Knopp, D., Application of immunological methods for the determination of environmental pollutants
    in human biomonitoring. A review, *Anal. Chim. Acta*, 311, 383–392, 1995.
14. Lowry, L. K., Rosenberg, J., and Fiserova-Bergerova, V., Biological monitoring III: measurements
    in urine, *Appl. Ind. Hyg.*, 4, F-11, 1989.
15. Rosenberg, J., Fiserova-Bergerova, V., and Lowry, L. K., Biological monitoring IV: measurements
    in urine, *Appl. Ind. Hyg.*, 4, 1-16, 1989.
16. Boeniger, M. F., Lowry, L. K., and Rosenberg, J., Interpretation of urine results used to assess
    chemical exposure with emphasis on creatinine adjustments: a review, *Am. Ind. Hyg. Assoc.
    J.*, 54, 615–627, 1993.
17. Ikeda, M., Ezaki, T., Tsukahara, T., Moriguchi, J., Furuki, K., Fukui, Y. et al., Bias induced by the
    use of creatinine-corrected values in evaluation of beta2-microglobin levels, *Toxicol. Lett.*, 145,
    197–207, 2003.
18. Droz, P. O., Biological monitoring I: sources of variability in human response to chemical exposure,
    Appl. Ind. Hyg., 4, F-20, 1989.
19. Rosenberg, J., Biological monitoring IX: concomitant exposure to medications and industrial
    chemicals, *Appl. Occup. Environ. Hyg.*, 9, 341–345, 1994.
20. Ogata, M., Fiserova-Bergerova, V., and Droz, P. O., Biological monitoring VII: Occupational
    exposures to mixtures of industrial chemicals, *Appl. Occup. Environ. Hyg.*, 8, 609–617, 1993.
21. Gutman, S., The role of food and drug administration regulation of in vitro diagnostic devices—
    applications to genetics testing, *Clin. Chem.*, 45, 746–749, 1999.
22. USDHHS, Medical devices: classification/reclassification; restricted devices; analyte specific
    reagents, Fed. Regist., 62, 62243–62260, 1997.
23. Biagini, R. E., Sammons, D. L., Smith, J. P., MacKenzie, B. A., Striley, C. A. F., Semenova,
    C. A. F. V., Steward-Clark, E. et al., Comparison of a multiplexed fluorescent covalent microsphere
    immunoassay and an enzyme-linked immunosorbent assay for measurement of human immunoglo-
    bulin G antibodies to anthrax toxins, *Clin. Diagn. Lab. Immunol.*, 11, 50–55, 2004.
24. Baker, S. E., Barr, D. B., Driskell, W. J., Beeson, M. D., and Needham, L. L., Quantification of
    selected pesticide metabolites in human urine using isotope dilution high-performance liquid chro-
    matography/tandem mass spectrometry, *J. Expo. Anal. Environ. Epidemiol.*, 10, 789–798, 2000.

25. Barr, D. B., Barr, J. R., Maggio, V. L., Whitehead, R. D., Sadowski, M. A., Whyatt, R. M., and Needham, L. L., A multi-analyte method for the quantification of contemporary pesticides in human serum and plasma using high-resolution mass spectrometry, *J. Chromatogr. B Analyt. Technol. Biomed. Life Sci.*, 778, 99–111, 2002.

26. Carabias-Martýnez, R., Garcýa-Hermida, C., Rodrýguez-Gonzalo, E., Soriano-Bravo, F., and Hernandez-Mendez, J., Determination of herbicides, including thermally labile phenylureas, by solid-phase microextraction and gas chromatography-mass spectrometry, *J. Chromatogr. A*, 1002, 1–12, 2003.

27. Bernstein, D. I., Biagini, R. E., Karnani, R., Hamilton, R., Murphy, K., Bernstein, C., Arif, S. A., Berendts, B., and Yeang, H. Y., In vivo sensitization to purified Hevea brasiliensis proteins in health care workers sensitized to natural rubber latex, *J. Allergy Clin. Immunol.*, 111, 610–616, 2003.

28. Biagini, R. E., Driscoll, R. J., Bernstein, D. I., Wilcox, T. G., Henningsen, G. M., Mackenzie, B. A., Burr, G. A., Scinto, J. D., and Baumgardner, E. S., Hypersensitivity reactions and specific antibodies in workers exposed to industrial enzymes at a biotechnology plant, *J. Appl. Toxicol.*, 16, 139–145, 1996.

29. Biagini, R. E., MacKenzie, B. A., Sammons, D. L., Smith, J. P., Striley, C. A., Robertson, S. K., and Snawder, J. E., Evaluation of the prevalence of antiwheat-, anti-flour dust, and anti-alpha-amylase specific IgE antibodies in US blood donors, *Ann. Allergy Asthma Immunol.*, 92, 649–653, 2004.

30. Biagini, R. E., Murphy, D. M., Sammons, D. L., Smith, J. P., Striley, C. A., and MacKenzie, B. A., Development of multiplexed fluorescence microbead covalent assays (FMCAs) for pesticide biomonitoring, *Bull. Environ. Contam. Toxicol.*, 68, 470–477, 2002.

31. Biagini, R. E., Sammons, D. L., Smith, J. P., Page, E. H., Snawder, J. E., Striley, C. A. F., and MacKenzie, B. A., Determination of serum IgG antibodies to Bacillus anthracis protective antigen in environmental sampling workers using a fluorescent covalent microsphere immunoassay, *Occup. Environ. Med.*, 61, 703–708, 2004.

32. Biagini, R. E., Schlottmann, S. A., Sammons, D. L., Smith, J. P., Snawder, J. C., Striley, C. A., MacKenzie, B. A., and Weissman, D. N., Method for simultaneous measurement of antibodies to 23 pneumococcal capsular polysaccharides, *Clin. Diagn. Lab. Immunol.*, 10, 744–750, 2003.

33. Biagini, R. E., Smith, J. P., Sammons, D. L., MacKenzie, B. A., Striley, C. A., Robertson, S. K., and Snawder, J. W., Development of a sensitivity enhanced multiplexed fluorescence covalent microbead immunosorbent assay (FCMIA) for the measurement of glyphosate, atrazine and metolachlor mercapturate in water and urine, *Anal. Bioanal. Chem.*, 379, 368–374, 2004.

34. Ahn, K. C., Watanabe, T., Gee, S. J., and Hammock, B. D., Hapten and antibody production for a sensitive immunoassay determining a human urinary metabolite of the pyrethroid insecticide permethrin, *J. Agric. Food Chem.*, 52, 4583–4594, 2004.

35. Lee, H. J., Shan, G., Ahn, K. C., Park, E. K., Watanabe, T., Gee, S. J., and Hammock, B. D., Development of an enzyme-linked immunosorbent assay for the pyrethroid cypermethrin, *J. Agric. Food Chem.*, 52, 1039–1043, 2004.

36. Shan, G., Huang, H., Stoutamire, D. W., Gee, S. J., Leng, G., and Hammock, B. D., A sensitive class specific immunoassay for the detection of pyrethroid metabolites in human urine, *Chem. Res. Toxicol.*, 17, 218–225, 2004.

37. Cho, Y. A., Kim, Y. J., Hammock, B. D., Lee, Y. T., and Lee, H. S., Development of a microtiter plate ELISA and a dipstick ELISA for the determination of the organophosphorus insecticide fenthion, *J. Agric. Food Chem.*, 51, 7854–7860, 2003.

38. Penalva, J., Puchades, R., Maquieira, A., Gee, S., and Hammock, B. D., Development of immunosensors for the analysis of 1-naphthol in organic media, *Biosens. Bioelectron.*, 15, 99–106, 2000.

39. Staimer, N., Gee, S. J., and Hammock, B. D., Development of a sensitive enzyme immunoassay for the detection of phenyl-beta-D-thioglucuronide in human urine, *Fresenius J. Anal. Chem.*, 369, 273–279, 2001.

40. Lohse, C., Jaeger, L. L., Staimer, N., Sanborn, J. R., Jones, A. D., Lango, J., Gee, S. J., and Hammock, B. D., Development of a class-selective enzyme-linked immunosorbent assay for mercapturic acids in human urine, *J. Agric. Food Chem.*, 48, 5913–5923, 2000.

41. Lee, J. K., Ahn, K. C., Park, O. S., Kang, S. Y., and Hammock, B. D., Development of an ELISA for the detection of the residues of the insecticide imidacloprid in agricultural and environmental samples, *J. Agric. Food Chem.*, 49, 2159–2167, 2001.

42. Biagini, R. E., Klincewicz, S. L., Henningsen, G. M., MacKenzie, B. A., Gallagher, J. S., Bernstein, D. I., and Bernstein, I. L., Antibodies to morphine in workers exposed to opiates at a narcotics manufacturing facility and evidence for similar antibodies in heroin abusers, *Life Sci.*, 47, 897–908, 1990.

43. Engvall, E. and Perlman, P., Enzyme-linked immunosorbent assay (ELISA). Quantitative assay of immunoglobulin G, *Immunochemistry*, 8, 871–874, 1971.

44. Striley, C. A., Biagini, R. E., Mastin, J. P., MacKenzie, B. A., and Robertson, S. K., Development and validation of an ELISA for metolachlor mercapturate in urine, *Anal. Chim. Acta*, 399, 109–114, 1999.

45. Dankwardt, A., Immunochemical assays in pesticide analysis, In *Encyclopedia of Analytical Chemistry*, Chichester, M. R., Ed., Wiley, New York, pp. 1–27, 2000.

46. Khan, M., Bajpai, V. K., Anasari, S. A., Kumar, A., and Goel, R., Characterization and localization of fluorescent Pseudomonas cold shock protein(s) by monospecific polyclonal antibodies, *Microbiol. Immunol.*, 47, 895–901, 2003.

47. Trout, D. B., Seltzer, J. M., Page, E. H., Biagini, R. E., Schmechel, D., Lewis, D. M., and Boudreau, A. Y., Clinical use of immunoassays in assessing exposure to fungi and potential health effects related to fungal exposure, *Ann. Allergy Asthma Immunol.*, 92, 483–491, 2004, quiz 492–484, 575

48. Biagini, R. E., Bernstein, I. L., Gallagher, J. S., Moorman, W. J., Brooks, S., and Gann, P. H., The diversity of reaginic immune responses to platinum and palladium metallic salts, *J. Allergy Clin. Immunol.*, 76, 794–802, 1985.

49. Biagini, R. E., Murphy, D. M., Sammons, D. L., Smith, J. P., Striley, C. A. F., and MacKenzie, B. A., Development of multiplexed fluorescence microbead immunosorbent assays (FMIAs) for pesticide biomonitoring, *Bull. Environ. Contam. Toxicol.*, 68, 470–477, 2002.

50. Hines, C. J., Deddens, J. A., Striley, C. A., Biagini, R. E., Shoemaker, D. A., Brown, K. K., Mackenzie, R., and Hull, B. A. D., Biological monitoring for selected herbicide biomarkers in the urine of exposed custom applicators: Application of mixed-effect models, *Ann. Occup. Hyg.*, 47, 503–517, 2003.

51. Mastin, J. P., Striley, C. A. F., Biagini, R. E., Hines, C. J., Hull, R. D., MacKenzine, B. A., and Robertson, S. K., Use of immunoassays for biomonitoring of herbicide metabolites in urine, *Anal. Chim. Acta*, 376, 119–124, 1998.

52. Hryhorczuk, D. O., Moomey, M., Burton, A., Runkle, K., Chen, E., Saxer, T., Slightom, J., Dimos, J., McCann, K., and Barr, D., Urinary *p*-nitrophenol as a biomarker of household exposure to methyl parathion, *Environ. Health Perspect.*, 6 (suppl. 6), 1041–1046, 2002.

53. Smith, P. A., Thompson, M. J., and Edwards, J. W., Estimating occupational exposure to the pyrethroid termiticide bifenthrin by measuring metabolites in urine, *J. Chromatogr. B. Analyt. Technol. Biomed. Life Sci.*, 778, 113–120, 2002.

54. DHHS(NIOSH), *Triazine Herbicides and Their Metabolites in Urine. Method 8315*, Government Printing Office, Washington, DC, 2003.

55. Fulton, R. J., McDade, R. L., Smith, P. L., Kienker, L. J., and Kettman, J. R. Jr., Advanced multiplexed analysis with the flowmetrix system, *Clin. Chem.*, 43, 1749–1756, 1997.

56. Iannone, M. A., Consler, T. G., Pearce, K. H., Stimmel, J. B., Parks, D. J., and Gray, J. G., Multiplexed molecular interactions of nuclear receptors using fluorescent microspheres, *Cytometry*, 44, 326–337, 2001.

57. Colinas, R. J., Bellisario, R., and Pass, K. A., Multiplexed genotyping of beta-globin variants from PCR-amplified newborn blood spot DNA by hybridization with allele-specific oligodeoxynucleotides coupled to an array of fluorescent microspheres, *Clin. Chem.*, 46, 996–998, 2000.

58. de Jager, W., Velthuis, H. T., Prakken, B. J., Kuis, W., and Rijkers, G. T., Simultaneous detection of 15 human cytokines in a single sample of stimulated peripheral blood mononuclear cells, *Clin. Diagn. Lab. Immunol.*, 10, 133–139, 2003.

59. Vignali, D. A., Multiplexed particle-based flow cytometric assays, *J. Immunol. Methods*, 243, 243–255, 2000.

60. Biagini, R. E., Schlottmann, S. A., Sammons, D. L., Smith, J. P., Snawder, J. C., Striley, C. A. F., MacKenzie, B. A., and Weissman, D. N., Method for simultaneous measurement of antibodies to 23 pneumococcal capsular polysaccharides, *Clin. Diagn. Lab. Immunol.*, 10, 744–750, 2003.

61. Biagini, R. E., Sammons, D. L., Smith, J. P., MacKenzie, B. A., Striley, C. A. F., Semenova, V., Steward-Clark, E. et al., Comparison of a multiplexed fluorescent covalent microsphere immunoassay (FCMIA) and an enzyme-linked immunosorbent assay (ELISA) for measurement of human IgG antibodies to anthrax toxins, *Clin. Diagn. Lab. Immunol.*, 11, 50–55, 2004.

62. Schweitzer, B., Wiltshire, S., Lambert, J., O'Malley, S., Kukanskis, K., Zhu, Z., Kingsmore, S. F., Lizardi, P. M., and Ward, D. C., Inaugural article: Immunoassays with rolling circle DNA amplification: a versatile platform for ultrasensitive antigen detection, *Proc. Natl Acad. Sci. U.S.A.*, 97, 10113–10119, 2000.

63. Wilson, A. B., McHugh, S. M., Deighton, J., Ewan, P. W., and Lachmann, P. J., A competitive inhibition ELISA for the quantification of human interferon-gamma, *J. Immunol. Methods*, 162, 247–255, 1993.

64. Aoyagi, K., Miyake, Y., Urakami, K., Kashiwakuma, T., Hasegawa, A., Kodama, T., and Yamaguchi, K., Enzyme immunoassay of immunoreactive progastrin-releasing peptide (31–98) as tumor marker for small-cell lung carcinoma: development and evaluation, *Clin. Chem.*, 41, 537–543, 1995.

65. Jones, G., Wortberg, M., Kreissig, S. B., Bunch, D. S., Gee, S. J., Hammock, B. D., and Rock, D. M., Extension of the four-parameter logistic model for ELISA to multianalyte analysis, *J. Immunol. Methods*, 177, 1–7, 1994.

66. Plikaytis, B. D., Turner, S. H., Gheesling, L. L., and Carlone, G. M., Comparisons of standard curve-fitting methods to quantitate Neisseria meningitidis group A polysaccharide antibody levels by enzyme-linked immunosorbent assay, *J. Clin. Microbiol.*, 29, 1439–1446, 1991.

67. Nix, B. and Wild, D., *Calibration Curve-Fitting*, Nature Publishing Group, New York, 2001.

68. Baud, M., Data analysis, mathematical modeling, In *Methods of Immunological Analysis*, Masseyeff, R., Ed., VCH, New York, 1993.

69. Davies, C., Concepts, In *The Immunoassay Handbook*, Wild, D., Ed., Nature Publishing Group, New York, pp. 78–110, 2001.

70. Davis, D., Zhang, A., Etienne, C., Huang, I., and Malit, M., Principles of curve fitting for multiplex sandwich immunoassays, Rev B Tech Note 2861. In. Bio-Rad Laboratories, Inc., Hercules, CA, 2002.

71. Quinn, C. P., Semenova, V. A., Elie, C. M., Romero-Steiner, S., Greene, C., and Li, H., Specific, sensitive, and quantitative enzyme-linked immunosorbent assay for human immunoglobulin G antibodies to anthrax toxin protective antigen, *Emerg. Infect. Dis.*, 8, 1103–1110, 2002.

72. Sheedy, C. and Hall, J. C., Immunoaffinity purification of chlorimuron-ethyl from soil extracts prior to quantitation by enzyme-linked immunosorbent assay, *J. Agric. Food Chem.*, 49, 1151–1157, 2001.

73. Rubio, F., Veldhuis, L. J., Clegg, B. S., Fleeker, J. R., and Hall, J. C., Comparison of a direct ELISA and an HPLC method for glyphosate determinations in water, *J. Agric. Food Chem.*, 51, 691–696, 2003.

74. MacKenzie, B. A., Striley, C. A., Biagini, R. E., Stettler, L. E., and Hines, C. J., Improved rapid analytical method for the urinary determination of 3,5,6-trichloro-2-pyridinol, a metabolite of chlorpyrifos, *Bull. Environ. Contam. Toxicol.*, 65, 1–7, 2000.

75. Strategic Diagnostics, Immunoassay Specificity, Technical Bulletin T2001, Strategic Diagnostics Inc., Newark, DE.

76. Yoshida, H., Imafuku, Y., and Nagai, T., Matrix effects in clinical immunoassays and the effect of preheating and cooling analytical samples, *Clin. Chem. Lab. Med.*, 42, 51–56, 2004.

77. Watts, C. D. and Hegarty, B., Use of immunoassays for the analysis of pesticides and some other organics in water samples, *Pure Appl. Chem.*, 87, 1533–1548, 1995.

78. Blondell, J., Epidemiology of pesticide poisonings in the United States, with special reference to occupational cases, *Occup. Med.*, 12, 209–220, 1997.

79. CDC, Third National Report on Human Exposure to Environmental Chemicals, Department of Health and Human Services Centers for Disease Control and Prevention, Atlanta, GA, 2005.

80. Richter, E. D. and Chlamtac, N., Ames, pesticides, and cancer revisited, *Int. J. Occup. Environ. Health*, 8, 63–72, 2002.

81. Fenske, R. A. and Lu, C., Determination of handwash removal efficiency: incomplete removal of the pesticide chlorpyrifos from skin by standard handwash techniques, *Am. Ind. Hyg. Assoc. J.*, 55, 425–432, 1994.

82. Lu, C. and Fenske, R. A., Dermal transfer of chlorpyrifos residues from residential surfaces: comparison of hand press, hand drag, wipe, and polyurethane foam roller measurements after broadcast and aerosol pesticide applications, *Environ. Health Perspect.*, 107, 463–467, 1999.

83. Le, H. T., Szurdoki, F., and Szekacs, A., Evaluation of an enzyme immunoassay for the detection of the insect growth regulator fenoxycarb in environmental and biological samples, *Pest Manag. Sci.*, 59, 410–416, 2003.

84. Perry, M. J., Christiani, D. C., Mathew, J., Degenhardt, D., Tortorelli, J., Strauss, J., and Sonzogni, W. C., Urinalysis of atrazine exposure in farm pesticide applicators, *Toxicol. Ind. Health*, 16, 285–290, 2001.

85. Perry, M., Christiani, D., Dagenhart, D., Tortorelli, J., and Singzoni, B., Urinary biomarkers of atrazine exposure among farm pesticide applicators, *Ann. Epidemiol.*, 10, 479, 2000.

86. Jaeger, L. L., Jones, A. D., and Hammock, B. D., Development of an enzyme-linked immunosorbent assay for atrazine mercapturic acid in human urine, *Chem. Res. Toxicol.*, 11, 342–352, 1998.

87. Lucas, A. D., Jones, A. D., Goodrow, M. H., Saiz, S. G., Blewett, C., Seiber, J. N., and Hammock, B. D., Determination of atrazine metabolites in human urine: development of a biomarker of exposure, *Chem. Res. Toxicol.*, 6, 107–116, 1993.

88. Shan, G., Wengatz, I., Stoutamire, D. W., Gee, S. J., and Hammock, B. D., An enzyme-linked immunosorbent assay for the detection of esfenvalerate metabolites in human urine, *Chem. Res. Toxicol.*, 12, 1033–1041, 1999.

89. Feng, P., Wratten, S., Horton, S., Sharp, C., and Logusch, E., Development of an enzyme linked immunosorbent assay for alachlor and its application to the analysis of environmental water samples, *Agric. Food Chem.*, 38, 159–163, 1990.

90. Driskell, W. J., Hill, R. H., Shealy, D. B., Hull, R. D., and Hines, C. J., Identification of a major human urinary metabolite of alachlor by LC–MS/MS, *Bull. Environ. Contam. Toxicol.*, 56, 853–859, 1996.

91. Ciucu, A., Bioelectrochemical methods for environmental monitoring, *Roum. Biotechnol. Lett.*, 7, 691–704, 2002.

92. Mulchandani, A. and Bassi, A. S., Principles and applications of biosensors for bioprocess monitoring and control, *Crit. Rev. Biotechnol.*, 15, 105–124, 1995.

93. Patel, P. D., Biosensors for measurements of analytes implicated in food safety: a review, *Trends Anal. Chem.*, 21, 96–115, 2002.

94. Belleville, E., Dufva, M., Aamand, J., Bruun, L., and Christensen, C. B., Quantitative assessment of factors affecting the sensitivity of a competitive immunomicroarray for pesticide detection, *Biotechniques*, 35, 1044–1051, 2003.

95. Ligler, F. S., ed., Special issue: Biosensors for indentification of biological warfare agents, *Biosen. Bioelectron.*, 14, 749–881, 2000.

96. Ligler, F. S. and Rowe Taitt, C. A., Eds., Optical biosensors: Present and future, Elsevier, 2002.

97. Svitel, J., Dzgoev, A., Ramanathan, K., and Danielsson, B., Surface plasmon resonance based pesticide assay on a renewable biosensing surface using the reversible concanavalin A monosaccharide interaction, *Biosens. Bioelectron.*, 15, 411–415, 2000.

98. Pogorelova, S. P., Bourenko, T., Kharitonov, A. B., and Willner, I., Selective sensing of triazine herbicides in imprinted membranes using ion-sensitive field-effect transistors and microgravimetric quartz crystal microbalance measurements, *Analyst*, 127, 1484–1491, 2002.

99. Dehlawi, M. S., Eldefrawi, A. T., Eldefrawi, M. E., Anis, N. A., and Valdes, J. J., Choline derivatives and sodium fluoride protect acetylcholinesterase against irreversible inhibition and aging by DFP and paraoxon, *J. Biochem. Toxicol.*, 9, 261–268, 1994.

100. Chester, G., Evaluation of agricultural worker exposure to, and absorption of, pesticides, *Ann. Occup. Hyg.*, 37, 509–523, 1993.

101. Barr, D. B. and Needham, L. L., Analytical methods for biological monitoring of exposure to pesticides: a review, *J. Chromatogr. B Analyt. Technol. Biomed. Life Sci.*, 778, 5–29, 2002.

102. Carabias-Martinez, R., Garcia-Hermida, C., Rodriguez-Gonzalo, E., Soriano-Bravo, F. E., and Hernandez-Mendez, J., Determination of herbicides, including thermally labile phenylureas, by solid-phase microextraction and gas chromatography-mass spectrometry, *J. Chromatogr. A*, 1002, 1–12, 2003.

12 Targeted and Non-Targeted Approaches for Detecting Genetically Modified Organisms

Farid E. Ahmed

CONTENTS

12.1 REGULATORY FRAMEWORKS FOR GENETICALLY MODIFIED (GM) CROPS AND DERIVED FOODS

There are two types of regulatory framework for foods derived from GM crops. First, is the horizontal process-based that focuses on the process of genetic modification, as for example, those enacted in the European Union (EU) and Australia. Second, is the vertical product-based that focuses on the resulting product characteristics, as for example, in the United States (U.S.) and Canada [1]. These two legislature processes are diametrically divergent and have often led to clashes and trade disputes [2]. Biotechnology has been readily accepted in medicine because of the tangible benefits enjoyed by consumers. However, food biotechnology, as it is commercialized today, has no obvious direct benefit to the consumer. The benefit is predominantly to the providing multinational companies and farmers, whereas any real or perceived risks are borne by consumers, resulting in a cautious stand in some countries in Europe. Other social factors, including different cultural attitudes toward food and agriculture, the lack of trust in regulatory institutions or the global agri-food industry, the reliability of scientific advice, and the economic interests of the European farmers have also contributed to Europe's resistance to food biotechnology even though the technology may have indirect potential benefits. For instance, an indirect environmental benefit could be reduced pesticide use. A social benefit could be the improvement of agriculture and food security in developing countries by using biotechnology to improve locally adopted crops and seeds as opposed to focusing on GM corn and soybean crops that work for U.S. farmers [3]. The right to choose is a central focus to the European stance against food biotechnology as the choice issue arises differently in different countries. For example, in Africa and other developing countries, multinational biotechnology companies may have little economic incentive to invest in seeds and germplasms that will not benefit poor farmers or solve local farming problems, making such innovations a public sector responsibility.

To date, in the U.S., there is a reliance on voluntary and proprietary testing that depends on compositional changes rather than on testing health effects as a result of the intake of GM crops. The optional use of labels identifying new GM foods has been the method of choice. However, a recent 2004 report by the National Academy of Sciences (NAS) jointly commissioned by the Food and Drug Administration (FDA), the U.S. Department of Agriculture, and the Environmental Protection Agency, all of which share responsibilities for regulating GM organisms and foods developed in the U.S., recognized that unintended compositional changes resulting from altera-tions—particularly by genetic engineering methods—should be assessed on a case-by-case basis. Futhermore, improved tracing and tracking methods should be implemented, when warranted, such as in specific populations of consumers or when unexplained clusters of adverse health effects occur [4]. This is a more stringent requirement than the voluntary system advocated by these govern-mental agencies just a few years ago. This is also in line with the recommendation of the United Nations Codex Alimentarious Commission that dictates a case-by-case premarket safety assess-ment of all GM foods that includes an evaluation of both direct and unintended health effects (e.g., environmental health risks that may indirectly affect human health) [5]. Although the codex principles do not have a binding effect on national legislation, they are referred to specifically in the Sanitary and Phytosanitary Agreement of the World Trade Organization (SPS Agreement) and can be used as a reference in case of trade disputes such as those brought by the U.S., Canada, and Argentina to the WTO against the EU de facto moratorium on GM crops in 2003 [2].

In the EU, the labeling and traceability of GM foods is embedded in three new EU regulations that went into effect April 18, 2004, (1829/2003, 1830/2003, and 65/2004) has been the preferred approach for responding to consumers' right of choice. Other countries (e.g., Canada, Australia, and Japan) have adopted legislations for GM foods that are either a variation on the U.S.'s or the EU's approach [1,6,7]. Monitoring is now required in the 64 countries and the EU that signed onto the adoption of the Cartagena Protocol on Biosafety in Nairobi, Kenya, in May 2000. The protocol puts into effect rules that govern the trade and transfer of GM commodity shipments and that allows governments to prohibit the import of GM food when concerns over safety exist [7]. These universal legislatures have made it imperative for governments, the food industry, crop producers, and testing laboratories to develop ways to accurately quantitate GM materials in crops, food, and food ingredients to ensure compliance with threshold levels for GM products [6].

12.2 SAMPLING

A sample is a collection of individual observations selected from a population according to specific sampling protocols. The goal of a good sampling procedure is to minimize unavoidable sampling error so that the sample is a representation of the entire population. In the case of GMO testing, the analysis is performed to gain information on the composition of a large body of target material (i.e., lots of kernels, ingredients, or final food products). However, a very small amount of material is subjected to this analysis. Therefore, a multiple sampling process employed to reduce the lot to a much smaller amount of material (i.e., the analytical sample) from the target material (i.e., lot) is key to the reliable analysis of foods and agricultural products [8].

Several organizations have developed sampling plans for raw bulk material such as seeds and grains, but not specifically for GMOs. The main parameters that should statistically be considered include lot size and uniformity, accepted risks (tolerances), and adopted testing methods. Parameters that need to be addressed include increment size, rate of increment sampling, and preparation of the sample prior to the analysis. Pragmatic aspects include available sampling facilities and costs [9]. As shown in Table 12.1, the various sampling strategies differ substantially in maximum allowable lot size, number of increments used, and number of kernels present [10]. Sampling is based on the assumption of random distribution so that the mean and standard deviation and the risk to both the producer and consumer can be estimated according to the binomial, the Poisson, or the hypergeometric distribution [8]. However, in the case of bulky material such as kernels, industrial activities tend to promote segregation during transportation and handling, leading to the conclusion that heterogeneity is much more likely than homogeneity with respect to GMO distribution in bulk commodities. Another factor that influences sample nomogeneity is the degree of lot uniformity. Because lot uniformity cannot be assessed on an a priori basis, all sampling approaches have recommended including more than one parameter per lot (see Table 12.1).

All approaches presented in Table 12.1 recognize that a systematic approach is prefered over a random sampling approach with sampling an flowing material occuring during lot loading or unloading considered to be the best choice. However, in a case when this cannot be achieved, sampling of static loads (e.g., large batches in silos or trucks) has also been performed. Moreover, the number of increments depends on the extent of heterogeneity of the GMO. As lot heterogeneity increases, the number of sample increments should increase. The lack of data on the expected distribution of real lots makes it impossible to establish objective criteria to address this issue, and research must be done to produce data that alleviates this uncertainty [8]. The recommended size of the laboratory sample varies among the various approaches [10].

Sampling plans for primary ingredients and complex food products are even more complicated because of the divergent nature of the target analyte, the concentration, and distribution in the product the production and packing process and distribution channels, these factors being some of the factors that preclude the possibility of standardizing sampling plans for primary ingredients and final food products [8]. All secondary sampling steps that are needed to produce the final samples of suitable working size do not constitute a problem provided that handling of the material, in terms of grinding and mixing, is carried out properly to minimize errors [11].

12.3 CERTIFIED REFERENCE MATERIAL

Certified reference materials (CRMs) are essential tools in quality assurance measurement and calibration processes. They are produced, certified, and used in accordance with relevant International Organization of Standardization (ISO) and Community Bureau of Reference (BCR) guidelines. Reference materials for molecules such as DNA and proteins have to behave similar to compared samples to ensure full applicability to standardization in the field (i.e., commutability). Various materials can be used for CRM for the detection of GMOs: matrix-based material produced from seeds, pure DNA, or protein standards. Pure DNA standards can be derived from genomic (g) or plasmid (p) DNA. Pure proteins can be extracted from grown seeds, or they can be produced by recombinant technology. At present, CRM are present only for the GMOs that are authorized in the EU and are available from the Institute of Reference Materials and Measurements at the Joint

TABLE 12.1
Comparison of Sampling Approaches for Grain Lots

Source	Bulk size(s)	Tolerance	Bulk sample	Laboratory sample	Increments	Increments size
ISTA	Varies according to species: 10,000–40,000 kg (max)	5%	1 kg	1 kg (approx. 3,000 maize kernels) (for analysis of contamination by other seed varieties)	One increment per 300–700 kg	Not indicated
USDA/GIPSA	Up to 10,000 bushels (approx. 254,000 kg), or 10,000 sacks if the lot is not loose	5%	Equivalent to laboratory sample	approx. 2.5 kg, but not less than 2 kg	3 cups or 1 cup per 500 bushels (approx. 12,000 kg)	1.25 kg
USDA/GIPSA, StarLink™	Follows general USDA/GIPSA guidelines		Minimum 3 times laboratory sample, in practice approx. 2.5 kg	2400 kernels		
ISO 13690	Up to 500,000 kg	Not indicated	Not indicated	> 1 kg (for kernels)	15–33 (static < 50,000 kg), "as many as possible" free-flowing	Stated as 0.2–0.5 kg, but in practice up to 5 kg
ISO 542	Up to 500,000 kg	Not indicated	100 kg	2.5–5 kg	15–33 for loose bulk (up to 500,000 kg)	Not specified
EU Dir.98/53	No limit if not separable, otherwise up to 500,000 kg	20%	30 kg (lot size 50,000 kg) 1-10 kg (lot size < 50,000 kg)	10 kg	Up to 100	0.3 kg
CEN	Up to 500,000 kg	Not indicated	20 times laboratory sample (i.e., 60 kg)	100,000 kernels	As for ISO 13690, except the number of depths per sample point not specified	Stated as 0.5 kg, but in practice up to 5 kg, if ISO 542 fully applied
FAO/WHO	Discussed, but not specified	Not indicated	Various proposals	Not discussed	Not indicated	Not indicated

Source: Adapted from Kay, S., *Comparison of sampling approaches for grain lots*, Report code EUR20134EN, Ispara, Italy, European Commission, Joint Research Center. Publication Office, 2002.

Research Center in Geel, Belgium, offered through the chemical company Fluka (Buchs, Switzerland) [12].

The suitability of these reference materials for the validation of DNA-based methods for detection by PCR, for example, has been questioned because

1. PCR can be used to quantify the GMO content on the basis of genomic equivalence (i.e., relative ratios of DNA molecules), whereas the CRMs are produced on the basis of weight equivalence. Because the number of DNA molecules in one unit of GM may differ from non-GM, a significant error may be introduced at this step.
2. To ensure homogeneity of CRMs, the materials employed in products are ground to minimize size variation. However, the grinding may degrade the DNA [13].
3. CRMs are only available in low, relative percentage concentrations (0%–5%), but the dynamic range of a method may include a 100% level.
4. CRMs are usually pure, single-ingredient matrices that may give the wrong value for the limit of detection (LOD) and, in particular, the limit of quantitation (LOQ) defined as the lowest quantities that can be reliably detected and quantified, respectively [14]. The use of plasmids containing DNA sequences of specific regions in GM maize and soybean lines or universal sequences found in various GM lines [e.g., cauliflower mosaic virus (CaMV 35S) promoter and neopalin synthesase (NOS) terminator] as reference molecules has also been reported [15].

Aspects such as ploidy or zygosity effect the selection and production of matrix-based reference material for calibration because this impacts DNA quantification. Moreover, effects such as DNA degradation, DNA quality, and length; the similarity in the behavior of the reference material used for calibration and method validation; and the DNA extracted from a field sample in PCR reactions all have important roles in the production of CRMs [16].

Matrix-based GMO CRMs have advantages over DNA- or protein-based CRMs (Table 12.2). However, the availability of the raw material for the production is restricted because of intellectual property right considerations. Since it is unlikely to find isogenic parental lines of non-GMO seeds in the near future because of cross contamination, it may be possible to produce a 0% matrix-based GMO CRM. Therefore, it is possible to employ plasmid DNA CRM because of the ease of its production and low cost. However, problems can be expected such as the topology of the reference plasmid (linearized versus circular or supercoiled); their stability and precision in very low concentrations; the absence of PCR inhibitors in plasmid compared to genomic DNA-CRM; the putative differential amplification effectiveness between pDNA and gDNA; or the variable ratio between transgenes and endogens in GM seeds [12].

12.4 METHODS FOR DETECTION OF GMOS OR THEIR MOLECULAR DERIVATIVES

There are two types of methods for GMO detection – targeted and non-targeted methods. Targeted methods detect a molecule (DNA, RNA or protein) that is specifically associated with or derived from the induced genetic modification by investigating defined constituents. These technologies employ genomic, transcriptomic, proteomic, or metabolomic methods. Non-targeted methods (also called profiling) use technologies that measure a wide range of parameters that are not defined prior to analysis [17].

12.4.1 TARGETED METHODS

The majority of targeted methods have focused on detecting DNA, though there are a few methods that have been developed for detecting protein or RNA. This occurs because

TABLE 12.2
Advantages and Disadvantages of Various Types of GMO CRMs

Type of CRM	Advantages	Disadvantages
Matrix GMO CRMs	Suitable for protein- and DNA-based methods	Different extractability (?)
		Large production needs
	Extraction covered	Low availability of raw material due to restricted use of seeds
	Commutability	Degradation
		Variation of the genetic background
Genomic DNA CRMs	Good calibrant	Large production needed
	Less seeds needed	Low availability of raw material due to restricted use of seeds
	Commutability	Variation of the genetic background
		Long-term stability (?)
Pure protein DNA CRMs	Less seeds needed	Commutability (?)
Plasmidic DNA CRMs	Easy to produce in large quantities	Plasmid topology
	Broad dynamic range	Discrepancies
		Commutability (?)

Source: From Trapmann, S., Corbisier, P., and Schimmel, H., In *Testing of genetically modified organisms in foods*, Ahmeded, F. E., Ed., Haworth Press, Binghamton, NY, 101–115, 2004. With permission.

1. DNA can be purified and multiplied exponentially by a technique such as PCR.
2. The multiplication of RNA and proteins is a more complicated and slower process.
3. DNA is a stable molecule, whereas RNA easily fragments when extracted from its normal environment unless it is stabilized by certain chaotropic reagents.
4. The stability of a protein varies and depends on the type of protein studied.
5. There usually exists a linear correlation between the GMO and nuclear DNA, whereas such correlation between the quantity of GMO and RNA or protein does not exist.
6. The genetic modification is currently accomplished at the nuclear (not mitochondrial or chloroplast) DNA levels.

12.4.1.1 Protein-Based Methods

These methods rely on a specific binding between the protein and the specific antibody that recognizes it. Then, the bound complex is detected in a chromogenic (color) reaction or in an isotopic (radioactive) reaction by a method called enzyme-linked immunosorbent assay (ELISA) [18]. Both direct double antibody (the preferred method) and indirect triple antibody sandwich ELISA methods have frequently been employed to detect and measure novel proteins produced by the GM varieties. These methods are applicable to the measurement of bivalent and polyvalent antigens, and they are referred to as a sandwich assay because the analyte is sandwiched between the solid phase antibody and the enzyme-labeled secondary antibody (direct double antibody) or a second antigen-binding antibody that is bound by enzyme-labeled antibody (indirect triple antibody) [9]. Because these methods require minimal processing of samples, they can be quickly completed. On the basis of typical concentrations of transgenic material in plant tissue (> 10 μg/tissue), the detection limits of protein immunoassays are in the range of 1% GMOs [18].

Various matrices can affect method performance. Similar to extraction efficiency, it is not necessary that methods be completely free from matrix effects provided that each sample contains a consistent amount of matrix that does not vary in composition and that the method is calibrated using a reference material of the same matrix [19]. To evaluate matrix effects, diluted non-GMO containing extracts should be used in the buffer for making the standard curve for the amounts of each GM-containing ingredient. If significant interference is observed at the selected matrix concentration (e.g., $>10\%–15\%$ inhibition or enhancement) or if the shape of the calibration curve changes, the standard should be spiked into an appropriate level of non-GM containing extract to guarantee accurate quantification [9].

Microwell plate, coated-tube ELISA, and lateral flow strip assays are presently the most common protein formats employed for GMO detection. The microplate or coated-tube ELISA formats are used as qualitative, semi-quantitative or qualitative assays, whereas the lateral flow strips are mostly qualitative. The nitrocellulose strip has an analyte-specific capture antibody immobilized on it. When the strip is inserted in a 0.5 ml Eppendorf tube containing the test solution, the solution moves upward and solubilizes the reporter antibody that then binds to the target analyte, forming an analyte-antibody complex. When the complex passes over the zone where the capture antibody has been immobilized, it binds to the antibody and produces a color band. The presence of two bands indicates a positive test for the protein of interest, whereas a single band indicates that the test was performed correctly, but the specific protein was not present (Figure 12.1). This is a fast test, taking about 10 min, that produces a qualitative or semi-quantitative answer, and it is suitable for field or on-site applications [18]. A limitation of this threshold technology is the difficulty of testing all available biotech events using a single test. Further development of this technology will require increasing the capacity to detect multiple traits in a single sample [19].

Several limitations exist for quantitative determination with protein-based methods

1. These methods cannot detect a genetic modification if the modified gene is inactive in the cells from which an analyte sample is taken, and they cannot be used to distinguish between GMOs modified to produce the same protein (e.g., authorized and unauthorized).
2. Because expression levels of introduced traits are tissue-specific and developmentally regulated, protein levels in unknown samples can hardly be compared to those present in the employed CRM.
3. An accurate measurement is only possible if sample matrices are identical to the reference material or if matched standard materials or standards that have been validated for the matrix are available.
4. Comparability with a given reference standard is similarly impaired when a specimen has been exposed to mechanical, thermal, or chemical treatment during product processing.
5. Because current immunological detection methods measure only one analyte, these methods can only be applied to food samples consisting of just one taxon [9].

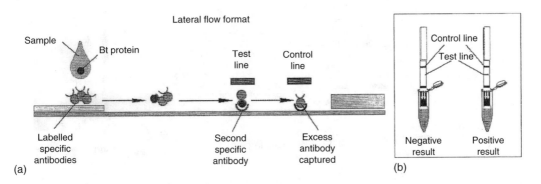

FIGURE 12.1 A schematic view of lateral flow strip assay format. Courtesy of Dean Layton, Envirrologix, Inc.

12.4.1.2 DNA-Based Methods

At present, the most commonly employed DNA-based methods involve amplification of a specific DNA with the PCR. In this technology, two short segments of synthetic DNA primers, each complementary to the DNA to be amplified, are needed. Amplification involves making two perfect copies of the original double-stranded (ds) DNA molecules. Once the cycle is completed, it can be repeated, and after 20 cycles the number of copies is $\sim 10^6$ higher than the first cycle. However, the number of cycles rapidly dwindles thereafter as a result of the exhaustion of available primers and nucleotides, and a plateau effect is reached, stopping amplification. The DNA can then be qualitatively detected, using agarose gel electrophoresis or quantitatively by high performance liquid chromatography (HPLC) or capillary electrophoresis, and the identity of amplified DNA can be further verified by DNA sequencing or other methods [20,21].

The choice of target sequence motif is the single most important factor controlling the specificity of the PCR reaction. Four types of PCR-based assays exist with various degrees of sensitivity and specificity (Figure 12.2). The first type includes screening methods where GMOs contain the transforming constructs CaMV 35S promoter or the *Agrobacterium tumefaciens* NOS terminator, and selectable marker genes coding for resistance to the antibiotics ampicillin (bla) or neomycin/kanamycin (npt II). However, the presence of these elements cannot identify the GMO, because CaMV can naturally occur in soil, and some DNA polymerases (e.g., Ampli Taq, Applied Biosystems) have been shown to contain amplifiable bla DNA. The second type includes gene specific methods where the gene of interest may be of a natural origin, but it is usually modified by truncation altered codon usage (e.g., phosphinotricin acetyltransferase (bar) gene or the synthetic truncation altered gene Cry IA (b)). Usually, a positive signal may allow identifying from which GMO the gene is derived. The third type includes construct-specific methods that target all junctions between adjacent elements of the gene construct (e.g., between the promoter and the gene of interest). A positive signal will appear only in the presence of GM-derived material although the gene construct may have been transformed into more than one GMO or may be used for future transformations (e.g., Bt 11 maize, Zeneca tomato, Mon 810 maize). The fourth type includes extent specific methods that target all junctions at the integration locus between host plant genome and the inserted recombinant DNA (e.g., Roundup Ready soybean). This method provides the highest specificity for the PCR method [14].

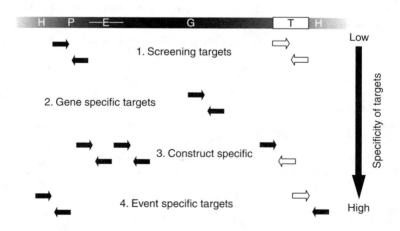

FIGURE 12.2 A schematic representation of a typical gene construct and four types of PCR-based assays, showing increasing specificity (from top to bottom). (From Holst-Jensen, A., Rønning, S. B., Løvseth, A., and Berdal, K. G., *Anal. Bional. Chem.*, 375, 2003. With permission.)

For quantitation, the copy number targeting the species-specific single copy gene of GM relative to genomic copies of the corresponding species is commonly used (e.g., soybean lectin gene for Soya-derived DNA). If the GM-specific target is inserted in a single copy, then quantitation is straightforward. For example, in Roundup Ready soybean, the full-length gene construct is inserted in a single copy number, and through backcrossing, the diploid GMO has been made homozygous. Therefore, each GM cell has a ratio of 1:1 of the GM target and the lectin gene. On the other hand, if the GM-specific target is inserted in more than one copy, quantification becomes uncertain. Moreover, heterozygosity and ploidy introduce additional uncertainty. Therefore, testing on single plants, tissues, or kernels/seeds/grains may be necessary.

There are generally three types of PCR-based quantifications – competitive, real-time, and multiplex testing. The first quantitative PCR tests were based on competitive PCR [22,23] where quantitation of the reference target is made by comparison to a competitor molecule with similar features and amplification ability, like the insert. When amplified together, standard and target DNA, compete with each other for available reagents, and the initial copy number is maintained at the end of the reaction. Following the PCR, the products are separated by gel electrophoresis that distinguishes the target DNA from the competitor by the size of the product. At the point of equilibrium (iv in Figure 12.3), the concentration of the internal strand and target are equal, and the GM content can be calculated from the ratio of standard/target $\times 100$ [23]. Although the method allowed for the detection of a GM content as little as 0.1%, it involved repeated pipettings of amplified DNA that increases the chance of contamination. Also, the process of running gels is time consuming (takes ~3 h). Therefore, this method was gradually replaced by the more sophisticated real-time PCR.

Real-time PCR allows for monitoring amplification of the products while the reaction is occurring, and it provides a larger dynamic range of amplification that leads to a higher sample throughput as well as to a more rapid quantification with reduction in the chance of cross-contamination [21]. Quantification can be done by either direct comparison of the Ct values (ΔCt) when the two fragments are amplified with the same frequency or by using a standard curve obtained using serial dilutions of the reaction products (Figure 12.4). Using the ΔCt method, two reactions of the same concentration of the unknown template DNA must be performed, one targeting a reference (R), and another targeting a GM-specific (G) sequence (black-stripped curves in Figure 12.4). Then $\Delta Ct = Ct_G - Ct_R$, and the formula $\%GM = (1/2)\Delta Ct \times 100\%$ estimates GM content.

When using a standard curve (regression line in Figure 12.4), the relative initial target copy number to Ct value allows for conversion of the Ct values of unknown samples to the initial copy number. The GM content can then be estimated by comparing the initial copy number of the two targets using the formula $\%GM = \Delta N_G/N_R \times 100\%$ [14]. Because most real-time PCR equipment on the market today has the capacity to simultaneously monitor multiple wavelengths, multiplication is recommended to simultaneously monitor many events. For a successful multiplex PCR assay, the relative concentration of the primers, the PCR buffer, the balance between the $MgCl_2$ and deoxynucleotide concentrations, the cycling temperatures, the amount of template DNA and Taq polymerase, and an optimal combination of annealing temperatures and buffer concentrations must be taken into account [20,21].

A multiplex SYBR Green I-based real-time PCR suitable for multiple GMO identification of specific amplification products for Maximizer 176, Bt 11, MON 810, GA21 maize, and GTS 40-3-2 soybean were developed employing melting curve analysis. The sensitivity levels were 1% for Maximizer 176 and GA 21 and 0.1% for GTS 40-3-2 soybean. Therefore, when further optimized, this convenient and cost effective (compared to specific FRET probes, molecular beacons, or Scorpion probes) SYBR Green method promises to provide an economic way for multiplex specific DNA quantification of several transgenic events [24]. Another multiplex PCR method for rapid screening of four transgenic maize (MON 810, Bt 11, Bt 176, and GA 21) and one transgenic soybean (Roundup Ready) gave a limit of detection of 0.25% [25]. A cheaper alternate to gel electrophoresis, was economically achieved using biotinylated probes immobilized on a membrane

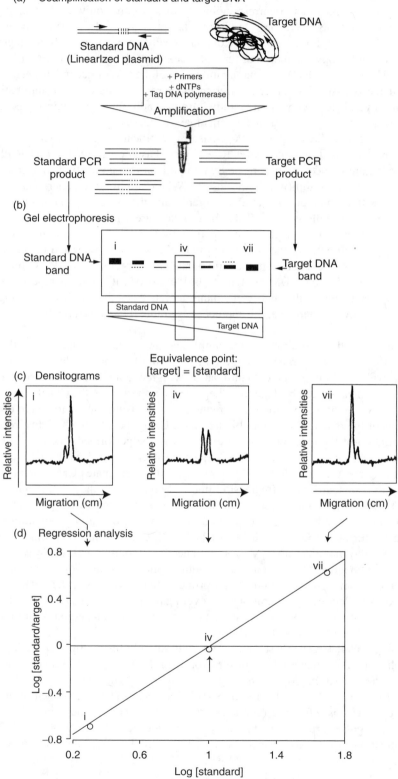

(a) Coamplification of standard and target DNA

Standard DNA
(Linearlzed plasmid)

Target DNA

+ Primers
+ dNTPs
+ Taq DNA polymerase

Amplification

Standard PCR
product

Target PCR
product

(b)
Gel electrophoresis

Standard DNA
band

Target DNA
band

Standard DNA

Target DNA

Equivalence point:
[target] = [standard]

(c) Densitograms

(d) Regression analysis

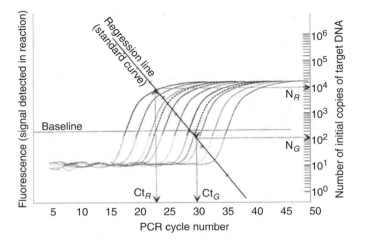

FIGURE 12.4 Real-time PCR quantification. Each amplification reaction is shown as a separate curved line. The fluorescence released in response to synthesis of the target amplicon is measured in real-time. The increase in amplification products in the initial PCR cycles is too small to be detected (dotted lines), but as the reaction proceeds, it enters a log linear phase (shown as solid lines). When the signal corresponds to a threshold line (the horizontal baseline), the corresponding cycle number for the reaction is scored (Ct value). Quantification can be carried out using a direct comparison of the Ct-values (ΔCt method) if the two targets are amplified with the same efficiency or by a comparison of copy numbers derived from a standard curve (standard curve method). In the latter method, a regression line relates initial target copy number to the Ct-value, allowing for conversion of the Ct values of the unknown sample into initial target copy numbers. In this figure, only one standard curve is shown. The two procedures, if carried out correctly, should give the same results. (From Holst-Jensen, A., RØnning, S. B., LØvseth, A., and Berdal, K. G., *Anal. Bional. Chem.*, 375, 2003. With permission.)

that hybridize with PCR products of amplified several target GM genes simultaneously and with subsequent calorimetric detection [26].

The LOD and LOQ, in addition to being method-specific, also depend on the sample being analyzed. It is nearly impossible to distinguish between these types of detection and quantification limits—the absolute limit (i.e., the lowest number of copies that must be present at the beginning of the first PCR cycle to obtain a probability of at least 95% detection correctly), the relative limit (i.e., the lowest relative percentage of GM event that can be detected or quantitated with certainty under optimal circumstances), and the practical limit (i.e., the limit applicable to the sample being assayed). It is important to distinguish between the LOD/LOQ of the test and report both values [27].

The major limitations for PCR-based detection of DNA derived from GMOs are access to information about PCR primers and a suitable DNA for analysis. For the majority of GMOs, there are currently no published primer pairs suitable for reliable amplification and quantitation because information describing the gene modification is proprietary to the biotech company. This also results in the unavailability of CRM, making method development and validation difficult. In addition, grinding, heating, acid/alkaline treatments, and other processes rapidly degrade DNA,

FIGURE 12.3 Schematic illustration of quantitative competitive PCR. Standard and target DNAs are co-amplified in the same reaction tube. (a) Following the PCR, the products are separated by gel electrophoresis that (b) distinguishes the standard DNA from the amplified target by the size of the product. At the equivalent point (iv), the starting concentrations of internal standard and target are equivalent. Densitometric analysis of the various bands (c) can be used to calculate the linear regression (d). (Modified from Hübner, P., Studer, E., and Lüthy, J., *Food Control*, 10, 1999. With permission.)

and refining effectively destroys DNA, leading to many processed products having very little low-quality GMO-derived DNA to access [14,20,21].

DNA extraction is important because the most limiting factor with respect to LOD/LOQ of the PCR test is the quality and quantity of template DNA [14]. Isolation of high quality DNA from plants is a challenge. Rigid cell walls must be dissolved through mechanical shearing and chemical means employed to release cell organelles. The tissues must be milled to release adequate quantities of DNA, but without excessive DNA degradation. Plant tissue may also contain secondary products that interfere with extraction of genomic DNA, and if these persist through subsequent steps, they may inhibit the PCR amplification. For example, polysaccharides and polyphenols (including tannins) may interfere with enzymatic reactions or even degrade the DNA. In a study that evaluated extraction methods for DNA from ground corn kernels spiked with 0.1% CBH351Starlink GM corn, six DNA extraction methods were evaluated for purity and fragment size using five commercial kits. Extraction efficiency was determined. The alcohol dehydrogenase gene (adh1) and the CBH351 (cry9C, 35S promoter) genes in the genomic DNA were determined by conventional end point and real-time PCR reactions. With every DNA extraction method employed, there were some extracted DNA samples where either the 35S promoter or cry9 (or both) were not detected by PCR, and the adh I gene was detected. Therefore, to choose an appropriate method for a particular plant or tissue type, extraction experiments must be carried out to determine the consistency of performance of trace amounts of biotechnology-derived genes. Additional purification steps may be required for certain plant genomic DNA prior to PCR amplification. Moreover, well-mixed samples need to be adequately ground to be a good representative of the entire sample [28].

12.4.2 Non-Targeted (Profiling) Approaches

Profiling approaches measure a wide range of parameters that are not defined prior to analysis. They have the advantage of being able to analyze the GM food as a whole. Genomics, transcriptomics, proteomics, and metabolomics have the potential of generating massive data that must be interpreted with caution and compared with reference data on compositional profiling and crop residues generated from crops known to be safe. These data, however, are lacking and must be developed to provide a basis for compositional equivalence and subsequent safety assessment [17,29].

12.4.2.1 Functional Genomics (Transcriptomics)

Functional genomics, or transcriptomics, refers to the study of direct expression of gene products [i.e., conversion of messenger (m) RNA into copy (c) DNA]. It has the potential to provide insight into routes that are relevant for the safety of crops and also into the natural variation in the expression of genes under different environmental conditions. In addition, this approach allows for the carrying out of expression profiles in specific plant tissues that are consumed by humans.

12.4.2.1.1   Differential Display (DD) Methods

DD has been employed to detect gene expression in GM plants. This method does not require prior knowledge of the inserted gene shows that employing targeted primer design can provide amplified transcriptional mRNA products of introduced plant gene sequences. However, because the method is labor intensive, it is unsuitable for routine expression studies on altered genes [30].

12.4.2.1.2   Microarray Technologies

These techniques have the advantage of parallel screening of a large number of known gene sequences for detecting differences in gene expression in tissues of variable origin. Two types of microarrays are available—DNA arrays that use cDNA probes immobilized spot-wise in an array fashion on solid surfaces with each spot containing numerous copies of the probe and oligo arrays that use oligonucleotide sequences (usually 20–25 bp long) instead, also spotted on various surfaces in an array fashion. Thereafter, mRNA is extracted from the sample of interest, reverse transcribed (RT) to produce a cDNA, and labeled either directly onto the backbone of the RNA molecule or indirectly during the RT process using fluorescent labels to allow easy detection by a confocal microscope. Sophisticated data analysis of the resulting pattern and relative intensities may reveal if the pattern can be attributed to a GM variety [29].

Other types of microarray systems include the electroarray system where the fluorescently labeled negative DNA fragments are guided to positively charged individual spots to increase the rate of hybridization events. Other systems that use gel-based DNA chips increase the surface of hybridization by using three-dimensional spot structure and pumping the fluorescently labeled fluid through microporous material to optimize hybridization conditions. Other systems called suspension arrays attach probes to uniquely-coated microsphere polystyrene beads (as opposed to gene-specific PCR products dotted or synthesized onto the surface for solid support) [9,17].

There are geometrical limitations for quantitatively detecting as little as 0.1% GMOs in solution because the total surface area of the chip must be compared to the small area of probe site on the chip. Another factor that limits widespread use of microarray in routine detection is the cost that may reach $500,000 just to establish a microarray facility [21]. To alleviate these problems, a signal visualization system that does not rely on fluorescence, but instead uses an enzyme-linked system that generates a dye visible to the naked eye, has been developed. The dimension of the spots where probes are attached to the chip are not a few microns as in conventional arrays, but rather, from 1 to 2 mm in diameter, hence the name Easy Read GMO Chip and are considerably more cost effective than the traditional microarray systems [31].

12.4.2.1.3 Biosensors

An affinity biosensor is a device that incorporates immobilized receptor molecules on a sensor surface that can reversibly detect nondestructively receptor ligand interactions (e.g., DNA, oligonucleotides, proteins). A stoichiometric binding event occurs in the sensor, and the associated physicochemical changes are detected by a transducer. For GMOs, two types of real-time biosensing transducers are currently used. There is the piezoelectric that is in the form of a quartz crystal microbalance (QCM), and surface plasmon resonance (SPR) [20].

In piezoelectric detection, an electric field applied to a material causes a minute mass change that can be electronically detected by the QCM. In a QCM system to detect Roundup Ready soybean CRM, single-stranded DNA probes are immobilized on the sensor surface of a QCM device and the hybridization between the immobilized probe and the target complementary sequence in solution is performed. The probe sequences are the 35S CaMV promoter and NOS terminator. The system can qualitatively detect the presence of GMOs, and it has the advantage of reusability of the sensor surface for more than 30 measurements [32].

SPR is based on transferring photons to a plasmon (a group of oscillating electrons on the metal surface). A thin film of differing dielectric (usually gold) is immobilized on the sensor chip. The surface is rinsed with fluid that contains a binding partner (e.g., GM sequence) to the surface-attached molecules (e.g., nucleotides). If the molecules from the fluid and on the surface of the chip are bound, the reflected light intensity is reduced, and this reduction can be measured [33]. Advantages of this technology are that it does not require highly purified samples, and detection can be carried out in one step in a real-time manner. A disadvantage, however, is that the sensitivity needs to be upgraded if it is needed to detect GM-foods at a sensitivity of 0.5% [20,21].

Electrochemical DNA sensors have also been developed [34] as well as DNA chips based on nanotechnology with Raman spectroscopic fingerprinting for DNA and RNA detection [35]. However, but these technologies also have to be further developed and their sensitivities increased to allow quantitative detection of GMOs present at low concentrations.

12.4.2.1.4 Proteomics

Proteomics is the large-scale study of protein properties to gain an understanding of gene function at the translational level. Unlike genomics, ultimately proteins undergo post-translational modifications (e.g., glycosylation and phosphorylation). Furthermore, protein cleavage and multi-protein complex formation occurs, and these processes might influence the functioning of protein molecules. Also, proteomics has no equivalent amplification mode (as for example by PCR), and the sensitivity of the method may be an issue in certain circumstances [36].

12.4.2.1.5 Plant Protein Separation and Identification

Currently, protein separation is largely based on two-dimensional gel electrophoresis (2-DE) where separation in the first dimension is by charge [isoelectric focusing (IEF)], and in the second dimension, according to the molecular weight, of the molecule by employing sodium dodecylsulfate polyacrylamide gel electrophoresis (SDA-PAGE). The combination of these two separation techniques allows for a resolution of up to 10,000 protein spots in a single 2-DE gel [37]. Various stains (e.g., Coomassie Brilliant Blue G and R dyes, silver staining, SYPRO fluorescent dyes, and cyanine-based dyes) have been employed for detection, followed by image analysis of the 2-DE gel. Proteins can then be identified by matching the images to the existing 2-DE databases. However, only a limited number of these databases are currently available [17].

A major advantage was achieved when it became possible to identify large polymer-like proteins by mass spectrometry (MS), coupled with a peptide mass fingerprinting strategy and MS-based peptide sequencing [29,36]. Employing large scale MS-based proteome studies revealed that two classes of proteins were underrepresented—hydrophobic proteins that are very difficult to keep in solution under conditions required by the IEF in the 2-DE gel, and low abundant proteins that are difficult to display on 2-DE gels because they are below the detection limits of the protein stains currently available. Furthermore, protein populations from a biological sample can show immense variability in protein expression levels with a dynamic range of up to 10^8 that makes parallel visualization of cell expressed proteins from such samples almost impossible [38].

Protein separation techniques that do not use gel electrophoresis have been tried. A liquid chromatography-based peptide separation strategy, named multidimensional protein identification technology (MudPIT), was developed [39]. The basis for this technology is the digestion of a complex mixture of proteins with a site-specific protease before separation with two independent liquid chromatographic (LC) separation systems. Nevertheless, this technique did not identify all rice proteins from three different tissues (leaf, root, and seed) [40]. Furthermore, because of a high number of peptides co-eluting from the chromatographic separation column into the mass spectrometer, the MudPIT technology did not seem to be suitable for differential analysis of a complex proteome. Additionally, MudPIT did not produce quantitative expression data of the identified proteins [20,36].

To overcome the above-described problems for protein identification and quantification, a new LC-MS technique called isotope-coded affinity tag (ICAT) was introduced. It allows for the comparison of two different protein products by making use of an affinity-based peptide-labeling step. In this method, a complex protein is digested with a sequence-specific protease. The cysteine-containing peptides are subsequently labeled with a regular ICAT reagent, and the peptides from a second sample are labeled with an ICAT reagent that contain a distinct number of heavy isotopes, making the affinity tag distinguishable from the regular ICAT. The labeled peptides are mixed, chromatographically separated, and directly eluted into an electrospray mass spectrometer. The corresponding peptides for the different samples are separated by the number of heavy isotopes times the number of cysteines in the peptides when the mass difference between the heavy and light isotope is unity. Because a mass spectrometer measures a mass-to-charge ratio, this correlation holds when the corresponding peptide ions are singly charged. For doubly charged particles, the value of the mass difference will be halved [41].

Protein expression profiling can also be analyzed by antibody arrays [42]. A key to this technology is the availability of highly purified and specific monoclonal antibodies. Although these have classically been generated by hybridoma technology, generation of these on a scale to cover the entire plant proteome will make the array exceedingly expensive. Antibodies can be economically and selectively produced in a high-throughput fashion against multiple plant antigens by phage selection of single chain Fv or Fab fragments (thereby called phage display) [43]. Phage display method offers several advantages over hybridoma technology

1. It can be used to create antibodies against immunogens as well as highly conserved antigens.
2. The introduction of a multiple spotting technique, that allows for multiplex screening of a thousand of Fabs or Fvs on a protein microarray representing the corresponding antigens, saves time and reagents [44].
3. The flexibility of phage display method allows it to be performed in vivo and in vitro.
4. Antibody fragments can be selected using inexpensive purification methods in the bacterium *Escherichia coli* [45].

Alternative strategies to phage display include ribosomal display or mRNA-protein fusion. Additionally, binding scaffolds, other than immunoglobulin domains, such as affibodies, fibronectin, lipocalin, or repeated domains have been employed [36]. Moreover, certain oligonucleotides (aptmers) can bind certain antigen proteins to antibodies at a low cost [46].

Antibody microarrays enable fast, easy, and parallel detection of a thousand samples with a single experiment. Originally, they were developed in the format of clinical ELISA microtiter plates with up to 96 wells [47]. Then, high density spotting of bacteria onto nitrocellulose filters ensued [48]. Optimization of this technique with the use of immobility solution chemistry and handling continued [44]. Although applications of antibody microarrays for specific investigation of plant proteome have not been widely reported, profiling of plants in response to environmental stresses could yield reliable targets for the development of plants for bioremediation or food production in adverse conditions [36]. Biosensors have also been used to screen proteins [49,50].

Proteomic approaches are crucial to an understanding of development, structure, and metabolism, and they are promising for the detection and understanding of unintended effects in GM crops when fully developed in the near future. Proteomics, combined with immunological detection methods, may also be useful for studying whether or not fusion proteins are produced from transgenic integration sites [17].

12.4.2.3 Metabolomics

The entire collection of metabolites in the cell is called a metabolome, and the science of measuring it is metabolomics. There are currently four types of metabolomic investigations: target compound analysis (i.e., analysis of specific compounds most directly affected by the modification); metabolic profiling (i.e., analysis of selected compounds for the same chemical group or compound linked by a known metabolic relationship); fingerprinting (i.e., rapid screening for sampling classification as, for example, by global analysis of spectroscopic data); and metabolomics (i.e., identification and quantification of as many individual compounds as possible across all compound classes). The main techniques for studying target compound analysis and metabolic profiling are chromatographic separations coupled with well-developed calibrations for specific analysis. For fingerprinting screening, methods like nuclear magnetic resonance (NMR), Fourier transform infrared spectroscopy (FTIR), or MS applied to crude extracts without prior separation steps are used. FTIR as a fingerprinting method has the advantage of ease of sample preparation, the speed of data acquisition, and a high degree of reproducibility [52]. Although the spectra are not necessarily interpreted as with other methods, subtle differences may be detected by chemometric methods, employing cluster analysis and soft independent modeling class analogy [53]. Metabolomics combines the first, second, and third elements listed above, aiming at accurate quantitative comparison and recognizing that extraction and measurement methods cannot be optimized for any single compound assessed [51]. Review of the different extraction methods used in metabolomics show that no direct comparison of performance regarding recovery and reproducibility has been made [51]. No single technique meets all the requirements for an ideal metabolomics measurement method. Gas chromatography (GC) provides high-resolution compound separation. However, most metabolites found in plant extracts are too involatile to be analyzed directly by GC. The compounds have to be first derivatized and converted to less polar, more volatile derivatives that may limit the utility of this technique when many samples need to be analyzed. GC combined with flame ionization detection (FID) [54] or, more recently, MS [55] have been used for routine detection – the latter methods allowing quantitation of minor peaks when shadowed by minor ones.

HPLC with ultraviolet (UV) detection has been used to identify compounds in complex profiles, but it often indicates the class of the compound rather than its exact identity [17]. The low sensitivity of LC/NMR spectrometry to LC/MS indicates that it will presently be used for structural characterization of the unknown rather than for comparative analysis of numerous samples [56].

Proton (^{1}H) NMR can detect any metabolite containing hydrogen, and their spectra of plant extracts are crowded not only by a large number of contributory compounds but also because of the low overall chemical shift dispersion and by spin-spin couplings. In ^{14}C NMR, the chemical shift dispersion is twenty times greater, and spin-spin interactions are removed by decoupling. However, the low sensitivity of ^{14}C NMR prevents its routine use with complex extracts [57].

Novel foods might be tested for substantial equivalence (i.e., equivalence to known safe food varieties) with respect to accumulated data of acceptable samples against a background of novel ones in the same way GC/MS and NMR data have been used for diagnosis of metabolic disorders. Pattern recognition methods can highlight abnormal samples against a background of normal ones, but this will require a large number of samples covering different genotypes, growing seasons and locations as well as stress conditions to model the acceptable class. These reference data should be generated from crops with an acceptable safety factor and used as crop specific benchmarks to compare with GM crops. At present, such reference data are not available, but should be generated. This will also require methods of data standardization across test laboratories [17].

With metabolomics, it may be easier to find some differences between two sets of samples than to convincingly prove that there are no differences. Nevertheless, as long as the differences are well defined, their importance can be assessed and any uncertainty lessened [17,29].

12.4.2.4   Near-Infrared Spectroscopic Methods

Non-targeted methods employing Near-Infrared Spectroscopic Methods (NIR) are commonly used at commodity food handling facilities throughout the world for nondestructive analysis of grains that will predict moisture, protein, oil, fiber, and starch. This is due to their combination of analytical power, rapidity, and ease of use as well as their nondestructive nature and low cost. Although NIR spectrometers are not precise enough to detect compounds such as DNA or protein at low concentration levels (parts per trillion), differences caused by large structural changes accompanying the GM modification might be measured [58]. A feasibility study was carried out on Roundup Ready soybean by a NIR Infratec 1220 spectrometer where whole grain samples flowed through a fixed path length. Locally Weighed Regression, using a database of ~8000 samples was 93% accurate for distinguishing the transgenic from the unmodified soy [59]. Another study was performed to create a calibration for discriminating between individual kernels of Bt and not Bt corn. It was possible to classify kernels with an average accuracy, in cross validation, of nearly 99% using partial least square discriminant analysis models [60].

TABLE 12.3
Summary of Methods Employed for Detection of Genetically Modified Organisms in Food

Parameter	Protein-based			DNA-based				Nontarget testing	
	ELISA	Lateral flow strips	Protein micro-arrays	QC-PCR	Real-time PCR	DNA micro-arrays	DNA[a] sensors	Metabolic profiling[b]	NIR
Ease of use	Moderate	Simple	Difficult	Difficult	Difficult	Difficult	Simple	Difficult	Simple
Needs special equipment	Yes	No	Yes	Yes	Yes	Yes	Yes	Yes	Yes
Sensitivity	High	High	Moderate	Very high	Very high	High	Moderate	Moderate-High	Moderate
Quantitative	Yes	No	No	Yes	Yes	No	No	Yes	No
Duration[c]	30–90 min	10 min	6 h[b]	2 days	1 day	2 days	2 days	2 days	1 hour
Cost/sample	U.S.$5	U.S.$2	U.S.$600	U.S.$350	U.S.$450	U.S.$600	U.S.$200	U.S.$500–4,000	$10
Provides quantitative results	Yes[d]	No	No	Yes	Yes[e]	Yes	No	Yes	No
Suitable for field test	Yes[d]	Yes	No	No	No	No	Yes	No	Yes
Employed mainly in	Test facility	Field testing	Academic[f] labs	Test facility	Test facility	Academic[f] labs	Field[f] testing	Research labs	Field[f] testing

[a] Applies also t0 protein chips.
[b] Includes GC, LC, MS, NMR or a combination thereof.
[c] Excluding time allotted for sample preparation.
[d] As in the antibody-coated tube format.
[e] With high precision.
[f] In development

The advantages of NIR are speed in that it takes <1 min to perform; sample preparation in that it is not required because the process uses whole kernels (~ 300 g) dropped into a measurement cell or flow through system; and cost in that it is relatively inexpensive. Among the disadvantages of this method, a major one includes the need for a large set of samples to generate spectra that are then used to predict the GM event. Therefore, this method cannot be more accurate than the reference method used to build the model. Second, calibrations need to be developed for each GMO to be predicted, and the accuracy of the method is limited because it does not detect a change in DNA or a single protein, instead, it detects much larger unknown structural changes that are introduced by the presence of the new DNA [58].

12.5 CONCLUSIONS

Table 12.3 summarizes the tests commonly employed for the detection of GMOs in foods and ingredients with respect to ease of use, requirements for specialized equipment, assay sensitivity, time to perform, quantitation, cost, and suitability for routine field testing.

Detailed characterization of the transformation events at the sequence level have been carried out for a large number of GMOs currently available. This allows for detailed genetic maps, including information essential for the development of event specific detection assays as well as details and the nature of insertion loci and potential unintended partial inserts and rearrangements of the host genomic DNA. However, the co-presence of event-specific sequence motifs observed in the so-called gene-stacked GMOs (i.e., the offspring of hybrids produced when two or more GMOs are crossed) will not be detected by event-specific detection and quantitation methods that do not distinguish between the stacked hybrid and parental GMOs [9].

A 100% identification of an ingredient stable in all varieties in terms of copy number per haploid genome and in terms of alleles based on DNA sequences unique to a single species has been achieved. These DNA sequences can serve as references for PCR-based quantitations. On the other hand, homozygous and heterozygous as well as diploid, triploid, and polyploidy lines will only give different qualitative estimates of the GMO by currently available detection methods. Moreover, it is unlikely that the DNA content per weight unit (for grains, for example) is the same when produced from different lines or when subjected to different handling conditions. It is believed that the analytical steps from sampling to extraction of analyte (whether protein or DNA) account for most measurement uncertainty of GMO content [9]. Multiplexing of quantitative analysis is a necessity in view of the increasing number of GMOs to be tested. Additionally, future developments in microarray technology hold promise for multiplex quantification [20].

Appropriate traceability and segregation systems may reduce the need for stringent testing strategies. In addition, measured GM content in processed material may not always reflect the actual GMO content in the raw material. This may affect the efficacy of the analytical control options in traceability systems. Threshold limits strongly influence the necessary separation practices aimed at meeting these limits. The lower the threshold, the stricter and more costly the separation practices will be. Moreover, the number of authorized and unauthorized events will affect the implementation of traceability [9].

REFERENCES

1. König, A., Cockburn, A., Cervel, R. W. R., Debruyne, E., Grafstroem, R., Hammerling, U., Knudsen, I. et al., Assessment of the safety of foods derived from genetically modified (GM) crops, *Food Chem. Toxicol.*, 42, 1047–1088, 2004.
2. Kinderlerer, J., The WTO complaint—Why now?, *Nature Biotechnol.*, 21, 735–736, 2003.
3. Taylor, M. R., Rethinking U.S. leadership in food biotechnology, *Nature Biotechnol.*, 21, 852–854, 2003.

4. NAS (National Academy of Sciences), *Safety of Genetically Engineered Foods: Approaches to Assessing Unintended Health Effects*, National Academy Press, Washington, DC, 2004.
5. Codex Alimentarius Commission, Joint FAO/WHO Foof Standard Programme, *Codex Ad Hoc Intergovernmental Task Force on Foods Derived from Biotechnology* (Codex, Yokohama, Japan) http://www.who.int/fsf/GMfood/codex_index.htm; http://www.codexalimentarius.net/ccfbt4/bt03_01e.htm (11–14 March, 2003).
6. Ahmed, F. E., Ed., *Testing of Genetically Modified Organisms in Foods*, Haworth Press, Binghamton, NY, 2004.
7. Gupta, J. A., Governing trade in genetically modified organisms: The Cartagena Protocol on Biosafety, *Environment*, 42, 22–23, 2000.
8. Paoletti, C., *Sampling for GMO analysis: The Euripean prespective Testing of Genetically Modified Organisms in Foods*, Haworth Press, Binghamton, NY, 2004.
9. Miraglia, M., Berdahl, K. G., Brera, C., Corbisier, P., Holst-Jensen, A., Kok, E. J., Marvin, H. J. P., et al., Detection and traceability of genetically modified organisms in the food production chain, *Food Chem. Toxicol.*, 42, 1157–1180, 2004.
10. Kay, S., *Comparison of sampling approaches for grain lots*, Report code EUR20134EN, Ispara, Italy, European Commission, Joint Research Center Publication Office, 2002.
11. Remund, K., Dixon, D. A., Wright, D. L., and Holden, L. R., Statistical considerations in seed purity testing for transgenic traits, *Seed Sci. Res.*, 11, 101–119, 2001.
12. Trapmann, S., Corbisier, P., and Schimmel, H., Reference material and standards, in *Testing of Genetically Modified Organisms in Foods*, Ahmed, F. E., Ed., Haworth Press, Binghamton, NY, pp. 101–115, 2004.
13. RØnning, S. B., Vaitilingom, M., Berdal, K. G., and Holst-Jensen, A., Event specific real-time quantitative PCR for genetically modified Bt11 maize (*Zea mays*), *Eur. Food Res. Technol.*, 216, 347–354, 2003.
14. Holst-Jensen, A., RØnning, S. B., LØvseth, A., and Berdal, K. G., PCR technology for screening and quantification of genetically modified organisms (GMOs), *Anal. Bional. Chem.*, 375, 985–993, 2003.
15. Kuribara, H., Shindo, Y., Matsuoka, T., Takubo, K., Futo, S., Aoki, N., Hirao, T. et al., New reference molecules for quantitation of genetically modified maize and soybean, *JOAC Int.*, 85, 1077–1089, 2002.
16. Corbisier, P., Trapmann, S., Ganceberg, D., Van Iwaarden, P. Y., Hannes, L., Catalani, P., Le Guern, L., and Schimmel, H., Effect of DNA fragmentation on the quantitation of GMO, *Eur. J. Biochem.*, 269 (suppl 1), 40, 2002.
17. Cellini, F., Ghesson, A., Colquhoun, I., Constable, A., Davies, H. V., Engel, K. H., Galehause, A. M. R. et al., Unintended effects and their detection in genetically modified crops, *Food Chem. Toxicol.*, 42, 1089–1125, 2004.
18. Ahmed, F. E., Protein-based methods: Elucidation of the principles, in *Testing of Genetically Modified Organisms in Foods*, Ahmed, F. E., Ed., Haworth Press, Binghamton, NY, pp. 117–146, 2004.
19. Stave, J. W., Protein-based methods: Case studies, in *Testing of Genetically Modified Organisms in Foods*, Ahmed, F. E., Ed., Haworth Press, Binghamton, NY, pp. 147–161, 2004.
20. Ahmed, F. E., DNA-based methods for GMOs detection: Historical developments and further perspectives, in *Testing of Genetically Modified Organisms in Foods*, Ahmed, F. E., Ed., Haworth Press, Binghamton, NY, pp. 221–253, 2004.
21. Fagan, J., DNA-based methods for detection and quantification of GMOs: Principles and standards, in *Testing of Genetically Modified Organisms in Foods*, Ahmed, F. E., Ed., Haworth Press, Binghamton, NY, pp. 163–220, 2004.
22. Hardegger, M., Brodmann, P., and Hermann, A., Quantitative detection of the 35S promoter and the NOS terminator using quantitative competitive PCR, *Eur. Food Res. Technol.*, 209, 83–87, 1999.
23. Hübner, P., Studer, E., and Lüthy, J., Quantitative competitive PCR for the detection of genetically modified organisms in food, *Food Control*, 10, 353–358, 1999.
24. Hernández, M., Rodríguez-Lázaro, D., Esteve, T., Prat, S., and Pla, M., Development of melting temperature-based SYBR Green I polymerase chain reaction methods for multiplex genetically modified organism detection, *Anal. Biochem.*, 323, 164–170, 2003.

25. Germini, A., Zanethi, A., Salati, C., Rossi, S., Forre, C., Schmid, S., Marchelli, R., and Fogher, C., Development of a seven-target multiple PCR for the simultaneous detection of transgenic soybean and maize in feeds and food, *J. Agr. Food Chem.*, 32, 3275–3280, 2004.

26. Su, W., Song, S., Long, M., and Liu, G., Multiplex polymerase chain reaction/membrane hybridization assay for detection of genetically modified organisms, *J. Biotechnol.*, 105, 227–233, 2003.

27. Berdal, K. G. and Holst-Jensen, A., Roundup Ready soybean event specific real-time quantitative PCR assay and estimation of the practical detection and quantification limits in GMO analysis, *Eur. Food Res. Technol.*, 213, 432–438, 2001.

28. Holden, M. J., Blasic, J. R. Jr., Bussjaeger, L., Kao, C., Shokere, L. A., Kendall, D. C., Freese, L., and Jenkins, G. R., Evaluation of extraction methodologies for corn kernel (*Zea mays*) DNA for detection of trace amounts of biotechnology-derived DNA, *J. Agr. Food Chem.*, 51, 2468–2474, 2003.

29. Ahmed, F. E., Other methods for GMO detection and overall assessment of the risks, in *Testing of Genetically Modified Organisms in Foods*, Ahmed, F. E., Ed., Haworth Press, Binghamton, NY, pp. 285–313, 2004.

30. Kok, E. J., van der Wal-Winnubst, E. N. W., Van Hoef, A. M. A., and Keijer, J., Differential display of mRNA, In *Molecular Microbial Ecology Manual*, Akkermans, A. D. L., van Elsas, J. D., and de Bruijn, F. J., Eds., Kluwer, Dordrecht, The Netherlands, pp. 1–10, 2001.

31. Fagan, J., DNA-based methods for detection and quantification of GMOs: principles and standards, In *Testing of Genetically Modified Organisms in Foods*, Ahmed, F. E., Ed., Haworth Press, Binghamton, NY, pp. 163–220, 2004.

32. Mannelli, I., Minunni, M., Tombelli, S., and Mascini, M., Quartz crystal microbalance (QCM) affinity biosensor for genetically modified organisms (GMOs) detection, *Biosens. Bioelectron.*, 18, 129–140, 2003.

33. Mariotti, E., Minunni, M., and Mascini, M., Surface plasmon resonance biosensor for genetically modified organism detection, *Anal. Chemica Acta.*, 453, 165–172, 2002.

34. Drummond, T. G., Hil, M. G., and Bantun, J. K., Electrochemical DNA sensors, *Nature Biotechnol.*, 21, 1192–1199, 2003.

35. Cao, Y. W. C., Jin, R., and Mirkim, C. A., Nanoparticle with Raman spectroscopic fingerprints for DNA and RNA detection, *Science*, 297, 1536–1540, 2003.

36. Kersten, B., Feilner, T., Angenendt, P., Giavalisco, P., Brenner, B., and Bürke, L., Proteomic approach in plant biology, *Curr. Proteomics*, 1, 131–144, 2004.

37. Klose, J. and Koblaz, U., Two-dimensional electrophoresis of proteins in an updated practical and complication for a functional analysis of the genome, *Electrphoresis*, 16, 1034–1059, 1995.

38. Corthals, G. L., Wasinger, V. C., Hochstrasser, D. F., and Sanchez, J. C., The dynamic range of protein expression. A challenge for proteomic research, *Electrophoresis*, 21, 1104–1115, 2000.

39. Lin, D., Alperts, A. J., and Yates, J. R. III, Multidimensional protein identification technology as an effective tool for proteomics, *Am. Genomics/Proteomics Technol.*, 38–47, Jul/Aug, 2001.

40. Koller, A., Washburn, M. P., Lange, B. M., Andon, N. L., Deciu, C., Haynes, P. A., Hays, L., and Schieltz, D., Proteomic survey of metabolic pathways in rice, *Proc. Natl. Acad. Sci.*, 99, 11969–11974, 2002.

41. Gygi, S. P., Rist, B., Greber, S. A., Turecek, F., Gelb, M. H., and Abersold, R., Quantitative analysis of complex protein mixtures using isotope-coded affinity tags, *Nature Biotechnol.*, 17, 994–999, 1999.

42. Kusnezow, W. and Hoheisel, J. D., Antibody microarrays: Promises and problems, *Biotechniques*, 33, S14–S23, 2002.

43. Halborn, J. and Carlsson, R., Automated screening procedure for high-throughput generation of antibody fragments, *Biotechniques*, 32 (suppl), 30–37, 2002.

44. Angenendt, P. and Glökler, J., Evaluation of antibodies and microarray coatings as a prerequisite for the generation of optimized antibody microarray, *Methods Mol. Biol.*, 278, 123–134, 2004.

45. Krebs, N., Rauchenberger, R., Reiffert, S., Rothe, C., Tesar, M., Thomassen, E., Cao, M., and Dreier, T., High-throughput generation and engineering of recombinant human antibodies, *J. Immunol. Methods*, 254, 67–82, 2001.

46. Schneider, D., Gold, L., and Platt, T., Selective enrichment of RNA species for tight binding of *Escherichia coli* rho factor, *FASEB J.*, 7, 201–207, 1993.

47. Engvall, E. and Perlman, P., Enzyme-linked immunosorbent assay (ELISA). Quantitative assay of immunoglobulin, *G. Immunochem.*, 8, 871–874, 1971.

48. de Wildt, R. M., Mundy, C. R., Gorick, B. D., and Tomlinson, I. M., Antibody arrays for high-throughput screening of antibody-antigen interactions, *Nature Biotechnol.*, 18, 989–994, 2000.
49. Wink, T., van Zuilen, S. J., Bult, A., and van Bennekom, W. P., Liposome-mediated enhancement of the sensitivity in immunoassays of proteins and peptides in surface plasmon resonance spectrometry, *Anal. Chem.*, 10, 827–832, 1998.
50. Lin, C. K., Tai, D. F., and Wu, T. Z., Discrimination of peptides by using a molecularly imprinted piezoelectric biosensor, *Chem. Eur. J.*, 9, 5107–5110, 2003.
51. Fiehn, O., Metabolomics—the link between genotypes and phenotypes, *Plant Mol. Biol.*, 48, 155–171, 2002.
52. Aharoni, A., de Vos, C. H. R., Verhoeven, H. A., Mahepoard, C. A., Kruppa, G., Bino, R., and Goodenowe, D. B., Nontargeted metabolome analysis by use of Fourier Trranfer ion cyclotron mass spectrometry, *OMICS*, 6, 217–234, 2002.
53. Kemsely, E. K., *Discriminant Analysis and Class Modeling of Spectroscopic Data*, Wiley, Chinchester, 1998.
54. Frenzel, Th., Miller, A., and Engel, K. H., A methodology for automated comparative analysis of metabolite profiling data, *Eur. Food Res. Technol.*, 216, 335–342, 2003.
55. Fiehn, O., Kopka, J., Dörman, P., Altmann, T., Trethenly, R. N., and Willmitzer, L., Metabolite profiling for plant functional genomics, *Nature Biotechnol.*, 118, 1157–1161, 2000.
56. Wolfender, J. L., Ndjoko, K., and Hostettman, K., The potential of LC-NMR in phytochemical analysis, *Phytochem. Anal.*, 12, 2–22, 2001.
57. Charlton, A., Allnutt, T., Holmes, S., Chroholm, J., Bean, S., Ellis, N., Mullineaux, P., and Oehlschlager, S., NMR profiling of transgenic pea, *Plant Biotechnol.*, 2, 27–35, 2004.
58. Roussel, S. A. and Cogdill, R. B., Near-infrared spectroscopic methods, in *Testing of Genetically Modified Organisms in Foods*, Ahmed, F. E., Ed., Haworth Press, Binghamton, NY, pp. 1425–1432, 2004.
59. Roussel, S. A., Hardy, C. L., Hurburgh, C. R. Jr., and Rippke, G. R., Detection of Roundup Ready soybean by near infrared spectroscopy, *Appl. Spec.*, 55, 1425–1432, 2001.
60. Cogdill, R. P., Hurburgh, C. R., Jr., and Rippke, G. R., Single-kernel maize analysis by near-infrared hyperspectral imaging, *Trans. ASAE*, 47, 311–320, 2004.

13 Bioanalytical Diagnostic Test for Measuring Prions

Loredana Ingrosso, Maurizio Pocchiari, and Franco Cardone

CONTENTS

13.1   INTRODUCTION

Prion diseases, or transmissible spongiform encephalopathies (TSEs), are fatal neurodegenerative disorders that include several human (Creutzfeldt-Jakob disease, CJD, variant Creutzfeldt-Jakob disease, vCJD, Gerstmann–Sträussler–Scheinker syndrome, GSS, fatal familial insomnia, FFI and sporadics fatal insomnia, sFI) and animal disorders (scrapie of sheep and goats, bovine spongiform encephalopathy, FSE, BSE and felina Spongiforn encephalopaty, FSE transmissible mink encephalopathies, TME, chronic wasting disease of cervids, CWD) [1].

Human prion diseases were recognized as independent nosological entities at the beginning of the 1920s. They are distributed worldwide with an incidence of about 1.5 cases per million people per year [2] and are classified as sporadic (sCJD), genetic, or acquired forms, depending on their etiology [3]. Affected patients present with a progressive neurological syndrome that usually lasts a few months, but longer durations have been reported, mostly in genetic cases [4]. The vast majority of cases are represented by the sporadic form, wheras the genetic forms account for about 5%–10% of TSEs and are all linked to mutations (insertion or deletions) of the PrP gene (*PRNP*) [5]. Acquired forms include a few hundred incidences of iatrogenic cases that are due to accidental transmission of infected materials either by peripheral (intramuscular injection of contaminated growth hormone of cadaveric origin) or by central (implant of cadaveric dura mater) routes [6]. Among acquired human TSEs, a disease called kuru, although now disappearing, retains a certain value for historical reasons, being the first human TSE shown to be transmissible to laboratory animals [2] and for its peculiar way of transmission. This disorder plagued a few tribes of Fore language in Papua New Guinea. It mostly affected women and children involved in ritual cannibalism practiced as a way of mourning. The dismissal of such practice during the last half of the last century resulted in a gradual disappearance of this disease. The third form of human acquired TSEs is related to the outbreak of BSE in the United Kingdom that is believed to be the cause of the appearance of a novel nosological entity called variant CJD (vCJD), affecting young adults [7]. These relatively new diseases (BSE and vCJD) have reaffirmed the importance of the oral route for TSE transmission both intra- and inter-species. Both in BSE and in vCJD, the highly suspected way of infection is the consumption of TSE-contaminated food. In particular, cows were subjected to a type of cannibalism as carcasses of TSE affected cows entered the rendering plants to be transformed into meat and bone meal (MBM) for production of high protein animal food [8], whereas in humans, the transmission of the disease most likely occurred by consumption of BSE-contaminated meat [9]. Variant CJD is almost exclusively confined to the U.K. ($n = 162$) and France ($n = 20$), but a few isolated cases have been officially confirmed in Ireland (since 2000), Italy (2002), Canada (2002), U.S.A. (2004), Saudi Arabia (2004), Japan (2005), Netherlands (2005), Portugal (2005), and Spain (2005). Variant CJD presents some peculiarities, consisting of an early age at onset, presence of early psychiatric symptoms, a long clinical duration, and distinct histopatological lesions with massive accumulation of PrP^{TSE} with the formation of "florid amyloid plaques" [10,11].

TSEs are caused by unconventional infectious agents called prions that are believed to be mainly, if not solely, composed by the pathological conformer (PrP^{TSE}) of the normal cellular prion protein (PrP^c). They are apparently devoid of any nucleic acids [1], thus some of the most powerful diagnostic methods such as polymerase chain reactions, antigen, or antibody titration, commonly used for the identification of other infectious agents, are inadequate for detection.

The central event in TSEs pathogenesis is the conformational switch of PrP^c into PrP^{TSE} (also called PrP^{Sc} after its first identification in rodents experimentally infected with the scrapie agent) that aggregates as rods or fibrils and accumulates into neural and often lymphoreticular cells [12,13]. Histological and biochemical lesions are restricted only to the CNS (spongiform encephalopathy) where there are the largest amounts of PrP^{TSE} and infectivity [1,3,14].

Experimental studies in rodent models and occasional observations in natural TSEs indicate that after peripheral inoculation, the infectious agent usually replicates in cells, particularly the follicular dendritic cells, of lymphoreticular tissues (spleen, lymph nodes, Peyer's patches) [15]. From this point, it enters local nerves and, following neural pathways, reaches the spinal cord or the brain stem, depending upon the site of infection and then colonizes the brain where replication continues until the development of clinical signs of disease and death of the host [16,17]. During this asymptomatic period, both the infectious agent (as measured by bioassay) and PrP^{TSE} (as measured by immunochemistry) are readily detectable in lymphoreticular tissues [18]. This fact has lead to the suggestion of measuring PrP^{TSE} in bioptic tonsil tissue for the diagnosis of scrapie in sheep [19] and, more recently, of vCJD [20–22]. The PrP^{TSE} diagnostic test in tonsils is unsuitable for other forms of human TSEs or for BSE because in these diseases, the level of PrP^{TSE} in tonsil tissues is either too low for detection or non-existent [23]. In Sporadic and genetic TSEs PrP^{TSE} accumulates mainly in the CNS, making difficult the development of early diagnostic methods based upon the detection of PrP^{TSE} in easily accessible tissues or body fluids. The increasing availability of highly sensitive detection methods [12] that detect PrP^{TSE} accumulation in lymphoid and in muscle tissue samples in sCJD patients [24] has partially modified views about the distribution of PrP^{TSE} in these patients. The diagnostic utility of the PrP^{TSE} detection in the muscles and spleen of sCJD patients is, however, still under examination as the positivity occurs only in a portion of the examined samples [24]. A recent report of PrP^{TSE} detection in the muscle of sheep and goats with experimental scrapie [25] and in hamsters and mice experimentally infected with adapted BSE or vCJD strains [26] shows how the comprehension of PrP^{TSE} distribution in tissues is continuously expanding also widening perspectives for future preclinical diagnosis of TSEs.

With the aim of developing preclinical diagnosis of TSE diseases in blood, experimental models of TSEs (rodents and small ruminants) have been used, and minute amounts of infectivity have been detected in whole blood, white blood cells, and plasma [27]. The transmissibility of BSE and natural scrapie to sheep through blood transfusion has also been demonstrated [28]. Novel findings on possible blood borne transmission in three vCJD patients who received a blood transfusion by individuals later diagnosed as having vCJD [29,30] has further raised concern in the matter of iatrogenic transmission possibly sustained by asymptomatic carriers of vCJD. The fact that PrP gene sequencing in one patient showed heterozygosity for methionine and valine at the polymorphic codon 129 indicates that, contrary to what was previously thought, a heterozygous subject can be infected by the vCJD agent, modifying what is the current perception of the epidemiological scenario of vCJD in the U.K. and the rest of the world as well [31]. The immunohistochemical examination of appendicectomy and tonsillectomy specimens for the presence of abnormal limphoreticular PrP^{TSE} deposition has given an estimated prevalence of 237 cases per million, in the U.K., i.e. 3808 subjects aged 10–30 might incubate vCJD [32]. In the absence of a screening method to identify asymtomatic individuals these might represent a reservoir for iatrogenic diffusion of infection [9,33–36].

These findings stimulated a great deal of research to improve techniques for the detection of PrP^{TSE} in blood [37–40] and in preclinical samples of experimental scrapie and BSE [41]. In this context, it is worthwhile to mention the Protein Misfolding Cyclic Amplification (PMCA) technique. This technique was first introduced by Saborio in 2001 [42]. It allows, using a mechanism conceptually similar to PCR (polymerase chain reaction), the serial amplification of undetectable amounts of PrP^{TSE} to an extent that is then easily revealed by western blot. PMCA has already been used to demonstrate the presence of PrP^{TSE} in blood samples from scrapie affected hamsters and, therefore, holds promise to become a very effective aid for an early, preclinical identification of PrP^{TSE} in natural TSE diseases [40].

13.2   DIAGNOSTIC METHODS: ROLE OF PrP AS DIAGNOSTIC MARKER OF PRION DISEASES

Irrespective of the mechanism that drives the PrP^c to PrP^{TSE} transition, and that still remains elusive [43], the formation and accumulation of pathological PrP occurs only in TSEs. Thus, the identification of PrP is a specific marker for these disorders. The pathological isoform retains the two glycosylation sites. The relative amount of variably glycosylated forms along with the size of PrP^{TSE} fragments obtained after digestion with proteinase K (See below), have been used to discriminate among different phenotype of TSE disease both in human and animals, contributing to a commonly accepted molecular classification of human disease [44].

Immunological methods for PrP^{TSE} detection begun in 1980 from identification of PrP^{TSE} in scrapie infected rodents, followed by the development of anti-PrP antibodies, and the development of immunological techniques for TSE diagnosis. Polyclonal, monoclonal, recombinant, Fab, and phagic antibodies have been produced, some of which are commercially available, some upon request of the producers, which other are more difficult to obtain [45–48].

With the availability of these antibodies, a large number of immunoassays have been developed and applied to the detection of PrP^{TSE} and TSE diagnosis. The current immunological methods for PrP^{TSE} diagnosis can be divided into two main groups. The first group includes the methods that require fixation of the tissue. The second group includes immunological assays that make use of non fixed tissues that should be reduced to a suspension (if not already available as fluids) before processing. All the methods, irrespective of the state of the starting material (fixed or not), require the tissue to be processed, to some extent, in order to eliminate contaminants that can reduce the sensitivity and the specificity of the test. These immunodiagnostic tests includes the following: western blot (WB); enzyme linked immunosorbent assay (ELISA); immunohistochemistry (IHC); histoblot (HB); paraffin-embedded tissue blot (PET); dissociation-enhanced lanthanide fluorescence immunoassay/conformation dependant immunoassay (DELFIA/CDI); and scanning for intensely fluorescent targets (SIFT).

This chapter describes each of these techniques, giving an idea of the logic behind the method, the procedures required for its execution, the sample preparation, the practical applications either for human/animal diagnosis, or research with an outlook at the advantages and pitfalls.

13.3   WESTERN BLOT (WB)

13.3.1   BASIC PRINCIPLES

The western blot is the best characterized and, by far, the most used method for PrP^{TSE} detection in unfixed tissues with a successful record of diagnostic and research applications [49]. For the execution of WB, the proteins extracted from the tissue under examination are first separated by SDS polyacrylamide gel electrophoresis and then electrotransferred to a nitrocellulose (or PVDF) membrane to make the relevant epitopes available for binding with a specific antibody. This solid-phase immunocomplex is then bound by a secondary antibody coupled to an enzyme able to catalyze a colorimetric or a chemiluminscent reaction for revealing the presence of the antigen.

The electrophoretic separation of proteins according to their molecular weights improves the sensitivity and the specificity of the immunoassay. The renders the WB one of the most powerful analytical tools available in the field of TSEs, giving good reason for the inclusion of this test into the club of the WHO (World Health Organization) reference methods for human TSEs diagnostic confirmation [50].

13.3.2 SAMPLING AND STORAGE OF SPECIMENS

Virtually all kinds of tissues or fluids can be analyzed by WB (provided they are not fixed) with the choice of the starting materials depending on the goal to be attained. For post-mortem diagnosis of human TSEs, the nervous tissue is the most suitable material because of its high content in PrP^{TSE}. A few cortex samples from distinct cerebral areas are usually sufficient for PrP^{TSE} detection of humans. Sporadic and genetic CJD requires the collection of a frozen hemi brain to include grey matter from frontal, parietal, temporal, occipital, and cerebellar cortical areas and the basal nuclei. These are selectively targeted in some uncommon TSE forms like fatal familial insomnia [50]. The spectrum of human TSEs encompasses a wide series of forms with distinct clinical and pathological features, and, occasionally, the cerebral cortex may be poor or not involved at all [44]. The formulation of a definite TSE diagnosis in these rarer forms is particularly relevant as this pool may hide the emergence of novel variants of the disease with unknown pathogenic properties.

The same rule should apply for BSE diagnosis. In nation-based surveillance programs, the brainstem is the material of choice. However, the recent description of the amyloidotic variant (BASE) of the disease [51], likely caused by a different infectious strain where the neocortex is more involved than the brainstem, confirms the validity of a multiple sampling strategy. This approach applies also with scrapie of sheep and goats. Although the brainstem is the preferred site for PrP^{TSE} accumulation, several combinations of infectious strains and host genotypes may be encountered with a heterogeneous distribution of PrP^{TSE} lesions in the brain [52]. At variance with BSE and sCJD, in scrapie-infected animals, many organs, particularly tissues of the lymphoreticular system, are loaded with PrP^{TSE} and can be employed for diagnostic purposes [53].

Although many studies showed that the incubation of samples for several days at room temperature from BSE-infected brains does not affect the detection of PrP^{TSE} by WB, this observation cannot be taken as a general rule. It is well known that the resistance to proteolytic degradation of PrP^{TSE} is strain dependent; therefore, a prompt storage of bioptic and autoptic specimens at subfreezing temperatures is desirable (if available, $-80°C$ is preferred).

13.3.3 SAMPLE PREPARATION

The preparation of the sample is a critical factor for the sensitivity and the specificity of the assay as it may affect the detectable concentration of PrP^{TSE} and the appearance of background noise. Protocols of different length and complexity are available to process TSE-infected tissues before WB analysis, allowing the desired level of analytical sensitivity.

A concise, high throughput preparation of CNS samples starts with fine homogenization (commonly at a weight/volume ratio of 10%) of the tissue in a buffered isotonic solution (Tris saline or PBS) that may also contain non-denaturing detergents (0.5%–1% of Nonidet P-40 or sodium deoxycholate or sarkosyl). The homogenate is usually cleared by a short centrifugation at low speed (5 min at $1000 \times g$) and finally digested with 50–100 micrograms of proteinase K for 1 h at 37°C (proteinase K degrades a significant portion of background proteins and cuts PrP^{TSE} at the N-terminus, yielding a heterogeneous mixture of truncated fragments whose doubly glycosylated isoform weight between 27 and 30 kDa, hence the term PrP27–30). The digestion may be stopped with protease inhibitors (100 μM PMSF) or by boiling in a gel loading buffer. This quick preparation functions very well if highly specific and sensitive immunological reagents are available for detection (as is the case with most TSEs). However, to improve the specificity and the analytical sensitivity, the pool of background proteins may be further reduced by the introduction of centrifugation and/or ultracentrifugation steps that take advantage of the insolubility of the pathological PrP in non-denaturing detergents and in salts (e.g., sodium chloride or the recently introduced sodium phosphotungstate) [12,54,55]. Very pure preparations of PrP^{TSE} are obtained combining a proteinase K digestion with prolonged extraction steps in detergents/salt solutions, but such materials are usually required only for biochemical studies [56,57].

13.3.4 PROCEDURE

Recently, the World Health Organization organized a comparative study, with the participation of several TSE laboratories, to assess the suitability of a selected group of brain homogenates from sporadic and variant CJD patients as standard reference reagents for human TSEs immunoassays. Most of the participants analyzed the samples by simple WB protocols. In spite of the differences between each procedure, a remarkably similar analytical sensitivity was observed, showing a variation below one order of magnitude between participant laboratories [58]. Although no official consensus has yet been reached on the standardization of WB for TSEs diagnosis, this study confirmed the WB as a rather robust system for PrP^{TSE} detection regardless of technical variations.

The protocol reported below is suitable for PrP^{TSE} detection in human and hamster samples [59], but it can be adapted to other species providing that anti-PrP antibodies recognize the PrP of that species.

Western Blot of Brain Tissue

Purified fractions (0.5–1 mg equivalents of fresh tissue are routinely used for PrP^{TSE} detection) are suspended in an equal volume of 2x gel loading buffers (Laemmli buffer or commercial proprietary preparations) to give a final volume of 15–25 μl and run in a disc-SDS-PAGE minigel (about 10 cm long and 1 mm thick) with a 12% resolving gel.

When the separation is complete, the gel is soaked in Towbin transfer buffer and layered onto a pre-wetted nitrocellulose filter for semi-dry electrotransfer of the proteins at 100 mA for 60 min at 4°C. After saturation of unoccupied sites with 3% non-fat dry milk in 10 mM Tris-HCl pH 8.0/0.05% Tween-20 (TBST buffer), membranes are incubated overnight with monoclonal antibody 3F4 (by far, the most used and best characterized among commercial antibodies that recognizes residues 109–112 of human and hamster PrPs, see Kascsak et al. [60]) diluted 1:2000 in TBST and incubated 1–2 h in an alkaline phosphatase-conjugated secondary antibody solution (1:5000 in TBST). Extensive washings (five changes of buffer, 5 min each) in TBST are performed after each of the three incubations to reduce background signals. A highly sensitive commercial chemiluminescence system (CDP-Star) is then applied to reveal PrP bands that are recorded on to sensitive films following manufacturer instructions.

13.3.5 APPLICATIONS

This technique offers good analytical sensitivity and possibly the highest specificity among biochemical TSE tests, in part due to the combination of an immunopositive chemiluminescent signal with a characteristic tri-banded appearance localized between 20 and 30 kDa (Figure 13.1). For these reasons, among European Union (EU)-validated tests for nation-wide BSE surveillance, WB tests are widely adopted either for screening or as a confirmatory test.

The western blot offers the advantage of recognizing different forms of PrP^{TSE} through the analysis of the molecular masses and the relative abundance of high, low, and non-glycosylated bands. These parameters characterize the so called PrP glycotype, a kind of PrP signature that varies among different forms of TSEs and therefore, is used to distinguish various forms of TSEs (e.g.,

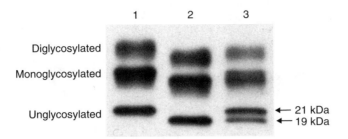

FIGURE 13.1 Western blot analysis of proteinase K treated PrPTSE extracted from the cerebral cortex of patients with sporadic Creutzfeldt-Jakob disease. The unglycosylated band has either a molecular mass of 21 kDa (type 1 PrP27-30, Lane 1) or 19 kDa (type 2 PrP27-30, Line 2). When the two types are present in the same cerebral area, a double, unglycosylated band can be seen (type 1 + 2, Lane 3). Courtesy of Dr P. Parchi, Università di Bologna, Italy.

Scrapie from BSE, Sporadic from variant (CJD) long before conclusive data from strain typing in laboratory rodents are available [61,62].

PrP^{TSE} glycotyping has been exploited for improving the classification of human TSEs [44,54,62]. In sporadic CJD, the two most frequent PrP^{TSE} glycotypes, type 1 and type 2a, each generated by a different cleavage site for proteinase K, are characterized by the unglycosylated fragment migrating either at 21 kDa (PrP^{TSE} type 1; Figure 13.1, lane 1) or at 19 kDa (PrP^{TSE} type 2; Figure 13.1, lane 2). The two glycotypes may be present simultaneously in the same cerebral area (Figure 13.1, lane 3). Recent work showed that this phenomenon is more frequent than previously thought and that the common WB methods, mostly relying on the use of the monoclonal antibody 3F4, demonstrate the presence of combined PrP^{TSE} glycotypes only when the two isoforms are deposited in similar amounts [63].

The combination of the two possible PrP^{TSE} glycotypes with the three possible genotypes of PrP at the polymorphic codon 129 (methionine homozygous, valine homozygous, or heterozygous) renders possible the molecular classification of human TSE disease in six distinct subgroups, each of which presents distinct clinical and pathological features [64]. The most common subtype (subtype 1) includes 129MM-1 and 129MV-1 and accounts for 60%–70% of all cases of sporadic prion diseases; over 95% of the patients of this group are 129MM-1, whereas the 129MV-1 patients are rare [65]. Patients in this group present the more classical presentation of sporadic CJD with a median disease duration of four months, clinical signs characterized by cognitive impairment, ataxia and the presence, as histological hallmarks, of fine spongiform degeneration, astrogliosis, and neuronal loss. The next most common type, type 2 (sCJD VV-2), is also called cerebellar or ataxic variant. Longer clinical duration and kuru plaques characterize the subtype 3 (sCJD MV-2), whereas subtype 4 (sCJD MM-2) and subtype 6 (sporadic familial insomnia) share the same 129 polymorphism and the same PrP^{TSE} glicotype while remaining phenotypically distinct [66]. Subtype 5 (sCJD VV-1) is extremely rare and related to an early onset of clinical signs.

Two dimensional-electrophoresis (2D-electrophoresis) allows a more detailed analysis of the PrP^{TSE} glycoform population (Figure 13.2) and the description of new atypical PrP^{TSE} glycotypes [67–70].

The WB offers some obvious limitations in comparison with other immunoassays (e.g., ELISA-type tests): relatively few samples can be processed in a single gel, the technique is time-consuming, cannot be completely automated, and requires experienced personnel for interpretation of results. However, because of the significant quantity of information it can provide, the WB (together with the immunohistochemistry) stands as a reference diagnostic test in the framework of human TSE surveillance activities.

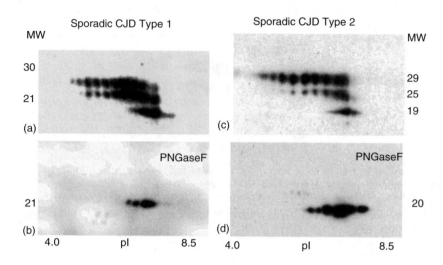

FIGURE 13.2 Two-dimensional western blot of proteinase K treated brain homogenates from MM-1 (a and b) and VV-2 (c and d) sporadic CJD cases. Trains of spots migrating around 25 kDa and 30 kDa represent the mono and the diglycosylated isoforms of PrP27-30. The unglycosylated isoforms with a molecular mass of 21 kDa and 19 kDa, respectively, in type 1 and type 2 PrP27-30 are composed of several spots that represent N-terminally truncated peptides with ragged ends (panels a and c). After PNGase F enzymatic treatment the number of spots at low molecular weight is increased by the addition of deglycosylated PrP isoforms (panels b and d). Courtesy of Dr G. Zanusso, Università di Verona, Italy.

13.4 ELISA

13.4.1 Basic Principles

The ELISA (enzyme-linked immunosorbent assay) test is based on the adsorption of the relevant antigen on the bottom of a multi-well polystyrene plate that is then filled with a solution containing a specific antibody linked (directly or through a second antibody) to an enzyme that catalyzes a colorimetric or chemiluminescent reaction. Binding of the antigen onto the plastic plate can be obtained by non specific interactions between the antigen and the surface of the well (direct ELISA) or by a selective capture by a specific antibody previously immobilized on the plastic well (sandwich ELISA).

13.4.2 Sampling and Storage of Specimens

Fresh or frozen samples are required for the ELISA analysis. Particular care should be applied to keep the specimen at or below freezing temperatures as tissue lysis may produce false negative results [71].

13.4.3 Procedure

The protocol for the application of the ELISA to animal TSEs has been subjected to continuous implementations. In the simplest version [72], the proteins contained in a proteinase K treated tissue homogenate (e.g., brain or spleen) are adsorbed onto the plastic well of a 96-well plate, and the bound proteins are then detected by sequential incubation with anti-PrP primary antibody and then with an enzyme-linked anti-Ig secondary antibody. To enhance the immunoreactivity of the adsorbed antigens and, consequently, to increase the analytical sensitivity, the adsorbed antigens can be treated with guanidinium thiocyanate (GdnSCN) at room temperature before the application of the first antibody solution [73]. Heat denaturation in the presence of this chaotropic salt allows the discrimination between normal and BSE brain homogenates without the need for proteolytic treatment [74].

This procedure (detailed below) takes advantage of the different conformations of PrP^c and PrP^{TSE}. In native conditions the relevant antigenic site of the folded PrP^{TSE} is hidden. Upon denaturation, however, it is exposed, yielding a strong increase in the immunoreactive signal.

This procedure bypasses the need for proteolytic removal of PrP^c and is also the basis of the conformation-dependent immunoassay (CDI, described later), a novel analytical test for TSEs with a traditional ELISA format but improved performance in terms of analytical sensitivity [75].

Enzyme Linked Immunosorbent Assay

The ELISA plate is coated with capture monoclonal antibody 6H4 [45] (overnight incubation at 4°C with 0.1 mL of a 1:100 dilution in carbonate buffer) and blocked (1 h at 37°C) with 0.2 mL per well of 10% dry milk diluted in RPB buffer (13.7 mM NaCl, 2.7 mM KCl, 1.4 mM KH$_2$PO$_4$, 8.1 mM Na$_2$HPO$_4$, pH 7.3) plus 0.01% Tween-20 prior to the addition of the brain homogenate. BSE-infected bovine brain samples are homogenized in 9 volumes of 0.32 M sucrose and clarified at 7000 × g for 5 in. One volume of the homogenate is diluted 1:2 in 0.2 M GdnSCN in RPB buffer and heated at 150°C for 10 min in a glass ampule. One hundred microliters of this suspension are added in triplicate to the wells of the precoated ELISA plate and incubated for 1 h at 37°C. After washes with RPB/Tween-20, the bound PrP is quantified by two successive incubations (1 h each, at 37°C). The first incubation consists of a 0.1 mL/well of a rabbit anti-PrP polyclonal serum (C15S or R#26) diluted 1:500 in RPB/Tween-20/3% dry milk. The second incubation is with 0.1 mL/well of an HRP-conjugated swine anti-rabbit antibody diluted 1:300 in PBS/Tween-20/3% dry milk. After the last incubation, the plates are washed with PBS/Tween-20 and the wells are filled with 0.2 mL of ABTS solution to detect the presence of PrP.

The chemiluminescent signal is registered through an automated plate reader.

13.4.4　Applications

Sensitive and specific ELISA protocols have been developed for the identification of PrP^{TSE} in brain and in other tissues, for diagnostic and research applications. Most of these methods have been optimized for animal TSEs [76–80] and are successfully used for nationwide BSE screening tests.

The emergence of the BSE epidemic pushed the academic and the industrial world to join forces on the development and the validation of fast, sensitive, and specific tests to implement the surveillance of the disease. In this effort, the ELISA format was the most successful test and several commercial kits are available on the market. Among these, a simple ELISA immunoassay based on a proprietary PrP^{TSE} extraction procedure and a proprietary polyclonal primary antibody has been validated by the European Union (Enfer Ltd), yielding diagnostic sensitivity and specificity figures equal to 100%. Although this test did not show the highest analytical sensitivity among the assays evaluated, it can return diagnostic answers in a few hours, and it permits simultaneously processing many samples.

In the same round of validation, the European Union evaluated two sandwich ELISA that offered a higher analytical sensitivity. One was developed by the French Commissariat a l'Energie Atomique and put on the market with the Bio-Rad Platelia-BSE brand. In this test, after a concentration step, proteinase K enzymatic treatment and denaturation treatments, the sample is transferred to in a microplate well previously layered with an anti PrP^{TSE} monoclonal antibody. Detection is obtained through chemiluminescence with a horseradish peroxidase-linked monoclonal antibody that recognizes a different epitope. This test can detect the presence of BSE contamination with the same analytical sensitivity of the mouse bioassay [77] and has the same sensitivity of other traditional diagnostic methods [78].

A slightly different chemiluminescent sandwich immunoassay for BSE screening has been proposed by Prionics AG. The Prionics Check LIA (luminescence immunoassay) includes protein-ase K treatment of brain homogenate before incubation with free monoclonal antibody and subsequent capture by a second immobilized monoclonal [79]. This test can be fully automated and is considerably faster than the Bio-Rad companion.

Although the EU validation procedure yielded outstanding results in terms of specificity, these three ELISA rapid tests that rely only on one criterion of positivity (i.e., the increase of the optical signal above an experimental cut-off level) may, in principle, yield false positive results driven, for example, by the presence of undigested PrP^c or by non PrP-specific binding of the detection antibodies. The problem of false positives may be particularly relevant when dealing with lysed samples as shown in a recent study from Japan [71]. For these reasons, in the common surveillance practice, a second confirmatory analysis, based on a different principle (such as western blot or immunohistochemistry), is recommended.

Although a CJD-dedicated ELISA protocol was proposed more than 15 years ago [73] for TSE diagnosis in humans, this assay is not included in the restricted selection of confirmatory tests for the diagnosis of human TSEs [50]. Why, in spite of the outstanding sensitivity and specificity reached by the latest versions of commercial BSE ELISA assays, is this test not used for human diagnosis?

One reason is that the speed and the high throughput offered by the ELISA, both essential for large scale animal screenings, represent no benefit versus other validated immunochemical techniques (western blot or immunohistochemistry) in human diagnosis where just a handful of samples are processed at a time. Also, as detailed above western blot and immunohistochemistry not only permit a highly reliable and sensitive detection of PrP^{TSE}, but also enable the classi-fication of PrP^{TSE} type. The ELISA format, however, may overcome this limitation by development of a two-phase test. In the first phase the presence of PrP^{TSE} is revealed through a C-terminal antibody able to recognize type 1 and type 2 PrP^{TSE}. Postive sample are then analyzed by an antibody that recognizes only the N-terminal portion of type 1 PrP^{TSE}. Because this fragment is removed by proteolytic treatment of PrP^{TSE} type 2, double positivity will suggest the presence of type 1 PrP^{TSE}, whereas a positive signal only at the first ELISA indicates type 2 PrP^{TSE}. This design, however, will not be able to discriminate between samples containing only type 1 PrP^{TSE} and those containing a mixture of type 1 and type 2 [58,63,81].

13.5 IMMUNOHISTOCHEMISTRY

13.5.1 BASIC PRINCIPLES

Immunohistochemistry (IHC) is a widely used histological technique that permits the revealing of an antigen in its native anatomical location. This procedure uses fixed tissue slices (5–7 micron thick) that are treated with a primary antibody that binds the relevant antigen and then are incubated with a secondary, enzyme linked, anti Ig-antibody that catalyses the in situ precipitation of an insoluble chromogen able to permanently mark the antigen.

Thanks to the large diffusion of the equipment (the basic instrumentation of a histology labora-tory is sufficient far enough) and the continuous technical improvements to increase the specificity and the sensitivity of the method, the IHC has become one of the reference methods for diagnostic confirmation in human and animal TSEs. Data accumulated since its introduction, together with those produced by histological analyses and PrP western blotting, combined to develop a highly refined pathological and biochemical classification of TSEs.

13.5.2 SAMPLING AND STORAGE OF SPECIMENS

Although PrP^{TSE} is notoriously resistant to degradation, tissue sampling and fixation should be ideally performed soon after death to avoid deterioration of the microarchitecture of the tissue. Fixation of tissue samples in 10% buffered formalin is the most popular way for the storage and preparation of specimens for immunohistochemical examination. Other fixatives, like Carnoy's solution, may be used as well [82].

In human TSEs, the WHO guidelines recommend, whenever possible, to perform the autopsy no later than 2–3 days after death and to fix the left brain hemisphere in formalin [50]. As pointed out in the western blot section, the topographical deposition of PrP^{TSE} varies in different human TSE forms and sometimes even within the same form. Immunostaining of a panel of brain areas prevent false negative assays and is also useful for identification of different forms of TSE. This heterogeneity also limits the usefulness of a biopsy in living patients that should be considered only when a treatable condition is hypothesized as an alternative diagnosis.

In BSE, similar rules apply for specimen collection and fixation (see the western blot paragraph). If only a limited portion of the brain is sampled, the medulla oblongata, is preferred.

13.5.3 PROCEDURE

The protocol for the immunohistochemical demonstration of PrP^{TSE} in TSE-affected tissues has been implemented over the years to deal with two related problems: the removal of PrP^c immunoreactivity and the exposure of PrP^{TSE} epitopes. A number of harsh denaturing and proteolytic treatments of tissue slices have been tested, and several combinations have been successfully used. A procedure with an optimized treatment schedule for brain tissue has been recently described, giving satisfactory signal-to-noise ratio even when the tissue has undergone overfixation [83].

Immunohistochemistry on Human Brain Tissue

Tissue (brain) blocks ($20 \times 30 \times 10$ mm) are fixed in 10% buffered formalin (4% formaldheyde in PBS) for a minimum of 4 days infectivity removed by immersion in 99% formic acid for 1 h and embedded in paraffin. Five micron thick sections are cut by microtome and after distension in water are collected onto silanized slides and deparaffinized in xylene.

Pre-treatments for antigen retrieval start with microwave cooking at 1000 W, three times for ten minutes each, 5 min treatment with 99% formic acid at 24°C, 2 h denaturation in 4 M GdnSCN at 4°C, and a final proteolytic digestion step with proteinase K at 5 μg/mL for 8 min at 24°C. After rinsing in water and an incubation step with a preimmune goat serum, PrP^{TSE}, immunostaining is performed by incubating the sample overnight at room temperature with the 3F4 primary antibody (diluted at 1:1000 in PBS) followed by a treatment for 10 min in 10% methanol and 3% hydrogen peroxide to block endogenous peroxidases followed by 25 min at room temperature with a prediluted biotinylated secondary antibody and then 25 min with streptavidin-linked horseradish peroxidase at room temperature (all the reagents are available as a commercial kit). PrP^{TSE} deposits are finally revealed with diaminobenzidine. After counterstaining with haematoxylin, the sections are mounted and finally ready for microscope examination.

13.5.4 APPLICATIONS

The immunohistochemical staining of tissues from TSE-suspected subjects is one of the earliest diagnostic applications of anti PrP^{TSE} antibodies, but in spite of this classicism, it is a solid frontline technique for the study and the diagnosis of TSEs. The reason for this success likely resides in the highly informative content of immunohistochemical analyses that, similarly to other immunological techniques, provide reliable information regarding the presence, the amount, and the gross tissue distribution of PrP^{TSE}. It also reveals the morphology and the cellular topology of PrP^{TSE} aggregates useful for the characterization of human and animal TSE [51,64,84,85].

For the confirmation of the clinical diagnosis of TSE, immunohistochemistry relies on the identification of Prp^{TSE} in the brain [1,3,14]. The availability of samples from several brain areas also supports the classification of different forms of TSEs [64], and the identification of rare forms characterized by specific immunostaining pattern in specific areas of the brain (like the so-called fatal insomnia, a MM-2 sCJD with prominent talamic involvement. Moreover, the possibility to obtain the microscopic resolution in the localization of PrP^{TSE} deposits makes immunohistochemistry the technique of choice to diagnose cases with rare, minute deposits of PrP^{TSE} that cannot be detected by other reference techniques such as western blot where a dilution effect may occur [86]. This is not a pure academic goal as the observation of highly unusual immunostaining patterns during surveillance programs contributes to the identification of novel, potentially dangerous, human and animal forms as already occurred with vCJD in the United Kingdom and bovine amyloidotic spongiform encephalopathy in Italy [7,51] (Figure 13.3).

In human TSEs, although a formally recognized univocal classification is still lacking, several typologies of PrP^{TSE}-positive immunostaining have been defined [64,83,85].

The so called synaptic staining consists in a diffuse, intense, and intracellular labeling of PrP^{TSE} [85] and is typically observed in the cerebellum and in the cerebral cortex of the majority of sporadic CJD cases (Figure 13.4 a).

Less common patterns of PrP deposition are observed in molecular and clinical subtypes of sporadic CJD [44,64,83,85,87,88]. In one third of classic sporadic CJD cases, typically bearing the 129 MM or MV PrP genotype and a PrP27–30 classified as type 1 at the western blot, a perivacuolar

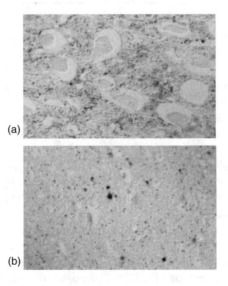

(a)

(b)

FIGURE 13.3 Immunohistochemical staining of PrP deposits in the obex of a cow with BSE (a, x 220) and in the olfactory tuberculum of a cow with BASE (b, x 250). Note the characteristic presence of amyloid plaques in BASE. Courtesy of Dr G. Zanusso, Università di Verona, Italy.

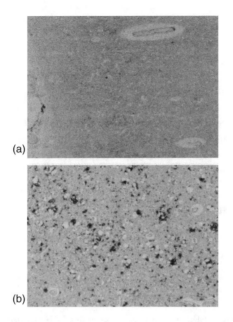

(a)

(b)

FIGURE 13.4 Immunohistochemistry for the prion protein in cerebral cortex samples from a MM-1 (a, x 100) and a MM-2 (b, x 100) sporadic CJD patient showing synaptic and perivacuolar staining patterns, respectively. Courtesy of Dr P. Parchi, Università di Bologna, Italy.

PrP^{TSE} positivity is observed in the cerebral cortex [Figure 13.4 b]. This pattern is also present in about half of sporadic CJD cases with type 2 PrP27–30 and 129 MM genotype.

PrP^{TSE} may aggregate to form amyloid plaques (10–50 µ large) that often are also visible after common haematoxylin-eosin histological staining and are characterized by red-green bire-fringence after Congo Red staining. These plaques may appear as single, isolated structures (unicentric plaque) or as irregular clusters that may partially overlap to form the so-called multicentric plaque that is typically observed in the brain of GSS patients [89]. A third, relevant class of plaques is made of a central PrP-positive nucleus surrounded by a corolla of vacuoles that gave rise to the definition of florid plaques. These structures are mostly seen in vCJD cases [11], even though a description of similar structures has been reported in other CJD forms [90,91].

Clearly emerging from this short summary of the principal immunohistochemical profiles (not to mention other, less common profiles that can be encountered) is that the spectrum of PrP^{TSE} immunostaining patterns is heterogeneous in humans. This issue, coupled with the potential for false positives caused by incomplete removal of PrP^c [85], implies that, for a correct diagnostic classification, the evaluation of immunohistochemical results should be accomplished by trained personnel. Immunohistochemistry is a robust technique that performs well on both recent and archived material from tissue collections and can be easily adapted to search for PrP^{TSE} in tissues other than the nervous system. Examination of lymphoreticular organs such as the spleen, tonsils, and lymphoreticular tissue distributed in the parenchyma of other organs may help to distinguish particular forms of TSEs where these tissues are primarily involved (e.g., vCJD) or to attempt an early preclinical diagnosis of these forms when peripheral infection is suspected. These possibilities have been recently exploited to provide an estimate of the number of individuals in the U.K. who may be incubating vCJD by studying surgically removed tonsillectomy and appendicectomy specimens (that are a specific site of PrP^{TSE} deposition in vCJD) stored in hospital tissue banks [21,32].

13.6 HISTOBLOT

13.6.1 BASIC PRINCIPLES

The histoblot (HB) technique was developed more than 10 years ago [92] to combine the anatomical and topographical information offered by immunohistochemistry with the analytical sensitivity of biochemical blotting methods. Fresh or frozen (unfixed) tissue slices are blotted by contact onto a piece of nitrocellulose membrane and, after a treatment to reduce the background (including PrP^c) and to unmask PrP^{TSE} epitopes, the membranes are immunostained to reveal PrP^{TSE}. The result is a natural size, monochromatic picture of PrP^{TSE} deposits in the framework of the anatomical section (Figure 13.5).

13.6.2 SAMPLING AND STORAGE OF SPECIMENS

The recommendations for sampling and storage of fresh or frozen samples for histoblots are the same as those described in the western blot section.

13.6.3 PROCEDURE

The original protocol of the histoblot [92] has been successfully applied by other authors (with only minor modifications) and is reported below.

Histoblot on Human or Animal Brain Tissue

Fresh frozen tissue blocks are cut by a cryostat set on a thickness of 8 μ, and the resulting slices are laid on a glass slide. Thawed sections are blotted (by contact transfer) by pressing the slide onto a piece of nitrocellulose membrane backed by two thick sheets of blotting paper pre-wetted in lysis buffer (0.5% Nonidet P-40/0.5% sodium deoxycholate/100 mM sodium chloride/10 mM EDTA/10 mM Tris-HCl pH 7.8). A slow rotation of the slide helps to avoid entrapment of bubbles during the blotting procedure. After transfer, the membrane is kept on the blotting paper for several minutes and then air dried and soaked for 1 h in 0.05% Tween-20/100 mM sodium chloride/10 mM Tris-HCl pH 7.8 (TBST buffer). To remove cellular PrP, the membrane is incubated for 18 h at 37°C in digestion solution (0.1% Brij 35/100 mM sodium chloride/10 mM Tris-HCl pH 7.8) containing proteinase K at 400 µg/mL (when applied to the hamster scrapie strain 263K) or 100 µg/ml (when applied to human CJD). After three washes in TBST buffer, the digestion is stopped by a 30 min-incubation in 3 mM PMSF/TBST. Exposure of antigenic PrP27-30 epitopes is achieved by soaking the membrane in a denaturation buffer (3 M GdnSCN/10 mM Tris-HCl pH 7.8) for ten minutes followed by three washes in a TBST buffer. For PrP immunostaining, the membrane is incubated in 5% non fat dry milk/TBST (30 min minimum, at room temperature) then for 18 h at 4°C with monoclonal antibody 3F4 diluted at 1:1000 (from ascitic fluid). A commercial kit containing alkaline phosphatase secondary antibody and chromogenic substrates is used for PrP revelation (Promega).

13.6.4 APPLICATIONS

Histoblot may provide a superior analytical sensitivity in comparison with immunohistochemistry; likely the result of a higher antigen retention offered by the nitrocellulose support and the absence

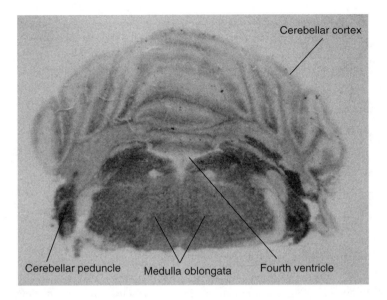

FIGURE 13.5 Histoblot of a coronal section of the hindbrain from a mouse infected with the 139A strain of scrapie. Diffuse PrP^{TSE} positivity is visible in the medulla oblongata as well as in the cerebellum. Courtesy of Drs. R. Nonno and M. A. Di Bari, Istituto Superiore di Sanità, Rome, Italy.

of harsh pre-treatment steps. It is, however, more laborious than conventional immunohistochemistry and more expensive in terms of consumables such as proteinase K and immunostaining products. It also requires fresh frozen material that is less practical to obtain and to handle than the fixed tissue.

An exhaustive collection of reference staining patterns, specific for each TSE form, is lacking. Whereas the absence of such a database hampers the application of the histoblot to the identification of TSE subtypes (particularly in humans), it does not limit the usage of this technique as a sensitive diagnostic and research tool. In fact, the histoblot has been applied to investigate the presence of PrP^{TSE} in the brain and in peripheral organs in BSE [93] and scrapie [93,94] affected animals. Histoblotting analysis of brains from experimental rodent models yields results of impressive realism: the strain-specific fingerprint produced by selective PrP^{TSE} targeting of cerebral areas can be appreciated at a glance. This property has been exploited to study the role of PrP^{TSE} production and accumulation in the pathogenesis of animal TSEs [95,96].

13.7 PARAFFIN-EMBEDDED TISSUE BLOT

13.7.1 BASIC PRINCIPLES

The paraffin-embedded tissue (PET) blot is an evolution of the histoblot, developed to overcome the problem of storage and transport of fresh samples allowing the usage of fixed tissues both from current and archival TSE cases [97].

In this original procedure, tissue samples undergo a three stage processing. The first stage is a typical histological procedure going from fixation to microtome cutting of paraffin embedded tissue slices that are then (second stage) deposited onto a nitrocellulose membrane then it is deparaffinized and rehydrated (three passages peculiar to the PET blot procedure). In the third stage, as described for the histoblot procedure, the membrane is extensively incubated with proteinase K, treated with denaturants, and the proteins are revealed by a colorimetric immunostaining reaction (Figure 13.6).

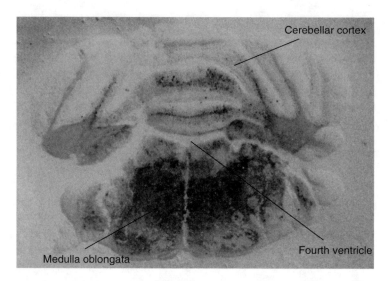

FIGURE 13.6 PET blot of a coronal section of the hindbrain of a mouse infected with mouse-adapted variant CJD. Intense PrP^{TSE} staining can be appreciated in the medulla oblongata, whereas the cerebellum appears less involved. Courtesy of Mr. M. Sbriccoli and Dr A. De Pascalis, Istituto Superiore di Sanità, Rome, Italy.

13.7.2 SAMPLING AND STORAGE OF SPECIMENS

Both CNS and lymphoreticular samples can be analyzed with this test. The recommendations for sampling and fixation of tissue specimens for PET blot are the same as those described in the immunohistochemistry section.

13.7.3 PROCEDURE

The original protocol has been developed on CNS tissue samples from human, sheep, bovine, and mouse TSEs [97].

The procedure has been subsequently improved to reveal minute PrP^{TSE} deposits in the lymphoreticular tissue of vCJD patients by the adoption of a highly sensitive detection system [98]. This protocol is reported below and can be adapted with excellent results on both CNS and lymphoreticular samples.

Paraffin-Embedded Tissue Blot on Human Tissues

Fresh tissue samples are fixed in formalin, inactivated by immersion in formic acid for 1 h, incubated again in formalin for 48 hs, included in paraffin and cut by a microtome (5–7 µm). Tissue slices are laid on a piece of wet nitrocellulose, dried, deparaffinized in xylene isopropanol, and rehydrated in graded alcohols. After washing in TBST (10 mM Tris-HCl pH 7.8/100 mM NaCl/0.05% Tween-20), proteinase K (25 µg/mL in 10 mM Tris-HCl pH 7.8/100 mM NaCl/0.1% Brij 35) is added, and the digestion is performed overnight at 55°C. This step eliminates cellular PrP (as well as other proteins) and fixes the undigested proteins to the filter. Membranes are washed several times in TBST, and protein epitopes are exposed by 10 min incubation in 3 M GdnSCN/10 mM Tris-HCl pH 7.8. After washes in TBST, the membrane is blocked for 30 min with 0.2% casein in TBST, incubated for 2 h with the primary antibody properly diluted (3F4 at

---

1:5000, Dako) in casein/TBST solution, and washed again as before. PrP^{TSE} deposits are then visualized using the amplified colorimetric detection system ABC-AmP (Vector laboratories) and alkaline phosphatase substrates NBT/BCIP (nitroblue tetrazolium/5-bromo-4-chloro-3-indolyl phosphate).

13.7.4 APPLICATIONS

The PET blot preserves the anatomical detail like other immuno-histological techniques (immunohistochemistry and histoblot), but it has the advantage of a superior analytical sensitivity likely the result of the particular combination of enhancement procedures that either allows it to retain a higher amount of PrP^{TSE} and/or makes the protein more accessible to the binding with the first antibody. The PET blot has been applied to the detection of PrP^{TSE} in the CNS of TSE-affected humans, sheep, cows, and mice and from preclinical BSE-infected cows [97] as well as to lymphoreticular tissues of vCJD patients [98]. In these studies, the PET blot showed higher analytical sensitivity and specificity versus the histoblot, the western blot, and the immunohistochemistry. These results sustain the need of a larger validation study able to compare the relative performances of all these tests in the light of a possible application of PET blotting to early TSE diagnosis.

In spite of the great advantages described above, the procedure is time consuming, requires a properly equipped laboratory (apparatuses for histology and protein biochemistry are needed), and can be only partially automated in its present format. It is, therefore, difficult to envision an application of the PET blotting to the field that would receive the highest benefit, i.e., large scale early diagnosis of animal TSEs. Yet, research applications and small scale screening tests that require a superior analytical sensitivity will take great advantage from the introduction of the PET blot.

13.8 DELFIA–CDI

13.8.1 BASIC PRINCIPLES

The DELFIA/CDI (dissociation-enhanced lanthanide fluorescence immunoassay/conformation dependant immunoassay) is an ELISA format assay (the antigen of interest adsorbed on a substrate is in contact with a solution containing a specific antibody linked to a detection system). In this specific test, the detection system is represented by time-resolved fluorescence (TRF). The TRF signal is obtained through a chelated lanthanide (commonly europium, Eu^{3+}, but terbium, Tb^{3+}, dysprosium, Dy^{3+} and samarium, Sm^{3+} can be used) that is coupled (directly or by a secondary antibody) to an anti-PrP^{TSE} detection antibody. This system exploits two characteristics of the chelated lanthanides that ensure an outstanding analytical sensitivity: a large Stoke's shift (i.e., the wavelength shift between the excitation and emission light is larger than 200 nm) and a long decay time. After UV excitation, chelated europium releases an orange flash that slowly decays and can be easily recorded after the background of surrounding fluorophores has been switched off. Europium fluorescence (or more properly, phosphorescence) is recorded by means of a time-resolved fluorometer during multiple excitation/delayed reading cycles per second [99]. This unprecedented analytical sensitivity can measure picograms of PrP^{TSE} per milliliter, and the need for specific equipment justifies the presentation of the DELFIA/CDI system as a separate entity from the ELISA.

The CDI [100] is a particular application of the DELFIA detection system. It allows detection of PrP^{TSE} even in the presence of an excess of PrP^c. Depending on the strain of TSE examined, some antigenic epitopes are exposed in normal PrP while they are hidden in the pathological isoform. When the sample is subjected to denaturation, hidden epitopes become exposed giving a rise in the signal. The difference observed is dependent of the amount of PrP^{TSE} present in the sample. The efficacy of the test is obviously dependent on the epitope chosen for antibody recognition; epitopes equally exposed in the normal and the pathological isoforms of PrP render this test ineffectual.

13.8.2 Sampling and Storage of Specimens

The DELFIA/CDI test has been applied for PrP^c measurement in unfixed blood fractions [101–105] and in cultured endothelial cells [106] as well as for the detection of PrP^{TSE} in fresh or frozen brain tissue from humans and various animal species [100,107–112]. Because the assay may demonstrate protease-sensitive isoforms of pathological PrP, particular care should be applied during storage and manipulation of all these unfixed materials to avoid sample degradation.

13.8.3 Procedure

The procedure of the DELFIA (and the CDI) is as simple as that of an ELISA assay. The main difference, arising from the usage of TRF, resides in the availability of a specific panel of reagents and a time-resolved fluorometer.

For the CDI application, two separate DELFIA measurements are performed on aliquots of the sample. One aliquot is processed with the protein denaturant guanidinium hydrochloride (denatured, D), and the other aliquot is not treated at all (native, N). The difference in fluorescence between denatured and native aliquots is proportional to the concentration of PrP^{TSE} in the sample, whereas the ratio between these two values gives an estimate of the accessibility of the epitope.

The D-N TRF differences from a series of normal samples are used to calculate a negative cut-off value. All the samples producing a D-N figure higher than this cut-off are regarded as TSE-positive.

The protocol of a recently published [112] conformation-dependent immunoassay, developed for CJD diagnosis in association with the sodium phosphotungstate PrP^{TSE} extraction method, is described below.

Conformation Dependent Immunoassay of Human Brain Tissue

Brain samples are homogenized in 4% sarkosyl/PBS and then diluted with 2 volumes of PBS. Proteinase K treatment (2.5–10 µg/mL for 60 min at 37°C) is performed to remove PrP^c. As this treatment removes protease-sensitive PrP^{TSE} as well, to improve the analytical sensitivity, the proteinase K may be omitted, and endogenous proteases should be inhibited. After microfuge clarification (500 × g for 5 min at room temperature) sodium phosphotungstate is added to 0.32% and MgCl$_2$ to 2.7 mM, the samples are incubated for one hour at 37°C and precipitated with a microfuge at 14000 × g for 30 min. The pellets are suspended in water with protease inhibitors (0.5 mM PMSF, 2 µg/mL aprotinin, 2 µg/mL leupeptin) and divided in two aliquots. One aliquot is not treated and analyzed as such (native, N), and the other is subjected to guanidinium hydrochloride (GdnHCl) denaturation (D) treatment.

GdnHCl is added to the resuspended pellet to a final concentration of 4 M and heated at 80°C for 5 min. The native and the denatured samples are loaded into a 96-well plastic plate either coated by overnight incubation with a capture anti-PrP monoclonal antibody (MAR 1, 10 µg/mL in sodium bicarbonate pH 8.6) in the indirect CDI, or simply activated with 0.2% glutaraldheyde for direct CDI. After a 2 h incubation and extensive washing with Tris buffered saline pH 7.8/0.05% Tween-20 (TBS-T), the wells are saturated by 1-hour incubation in 0.5% bovine serum albumin/6% sorbitol in TBS. The plates are then washed and incubated for 2 h with an europium-conjugated anti-PrP monoclonal antibody (3F4). After washing, the enhancement solution (Wallac Inc, Turku, Finland) is added, and the fluorescence signal of each well is analyzed in a time-resolved fluorescence reader. The difference in fluorescence between denatured and native samples is calculated to give an estimate of the concentration of PrP^{TSE} with reference to a cut-off D-N value obtained from negative samples.

13.8.4 APPLICATIONS

The CDI was the first application of the DELFIA system in the field of TSEs [100]. This original assay was developed to demonstrate that different TSE strains propagated in the golden Syrian hamster (a widely used experimental host for TSEs) associate with strain-specific conformations of the pathological PrP as deduced by the different accessibilities of the epitope recognized by 3F4 Ab (hamster PrP residues) [109–112].

In addition to the possibility to discriminate differently misfolded PrP^{TSE}s and to detect PrP^{TSE} in the presence of excess PrP^{c} (the assay allows to detect PrP^{TSE} traces mixed with a 3000 fold excess of PrP^{c}), the CDI provides an outstanding analytical sensitivity, comparable with that of transgenic mice bioassays of BSE or CJD for the detection of TSE infection [107,112]. In the latter study [112], the CDI showed a diagnostic sensitivity significantly higher than the immunohisto-chemistry, and it was proposed to take the place of this technique for routine histological human TSE diagnosis. The CDI is a high throughput technique. It is almost fully automated and can provide results in a handful of hours; therefore, its application in human diagnosis should be taken into serious consideration, but more extensive studies are required before IHC dismissal.

Caution is also suggested by a recent study completed on sheep experimentally infected with scrapie [111] showing that the CDI may miss a significant portion of western blot positive animals of a particular PrP genotype. This result is apparently in contrast with the results of a WHO validation study that suggested that the CDI has a 100% diagnostic sensitivity (percentage of affected people who tested positive) and a 50–100-fold larger analytical sensitivity in comparison with the western blot in sporadic and variant CJD samples [58]. These discrepancies could be explained coming back to the principle of this technique. The array of exposed epitopes in non-denatured PrP^{c} and PrP^{TSE} containing samples is strictly dependent both on the strain under examination and host genotype [100, 111]. In some conditions, like the scrapie model mentioned above [111], the relevant epitope may have similar accessibility both in native and denatured PrP^{TSE}, thus reducing the distinction between infected and non-infected samples. Such a limitation could have been already foreseen from the data reported in the first CDI paper [100] where it is shown that different hamster-adapted scrapie strains produce different D/N ratios. In other words, one of the strong points of the CDI (i.e., the possibility to discriminate between strains) turns out to be a weakness and the CDI may reveal poor sensitivity for the diagnostic screening of some TSEs. These difficulties may be overcome by preventive validation studies on a large number of samples of the different TSE forms involved and by the adoption of a panel of different antibodies, each recognizing different epitopes. However, even these strategies may not allow the identification of emerging mutant strains associated with novel PrP^{TSE} conformations that may arise in the field.

In conclusion, the CDI stands as one of the most powerful tools that may contribute to improving the diagnosis and the basic knowledge of TSEs, but as is recommended for other rapid screening tests, it should be used in combination with other confirmatory assays in diagnostic campaigns.

13.9 SIFT (SCANNING FOR INTENSELY FLUORESCENT TARGET)

13.9.1 BASIC PRINCIPLES

This advanced technique is based on fluorescence correlation spectroscopy (FCS) [113,114]. FCS recognizes single fluorescent molecules in solution as they pass between an exciting laser beam and the objective of a confocal microscope equipped with a single-photon counter. If the fluorochrome is coupled to an antibody raised against a given molecule, the FCS can be applied to the immuno-logical detection of this molecule in solution. The distinction between free and bound fluorophores is generally achieved by the measurement of the diffusion time of the target. Once optimized, the analysis is quickly performed and requires only a small volume of sample.

In this classical format, the FCS is of poor utility when the concentration of the analyte is scant in comparison with the fluorophore (fluorochrome plus antibody), a situation that may occur when the diagnostic method is applied to preclinical samples. To improve the analytical sensitivity of the assay for TSE diagnosis, a particular setting has been developed in consideration of the aggregated chemical state of PrP^{TSE} and the rarity of the target. In this setting, the large, heavy labeled PrP^{TSE} aggregates are actively searched by fast and continuous scanning of the sample under the focus of the microscope (scanning for intensely fluorescent targets, SIFT).

13.9.2 SAMPLING AND STORAGE OF SPECIMENS

The sample analyzed in the original and, thus far, the only report was cerebrospinal fluids collected according to the protocols adopted by the German National CJD Surveillance Unit [115]. In principle, suspensions or extracts properly prepared from other kinds of unfixed tissues may be analysed by this technique, provided that the tissues have been collected and stored to avoid proteolysis by following the principles detailed in the western blot section.

13.9.3 PROCEDURE

The procedure has been developed on control (non-CJD) human cerebrospinal fluid spiked with purified hamster PrP^{TSE} and applied to the detection of human PrP^{TSE} in the cerebrospinal fluid of sporadic CJD patients.

Scanning for Intensely Fluorescent Target in TSE Cerebrospinal Fluid

Two monoclonal antibodies 3F4 and 12F10 [116] that recognize distinct epitopes of the PrP^{TSE} are labeled with fluorescent green (Alexa 488, Molecular Probes) and red (Cy-5, Amersham) dyes, according to the manufacturer's instructions. Fluorescent antibodies are then added to 20 μl of the sample under examination (final concentration of antibodies 6 nM). This mixture is used to fill a 0.2×2 mm glass capillary that is sealed onto a glass support and analyzed for the presence of highly fluorescent particles by a confocal, dual color fluorescence microscope (Dual Color ConfoCor, Zeiss) equipped with a mobile positioning table that allows the scanning of the capillary at a speed of 1 mm/s. Fluorescent targets are identified by laser enlightenment of the sample solution at 633 nm (for Cy 5) and 488 nm (Alexa 488) and detected by a red and a green detector photocounter. The measurement time is 600 s, and the temperature of the sample is kept at 22°C. PrP^{TSE} aggregates are detected as intensely red-green fluorescent particles that peak above the background of free fluorescent antibodies. The photons registered by the detectors during a definite period of time (bin) can be plotted versus the number of bins to produce an intensity distribution graph that allows measuring of the signal derived from the target molecules.

13.9.4 APPLICATIONS

The analytical sensitivity and the specificity of dual color SIFT were experimentally measured with human cerebrospinal fluid mixed with purified amyloid PrP^{TSE}, yielding a detection limit of 2 pM that translates into a 20-fold higher sensitivity than the western blot. In this reported experimental setting, the SIFT allowed the detection, for the first time, of PrP^{TSE} in the cerebrospinal fluid of CJD patients, [114] and although not yet validated in other laboratories (the instrument is highly expensive), it remains one of the most promising diagnostic tools for TSEs.

13.10 CONCLUSIONS AND PERSPECTIVES

Diagnosis of TSE diseases in affected subjects does not pose major problems, particularly in humans where the sum of clinical signs, clinical records, strumental (EEG, MRI), biochemical (presence of 14-3-3 and tau proteins in cerebrospinal fluid), and genetic tests are usually sufficient to obtain a reliable diagnosis of possible sporadic, possible variant, or definite familial CJD. Post-mortem examination of the brain supplies the definite diagnosis of human TSE on the basis of the histological lesions, the immunohistochemical morphology of PrP deposits, and the biochemical definition of PrP^{TSE} glycotype.

However, when it comes to either preclinical diagnosis of healthy but at risk subjects or the evaluation of possibly infected materials such as blood or tissues intended for transplantation or therapy, prion diseases confront researchers with unprecedented difficulties that are tightly linked with basic research topics. These topics include the detailed definition of the structure of the infectious agent and the relationship between PrP^{TSE} and infectivity. Both, concern for public health and economic consequences deriving from the outbreak of bovine spongiform encephalopathy as well as the outrising of its human counterpart, that might spread horizontally through blood donations, have boosted the interest in the development of improved analytical methods to detect the marker protein. Although a highly sensitive screening test for the identification of TSE infectivity on easily available tissues (like blood or urine) [117,118] and in other relevant substrates (like food or biopharmaceuticals)[119] is not yet available, researchers have recently witnessed a number of significant advancements involving the development of immunological and non immunological detection systems [120–123] as well as original signal amplification strategies [124–126]. It is simple, therefore, to forecast that the combination of the most efficient systems from these areas will soon bring a new generation of bioanalytical diagnostic tests with an unprecedented sensitivity for measuring prions that hold promises for the realization of large screening surveys for the prevention and control of human and animal TSEs.

ACKNOWLEDGMENTS

We gratefully acknowledge Umberto Agrimi, Angela De Pascalis, Michele Di Bari, Romolo Nonno, Piero Parchi, Marco Sbriccoli, and Gianluigi Zanusso for providing high quality iconographic material.

We also thank Dr. Alessandra Garozzo and Marco del Re for the editorial assistance.

This work was partially supported by grants from the European Union (QLK2-CT-2002-81523; QLG3-CT-2002-81606; network of excellence Neuroprion) and from the Istituto Superiore di Sanità and the Italian Ministry of Health (Progetto di ricerca 1%/2003-4ANF; National Registry of Creutzfeldt-Jakob disease; Progetto di ricerca "Le encefalopatie spongiformi trasmissibili dell'uomo e degli animali causate da agenti infettivi non convenzionali").

REFERENCES

1. Collinge, J., Prion diseases of humans and animals: their causes and molecular basis, *Annu. Rev. Neurosci.*, 24, 519–550, 2001.
2. Ladogana, A., Puopolo, M., Croes, E. A., Budka, H., Jarius, C., Collins, S., Klug, G. M., et al., Mortality from Creutzfeldt-Jakob disease and related disorders in Europe, Australia, and Canada, *Neurology*, 64, 1586–1591, 2005.
3. Pocchiari, M., Prions and related neurological diseases, *Mol. Aspects Med.*, 15, 195–291, 1994.
4. Pocchiari, M., Puopolo, M., Croes, E. A., Budka, H., Gelpi, E., Collins, S., Lewis, V. et al., Predictors of survival in sporadic Creutzfeldt-Jakob disease and other human transmissible spongiform encephalopathies, *Brain*, 127, 2348–2359, 2004.

5. Kovacs, G. G., Puopolo, M., Ladogana, A., Pocchiari, M., Budka, H., van Duijn, C., Collins, S. J. et al., Genetic prion disease: The EUROCJD experience, *Hum. Genet.*, 118, 166–174, 2005.

6. Brown, P., Preece, M., Brandel, J. P., Sato, T., McShane, L., Zerr, L. I., Fletcher, A. et al., Iatrogenic Creutzfeldt-Jakob disease at the millennium, *Neurology*, 55, 1075–1081, 2000.

7. Will, R. G., Ironside, J. W., Zeidler, M., Cousens, S. N., Estibeiro, K., Alperovitch, A., Poser, S., Pocchiari, M., Hofman, A., and Smith, P. G., A new variant of Creutzfeldt-Jakob disease in the U.K., *Lancet*, 347, 921–925, 1996.

8. Wilesmith, J. W., Wells, G. A., Cranwell, M. P., and Ryan, J. B., Bovine spongiform encephalopathy: epidemiological studies, *Vet. Rec.*, 123, 638–644, 1998.

9. Ward, H. J., Everington, D., Cousens, S. N., Smith-Bathgate, B., Leitch, M., Cooper, S., Heath, C., Knight, R. S., Smith, P. G., and Will, R. G., Risk factors for variant Creutzfeldt-Jakob disease: a case-control study, *Ann. Neurol.*, 59, 111–120, 2006.

10. Will, R. G., Alperovitch, A., Poser, S., Pocchiari, M., Hofman, A., Mitrova, E., de Silva, R. et al., Descriptive epidemiology of Creutzfeldt-Jakob disease in six European countries, 1993–1995. EU collaborative study group for CJD, *Ann. Neurol.*, 43, 763–767, 1998.

11. Will, R. G., Zeidler, M., Stewart, G. E., Macleod, M. A., Ironside, J. W., Cousens, S. N., Mackenzie, J., Estibeiro, K., Green, A. J., and Knight, R. S., Diagnosis of new variant Creutzfeldt-Jakob disease, *Ann. Neurol.*, 47, 575–582, 2000.

12. Wadsworth, J. D., Joiner, S., Hill, A. F., Campbell, T. A., Desbruslais, M., Luthert, P. J., and Collinge, J., Tissue distribution of protease resistant prion protein in variant Creutzfeldt-Jakob disease using a highly sensitive immunoblotting assay, *Lancet*, 358, 171–180, 2001.

13. Head, M. W., Ritchie, D., Smith, N., McLoughlin, V., Nailon, W., Samad, S., Masson, S., Bishop, M., McCardle, L., and Ironside, J. W., Peripheral tissue involvement in sporadic, iatrogenic, and variant Creutzfeldt-Jakob disease: an immunohistochemical, quantitative, and biochemical study, *Am. J. Pathol.*, 164, 143–153, 2004.

14. Unterberger, U., Voigtlander, T., and Budka, H., Pathogenesis of prion diseases, *Acta Neuropathol. (Berl)*, 109, 32–48, 2005.

15. Mabbott, N. A. and Bruce, M. E., The immunobiology of TSE diseases, *J. Gen. Virol.*, 82, 2307–2318, 2001.

16. Baldauf, E., Beekes, M., and Diringer, H., Evidence for an alternative direct route of access for the scrapie agent to the brain bypassing the spinal cord, *J. Gen. Virol.*, 78, 1187–1197, 1997.

17. McBride, P. A., Schulz-Schaeffer, W. J., Donaldson, M., Bruce, M., Diringer, H., Kretzschmar, H. A., and Beekes, M., Early spread of scrapie from the gastrointestinal tract to the central nervous system involves autonomic fibers of the splanchnic and vagus nerves, *J. Virol.*, 75, 9320–9327, 2001.

18. Doi, S., Ito, M., Shinagawa, M., Sato, G., Isomura, H., and Goto, H., Western blot detection of scrapie-associated fibril protein in tissues outside the central nervous system from preclinical scrapie-infected mice, *J. Gen. Virol.*, 69, 955–960, 1998.

19. Schreuder, B. E., van Keulen, L. J., Vromans, M. E., Langeveld, J. P., and Smits, M. A., Tonsillar biopsy and PrPSc detection in the preclinical diagnosis of scrapie, *Vet. Rec.*, 124, 564–568, 1998.

20. Hill, A. F., Zeidler, M., Ironside, J., and Collinge, J., Diagnosis of new variant Creutzfeldt-Jakob disease by tonsil biopsy, *Lancet*, 349, 99–100, 1997.

21. Frosh, A., Smith, L. C., Jackson, C. J., Linehan, J. M., Brandner, S., Wadsworth, J. D., and Collinge, J., Analysis of 2000 consecutive U.K. tonsillectomy specimens for disease-related prion protein, *Lancet*, 364, 1260–1262, 2000.

22. Hilton, D. A., Ghani, A. C., Conyers, L., Edwards, P., McCardle, L., Penney, M., Ritchie, D., and Ironside, J. W., Accumulation of prion protein in tonsil and appendix: review of tissue samples, *BMJ*, 325, 633–634, 2002.

23. Hill, A. F., Butterworth, R. J., Joiner, S., Jackson, G., Rossor, M. N., Thomas, D. J., Frosh, A. et al., Investigation of variant Creutzfeldt-Jakob disease and other human prion diseases with tonsil biopsy samples, *Lancet*, 353, 183–189, 1999.

24. Glatzel, M., Abela, E., Maissen, M., and Aguzzi, A., Extraneural pathologic prion protein in sporadic Creutzfeldt-Jakob disease, *N. Engl. J. Med.*, 349, 1812–1820, 2003.

25. Andreoletti, O., Simon, S., Lacroux, C., Morel, N., Tabouret, G., Chabert, A., Lugan, S. et al., PrPSc accumulation in myocytes from sheep incubating natural scrapie, *Nat. Med.*, 10, 591–593, 2004.

26. Thomzig, A., Cardone, F., Kruger, D., Pocchiari, M., Brown, P., and Beekes, M., Pathological prion protein in muscles of hamsters and mice infected with rodent-adapted BSE or vCJD, *J. Gen. Virol.*, 87, 251–254, 2006.

27. Brown, P., Rohwer, R. G., Dunstan, B. C., MacAuley, C., Gajdusek, D. C., and Drohan, W. N., The distribution of infectivity in blood components and plasma derivatives in experimental models of transmissible spongiform encephalopathy, *Transfusion*, 38, 810–816, 1998.

28. Hunter, N., Foster, J., Chong, A., McCutcheon, S., Parnham, D., Eaton, S., MacKenzie, C., and Houston, F., Transmission of prion diseases by blood transfusion, *J. Gen. Virol.*, 83, 2897–2905, 2002.

29. Llewelyn, C. A., Hewitt, P. E., Knight, R. S., Amar, K., Cousens, S., Mackenzie, J., and Will, R. G., Possible transmission of variant Creutzfeldt-Jakob disease by blood transfusion, *Lancet*, 363, 417–421, 2004.

30. Peden, A. H., Head, M. W., Ritchie, D. L., Bell, J. E., and Ironside, J. W., Preclinical vCJD after blood transfusion in a PRNP codon 129 heterozygous patient, *Lancet*, 364, 527–529, 2004.

31. Hilton, D. A., Pathogenesis and prevalence of variant Creutzfeldt-Jakob disease, *J. Pathol.*, 208, 134–141, 2006.

32. Hilton, D. A., Ghani, A. C., Conyers, L., Edwards, P., McCardle, L., Ritchie, D., Penney, M., Hegazy, D., and Ironside, J. W., Prevalence of lymphoreticular prion protein accumulation in U.K. tissue samples, *J. Pathol.*, 203, 733–739, 2004.

33. Brown, P., Gibbs, C. J. Jr., Rodgers-Johnson, P., Asher, D. M., Sulima, M. P., Bacote, A., Goldfarb, L. G., and Gajdusek, D. C., Human spongiform encephalopathy: the National Institutes of Health series of 300 cases of experimentally transmitted disease, *Ann. Neurol.*, 35, 513–529, 1994.

34. Bruce, M. E., McConnell, I., Will, R. G., and Ironside, J. W., Detection of variant Creutzfeldt-Jakob disease infectivity in extraneural tissues, *Lancet*, 358, 208–209, 2001.

35. van Duijn, C. M., Delasnerie-Laupretre, N., Masullo, C., Zerr, I., de Silva, R., Wientjens, D. P., 1998, J. P. et al., Case-control study of risk factors of Creutzfeldt-Jakob disease in Europe during 1993–95. European Union (EU) Collaborative Study Group of Creutzfeldt-Jakob disease (CJD), *Lancet*, 351, 1081–1085, 1998.

36. Collins, S., Law, M. G., Fletcher, A., Boyd, A., Kaldor, J., and Masters, C. L., Surgical treatment and risk of sporadic Creutzfeldt-Jakob disease: a case-control study, *Lancet*, 353, 693–697, 1999.

37. Yakovleva, O., Janiak, A., McKenzie, C., McShane, L., Brown, P., and Cervenakova, L., Effect of protease treatment on plasma infectivity in variant Creutzfeldt-Jakob disease mice, *Transfusion*, 44, 1700–1705, 2004.

38. Cervenakova, L. and Brown, P., Advances in screening test development for transmissible spongiform encephalopathies, *Expert. Rev. Anti. Infect. Ther.*, 2, 873–880, 2004.

39. Minor, P. D., Technical aspects of the development and validation of tests for variant Creutzfeldt-Jakob disease in blood transfusion, *Vox Sang.*, 86, 164–170, 2004.

40. Castilla, J., Saá, P., and Soto, C., Detection of prions in blood, *Nat. Med.*, 9, 982–985, 2005.

41. Soto, C., Anderes, L., Suardi, S., Cardone, F., Castilla, J., Frossard, M. J., Peano, S. et al., Pre-symptomatic detection of prions by cyclic amplification of protein misfolding, *FEBS Lett.*, 579, 638–642, 2005.

42. Saborio, G. P., Permanne, B., and Soto, C., Sensitive detection of pathological prion protein by cyclic amplification of protein misfolding, *Nature*, 411, 810–813, 2001.

43. Weissmann, C., The state of the prion, *Nat. Rev. Microbiol.*, 2, 861–871, 2004.

44. Parchi, P., Castellani, R., Capellari, S., Ghetti, B., Young, K., Chen, S. G., Farlow, M. et al., Molecular basis of phenotypic variability in sporadic Creutzfeldt-Jakob disease, *Ann. Neurol.*, 39, 767–778, 1996.

45. Korth, C., Stierli, B., Streit, P., Moser, M., Schaller, O., Fischer, R., Schulz-Schaeffer, W. et al., Prion (PrPSc)-specific epitope defined by a monoclonal antibody, *Nature*, 390, 74–77, 1997.

46. Paramithiotis, E., Pinard, M., Lawton, T., LaBoissiere, S., Leathers, V. L., Zou, W. Q., Estey, L. A. et al., A prion protein epitope selective for the pathologically misfolded conformation, *Nat. Med.*, 9, 893–899, 2003.

47. Curin, V., Bresjanac, M., Popovic, M., Pretnar Hartman, K., Galvani, V., Rupreht, R., Cernilec, M., Vranac, T., Hafner, I., and Jerala, R., Monoclonal antibody against a peptide of human prion protein discriminates between Creutzfeldt-Jacob's disease-affected and normal brain tissue, *J. Biol. Chem.*, 279, 3694–3698, 2004.

48. Zou, W. Q., Zheng, J., Gray, D. M., Gambetti, P., and Chen, S. G., Antibody to DNA detects scrapie but not normal prion protein, *Proc. Natl Acad. Sci. U.S.A.*, 101, 1380–1385, 2004.

49. Head, M. W., Bunn, T. J., Bishop, M. T., McLoughlin, V., Lowrie, S., McKimmie, C. S., Williams, M. C. et al., Prion protein heterogeneity in sporadic but not variant Creutzfeldt-Jakob disease: U.K. cases 1991–2002, *Ann. Neurol.*, 55, 851–859, 2004.

50. WHO Manual for surveillance of human transmissible spongiform encephalopathies including variant Creutzfeldt-Jakob disease. 2003. World Health Organization, Communicable Disease Surveillance and Response.

51. Casalone, C., Zanusso, G., Acutis, P., Ferrari, S., Capucci, L., Tagliavini, F., Monaco, S., and Caramelli, M., Identification of a second bovine amyloidotic spongiform encephalopathy: molecular similarities with sporadic Creutzfeldt-Jakob disease, *Proc. Natl. Acad. Sci. U.S.A.*, 101, 3065–3070, 2004.

52. Begara-McGorum, I., Gonzalez, L., Simmons, M., Hunter, N., Houston, F., and Jeffrey, M., Vacuolar lesion profile in sheep scrapie: factors influencing its variation and relationship to disease-specific PrP accumulation, *J. Comp. Pathol.*, 127, 59–68, 2002.

53. Caplazi, P., O'Rourke, K., Wolf, C., Shaw, D., and Baszler, T. V., Biology of PrPsc accumulation in two natural scrapie-infected sheep flocks, *J. Vet. Diagn. Invest.*, 16, 489–496, 2004.

54. Cardone, F., Liu, Q. G., Petraroli, R., Ladogana, A., D'Alessandro, M., Arpino, C., Di Bari, M., Macchi, G., and Pocchiari, M., Prion protein glycotype analysis in familial and sporadic Creutz-feldt-Jakob disease patients, *Brain Res. Bull.*, 49, 429–433, 1999.

55. Zanusso, G., Ferrari, S., Cardone, F., Zampieri, P., Gelati, M., Fiorini, M., Farinazzo, A. et al., Detection of pathologic prion protein in the olfactory epithelium in sporadic Creutzfeldt-Jakob disease, *N. Engl. J. Med.*, 348, 711–719, 2003.

56. Silvestrini, M. C., Cardone, F., Maras, B., Pucci, P., Barra, D., Brunori, M., and Pocchiari, M., Identification of the prion protein allotypes which accumulate in the brain of sporadic and familial Creutzfeldt-Jakob disease patients, *Nat. Med.*, 3, 521–525, 1997.

57. Chen, S. G., Parchi, P., Brown, P., Capellari, S., Zou, W., Cochran, E. J., Vnencak-Jones, C. L. et al., Allelic origin of the abnormal prion protein isoform in familial prion diseases, *Nat. Med.*, 3, 1009–1015, 1997.

58. Minor, P., Newham, J., Jones, N., Bergeron, C., Gregori, L., Asher, D., van Engelenburg, F. et al., WHO working group on international reference materials for the diagnosis and study of transmissible spongiform encephalopathies. Standards for the assay of Creutzfeldt-Jakob disease specimens, *J. Gen. Virol.*, 85, 1777–1784, 2004.

59. Berardi, V., Cardone, F., Valanzano, A., Lu, M., Pocchiari, M., Preparation of soluble infectious samples from scrapie-infected brain. A new tool to study the clearance of transmissible spongiform encephalopathy agents during plasma fractionation. Transfusion 46, 652–658, 2006. doi: 10.1111/j. 1537–2995.2006.00763.x.

60. Kascsak, R. J., Rubenstein, R., Merz, P. A., Tonna-DeMasi, M., Fersko, R., Carp, R. I., Wisniewski, H. M., and Diringer, H., Mouse polyclonal and monoclonal antibody to scrapie-associated fibril proteins, *J. Virol.*, 61, 3688–3693, 1987.

61. Kuczius, T. and Groschup, M. H., Differences in proteinase K resistance and neuronal deposition of abnormal prion proteins characterize bovine spongiform encephalopathy (BSE) and scrapie strains, *Mol. Med.*, 5, 406–418, 1999.

62. Collinge, J., Sidle, K. C., Meads, J., Ironside, J., and Hill, A. F., Molecular analysis of prion strain variation and the aetiology of 'new variant' CJD, *Nature*, 383, 590–685, 1996.

63. Polymenidou, M., Stoeck, K., Glatzel, M., Vey, M., Bellon, A., and Aguzzi, A., Coexistence of multiple PrPSc types in individuals with Creutzfeldt-Jakob disease, *Lancet Neurol.*, 4, 805–814, 2005.

64. Parchi, P., Giese, A., Capellari, S., Brown, P., Schulz-Schaeffer, W., Windl, O., Zerr, I. et al., Kretzschmar H. Classification of sporadic Creutzfeldt-Jakob disease based on molecular and pheno-typic analysis of 300 subjects, *Ann. Neurol.*, 46, 224–233, 1999.

65. Gambetti, P., Kong, Q., Zou, W., Parchi, P., and Chen, S. G., Sporadic and familial CJD: Classification and characterisation, *Br. Med. Bull.*, 66, 213–239, 2003.

66. Parchi, P., Petersen, R. B., Chen, S. G., Autilio-Gambetti, L., Capellari, S., Monari, L., Cortelli, P., Montagna, P., Lugaresi, E., and Gambetti, P., Molecular pathology of fatal familial insomnia, *Brain Pathol.*, 8, 539–548, 1998.

67. Zanusso, G., Righetti, P. G., Ferrari, S., Terrin, L., Farinazzo, A., Cardone, F., Pocchiari, M., Rizzuto, N., and Monaco, S., Two-dimensional mapping of three phenotype-associated isoforms of the prion protein in sporadic Creutzfeldt-Jakob disease, *Electrophoresis*, 23, 347–355, 2002.

68. Pan, T., Colucci, M., Wong, B. S., Li, R., Liu, T., Petersen, R. B., Chen, S., and Gambetti, P., Sy M.S Novel differences between two human prion strains revealed by two-dimensional gel electrophoresis, *J. Biol. Chem.*, 276, 37284–37288, 2001.

69. Zanusso, G., Farinazzo, A., Prelli, F., Fiorini, M., Gelati, M., Ferrari, S., Righetti, P. G., Rizzuto, N., Frangione, B., and Monaco, S., Identification of distinct N-terminal truncated forms of prion protein in different Creutzfeldt-Jakob disease subtypes, *J. Biol. Chem.*, 279, 38936–38942, 2004.

70. Hill, A. F., Joiner, S., Wadsworth, J. D., Sidle, K. C., Bell, J. E., Budka, H., Ironside, J. W., and Collinge, J., Molecular classification of sporadic Creutzfeldt-Jakob disease, *Brain*, 126, 1333–1346, 2003.

71. Hayashi, H., Takata, M., Iwamaru, Y., Ushiki, Y., Kimura, K. M., Tagawa, Y., Shinagawa, M., and Yokoyama, T., Effect of tissue deterioration on postmortem BSE diagnosis by immunobiochemical detection of an abnormal isoform of prion protein, *J. Vet. Med. Sci.*, 66, 515–520, 2004.

72. Barry,, R. A., Kent, S. B., McKinley, M. P., Meyer, R. K., DeArmond, S. J., Hood, L. E., and Prusiner, S. B., Scrapie and cellular prion proteins share polypeptide epitopes, *J. Infect. Dis.*, 153, 848–854, 1986.

73. Serban, D., Taraboulos, A., DeArmond, S. J., and Prusiner, S. B., Rapid detection of Creutzfeldt-Jakob disease and scrapie prion proteins, *Neurology*, 40, 110–117, 1990.

74. Meyer, R. K., Oesch, B., Fatzer, R., Zurbriggen, A., and Vandevelde, M., Detection of bovine spongiform encephalopathy-specific PrP(Sc) by treatment with heat and guanidine thiocyanate, *J. Virol.*, 73, 9386–9392, 1999.

75. Wong, B. S., Green, A. J., Li, R., Xie, Z., Pan, T., Liu, T., Chen, S. G., Gambetti, P., and Sy, M. S., Absence of protease-resistant prion protein in the cerebrospinal fluid of Creutzfeldt-Jakob disease, *J. Pathol.*, 194, 9–14, 2001.

76. Gavier-Widen, D., Noremark, M., Benestad, S., Simmons, M., Renstrom, L., Bratberg, B., Elvander, M., and Segerstad, C. H., Recognition of the Nor98 variant of scrapie in the Swedish sheep population, *J. Vet. Diagn. Invest.*, 16, 562–567, 2004.

77. Deslys, J. P., Comoy, E., Hawkins, S., Simon, S., Schimmel, H., Wells, G., Grassi, J., and Moynagh, J., Screening slaughtered cattle for BSE, *Nature*, 409, 476–478, 2001.

78. Grassi, J., Comoy, E., Simon, S., Creminon, C., Frobert, Y., Trapmann, S., Schimmel, H., Hawkins, S. A., Moynagh, J., Deslys, J. P. et al., Rapid test for the preclinical postmortem diagnosis of BSE in central nervous system tissue, *Vet. Rec.*, 149, 577–582, 2001.

79. Biffiger, K., Zwald, D., Kaufmann, L., Briner, A., Nayki, I., Purro, M., Bottcher, S. et al., Validation of a luminescence immunoassay for the detection of PrP(Sc) in brain homogenate, *J. Virol. Methods*, 101, 79–84, 2002.

80. Grathwohl, K. U., Horiuchi, M., Ishiguro, N., and Shinagawa, M., Sensitive enzyme-linked immunosorbent assay for detection of PrP(Sc) in crude tissue extracts from scrapie-affected mice, *J. Virol. Methods*, 64, 205–216, 1997.

81. Puoti, G., Giaccone, G., Rossi, G., Canciani, B., Bugiani, O., and Tagliavini, F., Sporadic Creutzfeldt-Jakob disease: co-occurrence of different types of PrP(Sc) in the same brain, *Neurology*, 53, 2173–2176, 1999.

82. Giaccone, G., Canciani, B., Puoti, G., Rossi, G., Goffredo, D., Iussich, S., Fociani, P., Tagliavini, F., and Bugiani, O., Creutzfeldt-Jakob disease: Carnoy's fixative improves the immunohistochemistry of the proteinase K-resistant prion protein, *Brain Pathol.*, 10, 31–37, 2000.

83. Privat, N., Sazdovitch, V., Seilhean, D., LaPlanche, J. L., and Hauw, J. J., PrP immunohistochemistry: different protocols, including a procedure for long formalin fixation, and a proposed schematic classification for deposits in sporadic Creutzfeldt-Jakob disease, *Microsc. Res. Tech.*, 50, 26–31, 2000.

84. Bruce, M. E., Will, R. G., Ironside, J. W., McConnell, I., Drummond, D., Suttie, A., McCardle, L. et al., Transmissions to mice indicate that 'new variant' CJD is caused by the BSE agent, *Nature*, 389, 498–501, 1997.

85. Budka, H., Neuropathology of prion diseases, *Br. Med. Bull.*, 66, 121–130, 2003.

86. Bolea, R., Monleon, E., Schiller, I., Raeber, A. J., Acin, C., Monzon, M., Martin-Burriel, I., Struckmeyer, T., Oesch, B., and Badiola, J. J., Comparison of immunohistochemistry and two rapid tests for detection of abnormal prion protein in different brain regions of sheep with typical scrapie, *J. Vet. Diagn. Invest.*, 17, 467–469, 2005.

87. Bell, J. E., Gentleman, S. M., Ironside, J. W., McCardle, L., Lantos, P. L., Doey, L., Lowe, J. et al., Prion protein immunocytochemistry–U.K. five centre consensus report, *Neuropathol. Appl. Neurobiol.*, 23, 26–35, 1997.

88. Hauw, J. J., Sazdovitch, V., Privat, N., Seilhean, D., Kopp, N., Laupretre, N., Brandel, J. P., Deslys, J. P., Laplanche, J. L., and Alpérovitch, A., The significance of morula-type spongiform change in sporadic Creutzfeldt-Jakob disease. A study in 70 patients, *Neuropathol. Appl. Neurobiol.*, 25 (suppl. 1), 62, 1999.

89. Liberski, P. P., Bratosiewicz, J., Walis, A., Kordek, R., Jeffrey, M., and Brown, P., A special report I. Prion protein (PrP)—amyloid plaques in the transmissible spongiform encephalopathies, or prion diseases revisited, *Folia Neuropathol.*, 39, 217–235, 2001.

90. Kretzschmar, H. A., Sethi, S., Foldvari, Z., Windl, O., Querner, V., Zerr, I., and Poser, S., Iatrogenic Creutzfeldt-Jakob disease with florid plaques, *Brain Pathol.*, 13, 245–249, 2003.

91. Shimizu, S., Hoshi, K., Muramoto, T., Homma, M., Ironside, J. W., Kuzuhara, S., Sato, T., Yamamoto, T., and Kitamoto, T., Creutzfeldt-Jakob disease with florid-type plaques after cadaveric dura mater grafting, *Arch. Neurol.*, 56, 357–362, 1999.

92. Taraboulos, A., Jendroska, K., Serban, D., Yang, S. L., DeArmond, S. J., and Prusiner, S. B., Regional mapping of prion proteins in brain, *Proc. Natl Acad. Sci. U.S.A.*, 89, 7620–7624, 1992.

93. Kimura, K. M., Yokoyama, T., Haritani, M., Narita, M., Belleby, P., Smith, J., and Spencer, Y. I., In situ detection of cellular and abnormal isoforms of prion protein in brains of cattle with bovine spongiform encephalopathy and sheep with scrapie by use of a histoblot technique, *J. Vet. Diagn. Invest.*, 14, 255–257, 2002.

94. Heggebo, R., Press, C. M., Gunnes, G., Gonzalez, L., and Jeffrey, M., Distribution and accumulation of PrP in gut-associated and peripheral lymphoid tissue of scrapie-affected Suffolk sheep, *J. Gen. Virol.*, 83, 479–489, 2002.

95. Jeffrey, M., Martin, S., Barr, J., Chong, A., and Fraser, J. R., Onset of accumulation of PrPres in murine ME7 scrapie in relation to pathological and PrP immunohistochemical changes, *J. Comp. Pathol.*, 124, 20–28, 2001.

96. Yokoyama, T., Kimura, K. M., Ushiki, Y., Yamada, S., Morooka, A., Nakashiba, T., Sassa, T., and Itohara, S., In vivo conversion of cellular prion protein to pathogenic isoforms, as monitored by conformation-specific antibodies, *J. Biol. Chem.*, 276, 11265–11271, 2001.

97. Schulz-Schaeffer, W. J., Tschoke, S., Kranefuss, N., Drose, W., Hause-Reitner, D., Giese, A., Groschup, M. H., and Kretzschmar, H. A., The paraffin-embedded tissue blot detects PrP(Sc) early in the incubation time in prion diseases, *Am. J. Pathol.*, 156, 51–56, 2000.

98. Ritchie, D. L., Head, M. W., and Ironside, J. W., Advances in the detection of prion protein in peripheral tissues of variant Creutzfeldt-Jakob disease patients using paraffin-embedded tissue blotting, *Neuropathol. Appl. Neurobiol.*, 360–368, 2004.

99. http://las.perkinelmer.de/applicationssummary/applications/TRF-DELFIA.htm (last accessed September 2006).

100. Safar, J., Wille, H., Itzi, V., Groth, D., Serban, H., Torchia, M., Cohen, F. E., and Prusiner, S. B., Eight prion strains have PrP(Sc) molecules with different conformations, *Nat. Med.*, 4, 1157–1165, 1998.

101. MacGregor, I. and Drummond, O., Immunoassay of human plasma cellular prion protein, *Transfusion*, 41, 1453–1454, 2001.

102. Volkel, D., Zimmermann, K., Zerr, I., Bodemer, M., Lindner, T., Turecek, P. L., Poser, S., and Schwarz, H. P., Immunochemical determination of cellular prion protein in plasma from healthy subjects and patients with sporadic CJD or other neurologic diseases, *Transfusion*, 41, 441–448, 2001.

103. Fagge, T., Barclay, G. R., MacGregor, I., Head, M., Ironside, J., and Turner, M., Variation in concentration of prion protein in the peripheral blood of patients with variant and sporadic Creutzfeldt-Jakob disease detected by dissociation enhanced lanthanide fluoroimmunoassay and flow cytometry, *Transfusion*, 45, 504–513, 2005.

104. Bessos, H., Drummond, O., Prowse, C., Turner, M., and MacGregor, I., The release of prion protein from platelets during storage of apheresis platelets, *Transfusion*, 41, 61–66, 2001.

105. MacGregor, I., Hope, J., Barnard, G., Kirby, L., Drummond, O., Pepper, D., Hornsey, V. et al., Application of a time-resolved fluoroimmunoassay for the analysis of normal prion protein in human blood and its components, *Vox Sang.*, 77, 88–96, 1999.

106. Starke, R., Drummond, O., MacGregor, I., Biggerstaff, J., Gale, R., Camilleri, R., Mackie, I., Machin, S., and Harrison, P., The expression of prion protein by endothelial cells: a source of the plasma form of prion protein? *Br. J. Haematol.*, 119, 863–873, 2002.

107. Safar, J. G., Scott, M., Monaghan, J., Deering, C., Didorenko, S., Vergara, J., Ball, H. et al., Measuring prions causing bovine spongiform encephalopathy or chronic wasting disease by immunoassays and transgenic mice, *Nat. Biotechnol.*, 20, 1147–1150, 2002.

108. Bellon, A., Seyfert-Brandt, W., Lang, W., Baron, H., Groner, A., and Vey, M., Improved conformation-dependent immunoassay: suitability for human prion detection with enhanced sensitivity, *J. Gen. Virol.*, 84, 1921–1925, 2003.

109. Barnard, G., Helmick, B., Madden, S., Gilbourne, C., and Patel, R., The measurement of prion protein in bovine brain tissue using differential extraction and DELFIA as a diagnostic test for BSE, *Luminescence*, 15, 357–362, 2000.

110. Tremblay, P., Ball, H. L., Kaneko, K., Groth, D., Hegde, R. S., Cohen, F. E., DeArmond, S. J., Prusiner, S. B., and Safar, J. G., Mutant PrPSc conformers induced by a synthetic peptide and several prion strains, *J. Virol.*, 78, 2088–2099, 2004.

111. McCutcheon, S., Hunter, N., and Houston, F., Use of a new immunoassay to measure PrP Sc levels in scrapie-infected sheep brains reveals PrP genotype-specific differences, *J. Immunol. Methods*, 298, 119–128, 2005.

112. Safar, J. G., Geschwind, M. D., Deering, C., Didorenko, S., Sattavat, M., Sanchez, H., Serban, A. et al., Diagnosis of human prion disease, *Proc. Natl Acad. Sci. U.S.A.*, 102, 3501–3506, 2005.

113. Giese, A., Bieschke, J., Eigen, M., and Kretzschmar, H. A., Putting prions into focus: application of single molecule detection to the diagnosis of prion diseases, *Arch. Virol. Suppl.*, 16, 161–171, 2000.

114. Bieschke, J., Giese, A., Schulz-Schaeffer, W., Zerr, I., Poser, S., Eigen, M., and Kretzschmar, H., Ultrasensitive detection of pathological prion protein aggregates by dual-color scanning for intensely fluorescent targets, *Proc. Natl Acad. Sci. U.S.A.*, 97, 5468–5473, 2000.

115. Zerr, I., Pocchiari, M., Collins, S., Brandel, J. P., de Pedro, J., Cuesta, R. S., Knight, R. S., Bernheimer, H. et al., Analysis of EEG and CSF 14-3-3 proteins as aids to the diagnosis of Creutzfeldt-Jakob disease, *Neurology*, 55, 811–815, 2000.

116. Krasemann, S., Groschup, M. H., Harmeyer, S., Hunsmann, G., and Bodemer, W., Generation of monoclonal antibodies against human prion proteins in PrP0/0 mice, *Mol. Med.*, 2, 725–734, 1996.

117. Brown, P., Cervenakova, L., and Diringer, H., Blood infectivity and the prospects for a diagnostic screening test in Creutzfeldt-Jakob disease, *J. Lab. Clin. Med.*, 137, 5–13, 2001.

118. Seeger, H., Heikenwalder, M., Zeller, N., Kranich, J., Schwarz, P., Gaspert, A., Seifert, B., Miele, G., and Aguzzi, A., Coincident scrapie infection and nephritis lead to urinary prion excretion, *Science*, 310, 324–326, 2005.

119. Robinson, M. M., Transmissible encephalopathies and biopharmaceutical production, *Dev. Biol. Stand.*, 88, 237–241, 1996.

120. Yang, W. C., Schmerr, M. J., Jackman, R., Bodemer, W., and Yeung, E. S., Capillary electrophoresis-based noncompetitive immunoassay for the prion protein using fluorescein-labeled protein a as a fluorescent probe, *Anal. Chem.*, 77, 4489–4494, 2005.

121. Trieschmann, L., Navarrete Santos, A., Kaschig, K., Torkler, S., Maas, E., Schatzl, H., and Bohm, G., Ultra-sensitive detection of prion protein fibrils by flow cytometry in blood from cattle affected with bovine spongiform encephalopathy, *BMC Biotechnol.*, 5, 26, 2005.

122. Kneipp, J., Lasch, P., Baldauf, E., Beekes, M., and Naumann, D., Detection of pathological molecular alterations in scrapie-infected hamster brain by Fourier transform infrared (FT-IR) spectroscopy, *Biochim. Biophys. Acta*, 1501, 189–199, 2000.

123. Henry, J., Anand, A., Chowdhury, M., Cote, G., Moreira, R., and Good, T., Development of a nanoparticle-based surface-modified fluorescence assay for the detection of prion proteins, *Anal. Biochem.*, 334, 1–8, 2004.

124. Castilla, J., Saa, P., Hetz, C., and Soto, C., In vitro generation of infectious scrapie prions, *Cell*, 121, 195–206, 2005.

125. Grosset, A., Moskowitz, K., Nelsen, C., Pan, T., Davidson, E., and Orser, C. S., Rapid presymptomatic detection of PrPSc via conformationally responsive palindromic PrP peptides, *Peptides*, 26, 2193–2200, 2005.
126. Barletta, J. M., Edelman, D. C., Highsmith, W. E., and Constantine, N. T., Detection of ultra-low levels of pathologic prion protein in scrapie infected hamster brain homogenates using real-time immuno-PCR, *J. Virol. Methods*, 127, 154–164, 2005.

14 Environmental Applications of Immunoaffinity Chromatography

Annette Moser, Mary Anne Nelson, and David S. Hage

CONTENTS

14.1 INTRODUCTION

As awareness of the effects of even ultra-trace contamination in the environment grows, there is an increasing demand for better analytical methods in studying environmental samples. For instance, new regulations by agencies in North America [1] and Europe [2] have forced laboratories to adopt methods with lower limits of detection, greater specificity, and higher precision for environmental agents. The analysis of environmental samples is complicated by the fact that the selected technique must deal with a wide range of analytes and matrices. Immunoaffinity chromatography (IAC) is one method that has recently been used for this purpose.

IAC is a type of affinity chromatography where the stationary phase is an antibody or related binding agent. IAC has several features that make it attractive for the study of pollutants and environmental agents. For instance, the ability to couple IAC with other methods can lead to the development of multidimensional methods that are easily automated, sensitive, and specific with excellent reproducibility [3]. In addition, the use of antibodies allows affinity chromatography to be employed for the analysis of either a specific analyte or closely-related group of analytes [4].

Environmental samples can range from simple matrices like drinking water to more complex mixtures such as soil extracts or foods. The analytes in these samples range from small molecules (e.g., organic pesticides) to large pharmaceuticals and even proteins. In many of these samples, the analyte is present at trace or ultra-trace levels and in the presence of several interfering agents. As an example, pesticides found in ground and surface water can occur as either the parent compound or as degradation products [5–7]. In a recent study performed by the U.S. Geological Survey, it was shown that more than 95% of sampled streams and almost 50% of wells had at least one pesticide with low-level mixtures of several pesticides being the most common form of contamination [8]. These mixtures present a real challenge in the analysis of water samples because the substances they contain may have similar properties when examined by traditional analytical techniques.

The most common methods used for environmental testing are gas chromatography (GC), gas chromatography/mass spectrometry (GC/MS) [9–12], high performance liquid chromatography (HPLC) [13–17], and enzyme-linked immunosorbent assays (ELISAs) [18–21]. However, over the last decade, there has also been growing interest in the use of IAC for such work [22–28]. This has paralleled the increased use of ELISAs in environmental testing [29–31]. The use of IAC in this manner is often referred to as a chromatographic immunoassay or flow injection immunoanalysis (FIIA). Several authors have reviewed the use of such assays for environmental samples [32–34]. This chapter will discuss the various formats where antibodies have been employed for chromatographic immunoassays as well as the use of IAC in combination with other analytical methods.

The basic operation of IAC is relatively simple (see Figure 14.1). First, a column is prepared that contains antibodies or related ligands that are immobilized or adsorbed onto a solid support. Next, a sample containing solutes that can bind to these ligands is applied to the column under

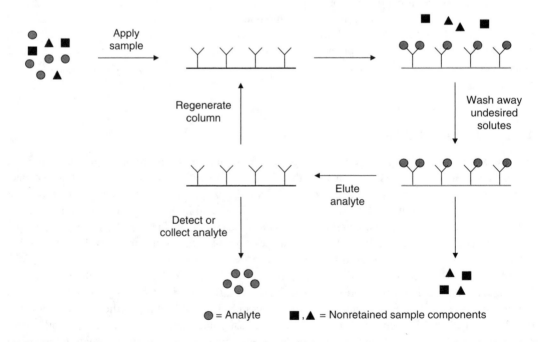

● = Analyte   ■,▲ = Nonretained sample components

FIGURE 14.1 Basic operation of IAC. A sample is applied to an antibody column under physiological conditions and compounds with an affinity for the immobilized antibodies bind. Elution is performed by disruption of the noncovalent forces that hold the analytes on the antibody column. After elution, the column is regenerated by reapplying the application buffer to renature the antibodies. If necessary, an additional step to separate the analytes may be performed after their elution from the antibody column.

conditions where strong association is obtained. This is typically done by using an application buffer that is at or near a physiological pH (i.e., pH 7–7.4). As a sample is applied to the column, the desired solutes are bound while other chemicals wash through non-retained. Later, a second buffer is applied to the column that causes the retained solutes to elute. These compounds are then collected for further analysis or monitored directly by an on-line detector. If desired, the IAC column can be placed back into the initial application buffer, allowing the antibody-based ligands to regenerate. Another sample can then be applied and the process repeated. In the following sections, the various components of this scheme will be examined in more detail (e.g., the antibodies, supports, and solvent conditions). Following this, the various formats where IAC has been used for environmental testing will be considered.

14.1.1 Antibody Structure and Production

The item that gives IAC its selectivity is the antibody or related ligand that is used as the stationary phase. An antibody, also known as an immunoglobulin, is simply a type of glycoprotein that is produced by the body's immune system in response to a foreign agent, or antigen. It has been estimated that the human body can produce antibodies for a million to a billion different foreign agents. As shown in Figure 14.2, the basic structure of a typical antibody (i.e., immunoglobulin G) consists of four polypeptides linked by disulfide bonds to form a Y- or T-shaped structure. The amino acids in the lower stem region (or F_c region) generally have the same sequence, but they show high variability at the two identical binding sites located at the upper ends of the antibody (the F_{ab} regions). In fact, it is this difference in amino acid composition at or near the binding sites that allows the production of antibodies with a variety of specificities and affinities to different chemical or biological substances that might enter the body.

One way to produce antibodies for a given agent is to inject a solution of the substance (or a conjugate of this agent and a large carrier) into an animal such as a rabbit or mouse. Samples of the animal's blood are then collected at specified intervals (typically a few weeks or months after injection) to obtain antibodies that have been produced against the foreign agent. This approach results in a heterogeneous mixture of antibodies that bind with a range of strengths and to various sites on the original injected agent or conjugate. These are known as polyclonal antibodies because they are produced by different cell lines within the body. Another approach for making antibodies is to isolate single antibody-producing cells and combine these with carcinoma cells to produce new

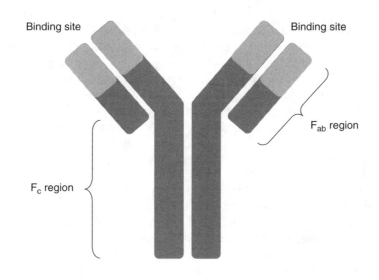

FIGURE 14.2 General structure of an antibody.

hybrid cells that can be grown in culture. These new cells are referred to as hybridomas, and their product is a single type of well-defined antibody known as a monoclonal antibody.

14.1.2 IAC Supports and Solvents

It can be seen from Figure 14.1 that several components are needed along with antibodies for IAC to work. These other items include a support to hold antibodies within the column, a means for attaching antibodies to this support, and solvents to apply and elute analytes from the column. Either low or high performance supports can be used in IAC methods. Examples of low performance materials used for this purpose include carbohydrate-related matrices and synthetic organic supports such as agarose, cellulose, acrylamide polymers, and polymethacrylate derivatives. High performance supports used in IAC include diol-bonded silica or glass beads, azalactone supports, and glycol-coated perfusion media.

In practice, both low and high performance methods have found use in environmental testing. Antibodies immobilized to low performance supports are mainly used for analyte concentration or extraction from a sample. This is due to the low cost and relative ease with which a low performance IAC column can be created and operated. The use of high performance IAC materials, giving rise to a method known as high performance immunoaffinity chromatography (HPIAC), is more common when producing automated systems for analyte detection. This is due to the increased stability of these latter supports as well as the greater precision inherent in HPLC methods.

There are numerous techniques for immobilizing antibodies and related ligands to low or high performance supports. For instance, antibodies can be directly coupled to many supports by reacting their free amine groups with materials activated by agents such as N,N'-carbonyl diimidazole, cyanogen bromide, and N-hydroxysuccinimide. A similar result is accomplished by using materials that have been treated to produce reactive epoxide or aldehyde groups on their surfaces. Antibodies and antibody fragments can also be immobilized through more site-selective methods. For example, the free sulfhydryl groups generated during the production of antibody F_{ab} fragments can be used to couple these fragments to surfaces activated by the divinylsulfone, epoxy, iodoacetyl/bromoacetyl, maleimide, TNB-thiol, or tresyl chloride/tosyl chloride methods. Intact antibodies can undergo site-selective immobilization by oxidizing the carbohydrate residues in their stem region with periodate to produce aldehyde groups that can then react with a hydrazide or amine-containing support [35].

Antibodies can also be placed onto IAC supports through non-covalent adsorption. One example is the conjugation of antibodies with biotin that can be used to bind these to a support containing immobilized streptavidin. Another approach for indirect immobilization involves adsorbing the antibody to a secondary ligand such as protein A or protein G. This makes use of the ability of both protein A and G to bind the stem region of many antibodies at a neutral pH, along with the ability of these ligands to later release the adsorbed antibodies when there is a decrease in pH. This method is attractive when high antibody activity is needed and/or when it is desirable to have frequent replacement of the antibodies on an IAC column. This allows good long-term reproducibility for the column binding capacity, but it does require the use of much larger amounts of antibody than direct immobilization methods.

Although the selection of an application buffer for an IAC column is usually straightforward (i.e., often being a neutral pH buffer), the choice of an appropriate elution solvent is not as simple. It is possible to use isocratic elution for some weak affinity antibodies [36], but this does not work for the high or moderate affinity antibodies used in the majority of IAC columns. Instead, the retained solutes must be eluted by changing the column conditions to lower the effective binding constant of the antibodies for the retained compound. This is often accomplished by using an acidic buffer (i.e., pH 1–3) in a step elution scheme. Another approach is to perform a gradient elution by gradually increasing the amount of a chaotropic agent, organic modifier, or denaturing agent in the mobile phase [37]. The proper choice of elution conditions is critical in analytical applications of IAC to provide rapid release of analyte from the column without causing permanent harm to the immobilized antibodies or support. This concern must currently be addressed on a case-by-case basis and is particularly important to consider when the same immunoaffinity column is to be employed for a large number of samples.

14.2 DIRECT DETECTION AND IMMUNOEXTRACTION

There are a variety of formats that can be used with IAC to analyze samples. The simplest approach is to use the scheme shown in Figure 14.1 to capture and elute an analyte, followed by either on-line or off-line detection. This is known as immunoextraction or immunoaffinity extraction. Other terms used to describe such a format are *direct detection* or the *on/off mode* of IAC. Immunoextraction is based on the injection of a sample onto an affinity column under conditions that promote specific binding between the analytes and immobilized ligand. As described in the previous section, if the ligand has strong binding (i.e., an association equilibrium constant greater than 10^5–10^6 M^{-1}), a change in column conditions will probably be needed to release analytes for detection. If the ligand has weaker binding, then analyte elution and detection may be possible under isocratic conditions [36]. Because of the high affinity and selectivity of antibody interactions, a high degree of molecular selectivity is obtained with IAC. This means that complex environmental samples can often be cleaned up or analyzed in a single step with antibody columns.

Combining IAC with other analytical techniques typically increases the selectivity of sample preparation while providing a means for discriminating between several chemicals captured by the same antibody column. There are two basic approaches used for this purpose. The first is off-line extraction where IAC is followed by fraction collection and subsequent analysis of the isolated substances by separate methods like GC, HPLC, or ELISA. The second approach is on-line extraction where the analyte is removed from a sample by an IAC column and passed directly onto a second method for measurement or detection. This latter approach generally involves coupling IAC to reversed-phase liquid chromatography (RPLC). However, there have been additional reports using IAC for on-line extraction with GC [38,39].

The choice of off-line or on-line extraction will depend on the overall goal of the analysis and the compatibility of this extraction with the second analysis method. For example, when IAC is combined with GC, an off-line mode is often used to go from the aqueous buffers used on the IAC column to a more volatile matrix that is compatible with GC. In this sense, the IAC column acts as a solid-phase extraction cartridge. However, when IAC extraction is used with RPLC, an on-line method might be preferred because of its higher precision and greater ease of automation. In either situation, the use of immunoextraction gives higher selectivity and less interference than traditional solid-phase extraction [39]. This occurs because most common solid-phase extraction cartridges retain analytes based on their polarity [40], a relatively non-specific property.

In addition to its simplicity, there are several other advantages to immunoextraction. For instance, the use of a high specificity antibody can allow direct detection for a particular analyte, whereas a more general group-specific antibody could be employed for the analysis of a broader class of compounds. Furthermore, when this approach is used as part of an HPLC system, measurements with precisions in the range of 1%–5% can generally be produced in as little as a few minutes. Good limits of detection can also be obtained through this technique, but will vary with the analyte and type of detection being used [41].

Immunoextraction is popular for sample pretreatment because it eliminates many of the extraction and derivatization steps required by traditional methods of sample preparation [5,42]. For instance, Figure 14.3 shows chromatograms obtained for groundwater samples with and without

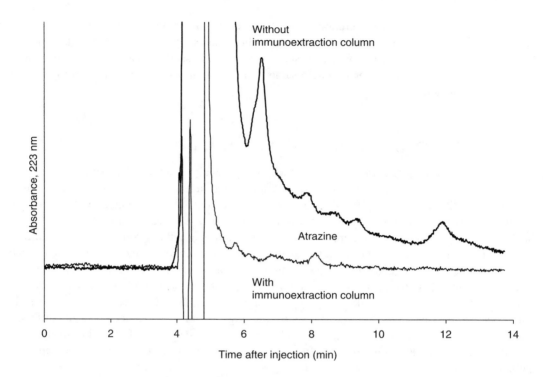

FIGURE 14.3 The analysis of atrazine in a groundwater sample using immunoextraction coupled with on-line RPLC. Without immunoextraction, a high sample background was observed on the RPLC column. However, when immunoextraction is used with RPLC, atrazine and related compounds could easily be detected at levels down to the low parts-per-billion range. (Adapted from Nelson, M. A., Gates, A., Dodlinger, M., and Hage, D. S., *Anal. Chem.*, 76, 805–813, 2004. With permission.)

immunoextraction. Without immunoextraction, the peak resulting from atrazine is masked by other sample components. However, the use of IAC for atrazine gives a well-defined peak [43].

Although the simplicity of immunoextraction makes it attractive for environmental analysis, it does have some limitations. When used for direct detection, one limitation is the need for a reasonably high analyte concentration to allow the measurement of the compound as it exits the IAC column. The main challenge here is to discriminate between the eluting analyte and the change in background signal that often occurs when the elution buffer is applied to the IAC column. For ultraviolet or visible absorbance detection, this requires that the analyte have a high molar absorptivity. Alternatively, fluorescent detection can be used with analytes that contain natural fluorophores or that have been derivatized with fluorescent tags through either pre-column or post-column methods [3]. Another limitation is that some antibodies are slow to release analytes, meaning a reconcentration step may be necessary prior to detection. This can be accomplished by coupling immunoextraction with another method such as RPLC, as discussed later in this chapter.

14.2.1 OFF-LINE IMMUNOEXTRACTION

Off-line immunoextraction is the easiest approach for combining IAC with another method. This typically involves the use of antibodies that are immobilized onto a low performance support and packed into a small disposable syringe or solid-phase extraction cartridge. After this column has been conditioned with the necessary application buffer or other solvents, the sample is applied, and any undesired sample components are washed away. An elution buffer is then passed through the support, and analyte fractions are collected. In some cases, these eluted fractions are analyzed directly, but they may also be dried down and reconstituted in a solvent that is more compatible with the method to be used for analyte quantitation. If needed, the collected solute fraction may be derivatized before it is examined by another technique. For instance, this last step can be used to improve a compound's detectability and/or volatility prior to separation and analysis by HPLC or GC [3].

As with any IAC method, off-line immunoextraction requires the availability of an antibody preparation that is selective for the desired analyte or group of analytes. If such antibodies are available, then immunoextraction offers the potential of a much greater specificity than traditional liquid–liquid or solid-phase extraction methods. However, it should be kept in mind that most antibodies will show some binding with compounds close to the desired analyte in structure. Ideally, this cross-reactivity should be evaluated for each immunoextraction support by performing binding and interference studies with any solutes or metabolites that are related to the analyte and that may be present in the samples of interest. However, even if several solutes do bind to the same IAC column, this will not present a problem as long as the analyte can be resolved or discriminated from these other compounds by the method used for quantitation. In fact, this cross-reactivity can be used as an advantage in that it may allow several structurally-related compounds to be examined simultaneously by a single method.

An important limitation of off-line immunoextraction is the need to transfer the extract to a second method. Because manual steps are generally used for both this transfer and the off-line extraction, the precision and recovery of the measurement may suffer. In addition, other steps (e.g., extraction into an organic solvent and evaporation of this solvent) may be needed before the IAC extract can be analyzed by HPLC, GC, or ELISA. This not only increases the analysis time, but it increases the cost of the assay and amount of reagents needed for sample pretreatment.

Off-line immunoextraction has been used in many environmental studies, including work with ochratoxins in baby food [44], deoxynivalenol in wheat [45], triazine biocides in seawater [46], and fumonisins in corn [47]. Typical limits of detection for these methods are in the ng/L range. One example of off-line immunoextraction has been its use in determining estrogens in wastewater [48]. In this method, wastewater was first treated by filtration and traditional solid-phase extraction. Next, the collected extracts were evaporated to dryness under nitrogen and reconstituted in a 5% solution of methanol in water. The reconstituted samples was then applied to an immobilized antibody column. Estrogens from the sample were bound to this column and later eluted with a 70% solution of methanol in water. This eluent was collected and evaporated to dryness, followed by reconstitution in 25% acetonitrile in water. The reconstituted analytes were analyzed by HPLC and detected by electrospray ionization mass spectrometry (ESI–MS) [48].

Off-line immunoextraction can also be used for solid samples. For example, polycyclic aromatic hydrocarbons in coral were extracted using supercritical fluid extraction followed by immunoaffinity column cleanup. The collected analytes were then separated and identified using GC/MS [49]. Using this method, a detection limit of 25 ng/g coral was obtained.

14.2.2 On-Line Immunoextraction

If immunoextraction is coupled directly with another method, this gives rise to a hyphenated technique such as immunoaffinity/reversed-phase liquid chromatography (IA–RPLC) or immunoaffinity/gas chromatography (IA–GC) [3]. This section will consider both these techniques.

14.2.2.1 Immunoaffinity Extraction Coupled with RPLC

The direct coupling of IAC columns to other analytical systems is an area that has seen rapid growth during the past decade. The most popular coupling scheme incorporating immunoextraction is IA–RPLC [3,50]. There are several reasons for this. One is that the elution buffer for an immunoaffinity column is a solvent which contains little or no organic modifier acts as a weak mobile phase for RPLC. This means that as a solute elutes from an IAC column to a RPLC column, it will tend to have strong retention on the reversed-phase support, leading to analyte reconcentration. This is valuable in dealing with analytes that have slow desorption from immunoaffinity columns that might make them too dilute to directly analyze.

The ease of incorporating immunoaffinity columns into an HPLC system makes this appealing as a means for automating immunoextraction methods and for reducing the amount of time required for sample pretreatment. In addition, the increased precision for on-line immunoextraction that comes with using high precision HPLC pumps and injection systems (giving more tightly controlled sample application and elution conditions) makes this combination an attractive alternative to off-line immunoextraction. In HPLC-based methods, the easiest approach for detecting analytes as they elute from immunoaffinity columns is to use on-line absorbance measurements. This works well for compounds present at moderate concentrations with relatively good chromophores in their structures.

In environmental testing, the on-line coupling of immunoextraction with RPLC has been used to create several methods with excellent reproducibility and fast analysis times. For instance, this approach has been used for the measurement of triazine herbicides and their degradation products at the parts-per-billion level in 6–12 min [42]. This same approach has been used with carbofuran [51] and carbendazim [52]. Low limits of detection are possible with this method because the IAC column is mass sensitive and responds to the moles rather than the concentration of a substance applied to the column. Thus, these columns can be optimized for work even at low parts-per-billion or parts-per-trillion levels [5].

A typical system used to perform on-line immunoextraction with RPLC is shown in Figure 14.4. Three solvents are often used in such a system: (1) an application buffer for the IAC column, (2) an elution buffer for this column, and (3) a mobile phase for the reversed-phase column. In this approach, the sample is first applied to the IAC column in the application buffer. Once binding of the analyte has occurred and non-retained components have been washed away, the IAC column is switched on-line with a reversed-phase precolumn, and the elution buffer is applied. This causes the analytes to dissociate from the IAC column; however, because this buffer also acts as a weak mobile phase for the reversed-phase precolumn, the analytes are trapped in this precolumn. This precolumn is later switched on-line with a second larger RPLC column and a mobile phase containing some organic modifier is applied, causing the elution and separation of the analytes based on their polarities.

A modified version of this system has been used for determining carbendazim in soil and lake water [52]. This was accomplished by using a high performance protein G column coupled to a reversed-phase analytical column through the use of a restricted access media trapping column. Prior to analysis, 20 μg of antibodies were loaded onto the protein G column to form an immunoaffinity support. A sample was then applied to this column, and the carbendazim was extracted. A stripping buffer containing 2% acetic acid was next pumped through the affinity column, causing elution of the adsorbed antibodies and carbendazim. The trapping column retained the carbendazim while the antibodies were washed through the system. The trapping column was then switched on-line with an analytical column and mass spectrometer. The limit of detection was 25 ng/L for carbendazim and the throughput was three samples per hour [52]. Other reports of on-line affinity columns in HPLC have been described for estrogens [53]; polycyclic aromatic hydrocarbons [54,55]; isoproturon [56]; phenylurea pesticides [56–59]; triazines [59]; aflatoxins [60]; *E. coli* [61]; and benzidene, dichlorobenzidene, aminobenzene, and azo dyes [62].

A unique application of immunoextraction has been its use to examine both a parent compound and its degradation products in environmental samples. Pathways for the degradation of herbicides can be of a chemical nature (e.g., hydrolysis or photodegradation) [6,7,63,64] or of biological origin (e.g., because of microbial action) [65–67]. The analysis of degradation products for environmental contaminants is complicated by several factors. First, these products generally occur at lower concentrations than the parent compound. Also, degradation of the parent compound often leads

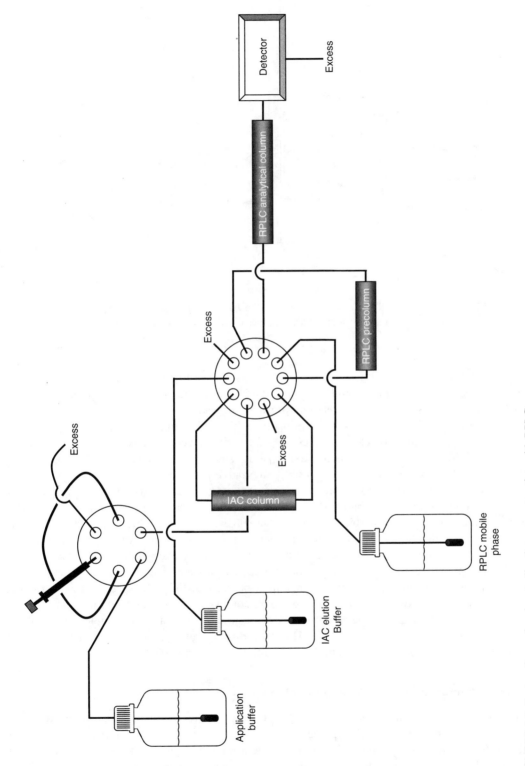

FIGURE 14.4 System for performing on-line immunoextraction with RPLC.

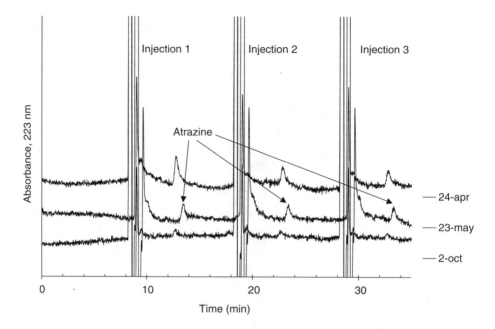

FIGURE 14.5 Results obtained by a field-portable device based on immunoextraction and RPLC for the triplicate analysis of triazines in river water at various times of the year. (Adapted from Nelson, M. A., Gates, A., Dodlinger, M., and Hage, D. S., *Anal. Chem.*, 76, 805–813, 2004. With permission.)

to a product that has a different polarity, activity, and ecological lifetime. Degradation analysis has been performed with on-line immunoextraction and RPLC in studies of atrazine in the environment. This has been demonstrated in the detection of primary degradation products for atrazine at levels extending down to the low parts-per-trillion range [5]. This same method has been used to determine the adsorption isotherms for these degradation products on adsorbents used for water treatment [68,69] and to examine the breakdown of atrazine by zero-valent iron (Fe^0) [70]. In this latter study, IA–RPLC was found to be a useful alternative to radiotracer studies.

Another application reported for on-line immunoextraction has been its use in the development of field-portable systems for herbicide detection. This approach has been used in the creation of a system for measuring triazine herbicides in water (see Figure 14.5) [43]. This same system has been adapted for use with other classes of herbicides such as 2,4-dichlorophenoxyacetic acid and related compounds. One advantage of this device is its ability to analyze a small amount of sample with only limited pretreatment. For instance, ground and surface water samples can be examined by simply filtering them through a 0.2 μm syringe filter before injection. The speed of this method is also quite good with an overall analysis time of 10 min or less per sample and a throughput of 5 min per sample. The lower limit of detection reported for atrazine with a 2 mL sample was 0.3 μg/L with a linear range extending up to 25 μg/L. However, this limit of detection and linear range can be adjusted by altering the amount of sample applied to the antibody column [43].

14.2.2.2 Immunoaffinity Extraction Coupled with GC

As noted earlier for immunoextraction and RPLC, the use of only traditional solid-phase extraction allows many sample components to proceed to a GC column where they can mask the peaks of interest. But immunoextraction provides more efficient sample cleanup and allows the analytes of interest to be more easily detected. Though it is not as common as on-line extraction with RPLC, there have been some studies investigating the use of on-line immunoextraction with GC.

For example, immunoextraction was coupled with GC for the measurement of atrazine in river water, wastewater, and orange juice [39]. Samples were first injected onto an immunoaffinity column and the retained solutes desorbed and collected by an on-line cartridge containing a styrene-divinyl benzene copolymer (PLRP-S). This cartridge helped to reconcentrate the analytes into a narrow band. Next, these compounds were desorbed from the PLRP-S cartridge with ethyl acetate and fed directly into a GC system where they were monitored by flame ionization or nitrogen phosphorus detectors. Prior to use on this system, all solvents were treated with a PLRP-S column to remove trace contaminants.

The recovery of triazines by this method was compared to that obtained using traditional solid-phase extraction. A spiked water sample containing 1 μg/L atrazine gave a recovery of 68% with traditional solid-phase extraction and 88% when using immunoextraction. At 100 ng/L atrazine, immunoextraction gave 96% recovery. The lower limit of detection for atrazine in this system was 170 pg when using a flame ionization detector and 15 pg when using a nitrogen phosphorous detector [39].

14.3 INDIRECT DETECTION METHODS

Chromatographic immunoassays can be created for the indirect detection of an analyte by observing how it reacts with a labeled binding agent or prevents a labeled analog of the analyte (sometimes referred to as the label) from interacting with an antibody [35]. In environmental testing, detection for these assays is usually accomplished by using an analyte analog that is labeled with a fluorescent group or enzymatic tag. This label is then monitored as the analog and analyte compete for antibody binding sites within a column or flow chamber. This approach is particularly useful for trace analytes that do not themselves produce a signal that can be directly detected.

One IAC method that uses indirect detection is a competitive binding immunoassay. This is the most common type of chromatographic immunoassay. It is based on the competition between the analyte in the sample and a fixed amount of a labeled analog of the analyte for binding sites on the immobilized antibodies. Other IAC formats using indirect detection are noncompetitive binding immunoassays. These include the homogeneous immunoassay, the one-site immunometric assay, and sandwich immunoassays. Although the sandwich immunoassay has not yet been used in environmental testing, it could potentially be used for looking at protein- and peptide-based herbicides.

14.3.1 COMPETITIVE BINDING IMMUNOASSAYS

A competitive binding immunoassay can be conducted using either an immobilized analog of the analyte or immobilized antibodies [32]. The first of these two approaches makes use of a small amount of labeled antibodies in solution for which the analyte and immobilized analog compete. The second method makes use of a labeled analog in solution that competes with the analyte for a limited number of immobilized antibodies on a support. This latter format is more cost-effective with regards to the antibodies (generally, the most expensive component of the assay) because these can often be reused for multiple assays. This section will consider three possible types of competitive binding immunoassays in flow-based systems: the simultaneous injection, sequential injection, and displacement methods.

14.3.1.1 Simultaneous Injection Assays

When using an immobilized antibody, there are two ways in which the sample and labeled analog can be applied to the column. The first of these formats is the simultaneous injection assay. In this approach, the analyte and labeled analog are mixed off-line and injected together onto the column where they compete for a small number of antibody binding sites. The amount of analyte in the sample is then determined by using a calibration curve that plots the amount of label that binds to or passes through the column versus the known concentration of analyte in injected standards.

The simultaneous injection assay has been used in several environmental studies, including a method developed for detecting the herbicide isoproturon. This made use of a column that contained immobilized antibodies able to bind this herbicide and an analog of isoproturon conjugated with the enzyme horseradish peroxidase. In this assay, samples were combined off-line with a fixed concentration of labeled isoproturon, and this mixture was injected onto the antibody column. After non-retained agents had been washed from the system, a substrate for the enzyme label was applied for detection. The limit of detection reported for isoproturon was 0.12 g/L, and the total run time was approximately 25 min [71].

A similar assay has been described for atrazine. In this method, the sample or standard was combined with an atrazine analog labeled with horseradish peroxidase. After injecting this mixture into a flow cell that contained immobilized antibodies for atrazine, the non-retained species were washed from the column and a substrate for the enzyme label was applied. The resulting fluorescent product was excited at 320 nm and detected at 405 nm, giving a measure of the labeled analog that was bound to the column. The run time was 20 min per sample, and the detection limit was 75 ng/L (see Figure 14.6) [72]. Other simultaneous injection chromatographic immunoassays have also been developed for atrazine [73,74].

An interesting variation of the flow-based competitive binding immunoassay involves the use of liposomes for detection [75–77]. This has given rise to a method known as flow-injection liposome immunoanalysis (FILIA). In this technique, a liposome containing a large amount of a fluorescent dye or other detectable marker is used to tag an analyte analog. This labeled analog is then allowed to compete with the analyte for immobilized antibodies in a flow cell. A detergent is later passed through the flow cell to lyse the retained liposomes and release their internal dye molecules or markers. These markers are measured downstream by an on-line detector and provide a signal inversely related to the amount of analyte in the original sample. When used to determine the concentration of alachlor, this technique had an analysis time of 6 min when operated at a flow rate of 450 μL/min and gave a limit of detection for alachlor of 5 μg/L when 25 μg of antibody were used to prepare the immunoaffinity column [75].

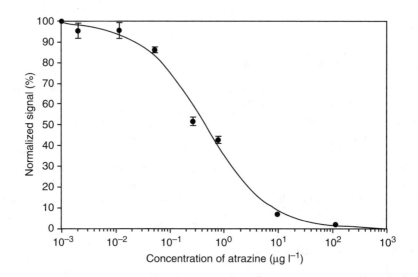

FIGURE 14.6 Simultaneous injection assay for the determination of atrazine in water. (From Gascon, J., Oubina, A., Ballesteros, B., Barcelo, D., Camps, F., Marco, M-P., Angel Gonzalez-Martinez, M., Morais, S., Puchades, R., and Maquieira, A., *Anal. Chim. Acta.*, 347, 149–162, 1997. With permission.)

A simultaneous injection assay can also be used in capillary electrophoresis or capillary electrochromatography [78]. For instance, anti-atrazine antibodies have been adsorbed within a C_8-modified capillary followed by injection of a mixture of fluorescent labeled atrazine and sample. A dissociation buffer was used to disrupt binding between the analyte and antibodies with the fluorescent analog then being detected.

14.3.1.2 Sequential Injection Assays

The second possible format for a competitive binding immunoassay in a chromatographic system is the sequential injection method. This technique differs from the simultaneous injection method in that the sample and label are now applied separately to the IAC column. The sample is injected first, followed by the label that now must bind to any sites that remain on the column. The bound analyte and label are then eluted together from the column, and the system is regenerated prior to injection of the next sample. One advantage of this format over the simultaneous injection method is that the label is not subject to matrix interferences because it is never in contact with the actual sample [79]. This is particularly important if a fluorescent tag is used for the label because it eliminates any quenching by sample components.

Sequential injection has been used to develop an automated immunoassay for atrazine in water and soil samples. In this approach, a flow cell containing immobilized antibodies was first rinsed with an equilibration buffer. A standard or sample was next pumped into this flow cell, followed by injection of a fixed amount of horseradish peroxidase-labeled atrazine. After a short rinsing step, a substrate for the enzyme label was added to the flow cell and allowed to react. The fluorescence of the generated product was then detected downstream using an excitation wavelength of 320 nm and an emission wavelength of 404 nm [80]. Other sequential injection assays for atrazine have also been published [81,82].

Another example of this approach has been reported for imazethapyr [77]. This assay was performed using a column that contained immobilized anti-imazethapyr antibodies. The sample was injected first onto the column, followed by injection of either a liposome conjugate of imazethapyr or imazethapyr containing a single fluorescein tag. In both cases, the amount of label bound to the column was determined with a fluorescence detector. When using the liposome conjugate as a label, a limit of detection of 0.5 ppb was achieved. This was approximately 1000-fold better than the results obtained when using the label that contained a single fluorescein tag. Other examples of the sequential injection assay for imazethapyr have also been reported [83,84].

Both the simultaneous and sequential injection formats can be used for either large or small analytes [80,85]. However, they differ in their analytical characteristics. For instance, the sequential injection assay provides a lower limit of detection and larger change in response than the simultaneous injection assay at low analyte concentrations. However, the simultaneous injection format is slightly easier to perform and has a calibration curve that covers a broader dynamic range.

14.3.1.3 Displacement Assays

The third possible format for a competitive binding immunoassay in chromatography is the displacement method [86]. This employs an immobilized antibody column that is first saturated with a labeled analog of the analyte. As a sample is applied to this column, any of the labeled analog that is momentarily free in solution will be displaced and eluted from the column. The amount of this displaced label is then detected, giving a response proportional to the amount of analyte in the sample. In many cases, several samples can be injected onto the column before regeneration provided that enough label remains for the production of a consistent, measurable signal.

The ability to use a single application of label for the injection of multiple samples is one advantage of this approach. Another is its speed because the displacement peak appears at or near the non-retained peak for the column. The fact that the signal increases with analyte concentration is also an advantage when compared to other competitive binding formats. However, care must be taken in selecting the optimal conditions for this approach to work properly. This includes the choice of an appropriate labeled analog, column size, and flow rate for this method.

One example of this format is an assay that was developed for detection of polychlorinated biphenyls (PCBs). In this study, an anti-PCB antibody was immobilized onto Emphaze beads and placed into a column. This column was then loaded with a fluorescent derivative of a PCB ((i.e., 2,3,5-trichlorophenoxy)propyl-Cy5). Any weakly bound label was then washed from the column. Samples were next injected onto the column, resulting in displacement of some of the remaining label. This label was then measured with a fluorescence detector using an excitation wavelength of 635 nm and an emission wavelength of 661 nm. The resulting calibration curve showed an increase in signal as the amount of PCBs in the sample was increased. The limit of detection was 4 ppb, and the linear range extended up to 20 µg/mL [87].

14.3.2 Noncompetitive Chromatographic Immunoassays

There are two types of noncompetitive binding immunoassays for environmental agents that have been performed in flow-based systems: the homogenous immunoassay and the one-site immunometric immunoassay.

14.3.2.1 Homogeneous Immunoassays

A *homog*eneous immunoassay is a method where the reaction between the antibody, analyte, and label occurs in solution. One way this can be performed by chromatography is to use a restricted access assay. This employs a random access media column that can separate a small, labeled analog from larger antibody-analog complexes based on their differences in size [88]. Such an assay has been used to screen plasma and water samples for atrazine, related *s*-triazines, and atrazine degradation products. This method used a fluorescein-labeled analog of atrazine and off-line incubation of the sample with the labeled analog and a small amount of anti-atrazine antibodies. This mixture was then introduced onto a restricted access media column. This column contained a support with a reversed-phase stationary phase within its pores that retained the labeled analog but not the excluded antibody-analog complexes. The amount of antibody-bound analog in the non-retained peak was then used as an indirect measure of analyte in the sample. This gave a detection limit of 20 pg/mL for atrazine and a throughput of 80 samples per hour [89,90].

A similar technique called immuno-supported liquid membrane extraction (immuno-SLM) has been reported [91]. In this technique, sample matrix effects are minimized by allowing the analyte to diffuse across a membrane to form antibody-analyte complexes (see Figure 14.7). Because the antibody is present in excess, both the free analyte and antibody-analyte complexes are present. Excess labeled analyte is then added to the mixture and complexes with the remaining antibodies are allowed to form. Free analyte (both labeled and unlabeled) is then separated from the complexes using a restricted access column. The labeled antibody-analyte complexes can then be detected and the amount of analyte in the sample measured. Using this method, atrazine in tap water, river water, and orange juice was determined in the range of 5–100 µg/L.

14.3.2.2 One-Site Immunometric Assays

A second type of non-competitive immunoassay is a one-site immunometric assay. Here, the sample is incubated with a known excess of labeled antibodies or F_{ab} fragments that can bind to the analyte. This mixture, now containing analytes bound to the labeled F_{ab} fragments or antibodies as well as excess labeled fragments and antibodies, is injected onto a column where an analog of the analyte has been immobilized. This column is used to extract the excess antibodies or F_{ab} fragments, allowing those that are bound to analyte to pass through non-retained. The labeled antibodies or F_{ab} fragments in this non-retained peak give a signal that is proportional to the

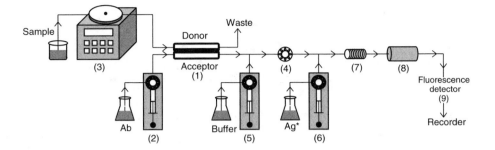

FIGURE 14.7 An immuno-SLM–FFIA system for atrazine. System components: (1) SLM unit, (2, 5, 6) automatic syringe pumps, (3) peristaltic pump, (4) manual injection valve, (7) mixing coil, (8) restricted access column, and (9) fluorescence detector. Solutions of anti-atrazine antibody (Ab) and labeled analog (Ag*) are injected using syringe pumps. The restricted access column allows the antibody-bound fractions of the analyte and labeled analog to pass through non-retained while the column binds to the free analyte or analog that remains in solution. (From Tudorache, M., Rak, M., Wieczorek, P. P., Jonsson, J. A., and Emneus, J., *J. Immunol. Meth.*, 284, 2004. With permission.)

amount of analyte in the original sample. Alternatively, the remaining amount of excess antibodies or F_{ab} fragments can be determined as they elute from the column.

Although the one-site immunometric assay has been used in only a few studies, it has several potential advantages over other chromatographic immunoassays. For instance, like a competitive binding immunoassay, this method is able to detect both small and large solutes. However, it gives a signal directly proportional to the amount of analyte in a sample. In addition, the fact that an immobilized analyte analog is used rather than an antibody in the column creates the possibility of employing a fairly wide range of elution conditions for column regeneration.

One disadvantage of this approach is that a different immobilized analog column must be used for each analyte of interest. In addition, highly active and pure labeled antibodies or F_{ab} fragments are needed when monitoring the non-retained fraction to provide a low background signal for label detection. This means that special precautions need to be taken when purifying these labeled agents to ensure that they do not lose a significant amount of their activity. It is also necessary to carefully monitor the stability of these binding agents during storage or use.

A one-site immunometric assay was developed to determine the concentration of terbutryn in drinking water [92]. In this method, unlabeled antibodies were incubated with the sample, and this solution applied to a surface containing immobilized terbutryn. The antibody complexes were detected by monitoring the change in refractive index on this surface. The surface was later regenerated by washing it with a 10 µg/L proteinase solution. The detection range for this assay was 15–200 µg/L [92].

One-site immunometric assays can be used to monitor the elution of specific analytes from other chromatographic columns. Using antibodies for this purpose is referred to as postcolumn immunodetection [3,93]. This technique involves taking the eluent from an HPLC column and combining it with a solution of labeled antibodies or F_{ab} fragments specific for the analyte of interest. This mixture is then allowed to react in a mixing coil and passed through a column that contains an immobilized analog of the analyte. The antibodies or F_{ab} fragments that are already complexed with the analyte will pass through this column and into a detector, yielding a signal proportional to the amount of analyte in the sample. The immunodetection column can later be washed with an elution buffer to dissociate the retained antibodies or F_{ab} fragments.

This type of post-column method was used to detect 2,4-dichlorophenoxyacetic acid (2,4-D) in water (see Figure 14.8) [94]. In this scheme, polyclonal antibodies conjugated with alkaline phosphatase were mixed with the sample and applied to a column containing immobilized 2,4-D.

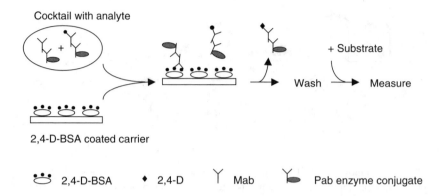

FIGURE 14.8 One-site immunometric assay for 2,4-D. In this system, an alkaline phosphatase label on the antibodies converts the substrate, p-aminophenylphosphate, to p-aminophenol that can be electrochemically detected. Abbreviations: BSA, bovine serum albumin; mab, monoclonal antibody; pab, alkaline phosphatase labeled antibody. (From Wilmer, M., Trau, D., Renneberg, R., and Spener, F., *Anal. Lett.*, 30, 515–525, 1997. With permission.)

Low detection limits were achieved by using p-aminophenylphosphate as a substrate for the enzyme label. The dephosphorylated product, p-aminophenol, was detected at $+350$ mV versus an Ag/AgCl reference electrode. The subsequent detection limit for 2,4-D was 0.1 μg/L.

14.4 FUTURE TRENDS AND DEVELOPMENTS

Although IAC is a relatively new technique in the environmental area, there are many possible advantages of using this method for such work. These advantages include the specificity of IAC and its ability to be directly used with a wide variety of samples. Other potential benefits include the simplicity of this approach for sample pretreatment and its ability to be used for either selective measurements or the study of a group of related compounds.

IAC in its present form clearly has the potential for widespread use in environmental testing. However, there is also a need for the improvement of IAC supports and development of new applications for these supports. For example, work is now being performed in the creation of improved supports for IAC such as porous glass beads or monolithic supports, elements that might increase sample throughput [95]. In addition, efforts continue in the coupling of immunoextraction with methods such as CE, mass spectrometry, and biosensors [96]. Another development expected to impact the use of IAC is the creation of micromachined analytical systems. This later area may open the possibility of using arrays of antibodies for the determination of a large panel of analytes. As these and other applications continue to be developed, IAC and related techniques should find even greater use in the analysis of environmental compounds.

REFERENCES

1. U.S. Environmental Protection Agency, *National Primary Drinking Water Standards*, 816-F-03-016, Washington, D.C, 2003.
2. Pesticides in ground and drinking water, *Off. J. Council Eur.Union*, 330, 32–54, 1998.
3. Hage, D. S., Survey of recent advances in analytical applications of immunoaffinity chromatography, *J. Chromatogr. B.*, 715, 3–28, 1998.
4. Stevenson, D., Immunoaffinity solid-phase extraction, *J. Chromatogr. B.*, 745, 39–48, 2000.
5. Rollag, J. G., Beck-Westermeyer, M. S., and Hage, D. S., Analysis of pesticide degradation products by tandem high-performance liquid chromatography, *Anal. Chem.*, 68, 3631–3637, 1996.

6. Korte, F., Konstantinova, T., Mansour, M., Ilieva, P., and Bogdanova, A., On the photodegradation of some unsaturated triazine derivatives with herbicide and bactericide activity, *Chemosphere*, 35, 51–54, 1997.

7. Mansour, M., Freicht, E. A., Behechti, A., and Scheunert, I., Experimental approaches to studying the photostability of selected pesticides in water and soil, *Chemosphere*, 35, 39–50, 1997.

8. Gilliom, R., Barbash, J., Kolpin, D., and Larson, S., Testing water quality for pesticide pollution, *Env. Sci. Tech.*, 33, 164A–169A, 1999.

9. Namiesnik, J. and Zygmunt, B., Selected concentration techniques for gas chromatographic analysis of environmental samples, *Chromatographia*, 56, S9–S18, 2002.

10. Zygmunt, B., Gas chromatographic determination of volatile environmental organic pollutants based on solvent-free extraction, *Chemia Inzyneria Ekologiczna*, 8, 973–980, 2001.

11. Eckenrode, B. A., Environmental and forensic applications of field-portable GC–MS: An overview, *J. Am. Soc. Mass. Spectrom.*, 12, 683–693, 2001.

12. Clement, R. E., Yang, P. W., and Koester, C. J., Enviromental analysis, *Anal. Chem.*, 73, 2761–2790, 2001.

13. Jedrzejczuk, A., Goralczyk, K., Czaja, K., Strucinski, P., and Ludwicki, J. K., High performance liquid chromatography: Application in pesticide residue analysis, *Roczniki Panstwowego Zakladu Higieny*, 52, 127–138, 2001.

14. Obrist, H., On-line solid phase extraction for HPLC analysis, *Chimia*, 55, 46–47, 2001.

15. Tribaldo, E. B., Residue analysis of carbamate pesticides in water, *Food Sci. Tech.*, 102, 537–570, 2000.

16. Boyd-Boland, A. A., SPME–HPLC of environmental pollutants, In *Applications of Solid Phase Microextraction*, Pawliszyn, J., Ed., Royal Society of Chemistry, Cambridge, pp. 327–332, 1999.

17. Pobozy, E., On-line concentration of trace elements for HPLC determination, *Chemia Analityczna*, 44, 119–135, 1999.

18. Linde, C. D. and Goh, K. S., Immunoassays (ELISAs) for pesticide residues in environmental samples, *Pest Outlook*, 6, 18–23, 1995.

19. Niessner, R., Immunoassays in environmental analytical chemistry: Some thoughts on trends and status, *Anal. Meth. Instr.*, 1, 134–144, 1993.

20. Jeannot, R., Trends in analytical methods for determination of organic compounds in the environment. Application on waters and soils matrixes, *Spectra Analyse*, 26, 17–24, 1997.

21. Nunes, G. S., Toscano, I. A., and Barcelo, D., Analysis of pesticides in food and environmental samples by enzyme-linked immunosorbent assays, *Trends Anal. Chem.*, 17, 79–87, 1998.

22. Aga, D. S. and Thurman, E. M., Environmental immunoassays: Alternative techniques for soil and water analysis, In *Immunochemical Technology for Environmental Applications*, Aga, D. S. and Thurman, E. M., Eds., American Chemical Society, Washington, pp. 1–20, 1997.

23. Bouzige, M. and Pichon, V., Immunoextraction of pesticides at the trace level in environmental matrixes, *Analusis*, 26, M112–M117, 1998.

24. Dankwardt, A. and Hock, B., Enzyme immunoassays for the analysis of pesticides in water and food, *Food Tech. Biotech.*, 35, 165–174, 1997.

25. Fitzpatrick, J., Fanning, L., Hearty, S., Leonard, P., Manning, B. M., Quinn, J. G., and O'Kennedy, R., Applications and recent developments in the use of antibodies for analysis, *Anal. Lett.*, 33, 2563–2609, 2000.

26. Groopman, J. D. and Donahue, K. F., Aflatoxin, a human carcinogen: Determination in foods and biological samples by monoclonal antibody affinity chromatography, *J. AOAC.*, 71, 861–867, 1988.

27. Harris, A. S., Wengatz, I., Wortberg, M., Kreissig, S. B., Gee, S. J., and Hammock, B. D., Development and application of immunoassays for biological and environmental monitoring, In *Multiple Stresses in Ecosystems*, Cech, J. J. Jr., Wilson, B. W., and Crosby, D. G., Eds., Lewis, Boca Raton, FL, pp. 135–153, 1998.

28. Hennion, M. C. and Barcelo, D., Strengths and limitations of immunoassays for effective and efficient use for pesticide analysis in water samples, *Anal. Chim. Acta.*, 362, 3–34, 1998.

29. Van Emon, J. M., Gerlach, C. L., and Bowman, K., Bioseparation and bioanalytical techniques in environmental monitoring, *J. Chromatogr. B.*, 715, 211–228, 1998.

30. Newman, D. J. and Price, C. P., Future developments in immunoassay, In *Principles and Practice of Immunoassay*, Price, C. P. and Newman, D. J., Eds., Macmillan, London, pp. 649–656, 1997.

31. Unger, K. K. and Anspach, B., Trends in stationary phases in high-performance liquid chromatography, *TrAC.*, 6, 121–125, 1987.
32. Zhi, Z-L. , Flow-injection immunoanalysis, a versatile and powerful tool for the automatic determination of environmental pollutants, *Lab. Rob. Automat.*, 11, 83–89, 1999.
33. Kramer, P., Franke, A., and Standfuss-Gabisch, C., Flow injection immunoaffinity analysis (FIIAA)—A screening technology for atrazine and diuron in water samples, *Anal. Chim. Acta.*, 399, 89–97, 1999.
34. Kraemer, P. and Schmid, R., Flow injection immunoanalysis (FIIA)—A new format of immunoassay for the determination of pesticides in water, *GBF Monographs*, 14, 243–246, 1991.
35. Hage, D. S. and Nelson, M. A., Chromatographic immunoassays, *Anal. Chem.*, 73, 198A–205A, 2001.
36. Pignatello, J. J., The measurement and interpretation of sorption and desorption rates for organic compounds in soil media, *Adv. Agron.*, 69, 1–73, 2000.
37. Green, T. M., Charles, P. T., and Anderson, G. P., Detection of 2,4,6-trinitrotoluene in seawater using a reversed-displacement immunosensor, *Anal. Biochem.*, 310, 36–41, 2002.
38. Mackert, G., Reinke, M., Schweer, H., and Seyberth, H. W., Simultaneous determination of the primary prostanoids prostaglandin E2, prostaglandin F2 alpha and 6-oxoprostaglandin F1 alpha by immunoaffinity chromatography in combination with negative ion chemical ionization gas chromatography-tandem mass spectrometry, *J. Chromatogr.*, 494, 13–22, 1989.
39. Dalluge, J., Hankemeier, T., Vreuls, R. J. J., and Brinkman, U. A. T., Online coupling of immunoaffinity-based solid-phase extraction and gas chromatography for the determination of s-traizines in aqueous samples, *J. Chromatogr. A.*, 830, 377–386, 1999.
40. Junk, G. A., Avery, M. A., and Richard, J. J., Interfaces in solid-phase extraction using C-18 bonded porous silica cartridges, *Anal. Chem.*, 60, 1347–1350, 1988.
41. Piehler, J., Brandenburg, A., Brecht, A., Wagner, E., and Gauglitz, G., Characterization of grating couplers for affinity-based pesticide sensing, *Appl. Optics*, 36, 6554–6562, 1997.
42. Thomas, D., Beck-Westermeyer, M., and Hage, D. S., Determination of atrazine in water using tandem high-performance immunoaffinity chromatography and reversed-phase liquid chromatography, *Anal. Chem.*, 66, 3823–3829, 1994.
43. Nelson, M. A., Gates, A., Dodlinger, M., and Hage, D. S., Development of a portable immunoextraction-reversed-phase liquid chromatography system for field studies of herbicide residues, *Anal. Chem.*, 76, 805–813, 2004.
44. Burdaspal, P., Legarda, T. M., Gilbert, J., Anklam, E., Apergi, E., Barreto, M., Brera, C. *et al.*, Determination of ochratoxin A in baby food by immunoaffinity column cleanup with liquid chromatography: Interlaboratory study, *J. AOAC Intl.*, 84, 1445–1452, 2001.
45. Cahill, L. M., Kruger, S. C., McAlice, B. T., Ramsey, C. S., Prioli, R., and Kohn, B., Quantification of deoxynivalenol in wheat using an immunoaffinity column and liquid chromatography, *J. Chromatogr. A.*, 859, 23–28, 1999.
46. Carrasco, P. B., Escola, R., Marco, M. P., and Bayona, J. M., Development and application of immunoaffinity chromatography for the determination of the triazine biocides in seawater, *J. Chromatogr. A.*, 909, 61–72, 2001.
47. de Girolamo, A., Solfrizzo, M., von Holst, C., and Visconti, A., Comparison of different extraction and clean-up procedures for the determination of fumonisins in maize and maize-based food products, *Food Add. Contam.*, 18, 59–67, 2001.
48. Ferguson, P. L., Iden, C. R., McElroy, A. E., and Brownawell, B. J., Determination of steroid estrogens in wastewater by immunoaffinity extraction coupled with HPLC-electrospray-MS, *Anal. Chem.*, 73, 3890–3895, 2001.
49. Thomas, S. D. and Li, Q. X., Immunoaffinity chromatography for analysis of polycyclic aromatic hydrocarbons in corals, *Env. Sci. Tech.*, 34, 2649–2654, 2000.
50. de Frutos, M. and Regnier, F. E., Tandem chromatographic-immunological analyses, *Anal. Chem.*, 65, 17A–20A, 1993. See also pp. 22A–25A3
51. Rule, G. S., Mordehai, A. V., and Henion, J., Determination of carbofuran by online immunoaffinity chromatography with coupled-column liquid chromatography/mass spectrometry, *Anal. Chem.*, 66, 230–235, 1994.
52. Bean, K. A. and Henion, J. D., Determination of carbendazim in soil and lake water by immunoaffinity extraction and coupled-column liquid chromatography-tandem mass spectrometry, *J. Chromatogr. A.*, 791, 119–126, 1997.

53. Farjam, A., Brugman, A. E., Lingeman, H., and Brinkman, U. A. T., On-line immunoaffinity sample pretreatment for column liquid chromatography: Evaluation of desorption techniques and operating conditions using an anti-estrogen immuno-precolumn as a model system, *Analyst*, 116, 8896–8981, 1991.

54. Bouzige, M., Pichon, V., and Hennion, M. C., Online coupling of immunosorbent and liquid chromatographic analysis for the selective extraction and determination of polycyclic hydrocarbons in water samples at the ng l^{-1} level, *J. Chromatogr. A.*, 823, 197–210, 1998.

55. Perez, S., Ferrer, I., Hennion, M. C., and Barcelo, D., Isolation of priority polycyclic aromatic hydrocarbons from natural sediments and sludge reference materials by an anti-fluorene immunosorbent followed by liquid chromatography and diode array detection, *Anal. Chem.*, 70, 4996–5001, 1998.

56. Delaunay-Bertoncini, N., Pichon, V., and Hennion, M. C., Comparison of immunoextraction sorbents prepared from monoclonal and polyclonal anti-isoproturon antibodies and optimization of the appropriate monoclonal antibody-based sorbent for environmental and biological applications, *Chromatographia*, 53, S224–S230, 2001.

57. Pichon, V., Chen, L., and Hennion, M. C., Online preconcentration and liquid chromatographic analysis of phenylurea pesticides in environmental water using a silica-based immunosorbent, *Anal. Chim. Acta.*, 311, 429–436, 1995.

58. Martin-Esteban, A., Fernandez, P., Stevenson, D., and Camara, C., Mixed immunosorbent for selective online trace enrichment and liquid chromatography of phenylurea herbicides in environmental waters, *Analyst*, 122, 1113–1117, 1997.

59. Ferrer, I., Hennion, M-C., and Barcelo, D., Immunosorbents coupled online with liquid chromatography/atmospheric pressure chemical ionization/mass spectrometry for the part per trillion level determination of pesticides in sediments and natural waters using low preconcentration volumes, *Anal. Chem.*, 69, 4508–4514, 1997.

60. Groopman, J. D., Trudel, L. J., Donahue, P. R., Marshak-Rothstein, A., and Wogan, G. N., High-affinity monoclonal antibodies for aflatoxins and their application to solid-phase immunoassays, *Proc. Natl Acad. Sci. U.S.A.*, 81, 7728–7731, 1984.

61. Bouvrette, P. and Luong, J. H. T., Development of a flow injection analysis (FIA) immunosensor for the detection of Escherichia coli, *Intl. J. Food Micro.*, 27, 129–137, 1995.

62. Bouzige, M., Legeay, P., Pichon, V., and Hennion, M. C., Selective on-line immunoextraction coupled to liquid chromatography for the trace determination of benzidine, congeners and related azo dyes in surface water and industrial effluents, *J. Chromatogr. A.*, 846, 317–329, 1999.

63. Stolpe, N. B. and Shea, P. J., Alachlor and atrazine degradation in Nebraska soil and underlying sediments, *Soil Sci.*, 160, 359–370, 1995.

64. Nelieu, S., Kerhoas, L., and Einhorn, J., Atrazine degradation by ozonization in the presence of methanol as scavenger, *Intl. J. Environ. Anal. Chem.*, 65, 297–311, 1996.

65. Boundy-Mills, H. L., de Souza, M. L., Mandelbaum, R. T., Wackett, L. P., and Sadowsky, M. J., The atzB gene of Pseudomonas sp. strain ADP encodes the second enzyme of a novel atrazine detection pathway, *App. Environ. Microbio.*, 63, 916–923, 1997.

66. Sadowsky, M. J., Wackett, L. P., de Souza, M. L., Boundy-Mills, K. L., and Mandelbaum, R. T., Genetics of atrazine degradation in Pseudomonas sp. strain ADP, *ACS Symp. Ser.*, 683, 88–94, 1998.

67. Shapir, N. and Mandelbaum, R. T., Atrazine degradation in subsurface soil by indigenous and introduced microorganisms, *J. Agric. Food Chem.*, 45, 4486–4491, 1997.

68. Hilts, B. A., Dvorak, B. I., Rodriguez-Fuentes, R., and Miller, J. A., GAC treatment of Lincoln's water: Implications of pretreatment and herbicides, *Proc. Conf. Am. Water Works Assoc.*, 1060–1066, 2000.

69. Ashley, J. M., Dvorak, B. I. and Hage, D.S, Activated carbon treatment of s-triazines and their metabolites, Paper presented at *Proc. Natl Conf. Environ. Eng.*, Chicago, 1998.

70. Singh, J., Zhang, T. C., Shea, P. J., Comfort, S. D., Hundal, L. S., and Hage, D. S., Transformation of atrazine and nitrate in contaminated water by iron-promoted processes, *Proc WEFTEC*, Dallas, Vol. 3, 1996.

71. Katmeh, M. F., Godfrey, A. J. M., Stevenson, D., and Aherne, G. W., Enzyme immunoaffinity chromatography—A rapid semi-quantitative immunoassay technique for screening the presence of isoproturon in water samples, *Analyst*, 122, 481–486, 1997.

72. Gascon, J., Oubina, A., Ballesteros, B., Barcelo, D., Camps, F., Macro, M-P., Angel Gonzalez-Martinez, M., Morais, S., Puchades, R., and Maquieira, A., Development of a highly sensitive

enzyme-linked immunosorbent assay for atrazine. Performance evaluation by flow injection immunoassay, *Anal. Chim. Acta.*, 347, 149–162, 1997.

73. Bjarnason, B., Bousios, N., Eremin, S., and Johansson, G., Flow injection enzyme immunoassay of atrazine herbicide in water, *Anal. Chim. Acta.*, 347, 111–120, 1997.

74. Turiel, E., Fernandez, P., Perez-Conde, C., Gutierrez, A. M., and Camara, C., Flow-through fluorescence immunosensor for atrazine determination, *Talanta*, 47, 1255–1261, 1998.

75. Reeves, S. G., Rule, G. S., Roberts, M. A., Edwards, A. J., and Durst, R. A., Flow-injection liposome immunoanalysis (FILIA) for alachlor, *Talanta*, 41, 1747–1753, 1994.

76. Rule, G. S., Palmer, D. A., Reeves, S. G., and Durst, R. A., Use of protein A in a liposome-enhanced flow-injection immunoassay, *Anal. Proc.*, 31, 339–340, 1994.

77. Lee, M., Durst, R. A., and Wong, R. B., Comparison of liposome amplification and fluorophor detection in flow-injection immunoanalyses, *Anal. Chim. Acta.*, 354, 23–28, 1997.

78. Ensing, K. and Paulus, A., Immobilization of antibodies as a versatile tool in hybridized capillary electrophoresis, *J. Pharm. Biomed. Anal.*, 14, 305–315, 1996.

79. Hage, D. S., Affinity chromatography: A review of clinical applications, *Clin. Chem.*, 45, 593–615, 1999.

80. Wittmann, C. and Schmid, R. D., Development and application of an automated quasi-continuous immunoflow injection system to the analysis of pesticide residues in water and soil, *J. Agric. Food Chem.*, 42, 1041–1047, 1994.

81. Kramer, P. and Schmid, R., Flow injection immunoanalysis (FIIA)—a new immunoassay format for the determination of pesticides in water, *Biosen Bioelectron*, 6, 239–243, 1991.

82. Kramer, P. M. and Schmid, R. D., Automated quasi-continuous immunoanalysis of pesticides with a flow injection system, *Pest Sci.*, 32, 451–462, 1991.

83. Lee, M. and Durst, R. A., Determination of imazethapyr using capillary column flow injection liposome immunoanalysis, *J Agric. Food Chem.*, 44, 4032–4036, 1996.

84. Lee, M., Durst, R. A., and Wong, R. B., Development of flow-injection liposome immunoanalysis (FILIA) for imazethapyr, *Talanta*, 46, 851–859, 1998.

85. Wittmann, C., Immunochemical techniques and immunosensors for the analysis of dealkylated degradation products of atrazine, *Intl. J. Environ. Anal. Chem.*, 65, 113–126, 1996.

86. Kronkvist, K., Loevgren, U., Svenson, J., Edholm, L-E., and Johansson, G., Competitive flow injection enzyme immunoassay for steroids using a post-column reaction technique, *J. Immunol. Methods*, 200, 145–153, 1997.

87. Charles, P. T., Conrad, D. W., Jacobs, M. S., Bart, J. C., and Kusterbeck, A. W., Synthesis of a fluorescent analog of polychlorinated biphenyls for use in a continuous flow immunosensor assay, *Bioconjugate Chem.*, 6, 691–694, 1995.

88. Onnerfjord, P. and Marko-Varga, G., Development of fluorescence based flow immunoassays utilising restricted access columns, *Chromatographia*, 51, 199–204, 2000.

89. Onnerfjord, P., Eremin, S. A., Emneus, J., and Marko-Varga, G., A flow immunoassay for studies of human exposure and toxicity in biological samples, *J. Mol. Recog.*, 11, 182–184, 1998.

90. Onnerfjord, P., Eremin, S. A., Emneus, J., and Marko-Varga, G., High sample throughput flow immunoassay utilising restricted access columns for the separation of bound and free label, *J. Chromatogr. A.*, 800, 219–230, 1998.

91. Tudorache, M., Rak, M., Wieczorek, P. P., Jonsson, J. A., and Emneus, J., Immuno-SLM–a combined sample handling and analytical technique, *J. Immunol. Methods*, 284, 107–118, 2004.

92. Bier, F. F., Jockers, R., and Schmid, R. D., Integrated optical immunosensor for s-triazine determination: Regeneration, calibration and limitations, *Analyst*, 119, 437–441, 1994.

93. Irth, H., Oosterkamp, A. J., Tjaden, U. R., and van der Greef, J., Strategies for online coupling of immunoassays to HPLC, *Trends Anal. Chem.*, 14, 355–361, 1995.

94. Wilmer, M., Trau, D., Renneberg, R., and Spener, F., Amperometric immunosensor for the detection of 2,4-dichlorophenoxyacetic acid (2,4-D) in water, *Anal. Lett.*, 30, 515–525, 1997.

95. Schuste, M., Wasserbauer, E., Neubauer, A., and Jungbauer, A., High speed immunoaffinity chromatography on supports with gigapores and porous glass, *Biosep*, 9, 259–268, 2000.

96. Willumsen, B., Christian, G. D., and Ruzicka, J., Flow injection renewable surface immunoassay for real time monitoring of biospecific interactions, *Anal. Chem.*, 69, 3482–3489, 1997.

15 Sol–Gel Immunoassays and Immunoaffinity Chromatography

Miriam Altstein and Alisa Bronshtein

CONTENTS

15.1 INTRODUCTION

The sol–gel process and its use in the encapsulation of biomolecules of various kinds have recently seen important developments. Since its first introduction over two decades ago, sol–gel encapsulation has opened intriguing new ways to immobilize biological materials that offer an immense potential for the design of a large variety of applications. A great number of biomolecules, including enzymes, antibodies (Abs) (monoclonal, polyclonal, recombinant, and catalytic), DNA, RNA, and live animal, plant, bacterial, and fungal cells as well as whole protozoa have been encapsulated and then tested and implemented as optical and electrochemical sensors as well as core components of diagnostic, chromatographic, and catalytic devices.

In the past decade, several comprehensive reviews have been written on the sol–gel process, especially on the encapsulation of the above biomolecules and living cells [1–7]. The main topics covered include detailed explanations of the sol–gel process itself, information on the entrapment of proteins via the sol–gel process and on various types of sol–gel-derived biocomposites, a description of the behavior of sol–gel-entrapped biomolecules (distribution, conformation, dynamics, accessibility, activity, kinetics, etc.), details of the applications of sol–gel based biocomposites (analytical, biomedical, biophysical, biosynthetic), and conjectures regarding the future development of sol–gel bio-immobilization.

Despite the large number and diverse nature of the biomolecules that have been, so far, entrapped in the sol–gel matrix, most of the recent reviews focus on the entrapment of enzymes, which represent the largest and most studied group of entrapped biomolecules [1,2,4,5]. Other biomolecules, including Abs, have drawn less attention, although the number of studies that have been carried out and the progress the field has achieved in the past decade as well as their potential for practical applications is immense. The present review focuses on entrapment of Abs and various antigens (Ags) in sol–gel matrices and their applications for immunoassay (IA) and immunoaffinity chromatography (IAC). For the convenience of the reader, a brief introduction to the sol–gel process, biomolecule entrapment, and the recent innovations in the field of biomolecule encapsulation in this matrix is presented. Detailed reviews covering the above issues as well as topics related to the properties of sol–gel encapsulated biomolecules, their conformation, dynamics, accessibility, reaction kinetics, stability, and the recent developments in their application are available [1,2,4,5], and the reader is referred to them for further information.

15.1.1 The Sol–Gel Process

The term sol–gel refers to a chemical process where metallic or semi-metallic alkoxide precursors or their derivatives form composites at moderate temperatures through a chemical reaction that involves hydrolysis followed by condensation–polymerization (Figure 15.1). Most sol–gels are silicon-based oxides, although other oxides such as aluminum silicates, titanium dioxide, zirconium dioxide, and many other oxide compositions are also employed [8–11]. Silica oxides (SiO_2) sol–gel matrices can be composed in a wide range of physical properties (e.g., porous texture, network structures, surface functionalities) and under a wide variety of processing conditions (e.g., ambient temperatures, moderate pH values, and short gelation time), making silica alkoxides the most favorable group of sol–gel precursors. These matrices may take the form of porous wet gels, ambigels, aerogels, xerogels, or organically modified sol–gels (termed ormosils). The resulting matrix has high surface area and porosity, inertness and stability to chemical and physical agents, and optical clarity in the visible and UV ranges.

15.1.2 Entrapment of Biomolecules in Sol–Gel Matrices

Inorganic gels have been studied for over a century. In the past two decades, however, the sol–gel process underwent a significant development when it turned into a generic methodology for the incorporation of bioactive molecules. Although the first report on the entrapment of enzymes appeared in the mid-1950s [12], it was only three decades later that the importance of this finding was realized by Avnir et al. who entrapped a series of enzymes in a silica-based matrix [13]. Since then, a large body of work has emerged that encompasses the entrapment of a wide variety of biological molecules and even whole cells [1,2,4,5,14–16].

To achieve successful entrapment in sol–gel matrices, it is necessary to maintain the active conformation of the biomolecule within the matrix, to optimize the configuration of the doped sol–gel to provide the utmost performance, and to ensure that the entrapped biomolecule gains access to substrates, ligands, or analytes. The preparation of biologically doped materials via the sol–gel process fulfills the above requirements because the encapsulation is based on the growth of the polymer chains around the biomolecule, thus minimizing its denaturation. The sol–gel process can be carried out in aqueous solutions where hydrolysis and condensation–polymerization conditions occur at room temperature and at ranges of pH and ionic strength that are favorable for the biomolecule. Moreover, unlike the process in other polymers, the formation of the polymer backbone (e.g., Si–O–Si in the polysiloxane-based polymers) does not involve intermediates that could

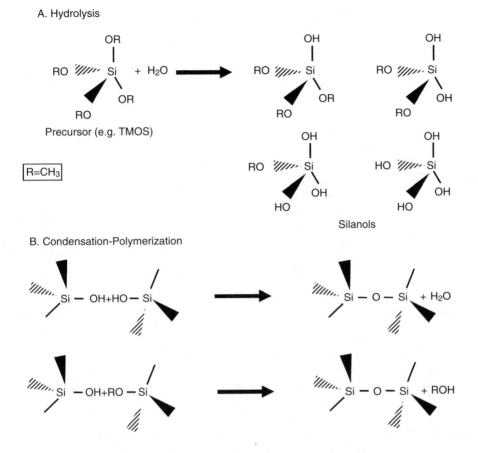

FIGURE 15.1 General scheme of the sol–gel process. TMOS: Tetramethoxysilane.

interact via a stable covalent bond with the protein being entrapped and potentially lead to its denaturation. Also, the pore size of the matrix can be controlled so that it is possible to obtain pores large enough to allow analytes, ligands, and substrates to reach the biomolecules, and the matrix can be modified to include residues and additives that modify the internal environment so as to improve bioactivity.

Entrapment of biomolecules is usually carried out according to the scheme shown in Figure 15.2. The precursor is first hydrolyzed (usually under acidic conditions that minimize the rate of siloxane condensation in the case of silicon oxides) to form an aqueous sol, then the hydrolyzed precursor is mixed with an aqueous solution of the biomolecules (in an appropriate buffer and at a pH that is suitable for bioactivity). The process results in condensation–polymerization of the hydrolyzed precursor followed by gelation of the aqueous sol to form a wet gel within which the biomolecules are entrapped. The initial gels have high water content and large pores, but over a period of time (days to weeks), further condensation occurs, strengthening the network. Further dehydration of the wet gel results in shrinkage of the polymer, collapse of its porous structure, and formation of a dry gel designated xerogel (Figure 15.2). As indicated above, most sol–gels are silica alkoxides, mainly because the hydrolysis and condensation rates of silica alkoxides are slow, enabling each step to be controlled independently and the kinetics to be optimized for specific needs.

Sol–gel biocomposites can be prepared from inorganic (silica or other metal or semi-metal) alkoxide precursors of the general formula $X(OR)_4$ where X is a metal or a semi-metal residue or

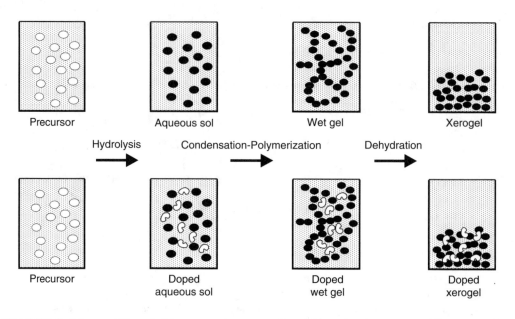

FIGURE 15.2 Scheme of the gelation process (upper panel) and biomolecule entrapment (lower panel).

from combined organic and inorganic materials of the general formula $YX(OR)_3$ where Y is an organic group such as ethyl or methyl, giving tetraethylorthosilicate (TEOS) or tetramethylsilane (TMOS). In addition, it is possible to encapsulate, together with the entrapped biomolecules, additives that can be beneficial for the stability and activity of the molecule. Such additives may be hydrophobic moieties, polymers (e.g., polyethylene glycol, glycerol, polyvinylimidazole, etc.), surfactants, liposomes, polysaccharides (dextran, cellulose, and chitosan), cofactors (e.g., redox modifiers), or even biological or synthetic additives. These compounds can be directly mixed with the sol–gel precursors before gelation to form hybrid organic–inorganic gels. The additives may affect the physical properties of the gel by changing its rigidity, mechanical stability, pore size, and optical or electrochemical clarity, and they may also affect the interactions of the gel with the entrapped biomolecules, thus providing higher overall activity. The nature of such gels and their effects on the entrapped biomolecules are discussed in detail in several reviews [1,4,5].

15.1.3 CHARACTERISTICS OF ENTRAPPED BIOMOLECULES

Progress in the encapsulation of biomolecules resulted in entrapment of a large variety of molecules such as enzymes, Abs, regulatory proteins, transport proteins, membrane-bound proteins, and nucleic acids [1,4,5]. Overall, these studies revealed that the entrapped biomolecules are surrounded by the porous gel network in a capsule- or cage-like manner. Depending on the nature and concentration of the entrapped molecule, the specific precursor and the additives used to form it may be homogenously dispersed in the pore solvent or adsorbed onto the silica in a variety of orientations, groups, or aggregates. A detailed review is presented by Jin and Brennan [1]. Entrapped biomolecules are not physically adsorbed or attached to the polymer texture although such interactions may occur naturally and could affect the activity. The interactions between the biomolecule and the inorganic, organic, or hybrid composite material and the additives in the gel determine the degree to which the biomolecule retains its native properties.

A variety of methods were employed to determine and characterize the bioactivity of entrapped biomolecules. They included measurements of activity by electrochemical and spectrophotometric methods, measurement of the absorbance changes that accompany ligand/substrate/Ag/Ab binding, use of fluorescence methods to probe ligand binding by regulatory proteins, and detailed studies of the interactions of Abs with their respective Ags. These studies revealed that the biomolecules are strongly encapsulated within the matrix and cannot diffuse out; the molecules retain their activity; their stability is enhanced; and they can react with compounds that diffuse into the highly porous matrix. Although the moderate temperatures and the mild hydrolysis and condensation–polymerization conditions that characterize the entrapment process allow the biomolecules to be entrapped without being denatured, conformational studies of entrapped molecules have revealed that conformational changes do occur in the course of the sol–gel entrapment. The enhanced stability of the entrapped biomolecules and the physical and chemical properties of the matrix are among the reasons for the attractiveness of the sol–gel approach to immobilization in general and to that of proteins in particular. Detailed information on the conformational stability of sol–gel-entrapped molecules and their dynamics, protein–gel interactions, response times for entrapped proteins, and their influence on the gelation process are all described by Jin and Brennan and by Pierre [1,5].

15.1.4 APPLICATIONS OF ENTRAPPED BIOMOLECULES

Doped silicate materials have been applied in a variety of fields. The activity in this field has been quite intense in recent years and has resulted in many biomaterials with diverse applications. Some of the applications of sol–gel-derived biomaterials include the successful immobilization of numerous biotechnologically important enzymes, the construction of optical and electrochemical biosensors for clinical, industrial, environmental and domestic use, enzymatic electrodes, stationary phases for affinity chromatography, generation of immunosorbents, solid-phase extraction (SPE) materials, controlled-release agents and solid-phase biosynthesis, construction of bioactive optical components, biocatalytic paints and films, preparation of biomatrices that can be applied for environmental studies and functional proteins, DNA and RNA biochips, etc. The ability to shape gels in a variety of formats (blocks, thin films, microarrays, columns, fibers, and powders) that are compatible with a variety of applications has turned out to be very important for industrial use, and a number of original designs based on these formats have extended the applicability of these encapsulated materials. Further details on some of the above applications are presented in [1,2,5] and in the references therein.

This review focuses on one group of biologically important molecules that have drawn the attention of many research groups: sol–gel-entrapped Abs and Ags for use in the development of IAs and IAC devices. The studies that have been performed in this connection and the applications of this approach are described below.

15.2 SOL–GEL IMMUNOASSAYS AND IMMUNOCHROMATOGRAPHY

15.2.1 SOLID PHASE IMMUNOASSAYS—GENERAL ASPECTS

Solid phase IAs are well-established methods in many fields. This powerful assay technology crosses discipline boundaries and is applied extensively as a research and diagnostic tool in both applied research and basic science as well as in medical, agricultural, and environmental studies. Solid-phase IAs are based on the fact that Abs or Ags can be immobilized on solid-phase matrices, thus enabling a simple and quick separation of free (unbound) analytes from the complexes immobilized on the solid surface. This simple feature made it possible to develop the method for quantitative detection of analytes and was, apparently, the most important reason for the rapid increase in its popularity and its wide application. Since the development of the first IA in the mid-1960s, hundreds of assays have been developed for native and synthetic molecules, and currently, they form one of the most generic diagnostic methods. The rapidity, sensitivity, simplicity, and cost-effectiveness of the method have made it an attractive tool, and it has been optimized for a large number of analytes. It also has been automated, and it now serves as a routine procedure in monitoring a wide variety of analytes in the fields of medical, environmental, and food sciences.

As indicated above, solid phase IAs are based, first and foremost, on the successful immobilization of Abs or Ags onto or within a suitable solid surface. The requirements for useful immobilization include high density of the immobilized molecule (high surface-to-volume ratio), high activity, good orientation of the adsorbed molecule, long-term stability under potentially adverse reaction conditions, good accessibility for analytes, rapid response time, and resistance to leaching or desorption. Several methods of immobilizing molecules on inorganic, organic, and polymeric surfaces have been reported; they include physical adsorption, covalent binding to the surface, entrapment in semi-permeable membranes, and micro-encapsulation in polymer microspheres or hydrogels (for a detailed review see [17]). Although some of the techniques have been successful, none of them are generic; namely, they could not be applied to a wide variety of molecules and had to be carefully optimized based on the chemical and physical properties of the entrapped molecule and the entrapping matrix. As a result, those techniques could be used, in most cases, for only a limited number of molecules and applications. Problems related to low surface loading, long preparation procedures, leaching and desorption of molecules, difficulties in controlling molecule orientation, and instability of the immobilized molecule resulted in the need to invest substantial efforts in the optimization of the immobilization protocol for each molecule, and this limited the applicability of IAs The emergence of the sol–gel technology, with all its advantages, and the reports on successful entrapment of biomolecules in such matrices opened the road to implementation of the method for the development of IAs in general and for those employed for food and environmental monitoring, in particular.

Immunochemical methods such as enzyme IA have become increasingly important during recent years for the determination of pesticides and other xenobiotics [18–23]. Many commercial kits for assaying pesticides are available, and several hundred assays have been described in the literature [24]. A major problem associated with the use of Abs in immunochemical assays for environmental or agricultural monitoring is their limited stability that can be a source of variation or deterioration in the test performance quality. Fruit and vegetable extracts as well as soil samples contain organic and inorganic substances that may directly interfere with Ab–Ag binding or impair it indirectly by denaturing the Ab or complexing with it, thus decreasing the efficiency of the assay and its monitoring capability. On top of the problems associated with matrix interference, some of the assays suffer from low reproducibility, slow reaction times, and high costs, and for many cases, they need to be used in on-site monitoring (in the field, packing houses, contaminated sites, etc.). Introduction of the sol–gel method for biomolecule entrapment, the unique nature of alkoxides polymer chemistry, and the advantages that are introduced by this method have opened up new possibilities for such applications and have enabled the development of simple, highly sensitive, highly reproducible, and cost-effective assays for off-line, on-line, and on-site use (that, in some cases, have proved to be significantly better than the standard existing ones).

A summary of the sol–gel based IAs developed in the course of the past two decades for environmental, agricultural, and medical applications is presented below.

15.2.2 Sol–Gel Based Solid Phase Immunoassays

Encapsulation of Abs and Ags in sol–gels has been well documented in the literature (see lists of references in Table 15.1 and Table 15.2), although less than a dozen IAs (assays that can be applied for monitoring real samples) have been developed in over a decade. Key issues in the development of a functional sol–gel based IA focused on optimization of the configuration of the doped sol–gel to provide the utmost performance, immobilization of biomolecules at a high density in a manner that will render them high activity, long term stability, good accessibility to analytes, rapid response time, and resistance to leaching. Indeed, those were the main issues that were addressed in most of the studies as indicated below. Early studies focused mainly on demonstrating the ability of

TABLE 15.1
Sol–gel Entrapped Antibodies Applied for Immunoassays

Entrapped Antibody	Immobilization Matrix	Sol–Gel Form	Method of Detection	Application	Tested Parameters	References
Anti-progesterone	TEOS/3-aminopropyl trimethoxysilane	Slurry suspension	Liquid scintillation counter	IA-POC	Binding affinity, leaching, effect of pH, entrapment capacity, comparison with ELISA	[29]
Anti-fluorescein	TMOS	Monolith	Spectrofluorimeter	IA-POC	Dose response, effect of gel aging, drying and storage on analyte binding and affinity	[28]
Anti-fluorescein	TEOS	Aerosol derived thin-films	Spectrofluorimeter	IA-POC	Binding activity as a function of storage time, leaching, non-specific binding, regenerability	[35]
Anti-dansyl	TEOS	Monolith	Spectrofluorimeter and time resolved fluorometry	IA-POC	Binding affinity, stability upon storage, mobility of entrapped Ab within the gel	[26]
Anti-TNT	TMOS	Monoliths & xerogels	Spectrofluorimeter	IA-POC	Dose and time response, comparison of binding in wet gels vs. xerogels, cross reactivity, stability	[25]
Anti-laminin	Aluminum isopropoxide	Thin-film capacitive electrochemical electrode	Voltametric analyzer	IA-POC	Optimization of binding conditions, time and dose response, binding capacity, non-specific binding, selectivity, reproducibility, comparison with SiO_2 sol–gel derived immunosensors	[34]
Anti-cortisol	TMOS	Monoliths & thin-films	Spectrofluorimeter	IA-POC	Binding (specific, non-specific), conformational changes of entrapped Ab, dose and time response binding, leaching	[27]
Anti-isoproturon (herbicide)	TMOS	Powder placed in a flow through spectrofluorimeter	Flow — injection Spectrofluorimeter	IA	Binding capacity, flow rate, leaching, recovery from spiked sea water and spiked food samples.	[36]
Fluorescein labeled anti-D dimer (a fibrin fragment)	TMOS	Coated tip of an optical fiber immunosensor & monoliths	Spectrofluorimeter	IA	Binding, LOD, dose response, aging, leaching, regenerability, determination of analyte in spiked human plasma and spiked whole blood samples	[31]

(continued)

TABLE 15.1 (Continued)

Entrapped Antibody	Immobilization Matrix	Sol–Gel Form	Method of Detection	Application	Tested Parameters	References
Anti-complement 3 (C3)	MTMOS/graphite	Squeezed paste in a form of an amperometric immunosensor	Electrochemical analyzer	IA	Optimization of operating conditions, sol–gel-graphite ratio, working potentials, sensitivity, reproducibility, non-specific adsorption, stability, detection range and LOD of C3, determination of C3 in real human serum samples	[32]
Anti-gentamicin (antibiotic)	TMOS/PEG	Slurry column in the form of a continuous flow displacement immunosensor	Spectrofluorimeter	IA	Effect of PEG on activity, non-specific binding, leaching, dose response, LOD column regeneration, stability, flow rate, recovery from spiked serum samples and real patient samples, validation with analytical methods	[33]
Anti-gentamicin (antibiotic)	Magnetite containing spherical silica (TMOS) nanoparticles	Magnetite-silica nanoparticles in solution	Spectrofluorimeter	IA	Binding capacity, recovery from spiked blood samples	[30][a]

The table divides the experiments into two groups: (A) experiments that proved the concept (POC) that sol–gel entrapped Abs can be used as immunoassay (IA) devices (such assays are termed IA-POC and involved binding of synthetic standard analytes); (B) experiments where the entrapped Ab was used to monitor analytes from spiked or real samples (such assays are termed IA). Abbreviations: IgG: immunoglobulin; LOD: Limit of detection; MTMOS: methyltrimethoxysilane; PEG: Polyethylene glycol; TEOS: Tetraethylorthosilicate; TMOS: Tetramethoxysilane.

[a] Unbound and bound analytes were separated with a magnet. In all other experiments, separation was by washing.

TABLE 15.2
Sol–Gel Entrapped Antigens Applied for Immunoassays

Entrapped Antigen	Immobilization Matrix	Sol–Gel Form	Method of Detection	Application	Tested Parameters	References
Rabbit-IgG	TEOS/hydroxypropyl cellulose/graphite	Thick-film amperometric electrode & electrochemical immunosensor	Voltmeter and ampermeter	IA-POC	Optimization of binding conditions, Ag loading, LOD, measurement range, different gel microstructures, reproducibility	[40]
Echinococcus granulosus Ag	TMOS	Gel cast in 96-well microplate	ELISA reader	IA	Detection of *Echinococcus granulosus* Abs in sera of infected patients	[16,38]
Leishmania whole cells	TMOS	Gel cast in a 96-well microplate	ELISA reader	IA	Characterization of the physical properties of the sol–gel matrix, ultrastructure of entrapped cells, storage detection of *Echinococcus granulosus* Abs in sera of infected patients, correlation with ELISA	[16]
Toxocara canis Ag	TEOS/polyvinyl alcohol	Sol–gel cast in 96-well microplates	Spectrophotometer (colorimetric signal)	IA	Determination of detection range, comparison of sol–gel based IA with conventional ELISA, detection of Abs in infected serum samples	[37]
CA19-9 (carbohydrate tumor marker)	Titanium isopropoxide (Ti based sol–gel)	Thick-film electrochemical electrode	Electrochemical analyzer	IA	Optimization of immunosensor preparation and working conditions, dose response analysis, selectivity, reproducibility, stability, measurements of CA19-9 in real serum samples from patients	[41]
Schistosoma japonicum Ag	TMOS/BSA/graphite	Fiber-like electrochemical electrode	Electrochemical and polarographic analyzer	IA	Binding conditions, non-specific adsorption, binding capacity, effect of incubation temperature, dose response, reproducibility, sensor stability, detection of Abs in infected serum samples	[42]

(continued)

TABLE 15.2 (Continued)

Entrapped Antigen	Immobilization Matrix	Sol–Gel Form	Method of Detection	Application	Tested Parameters	References
HCV and EBV derived peptide	Semicarbazide 3-aminopropyl trimethoxysilane	Slide microarrays	Fluorescence array scanner	IA	Non-specific binding, sensitivity, specificity, cross reactivity, reproducibility, comparison with ELISA, stability in storage, detection of HCV and EBV Abs in human serum	[39]

The table divides the experiments into two groups: (A) experiments that proved the concept (POC) that sol–gel entrapped Abs can be used as immunoassay (IA) devices (such assays are termed IA-POC and involve binding of Abs by the entrapped Ags); (B) experiments where the entrapped Ag was used to monitor Abs from spiked or real samples (such assays are termed IA).

Abbreviations: BSA: Bovine serum albumin; CA: Carbohydrate antigen; EBV: Epstein-Barr virus; HCV: hepatitis C virus; IgG: immunoglobulin; LOD: Limit of detection; TEOS: Tetraethylorthosilicate; TMOS: Tetramethoxysilane.

entrapped molecules to function within gels; later studies extended the scope of their analysis to more detailed characterization of the properties of entrapped biomolecules and even whole cells (such studies are referred to in the tables as proof of concept, POC IAs). More recently, studies have begun to focus on the implementation of sol–gel entrapment of molecules and whole cells for generation of sol–gel based IA devices for monitoring real samples (real IA).

A quick overview of the main issues that were addressed by the POC-IA studies reveals that they focused on the optimization of assay conditions (pH, sample volume, flow rate, effect of additives), kinetic parameters and thermodynamic analysis of Ab–Ag interactions (binding affinity, time kinetics), non-specific binding, cross reactivity and selectivity, reproducibility, regenerability, leaching, effects of gel drying and storage time on bioactivity (aging and stability), interaction of dopant and/or analyte with the matrix, effects of the dopants on pore size and of the pore size on bioactivity, conformation of the entrapped molecule and its mobility, and comparison of the sol–gel activity with that of other IA formats (e.g., enzyme linked immunosorbent assay, ELISA). The more applied studies extended their focus beyond the above to the examination of issues such as assay capacity, detection range, and detection limits relative to those of other IAs, ability to use sol–gel based IA for monitoring analytes in real samples, recovery, and validation against other analytical methods.

Regardless of their final purpose, whether POC or real IA, all of the above studies revealed that the entrapped biomolecules were active within the biogels and obeyed basic binding and kinetic rules (all Abs listed in Table 15.1), retained their specificity (e.g., [25]), were very stable after exposure to denaturing factors such as extreme pH values, high temperatures, organic solvents, etc. and retained their binding activity for several months (e.g., [25,26]). The biomolecules did not exhibit any detectable rotational reorientation (e.g., [26,27]), and their binding affinity was lower than that in solution (e.g., [26,28]) although, in some cases, the affinity did resemble that in solution (e.g., [29,30]). The non-specific binding of the assay could be reduced to a minimum by appropriate choice of matrix (e.g., [27]), the detection limit was within the necessary range for analytical or diagnostic purposes (e.g., [27,30–34]), and in most of the studies, aging did not affect the binding properties of the Abs (e.g., [26,31,35]).

Two main formats of sol–gel based IA were developed: those where Abs were immobilized/entrapped (Ab format, Table 15.1) and those where Ags were immobilized/entrapped (Ag format, Table 15.2). In each format, competitive and non-competitive assays were designed. In the first format, Abs were entrapped, and Ags that either exhibited self-fluorescence or were labeled with an enzyme, a fluorescent tag, or a radioactive isotope were tested for their binding to the entrapped Ab in the absence (non-competitive assay [26,28,31,34,35]) or the presence (competitive assay, [25,27,29,30,32,33,36]) of a free Ag. In the non-competitive assay, the decay of fluorescence was monitored; in the competitive assay, the decrease in enzymatic activity, fluorescence, or radioactivity (that resulted from an increase in the amount of free Ag that competed on binding to the entrapped Ab) was measured. In the second format, Ags were entrapped in the sol–gel matrix, and Abs of unknown samples (usually patient sera) were detected by a tagged secondary Ab conjugated either to an enzyme or to a fluorescent tag (non-competitive assay [16,37–39]), or by competition with a tagged primary Ab (competitive assay [40–42]).

Both IA formats employed a wide variety of entrapping matrices. Most assays were based just on silica alkoxides (TMOS or TEOS). Others used modified or derivatized alkoxides (3-aminopropyl trimethoxysilane, semicarbazide 3-aminopropyl trimethoxysilane, methyltrimethoxysilane) or silica alkoxides in combination with other additives (polyethylene glycol, PEG, hydroxypropyl cellulose, polyvinyl alcohol). Composites based on Al (aluminum isopropoxide) or Ti (titanium isopropoxide) were also employed. Graphite was added for generation of electrochemical electrodes. Both IA formats used a wide variety of sol–gel forms: slurry suspension, monoliths, aerosol-derived thin films, xerogels, gels cast in microplates, and a variety of electrochemical electrodes (thin-film capacitive electrodes, thick-film and fiber-like electrochemical electrodes, and amperometric electrodes). A whole array of immunosensors (optical, amperometric, and continuous flow) was generated as well as magnetite-containing spherical silica nano-particles and slide microarrays. The Ab format IA has been used to monitor environmental as well as medically related analytes, whereas the Ag format IA has been used only for medical applications. Examples of the different formats and their applications are presented below, and summarized in Table 15.1 and Table 15.2.

15.2.2.1 Sol–Gel Entrapped Ab Immunoassays (Ab Format)

The first study on entrapped Abs was reported in 1984 by Venton et al. [29]. It demonstrated the successful entrapment of anti-progesterone Abs in a polysiloxane copolymer prepared from a 3:1 mixture of TEOS and 3-aminopropyl trimethoxysilane. The study revealed that entrapped Abs retained their ability to bind Ag molecules with an apparent Ka equal to that of free antiserum (although only 50% of the entrapped Abs retained their activity). The study also revealed that the Abs did not leach out of the matrix, and they were stable at a wide range of pH values.

This study was followed almost a decade later by the work of Wang et al. [28] who entrapped anti-fluorescein Abs in a monolithic format and also confirmed the concept that sol–gel entrapped Abs retain their binding activity and affinity to their analytes (in this case, fluorescein). The group carried out a detailed quantitative examination of the influence of the encapsulation method and also of the aging, drying, and storage of the gel on the affinity constant of the Ab-analyte interaction. The study revealed that storage time and conditions affected the affinity of the sol–gel encapsulated Abs if the gel was kept dry, but the affinity could be retained if the samples were stored wet in the cold. Contrary to the findings of Venton et al. [29], encapsulation of antifluorescein resulted in a decrease in binding affinity by about two orders of magnitude compared with that obtained in solution.

Two other Abs were entrapped in a monolith sol–gel format: anti-TNT [25] and anti-dansyl [26]. The anti-TNT Abs were able to detect analytes at a ppm level, and they retained their ability to differentiate between TNT and other trinitroaromatic analogues. The sol–gel entrapped Abs exhibited better relative stability than Abs immobilized by means of surface attachment, and they proved to be highly stable upon exposure to denaturing conditions. Satisfactory activity and enhanced stability of the entrapped Abs was also demonstrated with the anti-dansyl Abs [26].

Despite the successful development of monolith-based IAs, such assays suffer from a limitation that could affect measurements of real samples because of the inherently long response time associated with the slow diffusion of the analyte. In order to overcome the above limitation and to improve the response time and accessibility of the analytes to the entrapped sensing molecule, thin film IAs were developed [27]. Zhou et al. entrapped anti-cortisol Abs in thin films and compared their activity with that of monolith-entrapped Abs, finding that both formats could be used as optical-sensing IAs. The data also indicated that the Abs retained their activity in both formats in a dose responsive manner, that the sensitivity of the assay was satisfactory (within the physiologically relevant range), and that the entrapped Abs showed only small changes in their secondary structure. A comparison between the sensing properties of the monolith and thin-film forms revealed that the thin films were more effective and gave improved accessibility of Ags to the encapsulated Abs, leading to a significant reduction in the required assay time. However, these advantages were gained at the expense of a decrease in signal intensity (although not a major one) and wider variability between assays. Another study was carried out by Jordan et al. [35] who applied the thin-film approach to the entrapment of anti-fluorescein Abs, and they quantified their response as a function of storage time. Although the study did not compare the time kinetics of thin-film reactions with those of monolith assays, it revealed that the Ab retained affinity for its hapten, that the binding was retained for over three months, and that the response time increased as a function of storage time (because of the particular structure of the device that was used, selecting Ab subpopulations whose behavior changed as a function of time). Partial regeneration of the device was possible for several cycles using a mild chaotropic reagent.

The thin-film sol–gel method was also applied to the development of capacitive label-free IAs with γ-alumina sol–gel entrapped Abs [34]. Capacitive immunosensors have been extensively studied as novel label-free IAs that offer high sensitivity and rapid testing that are used with ordinary instrumentation. Successful implementation of this method required formation of an electrically insulated film in order to allow the capacitive measurements and the formation of thin (nanometer size) films of the embedded Ab layer. In their study [34], Jiang et al. described the use of γ-alumina-based sol–gel (that is especially suitable for the formation of thin films because of the inherently high surface area of the matrix that enables significant reduction of the thickness of the formed films) for the development of a capacitive immunosensor. A multi-channel capacitive sol–gel-derived anti-human immunoglobulin (IgG) and anti-laminin IA was constructed on the basis of this system to illustrate the application. The device was able to measure Ags (human IgG and laminin) with the accuracy required for diagnostic analysis, it showed a low detection limit (lower than those of SiO_2 sol–gel-derived capacitive immunosensors or conventional ELISAs), demonstrated reproducible linear responses to the respective Ags, and was found to be selective and specific. This is the only example of a sol–gel matrix based on alumina alkoxide in this format.

Biosensors offer considerable advantages, especially when on-site monitoring is necessary. There are several formats of biosensors; the most common based on sol–gel technology are electrochemical immunosensors. Electrochemical immunosensors combine simple, portable, low-cost electrochemical systems with specific and sensitive IA procedures, representing a promising approach to clinical, biochemical, and environmental analyses. An example of a sol–gel amperometric immunosensor is presented in the study of Liu et al. [32]. The authors describe the generation of an amperometric immunosensor based on anti-complement C3 Ab entrapped in a mixture of sol–gel-BSA-graphite composite. The doped sol–gel paste (that was squeezed into a polyvinyl chloride tube) formed an amperometric immunosensor that was used in a competitive binding assay to detect C3 in human serum (with the aid of a C3-horse-radish peroxidase (HRP) labeled conjugate). The immunosensor was found to be sensitive in the range required for clinical analysis, stable, highly reproducible, and renewable.

Another application of sol–gel entrapped Abs is in the formation of immunosensors that can monitor analytes in real time. Flow-injection IAs or electrochemical immunoelectrodes offer many advantages for real time monitoring. Analysis can be performed under continuous flow and automated on-line performance in the laboratory or on-site without the need to use costly and complex instrumentation. This is especially important for environmental and forensic monitoring. Two studies address the implementation of sol–gel entrapped Abs for the formation of continuous-flow displacement immunosensors. In the first study [36], anti-isoproturon Abs were entrapped in TMOS that was crushed into a powder and applied in a flow-through spectrofluorimeter to monitor the herbicide isoproturon in spiked sea water and potato extracts. The method was found to be effective, rapid, and more sensitive than HPLC; matrix effects were minimized, the overall analysis was simple, and the immunosensor could be used for up to two months without changes in sensitivity. It is interesting to note that this assay is the only sol–gel based IA that was developed for environmental monitoring of real samples, highlighting the fact that although IAs are currently quite widely employed in environmental, forensic, and food-exposure studies [18–24], introduction and implementation of sol-gel based IAs to these fields is somewhat slow and lags behind that in the pharmaceutical and biomedical fields. This is very well reflected in view of the fact that, out of all the sol–gel IAs that have been developed, only two (the above study and the POC-IA on TNT [25]) are for environmental diagnostics. One possible explanation may lie in the need to develop Abs to diverse and sometimes lipophylic analytes that is not always a straightforward matter and to monitor diverse and complex matrices such as sediments, fatty foods, and crude acetonic extracts of fruits and vegetables, etc. that are much more complicated than body fluids.

A flow-injection IA was also introduced by Yang et al. [33] who used anti-gentamicin in a silicon sol–gel based matrix that contained PEG in the form of a slurry column. The IA was based on the competition between gentamicin and fluorescence-labeled gentamicin that competed for the entrapped Ab. The method was found to be efficient, the columns could be reused many times, non-specific binding was low, and it was possible to determine analytes in patients' serum in amounts close to those obtained with the commonly used IAs. A further verification of sol–gel entrapped anti-gentamicin Abs was presented by the same group [30] where the Abs were entrapped in magnetite-containing spherical nanoparticles and were used in a magnetic-separation IA system. The mechanically stable particles that are nanometer in size and magnetically separable were tested for their ability to bind fluorescent gentamicin in solution. The data revealed that the entrapped Abs retained their ability to bind gentamicin and that the magnetite-based IA was quantitative and behaved in a manner similar to that of IAs that used Abs in solution. The assay was also employed to analyze the recovery of spiked serum samples, and the method showed satisfactory recovery and reproducibility, thus indicating the high potential for the use of magnetite-containing spherical silica nanoparticles as improved biosensor devices in a micro-scale fluid system or in vivo for biomedical monitoring (e.g., drug release, drug metabolism, intake, etc.).

A very interesting application of the sol–gel method was reported by Grant and Glass [31] who entrapped fluorescent anti-D dimer Abs (a fibrin product resulting from the degradation of fibrin clots) at the tip of an optical fiber and used the device for monitoring the D-dimer analyte in spiked human plasma and whole blood. The study demonstrated the feasibility of application of the method in spiked PBS, human plasma, and whole blood. The sensor showed clinically relevant sensitivity, low leachability, some regenerability, and stability for about a month. Although several features (such as lifetime and regenerability) still need to be improved, the method offers an intriguing diagnostic application in monitoring thrombolysis in stroke patients undergoing thrombolytic therapy and the possibility of additional new intracatheter applications.

The sol–gel method was also used for the development of IAs and immunosensors where the Abs were not encapsulated in the matrix itself but rather adsorbed or covalently bound to the solid-phase surface. Although this topic is beyond the scope of this review, a few examples are listed. In two studies [43,44], the sol–gel method was employed to form a novel potentiometric immunosensor that comprised a two-dimensional sol–gel layer onto which nanoparticles and Abs (anti-diphtheria and anti-cortical hormone Abs) were adsorbed. In another study [45], a continuous-flow displacement immunosensor was formed from anti-RDX (hexahydro-1,3,5-trinitro-1,3,5-triazine) Abs covalently immobilized on TEOS. In other studies, a hybrid inorganic–organic composite based on polyvinyl alcohol and polysiloxane was used to covalently immobilize Ags obtained from the bacterium *Yersinia pestis* [46] and anti-2,4-D monoclonal Abs (MAbs) adsorbed via anti-mouse IgG onto a sol–gel treated glass capillary surface was used in a competitive chemi-luminescent reaction to develop a highly sensitive IA for the detection of the herbicide [47].

15.2.2.2 Sol–Gel Entrapped Ag Immunoassays (Ag Format)

The first study on the development of an IA of the Ag format was reported by Livage et al. in 1996 [16,38] (Table 15.2). An *Echinococcus granulosus* cyst fluid Ag was entrapped in a sol–gel matrix, cast in a 96-well microplate, and used for the detection of anti-*Echinococcus granulosus* Abs in sera of infected patients. The study proved the ability of the assay to detect relatively large molecules (Abs) with a sensitivity that did not significantly differ from that of the classical ELISA.

In another study [16], whole protozoa cells (*Leishmania*) were entrapped in a sol–gel matrix in a microplate format and were used to detect *Leishamnia* Abs in infected patients. The study revealed that the entrapment of whole-cell organisms within sol–gel matrices did not destroy their cellular organization and that the gelation process could be controled to ensure that the pores were large enough to accommodate a whole cell and to allow diffusion of large molecules such as IgGs in a manner that enabled detection of Abs in infected human patients.

A similar study was performed by Coelho et al. [37] with a *Toxocara canis* Ag entrapped in a hybrid sol–gel composite consisting of TEOS and polyvinyl alcohol (that enabled the Ag to be retained more strongly in the matrix by its covalent binding via glutaraldehyde). Detection of

Toxocara canis Abs in unknown samples was determined with a secondary anti-human IgG conjugated to HRP by means of a chromogenic substrate. The study revealed that the assay could detect Abs in infected patients and that the sensitivity of the assay was significantly better than that of conventional ELISA.

The first report on the formation of a disposable thick-film electrochemical immunosensor based on sol–gel technology was published by Wang et al. in 1998 [40]. The group entrapped an Ag (rabbit IgG) that competed with an analyte in solution for binding of Abs (anti-rabbit IgG) conjugated to a reporting enzyme. The study demonstrated that the Abs could readily access the embedded Ag despite the thick gel film, the immobilized Ag could still be recognized by the Ab, the sol–gel based immunoelectrode could be used in a competitive assay to detect Abs in solution, and the measurement range and detection limits of these electrodes compared favorably with those of other electrochemical immunosensors and of ELISAs.

Another example of a thick-film electrochemical electrode where the Ag was entrapped in a sol–gel matrix is described by Du et al. [41]. An electrochemical graphite electrode was developed, onto which a carbohydrate Ag (CA) 19-9 (a carbohydrate tumor marker) was immobilized with titanium sol–gel membranes. Detection of CA19-9 in unknown samples was determined on the basis of competition between the free and the entrapped Ags on an HRP-labeled CA19-9 Ab present in the sample solution. The immunosensor demonstrated good accuracy and acceptable selectivity, sensitivity, reproducibility, storage stability, and precision.

Another example of an electrochemical electrode is described by Zhong and Liu [42]. An electrochemical biosensor based on sol–gel was used to entrap an Ag of a human parasite, *Schistosoma japonicum*, for the detection of anti-*Schistosoma japonicum* Abs. Detection of Abs in unknown samples was determined on the basis of competition between an HRP-labeled anti-*Schistosoma japonicum* Ab and the Ab present in the sample solution. As in the entrapped Ab IA, the sensor exhibited excellent physical and electrochemical stability with a renewable external surface, low background, a wide range of working potentials, and a relatively long lifetime. The method also proved successful in providing a useful sensing device for directly monitoring the concentration of *Schistosoma japonicum* Abs in serum samples.

Sol–gel entrapped peptides were also used to make peptide arrays for highly sensitive and specific Ab-binding, high-throughput screening (HTS) IAs. Currently, the miniaturization of biological assays is gaining a lot of attention for advances in biological and medical research. The design of miniaturized devices for highly sensitive, specific, and simultaneous detection of multiple Abs from complex biological samples is of high importance. Over the past decade, the trends toward the development of miniaturized HTS assays have generated a great deal of interest in the scientific community. Microarrays have been employed for a variety of applications, especially in genomic studies, but their application to diagnostics has not yet been fully explored. In a study by Melnyk et al. [39], two hepatitis C virus (HCV)-derived peptides and an Epstein-Barr virus (EBV)-derived peptide were printed on a semicarbazide sol–gel layer for the detection of HCV and EBV Abs in human sera, and the amount of serum Ab was determined with a fluorescent secondary anti-human Ab. The assay displayed very high sensitivities and specificities for Abs directed toward several peptidic epitopes; it enabled detection in very small blood samples from infected individuals, and a comparison with standard ELISA demonstrated large gains in sensitivity and specificity.

15.2.3 Immunoaffinity Chromatography—General Aspects

Immunoaffinity chromatography is one of the most powerful techniques for single-step isolation and purification of individual compounds or classes of compounds from liquid matrices. IAC is based on the highly selective interaction between Abs and their Ags. Because of the high affinity and high selectivity of the Ag–Ab interaction, the method provides a high degree of molecular selectivity. IAC involves three steps: (a) preparation of the Ab matrix, followed by packing the sorbent as a column where the adsorption/desorption chromatography will be carried out; (b) binding the Ag to the Ab matrix; and (c) elution of the Ag. In the first step, Abs are immobilized onto a solid-phase matrix. After the preparation of the Abs-containing matrix, the Ag is bound and the contaminating macromolecules are removed by washing. In the last step, the Ab–Ag interaction is dissociated, and the Ag is released into the eluate. IAC can be performed off-line or on-line, and it can be coupled to a wide variety of separation techniques and analytical methods such as liquid chromatography (LC), gas chromatography (GC), and immunochemical analysis (e.g., ELISA).

IAC has been widely applied for over four decades for pharmaceutical and biomedical trace analysis and, in the more recent decades, for analysis of environmental contaminants and pesticide residues in occupational and environmental health monitoring, in forensic examinations, and in food safety analysis. The varied and complex matrices that serve as sources for analyte monitoring, the low concentrations of the analytes within the matrix, and the presence of compounds that interfere with the analytical method raised the need for a highly specific, quick, and cost-effective method of clean-up and concentration of the tested materials. This, together with the intensive research in the area of SPE that has led to the development of new formats and new sorbents and the drastic decrease in the use of traditional liquid–liquid extractions because of restrictions on some of the solvents (e.g., chlorinated organic compounds), has resulted in the emergence of IAC as a preferred method in trace analysis. To date, IAC has been successfully utilized for monitoring pesticides and other trace organics in environmental and food samples as well as for detecting drug metabolites and endogenous compounds in biological fluids in occupational exposure and clinical trials. Interestingly, the applications where the high potential of IAC for class-selective extractions, has been clearly shown, belong to the environmental field of analysis (see below item Section 15.2.4.1). The basic principles of IAC approaches, the new developments in this field, and their applications in clinical and environmental analysis have recently been reviewed in detail [21,22,48–51].

Successful employment of IAC requires special supports for immobilization of Abs. The supports must be (a) porous—to allow Ag penetration and to provide high-capacity support; (b) chemically and biologically inert—to minimize non-specific adsorption; (c) stable—to allow the use of denaturing reagents in the elution process; and (d) easily activated. To date, the usual approach to the production of bioaffinity chromatography devices has been based on covalent or affinity coupling of Abs (via streptavidin, protein A, or protein G) to solid supports, and the supports traditionally used in IAC include agarose, silica, cellulose, and synthetic polymers. This approach suffers from limitations such as loss of Ab activity upon coupling (because of poor control over protein orientation and conformation), low surface loading, potentially low mechanical stability (that prevents the on-line coupling of IAC columns with separation methods), difficulties in the loading of beads into narrow columns, difficulties in miniaturizing to very narrow columns, poor flexibility with certain proteins, long preparation time, low regenerability, and—most importantly—high cost. The recent advances in the development of monolithic columns have not been widely adapted to IAC as the method involves harsh chemical processes that are not compatible with biomolecules. The successful application of the sol–gel doping methodology to a wide variety of biomolecules, the ability to tailor the porosity of the sol–gel matrix, as well as all of the other above-mentioned long list of potential advantages of the sol–gel technique have stimulated the extension of the sol–gel studies to IAC applications for clinical, environmental, forensic, and food safety residue monitoring. A summary of the sol–gel based IAC applications of the past decade is presented below and in Table 15.3.

15.2.4 SOL–GEL BASED IMMUNOAFFINITY CHROMATOGRAPHY

Although entrapment of Abs in sol–gels has been widely studied (see above), only a few IAC applications (less than 20) have been reported in the literature, and even fewer studies have been

TABLE 15.3
Sol–Gel Entrapped Antibodies Applied for Immunoaffinity Chromatography

Entrapped Antibody	Immobilization Matrix	Sol–Gel Form	Method of Detection	Application	Tested Parameters	References
Anti-dinitro-benzene (environmental-contaminant)	TMOS TMOS/PEG	Slurry column	ELISA (colorimetric assay)	IAC-POC	Proof of binding, non-specific binding, comparison of binding capacity of whole serum vs. IgGs, optimization of sol–gel formats and additives, binding capacity, elution, reproducibility, leaching, comparison with other IAC methods	[52,58]
Anti-atrazine- (herbicide)	TMOS/PEG	Slurry column	ELISA (colorimetric assay)	IAC-POC	Binding, dose response, optimization of sol gel formats, capacity, elution, leaching, stability in storage, reproducibility	[54,57]
Anti-TNT	TMOS/PEG	Slurry column	ELISA (colorimetric assay)	IAC-POC	Binding, non-specific binding, dose response, comparison with other Ag–Ab binding assays (i.e., immunoprecipitation), leaching, stability, tolerance toward organic solvents, elution	[59]
Anti-1-nitro-pyrene- (environmental contaminant)	TMOS	Slurry column	ELISA or HPLC- (fluorescence detector)	IAC-POC	Proof of binding, non-specific binding, capacity, leaching, elution, reusability, aging and storage	[53]
Anti-pyrene (environmental contaminant)	TMOS	Slurry column	ELISA or HPLC (fluorescence detector)	IAC-POC	Proof of binding, non-specific binding, effects of surfactants, PEG and high-MW blockers on non-specific binding, capacity, cross reactivity, leaching, elution, reusability, aging and storage	[55,56]

(continued)

TABLE 15.3 *(Continued)*

Entrapped Antibody	Immobilization Matrix	Sol–Gel Form	Method of Detection	Application	Tested Parameters	References
Anti-2,4-D (herbicide)	TEOS	Slurry column	HPLC (UV detector)	IAC-POC	Binding capacity of wet gels vs. semi-dry gels and xerogels, flow rates, physical properties of doped vs. non doped gels; optimization of elution conditions, reusability, comparison with Ab–Ag binding in solution	[60]
Anti-1-nitro-pyrene (environmental contaminant)	TMOS	Slurry column	HPLC (fluorescence detector)	IAC	Mechanical stability, capacity, reusability, reproducibility, sample preparation compatible with sol–gel IAC, matrix interference and analyte recovery from spiked herb samples	[63]
Anti-pyrene (environmental contaminant)	TMOS	Slurry column inserted into an outdoor sample (rainwater) collector	HPLC (UV and fluorescence detector)	IAC	Binding capacity, effect of pH on binding, stability in storage at high temperatures, cross reactivity, analyte recovery from rain water	[62]
Anti-pyrene (environmental contaminant)	TMOS	Slurry column coupled on line to HPLC	HPLC (fluorescence detector)	IAC	Pressure stability of column, kinetics of Ag–Ab binding, column capacity, elution conditions, reusability, analysis of river samples	[64]
Anti-pyrene (environmental contaminant)	TMOS	Slurry column	HPLC (fluorescence detector)	IAC	Binding capacity, leaching, elution, specificity, matrix interference, comparison of sol–gel IAC with SPE, reusability, recovery from spiked and real urine samples	[65]
Anti-s-triazine (herbicide)	TMOS/glycerol	Slurry column	GC (NPD detector)	IAC	Ab immobilization efficiency, leaching, binding capacity, selectivity, non-specific binding, binding and elution conditions, reusability, recovery from spiked water and real soil samples, comparison with other SPE methods	[66]
Anti-bioallethrin-(insecticide)	TMOS/PEG	Slurry column	ELISA (colorimetric assay)	IAC	Determination of the analyte in spiked fruit and vegetable extracts	[69]

Anti-bisphenol A (environmental contaminant)	TMOS	Slurry column	HPLC (fluorescence detector)	IAC	Loading and elution conditions, binding capacity, recovery, regeneration, cross reactivity, determination of analyte in spiked and real samples (canned beverages and food)	[68]
Anti-morphine Anti-M3G Anti-M6G	TMOS	Slurry column	HPLC (fluorescence detector and laser induced fluorescence)	IAC	Comparison of sol–gel IAC with SPE extraction, capacity, recovery, determination of analytes in real blood samples	[67]
Anti-tumor IgG	TMOS/PEG/PVP/3-aminopropyl-trimethoxy-silane	Columns composed of glass fiber coated sol–gel membranes	HPLC (UV detector)	IAC	Non-specific binding, leaching, capacity, determination of tumor-associated antigens from patients' sera	[61]

Experiments are divided into two groups: (A) experiments that proved the concept (POC) that sol–gel entrapped Abs can be used as immunoaffinity chromatography (IAC) devices (such assays are termed IAC-POC and involve studies in which the Abs were used to purify standard analytes); (B) Experiments where the entrapped Abs were used to monitor analytes from spiked or real samples (such assays are termed IAC). Abbreviations: GC: Gas chromatography; HPLC: High pressure liquid chromatography; IgG: immunoglobulin; M3G: anti-morphine-3-β-D-glucuronide; M6G: anti-morphine-6-β-D-glucuronide; NPD: Nitrogen phosphorus detector; PEG: Polyethylene glycol; PVP: polyvinylpyrrolidone; SPE: solid phase extraction; TEOS: Tetraethylorthosilicate; TMOS: Tetramethoxysilane; TNT: 2,4,6-trinitrotoluene.

reported that describe the use of sol–gel based IAC for group-selective enrichment and recovery of small analytes from real samples (Table 15.3). As with sol–gel based IAs, some of the studies were designed as IAC-POC experiments, i.e., studies where the ability to use sol–gel entrapped Abs for IAC has been proven by using standard analytes [52–60]; other studies [61–69] employed the devices for clean-up and concentration of real samples. Most of the IAC experiments differed from those of the sol–gel based IAs described above by being focused on environmental analytes; only two [61,67] were concerned with compounds of clinical interest.

Several approaches have been used so far in the development of sol–gel based IAC devices. Most of the studies used crushed Ab-doped silica monoliths that were loaded into a column in the form of a slurry. Only one study [61] used a different format where glass fibers covered with Ab-doped sol–gel, as a support, were used for the affinity separation. Almost all studies used silica based composites (TEOS or TMOS) with or without PEG or glycine as the entrapping matrix; one study [61] used a mixture of TMOS, PEG, 3-aminopropyl trimethoxysilane, and polyvinylpyrrolidone (PVP). Almost all the columns were used as off-line devices. Only one study [64] used a slurry IAC column coupled to on-line HPLC via a reverse-phase pre-column. Eluted analytes from the on-line and off-line columns were monitored by ELISA, HPLC, or GC. A general scheme of a sol–gel based IAC process is depicted in Figure 15.3.

The development of a sol–gel based robust analytical device for IAC relies on having the entrapped Ab mimic or even be enhanced relative to its behavior in solution (with respect to its activity or stability to denaturation). In order to achieve this goal, it is important to optimize entrapping conditions, check binding, capacity and, most important, to compare the activity of the entrapped Abs with those in solution and with other IAC methods. These topics were indeed the main objective of most of the IAC-POC studies as indicated below. A quick overview of the main issues that were addressed by these studies reveals that they focused on: (a) optimization of gel formats and additives (wet gel, xerogels, presence of PEG, etc.) and mechanical stability of gel; (b)

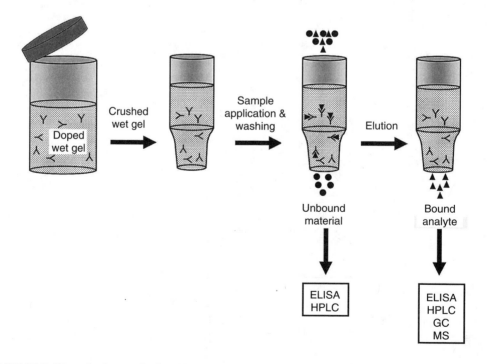

FIGURE 15.3 General scheme of sol–gel based IAC process. ELISA: Enzyme linked immunosorbent assay; GC: Gas chromatography; HPLC: High pressure liquid chromatography; MS: Mass spectrometry.

proof of binding; (c) optimal loading conditions; (d) non-specific binding and effect of surfactants on its reduction; (e) binding capacity; (f) cross reactivity/specificity (mainly in comparison with ELISA); (g) tolerance of entrapped Abs toward organic solvents; (h) elution conditions; (i) recovery; (j) leaching; (k) stability during storage and aging; (l) reproducibility; (m) reusability; and (n) comparison with other IAC methods and SPE. The applied studies extended their focus beyond the above and examined the ability of the IAC columns to clean up and concentrate analytes from complex food, environmental, and clinical samples. All of the studies proved, without doubt, that sol–gel entrapped Abs could serve as highly efficient, reproducible, stable, and reusable IAC devices as detailed below.

15.2.4.1 Sol–Gel Immunoaffinity Chromatography for Environmental, Forensic, and Occupational Monitoring

The first use of sol–gel entrapped Abs for IAC purposes was reported over a decade ago in 1994 by the group of Altstein et al. [52,58] who entrapped anti-dinitrobenzene (DNB) IgGs in a TMOS based sol–gel matrix. The study revealed that the entrapped Abs, in the form of a slurry column, retain their ability to bind analytes from solution. That study was followed by entrapment of anti-DNB antiserum in a TMOS/PEG sol–gel matrix [58], entrapment of anti-atrazine monoclonal Abs (MAbs) [54,57], and later of anti-trinitrotoluene (TNT) MAbs [59] and anti-pyrethroid MAbs [69] with the latter study focusing on employment of the sol–gel columns for IAC of the pyrethroid bioallethrin from food (crude acetonic extracts of tomatoes, cucumbers, and strawberries). Regardless of the entrapped Ab or tested analyte, the above studies revealed that the hydrophilic wet gel (in the form of a slurry column) with a TMOS:aqueous ratio of 1:8, enriched with 10% PEG, was the preferred working format, the entrapped Abs retained their ability to bind analytes from solution in a dose-dependent manner within a time frame that did not greatly differ from that in solution, the analytes could be eluted at high recoveries (86%–100%) with either organic solvents (ethanol, acetone, acetonitrile) or highly basic or acidic buffers, and the analytes did not adhere non-specifically to the matrix. The studies also revealed that the Abs did not leach from the column even under extreme elution conditions, the column could be stored for several months without losing activity, and the assays were highly reproducible. The assays were equally effective with either polyclonal Abs (PAbs) or protein A-purified IgGs (that eliminated the need to purify IgGs from the whole antiserum), and they could be carried out with PAbs, MAbs, or even hybridoma culture fluid (without the need to purify the MAbs from the tissue culture medium). Interestingly, the sol–gel IAC columns exhibited binding capacities that were either significantly higher than or equal to those obtained with protein A-agarose coupled Abs.

Almost in parallel to the above studies, Zhulke et al. [53] reported in 1995 the successful entrapment of anti-1-nitropyrene Abs in a TMOS-based sol–gel matrix. The study addressed similar questions to those listed above and further proved the feasibility of the concept, namely, that sol–gel entrapped Abs could serve as successful IAC devices. Zhulke's study was followed by additional reports from the same group who examined the properties of sol–gel entrapped anti-pyrene Abs [55,56]. These studies revealed, once again, that entrapped Abs bound analytes, that the Abs did not leach from the matrix, the columns could be regenerated and reused, aging did not affect the binding properties, and the method was highly efficient and yielded excellent analyte recoveries. Unlike the analytes that were tested by Altstein et al. (e.g., [57,59]) that did not adsorb non-specifically to the matrix, members of the polycyclic aromatic hydrocarbons (e.g., pyrene and related compounds [56]) did adsorb to the matrix and non-specific binding to the columns slowly increased during storage. A variety of approaches (e.g., modification of the sol–gel composite with PEG, addition of nonionic surfactants, or combination of a high-molecular-weight blocking agent with a surfactant) were found to markedly reduce non-specific adsorption and reduce cross reactivity [56].

From the late 1990s, the POC-IAC reports showed increasing attention to applied studies where the method was employed for monitoring real environmental, occupational, and food samples.

Several such reports were published, most of them from the group of Niessner and coworkers [62–66,68]. Application of the sol–gel based IAC method for recovery of analytes from real samples included entrapment of the following Abs: (a) anti-*s*-triazine for detection of triazine herbicides in water and soil samples [66]; (b) anti-1-nitropyrene for detection of the analyte from herb samples [63]; (c) anti-bisphenol A for clean-up of the analyte from canned beverages, fruits, and vegetables [68]; and (d) anti-pyrene for detection of polycyclic aromatic hydrocarbons (e.g., pyrene) and their metabolites in spiked and real urine [65], river water [64], and rain water [62] samples. The study described by Scharnweber, et al. [62] employed a device that was used outdoors and was intended to calculate the penetration of chemical compounds deposited from the atmosphere into the soil. Beyond the successful application of monitoring analytes from real samples, the study demonstrated the ability of the IAC device to withstand the harsh conditions that are met outdoors in on-site field experiments. All of the above studies revealed that the columns could effectively cleanup analytes from complex, spiked samples, and they could significantly reduce the matrix interference that could affect downstream chemical or immunochemical analyses.

Although most of the above studies confirmed what had already been reported and proven with regard to the successful entrapment of Abs in sol–gel matrices, several of them revealed drawbacks and problems with the method. For example, in the study of Braunrath and Cichna [68], the cross-reactivity pattern of the IAC columns was significantly higher than that obtained in ELISA. In another study [70], a few compounds that could be detected by Abs in solution could not interact with the sol–gel entrapped Abs. Furthermore, in a study by Cichna et al. [64], it was shown that the selectivity of the sol–gel IAC column was comparable with that of a conventional reverse-phase (RP-8) column, i.e., the sol–gel IAC approach did not show any advantage over the RP-8. In most cases, however, sol–gel IAC resulted in higher analyte recovery compared with that obtained with other SPE methods (e.g., [65–67]). Also, in some studies, the IAC column did not remove all interfering matrix components. This was obvious, especially when the samples were monitored by standard chemical analytical methods (rather than ELISA or electrochemical methods), and there was a need for further optimization of the purification procedure [63]. In some studies, the binding capacity of the entrapped Ab was lower than that obtained in solution (e.g., [66]), but in most cases, it was high enough to retain the analyte on the IAC support for further analysis.

15.2.4.2 Sol–Gel-Based Immunoaffinity Chromatography for Clinical Monitoring

Although immunological assays and IAC have been in regular use for many years in pharmaceutical, biomedical, and clinical research, sol–gel based IAC has not been widely implemented, and the method has been employed in only two cases: for monitoring morphine and its phase II metabolites in blood samples taken from heroin victims and heroin consumers [67] and for monitoring anti-tumor IgGs from patient sera [61]. In the first study, anti-morphine, anti-morphine-3-β-D-glucuronide (M3G), and anti-morphine-6-β-D-glucuronide (M6G) Abs were entrapped in TMOS and were used in combination with laser-induced fluorescence coupled to HPLC to monitor the above compounds in blood samples. The method enabled the detection of low analyte concentrations, and it could be applied without interference to complex matrices such as post-mortem blood samples.

The second study, by Zusman and Zusman [61], introduced a new IAC format, i.e., gel fiberglass membranes. The membranes were comprised of glass fibers covered with oxysilane to provide a sol–gel-glass matrix. A thin-layer of Abs trapped in the gel glass during its preparation was deposited on the surface of a glass lattice to form a sol–gel membrane with the entrapped Abs. The gel-fiberglass (GFG) membranes were used to form GFG columns that were assembled from a series of 20–30 membranes. The columns, containing anti-tumor IgGs, were used to isolate a variety of proteins. The IAC application involved the isolation of tumor-associated Ags from sera of cancer patients. As in all other cases, the sol–gel entrapped Abs were very stable and could be stored for several months at room temperature, and the columns were found to be highly effective in enabling the isolation of large quantities of proteins, mainly as a result of the large active area of the membranes.

15.3 FUTURE PROSPECTS AND CONCLUDING REMARKS

As described above, sol–gel derived biocomposites offer a series of significant advantages over other immobilization matrices or methods for the development of advanced analytical devices. Sol–gel derived materials can be based on a wide range of compositions and can be used to entrap a large number of different biomolecules. No other immobilization method provides such a generic and flexible approach where both the nature of the matrix and its interactions with the entrapped molecule can be so well controlled. The physical and chemical properties of the sol–gel, its amenability to modifications of the properties of the composite, the high biomolecule content that can be loaded onto the gel, and the improved properties of the entrapped biomolecules (e.g., high stability) present a unique combination with an immense application potential. The ability to use sol–gel biocomposites for the applications discussed above has emerged from a huge number of studies in this field, most of which were performed in the past 10–15 years. Overall, these studied have yielded useful insights into the fundamental factors that control the behavior of entrapped biomolecules and have thereby provided guidance in the development of improved materials and processing methods that enable the activity of entrapped Abs to be maintained for the above-mentioned applications.

Although sol–gel derived biocomposites have been shown to be useful in many analytical applications, some issues are still unresolved and need to be further studied. For example, material properties still need to be improved to reduce cracking, age-linked shrinkage, pore collapse, and phase separation. There is still ample space for significant improvement in the physical and chemical parameters of precursors and reaction conditions (e.g., nature of precursors and of additives hydrolysis ratio, nature and presence of solvents, condensation kinetics, etc.), and there is a need to improve organic–inorganic composite materials to gain a better bioactivity of the entrapped molecules. There is a need for better understanding of the protein (or any other entrapped molecule)-silica and analyte-silica interactions (caused by electrostatic, hydrogen, or hydrophobic interactions). Studies clearly indicate that polymer-protein interactions can be used advantageously to maximize the stability and function of some proteins. However, at present, there is very limited information available on the mechanisms by which other proteins may be stabilized. Studies along these lines should be extended, aiming at the improvement of bioactivity and stability of the entrapped molecules. Such interactions are also disadvantageous because they cause analytes to adsorb non-specifically to the matrix. The nature of such interactions should be further studied in order to find simple ways to minimize or overcome the problem. Furthermore, there is still a need to be able to optimize the physical properties of entrapped sol–gels (with respect to size, shape, pore size, etc.) so as to achieve better performance (e.g., faster diffusion of analyte) without losing activity, especially in the case of on-line biosensors. In addition, it is necessary to find ways to scale up the entrapment process and, most important, to find entrapment conditions under which the biomolecules will be compatible with prolonged use.

Some aspects of these needs are currently being addressed. Measures being examined to improve the bioactivity of entrapped molecules include changes in the physical and chemical parameters of precursors and reaction conditions (e.g., the nature of the precursor and of additives, hydrolysis ratio, the presence or absence of various solvents, condensation kinetics, etc.) and the use of different combinations of these factors and an introduction of improved organic–inorganic composites. Development of novel advanced materials is also underway. Adoption of combinatorial approaches in combination with high-throughput material characterization may enable the discovery of optimal biocomposites in a rapid screening process to select the most suitable matrices for a given application. Integration of molecular imprinting methods, controlled pore architectures, micro fabrication, nanotechnology and other novel methods (for review see [4]) may also lead to the development of new sol–gel based tools applicable for IAC, immunosensing, and a wide variety of other applications.

In the current review, we focused on two applications of sol–gel entrapped biomolecules: solid phase IAs and IAC. The entrapment of Abs in a sol–gel matrix in a simple one-step procedure that maintains their binding capacity, enhances their stability, enables dissociation of the analyte from the Ab at high recoveries, and allows no leaching, offers many advantages and opens the way for the development of highly selective biosensors/immunosensors for applications in immunochemical detection methods and in IAC. It is envisioned that the next decade will see much faster progress in sol–gel applications in the above and many other fields of research, and this will lead to the emergence of new and improved devices that will integrate the above-mentioned new approaches and will help to facilitate simple, highly reproducible, environmentally friendly, and cost-effective environmental, forensic, agricultural, and clinical monitoring.

ACKNOWLEDGMENTS

The authors wish to acknowledge The United States Environmental Protection Agency through its Office of Research and Development for funding and managing some of the research described in this review under contract # 68-D-99-011 to Battelle with a subcontract # 175199 Task 2B to M.A. at the Volcani Institute, Israel. We would also like to thank Mr. Oran Dan for his help in the organization of the references.

REFERENCES

1. Jin, W. and Brennan, J. D., Properties and applications of proteins encapsulated within sol–gel derived materials, *Anal. Chim. Acta.*, 461, 1–36, 2002.
2. Gill, I., Bio-doped nanocomposite polymers: Sol–gel bioencapsulates, *Chem. Mater.*, 13, 3404–3421, 2001.
3. Livage, J., Coradin, T., and Roux, C., Encapsulation of biomolecules in silica gels, *J. Phys.: Condens. Matter*, 13, R673–R691, 2001.
4. Gill, I. and Ballesteros, A., Bioencapsulation within synthetic polymers (Part 1): Sol–gel encapsulated biologicals, *Trends Biotechnol.*, 18, 282–296, 2000.
5. Pierre, A. C., The sol–gel encapsulation of enzymes, *Biocatal. Biotransfor.*, 22, 145–170, 2004.
6. Gill, I. and Ballesteros, A., Encapsulation of biologicals within silicate, siloxane, and hybrid sol–gel polymers: an efficient and generic approach, *J. Am. Chem. Soc.*, 120, 8587–8598, 1998.
7. Avnir, D. and Braun, S., *Biochemical Aspects of Sol–Gel Science and Technology*, Kluwer Academic Publishers, Boston, MA, 1996.
8. Chen, X. and Dong, S., Sol-gel-derived titanium oxide/copolymer composite based glucose biosensor. *Biosensor Bioelection.*, 18, 999–1004, 2003.
9. Chen, X. H., Hu, Y. B., and Wilson, G. S., Glucose microbiosensor based on alumina sol–gel matrix/electropolymerized composite membrane, *Biosens. Bioelectron.*, 17, 1005–1013, 2002.
10. Hsu, A. F., Foglia, T. A., and Shen, S., Immobilization of *Pseudomonas cepacia* lipase in a phyllo-silicate sol-gel matrix: effectiveness as a biocatalyst. *Biotech. App. Biochem.*, 31, 179–183, 2000.
11. Liu, B. H., Cao, Y., Chen, D. D., Kong, J. L., and Deng, J. Q., Amperometric biosensor based on a nanoporous ZrO2 matrix, *Anal. Chim. Acta*, 478, 59–66, 2003.
12. Dickey, F. H., Specific adsorption, *J. Phys. Chem.*, 59, 695–707, 1955.
13. Braun, S., Rappoport, S., Zusman, R., Avnir, D., and Ottolenghi, M, Biochemically active sol–gel glasses—The trapping of enzymes, *Mater. Lett.*, 10, 1–5, 1990.
14. Avnir, D., Braun, S., Lev, O., and Ottolenghi, M., Enzymes and other proteins entrapped in sol–gel materials, *Chem. Mater.*, 6, 1605–1614, 1994.
15. Brinker, C. J. and Scherer, G. W., *Sol–Gel Science: The Physics and Chemistry of Sol–Gel Processing*, Academic Press, Boston, MA, 1990.

16. Livage, J., Roux, C., DaCosta, J. M., Desportes, I., and Quinson, J. F., Immunoassays in sol–gel matrices, *C.R. Acad. Sci. Paris II*, 7, 45–51, 1996.
17. Tijssen, P., The immobilization of immunoreactants on solid phases, In *Practice and Theory of Enzyme Immunoassays*, Burdon, R. H. and Knippenberg, P. H., Eds.,, Elsevier, Amsterdam, pp. 297–329, 1985.
18. Hock, B., Enzyme immunoassays for pesticide analysis, *Acta. Hydrochim. Hydrobiol.*, 21, 71–83, 1993.
19. Meulenberg, E. P., Mulder, W. H., and Stoks, P. G., Immunoassays for pesticides, *Environ. Sci. Technol.*, 29, 553–561, 1995.
20. Sherry, J. P., Environmental chemistry—The immunoassay option, *Crit. Rev. Anal. Chem.*, 23, 217–300, 1992.
21. Van Emon, J. M., Immunochemical applications in environmental science, *J. AOAC. Int.*, 84, 125–133, 2001.
22. Van Emon, J. M., Immunoassay methods: EPA evaluations, In *Immunochemical Methods for Environmental Analysis*, 442, Van Emon, J. M. and Mumma, R. O., Eds., American Chemical Society Symposium Series, Washington, DC. pp. 58–64, 1990.
23. Van Emon, J. M., Gerlach, C. L., and Johnson, J. C., In *Environmental Immunochemical Methods: Perspectives and Applications*, Van Emon, J. M., Gerlach, C. L., and Johnson, J. C., Eds., *ACS Symposium Series*, Vol. 646, ACS, Washington DC, 1996.
24. Gabaldon, J. A., Maquieira, A., and Puchades, R., Current trends in immunoassay-based kits for pesticide analysis, *Crit. Rev. Food Sci. Nutrition*, 39, 519–538, 1999.
25. Lan, E. H., Dunn, B., and Zink, J. I., Sol–gel encapsulated anti-trinitrotoluene antibodies in immunoassays for TNT, *Chem. Mater.*, 12, 1874–1878, 2000.
26. Doody, M. A., Baker, G. A., Pandey, S., and Bright, F. V., Affinity and mobility of polyclonal anti-dansyl antibodies sequestered within sol–gel-derived biogels, *Chem. Mater.*, 12, 1142–1147, 2000.
27. Zhou, J. C., Chuang, M. H., Lan, E. H., Dunn, B., Gillman, P. L., and Smith, S. M., Immunoassays for cortisol using antibody-doped sol–gel silica, *Anal. Chem.*, 14, 2311–2316, 2004.
28. Wang, R., Narang, U., Prasad, P. N., and Bright, F. V., Affinity of antifluorescein antibodies encapsulated within a transparent sol–gel glass, *Anal. Chem.*, 65, 2671–2675, 1993.
29. Venton, D. L., Cheesman, K. L., Chatterton, R. T., and Anderson, T. L., Entrapment of a highly specific antiprogesterone antiserum using polysiloxane copolymers, *Biochim. Biophys. Acta.*, 797, 343–347, 1984.
30. Yang, H. H., Zhang, S. Q., Chen, X. L., Zhuang, Z. X., Xu, J. G., and Wang, X. R., Magnetite-containing spherical silica nanoparticles for biocatalysis and bioseparations, *Anal. Chem.*, 76, 1316–1321, 2004.
31. Grant, S. A. and Glass, R. S., Sol–gel-based biosensor for use in stroke treatment, *IEEE Trans. Biomed. Eng.*, 46, 1207–1211, 1999.
32. Liu, G. D., Zhong, T. S., Huang, S. S., Shen, G. L., and Yu, R. Q., Renewable amperometric immunosensor for Complement 3 assay based on the sol–gel technique, *Fresenius J. Anal. Chem.*, 370, 1029–1034, 2001.
33. Yang, H. H., Zhu, Q. Z., Qu, H. Y., Chen, X. L., Ding, M. T., and Xu, J. G., Flow injection fluorescence immunoassay for gentamicin using sol–gel-derived mesoporous biomaterial, *Anal. Biochem.*, 308, 71–76, 2002.
34. Jiang, D., Tang, J., Liu, B., Yang, P., and Kong, J., Ultrathin alumina sol–gel-derived films: allowing direct detection of the liver fibrosis markers by capacitance measurement, *Anal. Chem.*, 75, 4578–4584, 2003.
35. Jordan, J. D., Dunbar, R. A., and Bright, F. V., Aerosol-generated sol–gel-derived thin films as biosensing platforms, *Anal. Chim. Acta*, 332, 83–91, 1996.
36. Pulido-Tofino, P., Barrero-Moreno, J. M., and Perez-Conde, M. C., Sol–gel glass doped with isoproturon antibody as selective support for the development of a flow-through fluoroimmunosensor, *Anal. Chim. Acta*, 429, 337–345, 2001.
37. Coelho, R. D., Yamasaki, H., Perez, E., and de Carvalho, L. B., The use of polysiloxane/polyvinyl alcohol beads as solid phase in IgG anti-Toxocara canis detection using a recombinant antigen, *Mem. Inst. Oswaldo Cruz.*, 98, 391–393, 2003.

38. Roux, C., Livage, J., Farhati, K., and Monjour, L., Antibody-antigen reactions in porous sol–gel matrices, *J. Sol–Gel Sci. Technol.*, 8, 663–666, 1997.

39. Melnyk, O., Duburcq, X., Olivier, C., Urbes, F., Auriault, C., and Gras-Masse, H., Peptide arrays for highly sensitive and specific antibody-binding fluorescence assays, *Bioconjugate Chem.*, 13, 713–720, 2002.

40. Wang, J., Pamidi, P. V. A., and Rogers, K. R., Sol–gel-derived thick-film amperometric immuno-sensors, *Anal. Chem.*, 70, 1171–1175, 1998.

41. Du, D., Yan, F., Liu, S. L., and Ju, H. X., Immunological assay for carbohydrate antigen 19-9 using an electrochemical immunosensor and antigen immobilization in titania sol–gel matrix, *J. Immunol. Methods*, 283, 67–75, 2003.

42. Zhong, T. S. and Liu, G., Silica sol–gel amperometric immunosensor for *Schistosoma japonicum* antibody assay, *Anal. Sci.*, 20, 537–541, 2004.

43. Tang, D. Q., Tang, D. Y., and Tang, D. P., Construction of a novel immunoassay for the relationship between anxiety and the development of a primary immune response to adrenal cortical hormone, *Bioprocess. Biosyst. Eng.*, 27, 135–141, 2004.

44. Tang, D., Yuan, R., Chai, Y., Liu, Y., Dai, J., and Zhong, X., Novel potentiometric immunosensor for determination of diphtheria antigen based on compound nanoparticles and bilayer two-dimensional sol–gel as matrices, *Anal. Bioanal. Chem.*, 381, 674–680, 2005.

45. Holt, D. B., Gauger, P. R., Kusterbeck, A. W., and Ligler, F. S., Fabrication of a capillary immuno-sensor in polymethyl methacrylate, *Biosens. Bioelectron.*, 17, 95–103, 2002.

46. Barros, A. E. L., Almeida, A. M. P., Carvalho, L. B., and Azevedo, W. M., Polysiloxane/PVA-glutaraldehyde hybrid composite as solid phase for immunodetection by ELISA, *Braz. J. Med. Biol. Res.*, 35, 459–463, 2002.

47. Dzgoev, A. B., Gazaryan, I. G., Lagrimini, L. M., Ramanathan, K., and Danielsson, B., High-sensitivity assay for pesticide using a peroxidase as chemiluminescent label, *Anal. Chem.*, 71, 5258–5261, 1999.

48. Weller, M. G., Immunochromatographic techniques—a critical review, *Fresenius J. Anal. Chem.*, 366, 635–645, 2000.

49. Hennion, M. C. and Pichon, V., Immuno-based sample preparation for trace analysis, *J. Chromatogr. A.*, 1000, 29–52, 2003.

50. Delaunay-Bertoncini, N. and Hennion, M. C., Immunoaffinity solid-phase extraction for pharmaceutical and biomedical trace-analysis-coupling with HPLC and CE-perspectives, *J. Pharm. Biomed. Anal.*, 34, 717–736, 2004.

51. Van Emon, J. M. and Lopez Avila, V., Immunoaffinity extraction with on-line liquid chromatography mass spectrometry, In *Environmental Immunochemical Methods, 646*, Van Emon, J. M., Gerlach, C. L., and Johnson, J. C., Eds., ACS, Washington, DC, pp. 74–88, 1996.

52. Aharonson, N., Altstein, M., Avidan, G., Avnir, D., Bronshtein, A., Ottolenghi, M., Rottman, C., and Turniansky, A., Recent developments in organically doped sol–gel sensors: a micron-scale probe; successful trapping of purified polyclonal antibodies; solutions to the dopant-leaching problem, *Better Ceram. Though Chem.*, 246, 519–530, 1994.

53. Zuhlke, J., Knopp, D., and Niessner, R., Sol–gel glass as a new support matrix in immunoaffinity chromatography, *Fresen. J. Anal. Chem.*, 352, 654–659, 1995.

54. Turniansky, A., Avnir, D., Bronshtein, A., Aharonson, N., and Altstein, M., Sol–gel entrapment of monoclonal anti-atrazine antibodies, *J. Sol–Gel Sci. Technol.*, 7, 135–143, 1996.

55. Cichna, M., Knopp, D., and Niessner, R., Immunoaffinity chromatography of polycyclic aromatic hydrocarbons in columns prepared by the sol–gel method, *Anal. Chim. Acta*, 339, 241–250, 1997.

56. Cichna, M., Markl, P., Knopp, D., and Niessner, R., Optimization of the selectivity of pyrene immuno-affinity columns prepared by the sol–gel method, *Chem. Mater.*, 9, 2640–2646, 1997.

57. Bronshtein, A., Aharonson, N., Avnir, D., Turniansky, A., and Altstein, M., Sol–gel matrices doped with atrazine antibodies: atrazine binding properties, *Chem. Mater.*, 9, 2632–2639, 1997.

58. Bronshtein, A., Aharonson, N., Turniansky, A., and Altstein, M., Sol–gel-based immunoaffinity chromatography: application to nitroaromatic compounds, *Chem. Mater.*, 12, 2050–2058, 2000.

59. Altstein, M., Bronshtein, A., Glattstein, B., Zeichner, A., Tamiri, T., and Almog, J., Immunochemical approaches for purification and detection of TNT traces by antibodies entrapped in a sol–gel matrix, *Anal. Chem.*, 73, 2461–2467, 2001.

60. Vazquez-Lira, J. C., Camacho-Frias, E., Pena-Alvarez, A., and Vera-Avila, L. E., Preparation and characterization of a sol–gel immunosorbent doped with 2,4-D antibodies, *Chem. Mater.*, 15, 154–161, 2003.

61. Zusman, R. and Zusman, I., Glass fibers covered with sol–gel glass as a new support for affinity chromatography columns: a review, *J. Biochem. Biophys. Methods*, 49, 175–187, 2001.

62. Scharnweber, T., Knopp, D., and Niessner, R., Application of sol–gel glass immunoadsorbers for the enrichment of polycyclic aromatic hydrocarbons (PAHs) from wet precipitation, *Field Anal. Chem. Technol.*, 4, 43–52, 2000.

63. Spitzer, B., Cichna, M., Markl, P., Sontag, G., Knopp, D., and Niessner, R., Determination of 1-nitropyrene in herbs after selective enrichment by a sol–gel-generated immunoaffinity column, *J. Chromat. A.*, 880, 113–120, 2000.

64. Cichna, M., Markl, P., Knopp, D., and Niessner, R., On-line coupling of sol–gel-generated immunoaffinity columns with high-performance liquid chromatography, *J. Chromat. A.*, 919, 51–58, 2001.

65. Schedl, M., Wilharm, G., Achatz, S., Kettrup, A., Niessner, R., and Knopp, D., Monitoring polycyclic aromatic hydrocarbon metabolites in human urine: Extraction and purification with a sol–gel glass immunosorbent, *Anal. Chem.*, 73, 5669–5676, 2001.

66. Stalikas, C., Knopp, D., and Niessner, R., Sol–gel glass immunosorhent-based determination of *s*-triazines in water and soil samples using gas chromatography with a nitrogen phosphorus detection system, *Environ. Sci. Technol.*, 36, 3372–3377, 2002.

67. Hupka, Y., Beike, J., Roegener, J., Brinkmann, B., Blaschke, G., and Kohler, H., HPLC with laser-induced native fluorescence detection for morphine and morphine glucuronides from blood after immunoaffinity extraction, *Int. J. Legal Med.*, 119, 121–128, 2005.

68. Braunrath, R. and Cichna, M., Sample preparation including sol–gel immunoaffinity chromatography for determination of bisphenol A in canned beverages, fruits and vegetables, *J. Chromatogr. A.*, 1062, 189–198, 2005.

69. Kaware, M., Bronshtein, A., Safi, J., Van Emon, J. M., Chuang., J. C., Hock, B., Kramer, K. and Alstein, M. Enzyme-linked immunosorbent assay (ELISA) and sol–gel-based immunoaffinity purification (IAP) of the pyrethroid bioallethrin in food and environmental samples. *J. Agric. Food Chem.* 54, 6482–6492, 2006.

70. Zhu, Q. Z., Degelmann, P., Niessner, R., and Knopp, D., Selective trace analysis of sulfonylurea herbicides in water and soil samples based on solid-phase extraction using a molecularly imprinted polymer, *Environ. Sci. Technol.*, 36, 5411–5420, 2002.

16 Electrochemical Immunoassays and Immunosensors

Niina J. Ronkainen-Matsuno, H. Brian Halsall,
and William R. Heineman

CONTENTS

16.1 INTRODUCTION

Modern society releases numerous man-made chemicals into the environment every day. The emissions, from agriculture, industry, and municipalities, end up in the air, soil, surface, and ground water. Streams and rivers transport herbicides, insecticides, fungicides, halogenated aromatic hydrocarbons, and other trace contaminants into the oceans and large lakes where they cause widespread environmental damage and possible public health problems. Humans are usually exposed to pollutants by consuming contaminated water [1], but they can ingest them at any point in the food chain. Regulatory agencies have created guidelines to minimize public exposure to harmful pollutants and to protect public health [2,3]. The European Union Drinking Water Directive regulates the maximum admissible concentration of pesticides to

0.1 µg/L (0.1 ppb) for an individual pesticide and 0.5 µg/L (0.5 ppb) for total pesticides [4]. The maximum allowed concentration of any member for the s-triazine class of pesticides in drinking water is 3 µg/L in the United States [5]. A result of these low limits is that sensitive analytical methods are required to routinely monitor pollutant levels in the environment. Immunoassays (IAs) are one of the more promising techniques for sensitively and inexpensively monitoring pollutants in the environment. Some environmental IA kits are commercially available products such as Millipore and Ohmicron [6,7].

The electrochemical immunoassays (ECIAs) that depend upon voltammetric or amperometric detection have an important place among modern analytical methods in such areas as clinical analysis, process analysis, agriculture, and food control [8]. However, the full potential and wide use of environmental ECIAs has not yet been realized. Most applications of ECIA have come from research laboratories [6] because the difficulty and high initial cost of developing a new, widely accepted IA has impeded the use of IAs in environmental laboratories. However, once developed, many IA formats permit the simultaneous analysis of multiple samples that improve efficiency and make the assays relatively fast and cost-effective. In the past, the limited availability of antibodies (Abs), mainly produced by university and corporate laboratories, to common pollutants has slowed the development of IAs for environmental applications. The production of Abs against low molecular weight analytes of environmental interest with molecular weights below 1000 g/mol is more challenging and often requires coupling the analyte to a carrier protein via a spacer molecule before an immune response is provoked in the host animal [9]. More recently, several manufacturers of agrochemicals have started producing Abs to their products in case of chemical pollution [6,7].

The advantages of IAs such as high specificity, small required sample volumes, high sample throughput with simultaneous analysis of multiple samples, reduced sample preparation, reduction in the use of chemicals and production of waste, and ease of automation, far outweigh their limitations, making the IAs an attractive alternative to the more conventional analytical methods like gas chromatography (GC) and liquid chromatography (LC) [8]. IAs and biosensors are relatively easy to use once fully developed and optimized, and the overall cost of analysis is reduced by a decrease in waste disposal, chemical use, expensive instruments, and maintenance. Environmental IAs are well suited to mapping out contaminated sites and for on-site monitoring [9], and they can be used as a rapid, qualitative screening technique to complement other analytical methods in environmental laboratories [6]. IAs can be used to rapidly determine the absence of a certain analyte on-site, helping to reduce the number of environmental samples that require more detailed analysis in a laboratory. For example, enzyme-linked immunosorbent assay (ELISA) IA kits have been used to complement the analysis by high performance liquid chromatography (HPLC) of imazapyr and triclopyr herbicides in water samples from forest watersheds [10]. Water, perhaps the most common environmental sample matrix, is well suited to analysis by IA [11,12]. On-site sample screening decreases the time between sampling and measurement results for harmful analytes and, therefore, enables faster decision making about the future actions to be taken.

The four key factors involved in the design of a sensitive IA are the format of the assay, the type of label used, the method of detection, and the minimization of nonspecific binding (NSB) [13]. The key limitation that must be overcome is the NSB of the label to surfaces, which significantly erodes the sensitivity of the IA. NSB blockers such as the nonionic surfactant Tween 20 and proteins such as bovine serum albumin (BSA), gelatin, or casein are commonly used to minimize NSB, but tailored surfaces, chemically designed and constructed to minimize NSB, are being explored.

This chapter focuses on the applications of ECIA in environmental monitoring. In addition to traditional IAs, immunosensors with Ab or antigen (Ag) directly immobilized at the electrode surface seem promising for on-site environmental analysis. A competitive IA for the widely used pesticide 2,4-dichlorophenoxyacetic acid (2,4-D) in a bienzyme substrate-recycling biosensor will be described [14]. A sensitive electrochemical IA for the determination of a common herbicide, atrazine, will be used as an example of the application of the capillary enzyme IA with electrochemical detection to water monitoring [12]. A separation-free bienzyme IA with amperometric detection has been developed to measure the herbicide chlorsulfuron [15]. The successful application of electrochemical immunosensors to the detection of polycyclic aromatic hydrocarbons (PAHs) will also be described [9,16,17,18].

16.2 ENZYME IMMUNOASSAYS

Enzyme immunoassays (EIAs) were first introduced by Engvall and Perlmann [19] and van Weeman and Schuurs [20] as an alternative to radioimmunoassays (RIAs) in 1971. EIAs measure the activity of the enzyme label in generating an electroactive product rather than the radioactive label used in RIA. EIAs are safer than RIAs, though their sensitivity and cost are often comparable [21,22]. Enzymes can be highly selective for their given substrate and provide a large signal amplification as a result of a high turn-over rate that allows the detection of very low concentrations of the analyte. However, the activity of the enzyme labels is affected by reaction conditions that have to be controlled during the detection step [23].

The most commonly used enzyme labels in EIAs are alkaline phosphatase (ALP), β-galactosidase (β-Gal), horseradish peroxidase (HRP), and glucose oxidase (GOx) [8,23–26]. GOx has a lower activity than the other enzyme labels and is typically used in amperometric IAs where the product is directly detected [27,28].

There are two main EIA formats: homogeneous and heterogeneous [8,29,30,31]. Homogeneous assays that have no separation step benefit from speed and procedural simplicity, but have poorer limits of detection. They are also more susceptible to interference by other species in the sample than are RIAs or EIAs of other formats [22]. Heterogeneous assays include a physical separation step to isolate the Ab–Ag complex from the unbound constituents, followed by a wash step to remove any unbound materials. The separation step in a heterogeneous assay complicates the procedure, but it enables a substantially better limit of detection. Homogeneous and heterogeneous EIAs can be done competitively or noncompetitively. Competitive heterogeneous and sandwich EIAs are often referred to as ELISA because the Ab or the Ag is immobilized on a solid surface.

16.3 ELECTROCHEMICAL DETECTION

Electrochemistry is a surface technique that offers certain advantages for detection in IAs. Therefore, it is not strongly dependent on the reaction volume, and very small volumes can be used for detection. Electrochemical detection can be used to achieve low detection limits with little or no sample preparation, and atto- and zeptomole-detecting ECIAs have been constructed [14,32]. In homogeneous assays that have no separation step to isolate the Ab–Ag complex from the unbound assay constituents, electrochemical detection is not affected by sample components such as chromophores, fluorophores, and particles that often interfere with spectroscopic detection. Therefore, electrochemical measurements can be made on colored or turbid samples such as whole blood without interference from fat globules, red blood cells, hemoglobin, and bilirubin [33,34]. The nonspecific oxidation or reduction of other compounds that may be in the sample can usually be prevented by carefully selecting the potential for the detection step.

A typical electrochemical detection cell consists of a working electrode made of solid, conductive materials such as platinum, gold, or carbon; a reference electrode made of silver, coated with a layer of silver chloride (Ag/AgCl); and a platinum wire auxiliary electrode. These electrodes can easily be miniaturized, so diameters on the order of micrometers are common, and nanometer sizes have been demonstrated [35,36,37]. Very small sample volumes (on the order of microliters) are required for detection with such small electrodes because of their small surface areas, which is a significant advantage when the sample sizes are limited [25,38]. In contrast, most conventional methods used to monitor chemical pollutants in the environment such as HPLC and GC require several milliliters of the sample. Furthermore, electrochemical detectors and their required control instrumentation can be easily miniaturized at a relatively low cost by micromachining, making possible the manufacture of field-portable instruments for ECIA. Because the limiting current in voltammetry is temperature-dependent, the detection cell should be maintained at a constant temperature for analyzing calibrants and samples in order to obtain accurate and precise results [39].

Screen-printed electrodes (SPEs) have gained popularity as the working electrode in electrochemical biosensors and IAs because of their low cost and the ease and speed of mass production using thick film technology. These advantages allow disposable SPEs to be used in immunochemical sensors [15,40,41,42].

Regenerating the sensing surface for a subsequent measurement in a biosensor is typically very time-consuming and not reproducible [43]. This is a problem particularly in immunosensors where dissociating an Ab–Ag complex is difficult because of the high affinity constant of the immunocomplex. The regeneration conditions can also damage and release the immunoreagent bound to the surface of the transducer [43].

Electrochemical immunosensors can be grouped into those with amperometric, potentiometric, and conductiometric transducers. In amperometry, a fixed potential is applied to the sample and the current, due to the resulting redox processes, is measured with time [44]. Voltammetry is like amperometry except that the potential is scanned over a range as the current is measured. In potentiometry, the potential is measured at zero current. Amperometric detection often yields the lowest detection limit [31].

Various types of amperometric detection have been used with IAs [8,33]. Advantageously, the fixed potential during amperometric detection results in a negligible charging current, minimizing the background signal, and, in turn, affecting the limit of detection. In addition, hydrodynamic amperometric techniques can provide significantly enhanced mass transport to the electrode surface [44,45], for example, when the electrode moves with respect to the solution by rotating or vibrating [46,47] or in flow conditions where the sample solution passes the stationary electrodes [12,45,48]. Electrochemical detection in flow conditions can be applied to environmental monitoring and industrial processes more easily than steady-state batch systems because the flow conditions allow easier solution changes in multi-step assay procedures, and they are ideal for on-line monitoring.

16.4   ENZYME LABELS AND SUBSTRATES FOR ECIAs

The enzyme label chosen for the ECIA should have a high catalytic activity for the corresponding substrate. The redox active product that is formed by the enzyme catalysis should have a low redox potential to minimize interference from other components in the sample matrix, and the substrate should be electroinactive at the measuring potential to keep the background signal low [49]. It is not necessary to remove oxygen from the sample if the observed reaction is an oxidation occurring between $+200$ and $+900$ mV [31]. The lower end of the range is more desirable because the more positive values may result in solvent breakdown [31]. Several enzyme/substrate/product combinations satisfy the above requirements and are used in ECIAs and immunosensors (Table 16.1).

16.4.1   ALKALINE PHOSPHATASE (ALP)

ALP is a commonly used enzyme label that hydrolyzes orthophosphate from a wide variety of phosphate esters under alkaline conditions [50] with an optimum activity around pH 10. ALP has a high turnover number.

TABLE 16.1
Some Commonly Used Enzyme Labels and Substrates for Electrochemical Immunoassays and Immunosensors

Enzyme	Substrate	Electroactive Product	Oxidation Potential[a]	Reference
Alkaline phosphatase (ALP)	Phenyl phosphate	Phenol	+870 mV with C paste	[32, 33, 44]
Alkaline phosphatase (ALP)	p-aminophenyl phosphate (PAPP)	p-aminophenol (PAP)	+290 mV with Au	[8, 33, 52]
Alkaline phosphatase (ALP)	α-naphthyl phosphate	α-naphthol	+400 mV with C based	[53–55]
Alkaline phosphatase (ALP)	Hydroquinone diphosphate (HQDP)	Hydroquinone (HQ)	+63 mV with C based +135 mV with Pt +21 mV with Au	[56]
β-galactosidase (β-Gal)	p-aminophenyl-β-D-galactopyranoside (PAPG)	p-aminophenol (PAP)	+290 mV with Au RDE	[49, 57]
Glucose oxidase (GOx)	β-D-Glucose	H_2O_2	+50 mV with C ink SPE	[15]

RDE, rotating disk electrode; SPE, screen-printed electrode.

[a] All oxidation potentials are vs. Ag/AgCl reference electrode.

16.4.1.1 Phenyl Phosphate

The substrate-product system used in early ECIA development was phenyl phosphate/phenol [32,33,44]. Phenol is detected by oxidation at about +870 mV vs. Ag/AgCl, a potential where the substrate is not oxidized. Although this system has been used successfully, detecting phenol has three disadvantages. First, at concentrations of phenol above 40 μM, the oxidation product polymerizes at the electrode surface, fouling the electrode, so assay conditions must be carefully controlled to limit phenol production [8,33]. Second, the high detection potential required for phenol may compromise the detection limits for certain detection methods as a result of the high background current. The noise in electrochemical detection typically increases with applied potential. Finally, the chemical irreversibility prevents phenol detection with interdigitated array electrodes for redox cycling. To circumvent these limitations, researchers have developed p-aminophenyl phosphate (PAPP)/p-aminophenol (PAP) as a substrate-product pair [25,46,51].

16.4.1.2 p-Aminophenyl Phosphate

ALP hydrolyses PAPP to an electrochemically active product, PAP, that can be oxidized at a low potential of +290 mV vs. Ag/AgCl to p-quinone imine (PQI) [8,33] (Reaction 16.1). The conversion of PAP to PQI is electrochemically reversible. However, PAPP is less stable than other substrates such as p-aminophenyl-β-D-galactopyranoside, and PAP is susceptible to air oxidation and is unstable at pH 10 in all commonly used buffer systems, reducing the sensitivity of the assay under alkaline buffer conditions [52]. In addition, using ALP with PAPP inconveniently requires two buffers for most IA types: a pH 7.4 assay buffer and pH 9.0 detection buffer for high ALP activity. Finally, PAPP is only available from a few commercial sources and with variable quality.

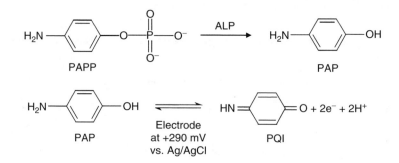

REACTION 16.1 Alkaline phosphatase (ALP) catalyzes the conversion of *p*-aminophenyl phosphate (PAPP) to *p*-aminophenol (PAP) and phosphate (not shown). The catalytic conversion is followed by the oxidation of PAP to *p*-quinone imine (PQI) at +290 mV at the working electrode.

16.4.1.3 α-Naphthyl Phosphate

ALP has been used as an enzyme label with α-naphthyl phosphate as the substrate. ALP catalyzes the dephosphorylation of α-naphthyl phosphate producing α-naphthol that can be oxidized at carbon-based electrodes (graphite-based SPEs and glassy carbon) at +400 mV vs. Ag/AgCl reference electrode [53,54,55]. The enzyme-substrate pair has been used in ECIAs combined with flow injection analysis [53], chronoamperometry [54], and SPEs [55] (Reaction 16.2).

16.4.1.4 [[(4-Hydroxyphenyl)Amino]Carbonyl]Cobaltocenium Hexafluorophosphate

The phosphoric acid ester of [[(4-hydroxyphenyl)amino]carbonyl]cobaltocenium hexafluorophosphate has also been used as the substrate for ALP in ECIAs [43]. The anionic substrate (S^-) is converted enzymatically into a cationic phenol derivative (P^+) by hydrolysis at pH 9.0 near the electrode surface [43]. The positively charged P^+ can be easily accumulated in a polyanionic Nafion coating on a SPE and determined by cyclic voltammetry (Reaction 16.3).

16.4.1.5 Hydroquinone Diphosphate (HQDP)

HQDP is another recently published alternative substrate for ALP that does not produce electrode fouling in ECIAs, even with repeated biosensor use [56]. HQDP also shows high hydrolytic stability and produces significantly larger amperometric responses than α-naphthyl phosphate

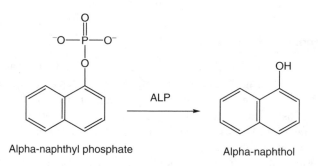

REACTION 16.2 Alkaline phosphatase (ALP)-catalyzed cleavage of α-naphthyl phosphate to form α-naphthol and phosphate (not shown).

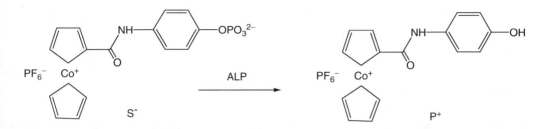

REACTION 16.3 Alkaline phosphatase (ALP)-hydrolyzed cleavage of a phosphate from anionic substrate (S^-) to form a cationic phenol derivative (P^+) and phosphate (not shown).

and phenyl phosphate with ECIA [56]. ALP hydrolyzes HQDP to hydroquinone (HQ) and two phosphates. HQ undergoes reversible oxidation that involves a two-electron transfer and deprotonation to produce benzoquinone (BQ) (Reaction 16.4).

The optimal potential for the reversible oxidation of HQ is $+63$ mV when using carbon, $+135$ mV with Pt, and $+21$ mV with a gold working electrode [56]. The oxidation potential of HQ is so low that interference from the oxidation of biological non-target analytes such as ascorbate is very unlikely. Cyclic voltammetry (CV) has shown that HQ does not passivate electrodes in the neutral or alkaline solutions that are optimal for ALP. Furthermore, the amperometric signal due to HQ oxidation can be amplified by recycling the quinone with GOx. The only major limitation of using HQDP as a substrate for ALP is that HQDP cannot be purchased from commercial sources. However, HQDP can be prepared from HQ in relatively few steps and in high yield.

16.4.2 β-Galactosidase (β-Gal)

PAPG, *p*-aminophenyl-β-D-galactopyranoside, has been used as a substrate in ECIA [49]. The enzyme, β-Gal, cleaves PAPG, yielding galactose and PAP, the same electrochemically active product as with ALP and PAPP [57] (Reaction 16.5).

β-Gal works optimally at pH 7.0–7.25 [49] and is, therefore, more suitable for homogeneous assays than ALP. Overall, PAPG is more stable than PAPP. For example, the amperometric response of 5 mM PAPG did not change in four hours at pH 7 [49]. The product, PAP, is also more stable at near neutral pH, and a less than 12% decrease in amperometric response was seen after four hours when PAP was measured at pH 7 [49]. Finally, PAPG is commercially available from several suppliers in the United States and is less expensive than PAPP.

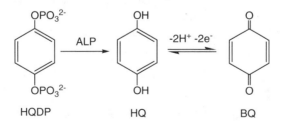

REACTION 16.4 Alkaline phosphatase (ALP)-hydrolyzed cleavage of two phosphates from hydroquinone diphosphate (HQDP) to produce hydroquinone (HQ) and phosphate (not shown). HQ is oxidized to benzoquinone (BQ).

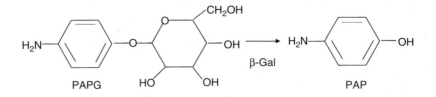

REACTION 16.5 β-galactosidase (β-gal)-catalyzed cleavage of *p*-aminophenyl-β-D-galactopyranoside (PAPG) to form *p*-aminophenol (PAP) and galactose (not shown).

16.5 OVERCOMING NONSPECIFIC BINDING (NSB)

NSB involves the adsorption of conjugated enzyme or other labels used for IA to materials other than the analyte. This phenomenon that increases the background signal is the major determinant of the detection limit of the ECIA. However, NSB can be reduced with blockers such as a nonionic surfactant, Tween 20, or protein blockers such as BSA. Other commonly used NSB blockers include gelatin [58], casein [59], and proprietary products from Pierce, Molecular Probes, and others. Much effort can be expended in minimizing NSB in IAs.

NSB has to be considered for the minimum detection limit of a non-competitive assay because the enzyme-labeled reagent antibody (Ab*) is usually present in excess to drive the Ab*–Ag binding reaction to completion. As the concentration of Ab* increases, the amount that is non-specifically bound also increases. In the absence of analyte, this results in a signal that is determined from a blank or zero dose analysis [8,13]. The blank signal generally determines the concentration detection limit of an ECIA, and it is important to take steps to minimize it.

With plastic surfaces such as polystyrene, there is a charge-interaction component, but the adsorption process is usually dominated by hydrophobic interactions [32]. The interactions are entropically driven and are commonly physically blocked by surface treatments such as a mixture of BSA and a detergent such as Tween 20 [60]. Sulfonate ion-pairing reagents can reduce NSB on positively charged surfaces [13,32] and 0.05% Tween-20 and 1.0% BSA have been added to Tris buffer to curb NSB with bead-based IAs [46,47]. The additives resulted in a 13-fold reduction in the detection limit in ECIA compared to unblocked assays [60].

NSB blocking agents should be avoided in solutions that are in contact with the electrodes because the blockers may adsorb onto the electrode surface, fouling it [61,62].

Polyethylene glycol (PEG) immobilized on glass surfaces and present in reaction buffers has previously been shown to significantly reduce NSB [63]. Competition between the solution and solid phases was used to reduce NSB of the enzyme label to surfaces. The immobilized PEG was derivatized with adipic acid dihydrazide that then reacted with oxidized immunoglobulin G, creating the immobilized primary Ab layer. The NSB decreased as the concentration of PEG increased in the reaction buffer to 4% v/v, beyond which the NSB began to increase. More recently, self-assembling monolayers of oligo(ethyleneglycol) [64–66] and dextran layers [67] have been used to great effect in the prevention of NSB on sensor surfaces.

16.6 IMMUNOASSAYS AND BIOSENSORS FOR ENVIRONMENTAL MONITORING

Representative examples of environmental applications are listed in Table 16.2 and discussed in the following sections.

TABLE 16.2
Selected Environmental Applications of Electrochemical Immunoassays and Immunosensors

Analyte	Label/Substrate	Detection Scheme	Limit of Detection and/or Range	Reference
Aroclor 1260 (PCB)	ALP/α-naphthyl phosphate	SPE	0.01–10 µg/mL	[55]
Atrazine	ALP/PAPP	Capillary-FIA–EC	0.10–10.0 µg/L	[12]
Benzo[a]pyrene (BaP)	Label free	Capacitance immunosensor	0.1–5 µM	[17]
Chlorsulfuron	HRP, GOx/glucose	SPE	0.01–1 ng/mL	[15]
2,4 Dichlorophenoxy-acetic acid (2,4-D)	ALP/ PP	Bi-enzymatic recycling biosensor	0.001–1000 µg/L	[14]
2,4 Dichlorophenoxy-acetic acid (2,4-D)	ALP	Nafion film coated SPE	0.01 µg/L, 1–100 µg	[43]
Phenanthrene	ALP/PAPP	SPE	0.8 ng/mL, 5–45 ng/mL	[16]
Polychlorinated biphenyls (PCB)	HRP/ferrocene acetic acid	FIA–EC	0.1 µg/mL, 0.1–50 µg/mL	[81]

ALP, alkaline phosphatase; FIA–EC, flow-injection analysis with electrochemical detection; GOx, glucose oxidase; HRP, horseradish peroxidase; PAPP, p-aminophenyl phosphate; PP, phenyl phosphate; SPE, screen printed electrode.

16.6.1 2,4-DICHLOROPHENOXY ACETIC ACID

A commonly used chlorinated phenoxyacetic acid herbicide, 2,4-dichlorophenoxyacetic acid (2,4-D), is typically detected in the environment by either HPLC or GC. The analyte has to be derivatized prior to the GC detection step. This procedure requires large sample volumes and has long analysis times per sample. The structure of 2,4-D is shown below.

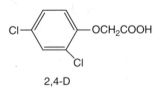

2,4-D

A bienzyme substrate-recycling biosensor coupled to flow injection analysis (FIA) was developed to detect 2,4-dichlorophenoxyacetic acid based on ECIA [14]. The low-cost IA detected 0.1–10 µg/L of 2,4-D. The amperometric biosensor consisted of Clark-type electrodes covered with a poly(vinyl alcohol) membrane that co-entrapped tyrosinase and the quinoprotein glucose dehydrogenase. Tyrosinase and glucose dehydrogenase convert catechol to o-quinone, and o-quinone to catechol, respectively (Figure 16.1). ALP dephosphorylates phenyl phosphate to phenol outside the flow cell, and the phenol is then oxidized in the sensor membrane inside the flow system via catechol to o-quinone by tyrosinase. Oxygen consumption associated with the tyrosinase reaction was monitored electrochemically, and the signal was amplified up to 350-fold as a result of the continuous substrate cycling.

A competitive EIA format was chosen for this assay because the 2,4-D analyte is too small to accommodate two antibodies simultaneously as required by the more sensitive sandwich IA. Small analytes (less than 1000 Daltons or so) such as 2,4-D typically only have one binding site (antigenic determinant) for the Ab. An IA using competitive binding is characterized by noncovalent, reversible binding of the analyte (2,4-D) to an immobilized antibody (mouse monoclonal Ab against 2,4-D) during incubation in a microtiter plate.

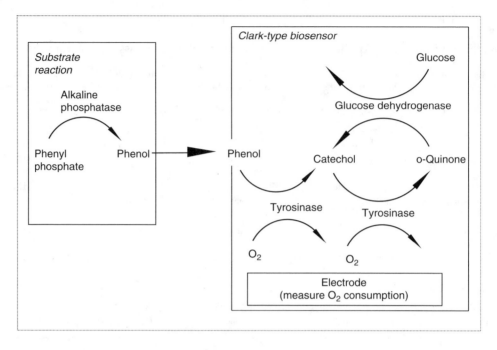

FIGURE 16.1 A bienzyme substrate-recycling biosensor coupled to flow injection analysis for 2,4-dichloro-phenoxyacetic acid (2,4-D).

The analyte, 2,4-D, competes with the labeled analyte 2,4-D-ALP conjugate for a limited number of Ab binding sites on the surface of wells in a microtiter plate.

The bienzyme substrate-recycling biosensor was operated in a flow injection analysis (FIA) system, providing hydrodynamic conditions for detection. FIA involves injecting sample into a mobile phase that carries the sample through the biosensor (convection is the main method of mass transport) to the working electrode. FIA has advantages that include short detection times, simple and inexpensive instrumentation, and the ability to analyze microliter sample volumes. However, data acquisition is not continuous, but occurs at discrete time periods during the enzyme substrate incubation step.

The biosensor had two enzyme membrane-covered Clark-type electrodes where the tyrosinase and glucose dehydrogenase-containing enzyme membranes were sandwiched between a dialysis membrane and a polypropylene membrane (Figure 16.1). Both modified electrodes were in contact with a flow channel between them. The signal was measured at -600 mV vs. Ag/AgCl by two potentiostats. As the amount of analyte increased, more oxygen was consumed by the tyrosinase-catalyzed reaction in the biosensor membrane, and the electrode response decreased.

Dequaire et al. have developed an immunomagnetic electrochemical sensor for 2,4-D based on a Nafion film-coated SPE in a microwell that uses a competitive IA format [43]. A limited number of anti-2,4-D Ab-coated magnetic beads were used as the capture surface for which 2,4-D and 2,4-D labeled with ALP competed. The bead-bound ALP activity was electrochemically determined by the amount of cationic product that accumulated in the polyanionic Nafion film during 30 min prior to performing CV between 0 and 0.65 V vs. Ag/AgCl. The biosensor had a detection limit of 0.01 µg/L of 2,4-D. The sensor also worked with untreated river water samples from three locations in central France spiked with 1 or 100 µg of 2,4-D. No significant matrix effects were observed in the river water.

16.6.2 ATRAZINE

Atrazine (2-chloro-4-(ethylamino)-6-(isopropylamino)-s-triazine) is a popular selective herbicide in the triazine family, and is used to control broadleaf and grassy weeds in various crops, including corn, sugarcane, pineapple, sorghum, and Christmas trees. It is one of the most commonly used pesticides in the world and is often considered as an indicator of the level of pesticide pollution in water systems. The molecular structure of atrazine is shown below:

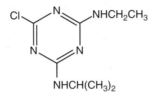

The standard analytical techniques used by environmental laboratories for detecting atrazine in soil and water samples are GC using nitrogen–phosphorus detectors [68] or GC-MS [69], HPLC with UV-vis detection [70] or HPLC-MS [71], and capillary zone electrophoresis [72]. These techniques usually require extensive pretreatment of the samples to achieve the required sensitivities, and they must be performed by skilled technicians. They can also require long analysis times and solvents that require disposal, increasing the operating costs. Therefore, ECIA and electrochemical immunosensors can be portable, sensitive, and inexpensive alternatives to standard laboratory analytical methods for detecting atrazine.

Some of the Abs against atrazine were found to cross-react with other triazine herbicides [73]. This is an advantage when samples are screened for the triazines as a class of herbicide. However, cross-reactivity is a serious limitation in an IA where specificity for one analyte is required. Goodrow et al. have produced more selective Abs against atrazine and other triazines, making possible the specific identification and quantitation of these potential pollutants [74].

16.6.3 CHLORSULFURON

Chlorsulfuron (1-(2-chlorophenylsulfonyl)-3-(4-methoxy-6-methyl-1,3,5-triazin-2-yl)urea) is a compound from the sulfonylurea class of herbicides that are widely used in agriculture. The molecular structure of chlorsulfuron is shown below.

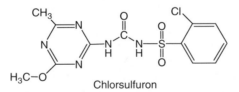

Chlorsulfuron

Clorsulfuron is usually detected off-site using GC or HPLC with appropriately prepared samples. Combining IAs with electrochemical detection can provide a simple and relatively rapid method for on-site chlorsulfuron determination.

Screen printing technology is versatile, and it can be used to produce electrodes containing various types of modifiers, catalysts, ligands, and biological molecules that are included in the base ink being deposited in a film of controlled pattern and thickness on an inert substrate. The electrochemical properties of the SPE can be adjusted by altering the ink formulation. SPEs have become popular in IAs for herbicides because the label or receptor molecules can be coated on the working electrode surface, thereby enhancing the catalytic signal produced by the reaction products [15].

Dzantiev et al. used SPEs containing HRP in the carbon ink of the working electrode as the detector in a competitive IA with a GOx label to determine chlorsulfuron [15]. A nylon membrane with immobilized anti-chlorsulfuron Abs was clamped to the electrode using a spring-loaded clip. Free chlorsulfuron in the sample and chlorsulfuron-GOx conjugates competed for available binding sites on the membrane. Hydrogen peroxide was produced by the GOx-conjugate when glucose was added. The hydrogen peroxide was reduced by the peroxidase, causing an electrical current change at the electrode that was measured to determine chlorsulfuron in the sample. A potential of $+50$ mV vs. Ag/AgCl was applied to the SPE during detection. Catalase was added to the buffered glucose solution to scavenge H_2O_2 formed by any unbound chlorsulfuron-GOx conjugate, thereby reducing the background and eliminating the need for a washing step. The nylon membranes where the assay steps were performed were discarded after each measurement while using the same SPE for all measurements. Reusing the electrode resulted in a high reproducibility for the measurements. The clever assay design helped to overcome the inter-electrode variation in geometry and thickness of the HRP-containing graphite layer that results in poor inter-electrode precision, a common problem with SPEs.

The total assay time was only 15 minutes (using previously prepared membranes) with the electrochemical measurements requiring less than 2 minutes [15]. The authors prepared a multi-well holder for ten electrodes that decreased the total assay time because complete testing of ten chlorsulfuron samples could be completed in under 35 minutes. The speed and relatively low cost make the assay an attractive alternative to traditional methods. Chlorsulfuron was detected at levels between 0.01 and 1 ng/mL.

16.6.4 POLYCYCLIC AROMATIC HYDROCARBONS (PAHs)

PAHs are a class of compounds consisting of two or more fused aromatic rings. PAHs are always found as a mixture of individual compounds that include over 100 chemicals. The majority of PAHs enter the environment through the incomplete combustion of fossil fuels such as crude oil or coal, and they can be found in many matrices including the air, water, soil, and food. The standard laboratory methods for the detection of PAHs include liquid/liquid extraction, solid phase extraction, and supercritical fluid extraction in combination with HPLC, LC, or GC separation. The common detection methods in PAH determination include ultraviolet absorbance or fluorescence for HPLC and mass spectrometry with GC [9]. All of these extraction, separation, and detection methods are labor-intensive, time-consuming, and costly.

Benzo[a]pyrene (BaP) is a PAH often found in cigarette smoke, heavily polluted air, soil, charbroiled meat, and water. It is highly toxic and carcinogenic; therefore, it is important in environmental monitoring [9,16,75] and clinical analysis [76]. The molecular structure of BaP is shown below.

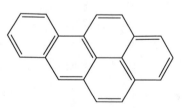

BaP is only one of the over 100 PAHs that are formed by burning fossil fuels and organic substances in garbage, food, or cigarettes. The U.S. Environmental Protection Agency (EPA) has identified 17 harmful PAHs that are of interest in water, and it has established limits of detection for them. PAH levels in drinking water range from 1 ng/L to 11 μg/L [9]. The standard allowed value of PAHs in drinking water released by the EU is 0.2 μg/L.

Many PAHs are very similar in their molecular structure, molecular weight, lack of side chain groups, and electron density, making the production of Abs that are specific for only one PAH compound impossible [16]. All Abs raised against PAHs show cross-reactivity toward other PAHs, and they are usually characterized by cross reactivity patterns. Therefore, it is common to develop

IAs for the determination of the overall PAH concentration or use one compound, often BaP, as a representative marker for the presence of PAHs.

Liu et al. have developed a capacitance immunosensor for BaP that is easy to prepare and has a rapid detection response [17]. These capacitance immunosensors use Abs against the analyte as the bio-recognition element to directly monitor the Ab–Ag binding if the analyte is present in the sample. Metals [77,78] and semiconductors [79] have been used for capacitive immunosensors. The main advantage of using capacitance measurements to monitor the Ab–Ag interaction is the opportunity for direct, label-free electrochemical monitoring. The capacitance of electrolytic capacitors depends on the thickness of a dielectric layer on the metal surface of the working electrode and the dielectric properties of the Ag. The authors modified a gold electrode surface with a self-assembled monolayer (SAM) of cystamine to allow a monoclonal Ab to be easily immobilized on the electrode surface [17]. They also used a second format where the Ag was immobilized on the modified electrode surface. The capacitance change that occurred during protein binding or release was monitored using linear sweep voltammetry. The charging current decreased with increasing layers on the electrode surface.

Liu et al. have also developed a label-free electrochemical immunosensor for PAHs that is based on the CV of ferricyanide, $Fe(CN)_6^{3-/4-}$ [18]. The authors modified a gold electrode surface with an organic SAM of a thioctic acid-pyrene conjugate to which they bound monoclonal Abs (mAb10c10) in concentrations between 2 and 96 ppm. The hydrophilicity of the electrode surface decreased upon modification of the sensor surface with the PAH antigen, resulting in a measurable decrease in the Faradaic current of the $Fe(CN)_6^{3-/4-}$ redox couple caused by decreased redox kinetics. The preliminary results for a nonoptimized sensor gave a detection range from 0.1 to 5 µM for BaP using a competitive inhibition of the Ab-hapten association.

Disposable biosensors have many advantages such as avoiding the effects of contamination by the sample components and eliminating the loss of signal as a result of fouling of the electrode with repeated use. Fähnrich et al. have developed a disposable amperometric immunosensor for phenanthrene on SPEs [16]. The molecular structure of phenanthrene is shown below.

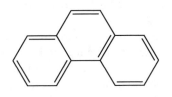

The working surfaces of disposable carbon SPEs were coated with the antigen as a conjugate of phenanthrene-9-carboxaldehyde coupled to BSA via adipic acid dihydrazide [16]. Monoclonal mouse anti-phenanthrene Abs conjugated with ALP were used to capture possible PAHs. Then amperometric detection at $+300$ mV (vs. Ag/AgCl) was used to observe the PAP generated from PAPP substrate by the ALP enzyme labels. The linear detection range of phenanthrene was from 5 to 45 ng/mL [16]. The limit of detection (LOD) of phenanthrene was 0.8 ng/ml (800 ppt) with the most sensitive indirect co-exposure competition assay format whereas indirect competition and indirect displacement assays had a LOD of 2 ng/mL. Anthracene and chrysene showed strong cross-reactivity when the cross-reactivities of 16 PAHs were tested [16]. A slight loss in sensitivity was observed when the immunosensor performance was tested in tap water and river water. Phenanthrene could be detected with a LOD of 5.0 ng/mL and 6.3 ng/mL in river water and tap water, respectively.

The biosensor was further improved by changing the sample cell design so that only a single drop of sample (100 µL) is required [80]. The model system gave a LOD of 1.4 ng/mL for phenanthrene and a linear range of 2–100 ng/mL. The single drop analysis format of the PAH biosensor was also applied to samples of sea, river, and tap water spiked with phenanthrene.

16.6.5 POLYCHLORINATED BIPHENYLS

PCBs are complex mixtures of up to 209 variably chlorinated biphenyls that are often referred to as Aroclors, followed by a number (i.e., Aroclor 1260: 12 carbon atoms and 60% chlorination). PCBs are solids or oily liquids, and some may be found as a vapor in the air.

PCBs have been classified as serious environmental pollutants, and the congeners are moderately to highly toxic and represent a serious health risk. Exposure to PCBs can result in acne-like skin conditions in adults as well as neurobehavioral and immunological changes in children. PCBs also cause cancer in animals. Although their production has been banned in the United States since 1977, PCBs are still found throughout the environment, and they are still used as coolants or lubricants in certain capacitors and transformers because of their insulating properties. Therefore, the environmental PCB burden may increase with time, and strict monitoring is required. Environmental agencies have placed much effort into monitoring and cleaning up PCB-contaminated sites. Gas chromatographs equipped with an electron capture detector or a mass spectrometer have been commonly used to determine PCBs, but this method is costly and not portable. Millipore, Ensys, and Ohmicron market portable PCB test kits that are based on immunochemistry with spectrophotometric determination of the enzymatic product.

Del Carlo et al. have reported an enzyme IA with amperometric FIA for determining PCBs [81], also using disposable SPEs to immunochemically detect PCBs [55]. The competitive microtitre plate-based ELISA for Aroclor 1260 was coupled with amperometric detection [81]. The assay uses HRP as a label with ferrocene acetic acid substrate. A PCB-gelatin conjugate was used to immobilize the antigen on the solid surface of the microtiter wells. Bound and free analyte competed for the anti-PCB IgG upon introduction of the sample into the wells. Finally, a HRP-labeled 2° Ab was added to complete the assay, and the enzymatically-produced ferricinium ion was determined by measuring the peak currents using amperometric FIA hydrodynamic voltammetry. The authors used 1% milk and 0.05% Tween 20 as NSB blockers in carbonate buffer [81]. The overall assay time was 2.5 hours because of the lengthy incubation steps. The calibration curve for the IA lay between 0.1 and 50 μg/mL. The LOD was estimated to be 0.1 μg/mL with a signal about 10% below the blank. The within-assay reproducibility between six measurements at each concentration was less than 10% for the entire range [81].

Del Carlo et al. extended the above work for immunochemical PCB determination with disposable SPEs [55] as the basis for field use with electrochemical detection. Disposing of the electrodes after each sample analysis rules out electrode fouling over time. The detection system for SPEs is also much simpler and cheaper to acquire than a FIA apparatus. ALP was used as the enzyme label with α-naphthyl phosphate as the substrate in the competitive IA, and the α-naphthol was oxidized at a graphite-based SPE at +400 mV vs. Ag pseudo-reference electrode.

Enzyme-linked immunofiltration assay (ELIFA™), a membrane-based ELISA, was performed in 96 flow-through wells containing the membrane using a commercial ELIFA™ system [55]. The IA reagents were forced through the membrane by suction. When large amounts of BSA–PCB conjugate were immobilized in the membrane and the ELIFA performed the format compared well to the polystyrene surface of microtiter plates. This device also helped to overcome the limited diffusion of the reagent to the solid phase.

Aroclor 1260 was measured in the range of 0.01–10 μg/mL [55]. The main limitation of the assay was a high background signal that was likely due to a cross reaction of the anti-PCB IgG with the BSA molecule in the BSA–PCB conjugate. As suggested by the authors, this may be minimized by changing the conjugating molecule used in the immobilization step.

16.7 CONCLUSION

The presence of pesticides, herbicides, PCBs, and PAH compounds in the environment can cause widespread environmental damage and possible public health problems. This has led to the development of sensitive screening methods such as electrochemical IAs and biosensors for these harmful pollutants. The advantages of IAs, such as their high specificity, use of small sample volumes, high sample throughput with simultaneous analysis of multiple samples, reduced sample preparation, reduction in the use of chemicals and production of waste, and ease of automation, far outweigh their limitations, making IAs an attractive alternative to the more conventional analytical methods in environmental monitoring. The commercial availability of inexpensive IA kits and antibodies to common pollutants is likely to increase the use of IAs in the future. Electrochemical detection can be used to achieve low detection limits in very small samples with little or no sample preparation. The assay format, the type of enzyme substrate pair used, and the ability to minimize NSB are important factors in the optimization of ECIA methods. Advances in electrochemical detection methods will continue to improve the limits of detection for ECIAs and immunosensors.

REFERENCES

1. Stevens, J. B. and Swackhamer, D. L., Environmental pollution. A multimedia approach to modeling human exposure, *Environ. Sci. Technol.*, 23, 1180–1186, 1989.
2. Lijinsky, W., Environmental cancer risks—real and unreal, *Environ. Res.*, 50 (2), 207–209, 1989.
3. www.foodsatty.gov, Gateway to Government Food Satty Information.
4. E.U. Directive 80/778/EEC of 15 July 1980. Commission of the European Union.
5. U.S. EPA Agency, National survey of pesticides in drinking water wells, phase II report, EPA 570/9-91-020, National Technical Information Service, Springfield, VA, 1992.
6. Knopp, D., Immnoassay development for environmental analysis, *Anal. Bioanal. Chem.*, 385, 425–427, 2006.
7. Rubio, F., Veldhuis, L. J., Clegg, B. S., Fleeker, J. R., and Hall, J. C., Comparison of a direct ELISA and an HPLC method for glyphosate determinations in water, *J. Agric. Food Chem.*, 51, 691–696, 2003.
8. Ronkainen-Matsuno, N. J., Thomas, J. T., Halsall, H. B., and Heineman, W. R., Electrochemical immunoassay moving into the fast lane, *Trends Anal. Chem.*, 21, 213–225, 2002.
9. Fahnrich, K. A., Pravda, M., and Guilbault, G. G., Immunochemical detection of polycyclic aromatic hydrocarbons, *Anal. Lett.*, 35 (8), 1269–1300, 2002.
10. Fischer, J. B. and Michael, J. L., Use of ELISA immunoassay kits as a complement to HPLC analysis of Imazapyr and Triclopyr in water samples from forest watersheds, *Bull. Environ. Contam. Toxicol.*, 59, 611–618, 1997.
11. Ruppert, T., Weil, L., and Niesser, R., Influence of water contents on an enzyme immunoassay for triazine herbicides, *Vom Wasser.*, 78, 387–401, 1992.
12. Jiang, T., Halsall, H. B., and Heineman, W. R., Capillary enzyme immunoassay with electrochemical detection for the determination of atrazine in water, *J. Agr. Food Chem.*, 43, 1098–1104, 1995.
13. Wittstock, G., Jenkins, S. H., Halsall, H. B., and Heineman, W. R., Continuing challenges for the immunoassay field, *Nanobiology*, 4, 153–162, 1998.
14. Bauer, C. G., Eremenko, A. V., Ehrentreich-Forster, E., Bier, F. F., Makower, A., Halsall, H. B., Heineman, W. R., and Scheller, F. W., Zeptomole-detecting biosensor for alkaline phosphatase in an electrochemical immunoassay for 2,4-dichlorophenoxyacetic acid, *Anal. Chem.*, 68 (15), 2453–2458, 1996.
15. Dzantiev, B. B., Yazynina, E. V., Zherdev, A. V., Plekhanova, Y. V., Reshetilov, A. N., Chang, S.-C., and McNeil, C. J., Determination of the herbicide chlorsulfuron by amperometric sensor based on separation-free bienzyme immunoassay, *Sensor Actuat. B*, 98, 254–261, 2004.
16. Fahnrich, K. A., Pravda, M., and Guilbault, G. G., Disposable amperometric immunosensor for the detection of polycyclic aromatic hydrocarbons (PAH's) using screen-printed electrodes, *Biosens. Bioelectron.*, 18 (1), 73–82, 2003.
17. Liu, M., Rechnitz, G. A., Li, K., and Li, Q. X., Capacitive immunosensing of polycyclic aromatic hydrocarbon and protein conjugates, *Anal. Lett.*, 31, 2025–2038, 1998.
18. Liu, M., Li, Q. X., and Rechnitz, G. A., Gold electrode modification with thiolated hapten for the design of amperometric and piezoelectric immunosensors, *Electroanal.*, 12, 21–26, 2000.

19. Engvall, E. and Perlmann, P., Enzyme-linked immunosorbent assay (ELISA). Quantitative assay of immunoglobulin G, *Immunochemistry*, 8, 871–874, 1971.
20. Van Weeman, B. K. and Schuurs, A. H. W. M., Immunoassay using antigen-enzyme conjugates, *FEBS Lett.*, 15, 232–235, 1971.
21. Chard, T., *An Introduction to Radioimmunoassay and Related Techniques*, Elsevier, Amsterdam, 1987.
22. Wisdom, G. B., Enzyme-immunoassay, *Clin. Chem.*, 22, 1243–1255, 1976.
23. Tijssen, P., *Practice and Theory of Enzyme Immunoassays*, Elsevier, Amsterdam, 1985.
24. O'Sullivan, M. J., In *Practical Immunoassay*, Butt, W. R., Ed., Marcel Dekker, New York, pp. 37–69, 1984.
25. Thomas, J. T., Ronkainen-Matsuno, N. J., Farrell, S., Halsall, H. B., and Heineman, W. R., Microdrop analysis of a bead-based immunoassay, *Microchem. J.*, 74, 267–276, 2003.
26. Sauer, M. J., Foulkes, J. A., and Morris, B. A., *Immunoassays in Food Analysis*, Elsevier, London, 1985.
27. Tsuji, I., Eguchi, H., Yasukouchi, K., Unoki, M., and Taniguchi, I., Enzyme immunosensors based on electropolymerized polytyramine modified electrodes, *Biosens. Bioelectron.*, 5 (2), 87–101, 1990.
28. Robinson, G. A., Hill, H. A. O., Philo, R. D., Gear, J. M., Rattle, S. J., and Forrest, G. C., Bioelectrochemical enzyme immunoassay of human choriogonadotrophin with magnetic electrodes, *Clin. Chem.*, 31, 1449–1452, 1985.
29. Heineman, W. R. and Halsall, H. B., Strategies for electrochemical immunoassay, *Anal. Chem.*, 57, 1321A–1331A, 1985.
30. Lunte, C. E., Heineman, W. R., and Halsall, H. B., Electrochemical enzyme immunoassay, *Curr. Separations*, 8, 18–22, 1987.
31. Halsall, H. B. and Heineman, W. R., Electrochemical immunoassay: an ultrasensitive method, *J. Int. Fed. Clin. Chem.*, 2, 179–187, 1990.
32. Jenkins, S. H., Heineman, W. R., and Halsall, H. B., Extending the detection limit of solid-phase electrochemical enzyme immunoassay to the attomole level, *Anal. Biochem.*, 168 (2), 292–299, 1988.
33. Wijayawardhana, C. A., Halsall, H. B., and Heineman, W. R., In *Electroanalytical Methods of Biological Materials*, Brajter-Toth, A. and Chambers, J. Q., Eds., Marcel Dekker, New York, pp. 329–365, 2002.
34. Yao, H., Jenkins, S. H., Pesce, A. J., Halsall, H. B., and Heineman, W. R., Electrochemical homogeneous enzyme immunoassay of theophylline in hemolyzed, icteric, and lipemic samples, *Clin. Chem.*, 39, 1432–1434, 1993.
35. Wightman, R. M. and Wipf, D. O., In *Electroanalytical Chemistry*, Bard, A. J., Ed., Marcel Dekker, New York, pp. 267–353, 1989.
36. Wightman, R. M., Microvoltammetric electrodes, *Anal. Chem.*, 53 (9), 1125A–1130A, 1981.
37. Heinze, J., Electrochemistry with ultramicroelectrodes, *Angew. Chem. Int. Ed. Engl.*, 32, 1268–1288, 1993.
38. Farrell, S., Ronkainen-Matsuno, N. J., Halsall, H. B., and Heineman, W. R., Bead-based immunoassays with microelectrode, *Anal. Bioanal. Chem.*, 379, 358–367, 2004.
39. Kissinger, P. T. and Heineman, W. R., Cyclic voltammetry, *J. Chem. Educ.*, 60 (9), 702–706, 1983.
40. Rippeth, J. J., Gibson, T. D., Hart, J. P., Hartley, I. C., and Nelson, G., Flow-injection detector incorporating a screen-printed disposable amperometric bionsensor for monitoring organophosphate pesticides, *Analyst*, 122, 1425–1429, 1997.
41. Hu, T., Zhang, X.-E., and Zhang, Z.-P., Disposable screen-printed enzyme sensor for simultaneous determination of starch and glucose, *Biotechnol. Techn.*, 13 (6), 359–362, 1999.
42. Zen, J.-M., Chung, H.-H., and Kumar, A. S., Flow injection analysis of hydrogen peroxide on copper-plated screen-printed carbon electrodes, *Analyst*, 125, 1633–1637, 2000.
43. Dequaire, M., Degrand, C., and Limoges, B., An immunomagnetic electrochemical sensor based on a perfluorosulfonate-coated screen-printed electrode for the determination of 2,4 dichlorophenoxyacetic acid, *Anal. Chem.*, 71 (13), 2571–2577, 1999.
44. Kissinger, P. T. and Heineman, W. R., *Laboratory Techniques in Electroanalytical Techniques*, Marcel Dekker, New York, 1996.
45. Trojanowicz, M., Szewczynska, M., and Wcislo, M., Electroanalytical flow measurements-recent advances, *Electroanalysis*, 15 (5–6), 347–365, 2003.

46. Wijayawardhana, C. A., Purushothama, S., Cousino, M. A., Halsall, H. B., and Heineman, W. R., Rotating disk electrode amperometric detection for a bead-based immunoassay, *J. Electroanal. Chem.*, 468, 2–8, 1999.

47. Wijayawardhana, C. A., Halsall, H. B., and Heineman, W. R., Micro volume rotating disk electrode (RDE) amperometric detection for a bead-based immunoassay, *Anal. Chim. Acta.*, 399, 3–12, 1999.

48. Puchades, R. and Maquieira, A., Recent developments in flow injection immunoanalysis, *Crit. Rev. Anal. Chem.*, 26 (4), 195–218, 1996.

49. Masson, M., Liu, Z., Haruyama, T., Kobatake, E., Ikariyama, Y., and Aizawa, M., Immunosensing with amperometric detection, using galactosidase as label and p-aminophenyl-β-D-galactopyranoside as substrate, *Anal. Chim. Acta.*, 304, 353–359, 1995.

50. Aslam, M. and Dent, A., *Bioconjugation*, Groves Dictionaries, Inc., New York, 1998.

51. Tang, H. T., Lunte, C. E., Halsall, H. B., and Heineman, W. R., p-Aminophenyl phosphate: an improved substrate for electrochemical enzyme immunoassay, *Anal. Chim. Acta.*, 214, 187–195, 1988.

52. Thompson, R. Q., Porter, M., Stuver, C., Halsall, H. B., Heineman, W. R., Buckley, E., and Smyth, M. R., Zeptomole detection limit for alkaline phosphatase using 4-aminophenylphosphate, amperometric detection, and an optimal buffer system, *Anal. Chim. Acta.*, 271, 223–229, 1993.

53. Cardosi, M., Birch, S., Talbot, J., and Phillips, A., An electrochemical immunoassay for Clostridium perfringens phospholipase C, *Electroanalysis*, 3 (3), 169–176, 1991.

54. Athey, D., Ball, M., and McNeil, C. J., Avidin-biotin based electrochemical immunoassay for thyrotropin, *Ann. Clin. Biochem.*, 30, 570–577, 1993.

55. Del Carlo, M., Lionti, I., Taccini, M., Cagnini, A., and Mascini, M., Disposable screen-printed electrodes for the immunochemical detection of polychlorinated biphenyls, *Anal. Chim. Acta.*, 342, 189–197, 1997.

56. Wilson, M. and Rauh, R., Hydroquinone diphosphate: an alkaline phosphatase substrate that does not produce electrode fouling in electrochemical immunoassays, *Biosens. Bioelectron.*, 20, 276–283, 2004.

57. Thomas, J. H., Kim, S. K., Hesketh, P. J., Halsall, H. B., and Heineman, W. R., Bead-based electrochemical immunoassay for bacteriophage MS2, *Anal. Chem.*, 76, 2700–2707, 2004.

58. Kato, K., Umedo, Y., Suzuki, F., and Kosaka, A., Improved reaction buffers for solid-phase enzyme immunoassay without interference by serum factors, *Clin. Chim. Acta.*, 102 (2–3), 261–265, 1980.

59. Vogt, R. F. Jr., Phillips, D. L., Henderson, L. O., Whitfield, W., and Spierto, F. W., Quantitative differences among various proteins as blocking agents for ELISA mictotiter plates, *J. Immunol. Methods*, 101 (1), 43–50, 1987.

60. Kaneki, N., Xu, Y., Kumari, A., Halsall, H. B., Heineman, W. R., and Kissinger, P. T., Electrochemical enzyme immunoassay using sequential saturation technique in a 20-uL capillary: digoxin as a model analyte, *Anal. Chim. Acta.*, 287, 253–258, 1994.

61. Centonze, D., Guerrieri, A., Malitesta, C., Palmisano, F., and Zambonin, P. G., Interference-free glucose sensor based on glucose oxidase-immobilized in an overoxidized non-conducting polypyrrole film, *Anal. Bioanal. Chem.*, 342 (9), 729–733, 1991.

62. Djane, N.-K., Armalis, S., Ndung'u, K., Johansson, G., and Mathiasson, L., Supported liquid membrane coupled on-line to potentiometric stripping analysis at a mercury-coated reticulated vitreous carbon electrode for trace metal determinations in urine, *Analyst*, 123, 393–396, 1998.

63. Kumari, A., Use of (poly)ethylene glycol modified solid supports in electrochemical immunoassay, M.S. Thesis. University of Cincinnati, Cincinnati, 1991.

64. Hodneland, C. D., Lee, Y.-S., Min, D.-H., and Mrksich, M., Selective immobilization of proteins to self-assembled monolayers presenting active site-directed capture ligands, *Proc. Natl Acad. Sci. U.S.A.*, 99, 5048–5052, 2002.

65. Chen, C. S., Mrksich, M., Huang, S., Whitesides, G. M., and Ingber, D. E., Micropatterned surfaces for control of cell shape, position, and function, *Biotechnol. Prog.*, 14, 356–363, 1998.

66. Mrksich, M. and Whitesides, G. M., Interactions of SAMs with proteins, *Biophys. Biomol. Struct.*, 25, 55–78, 1996.

67. Johnsson, B., Lofas, S., and Lindquist, G., Immobilization of proteins to a carboxymethyldextran-modified gold surface for biospecific interaction analysis in surface plasmon resonance sensors, *Anal. Biochem.*, 198, 268–277, 1991.

68. Sabik, H. and Jeannot, R. J., Determination of organonitrogen pesticides in large volumes of surface water by liquid-liquid and solid-phase extraction using gas chromatography with nitrogen-phosphorus detection and liquid chromatography with atmospheric pressure chemical ionization mass spectrometry, *Chromatogr. A.*, 818 (2), 197–207, 1998.

69. Quintana, J., Marti, I., and Ventura, F., Monitoring of pesticides in drinking and related waters in NE Spain with a multiresidue SPE-GC-MS method including an estimation of the uncertainty of the analytical results, *J. Chromatogr. A.*, 938 (1–2), 3–13, 2001.

70. Carabias-Martinez, R., Rodriquez-Gonzalo, E., Herrero-Hernandez, E., Sanchez-San Roman, F. J., and Flores, M. G. P., Determination of herbicides and metabolites by solid-phase extraction and liquid chromatography evaluation of pollution due to herbicides in surface and groundwaters, *J. Chromatogr. A.*, 950 (1–2), 157–166, 2002.

71. Barcelo, D., Durand, G., Bouvot, V., and Nielen, M., Use of extraction disks for trace enrichment of various pesticides from river water and simulate, *Environ. Sci. Technol.*, 27, 271–277, 1993.

72. Penmetsa, K. V., Leidy, R. B., and Shea, D., Herbicide analysis by micellar electrokinetic chromatogrphy, *J. Chromatogr. A.*, 745 (1), 201–208, 1996.

73. Giersch, T., A new monoclonal antibody for the sensitive detection of atrazine with immunoassay in microtiter plate and dipstick format, *J. Agric. Food Chem.*, 41, 1006–1011, 1993.

74. Wortberg, M., Goodrow, M. H., Gee, S. J., and Hammock, B. D., Immunoassay for simazine and atrazine with low cross-reactivity for propazine, *J. Agric. Food Chem.*, 44, 2210–2219, 1996.

75. Monarca, S., Pasquini, R., Scassellati Sforzolini, G., Savino, A., Bauleo, F. A., and Angeli, G., Environmental monitoring of mutagenic/carcinogenic hazards during road paving operations with bitumens, *Int. Arch. of Occ. and Env. Health*, 59 (4), 393–402, 1987.

76. Li, D., Firozi, P. F., Wang, L.-E., Bosken, C. H., Spitz, M. R., Hong, W. K., and Wei, Q., Sensitivity to DNA damage induced by benzo(a)pyrene diol epoxide and risk of lung cancer: a case-control analysis, *Cancer Res.*, 61, 1445–1450, 2001.

77. Souteyrand, E., Martin, J. R., and Martelet, C., Direct detection of biomolecules by electrochemical impedance measurements, *Sens. Act. B.*, 20 (1), 63–69, 1994.

78. Berggren, C. and Johansson, G., Capacitance measurement of antibody-antigen interactions in a flow system, *Anal. Chem.*, 69, 3651–3657, 1997.

79. Gebbert, A., Alvarez-Icaza, M., Stocklein, W., and Schmid, R. D., Real-time monitoring of immunochemical interactions with a tantalum capacitance flow-through cell, *Anal. Chem.*, 64, 997–1003, 1992.

80. Moore, E. J., Kreuzer, M. P., Pravda, M., and Guilbault, G. G., Development of a rapid single-drop analysis biosensor for screening of phenanthrene in water samples, *Electroanalysis*, 16 (20), 1653–1659, 2004.

81. Del Carlo, M. and Mascini, M., Enzyme immunoassay with amperometric flow-injection analysis using horseradish peroxidase as a label. Application to the determination of polychlorinated biphenyls, *Anal. Chim. Acta.*, 336, 167–174, 1996.

17 Biosensors for Environmental Monitoring and Homeland Security

Kanchan A. Joshi, Wilfred Chen, Joseph Wang, Michael J. Schöning, and Ashok Mulchandani

CONTENTS

17.1 INTRODUCTION

Organophosphorus compounds (OPs) are widely used as pesticides, insecticides, and chemical warfare agents (CWAs) [1–3]. These neurotoxic compounds irreversibly inhibit the enzyme acetyl-cholinesterase (AChE), essential for central nervous system function in humans and insects, resulting in the buildup of the neurotransmitter acetylcholine that interferes with muscular responses, produces serious symptoms in vital organs, and eventually causes death [1–5].

Approximately 91 million pounds of organophosphate insecticides were used in the United States in 1999, the last year for which data are available [6]. The extensive use of these pesticides has resulted in the widespread presence of trace amounts in surface and ground waters across the United States [7]. Therefore, rapid, sensitive, selective, and reliable determination of OPs is necessary in order to take immediate action. The recent increase in terrorist activities and homeland security concerns has further aggravated this need. Additionally, any process that may be adopted to treat the large volume of wastewater generated at both the producer and consumer levels and the mandated destruction of the stockpile of CWAs that belong to the OP group requires analytical tools to properly monitor and control the process.

Current analytical techniques such as gas and liquid chromatography [8,9] are very sensitive and reliable, but they cannot be easily performed in the field. These techniques are time consuming and expensive, and must to be performed by highly-trained technicians. Bioanalytical methods such as immunoassays and inhibition of cholinesterase activity for OP determination have also been

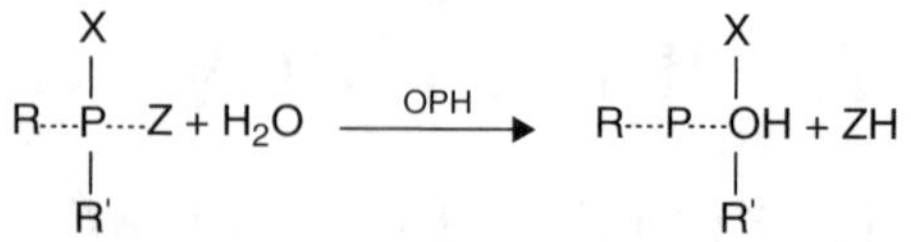

$$\begin{array}{c} X \\ | \\ R\cdots P\cdots Z + H_2O \end{array} \xrightarrow{\text{OPH}} \begin{array}{c} X \\ | \\ R\cdots P\cdots OH + ZH \\ | \\ R' \end{array}$$

FIGURE 17.1 Reaction scheme for the OPH-catalyzed hydrolysis of organophosphorus compounds (OPs). X is oxygen or sulfur; R is an alkoxy group ranging in size from methoxy to butoxy; R', is an alkoxy or phenyl group; and Z is a phenoxy group, a thiol moiety, a cyanide, or a fluorine group. (From Mulchandani, A., Chen, W., Mulchandani, P., Wang, J., and Rogers, K. R., *Biosens. Bioelectron.*, 16, 225, 2001. With permission.)

reported [8]. Despite the advantages of immunoassay techniques, these methods may require long analysis time (1–2 h) and extensive sample handling (a large number of washing steps); therefore, they are unsuitable for online monitoring of detoxification processes. Biosensing analytical devices based on the AChE inhibition have been reported [10–28]. Although sensitive, AChE inhibition-based biosensors have limitations of poor selectivity, tedious and time-consuming protocol, and single use.

Organophosphorus hydrolase (OPH) is an organophosphotriester hydrolyzing enzyme first discovered in soil microorganisms *Pseudomonas diminuta* MG and *Flavobacterium* sp. The enzyme has broad substrate specificity and is able to hydrolyze a number of OP pesticides such as paraoxon, parathion, coumaphos, diazinon, dursban, methyl parathion, and the CWAs sarin and soman [29–32]. As shown in Figure 17.1, OPH-catalyzed hydrolysis of the P–O, P–F, P–S, or P–CN bonds of OP compounds generates an acid and an alcohol that, in many cases, is chromophoric and/ or electroactive. This enzyme reaction can be combined with a variety of transduction schemes to construct simple biosensors for the direct and rapid determination of OPs. For example, OPH can be combined with potentiometric transducers such as a pH electrode or a field effect transistor or a pH-indicator dye to quantify protons produced that can be then correlated to the OP concentration. Similarly, OPH can be combined with an optical transducer to monitor the production of chromophoric products such as *p*-nitrophenol (PNP) produced during the hydrolysis of OPs such as paraoxon, parathion, methyl parathion, or chlorferon from the hydrolysis product of coumaphos. OPH can be integrated with an amperometric transducer to monitor the oxidation current of hydrolysis products such as PNP and thiols. The amperometric transduction systems include thick-film screen printed carbon, carbon paste, and multiwalled carbon nanotubes electrodes.

Organophosphorus hydrolase purification requires a costly and time-consuming effort. To alleviate this, this team genetically engineered microorganisms such as *Escherichia coli* and *Moraxella* sp. to express OPH on their cell surface, and combined the new biological sensing element with potentiometric, optical, and amperometric transducers to make microbial biosensors. The surface expression of the enzyme considerably improved the sensitivity, stability, and response time of the biosensors. Additionally, this group combined purified OPH with microorganisms that degrade/metabolize/oxidize PNP, and genetically engineered the OPH activity in these PNP degraders for construction of highly sensitive and selective biosensors for PNP-substituted OPs such as parathion, methyl parathion, paraoxon, fenitrothion, and O-ethyl-O-*para*-nitrophenyl phenylphosphonothioate (EPN).

In order to differentiate/discriminate among OP compounds, this group combined the OPH-modified potentiometric and amperometric transducers in series of a flow-injection analyzer. Combining the two transduers differentiated *p*-nitrophenyl substituted OP from the others. Furthermore, it combined precolumn enzyme-catalyzed hydrolysis electrophoretic separation in a microcapillary column of a lab-on-a-chip with a contactless conductivity detector for a complete fingerprint of the target. The attractive analytical features of these devices make them particularly well-suited for various environmental and defense scenarios.

The following presents a brief overview of these different biosensor configurations.

17.2 OPH-BASED POTENTIOMETRIC ENZYME ELECTRODES

A pH electrode, modified with an immobilized purified OPH layer that was created by cross-linking OPH with bovine serum albumin and glutaraldehyde, formed the basic element of this very simple enzyme electrode. Other components of the system included a pH meter, a measuring cell placed on a magnetic stirrer for mixing, and a chart recorder (Figure 17.2). The sensor signal and response time were optimized with respect to the buffer pH, ionic concentration of buffer, temperature, and units of OPH immobilized using paraoxon as substrate. The best sensitivity and response times were obtained using a sensor constructed with 500 IU of OPH and operating in pH 8.5, 1 mM 4-(2-hydroxyethyl-1-piperazineethanesulfonic acid (HEPES) buffer supplemented with 100 mM NaCl and 0.05 mM cobalt chloride at 20°C. Operating under these conditions, the biosensor was able to detect as low as 2 µM of paraoxon, ethyl parathion, methyl parathion, and diazinon in approximately 2 min with very good accuracy and selectivity against other non-organophosphate pesticides such as simazine, triazine, atrazine, sevin, and sutan. The biosensor was completely stable for at least one month when stored in pH 8.5, 1 mM HEPES + 100 mM NaCl buffer at 4°C. This biosensor was used to measure paraoxon, parathion, and methyl parathion in simulated samples and showed a high correlation ($r^2 = 0.998$) to a spectrophotometric enzymatic assay [33].

A transducer, consisting of a pH sensitive capacitive electrolyte insulator semiconductor (EIS), replaced the glass pH electrode in order to firstly miniaturize the potentiometric enzyme biosensor, and secondly to eliminate mechanical instability of the glass [34]. The EIS transducer was made of a layer sequence of $Al/p\text{-}Si/SiO_2$ with Ta_2O_5, Al_2O_3 or Si_3N_4 as the pH-sensitive layer. To obtain the enzyme biosensor, OPH was immobilized on top of the pH-sensitive layer structure. Figure 17.3 shows the schematic of the biosensor fabrication. Typical sensor characteristics such as pH sensitivity, sensitivity toward paraoxon, reversibility of the sensor signal, response time, detection limit, long-term stability, and selectivity to various pesticides were determined to be comparable to the OPH-modified glass pH electrode.

17.3 OPH-BASED OPTICAL BIOSENSORS

Two different optical-based biosensors were constructed. The first was based on the measured decrease in the fluorescence intensity of the fluorescein isothiocyanate (FITC) covalently conjugated to the enzyme. The method employed poly(methyl methacrylate) beads to which the

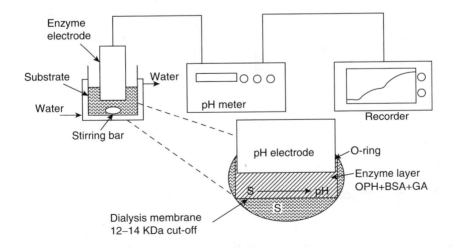

FIGURE 17.2 Schematic of potentiometric OPH-based biosensor. (From Mulchandani, P., Mulchandani, A., Kaneva, I., and Chen, W., *Biosens. Bioelectron.*, 14, 77, 1999. With permission.)

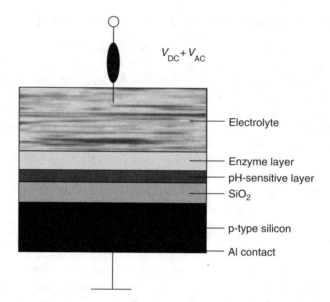

FIGURE 17.3 Schematic of capacitive electrolyte insulator semiconductor (EIS) biosensor with immobilized enzyme layer on a pH-sensitive transducer structure. (From Schoning, M. J., Mulchandani, P., Chen, W., and Mulchandani, A., *Sens. Actuators, B*, 91, 92, 2003. With permission.)

FITC-labeled enzyme was adsorbed. Analytes were then measured using a commercially available microbead fluorescence analyzer (KinExA, Sapidyne Inc., Boise, ID, U.S.A.). Figure 17.4 shows the schematic of the biosensor. This assay was based on a substrate-dependent change in pH at the local vicinity of the enzyme. Similar to the OPH-modified pH electrode biosensor, the sensitivity of

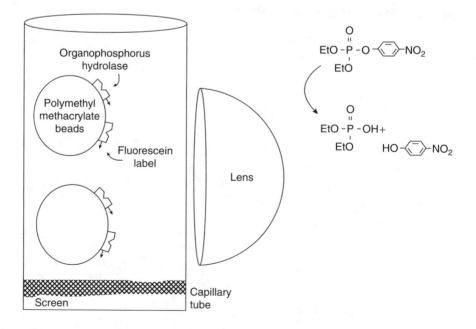

FIGURE 17.4 Schematic of the organophosphate (OP) assay in the KinExA fluorescence analyzer. (From Roger, K. R., Wang, Y., Mulchandani, A., Mulchandani, P., and Chen, W., *Biotechnol. Prog.*, 15, 517, 1999. With permission.)

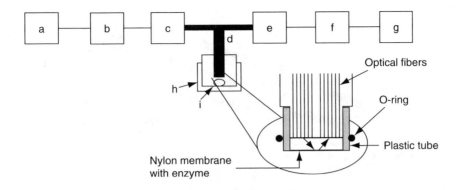

FIGURE 17.5 Schematic of organophosphorus hydrolase (OPH)-modified fiber-optic biosensor. (From Mulchandani, A., Pan, S., and Chen, W., *Biotechnol. Prog.*, 15, 130, 1999. With permission.)

this biosensor was an inverse function of the ionic strength of the assay buffer. The dynamic concentration range for the assay extended from 25 to 400 μM for paraoxon with a detection limit of 8 μM. Organophosphorus insecticides measured using this technique included parathion, methyl parathion, dursban, fensulfothion, crotoxyphos, diazinon, mevinphos, dichlorvos, and coumaphos. The technique was used to measure coumaphos in biodegradation samples of cattle dip wastes and showed a high correlation ($r^2 = 0.998$) to a HPLC method [35].

In the second optic-based biosensor format, the analysis was based on the relationship between the amount of OP hydrolyzed and the amount of chromophoric product formed (quantified by measuring the absorbance at the λ_{max} of the product) by enzyme catalyzed hydrolysis. Figure 17.5 shows the fiber-optic biosensor set-up. Operating at optimized analytical conditions (pH 9, 30°C, and 123 IU of immobilized OPH), the biosensor was able to detect as low as 2 μM of paraoxon and parathion and 5 μM of coumaphos, selectively, without interference from carbamates and triazines in less than 2 min. The biosensor response had excellent reproducibility, and the immobilized enzyme was stable for more than one month when stored at 4°C in the analytical buffer [36].

An advantage of the present sensor over other biosensors based on the pH change is the use of higher ionic strength buffer used in the analysis. The use of higher ionic strength buffer allows the enzyme to function at its maximum activity over the complete duration/range of the assay rather than just at the start. Additionally, the use of higher ionic strength buffer precludes the need to adjust the sample pH to that of the analytical buffer.

17.4 OPH-BASED AMPEROMETRIC ENZYME ELECTRODES

Hydrolysis of certain OP pesticides such as parathion, methyl parathion, paraoxon, EPN, and fenitrothion generates PNP that is electroactive. PNP can be electrochemically oxidized at the anode. The oxidation current, measured at a fixed potential using a potentiostat, is directly proportional to the concentration of PNP formed. This group constructed several amperometric enzyme electrodes for OP determination. In the first format (Figure 17.6), screen printed thick-film carbon electrodes were modified by OPH that was deposited on the electrode in a Nafion film. Hydrodynamic voltammetry investigations identified the oxidation potential of PNP to be 0.85 V versus Ag/AgCl reference electrode and 1080 IU of OPH to be optimum for the biosensor. The amperometric signals were linearly proportional to the concentration of the hydrolyzed paraoxon and methyl parathion substrates up to 40 and 5 μM, showing detection limits of 9×10^{-8} and 7×10^{-8} M, respectively. Such detection limits were substantially lower compared to the

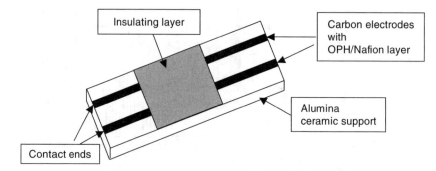

FIGURE 17.6 Schematic of organophosphorus hydrolase (OPH)-modified screen printed thick-film electrode. (From Mulchandani, A., Mulchandani, P., Chen, W., Wang, J., and Chen, L., *Anal. Chem.*, 71, 2246, 1999. With permission.)

$(2-5) \times 10^{-6}$ M values for OPH-based potentiometric and fiber-optic devices. The high sensitivity was coupled to a faster and simplified operation with the potential for field-deployment [37].

In another format, an OPH-based amperometric biosensor for remote sensing was constructed. This remote-electrochemical sensor consisted of a PVC-housing tube connected to a 16-m-long shielded cable via three-pin, environmentally sealed, rubber connectors. The assembly included the OPH-modified carbon-paste working electrode, a Ag/AgCl reference electrode (BAS, Model RE-4), and a platinum wire counter electrode. Two female coupling connectors fixed with epoxy in the PVC tube served for mounting the working and reference electrodes; brass screws, located within these connectors, provided the electrical contact. The carbon paste electrode was prepared by thoroughly hand mixing 120 mg mineral oil and 180 mg graphite powder. The resulting paste was firmly packed into the electrode cavity (3 mm diameter, 1 mm depth) of a 4-cm-long Teflon sleeve. Electrical contact (to the inner part of the paste) was established via a stainless steel screw, contacting the brass screw within the female connector on the PVC housing. The paste surface was smoothed on a weighing paper. The enzyme OPH was immobilized by casting a 10-µl droplet, containing 5 µl (of 108 IU/ml) OPH and 5 µl Nafion (in 1% ethanol) onto the carbon surface, allowing the solvent to evaporate. Operating in chronoamperometric mode (from open circuit to $+0.85$ V), the response of the biosensor was linear in the range of 4.6–46 µM for paraoxon and up to 5 µM for methyl parathion. The lower detection limits of the biosensor for paraoxon and methyl parathion were 0.9 and 0.4 µM, respectively. An important characteristic of a remote sensor is its rapid response to sudden concentration changes with no carry over between samples (Figure 17.7). High stability (Figure 17.8) is another important characteristic of the in situ sensor [38].

Monitoring and controlling the destruction of CWAs stockpiles and OP insecticides in industrial and agricultural waste waters is important because of the large quantity of these compounds produced and consumed. A flow-injection amperometric biosensor (FIAB) particularly suitable for such an application was developed [39]. The schematic drawing is shown in Figure 17.9. The biosensor incorporated an immobilized enzyme reactor that contained OPH covalently immobilized on activated aminopropyl-controlled pore glass beads and an electrochemical flow-through detector, containing a carbon paste working electrode, a silver/silver chloride reference electrode, and a stainless steel counter electrode. The OPH catalyzed the hydrolysis of organophosphate with nitrophenyl substituent to generate PNP that was then electrochemically detected downstream at the carbon paste electrode poised at 0.9 V versus the reference electrode. The amperometric response of the biosensor was linear up to 120 and 140 µM with lower detection limits of 20 and 20 nM for paraoxon and methyl parathion, respectively. The response was very reproducible (Residual standard deviation (RSD) 2%, $n = 35$) and stable for over one month when the

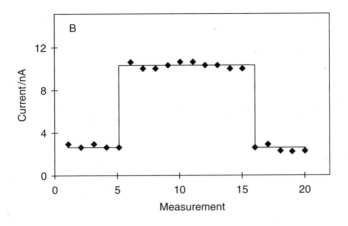

FIGURE 17.7 Response of the remote biosensor to changes in methyl parathion concentration from 4.6 to 23 μM and back to 4.6 μM. (From Wang, J., Chen, L., Mulchandani, A., Mulchandani, P., and Chen, W., *Electroanalysis*, 11, 866, 1999. With permission.)

immobilized enzyme column was stored at 4°C. Each assay took about 2 min, giving a sample throughput of 30 h^{-1}. The applicability of the biosensor to monitor paraoxon and methyl parathion was demonstrated in distilled water and simulated well water.

Although very sensitive, the selectivity of the above amperometric biosensors, based on electrochemical oxidation of PNP, was compromised by other phenols [40] and thus not suitable for most practical environmental applications. To improve the selectivity, this group used a carbon nanotube-modified (CNT) transducer for electrooxidation of PNP produced from the OPH catalyzed hydrolysis of PNP-substituted OPs. The ability of CNT-modified electrodes to promote the oxidation of phenolic compounds (including the PNP product of the OPH reaction) and to minimize surface fouling associated with such oxidation processes paved the way to the new OPH-CNT amperometric biosensor. As shown in the cyclic voltammograms in Figure 17.10 and Figure 17.11, compared to an unmodified electrode, the CNT-modified electrode showed enhancement in current and, consequently in, the sensitivity, and it also showed decreased surface

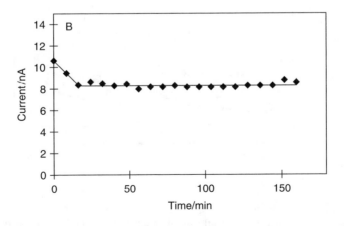

FIGURE 17.8 Stability of the organophosphorus hydrolase (OPH)-modified remote electrochemical sensor response to 7.5 μM methyl parathion. (From Wang, J., Chen, L., Mulchandani, A., Mulchandani, P., and Chen, W., *Electroanalysis*, 11, 866, 1999. With permission.)

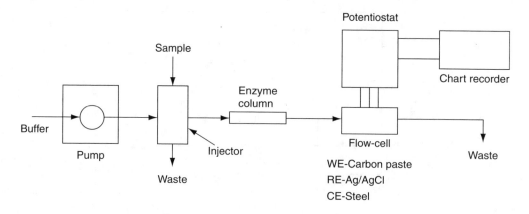

FIGURE 17.9 Schematic drawing of the flow-injection amperometric biosensor configuration. (From Mulchandani, P., Chen, W., and Mulchandani, A., *Environ. Sci. Technol.*, 35, 2562, 2001. With permission.)

passivation for PNP. The optimization and advantages of the CNT-based OPH amperometric biosensor were reported in Deo et al. [41].

In another approach to improve the OPH-modified amperometric biosensor selectivity, this group incorporated PNP mineralizing bacterium, *Arthrobacter* sp., along with purified OPH to construct a hybrid amperometric biosensor. The biocatalytic layer was prepared by co-immobilizing *Arthrobacter* sp. JS443 and OPH on a carbon paste electrode. OPH catalyzed the hydrolysis

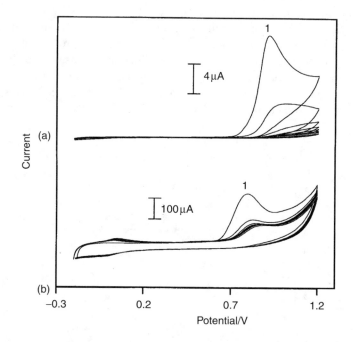

FIGURE 17.10 Cyclic voltammograms showing 10 repetitive scans of 1 mM *p*-nitrophenol at unmodified (a), chemical vapor deposition produced multi-wall carbon nanotube (MWNT-CVD)-modified (b) glassy carbon (GC) electrodes. The first scans in (a) and (b) are indicated as 1. Scan rate, 50 mV/s. Electrolyte, phosphate buffer (0.05 M, pH 7.4). (From Deo, R. P., Wang, J., Block, I., Mulchandani, A., Joshi, K. A., Trojanowicz, M., Scholz, F., Chen, W., and Lin, Y., *Anal. Chim. Acta*, 530, 185, 2005. With permission.)

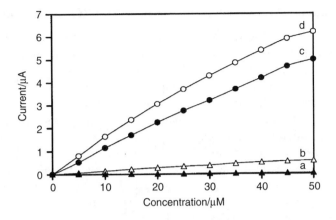

FIGURE 17.11 Calibration plots resulting from amperometric measurement of 5 μM successive additions of *p*-nitrophenol at various electrodes: (a) bare glassy carbon (GC); (b) ARC produced multi-wall carbon nanotube (MWNTARC); (c) chemical vapor deposition produced single-wall carbon nanotube (SWNT-CVD); and (d) chemical vapor deposition produced multi-wall carbon nanotube (MWNT-CVD)-modified GC electrodes. Electrolyte, phosphate buffer (0.05 M, pH 7.4); stirring rate, ~300 rpm; operating potential, +0.85 V. Data points are average of duplicate measurements. (From Deo, R. P., Wang, J., Block, I., Mulchandani, A., Joshi, K. A., Trojanowicz, M., Scholz, F., Chen, W., and Lin, Y., *Anal. Chim. Acta*, 530, 185, 2005. With permission.)

of organophosphorus pesticides with a *p*-nitrophenyl substituent such as paraoxon and methyl parathion to release the PNP that was oxidized by the enzymatic machinery of *Arthrobacter* sp. JS443 to carbon dioxide through electroactive intermediates 4-nitrocatechol and 1,2,4-benzenetriol as shown in Figure 17.12. The oxidization current of the intermediates was measured and correlated to the concentration of organophosphates. The best sensitivity and response time were obtained using a sensor constructed with 0.06 mg dry weight of cell and 965 IU of OPH, operating at 400 mV applied potential (versus Ag/AgCl reference) in 50 mM citrate-phosphate pH 7.5 buffer at room temperature. Using these conditions, the biosensor measured as low as 2.8 ppb (10 nM) of paraoxon and 5.3 ppb (20 nM) of methyl parathion without interference from phenolic compounds, carbamate pesticides, triazine herbicides, and organophosphate pesticides that do not have the *p*-nitrophenyl substituent. The biosensor had excellent operational lifetime stability with no decrease in response for more than 40 repeated uses over a 12 h period when stored at room temperature, whereas its storage life was approximately two days when stored in the operating buffer at 4°C [42].

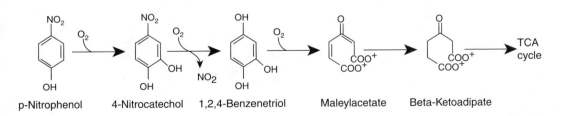

FIGURE 17.12 Proposed pathway for *p*-nitrophenol (PNP) degradation by *Arthrobacter* sp. (From Lei, Y., Mulchandani, P., Chen, W., Wang, J., and Mulchandani, A., *Biotechnol. Bioeng.*, 85, 706, 2004. With permission.)

17.5 OPH-BASED DUAL POTENTIOMETRIC–AMPEROMETRIC BIOSENSOR

Although the potentiometric biosensor favorably responds to all OP compounds, reflecting the pH changes associated with the OPH activity, the amperometric device displays well-defined signals only toward OP substrates (pesticides) liberating the oxidizable PNP product. Therefore, simultaneous/combined use of the potentiometric and amperometric transducers in connection to the same OPH enzyme would enhance the information content, and would provide discrimination between various subclasses of OP compounds. To demonstrate this hypothesis, this group constructed a miniaturized dual, potentiometric-amperometric, flow-injection OPH-based biosensor. Both transducers were prepared by a thin-film fabrication technology, and they rapidly and independently responded to sudden changes in the level of the corresponding OP compound with no apparent cross-reactivity even when operated in series (Figure 17.13). Typical amperometric and potentiometric peaks recorded simultaneously with the combined OPH biosensor system for the injection of solutions of various OP compounds is shown in Figure 17.14. The effect of relevant experimental variables like capacitance and applied potential, buffer concentration, and flow rate were evaluated and optimized. The dual system holds great promise for field screening of OP neurotoxins in connection to various defense and environmental scenarios [43]. The multiple-transduction concept could be extended for increasing the information content of other class-enzyme biosensor systems.

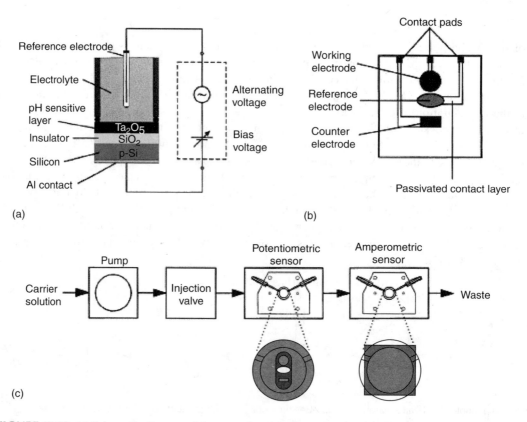

FIGURE 17.13 (a) Schematic diagram of the potentiometric electrolyte insulator semiconductor (EIS) sensor system. (b) Design of the thin-film amperometric gold electrode. (c) Schematic diagram of the dual-biosensor flow-injection system. (From Wang, J., Krause, R., Block, K., Musameh, M., Mulchandani, A., Mulchandani, P., Chen, W., and Schöning M. J., *Anal. Chim. Acta*, 469, 197, 2002. With permission.)

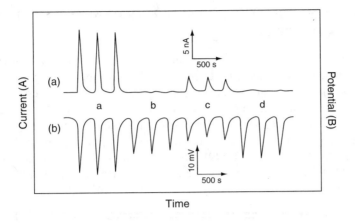

FIGURE 17.14 Simultaneous amperometric (A) and potentiometric (B) measurements of (a) 50 μM para-oxon; (b) 100 μM dichlorvos; (c) 200 μM parathion; and (d) 200 μM diazinon. Carrier solution, phosphate buffer (0.5 mM, pH 9.0)/10 mM KCl; flow rate, 1.8 ml min^{-1}; injection loop, 1000 μl; operating potential (A), +0.75 V; constant capacitance (B), 22 nF. (From Wang, J., Krause, R., Block, K., Musameh, M., Mulchandani, A., Mulchandani, P., Chen, W., and Schöning M. J., *Anal. Chim. Acta*, 469, 197, 2002. With permission.)

17.6 MICROBIAL BIOSENSORS

The biosensors described above were based on purified OPH. Purifying OPH is a laborious, time consuming, and costly effort. To avoid this, this group genetically engineered microorganisms such as *E. coli* and *Moraxella* sp. to express OPH on the cell surface. The expression of OPH on the surface of the cell alleviates the mass transport resistance as a result of the cell walls experienced when the cells expressing the enzyme in the cytoplasm or periplasm are used. The targeting of OPH on the cell surface of *E. coli* was achieved by using the Lpp-OmpA anchor system, whereas the ice-nucleation protein anchor derived from *Pseudomonas syriange* was used to display OPH on the cell surface of *Moraxella* sp. The targeting of OPH on the cell surface afforded a seven-fold increase in the parathion hydrolysis rate compared to cells expressing a similar amount of OPH intracellularly and a significantly improved stability [44,45]. Cells expressing OPH on the cell surface were combined with potentiometric, optical, and amperometric transducers to construct microbial electrodes [46–48]. When compared to the enzyme-based biosensors, the microbial biosensors had comparable sensitivity, selectivity, and operational and storage lifetimes.

Recently, this group genetically engineered a PNP degrader *Pseudomonas putida* JS444 to endow it with OPH activity on the cell surface and combined it with the Clark-dissolved oxygen electrode to build a simple microbial biosensor for sensitive and selective determination of PNP-substituted OPs [47]. Surface-expressed OPH catalyzed the hydrolysis of organophosphorus pesticides with a *p*-nitrophenyl substituent such as paraoxon, methyl parathion, and parathion to release PNP that was oxidized by the enzymatic machinery of *Pseudomonas putida* JS444 to carbon dioxide while consuming oxygen. The oxygen consumption was measured and correlated to the concentration of organophosphates. The sensor signal and response time were optimized with 0.086 mg dry weight of cell and operated in 50 mM pH 7.5 citrate-phosphate buffer with 50 μM CoCl$_2$ at room temperature. When operated at optimized conditions, the biosensor measured as low as 55 ppb of paraoxon, 53 ppb of methyl parathion, and 58 ppb of parathion without interference from most phenolic compounds and other commonly used pesticides such as atrazine, coumaphos, sutan, sevin, and diazinon. The operational life of the microbial biosensor was approximately five days when stored in the operating buffer at 4°C.

17.7 LAB-ON-A-CHIP DEVICES FOR SEPARATION AND DETECTION OF OP NERVE AGENTS

Miniaturized analytical devices based on microfluidic chips are attracting growing interest as a result of their enhanced performance, faster response, high degree of integration, and substantial reduction in the overall weight and size. An on-chip separation/detection microsystem can meet the requirements of on-site monitoring or rapid field detection of CWAs. A miniaturized analytical system for separating and detecting toxic organophosphate nerve agent compounds based on the coupling of a micromachined capillary electrophoresis chip with a thick-film amperometric detector was developed by Wang et al. [49]. The separation was based on micellar electrokinteic chromatography (MEKC) followed by sensitive electrochemcial detection of the OP nerve agents.

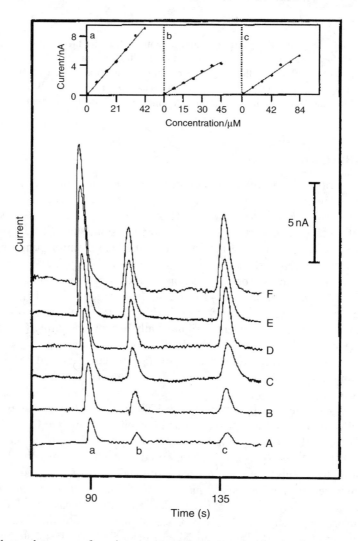

FIGURE 17.15 Electropherograms for mixtures containing increasing levels of paraoxon (a), methyl parathion (b), and fenitrothion (c) in steps of 7.1×10^{-5}, 7.5×10^{-5}, and 1.4×10^{-4} M, respectively (a–f). Also shown (as insets) are the resulting calibration plots. Separation buffer, 20 mM MES (pH 5.0) containing 7.5 mM sodium dodecyl sulfate (SDS); separation voltage, $+2000$ V; injection voltage, $+1500$ V; injection time, 3 s; detection potential, -0.5 V (versus Ag/AgCl wire) at bare carbon screen-printed electrode. (From Wang, J., Chatrathi, M. P., Mulchandani, A., and Chen, W., *Anal. Chem.*, 73, 1804, 2001. With permission.)

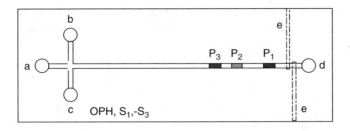

FIGURE 17.16 Layout of the capillary electrophoresis (CE) microchip electrophoretic system with the contactless conductivity detector for enzymatic assays of organophosphate (OP) nerve agents. Running buffer reservoir (a), unused reservoir (b), and sample reservoir (c) containing the substrates S_1–S_3 along with organophosphorus hydrolase (OPH), separated enzyme reaction products (P_1–P_3); outlet reservoir (d); aluminum sensing electrodes (e). (From Wang, J., Chen, G., Muck, A., Chatrathi, M. P., Mulchandani, A., and Chen, W., *Anal. Chim. Acta*, 505, 183, 2004. With permission.)

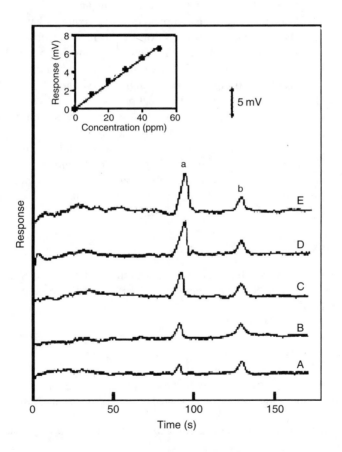

FIGURE 17.17 Electropherograms for mixtures containing increasing levels of methyl parathion (a) in steps of 10 ppm (a–e) and in the presence of 20 mg/l paraoxon (b). Also shown (inset) is the resulting calibration plot based on the reaction of organophosphorus hydrolase (OPH) with the organophosphate (OP) nerve agents in the conditions of: separation voltage, 1000 V; injection voltage, 1000 V; injection time, 1 s; frequency, 200 kHz; peak-to-peak amplitude, 10 V; sinusoidal waveform; nerve agents concentration, 50 mg/l each; OPH activity, 80 U/ml; running buffer, MES/HIS (5 mM, pH 6.1). (From Wang, J., Chen, G., Muck, A., Chatrathi, M. P., Mulchandani, A., and Chen, W., *Anal. Chim. Acta*, 505, 183, 2004. With permission.)

Factors influencing the on-chip separation and detection processes such as applied potential and composition of the running buffer were optimized. Using a 2-(N-morpholino) ethanesulfonic acid (MES) buffer (20 mM, pH 5.0) running buffer, a 72-mm long separation channel, and a separation voltage of 2000 V, baseline resolution was observed for paraoxon, methyl parathion, fenitrothion, and ethyl parathion in 140 s. A typical electropherogram for sample mixtures containing increasing levels of OP insecticides is shown in Figure 17.15. The miniaturization and speed advantages were coupled to submicromolar detection limits and good precision. Applicability of the system for the detection of river water spiked with OP compounds was demonstrated. Further improvement in the sensitivity of monitoring for various electroactive OP warfare agents as desired for various defense and environmental applications can be brought about by on-chip preseparation and preconcentration.

Combining capillary electrophoresis (CE) with the selectivity and amplification provided by enzymes can lead to enhanced analytical features and versatility of microchip devices. Recently, this group developed an on-chip enzymatic assay for screening OP nerve agents based on a precolumn reaction of OPH, electrophoretic separation by CE of the phosphonic acid products, and their contactless-conductivity detection [50]. Figure 17.16 shows the layout of the microchip enzymatic assay. Factors affecting the enzymatic reaction and the separation and detection processes were assessed and optimized. The complete bioassay required 1 min of the OPH reaction along with 1–2 min for the separation and detection of the reaction products. Operating under optimized conditions, the response was linear with detection limits of 5 and 3 ppm for paraoxon and methyl parathion, respectively (Figure 17.17). Compared to conventional OPH-based biosensors, the OPH-biochip was able to differentiate between the individual OP substrates. The attractive behavior of the new OPH-based biochip indicates great promise for field screening of OP pesticides and nerve agents. The study also demonstrated, for the first time, the suitability of the contactless-conductivity detection for on-chip monitoring of enzymatic reactions.

In conclusion, various biosensor devices have been developed for the selective, sensitive, direct, fast, and cost-effective detection and screening of OP compounds intermittently or continuously in the field. Such sensors can be extremely important tools for homeland security and environmental monitoring in order to verify the presence and identity of these dangerous chemicals.

ACKNOWLEDGMENTS

This work was supported by grants from the U.S. Environmental Protection Agency (R8236663-01-0, R82816001-0), the National Science Foundation (BES 9731513), and the U.S. Department of Agriculture (99-35102-8600).

REFERENCES

1. Compton, J. A., *Military Chemical and Biological Agents*, Telford Press, Caldwell, NJ, 1988.
2. Food and Agricultural Organization of the United Nations (FAO), Rome, *FAO Product Yearbook*, 43, 320, 1989.
3. US Department of Agriculture, *Agricutural Statistics*, US Government Printing Office, Washington, DC, p. 395, 1992.
4. Donarski, W. J., Dumas, D. P., Heitmeyer, D. P., Lewis, V. E., and Raushel, F. M., Structure–activity relationships in the hydrolysis of substrates by the phosphotriesterase from *Pseudomonas diminuta*, *Biochemistry*, 28, 4650, 1989.
5. Chapalamadugu, S. and Chaudhry, G. S., Microbiological and biotechnological aspects of metabolism of carbamates and organophosphates, *Crit. Rev. Biotechnol.*, 12, 357, 1989.
6. US EPA, *Pesticide Market Estimates: Usage. 1998–1999*, http://www.epa.gov/oppbead1/pestsales/99pestsales/usage1999_3.html

7. Gilliom, R. J., Barbash, J. E., Kolpin, D. W., and Larson, A. J., Testing water quality for pesticide pollution, *Environ. Sci. Technol.*, 133, 164, 1999.
8. Sherma, J., Pesticides, *Anal. Chem.*, 65, R40, 1993.
9. Yao, S., Meyer, A., Henze, G., and Fresnius, J., Comparison of amperometric and UV-spectrophotometric monitoring in the HPLC analysis of pesticides, *Anal. Chem.*, 339, 207, 1991.
10. Palchetti, I., Cagnini, A., Del Carlo, M., Coppi, C., Mascini, M., and Turner, A. P. F., Determination of acetylcholinesterase pesticides in real samples using a disposable biosensor, *Anal. Chim. Acta*, 337, 315, 1997.
11. Diehl-Faxon, J., Ghindilis, A. L., Atanasov, P., and Wilkins, E., Direct electron transfer based tri-enzyme electrode for monitoring of organophosphorus pesticides, *Sens. Actuators, B*, 35–36, 448, 1996.
12. La Rosa, C., Pariente, F., Hernandez, L., and Lorenzo, E., Determination of organophosphorus and carbamic pesticides with an acetylcholinesterase amperometric biosensor using 4-aminophenyl acetate as substrate, *Anal. Chim. Acta*, 295, 273, 1994.
13. Martorell, D., Céspedes, F., Martínez-Fáregas, E., and Alegret, S., Amperometric determination of pesticides using a biosensor based on polishable graphite-epoxy biocomposite, *Anal. Chim. Acta*, 290, 343, 1994.
14. Skladal, P., Determination of organophosphate and carbamate pesticides using cobalt phthalocyanine-modified carbon paste electrode and a cholinesterase enzyme membrane, *Anal. Chim. Acta*, 252, 11, 1991.
15. Marty, J. -L., Sode, K., and Karube, I., Biosensor for detection of organophosphate and carbamate insecticides, *Electroanalysis*, 4, 249, 1992.
16. Palleschi, G., Bernabei, M., Cremisini, C., and Mascini, M., Determination of organophosphorus insecticides with a choline electrochemical biosensor, *Sens. Actuators, B*, 7, 513, 1992.
17. Skladal, P. and Mascini, M., Sensitive detection of pesticides using amperometric sensors based on cobalt phthalocyanin-modified composite electrodes and immobilized cholinesterase, *Biosens. Bioelectron.*, 7, 335, 1992.
18. Mionetto, N., Marty, J. -L., and Karube, I., Acetylcholinesterase in organic solvents for the detection of pesticides: biosensor application, *Biosens. Bioelectron.*, 9, 463, 1994.
19. Trojanowicz, M. and Hitchman, M. L., Determination of pesticides using electrochemical biosensors, *Trends Anal. Chem.*, 15, 38, 1996.
20. Kumaran, S. and Tranh-Minh, C., Determination of organophosphorus and carbamate insecticides by flow injection analysis, *Anal. Biochem.*, 200, 187, 1992.
21. Tran-Minh, C., Pandey, P. C., and Kumaran, S., Studies of acetylcholine sensor and its analytical application based on the inhibition of cholinesterase, *Biosens. Bioelectron.*, 5, 461, 1990.
22. Chuna Bastos, V. L. F., Chuna Bastos, J., Lima, J. S., and Castro Faria, M. V., Brain acetylcholinesterase as an in vitro detector of organophosphorus and carbamate insecticides in water, *Water Res.*, 25, 835, 1991.
23. Kumaran, S. and Morita, M., Application of a cholinesterase biosensor to screen for organophosphorus pesticides extracted from soil, *Talanta*, 42, 649, 1995.
24. Dzyadevich, S. V., Soldatkin, A. P., Shul'ga, A. A., Strikha, V. I., and El'skaya, A. V., Conductometric biosensor for determination of organophosphorus pesticides, *J. Anal. Chem.*, 49, 874, 1994.
25. Rogers, K. R., Cao, C. J., Valdes, J. J., Eldefrawi, A. T., and Eldefrawi, M. E., Acetylcholinesterase fiber-optic biosensor for detection of acetylcholinesterases, *Fund. Appl. Toxicol.*, 16, 810, 1991.
26. Hobel, W., Polster, J., and Fresenius, J., Fiber optic biosensor for pesticides based on acetylcholine esterase, *Anal. Chem.*, 343, 101, 1992.
27. Garcia de Maria, C., Munoz, T. M., and Townhend, A., Reactivation of an immobilized enzyme reactor for the determination of acetylcholinesterase inhibitors. Flow injection determination of paraoxon, *Anal. Chim. Acta*, 295, 287, 1994.
28. Moris, P., Alexandre, I., Roger, M., and Remacle, J., Chemiluminescence assay of organophosphorus and carbamate pesticides, *Anal. Chim. Acta*, 302, 53, 1995.
29. Munnecke, D. M., Enzymatic detoxification of waste organophosphate pesticides, *J. Agric. Food Chem.*, 28, 105, 1980.
30. Dumas, D. P., Wild, J. R., and Raushel, F. M., Diisopropylfluorophosphate hydrolysis by a phosphotriesterase from *Pseudomonas diminuta*, *Biotech. Appl. Biochem.*, 11, 235, 1989.

31. Dumas, D. P., Caldwell, S. R., Wild, J. R., and Raushel, F. M., Purification and properties of the phosphotriesterase from *Pseudomonas diminuta*, *J. Biol. Chem.*, 33, 19659, 1989.

32. Dumas, D. P., Durst, H. D., Landis, W. G., Raushel, F. M., and Wild, J. R., Inactivation of organophosphorus nerve agents by the phosphotriesterase from *Pseudomonas diminuta*, *Arch. Biochem. Biophys.*, 227, 155, 1990.

33. Mulchandani, P., Mulchandani, A., Kaneva, I., and Chen, W., Biosensor for direct determination of organophosphate nerve agents. 1. Potentiometric enzyme electrode, *Biosens. Bioelectron.*, 14, 77, 1999.

34. Schoning, M. J., Mulchandani, P., Chen, W., and Mulchandani, A., A capacitive field-effect sensor for the direct determination of organophosphorus pesticides, *Sens. Actuators, B*, 91, 92, 2003.

35. Rogers, K. R., Wang, Y., Mulchandani, A., Mulchandani, P., and Chen, W., Organophosphorus hydrolase-based fluorescence assay for organophosphate pesticides, *Biotechnol. Prog.*, 15, 517, 1999.

36. Mulchandani, A., Pan, S., and Chen, W., Fiber-optic biosensor for direct determination of organophosphate nerve agents, *Biotechnol. Prog.*, 15, 130, 1999.

37. Mulchandani, A., Mulchandani, P., Chen, W., Wang, J., and Chen, L., Amperometric thick-film strip electrodes for monitoring organophosphate nerve agents based on immobilized organophosphorus hydrolase, *Anal. Chem.*, 71, 2246, 1999.

38. Wang, J., Chen, L., Mulchandani, A., Mulchandani, P., and Chen, W., Remote biosensor for in-situ monitoring of organophosphate nerve agents, *Electroanalysis*, 11, 866, 1999.

39. Mulchandani, P., Chen, W., and Mulchandani, A., Flow injection amperometric enzyme biosensor for direct determination of organophosphate nerve agents, *Environ. Sci. Technol.*, 35, 2562, 2001.

40. Mulchandani, A., Chen, W., Mulchandani, P., Wang, J., and Rogers, K. R., Biosesnors for direct determination of organophosphate pesticides, *Biosens. Bioelectron.*, 16, 225, 2001.

41. Deo, R. P., Wang, J., Block, I., Mulchandani, A., Joshi, K. A., Trojanowicz, M., Scholz, F., Chen, W., and Lin, Y., Determination of organophosphate pesticides at a carbon nanotube/organophosphorus hydrolase electrochemical biosensor, *Anal. Chim. Acta*, 530, 185, 2005.

42. Lei, Y., Mulchandani, P., Chen, W., Wang, J., and Mulchandani, A., Whole cell-enzyme hybrid amperometric biosensor for direct determination of organophosphate nerve agents with p-nitrophenyl substituents, *Biotechnol. Bioeng.*, 85, 706, 2004.

43. Wang, J., Krause, R., Block, K., Musameh, M., Mulchandani, A., Mulchandani, P., Chen, W., and Schöning, M. J., Dual amperometric-potentiometric biosensor detection system for monitoring organophosphate neurotoxins, *Anal. Chim. Acta*, 469, 197, 2002.

44. Richins, R., Kaneva, I., Mulchandani, A., and Chen, W., Biodegradation of organophosphorus pesticides by surface-expressed organophosphorus hydrolase, *Nat. Biotechnol.*, 15, 984, 1997.

45. Mulchandani, A., Kaneva, I., and Chen, W., Detoxification of organophosphate nerve agents by immobilized *Escherichia coli* with surface-expressed organophosphorus hydrolase, *Biotechnol. Bioeng.*, 63, 216, 1999.

46. Mulchandani, A., Mulchandani, P., Kaneva, I., and Chen, W., Biosensor for direct determination of organophosphate nerve agents using recombinant *Escherichia coli* with surface-expressed organophosphorus hydrolase. 1. Potentiometric electrode, *Anal. Chem.*, 70, 4140, 1998.

47. Mulchandani, A., Kaneva, I., and Chen, W., Biosensor for direct determination of organophsophate nerve agents using recombinant *Escherichia coli* with surface-expressed organophosphorus hydrolase. 2. Fiber-optic microbial biosensor, *Anal. Chem.*, 70, 5042, 1998.

48. Lei, Y., Mulchandani, P., Chen, W., and Mulchandani, A., Direct determination of p-nitrophenyl substituent organophosphorus nerve agents using recombinant *Pseudomonas putida* JS444-modified Clark oxygen electrode, *J. Agric. Food Chem.*, 53, 524, 2005.

49. Wang, J., Chatrathi, M. P., Mulchandani, A., and Chen, W., Capillary electrophoresis microchips for rapid separation and detection of organophosphate nerve agents, *Anal. Chem.*, 73, 1804, 2001.

50. Wang, J., Chen, G., Muck, A., Chatrathi, M. P., Mulchandani, A., and Chen, W., Microchip enzymatic assay of organophosphate nerve agents, *Anal. Chim. Acta*, 505, 183, 2004.

18 Bioarrays: Current Applications and Concerns for Developing, Selecting, and Using Array Technology

Joany Jackman

CONTENTS

18.1 BACKGROUND OF ARRAY TECHNOLOGY

18.1.1 DEFINING ARRAYS

Assay multiplexing is a means to provide the maximum amount of information about a sample while conserving reagents, conserving sample, and increasing the speed of analysis. Microarray technology is assay multiplexing on steroids. Simply put, microarrays are a logical presentation of biological molecules laid out on a solid planar platform, generally a glass slide [1]. The resulting presentation is referred to as an array or as it is sometimes called a chip. In this format, individual molecules are separated from one another, allowing the user to interrogate each molecule independently but simultaneously. As a result, it is possible to simultaneously obtain relational information about hundreds and thousands of different events. No other technique provides such a data rich environment in a single test. The types of molecules that may be presented or arrayed are discussed in more detail below include DNA oligonucleotides, cDNAs, peptides, or sugar molecules, although each array contains multiples of only one type of molecule.

Microarray technology revolutionized the way that genomic information is used and analyzed. A full history of microarrays can be found in reviews by Venkatasubbarao [2] and Southern [3]. Microarray technology for the purpose of genetic analysis was conceived about 10 years ago. Whereas Southern was the first to print microarrays on glass slides, recognizing the potential value of microarrays for multiple probes in parallel, it was Brown's group who first reduced to practice the concept of simultaneously monitoring 1000 genes for alterations in changes of gene expression using cDNA arrays as capture probes [4,5]. One way of understanding how microarrays work would be to compare them to older, more well-known technologies from which the concept of microarrays was conceived. In some sense, microarrays can be related to Southern blot and associated dot blot technologies (Figure 18.1). Researchers using dot blots or Southern blots use a labeled probe to determine the presence of a gene or amount of a transcript in a variety of targets arrayed in columns or in 96 well format. The definition of a target is application dependent. Targets can be derived from a single subject at various times or conditions or derived from multiple subjects. The purpose of these types of experiments is to compare a single gene or sequence variant in a variety of targets. In this way, the targets are arrayed when presented to the probe. The probe is labeled using fluourescent, radioactive, or enzymatic tags. Probes are used singly or in small groups to interrogate the target sequence. Detection occurs when the labeled probe is hybridized or captured out of solution by its complimentary target sequence.

In contrast to the Southern blot technology, microarrays present or array a number of different probes rather than targets. The target is applied to the array, and in this manner, a single sample is interrogated by hundreds or thousands of probes. When bound to a microarray, the probes, now called capture probes, differ from probes used by Southern blots in that they are unlabeled. Instead, the target molecules are labeled using, in most cases, fluorescent tags although electrochemical

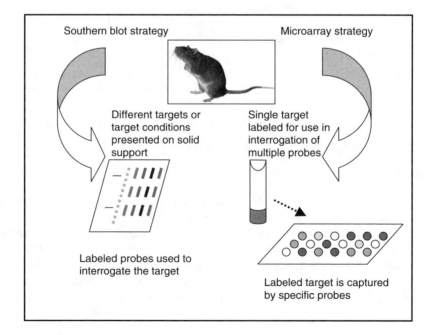

FIGURE 18.1 Comparison of microarray technology and southern blots.

methods exploiting ruthidium chemistry and other methods that generate observable signals have been used in a few systems. The target is captured by a subset of probes linked to a substrate. The binding is quantifiable. Detection occurs when the labeled target is retained by the surface-bound probe. Tens, hundreds, or thousands of probes can be interrogated against a single target at once. As a result, a more complete picture of a single target organism is obtained by defining the relationship of gene transcripts to one another or the appearance of specific genetic polymorphisms to one another. In contrast, Southern blot technology describes the behavior of a single gene in a population, a pathway, or an environment. Both approaches are valid, but they reflect different viewpoints, addressing similiar scientific questions. Microarray technology dramatically improves the ability of researchers to study genes in functional groups or biochemical pathways rather than in isolation.

Labeling materials and methods used in microarray technology have been taken directly or adapted from other methods used to label specific molecules in other technologies such as ELISA and PCR assays. Many of these have already been described in detail by other authors in the volume (Heineman, Biagini, Cooper, Dill, Mulchandani, and Nie). As should be noted from this list of related methods, almost any labeling method can be adapted or applied to microarray technology.

18.1.2 Array Formats

Arrays are generally printed directly onto glass slides and are referred to as planar or two-dimensional arrays (2-D). Three-dimensional (3-D) arrays refer to glass slides modified to contain hydrogel (protein arrays) or polyacrylamide (gene arrays) layers into which probes are introduced. Microarrays are grouped by type and density of arrayed molecules: low density contain tens of individual elements, medium density contain hundreds of elements, and high density arrays contain thousands of elements. High density arrays such as the GeneChip Human Genome array (Affymetrix) contain over 54,000 elements and evaluate over 1 million different binding events on a single array [6] although high density arrays containing between 10,000 and 25,000 elements are a more common size (Amersham, Agilent, Combimatrix).

Bead-based microarrays are notable as an alternative array format to the planar array [7,8]. In one version, the arrayed molecules are placed on polymer bead surfaces that mix in solution [9]. Whereas the beads are a solid surface, the array itself is not fixed in space. The visual array does not exist in real time, but it can only be visualized in virtual space through the use of a flow cytometer and computer. It should be noted, therefore, that these types of arrays, although fast and capable of imaging up to 100 different beads, do not offer the benefit of physical separation of assayed materials as in the glass slide format. Generally, this type of assay format is limited to about 100 simultaneous tests. In another version of the bead based array, the molecules are attached to beads, and the beads are then arrayed on the tips of optical fibers or planar surfaces [10–12]. These bead-based systems are able to simultaneously array thousands of surface bound beads and do offer the advantage of physical separation of probes.

Tissue microarrays (TMAs) use pieces of various tumor tissues on an array as targets. These small tissue slices are logically patterned onto glass slides and probed to obtain profiles of gene expression [13]. Although they are also considered microarrays, tissue arrays and cell arrays are substantially different in set up and analysis from other types of arrays mentioned here as both these array types represent arrays of surface captured targets rather than individual probes. Hence, TMAs are more related to Southern blot technology with the exception that the organ specific architecture of the target is maintained. Frequently, biochemical pathways (cell-based arrays (CBAs)) or biomarkers (TMAs) initially described or discovered using DNA or protein arrays can subsequently be rapidly screened against populations of tumors and metabolic responses. Of concern in this particular technology is the issue of uniformity and reproducibility. TMAs and CBAs are substantially different in their set up and handling. Because TMAs and CBAs are derived from complex biological materials, organ and tumor microheterogenity or cell population variations occur naturally and present a greater quality control concern than for synthetic arrays. More detailed discussions of this unique variant of microarray technology are addressed in reviews by Fedor [14], Wheeler [15], and Kallioniemi [16].

Although it is clear that microarray technology can revolutionize approaches to medicine and improve understanding of complex biological processes, it is important to note that microarray technology remains primarily a research tool at this time. The challenges, to date, have been overcoming the quality control issues in microarray manufacture, simultaneous validation of multiple probes that might be present on a single microarray, and understanding and distinguishing the important features from among a background of hundreds or thousands of datapoints. In January of 2005, the first microarray to be considered for U.S. Food and Drug Administration (U.S. FDA) approval as a diagnostic test was introduced by Roche Molecular Systems [17]. Whereas technically this is a microarray, it is limited in size and scope. The AmpliChip CYP450 genotyping test evaluates subjects for the presence of variation in only two P450 genes, CYP2D6 and CYP2C19, accounting for 33 distinct allelic patterns [18]. Although limited in the number of polymorphisms that can be simultaneously assessed, the impact of advancing microarray technology into the treatment and diagnosis of patients is notable. Microarrays have been recognized to be a key technology for the practical application of pharmocogenomics and toxicogenomics and have enabled the concept of individualized medicine as a practicable idea [19,20].

18.2   MICROARRAY ADVANTAGES AND DISADVANTAGES

18.2.1   ADVANTAGES

The advantages of microarrays are measured in sheer numbers—the number of different genes simultaneously quantified, the number of different single nucleotide polymorphisms (SNPs) simultaneously analyzed, the number of biochemical pathways evaluated, or the number of binding sites available (Table 18.1). As a general note, at the time of this writing, over 140 manufacturers are involved in production of microarrays, microarray readers, or software for analysis of microarrays (Table 18.2 through Table 18.4). Hence, the lists provided are not meant to be comprehensive, but to represent a cross section of suppliers. Two websites, www.biocompare.com and www.functionalgenomics.org.uk/sections/resources maintain up-to-date lists of commercial suppliers (Table 18.4). As this field is expanding, new manufacturers, products, and users are adding

---

TABLE 18.1
Microarray Balance Sheet

Advantages
Comparison large numbers of genes or protein simultaneously
Ability to profile multiple polymorphisms in a single individual, tissue, or cell line
Conservation of analyte
Enables data driven approaches for discovery of natural associations

Disadvantages
Costly investment in printed arrays, imagers, and/or print facilities
May be less rapid than other genetic or protein-based tests
Reproducibility and sensitivity difficult to measure
Data management more challenging
Quality metrics are evolving

TABLE 18.2
Selected Sources of Commercial Printed Microarrays

Company	Website	Target[a]	Types	Uses	Species
Agilent	www.chem.agilent.com	N	Spotted	Gene expression, re-sequencing, genome profiling	Arabidopsis, Human, Mouse, Magneporta grisea, Rat, Rhesus Monkey, Xenopus, Zebrafish
Combimatrix	www.combimatrix.com	N	In situ	Genome profiling, disease profiling	Human, DogMouse, Rat, Fish, E.coli, Fungus, Cotton, Malaria, Virus
Glycominds	www.glycochip.com	G	Spotted	Antibody profiling, infectious disease profiling	Mammalia, Bacteria (Various)
Invitrogen	www.invitrogen.com	P, QC	Spotted	Genome profiling, quality assurance	Human, Mouse, Yeast
Affymetrix	www.affymetrix.com	N, QC	In situ	Sequencing, gene expression, SNP, quality assurance	Human, Mouse
ArrayIt	www.arrayit.com	N, P, QC	Spotted	Gene expression, transcription profiling, quality assurance	Human, Mouse, Rat, Arabidopsis
GE Healthcare (formerly Amersham Biosciences)	www5.amershambiosciences.com/aptrix/upp01077.nsf/Content/microarrays_introduction_page	N	Spotted	Genomic profiling, targeted gene profiling	Human, Mouse, Rat
Procognia	www.procognia.com/index	G, P	Spotted	Protein folding dependent arrays, drug screening, glycomic profiling	Human, Mouse
Stratagene	www.stratagene.com	QC	Spotted	Reference RNA (MAQC control)	Human, Mouse, Rat
Plexigen	www.plexigen.com	N, P	Spotted	Antigen/antibody profiling, genomic profiling	Cattle, Swine, Bacteria, Virus
Allied Biotech	www.alliedbiotech.com	P	Spotted	Pathway profiling, disease profiling	Human, Mouse, Rat, Virus
Super Array	www.superarray.com	N	Spotted	Biochemical/disease pathway profiling	Human, Mouse, Rat
Sigma	www.sigma.com	T, P	Spotted	Tumor profiling, cell signaling pathways	Human, Mouse, Rat
Illumina	www.illumina.com	n	Bead	SNP, Gene expression	Human, Mouse
Panomics	www.panomics.com	P, N	Spotted	Antibody array	Human, Mouse
Biogenex	www.biogenex.com	T	—	Tumor/disease profiling, organ profiling	Human, Mouse, Rat
Biochain	www.biochain.com	N, P, T	Spotted	Fetal tissue profiling, tumor profiling, gene expression, protein expression	Human

[a] N, Nucleic acid; P, Protein; G, Glycan; T, Tissue/cell; QC, Quality control.

TABLE 18.3
Selected Manufacturer of Bioarray Readers

Device/Company	Website	Type	Features	Processing
Agilent	www.agilent.com	Laser Scanning	2 Color	Serial
Axon (Molecular devices)	www.moleculardevices.com/pages/instruments/microarray_main.html	Laser Scanning	2 Color and 4 color	Serial
Perkin elmer	www.las.perkinelmer.com	Laser Scanning	Up to 5 color	Serial
Aurora photonics	www.auroraphotonics.com	Flat field illumination	2 Color and 4 color, thermal stage	Parallel
Nanogen	www.nanogen.com/technologies/microarray	Flat field illumination	2 Color	Parallel
Affymetrix	www.affymetrix.com	Laser Scanning	2 Color	Serial
Tecan	www.tecan.com/platform/apps/product	Laser Scanning	2 Color and 4 color	Serial
Genomic solutions	www.genomicsolutions.com	Laser Scanning	2 Color and 4 color	Serial
Arrayit	www.arrayit.com	Laser scanning	2 Color and 6 color	Serial
Luminex	www.luminexcorp.com/01_xMAPTechnology/01_Products/01_Instuments	Flow cytometer	1 Color detection/2 colors probe address	Parallel

technology so quickly that web-based resources should be used and compared to supplement selection of the most appropriate versions of this technology.

Most microarray applications use high density arrays with thousands of elements. High density arrays are advantageous in that it is now possible to obtain large amounts of relational information about a single target quickly. In fact, for the first time, it is possible to obtain a complete picture of an individual's transcriptional status. However, the challenges to this approach are additive. To the typical challenges of gene expression analysis are added the challenge of data handling and data mining on a massive level. If pair-wise comparisons between 1000 different genes are performed, the number of potential data comparisons can approach a million. Manual analysis of high and medium density microarray results is no longer possible. Therefore, data analysis programs supplied as part of microarray readers or developed independently by bioinformatics sources are a requirement for use of microarrays. Together, these technologies permit users to apply methods of scientific investigation where data drive the interpretation via natural association in large data sets rather than the more classical methods that are hypothesis driven. This is often referred to in the microarray community as unsupervised (data driven) and supervise (hypothesis driven) approaches to data analysis.

Use of unsupervised discovery approaches do not obviate the requirement for good experimental design and a clear biological question. The idea of unsupervised data analysis is that instead of developing the support for a pre-conceived hypothesis, researchers can allow the data to drive natural associations between complex biological phenomenon as a means of hypothesis development. When performed correctly, it is assumed that the data set contains all the necessary results for an unbiased interpretation. In this manner, microarray technologies have permitted researchers to investigate biological pathways and multi-gene disease models in a manner not previously possible. Methods to normalize signal, control samples, expression levels, and data mining have been the subject of numerous publications and technical reviews, and they are important considerations before designing or even working with the first microarray [21–23]. Improved imaging processing

TABLE 18.4
Selected Microarray Software: Commercial and Non Commercial Sources

Program	Features	Source	Web Site
Rosetta Resolver	Data management	Rosetta Biosoftware	http://www.rosettabio.com/default.htm
	Gene expression analysis		
	Data sharing		
QRI	Signal processing software	Vialogics	www.vialogy.com
	Gene expression analysis		
Geo	Gene expression databrowser	NCBI	www.ncbi.nlm.nih.gov/geo
Matchminer	Coordinates gene names	NCI	www.discover.nci.nih.gov/matchminer
GeneSight	Supervised and unsupervised analysis	Biodiscovery	www.biodiscovery.com
TM4	Statistical analysis suite	TIGR	www.tigr.org/software/microarray.shtml
MAPS	Database tool and statistical analysis	NIEHS	www.dir.niehs.nih.gov/microarray/software/maps
Genespring	Statistical analysis	Agilent	www.chem.agilent.com
Nimblescan, Array Start	Statistical tools	Nimblegen	www.nimblegen.com/products/software/index.html
ArrayScribe	Design tool	Nimblegen	www.nimblegen.com/products/software/index.html
GeneTraffic	Database tool and statistical analysis	Stratagene	www.stratagene.com
Acuity	Database tool and statistical analysis	Molecular Devices	www.axon.com
GEArray Expression Analysis Suite	Web based subscription to analysis tools	Superarray	www.superarray.com/support_software.php
GTYPE/GSEQ	Analysis tools for genotyping and sequence analysis	Affymetrix	www.affymetrix.com/products/software
VersArray	Image analysis	Biorad	www.biorad.com
Cluster/TreeView/ SAM	Supervised and unsupervised analysis programs	Eisen laboratory	genome-www5.stanford.edu/resources/restech.shtml
Freeware Site	Various tools: image, database, cluster analysis		www.nslij-genetics.org/microarray/soft.html

programs and imagers are being developed in order to extract more information concerning differential gene expression where expression levels between the control sample and test sample vary by as little as two-fold.

18.2.2 DISADVANTAGES

The primary disadvantages of microarrays are cost, speed, and sensitivity (Table 18.1). Generally, the larger the number of different elements, the higher the cost and the slower the analysis. While relative cost is certainly less than if each gene was analyzed individually, establishment of the hardware and software is more costly than other types of genetic analysis technology such as PCR or protein assays such as ELISAs. Inclusive of sample preparation, target detection, and assay analysis, microarrays require significantly more time than real time PCR or rapid lateral flow assays. Without the use of PCR for sample preparation of rare genetic targets, microarrays are not as sensitive for genetic detection as PCR alone. Cost of the microarray itself impacts other areas of microarray technology, including reproducibility and validation. High density arrays composed of oligonucleotides can cost more than $1000 per array. With minor exception, microarrays cannot be stripped of signal and reused. Combimatrix™ Custom and 4×2 arrays can be stripped up to four times as long as the oligonucleotide capture probes do not exceed 40 mers in length. Even so, as per the manufacturer's web site, stripped microarrays may show a reduction in signal following stripping (www.combimatrix.com/support_faq4X2.htm). A second microarray to be excepted from the no stripping rule is the MAGIChip™ [24]. The MAGIChip™, a gel-based, spotted 21 mer synthetic oligonucleotide array designed specifically for rapid discrimination of pathogenic bacteria (Argonne National Laboratory) can be stripped and reused up to 50 times with no loss of signal and no sample memory effect, thus bringing the cost per use to less than $10 [25,26]. This novel microarray uses a proprietary method to create non continuous polyacrylamide patterns to create gel pads. Using a quartz mask, the gel matrix, when affixed in a specific geometry using photopolymerization, creates a low cost platform where proteins, DNA, or RNA can be added. Oligonucleotides are deposited and chemically linked to the matrix in such a manner that duplexes created during a hybridization can be dissociated readily without removal of the single stranded capture probe. The resulting product has the benefit of physically limiting spot geometry and creating a reusable microarray format that reduces per sample cost and simplifies issues relating to reproducibility and assay variation [27].

Costs associated with the introduction of microarray technology are not limited to the purchase of microarray chips alone. To image a microarray requires the purchase of a microarray scanner/-imager and appropriate software as mentioned previously (Table 18.3 and Table 18.4). Not all microarray manufacturers require the user to purchase a specific reader; however, some manufacturers do not guarantee or support the results of their microarray platforms with different imagers. Optical imagers generally fall into two categories: laser scanning readers and flat field illumination imagers. The first type of imager interrogates and acquires data on each spot individually in sequential fashion. The second type uses microscope-like optics to simultaneously capture the signal from all microarray elements in parallel. Unlike most fluorescent microscopes, the illumination occurs evenly across the field of view rather than falling off gradually from the center of the field of view. The disadvantage of parallel type imagers as compared to sequential type imagers is that higher amounts of background fluorescence may be observed as a result of the illumination occuring in areas between pads as well as within the pads. This fluorescence can be removed or reduced by post processing either the image or extracted data. Other types of microarray readers that measure non optical output are under development and some are not yet fully commercialized. Devices such as the Bead Array Counter (BARC) measures magnetic fields when a specific capture event is observed [28,29].

Except for specific limitations referenced by each manufacture, generally any type of microarray that can be imaged by one type of reader can be imaged by another device with a comparable scan area or field of view. This group compared flat field illumination imagers (Aurora Photonics) and found them to display sensitivity comparable to the laser scanning type of readers (GenePix). Although the laser scanning readers were much earlier to market and have been applied to a greater number of applications, the new flat field illumination imagers are one half to one quarter the price of laser scanning readers and, because of the absence of moving parts, have few maintenance issues.

No matter what type of reader is purchased, it is important to make sure that the imager comes with software that supports the export of data to multiple common platforms, minimally TIF or JPEG, in addition to its own formats (Table 18.3). Genepix software requires GAL file types for analysis of data obtained by other programs (www.moleculardevices.com/pages/instrument/gn_genepix4100.html). Whereas vendor specific software and algorithms come with devices, after market software programs have flourished as new methods for handling, analyzing and interpreting microarray data arise.

Finally, if the user intends to create microarrays on site, a microarray printer or spotter will be needed (Table 18.3). The cost of these devices varies tremendously with the amount of robotics and throughput required. Costs generally range from $25,000 to >$100,000. Additional cost may be included to place these devices in an appropriate environment. Random dust present during printing results in spotty background that cannot be removed by any washing protocols and can obscure underlying signals. In this group's experience, even the relatively low cost ARRAYIT printers require environmental conditions that may not be commonly available in research laboratory areas. It found it necessary to place the printers in humidity-controlled CLASS 100 clean rooms with specialized equipment to eliminate vibration in order to print microarrays with demonstrated quality and reproducibility. Extensive methodology has been described in the literature for assessing the quality of microarray components prior to printing [30–32] and for evaluating the spot quality after printing [33–35]. The simplest methods involve the co-deposition or staining of microarray products with fluorescent dyes [36,37]. There are several good resources for chip manufacture protocols and troubleshooting methods [38], (Table 18.5). However, for the small laboratory or laboratory that only intends to print a few custom arrays, it is likely that for the cost required to purchase and maintain a print facility, it will be as cost effective to obtain arrays from a commercial source despite the relatively high cost per array. Printing at core facilities can be a less expensive alternative, but the results are highly dependent upon the experience, equipment, and quality assurance methods employed at each site. Tools for evaluating the quality of microarrays

TABLE 18.5
Selected Web Sites of Interest

MIAME	Minimum information about a microarray experiment	www.med.org/workgroups/MIAME/miame.html
MGED	Microarray gene expression data and ontology tool	www.mged.org
MAQC	Microarray quality control	www.fda.gov/nctr/science/centers/ toxicoinformatics/maqc
Biocompare	Resource for identifying commercial sources	www.biocompare.com
Functional genomics	European Science Foundation site for gene array and protein array information	www.functionalgenomics.org.uk/sections/resources
Array track	Freeware for data management and analysis	www.fda.gov/nctr/science/centers/ toxicoinformatics/ArrayTrack/index.htm
MicroGen	Freeware for data management and analysis	www.bioinformatics.polimi.it/MicroGen
TIGR	Microarray protocols/resources	www.tigr.org/tdb/microarray
The Brown Lab	Microarray protocols/resources	www.cmgm.stanford.edu/pbrown/protocols/
TMA Database	Tissue based microarray data management tool	www.bioinfo.itc.it/TMA/home.htm
dCHIP software	Supervised and unsupervised gene clustering analysis	www.biosun1.harvard.edu/complab/dchip
Functional glycomics	Glycan array, glycan profiling and glycan structure	www.functionalglycomics.org/static/consortium /main.shtml and www.ccrc.uga.edu
Bacterial carbohydrate structure database	Glycans associated with microbes	www.glyco.ac.ru/bcsdb/ start.shtml
3D Glycan database	Glycan structures	www.cermav.cnrs.fr/lectines

are discussed at the end of this chapter. Before designing microarrays, individuals should preview the designs available (Table 18.2) and, in particular, the controls incorporated into commercial chips of similar design for anticipated needs. Although the controls described may not be directly applicable to the desired chip design, they should highlight the types of error checking needed to interpret the array data [23]. Prior to the introduction of flat field illumination imagers, $200,000 was considered a reasonable starting investment for the hardware, software, and disposables required to implement this technology [39].

18.2.3 GETTING STARTED: MICROARRAY SELECTION

For the new user, the first decision to be made is the type and density of microarray to be used. This assumes that the manufacturer of a particular array format is less important to the experimental outcome than are the probes present on the array. This is a point that will be revisited in detail when quality metrics are discussed later in this chapter.

18.2.3.1 Genetic Arrays

By far, the most common application of microarray technology involves interrogation of nucleic acid targets on planar arrays. The first applications used arrays to monitor whole genomes for gene expression differences [5]. Target mRNA from normal and test materials were labeled using different fluorescent dyes, and they were applied to glass slides printed with 1000+ unknown cDNAs derived from transformed human leukocytes. The resulting signal was imaged using excitation filters to distinguish the differences in labeling that defined each of the two RNA target mixtures. Practical applications of this methodology include gene profiling as a means of understanding toxicological mechanisms and comparative gene profiling between tumor and normal tissue [40–42]. Microarrays have been developed to profile SNPs in a single gene and in a subset of ancestral genes located on multiple chromosomes in humans and between species [43,44]. Generally, SNP arrays work by using a ratiometric approach to probe hybridization comparisons. One probe contains a cannonical sequence that is an exact match to a particular target. This is called the perfect match (PM). Other probes are designed to contain single base mismatches (MM). A ratio derived from the signal produced by the PM relative to the MM that is greater than 1 means that the best match to the target contained the same sequence as the PM probe. A ratio less than 1 of the PM:MM means that the mismatch contains the same sequence as the target, and the MM defines the sequence in the target sample [45]. Re-sequencing microarrays have also been developed to entire genomic regions where multiple mutations (transversions, deletions, insertions, etc.) may occur [43]. A ratiometric approach is used to distinguish between overlapping probes containing a single base change for each of four possible bases. Re-sequencing arrays have been reported to have 100 times the accuracy of gel base techniques because of the high redundancy of information at each base [46]. Re-sequencing chips and variants thereof are especially useful in rapid typing of viruses [47] and for monitoring viral re-assortment [48–50]. Other ratiometric approaches combine SNP analysis and re-sequencing arrays to address evolutionary distances between organisms using genetic markers in ribosomal RNA for bacterial identification using microarrays. These types of approaches are useful in identifying members of bacterial communities even from complex environments. Organisms that cannot be cultured can be identified as present based on their genetic identification in mixtures. Oligonucleotides SNPs arrays have been applied to conserved genes for quantitative trait loci (QTL) mapping as a means of understanding the link between phenotypic and genetic differences and for studies of gene evolution [51]. There are fundamental differences in preparation of target materials and controls used for each of these applications of nucleic acid arrays. Some approaches require conversion of RNA into cDNA [41], and others use direct capture of RNA labeled in vitro [25].

Commercial gene arrays can be spotted or synthesized in situ on glass slides. Spotted arrays are most common in both commercial and core facility-based manufacture. Spotted arrays use nucleic acids, either synthetic or derived from amplified cDNA libraries. Oligonucleotides between 20 and 70 base pairs have been successfully used as capture probes. cDNA arrays are prepared from a cDNA library by the laboratory personnel from commercial cDNA cloning kits or obtained from commercial libraries (www.biocompare.com/jump/176/cDNA-Libraries.html). Individual cDNA clones are amplified then spotted onto commercial glass slides modified with amino silane (silanated) or aldehyde coatings (silylated) to permit covalent cross linking of the DNA to the substrate. Whereas many of the commercial sources listed in Table 18.2 will produce custom arrays, the user base for commercial off the shelf (COTS) cDNA and oligonucleotide arrays directed to specific species or tissues has grown so rapidly that companies now provide standardized arrays for whole genome analysis. This eliminates the need to generate cDNA libraries, and it often provides a better platform to compare results with other users. As previously mentioned, planar printed arrays can be 2-D or 3-D. There are several benefits of three-dimensional arrays. For gene arrays, the novel 3-D surface chemistry is comprised of a long-chain, hydrophilic polymer containing amine-reactive groups that are coated over the surface of the slide (Amersham). This polymer is covalently crosslinked to itself and to the surface of the slide. The crosslinked polymer, combined with end-point attachment, orients the immobilized DNA and holds it away from the surface of the slide. This combination makes the DNA more readily available for hybridization and also eliminates the need for inert sequence spacers to reduce steric effects. Additionally, the hydrophilic nature of the polymer reduces background in the final product. Three-dimensional arrays have dynamic range of greater than three logs, much greater than typically observed for two-dimensional arrays. The MAGIChip gel pad array is a variation on continuous gel 3-D arrays where coating is applied as spatially restricted areas referred to as gel pads. As a result, spotting and any subsequent spread of printed oligonucleotides is physically restricted to the gel pad as is probe distribution. As a result, the analysis area and geometry between spots does not vary. This group noted that the performance parameters of oligonucleotides have similar probe specificity on gel pad arrays as on two-dimensional arrays though hybridization on gel pad arrays can be faster for the same application (less than one hour). A newer version of the gel pad 3-D array technology uses a co-deposition method spotting the polymer and the oligomer at the same time and is referred to as a gel drop array (Akonni). The benefit of this technology is that these chips can be spotted more rapidly, require less reagent, and are lower cost to produce. At equilibrium conditions as occur with overnight hybridization, the gel drop or co-polymerization chip performs similarly to the gel pad chips (Jackman, pers. comm.). However, at non equilibrium conditions as defined by shorter hybridization times (less than one hour), the gel drop chips have very dissimilar probe performance. Therefore, longer hybridzation times or higher sample concentrations were needed to permit comparisons between experiments.

In situ synthetic arrays are generally only a commercial product because of the cost of the equipment and the expertise required to create chips. In this case, oligonucleotides are not synthesized prior to printing, but the phosphoramidite chemistry is carried out directly on the chip in each location so that each probe is built on the substrate. Affymetrix and Combimatrix both use in situ synthesis methods with slight differences. Affymetrix uses patterned light to direct sequence specific synthesis via classical combinatorial chemistry. Combimatrix pairs combinatorial chemistry with complementary metal-oxide semiconductor technology. Electrical current is used to address synthesis to specific array elements. Microelectronics direct the coordinated synthesis of thousands of oligonucleotides simultaneously. Both companies produce standardized arrays against a number of species and of different levels of complexity as well as offering services to make custom arrays, giving users a broad spectrum of targets to investigate.

18.2.3.2 Protein Arrays

Protein arrays are a relatively new technology, and they have substantially more quality control issues relating to their use [52]. Generally, these arrays are used to profile differential protein expression between tumor and normal tissue and between disease and non-disease states; they are also used to look for organ-specific differences in protein expression (Biochain, Sigma). Whole organism protein/peptide chips specific for infectious organisms or inflammatory responses are also available (Combimatrix; Plexigen). Detection is achieved by one of two common methods. One method uses an antibody sandwich capture detection format (ELISA-like). The variation from a standard ELISA is that multiple probes are spotted into a single well or membrane [53]. Membrane arrays may use a glass slide support depending on the manufacturer so that the physical format can be introduced and used by many of the same readers. ELISA plates have beeen used to array antibodies, however, the number of spots per well is space constricted, and therefore, these arrays are relatively limited in the number of targets that they can interogate. RayBiotech and Novagen have produced multiwell arrays with a focused target content. Novagen arrays target up to 12 cytokines simultaneously. Each capture probe is represented in four spots, and controls are included in each well. Larger arrays are printed on modified glass slides or slides laminated with nitrocellulose or other membrane (Panomics, Invitrogen). These arrays are composed of antibodies as capture surfaces or capture peptides synthetically derived from proteins expressed and manufactured from cDNA libraries [54–56]. One complication of cDNA library-generated proteins is that correct folding of proteins during manufacture and maintenance of peptide folding following printing is needed for specific and accurate target binding [57].

Expressed peptide arrays have been used to study protein–protein interactions [58]. Protein arrays can be used for antibody profiling performed using peptide arrays in order to look for autoantigens in cases where autoimmune disease is suspected or for allergen reactivity on a large scale [59,60]. Capture of antibody target can be disclosed by labeled antibodies directed against the Fc region of human immunoglobins. In other applications, peptides cloned into cDNA libraries are tagged with small ligands such as *his* sequence tags to enable them to be affixed to slide substrates without excessive exposure of the peptides to linkage chemistries that may damage the proteins [61,62]. A novel protein array application that models some characteristics of cell or tissue base microarrays (see below) is the use of cell lysates as capture probes for protein arrays [63]. Variations of these applications have been conceived where cells introduced to nucleic acid (recombinant DNA or silencing RNA) or gene knock out cell lines are prepared for spotting onto microarrays [64,65]. The idea is that either over-expression proteins or elimination of specific pathways via siRNA will result in a pattern changes reflected in the array. Captured targets can be visualized by differential labeling (Figure 18.2) or using ELISA sandwich detection methods with labeled secondary antibodies. As with any formats using antibody capture, antibody specificity is always a consideration such that it is suggested that multiple antibodies (at least two) are used to reflect the capture of a specific protein [65]. If the comparison is made between proteins derived from two conditions such as normal and cancerous tissue or normal and treated tissue, the method most often used is differential labeling. As previously described, each condition is labeled with a separate amino reactive dye, generally Cy3 or Cy5. After labeling and removal of unincorporated dye, the two sets of proteins are mixed prior to application and capture on the protein array [66]. Differences in patterns and levels of proteins can be readily distinguished by differences in fluorescent emission using a series of filters. Notably, for any assays where a comparison of two dye labels is being used, controls for dye bias should be included. Dye switching experiments where the labels of each condition are reversed and the experiment is repeated to ensure that the observed differences are due to differences in the sample rather than differences in dye incorporation.

Protein arrays have all the complexities associated with gene arrays plus an additional level of complexity related to the capture target and capture molecules [67]. Quality control issues for protein arrays are more substantial than just concerns about probe position and printing consistency.

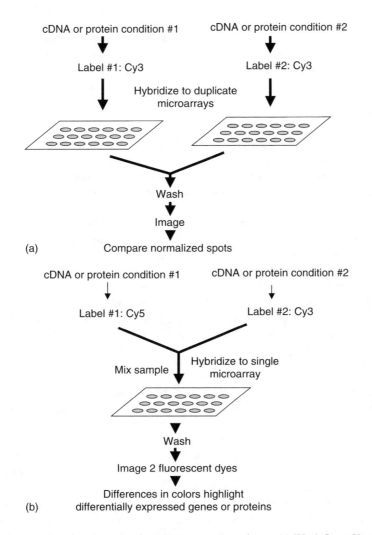

cDNA or protein condition #1 cDNA or protein condition #2

Label #1: Cy3   Label #2: Cy3

Hybridize to duplicate
microarrays

Wash

Image

(a)   Compare normalized spots

cDNA or protein condition #1 cDNA or protein condition #2

Label #1: Cy5   Label #2: Cy3

Mix sample   Hybridize to single
microarray

Wash

Image 2 fluorescent dyes

(b)   Differences in colors highlight
differentially expressed genes or proteins

FIGURE 18.2 Sample processing strategies for microarray comparisons. (a) Work flow: Single color analysis and labeling. (b) Work flow: Two color analysis and labeling.

By comparison, the binding of a nucleic acid to its complement is less complex and better understood than associations between proteins. Protein arrays have inherent issues with epitope availability based on protein modification, and protein secondary and tertiary structure; complications introduced from linkage of fluorescent labels through haptens to the protein; and orientation and protein stability when immobilized to the chip [65]. Comparison of the same sample on protein and gene arrays may not agree as a result of differences in translation and transcription control. Therefore, there is no requirement despite underlying expectations that gene expression and protein expression data agree. Hydrogel or polyacrylamide three-dimension chips may enhance protein stability and conformation because these surfaces maintain a protective water layer [68]. Passive absorption of a protein to its matrix does not control for changes in protein orientation or for concentration issues. Hence, active binding methods where attachment is regulated through a specific binding event or epitope tag has advantages. In vitro translated cDNA library arrays are preferred for use in evaluation changes in protein expression because the protein is fixed to the capture surface through its peptide tag [61,62]. Whereas capture by antibodies is fairly well characterized, there are differences between capture affinities among individual antigens. The breadth of hybridization requirements is amplified when considering chips constructed of large numbers of purified proteins or peptide to evaluate protein–protein interactions. The reason being that unlike nucleic acid arrays, the biochemical requirements for correct peptide folding and target binding may be very different between proteins. Regardless of the potential difficulties associated with protein chips, their utility for proteomics applications and disease diagnosis is evident from the large number of disease or tissue specific arrays currently available (Table 18.2).

18.2.3.3 Glycomic Arrays

By comparison to other array technologies, glycan arrays and glycomic applications are still in their infancy. Yet, just as biologists before the twenty-first century entered a post genomic era, scientists have entered the post-proteomic era in the twenty-first century. There is wide recognition that the sugars that decorate proteins are actually important signaling molecules for tissue formation, protein, protein interactions, and pathogen recognition [65,69]. Wang et al. were the first to recognize the value of microarrays for carbohydrate profiling for identification of infectious organisms [70]. This observation was quickly followed by improvements to the surface chemistry for arraying glycans on glass slides [71,72]. As yet, glycan microarrays have found limited application as most glycomics and glycome research is still focused on the regulation of the transcription of the genes involved in glycome pathways or observed changes in the glycomes during disease progression [73]. However, glycan arrays may develop into a valuable diagnostic screening for viral and bacterial virulence because pathogen adhesion appears to be, in part, regulated by surface sugars. Glycan arrays can be used to improve and confirm the glycosylation events predicted by gene profiling arrays [72,74].

18.2.3.4 Tissue-Based Arrays

Tissue-based microarrays (TMA) are a means of simultaneously profiling a particular genetic or proteomic biomarker among a large number of conditions [75]. In this application, tissue sections (or cells) derived from cancerous and normal tissue are sliced into thin sections onto a solid support. Blocks of tissues are cut from fresh and archival materials fixed by various methods [76]. Just as protein and genetic arrays can be used for biomarker discovery, tissue arrays can provide a large amount of information concerning normal and abnormal presentation, or they can provide population distribution information of a particular biomarker. Specimens can be derived from archival materials or tissue banks although the method of fixation determines the types of analysis that can be performed. Archival formalin fixed tissue can be used for genomic applications involving DNA; however, RNA may be destroyed by these older procotols for fixation, eliminating use of these tissues for some proteins [77]. Frozen sections and newer fixation methods do not suffer from these deficiencies and can be used to profile gene expression or protein expression. In this way, TMA can be used for rapid evaluation of potential biomarkers discovered in the application of other microarray approaches. New methods have been developed to help overcome some of the technical difficulties in printing these arrays [78]. Unlike nucleic acid and protein arrays where the reagents used to assemble the array are well characterized and relatively homogeneous, tissue-based arrays are inherently less defined. Even though small tissue sections are used, heterogeneity within tumors produces great variations among targets in a sample compared to extracted homogenized materials used to construct other arrays [76]. In addition, whereas there are many tumors of a similar type, there is a limited supply of tumor tissue from a particular individual, source, or isolated node, suggesting that exact comparisons between print groups may be difficult.

18.2.4 ANALYSIS OF MICROARRAYS

18.2.4.1 Data Management

Once the biological question that needs to be addressed is identified, and the microarray for use has been designed or selected, one critical step is needed before initiating experiments or even finalizing the printing of an array if a commercial product is not being used. One must determine how to manage and analyze the data.

Whereas microarrays have the advantage of rapidly providing a large amount of high quality information, microarrays are also extremely sensitive to biases produced by artifacts. This means that experimental variations that do not appear to effect other assays can produce confounding effects in microarray experiments that may be interpreted as true measurements. Poor management of the experimental processes, poor or inadequate control design, and insufficient sampling can lead to erroneous results. Microarray analysis is clearly a case of "garbage in, garbage out." Hence, analysis of microarray and quality metrics have been the subject of significant discussion in the microarray community. Luckily, governmental organizations such as the U.S. FDA, the National Institute of Standards (United States), and others in the European Scientific Community have recognized the scientific power of microarray technology, and they have been a mobilizing support for microarray efforts in this area. Web-based FDA freeware, Array Track, is available for use by all researchers for the storage, analysis, and interpretation of microarray data in a common format that is standardized. (www.fda.gov/nctr/science/centers/toxicoinformatics/Array-Track/index.htm) [79]. Likewise, MicroGen (www.bioinformatics.polimi.it/MicroGen/) is another freeware product developed for storage of microarray information [80]. Be forewarned; the bulk of the current papers and efforts involved in standardization and quality metrics occur in the area of gene expression type microarrays. Other applications such as SNP analysis, protein arrays, and glycomic applications discussed above are not at the same level of maturity for diagnostic use, although many of the current tools clearly could be adopted by these user communities with modification. A new tissue-based microarray data management tool, TMABoost, has been put forward as a means of supporting biomarker discovery and analysis, and it is dedicated to the intricacies of tissue microarray platforms (www.bioinfo.itc.it/TMA/home.htm).

In order to compare results across laboratories, the microarray gene expression community has derived an agreed standard of necessary information that needs to be captured for microarrays [81]. MIAME describes the essential information that needs to be captured for data presentation and publication (Figure 18.3). This includes information on the design of the experiment, the array, analysis methods, normalization methods, imaging parameters, and microarray hybridization parameters. New users can direct the capture of this information by using the handy checklist provided on the MIAME website. Adoption of this standard permits sharing of data and information across laboratories and platforms. However, one gap that still needs to be addressed is a lack of common language between community members regarding gene ontology and expression quantification [82]. The omission of a standard unit of measure for gene expression data is a particular hinderance [83,84]. Even though it was demonstrated that for well-controlled calibrated RNAs, it is possible to obtain an exact value for mRNA binding, most samples presented to microarrays are not as well characterized, and therefore, other methods of comparison that are more qualitative still stand as the standard by which microarray data are compared between laboratories or platforms [85–88]. Some of the lack of consistency across microarray platforms is due to the bias of specific gene ontologies

1. Experimental design: the set of hybridization experiments as a whole
2. Array design: each array used and each element (spot, feature) on the array
3. Samples: samples used, extract preparation and labeling
4. Hybridizations: procedures and parameters
5. Measurements: images, quantification and specifications
6. Normalization controls: types, values and specifications

FIGURE 18.3 MIAME standardization of microarray experimentation. (Brazma, A., *Nat. Genet.*, 29, 365–371, 2001. With permission.)

represented by various manufacturer [89,90]. Hence, the MAIME standard has been extended by the development of an ontology tool (www.mged.sourceforge.net/ontologies/index.php) to assist in the standardization of gene annotations use to interpretate microarray outputs [91].

18.2.4.2 Quality Metrics: Designing Controls and Defining Error

Even in unsupervised data analysis approaches (hypothesis generating, data driven), understanding the error within the data sets and mapping the variation in the observed signal must be approached with rigor to avoid reporting artifactual gene clustering or bogus metabolic pathway involvement [92]. It is clear that not all variation in signal is due to biological variation. Variation other than biological variation includes, but is not limited to, differences in sample preparation, differences in hybridization efficiency, variations in chip printing, variations in probe hybridization intensity based on location, and variations in the equipment. For microarrays that can only be used once, validating the array must be performed during the assay with appropriate controls, either incorporated as fluorescent molecules into the chip or evaluated using spiked external controls [34,37,93]. In addition, at least one company has developed microarrays for performing quality assessment of materials prior to hybridization (Affymetrix Test Array, Table 18.1). In many cases, however, agreement with other methods in particular QRT–PCR or capillary electrophoresis are used to confirm the quality of the materials prior to hybridization or validate following hybridization [85,94,88].

No matter the type of bioarray employed, users should incorporate the means to define reproducibility of the experiment. There are two major types of replication that should be considered: biological and technical [95,96]. Biological replication describes the ability of the assay to achieve the same values from the same starting materials. The amount and type of biological replication include evaluating sampling methods to determine within sample and between sample variations to name a few considerations. The biological replicate can be defined as one derived from an individual, a tumor, tissue, or aliquots from a single extract. Technical replication includes the effects of the methods of the sampling technique and sample processing, instrumental variation, computational, and interpretative methodologies [97]. Standardization of sample handling is critical to the outcome of microarray experiments and to one's ability to share data across laboratories and platforms [19]. Because microarray technology is relatively new and part of its value lies in its extreme sensitivity to detect low level genes, small inconsequential differences in sampling can produce large differences in the outcome observed. Validation of results across platforms and across analysis methods has been raised as an important consideration in the decision to identify microarray results with biomarker or pathway discovery [85,98–100].

In one approach used by Tan, microarrays from three manufacturers were hybridized with a common sample set [101]. The agreement across platforms could only be described as dismal; differential expression was evident on all platforms for only four genes derived from a subset of over 2,000 genes common to all three platforms. As might be expected, this analysis generated a robust discussion in the microarray community. The apparent lack of reproducibility or agreement in microarray experiments has led to a flurry of effort to develop standards for microarray use and analysis. The U.S. FDA, concerned about the reliability of this new enabling technology extensively used in its programs in toxicogenomics and pharmacogenomics, conducted its own review of the dataset generously supplied by the original authors [19,102]. The resulting papers highlight the need for users to examine the intra-platform consistency and to establish rigorous standards for data analysis [95]. By far, the most important notation by the U.S. FDA concerning the Tan study was that the methods used in Tan's study were both statistically correct and employed common current practices of the microarray community. Hence, the MAQC project was established. The purpose of this multicenter pilot study is to develop vendor-independent quality assessment tools for use by researchers employing microarrays. This tool includes establishment of control mRNAs and other best practices for microarray analysis. The final guidance document is anticipated to be e-published in 2007, although platform specific quality guidance documents are currently available on the MAQC website (see Table 18.5). The MAQC website is primarily devoted to descriptions of gene expression arrays. Subsequent studies have examine inter-platform agreements between microarray types (Affymetrix and Illumina), and they found a high level of agreement between bead-based and in situ microarrays [85]. There are, as yet, few efforts of this magnitude devoted to the guidance in the use of other microarrays for protein and glycomic profiling.

Typically, samples derived from the same material or a single sample batched processed and subsequently hybridized on multiple arrays are used to establish baseline values in microarray experiments and to address biological and technical replication consistency. A number of companies are manufacturing quality control chips. It is important to note that in experiments where two dyes are used, dye switching experiments have been recommended to account for the inherent bias of specific imagers for a specific dye preference. Failure to perform such analysis could lead to an artifactual underestimation of gene induction as a result of the difference in scanner efficiencies for each dye [103]. The methods to carry out and evaluate these experiments have generated some controversy concerning if dye swapping experiments may uncover another type of error, gene specific dye bias [104,105]. In the instance of dye switching experiments in particular, it is suggested to keep the laser power relatively low (30%) to prevent photobleaching and to adjust photomultiplier tube (PMT) gain settings to balance the dye channels and reduce bias [21]. However, prior to attempting this, it is critical to understand the effect of PMT gain on the reproducibility of measurements. Shi et al. described a method for obtaining optimum PMT settings for laser scanners such that PMT settings require little, if any, adjustments and are operated in the linear portion of the calibration curve [102]. If possible, all microarrays to be compared within an experiment should be scanned with the same calibrated settings in order to reduce inter-chip reproducibility. Second, the location of probes must be taken into account when normalizing probes, even when using laser scanning devices [106]. Intra-chip probe comparisons often show differences based on the group of probes printed together. Global methods (mean, Lowess) for normalization of signal have been improved when the printing group is taken into account (print tip normalization), and they seem to provide a better means to evaluate true expression difference [21,107]. Third, methods to normalize between microarray variations need to be considered because at least one of these methods requires pre-planning before hybridization occurs. Three common methods of normalization are the use of housekeeping genes whose expression is considered not to change; the use of spiking or doping external control RNA into each hybridization; or a comparison of the total intensity within the dataset. Each of these methods has its own caveats. In reality, there are very few instances where housekeeping genes are conclusively known not to change [108,109]. Global normalization to the total intensity assumes that, whereas individual gene expression can change within a sample, the total number of transcripts is essentially the same [110]. The addition of spiked nucleic acid assumes that the behavior of external controls spiked into the sample reflect the behavior of all the probes on the chip, independent of concentration. This is not to say that these methods cannot be used; it is just to note the potential for experimental bias.

18.2.4.3 Imaging and Extraction of Data

At this stage, the microarray platform has been selected and appropriate controls incorporated into the array or into the experiment; the array is hybridized according to protocols, either established by the manufacturer or by other users (for example, www.tigr.org/tdb/microarray or www.cmgm.stanford.edu/pbrown/protocols). Images of washed microarrays are collected using calibrated instruments. Before analysis of the data begins in earnest, there is another step to consider. Whereas microarrays are developed for high throughput applications, the images generated by high throughput methods must also undergo quality assessment prior to extraction of intensity data. Image generation can be automated, and most manufacturers of scanners include an automated program for setting and aligning an area from which data will be captured. However, a number of after market software solutions are also available. The process of alignment is referred to as gridding. Programs used for gridding vary in the way that spots are localized and in the alignment of the grid [111–113]. Spot localization is referred to as spot registration. Defects in the microarrays can lead to faults in grid alignment. As previously mentioned, microarrays can attract dust that will produce a fluorescent signal, confounding automated spot registration. Usually, fluorescence because of dust, is local, and if it occurs near or over a spot, the entire spot must be eliminated from the analysis. Residual fluorescent background is not specific to in-house printed arrays [114]. Spotted arrays can exhibit other deformities that would make the entire array a candidate for elimination from the analysis. Spots with deformed signal (donuts, partial circles, asymmetric spots) are indicators of poor print quality, print pin artifacts or damage to the array. Ghost images are empty pads in areas of high background, and they are indicative of chip production problems where surface modifications results in non-specific fluorescence around the spot. In addition, array printers vary in performance, and in some cases, the spots do not perfectly align. This group found that some printers need to be placed on isolator tables to prevent vibrations from interfering with the uniformity of the printing process. Automated gridding programs that do not adjust spot registration when alignment issues arise result in poor intensity data even though the quality of the data imbedded in the image is quite good. This group found that a well-trained user, unrestricted by time, performs as well as or better than auto-mated gridding programs. It has conducted studies comparing the results of data analysis and gridding performance of well-trained and naïve users. In these studies, 40% of the errors in the data acquisition were the results of poor grid placement (Jackman, pers. comm.). Before using any gridding program, the following tests or proficiency testing should be considered. First, compare the results of manual alignments to determine the user's ability to reproducibly align the image on high quality microarrays with no obvious defects. This will provide a baseline evaluation to determine the variation for a particular user. Second, compare data and variation in signal among controls using an automated program and a manual alignment for a high quality chip. Third, using an image of poor quality because of hybridization artifacts or microarray defects (dust, high localized backgrounds, uneven spotting pattern), conduct the same analysis to characterize the program performance on less than optimal arrays. This should assist the user in understanding the deviations for which automated gridding programs should not be used. If the deviation is dramatic enough, manual alignments may not help in extracting additional quality data. Also, it is important to remember that, for comparative studies between spot intensities, images must be acquired below pixel saturation. Presence or absence measurements can be made on extreme images that are black and white, but comparisons of expression levels or SNPs require shades of gray. If too many spots on an image are at maximum intensity, representing complete saturation, then it will not be possible to obtain a true ratiometric evaluation between spots. In these cases, the image should not be used for data extraction, and the light source or acquisition time should be lowered before the microarray is re-scanned. The determination of "too many" is somewhat dependent on the number of microarray elements, though generally, if 0.1%–1% of the total spot intensities are saturated, the data are not accepted.

18.2.4.4   Data Analysis

Once the data is extracted from the image and normalized, it can be subjected to a number of statistical analysis and have been the subject of several recent reviews [115–118]. Because of the breath and specificity of the applications, it is not possible to review all the permutations of analysis methods used to assess microarrays within this chapter. However, it has been mentioned that one of the great strengths of microarrays is in the application of unsupervised data analysis methods. Gene clustering approaches can be supervised (user defined) or unsupervised [119]. Unsupervised approaches permit users to combine data from multiple experiments and then allow clustering or organization of the data based on similarity of expression. There are three types or models of unsupervised methods, and they are listed in the order of their application: hierarchial clustering

[120], K clustering [121], and self-organizing maps [122]. Hierarchial clustering, also sometimes known as Eisen clustering after its creator, displays genes grouped by their common expression characteristics. No assumptions are made about the number of clusters the data represent so that the number of clusters that will be displayed is unknown. K clustering also groups of genes by their common expression characteristics with the caveat that the user defines the acceptance criteria for number of clusters and then the algorithm partitions genes of similar expression profiles into groups. Self-organizing maps are the most rigorous method of unsupervised analysis as the analysis describes groups of genes not only on the basis of their expression profiles but also on the relativeness of similarly expressed genes by relatedness. Self-organizing maps are computationally driven by neural networks of which there are many versions. Whereas self organizing maps are recognized as important discovery tools, the relevance of these analysis tools improves significantly with sample size.

18.3 SUMMARY

In summary, bioarrays are an expanding tool for gene discovery, bioinformatics, and drug development. Whereas appropriate caution must be employed in their generation, use, and interpretation, the benefits of microarray technology far outweigh any concerns for their use. Expanded field application of arrays to food safety and diagnostics will improve the robustness of the interpretation and likely reduce the associated costs as well as be an enabling factor in genetic testing and personalized medicine [20]. The information presented here is an overview of all microarray technologies. For more detailed information about specific microarray type or applications to field such as diagnostics, a number of excellent reviews concerning protein bioarrays [65,123,52,124], nucleic acid arrays and their statistical analysis [116,125], glycan arrays [69,126], and tissue-based arrays [75] are encouraged reading as well as current reviews concerning arrays as diagnostics [127–130] and for food and nutrition applications [131,132].

REFERENCES

1. Maskos, U. and Southern, E., A novel method for the parallel analysis of multiple mutations in multiple samples, *Nucleic Acids Res.*, 21 (9), 2269–2270, 1993.
2. Venkatasubbarao, S., Microarrays—status and prospects, *Trends Biotechnol.*, 22 (12), 630–637, 2004.
3. Southern, E., DNA microarrays: history and overview, in *DNA Arrays*, Rampal, J. B., Ed., Vol. 170, Humana Press, New York, pp. 1–16, 2001.
4. Schena, M., Shalon, D., Davis, R. W., and Brown, P. O., Quantitative monitoring of gene expression patterns with a complementary DNA microarray, *Science*, 270, 467–470, 1995.
5. Schena, M., Shalon, D., Heller, R., Chai, A., Brown, P. O., and Davis, R. L., Parallel human genome analysis: microarray-based expression monitoring of 1000 genes, *PNAS*, 93, 10614–10619, 1996.
6. Aris, V. M., Cody, M. J., Cheng, J., Dermody, J. J., Soteropoulos, P., Recce, M., and Tolias, P. P., Noise filtering and nonparametric analysis of microarray data underscores discriminating markers of oral, prostate, lung, ovarian and breast cancer, *BMC Bioinformatics*, 5 (1), 185–194, 2004.
7. Spiro, A., Lowe, M., and Brown, D., A bead-based method for multiplexed identification and quantitation of DNA sequences using flow cytometry, *Appl. Environ. Microbiol.*, 66 (10), 4258–4265, 2000.
8. Gunderson, K. L., Kruglyak, S., Graige, M. S., Garcia, F., Kermani, B. G., Zhao, C., Che, D. et al., Decoding randomly ordered DNA arrays, *Genome Res.*, 14 (5), 870–877, 2004.
9. Pickering, J. W., McMillin, G. A., Gedge, F., Hill, H. R., and Lyon, E., Flow cytometric assay for genotyping cytochrome p450 2C9 and 2C19: comparison with a microelectronic DNA array, *Am. J. Pharmacogenomics*, 4 (3), 199–207, 2004.
10. Walt, D., Bead-based fiber-optic arrays, *Science*, 287 (5452), 451–452, 2000.

11. Fan, J. B., Hu, S. X., Craumer, W. C., and Barker, D. L., Bead array-based solutions for enabling the promise of pharmacogenomics, *Biotechniques*, 39 (4), 583–588, 2005.

12. Kuhn, K., Baker, S. C., Chudin, E., Lieu, M. H., Oeser, S., Bennett, H., Rigault, P., Barker, D., McDaniel, T. K., and Chee, M. S., A novel, high-performance random array platform for quantitative gene expression profiling, *Genome Res.*, 14 (11), 2347–2356, 2004.

13. Kononen, J., Bubendorf, L., Kallioniemi, A., Barlund, M., Schraml, P., Leighton, S., Torhorst, J., Mihatsch, M. J., Sauter, G., and Kallioniemi, O. P., Tissue microarrays for high-throughput molecular profiling of tumor specimens, *Nat. Med.*, 4 (7), 844–847, 1998.

14. Fedor, H. L. and De Marzo, A. M., Practical methods for tissue microarray construction, *Methods Mol. Med.*, 103, 89–101, 2005.

15. Wheeler, D. B., Carpenter, A. E., and Sabatini, D. M., Cell microarrays and RNA interference chip away at gene function, *Nat. Genet.*, 37, S25–S30, 2005.

16. Kallioniemi, O. P., Wagner, U., Kononen, J., and Sauter, G., Tissue microarray technology for high-throughput molecular profiling of cancer, *Hum. Mol. Genet.*, 10 (7), 657–662, 2001.

17. Pizzi, R., The challenges of pharmocogenomic tests: has personalized medicine arrived? *Clin. Lab. News*, 31, 1–3, 2005.

18. Jain, K. K., Applications of the amplichip, *Mol. Diagn.*, 9, 119–127, 2005.

19. Shi, L., Tong, W., Goodsaid, F., Frueh, F. W., Fang, H., Han, T., Fuscoe, J. C., and Casciano, D. A., QA/QC: challenges and pitfalls facing the microarray community and regulatory agencies, *Expert Rev. Mol. Diagn.*, 4 (6), 761–777, 2004.

20. Mansfield, E., Genetic testing and personalized medicine, *Preclinica*, 1, 155–158, 2003.

21. Leung, Y. F. and Cavalieri, D., Fundamentals of cDNA microarray data anlaysis, *Trends Genet.*, 10, 649–659, 2003.

22. Nadon, R. and Shoemaker, J., Statistical issues with microarrays: processing and analysis, *Trends Genet.*, 18, 265–271, 2002.

23. Hegde, P., Qi, R., Abernathy, K., Gay, C., Dharap, S., Gaspard, R., Hughes, J. E., Snesrud, E., Lee, N., and Quackenbush, J., A concise guide to cDNA microarray analysis, *Biotechniques*, 29 (3), 548–550, 2000, see also pp. 552–554 and p. 556.

24. Zlatanova, J. and Mirzabekov, A., Gel immobilizd microarrays of nucleic acids and proteins, in *Methods in Molecular Biology*, Rampal, J. B., Ed., Vol. 170, Humana Press, New Jersey, pp. 17–38, 2001.

25. Bavykin, S. G., Lysov, Y. P., Zakhariev, V., Kelly, J. J., Jackman, J., Stahl, D. A., and Cherni, A., Use of 16S rRNA, 23S rRNA, and gyrB gene sequence analysis to determine phylogenetic relationships of Bacillus cereus group microorganisms, *J. Clin. Microbiol.*, 42 (8), 3711–3730, 2004.

26. Theodore, M. L., Jackman, J., and Bethea, W. L., Counterproliferation with advanced microarray technology, *Johns Hopkins APL Tech. Digest.*, 25 (1), 38–43, 2004.

27. Proudnikov, D., Timofeev, E., and Mirzabekov, A., Immobilization of DNA in polyacrylamide gel for the manufacture of DNA and DNA-oligonucleotide microchips, *Anal. Biochem.*, 259 (1), 34–41, 1998.

28. Baselt, D. R., Lee, G. U., Natesan, M., Metzger, S. W., Sheehan, P. E., and Colton, R. J., A biosensor based on magnetoresistance technology, *Biosens. Bioelectron.*, 13 (7–8), 731–739, 1998.

29. Lee, G. U., Chrisey, L. A., and Colton, R. J., Direct measurement of the forces between complementary strands of DNA, *Science*, 266 (5186), 771–773, 1994.

30. Grissom, S. F., Lobenhofer, E. K., and Tucker, C. J., A qualitative assessment of direct-labeled cDNA products prior to microarray analysis, *BMC Genomics*, 6 (1), 36, 2005.

31. Rickman, D. S., Herbert, C. J., and Aggerbeck, L. P., Optimizing spotting solutions for increased reproducibility of cDNA microarrays, *Nucleic Acids Res.*, 31 (18), e109, 5221–5461, 2003.

32. Boa, Z., Ma, W. L., Hu, Z. Y., Rong, S., Shi, Y. B., and Zheng, W. L., A method for evaluation of the quality of DNA microarray spots, *J. Biochem. Mol. Biol.*, 35 (5), 532–535, 2002.

33. Shearstone, J. R., Allaire, N. E., Getman, M. E., and Perrin, S., Nondestructive quality control for microarray production, *Biotechniques*, 32 (5), 1051–1052, 2002, see also p. 1054 and pp. 1056–1057.

34. Hessner, M. J., Wang, X., Khan, S., Meyer, L., Schlicht, M., Tackes, J., Datta, M. W., Jacob, H. J., and Ghosh, S., Use of a three-color cDNA microarray platform to measure and control support-bound probe for improved data quality and reproducibility, *Nucleic Acids Res.*, 31 (11), e60, 2705–2974, 2003.

35. Diehl, F., Beckmann, B., Kellner, N., Hauser, N. C., Diehl, S., and Hoheisel, J. D., Manufacturing DNA microarrays from unpurified PCR products, *Nucleic Acids Res.*, 30 (16), e79, 3497–3642, 2002.

36. Battaglia, C., Salani, G., Consolandi, C., Bernardi, L. R., and De Bellis, G., Analysis of DNA microarrays by non-destructive fluorescent staining using SYBR green II, *Biotechniques*, 29 (1), 78–81, 2000.

37. Hessner, M. J., Singh, V. K., Wang, X., Khan, S., Tschannen, M. R., and Zahrt, T. C., Utilization of a labeled tracking oligonucleotide for visualization and quality control of spotted 70-mer arrays, *BMC Genomics*, 5 (1), 12, 2004.

38. Bowtell, D. and Sambrook, J., *DNA Microarrays*, Cold Spring Harbor Laboratory Press, Cold Spring Harbor, NY, 2003.

39. Sender, A. J., Chipping away at affy, *Genome Technol.*, 72–78, October, 2002.

40. Gershon, D., DNA microarrays: More than gene expression, *Nature*, 437, 1195–1198, 2005.

41. Vrana, K. E., Freeman, W. M., and Aschner, M., Use of microarray technologies in toxicology research, *Neurotoxicology*, 24 (3), 321–332, 2003.

42. Golub, T. R., Slonim, D. K., Tamayo, P., Huard, C., Gaasenbeek, M., Mesirov, J. P., Coller, H. et al., Molecular classification of cancer: class discovery and class prediction by gene expression monitoring, *Science*, 286 (5439), 531–537, 1999.

43. Hacia, J. G., Resequencing and mutational analysis using oligonucleotide microarrays, *Nat. Genet.*, 21, 42–48, 1999.

44. Raitio, M., Lindroos, K., Laukkanen, M., Pastinen, T., Sistonen, P., Sajantila, A., and Syvanen, A. C., Y-chromosomal SNPs in Finno-Ugric-speaking populations analyzed by minisequencing on microarrays, *Genome Res.*, 11 (3), 471–482, 2001.

45. Liu, W. T., Mirzabekov, A. D., and Stahl, D. A., Optimization of an oligonucleotide microchip for microbial identification studies: a non-equilibrium dissociation approach, *Environ. Microbiol.*, 3 (10), 619–629, 2001.

46. Drmanac, S., Kita, D., Labat, I., Hauser, B., Burczak, J., and Drmanac, R., Accurate sequencing by hybridization for DNA diagnostics and individual genomics, *Nat. Biotechnol.*, 16, 54–58, 1998.

47. Laassri, M., Chizhikov, V., Mikheev, M., Shchelkunov, S., and Chumakov, K., Detection and discrimination of orthopoxviruses using microarrays of immobilized oligonucleotides, *J. Virol. Methods*, 112 (1–2), 67–78, 2003.

48. Cherkasova, E., Laassri, M., Chizhikov, V., Korotkova, E., Dragunsky, E., Agol, V. I., and Chumakov, K., Microarray analysis of evolution of RNA viruses: evidence of circulation of virulent highly divergent vaccine-derived polioviruses, *Proc. Natl Acad. Sci. U.S.A.*, 100 (16), 9398–9403, 2003.

49. Lin, B., Vora, G. J., Thach, D., Walter, E., Metzgar, D., Tibbetts, C., and Stenger, D. A., Use of oligonucleotide microarrays for rapid detection and serotyping of acute respiratory disease-associated adenoviruses, *J. Clin. Microbiol.*, 42 (7), 3232–3239, 2004.

50. Ivshina, A. V., Vodeiko, G. M., Kuznetsov, V. A., Volokhov, D., Taffs, R., Chizhikov, V. I., Levandowski, R. A., and Chumakov, K. M., Mapping of genomic segments of influenza B virus strains by an oligonucleotide microarray method, *J. Clin. Microbiol.*, 42 (12), 5793–5801, 2004.

51. Zeng, Z. B., QTL mapping and the genetic basis of adaptation, *Genetica*, 123, 25–37, 2005.

52. Huang, R. P., Protein arrays, an excellent tool in biomedical research, *Front. Biosci.*, 8, d559–d576, 2003.

53. Huang, R. P., Detection of multiple proteins in an antibody-based protein microarray system, *J. Immunol. Methods*, 255 (1–2), 1–13, 2001.

54. Angenendt, P., Nyarsik, L., Szaflarski, W., Glokler, J., Nierhaus, K. H., Lehrach, H., Cahill, D. J., and Lueking, A., Cell-free protein expression and functional assay in nanowell chip format, *Anal. Chem.*, 76 (7), 1844–1849, 2004.

55. He, M. and Taussig, M. J., Discern array technology: a cell-free method for the generation of protein arrays from PCR DNA, *J. Immunol. Methods*, 274 (1–2), 265–270, 2003.

56. Sreekumar, A., Nyati, M. K., Varambally, S., Barrette, T. R., Ghosh, D., Lawrence, T. S., and Chinnaiyan, A. M., Profiling of cancer cells using protein microarrays: discovery of novel radiation-regulated proteins, *Cancer Res.*, 61 (20), 7585–7593, 2001.

57. Nakayama, M. and Ohara, O., A system using convertible vectors for screening soluble recombinant proteins produced in Escherichia coli from randomly fragmented cDNAs, *Biochem. Biophys. Res. Commun.*, 312 (3), 825–830, 2003.

58. Kukar, T., Eckenrode, S., Gu, Y., Lian, W., Megginson, M., She, J. X., and Wu, D., Protein microarrays to detect protein-protein interactions using red and green fluorescent proteins, *Anal. Biochem.*, 306 (1), 50–54, 2002.

59. Robinson, W. H., DiGennaro, C., Hueber, W., Haab, B. B., Kamachi, M., Dean, E. J., and Fournel, S., et al. Autoantigen microarrays for multiplex characterization of autoantibody responses, *Nat. Med.*, 8, 295–301, 2002.

60. Feng, Y., Ke, X., Ma, R., Chen, Y., Hu, G., and Liu, F., Parallel detection of autoantibodies with microarrays in rheumatoid diseases, *Clin. Chem.*, 50 (2), 416–422, 2004.

61. Busso, D., Kim, R., and Kim, S. H., Expression of soluble recombinant proteins in a cell-free system using a 96-well format, *J. Biochem. Biophys. Methods*, 55 (3), 233–240, 2003.

62. Lue, R. Y., Chen, G. Y., Zhu, Q., Lesaicherre, M. L., and Yao, S. Q., Site-specific immobilization of biotinylated proteins for protein microarray analysis, *Methods Mol. Biol.*, 264, 85–100, 2004.

63. Yan, F., Sreekumar, A., Laxman, B., Chinnaiyan, A. M., Lubman, D. M., and Barder, T. J., Protein microarrays using liquid phase fractionation of cell lysates, *Proteomics*, 3 (7), 1228–1235, 2003.

64. Paweletz, C. P., Charboneau, L., Bichsel, V. E., Simone, N. L., Chen, T., Gillespie, J. W., Emmert-Buck, M. R., Roth, M. J., Petricoin, E. F. III, and Liotta, L. A., Reverse phase protein microarrays which capture disease progression show activation of pro-survival pathways at the cancer invasion front, *Oncogene*, 20 (16), 1981–1989, 2001.

65. Howbrook, D. N., van de Valk, A. M., O'Shaughnessy, M. C., Sarker, D. K., Baker, S. C., and Lloyd, A. W., Developments in microarray technologies, *Drug Discov. Today*, 8 (14), 642–650, 2003.

66. Haab, B. B., Methods and application of antibody microarray in cancer research, *Proteomics*, 3, 416–422, 2003.

67. Kricka, L. J. and Master, S. R., Validation and quality control of protein microarray-based analytical methods, *Methods Mol. Med.*, 114, 233–255, 2005.

68. Rubina, A. Y., Dementieva, E. I., Stomakhin, A. A., Darii, E. L., Pan'kov, S. V., Barsky, V. E., Ivanov, S. M., Konovalova, E. V., and Mirzabekov, A. D., Hydrogel-based protein microchips: Manufacturing, properties, and applications, *Biotechniques*, 34 (5), 1008–1014, 2003. see also pp. 1016–1020 and p. 1022.

69. Raman, R., Raguram, S., Venkataraman, G., Paulson, J. C., and Sasisekharan, R., Glycomics: an integrated systems approach to structure-function relationships of glycans, *Nat. Methods*, 2 (11), 817–824, 2005.

70. Wang, D., Liu, S., Trummer, B. J., Deng, C., and Wang, A., Carbohydrate microarrays for the recognition of cross-reactive molecular markers of microbes and host cells, *Nat. Biotechnol.*, 20 (3), 275–281, 2002.

71. Willats, W. G., Rasmussen, S. E., Kristensen, T., Mikkelsen, J. D., and Knox, J. P., Sugar-coated microarrays: a novel slide surface for the high-throughput analysis of glycans, *Proteomics*, 2 (12), 1666–1671, 2002.

72. Blixt, O., Head, S., Mondala, T., Scanlan, C., Huflejt, M. E., Alvarez, R., Bryan, M. C. et al., Printed covalent glycan array for ligand profiling of diverse glycan binding proteins, *Proc. Natl Acad. Sci. U.S.A.*, 101 (49), 17033–17038, 2004.

73. Bragonzi, A., Worlitzsch, D., Pier, G. B., Timpert, P., Ulrich, M., Hentzer, M., Andersen, J. B., Givskov, M., Conese, M., and Doring, G., Nonmucoid pseudomonas aeruginosa expresses alginate in the lungs of patients with cystic fibrosis and in a mouse model, *J. Infect. Dis.*, 192 (3), 410–419, 2005.

74. Kawano, S., Hashimoto, K., Miyama, T., Goto, S., and Kanehisa, M., Prediction of glycan structures from gene expression data based on glycosyltransferase reactions, *Bioinformatics*, 21 (21), 3976–3982, 2005.

75. Watanabe, A., Cornelison, R., and Hostetter, G., Tissue microarrays: applications in genomic research, *Expert Rev. Mol. Diagn.*, 5 (2), 171–181, 2005.

76. Hoos, A. and Cordon-Cardo, C., Tissue microarray profiling of cancer specimens and cell lines: Opportunities and limitations, *Lab. Invest.*, 81 (10), 1331–1338, 2001.

77. Paik, S., Kim, C. Y., Song, Y. K., and Kim, W. S., Technology insight: application of molecular techniques to formalin-fixed paraffin-embedded tissues from breast cancer, *Nat. Clin. Pract. Oncol.*, 2 (5), 246–254, 2005.

78. Chen, N. and Zhou, Q., Constructing tissue microarrays without prefabricating recipient blocks: a novel approach, *Am. J. Clin. Pathol.*, 124 (1), 103–107, 2005.

79. Tong, W., Cao, X., Harris, S., Sun, H., Fang, H., Fuscoe, J., Harris, A. et al., ArrayTrack-supporting toxicogenomic research at the U.S. Food and Drug Administration National Center for Toxicological Research, *Environ. Health Perspect.*, 111 (15), 1819–1826, 2003.

80. Burgarella, S., Cattaneo, D., Pinciroli, F., and Masseroli, M., MicroGen3: A MIAME compliant web system for microarray experiment information and workflow management, *BMC Bioinformatics*, 6 (suppl. 4), S6, 2005.

81. Brazma, A., Minimum information about a microarray experiment (MAIME)-toward standards for microarray data, *Nat. Genet.*, 29, 365–371, 2001.

82. Reid, B., MAIME or Bust: realizing the promise of microarray data, *Preclinica*, 1, 151–152, 2003.

83. Hekstra, D., Taussig, A. R., Magnasco, M., and Naef, F., Absolute mRNA concentrations from sequence-specific calibration of oligonucleotide arrays, *Nucleic Acids Res.*, 31 (7), 1962–1968, 2003.

84. Townsend, J. P. and Hartl, D. L., Bayesian analysis of gene expression levels: statistical quantification of relative mRNA level across multiple strains or treatments, *Genome Biol.*, 3 (12), Research0071.1–Research0071.16, 2002.

85. Barnes, M., Freudenberg, J., Thompson, S., Aronow, B., and Pavlidis, P., Experimental comparison and cross-validation of the Affymetrix and Illumina gene expression analysis platforms, *Nucleic Acids Res.*, 33 (18), 5914–5923, 2005.

86. van Ruissen, F., Ruijter, J. M., Schaaf, G. J., Asgharnegad, L., Zwijnenburg, D. A., Kool, M., and Baas, F., Evaluation of the similarity of gene expression data estimated with SAGE and Affymetrix GeneChips, *BMC Genomics*, 6, 91, 2005.

87. Irizarry, R. A., Warren, D., Spencer, F., Kim, I. F., Biswal, S., Frank, B. C., Gabrielson, E. et al., Multiple-laboratory comparison of microarray platforms, *Nat. Methods*, 2 (5), 345–350, 2005.

88. Larkin, J. E., Frank, B. C., Gavras, H., Sultana, R., and Quackenbush, J., Independence and reproducibility across microarray platforms, *Nat. Methods*, 2 (5), 337–344, 2005.

89. Draghici, S., Khatri, P., Shah, A., and Tainsky, M. A., Assessing the functional bias of commercial microarrays using the onto-compare database, *Biotechniques*, suppl. 55–61, March 2003.

90. Zhong, S., Tian, L., Li, C., Storch, K. F., and Wong, W. H., Comparative analysis of gene sets in the Gene Ontology space under the multiple hypothesis testing framework, *Proc. IEEE Comp. Syst. Bioinform. Conf.*, 425–435, 2004.

91. Whetzel, P. L., Parkinson, H., Causton, H. C., Fan, L., Fostel, J., Fragoso, G., Game, L. et al., The MGED Ontology; a resource for semantics-based description of microarray experiments, *Bioinformatics*, 22, 866–873, 2006.

92. van Hijum, S. A., de Jong, A., Baerends, R. J., Karsens, H. A., Kramer, N. E., Larsen, R., den Hengst, C. D., Albers, C. J., Kok, J., and Kuipers, O. P., A generally applicable validation scheme for the assessment of factors involved in reproducibility and quality of DNA-microarray data, *BMC Genomics*, 6 (1), 77, 2005.

93. External RNA Controls Consortium, Proposed methods for testing and selecting the ERCC external RNA controls, *BMC Genomics*, 6, 150, 2005.

94. Dallas, P. B., Gottardo, N. G., Firth, M. J., Beesley, A. H., Hoffmann, K., Terry, P. A., Freitas, J. R., Boag, J. M., Cummings, A. J., and Kees, U. R., Gene expression levels assessed by oligonucleotide microarray analysis and quantitative real-time RT–PCR—how well do they correlate? *BMC Genomics*, 6 (1), 59, 2005.

95. Imbeaud, S. and Auffray, C., 'The 39 steps' in gene expression profiling: critical issues and proposed best practices for microarray experiments, *Drug Discov. Today*, 10 (17), 1175–1182, 2005.

96. Altman, N., Replication, variation and normalisation in microarray experiments, *Appl. Bioinformatics*, 4 (1), 33–44, 2005.

97. Lee, M. L., Kuo, F. C., Whitmore, G. A., and Sklar, J., Importance of replication in microarray gene expression studies: statistical methods and evidence from repetitive cDNA hybridizations, *Proc. Natl Acad. Sci. USA*, 97 (18), 9834–9839, 2000.

98. Yuen, T., Wurmbach, E., Pfeffer, R. L., Ebersole, B. J., and Sealfon, S. C., Accuracy and calibration of commercial oligonucleotide and custom cDNA microarrays, *Nucleic acids Res.*, 30 (10), e48, 2097–2260, 2002.

99. Marshall, E., Getting the noise out of gene arrays, *Science*, 306 (5696), 630–631, 2004.

100. Borden, E. C., Baker, L. H., Bell, R. S., Bramwell, V., Demetri, G. D., Eisenberg, B. L., Fletcher, C. D. et al., Soft tissue sarcomas of adults: state of the translational science, *Clin. Cancer Res.*, 9 (6), 1941–1956, 2003.

101. Tan, P. K., Downey, T. J., Spitznagel, E. L. Jr., Xu, P., Fu, D., Dimitrov, D. S., Lempicki, R. A., Raaka, B. M., and Cam, M. C., Evaluation of gene expression measurements from commercial microarray platforms, *Nucleic Acids Res.*, 31 (19), 5676–5684, 2003.

102. Shi, L., Tong, W., Su, Z., Han, T., Han, J., Puri, R. K., Fang, H. et al., Microarray scanner calibration curves: characteristics and implications, *BMC Bioinformatics*, 6 (suppl. 2), S11, 2005.

103. Shi, L., Tong, W., Fang, H., Scherf, U., Han, J., Puri, R. K., Frueh, F. W. et al., Cross-platform comparability of microarray technology: intra-platform consistency and appropriate data analysis procedures are essential, *BMC Bioinformatics*, 6 (suppl. 2), S12, 2005.

104. Martin-Magniette, M. L., Aubert, J., Cabannes, E., and Daudin, J. J., Evaluatation of the gene specific dye bias in cDNA microarray experiments, *Bioinformatics*, 21 (9), 1995–2000, 2005.

105. Dobbin, K. K., Shih, J. H., and Simon, R. M., Comment on 'Evaluation of the gene-specific dye bias in cDNA microarray experiments', *Bioinformatics*, 21 (12), 2803–2804, 2005.

106. Balazsi, G., Kay, K. A., Barabasi, A. L., and Oltvai, Z. N., Spurious spatial periodicity of co-expression in microarray data due to printing design, *Nucleic Acids Res.*, 31 (15), 4425–4433, 2003.

107. Schuchhardt, J., Beule, D., Malik, A., Wolski, E., Eickhoff, H., Lehrach, H., and Herzel, H., Normalization strategies for cDNA microarrays, *Nucleic Acids Res.*, 28 (10), e47, 2019–2206, 2000.

108. Thellin, O., Zorzi, W., Lakaye, B., De Borman, B., Coumans, B., Hennen, G., Grisar, T., Igout, A., and Heinen, E., Housekeeping genes as internal standards: use and limits, *J. Biotechnol.*, 75 (2–3), 291–295, 1999.

109. Lee, P. D., Sladek, R., Greenwood, C. M., and Hudson, T. J., Control genes and variability: absence of ubiquitous reference transcripts in diverse mammalian expression studies, *Genome Res.*, 12 (2), 292–297, 2002.

110. Duggan, D. J., Bittner, M., Chen, Y., Meltzer, P., and Trent, J. M., Expression profiling using cDNA microarrays, *Nat. Genet.*, 21 (suppl. 1), 10–14, 1999.

111. Galinsky, V. L., Automatic registration of microarray images II. Hexagonal grid, *Bioinformatics*, 19 (14), 1832–1836, 2003.

112. Galinsky, V. L., Automatic registration of microarray images. I. Rectangular grid, *Bioinformatics*, 19 (14), 1824–1831, 2003.

113. Bajcsy, P., Gridline: automatic grid alignment in DNA microarray scans, *IEEE Trans. Image Process.*, 13 (1), 15–25, 2004.

114. Martinez, M. J., Aragon, A. D., Rodriguez, A. L., Weber, J. M., Timlin, J. A., Sinclair, M. B., Haaland, D. M., and Werner-Washburne, M., Identification and removal of contaminating fluorescence from commercial and in-house printed DNA microarrays, *Nucleic Acids Res.*, 31 (4), e18, 1119–1373, 2003.

115. Reimers, M., Statistical analysis of microarray data, *Addict. Biol.*, 10 (1), 23–35, 2005.

116. Armstrong, N. J. and van de Wiel, M. A., Microarray data analysis: from hypotheses to conclusions using gene expression data, *Cell Oncol.*, 26 (5–6), 279–290, 2004.

117. Churchill, G. A., Using ANOVA to analyze microarray data, *Biotechniques*, 37 (2), 173–175, 2004, see also p. 177.

118. Liu, D. K., Yao, B., Fayz, B., Womble, D. D., and Krawetz, S. A., Comparative evaluation of microarray analysis software, *Mol. Biotechnol.*, 26 (3), 225–232, 2004.

119. D'haeseleer, P., How does gene expression clustering work?, *Nat. Biotechnol.*, 23, 1499–1501, 2005.

120. Eisen, M. B., Spellman, P. T., Brown, P. O., and Botstein, D., Cluster analysis and display of genome-wide expression patterns, *Proc. Natl Acad. Sci. U.S.A.*, 95 (25), 14863–14868, 1998.

121. De Smet, F., Mathys, J., Marchal, K., Thijs, G., De Moor, B., and Moreau, Y., Adaptive quality-based clustering of gene expression profiles, *Bioinformatics*, 18 (5), 735–746, 2002.

122. Kohonen, T., *Self-Organizing Maps*, Springer, Berlin, 1995.

123. Bertone, P. and Snyder, M., Advances in functional protein microarray technology, *FEBS J.*, 272 (21), 5400–5411, 2005.
124. Lueking, A., Cahill, D. J., and Mullner, S., Protein biochips: a new and versatile platform technology for molecular medicine, *Drug Discov. Today*, 10 (11), 789–794, 2005.
125. Allison, D. B., Cui, X., Page, G. P., and Sabripour, M., Microarray data analysis: from disarray to consolidation and consensus, *Nat. Rev. Genet.*, 7 (1), 55–65, 2006.
126. Feizi, T., Fazio, F., Chai, W., and Wong, C. H., Carbohydrate microarrays—a new set of technologies at the frontiers of glycomics, *Curr. Opin. Struct. Biol.*, 13 (5), 637–645, 2003.
127. Geschwind, D. H., DNA microarrays: translation of the genome from laboratory to clinic, *Lancet Neurol.*, 2 (5), 275–282, 2003.
128. Calvo, K. R., Liotta, L. A., and Petricoin, E. F., Clinical proteomics: from biomarker discovery and cell signaling profiles to individualized personal therapy, *Biosci. Rep.*, 25 (1–2), 107–125, 2005.
129. Simon, R., Roadmap for developing and validating therapeutically relevant genomic classifiers, *J. Clin. Oncol.*, 23 (29), 7332–7341, 2005.
130. Giltnane, J. M. and Rimm, D. L., Technology insight: identification of biomarkers with tissue microarray technology, *Nat. Clin. Pract. Oncol.*, 1 (2), 104–111, 2004.
131. Spielbauer, B. and Stahl, F., Impact of microarray technology in nutrition and food research, *Mol. Nutr. Food Res.*, 49 (10), 908–917, 2005.
132. Kuiper, H. A., Kok, E. J., and Engel, K. H., Exploitation of molecular profiling techniques for GM food safety assessment, *Curr. Opin. Biotechnol.*, 14 (2), 238–243, 2003.

19 Microelectrode Protein Microarrays

*Kilian Dill, Andrey L. Ghindilis, Kevin R. Schwarzkopf,
H. Sho Fuji, and Robin Liu*

CONTENTS

19.1   DISCUSSION OF THE CMOS CHIP

19.1.1   CHIP CONSTRUCTION AND ELECTRONICS

Protein arrays have been in existence since the 1980s. At that time, they resulted from spotting of proteins onto porous membranes such as nitrocellulose. Considerably later, nonporous materials such as plastic or derivatized glass were used as a matrix or as support for the spotted proteins. These arrays were used to measure protein–protein interactions, oligonucleotide–protein interactions, receptor-drug interactions, and drug discovery or to quantitate cell components present using antibodies for the capture mechanism. In most cases, a fluorescence label was used to detect and quantitate the protein or small molecule that was captured.

The most common method of detection used, especially in the early days, was based upon fluorescence emission of a reporter molecule. Additionally, other methods such as luminescence were used. This chapter details a new method used for the detection of captured proteins: the use of redox enzyme amplified electrochemical detection.

A key aspect of this new technology platform is a semiconductor-based microarray that allows for the manufacture of high density microelectrode arrays that vary from a density of 1000 to greater than 100,000 electrodes/cm^2. Using active circuit elements in the design permits the selection and parallel activation of individual electrodes in the array to perform in-situ oligonucleotide synthesis of customized content on the chip. Proprietary hardware and software developed by CombiMatrix controls the chip. To date, CombiMatrix produces the 1K chip and has just released the 12K chip. Most of the data produced for this chapter details work using the 1 K chips.

The CombiMatrix microarray chip is a silicon-integrated circuit manufactured using a commercial mixed signal complimentary metal oxide semiconductor (CMOS) process. Depending on the electrode density of the chip, the minimum feature size of the CMOS manufacturing process is between 0.6 μ and 1.0 μ and uses two to three levels of interconnect metal. Following the conventional CMOS process, several additional post-process steps are performed at the wafer level to fabricate and expose the platinum metal (Pt) that is the top electrode layer. These steps include the deposition of Pt with an appropriate adhesion layer (either a titanium–tungsten alloy or a titanium/-titanium nitride/titanium metal stack), patterning of the deposited metal layer, deposition of a silicon nitride dielectric film over the surface of the chip, and the patterning of the silicon nitride film to expose the features of the active electrode. Typical electrode sizes range from 44 μ to 92 μ in diameter.

19.1.2   HOW THE CHIP WORKS

The CMOS integrated circuit technology creates active circuit elements and digital logic on the chip that allows complex functions to be implemented. These include a high speed digital interface to communicate to the chip, data writing and reading from the electrode array, setting of appropriate electrical conditions at the electrode to perform in situ oligonucleotide synthesis, and monitoring or detecting of the signal from the electrode array. Figure 19.1 illustrates the architecture and layout of the CombiMatrix CustomArray 12 k chip. This design uses a Serial Peripheral Interface (SPI) to minimize the number of external electrical connections required to communicate to the chip. The SPI is a synchronous serial interface using clock and data pins that allow the control of multiple devices on a communication bus line. In this case, thirteen signal lines allow for both the selection and appropriate voltage/current control of the electrodes and the readout of selective parameters during detection. The large contact pads located along the side of the chip are used for direct electrical connection to the chip by pogo pin connectors. A 56×224 array of electrodes is located in the center of the chip, providing a total of 12,544 spots for the generation of oligonucleotide probes. Specific circuits to control the row and column selection of electrodes in the array are located along the periphery of the electrode matrix. Fiducial marks to assist in the optical analysis of the chip such as stitching of images and auto-templating of the electrode array are also included on the chip.

Each electrode is fabricated within a unit cell of circuit elements that allows precise control of the electrical characteristics of the electrode. A 3-bit memory element within each unit cell can set the electrode to one of several externally controlled voltage lines, an internal current source control circuit, an internal current sink control circuit, or a read-back line that allows external monitoring of the electrode parameters. Figure 19.2 shows a magnified image of the unit cell and associated electrode.

The surface condition, both physical and chemical, of the electrode is important in the performance of this microarray device. For the physical aspect, having a smooth topology without steep angled features is necessary to allow uniform dielectric film removal during the post-processing steps and uniform coating and synthesis of the oligonucleotide. A spin-on-glass etch-back process is performed as parts of the final interconnect metal layer dielectric deposition to planarize

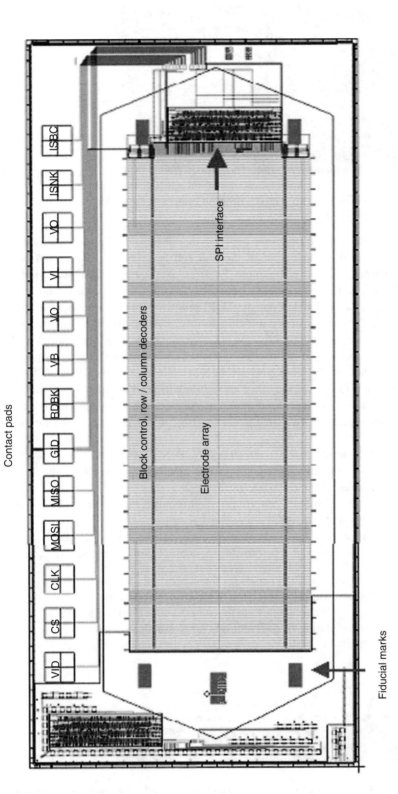

FIGURE 19.1 Chip architecture of CustomArray 12k chip.

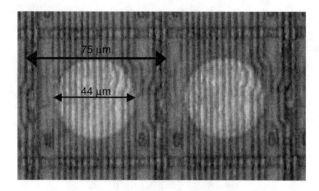
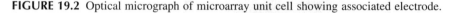

FIGURE 19.2 Optical micrograph of microarray unit cell showing associated electrode.

the chip topology. Also, the electrical connection between the CMOS aluminum layer and the platinum electrode metal is placed at the corner of the unit cell to keep this nonplanar feature outside of the active electrode area of exposed platinum. Figure 19.3 is a cross-sectional SEM of the electrode area, showing the nearly planar electrode topography. Additional planarization processes such as the inclusion of a chemical mechanical polish (CMP) following the interconnect metalliza- tion step can further improve the flatness of the top chip surface. For the chemical aspect, the Pt electrode surface needs to be as contamination free as possible. Therefore, the preparation of the electrode's surface is important before the microarray synthesis. The chip handling protocols and assembly processes used to package the chip prevent the direct contact or handling of the top surface of the chip, particularly the electrode array area. In addition, the exposure of the chip surface to any volatile contaminants, such as outgassing components from epoxies or other adhesives and encapsulants typically used in the packaging of integrated circuit devices, is minimized.

Following the packaging of the microarray, the chip surface is plasma, and it is chemically cleaned before the application of a reaction layer coating. This coating layer facilitates the attach- ment and synthesis of biomolecules to the electrode surface.

A diagram of a unit cell for CMOS-based CombiMatrix 98001 chip 1 K is shown in Figure 19.4 [1] with details concerning the intricate microcircuitry. The 1 K chips contain 1024 electrodes that are available for various experiments. This is the immunochemical detection platform that was introduced by CombiMatrix Corporation in the form of a silicon chip microarray. Unlike many

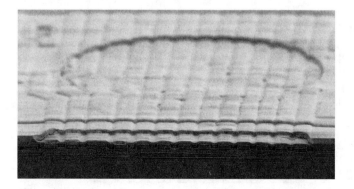

FIGURE 19.3 Cross-sectional SEM of electrode area showing underlying metal interconnect lines and nearly planar metal surface.

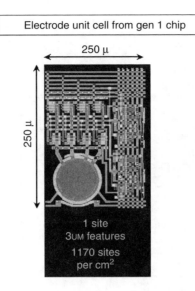

FIGURE 19.4 Diagram of an electrode unit cell for the CombiMatrix 98001 chip. The electrode surface is ~100 microns in diameter and the electrode density is ~1,000 electrodes per sq cm.

other microarray systems that are commercially available, this system is based upon existing technology using a silicon chip. All the electrodes on this chip are individually addressable so that unique chemistries may be carried out at each individual site based upon a potential setting and current flow. Conversely, because each site is individually addressable, the electrode surface may also be monitored for changes in current or potential emanating from the sample at that site.

19.1.3 Surface Components and Coating

The CMOS surface of the CombiMatrix chip is composed of silicon nitride with the electrode composed of a clean platinum surface. Many commercial electrode systems may use gold as this surface provides reactive metal for sulfur-based biomolecule deposition. Monolayers and proteins containing sulfhydryls or disulfides may be chemically attached. Platinum is a somewhat different metal and biomaterials may not readily bond or stick to it. Moreover, DNA synthesis procedures require a matrix that is biofriendly, contains hydroxyl groups, and can withstand the organic media used—the acid conditions generated for the synthesis and the aqueous media required for biological samples. The membrane/matrix support cannot come off the chip or deform under these conditions.

To this end, this group has tested a large number of membranes that meet the above-mentioned conditions. Some are being used commercially and patents have been filed.

19.2 SYNTHESIS CHEMISTRY

19.2.1 DNA Synthesis

DNA can readily be synthesized on the CombiMatrix chip using commercially available reagents except for the generation of protons [2]; they are generated electrochemically. Another crucial point in all these syntheses is that the matrix above the electrodes must contain free OH groups that are usually attached to a carbon backbone. It is on these OH groups (covalent attachment) that DNA synthesis takes place.

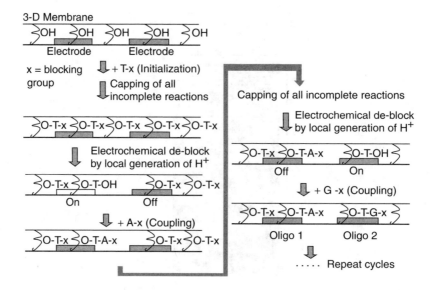

FIGURE 19.5 DNA synthesis scheme.

As previously discussed, a biofriendly matrix is placed uniformly above the chip. This matrix must be stable to the various organic solvents that are used in DNA syntheses as well as aqueous samples containing salts that are used for protein and DNA hybridization and sample detection.

For DNA synthesis, the entire three-dimensional membrane is initialized with dT dimethoxytrityl (DMT) and the remaining unreacted, free hydroxyls are capped with acetic anhydride. This means that dT(DMT) reacts with every free hydroxyl present, whether or not it is over the membrane. The blocking group is then removed by turning on selective electrodes; only those electrodes turned on will lose the DMT group (Figure 19.5 and [2]) in the presence of acid (H^+) that is produced. As long as the acid is contained within the area of the electrode (virtual flask), the DMT groups that will be removed are those confined to the area of the electrode (Figure 19.5). This leaves the DMT protecting groups unaltered on all other dT groups (not near the electrode).

The electrodes that now contain free hydroxyl groups are ready for the next step in the synthesis process. The next activated nucleotide reagent is introduced, and it reacts with the free hydroxyl groups in these solution conditions. The chip is washed, capped, and oxidized to stabilize the central phosphorous atom. The process continues with the deprotection of certain electrodes and a coupling step. The regents used in this process can be purchased off the shelf except for the deblocking solution. In this way, unique oligomers of 100 oligomers or higher can be synthesized at each electrode. Typically, 15 oligomers have been found to be sufficient for use.

19.2.2 OTHER REACTIONS

A host of electrochemical reactions can be performed on the chip. One of the earlier reactions carried out was the electrochemical biotinylation and fluoresceination of the matrix above the electrode on the chip. The reaction scheme for biotin is shown below with the appropriate reaction conditions (Figure 19.6). There is a question of whether the reaction is a carbon insertion or an esterfication process [3,4].

The biotin is incorporated onto the membrane and detected by the addition of fluorescein-labeled streptavidin (F-SA). F-SA cannot be detected unless biotin is covalently attached as shown in Figure 19.7 [3]. The figure shows that with increasing electrode on-time, biotinylation

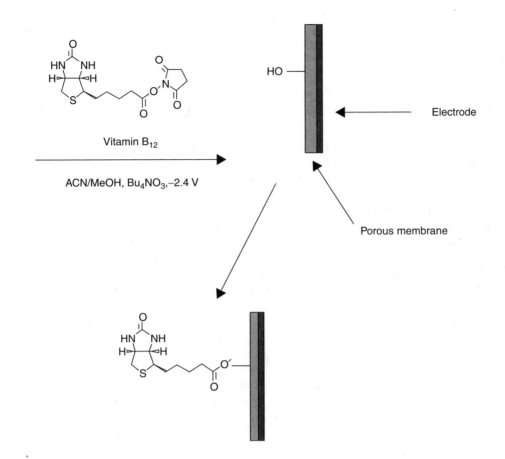

Vitamin B$_{12}$

ACN/MeOH, Bu$_4$NO$_3$,−2.4 V

HO

Electrode

Porous membrane

FIGURE 19.6 Electrochemical biotinylation process on the membrane surface.

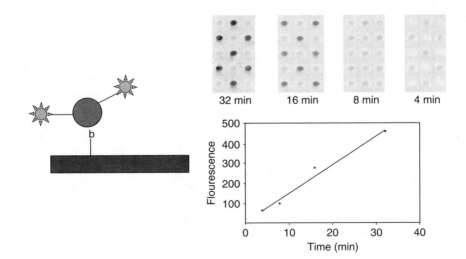

32 min   16 min   8 min   4 min

FIGURE 19.7 Biotinylation of the matrix above the electrode.

is also increased based upon the quantity of F-SA that can be detected. This is just one example of electrochemical reactions performed at the electrode surface.

19.3 ATTACHMENT OF ANTIBODIES TO THE CHIP

19.3.1 Antibody Labeling

For the antibody self-assembling process, the antibodies must be labeled with a DNA sequence that is complementary to the sequence that is produced at the specific electrode. The labeling procedure can be done using two methods (see below). Both use the dual linking reagent, Succinimidye-4-(maleimidomethyl) cyclohexane-1-carbonylate (SMCC); the reactants N-hydroxy succinimide need an amino group on one molecule and a free SH group on a second molecule. The amino labeling reagent SMCC, an N-hydroxy succinimide (NHS) ester, is very reactive and must be used for labile amino groups in step one of any cross-linking procedure.

The structure of an unmodified antibody is shown in Figure 19.8 [1,5]. The connections between the heavy chains and heavy and light chains are disulfide bonds. The vast majority of antibodies do not contain free sulfhydryl (cysteine) on the surface of the macromolecule. However, most do contain free amino groups (ε-amino group of lysine) on the surface.

One way to label the protein for self-assembly is to reduce the heavy chain-heavy chain disulfide bonds and to use the free sulfhydryls produced for the labeling chemistries. The labeling is simple, but the antibody could be damaged or left with only a partial reduction of the disulfide bonds. Moreover, the single chain antibody may not bind antigen as well as the intact antibody. The antibody must be reduced and used within hours of preparation.

The SMCC-NHS ester is reacted with the amino group on the $3'$-end of the complementary oligomer that can be purchased. The reaction is purified via a spin column (Bio Rad) and the maleimide group of the labeled oligonucleotide is ready to react with the sulfhydryls of the reduced antibody (Figure 19.8 and Figure 19.9).

The simpler method of labeling antibodies is to use the amino groups on the protein and cross-link these to the sulfhydryl groups on the oligonucleotide. The antibody is first treated with SMCC that will react with the free amino groups on the antibody. A controlled amount of SMCC is used so that about three lysines per antibody that have been labeled with the dual-functional reagent should be obtained. The mixture is purified through a P6 spin column to removed low molecular weight by-products. An oligonucleotide labeled with an extender (eighteen polyethylene glycol groups) at the 3-prime end (commercially available linked to this end). This extender contains a disulfide unit

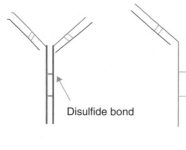

Intact antibody   Sulfhydryl tagging

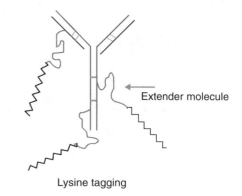

Lysine tagging

FIGURE 19.8 DNA labeling chemistries for antibodies.

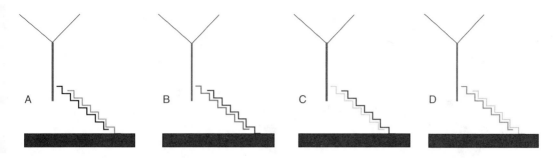

FIGURE 19.9 Antibodies self-assembled to complementary arrays.

(–R-S-S-R′) that must first be reduced with Cleland's reagent. This reaction mixture is purified using a Bio Rad spin column and added to the SMCC-treated antibody. Material is allowed to react and finally purified on a P-30 spin column.

19.3.2 SELF-ASSEMBLY

The unique feature about preparing antibody labeled chips is that the current protocol allows self-assembly. That is, the antibodies that are meant for various chip sectors can all be mixed together and system hybridized for self-assembly. This is usually done in higher salt concentrations and at slightly elevated temperatures. Typically, 40 degrees for 1/2 h in a 2×phosphate buffered saline, tween (PBST) buffer concentration is used. Those not hybridized to the chip surface were washed away with a 2×PBST buffer.

The oligomers were not randomly chosen, but rather, extracted from CombiMatrix's probe selection processes programs so that each probe has a similar melting point and is substantially different in sequence. In addition, it was established that the probes would not contain any significant secondary structure.

19.3.3 ASSAY FORMATS

There are two assay formats that can be used in this system: sandwiched-based assay or competitive assay (Figure 19.10).

In the sandwich assay format, an antibody is used to capture antigens on the chip surface; which often may be a purified monoclonal antibody that has a very high affinity for the analyte. The signal antibody (usually a polyclonal antibody) is labeled with a fluorescence marker or enzyme to provide evidence that the analyte has indeed been captured. Often, this is a mixture of high affinity antibodies containing binding sites for various epitopes on the analyte surface; it is crucial that these Ab-analyte binding sites do not overlap with the analyte binding site provided by the capture antibody. This is the preferred assay format as the signal increases and the quantity of captured-analyte increases. In the sequential assay format, the Hooke effect never comes into play.

The competitive assay format is an indirect method as the signal is greatest when the nonlabeled analyte is present. A known quantity labeled analyte is used, and it is titrated with a true or nonlabeled analyte that is found in samples. In other words, the nonlabeled material competes with the labeled species for the antibody binding site. Therefore, the signal decreases when the sample containing the nonlabeled analyte increases. Here, the assay is never quite as sensitive as the direct sandwich-based method, and the dynamic range (detection region) is usually smaller and limited to the sigmoidal detection region [5].

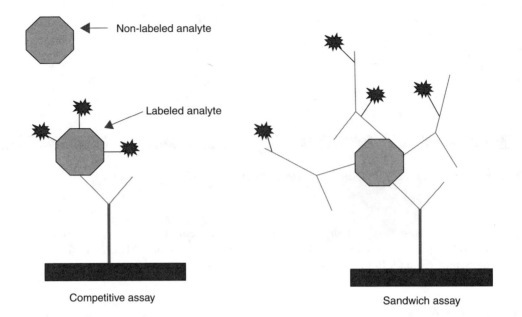

FIGURE 19.10 Various immunoassay formats.

19.4 DETECTION METHODS

19.4.1 FLUORESCENCE

Numerous methods can be used for the detection process in immunoassays. Many optical methods are not sensitive enough to be employed. However, methods such as luminescence or fluorescence are traditionally used in these assays. They do require a label on the protein and a rather large instrument for excitation and detection at discrete frequencies. The fluorescent detection method was initially deployed to show that immunoassays can indeed be used on the microarray format. Because this was not the focus of the current chapter, we shall show but a few examples that fluorescence-based assays can indeed be used [3].

Results for the fluorophore-based sandwich immunoassays for selected analytes are shown in Figure 19.11. The antibodies were labeled with different fluorophores (Cy5 and Texas Red). The right lane on each chip is the control. The analytes range from proteins to much larger spores.

Figure 19.12 shows the results for the standard detection curves. The immunosandwich assay standard curve is shown for the detection of BG spores. The competitive curve is for F-BSA-b (fluorescein-bovine serum albumin (BSA) was used as the analyte).

19.4.2 ELECTROCHEMISTRY

Another detection method that can be used is an electrochemical-based system. In this case, a change in the oxidation state of a solution component is measured. Literature methods using various enzymes to enhance/detect analytes in immunoassays systems have been published [1,4,6–8].

For this redox-based enzyme amplification electrochemical system, horseradish peroxidase was used. Horseradish peroxidase is an oxidoreductase that catalyzes the oxidation of a substrate (such as OPD and related compounds) while using peroxide as the electron acceptor. The molecular weight of HRP is about 30,000 daltons, and the enzyme turnover rate is very high. For this sandwich immunoassay-based system (Figure 19.13), a streptavidin-HRP conjugate is bound to

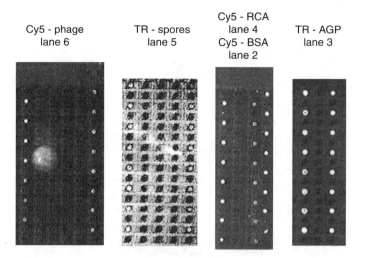

FIGURE 19.11 Fluorescence-based immunoassays using the 1K CombiMatrix chip. The single antibody for each analyte is tagged with a different fluorophore.

the immunocomplex through the biotinylated antibody. Alternatively, the reporter antibody may be directly conjugated with HRP and, hence, not require the biotin labeling procedure.

To show the feasibility of using one of the ~ 100-micron electrodes on the CombiMatrix BioChip for the electrochemical chemical detection of redox reactions in solution, a chip is immersed into a 500 μL solution containing substrate, buffer, and enzyme (HRP). The enzymatic events could be monitored as a function of potential (V) and current (I). Figure 19.14 shows the voltage/current measurements for the HRP reaction (catechol as substrate) as a function of time. A single 92 μ electrode on the chip was used to collect the data from reactions ongoing in the 500 μL solution. The measurements were made of the solution that contained buffer and substrate, with and without enzyme present. The voltage was set relative to the platinum electrode, and the resulting current was measured with and without HRP present. Clearly, a noticeable difference is achieved when HRP is present as the enzymatic reaction is monitored. The results give two important pieces of information: (1) A potential may be chosen to maximize the differential

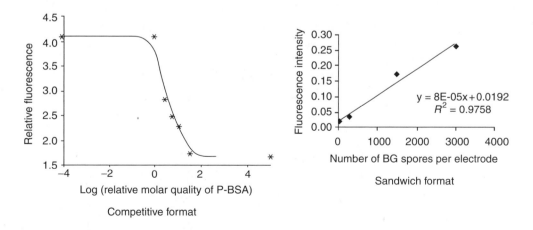

FIGURE 19.12 Result from two different immunoassays formats. The left is for fluorescein and the right is for BG spores.

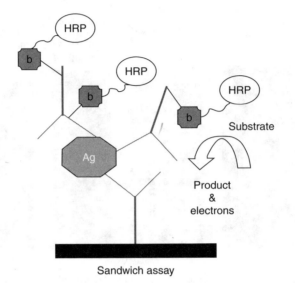

Sandwich assay

FIGURE 19.13 A depiction of a sandwich immunoassay format on the chip using the e-chem detection system.

current with and without enzyme present, and (2) a single microelectrode may be used to monitor enzymatic events surrounding the electrode.

The results presented in Figure 19.15 show that the current generated from the redox reaction may be monitored as a function of time with an optimized voltage setting.

Again, a chip microelectrode is used to monitored events occurring in solution. Clearly, the current rises with time (from −10 nA to −60 nA) as the enzyme converts the substrate to product. A steady state is reached after three seconds, indicating that a short waiting period may be optimum for data collection.

This group has shown that the microarray can be used to monitor enzymatic redox reactions in a solution surrounding the chip. Can the microelectrode monitor events based upon chemistries

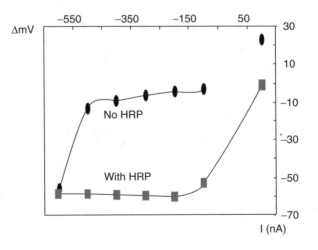

FIGURE 19.14 Monitoring of the ongoing enzymic solution reaction by a chip microelectrode. The chip potential is varied and the current is measured. Catechol was used as a substrate in this system.

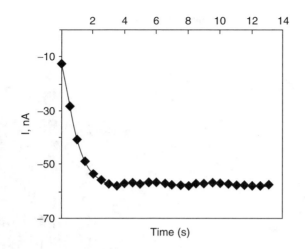

FIGURE 19.15 Monitoring the current at one electrode as a function of ongoing enzymatic activity of horseradish peroxidase on the substrate (catechol).

ongoing at a specific site without loss of signal because of diffusion or bleed over from events occurring at other electrodes? To show that the electrochemical detection may be feasible for chemistries at selected electrodes, it has devised an immunoassay scheme that uses selected electrodes to monitor immunocomplexes that develop at those sites (Figure 19.13). In this case, a signal is produced if the immunocomplex is completed and the redox signal can be monitored. If the group is able to chaperon selected antibodies to specific predetermined locations, it should be able to determine if immunoassay detection is feasible and if the chemistries can be localized to the selected electrodes in question. It should be noted that a similar enzyme amplified electrochemical detection system using a single flow cell provided excellent electrochemical data for immunoassays [9].

Figure 19.16 shows some of the electrochemical data for the detection of the immunoassay complexes. In this case, the biotinylated reporter antibody contained bound SA-HRP. The analytes present were in saturated conditions, and the addition order of the immunocomplex components was sequential. Clearly, the two analytes were detected according to the sandwich assay scheme depicted in Figure 19.13. The data were collected as the enzymic reaction reached steady state conditions. The statistics for the data in Figure 19.16 are given on the left-hand side of that figure. The standard deviation for α1-acid glycoprotein (AGP) was found to be 16.4%, and those for the detection of M13 phage were 7.8%. Typically, a standard deviation of 10% or less is observed for a given analyte concentration.

How many electrodes are needed for good statistics in obtaining accurate data for analyte concentrations? Traditionally, several rows of electrodes are used, but are these necessary? Figure 19.17 shows the standard deviation for a phage assay as a function of the number of electrodes used. Clearly, a minimum of ten sample repetitions (electrodes) for the same analyte concentration is required for the best results. Therefore, using two rows of electrodes (comprising sixteen electrodes) would be adequate for the analysis of a single analyte concentration.

A number of additional parameters enable the optimization of the immunoassays. One of those parameters is analyte incubation time. This is a crucial parameter as it allows the signal output to be optimized. Figure 19.18a and b show the results for the signal output in the immunoassays for ricin (RCA) and phage as a function of concentration and time. For the best results, sixty minutes (statistics and maximum signal) is optimum; however, for quicker results, a shorter incubation time (twelve minutes or less) can be used.

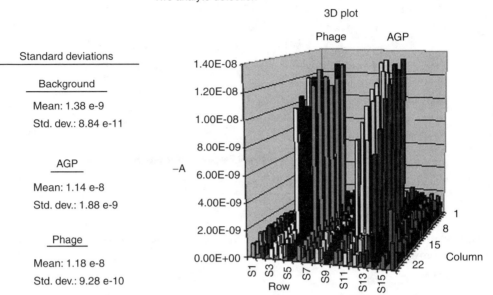

FIGURE 19.16 The three-dimensional plot for results from a sandwich immunoassay detection of AGP and Phage using the e-chem detection system. Mean values and standard deviations for background and each analyte are given on the left-hand side of the figure.

Figure 19.19 shows that a host of different antibodies may be self-assembled on a biochip to perform numerous multiplexed immunoassays on a single chip. The results for five analytes are presented in two-dimensional and three-dimensional plots. To date, a chip that contained up to fifteen unique antibodies has been constructed. The intensities (signal output in Figure 19.19) are not the same for all analytes as differing concentrations were used. Additionally, the assay performance for each analyte will differ based upon the affinity constant of each antibody-analyte pair. AGP concentration was just above the LOD conditions.

Standard curves and limits of detection (LODs) for these analytes are given in Figure 19.20 and Figure 19.21. The results show that the system is indeed very sensitive (attomole levels), and the

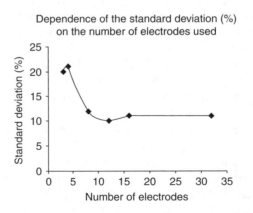

FIGURE 19.17 Dependence of the standard deviation (%) on the number of electrodes used in the calculation. Analyte used was phage under saturation conditions with a one-hour incubation period.

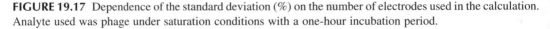

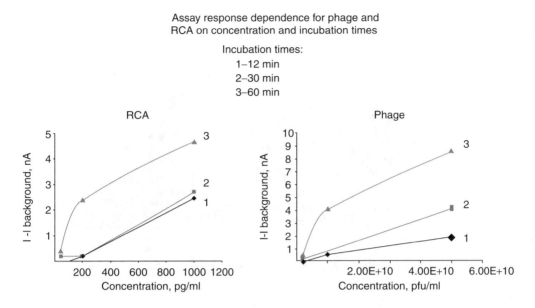

FIGURE 19.18 Assay response as a function of analyte concentration and incubation time. Analytes used were phage and ricin (RCA).

system can detect an analyte over a range of at least four logs. The results and detection limits for AGP and ricin are similar to those published for RCA and AGP using light addressable potentiometric sensor (LAPS) detection [10,11]. Because this platform is still under development, improvements in LODs are still in the offing. The LOD value for phage is far from optimal, and this may be because the antibody affinity constant for the phage is poor, and, more than likely, the use of the identical monoclonal antibody for the capture and detection of phage is far from optimum.

The analyte standard curves shown in Figure 19.20 and Figure 19.21 were derived from data accumulated from at least five different chips. The reproducibility of the data then relies on intrachip and interchip variations. The intrachip standard deviations are typically less than 10% for a given analyte concentration; this, of course, depends upon the number of data points (electrodes) per analyte. Typically, an onboard standard is used to normalize chip variations, and, in this format, chip-to-chip variations were found to be much less than 5%.

Not presented in Figure 19.22 is the LOD data for BG spores. These data are presented in Figure 19.23. From the number of spores placed on the chip, the area covered by these spores, the fraction of the total area that the electrodes encompass, and the signal output, on average, one spore per electrode can be detected. See Figure 19.23 for details. The high degree of sensitivity results from the large number of identical epitopes that are likely found on the spore coat protein (large number of signal antibodies is bound).

19.5 NEW CHIP DEVELOPMENTS

19.5.1 INTEGRATED MICROFLUIDIC BIOCHIPS

The research on integrated microfluidics lab-on-a-chip technology involves the development of miniaturized devices, miniaturized systems, and new applications related to the handling of fluids (both liquids and gases). Several researchers have developed devices allowing for performance of multistep assays using complicated channel networks, whereas pumps, valves, and detectors were

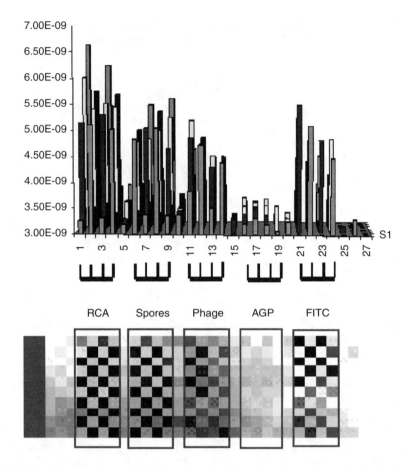

FIGURE 19.19 Signal output from a multiplexed immunoassay for five analytes. Both the two-dimensional and three-dimensional plots are provided. Analyte incubation was for a one-hour duration.

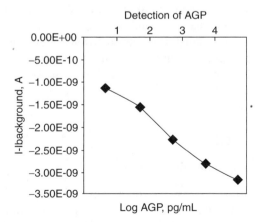

FIGURE 19.20 The standard curve for the detection of AGP.

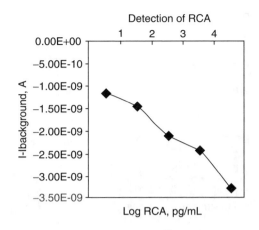

FIGURE 19.21 Standard curve for the detection of ricin (RCA).

left off-chip and were built into the desktop test station [12]. Others argued for integrating all functional components into the chip and preferred portable solutions [13]. The latter efforts led to ingenious demonstrations of on-chip valving [14] and pumping schemes [15] in an attempt to depart from traditional Microelectro-mechanical Systems (MEMS) approaches that are complicated in fabrication and expensive. Most of the work has been directed toward the integration of DNA amplification with capillary electrophoresis (CE) [16]. This would allow one to automate DNA and immunoassays and reduce costs. It would also allow one to produce a handheld platform.

A self-contained and fully integrated microfluidic biochip device that integrates microfluidics with an enzyme amplified electrochemical detection-based microarray for protein (immunoassays) and DNA detection has recently been developed. The device (as shown in Figure 19.24) consists of a plastic fluidic cartridge and a CombiMatrix microarray chip. The plastic cartridge includes a reaction chamber for Ag-Ab binding or DNA hybridization, a number of reagent storage chambers, and micropumps.

The operation of the biochip device is as follows. In the case of immunoassay, a biological sample solution containing antigens is loaded in the microarray chamber. Other solutions, including a wash buffer, a secondary antibody solution, a streptavidin-HRP conjugate solution, and an electrochemical detection buffer mixture are separately loaded in other storage chambers. The biochip device is then inserted into an instrument that provides electrical power for pumping, hybridization heating, and electrochemical signal readout. The process starts with a one-hour hybridization at 37°C in the microarray chamber. The washing buffer is then pumped through the microarray chamber using an integrated electrochemical micropump to wash the array. The secondary antibody solution is subsequently pumped into the microarray chamber followed by a

1. AGP: 5 pg/ml. 8 amol in 50 μl volume

2. RCA: 300 pg/ml. 250 amol in 50 μl volume

3. Phage*: 3.75 e8pfu/ml. ~1 million pfu in 40 μl volume

Dynamic detection range is 4 logs

* Identical monoclonal antibody was used for capture and detection

FIGURE 19.22 Limits of detection determined for the three analytes. The dynamic range for the analytes is also presented.

Detection of single spores

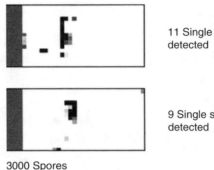

11 Single spores
detected

9 Single spores
detected

3000 Spores
5% Electrode area on chip
150 Spores/500 electrodes
~12 Electrodes with single spore detection

FIGURE 19.23 The detection of a single spore.

thirty-minute incubation allowing the secondary Ab to bind with the Ag on the assay surface. Once the incubation is completed, the washing buffer is pumped through the microarray chamber washing the array surface. The streptavidin-HRP conjugate solution is then introduced into the microarray chamber, and the solution/chip is incubated for thirty minutes at room temperature. A final wash is implemented in the microarray chamber before the introduction of the electro-chemical detection buffer mixture. The electrochemical hybridization signals corresponding to the redox-reaction amplified by the enzymes (HRP) that are bound to the immunocomplex on the surface of probe electrodes are detected on the chip and recorded by the instrument.

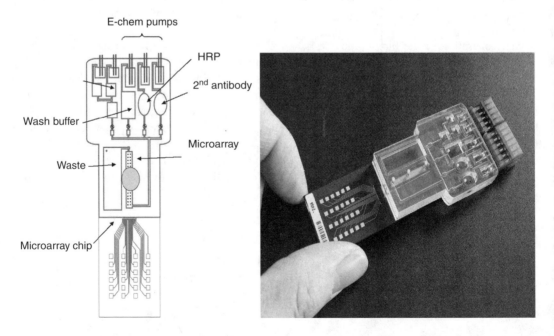

FIGURE 19.24 (A) Schematic of the microfluidic biochip device. (B) Photograph of the integrated device that consists of a plastic fluidic cartridge and a Combimatrix's CustomArray™ chip.

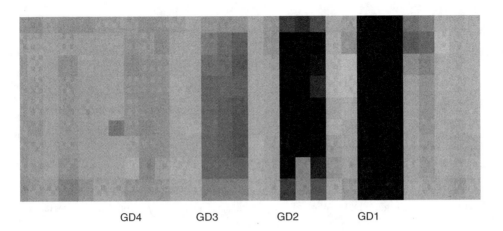

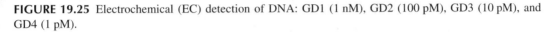

FIGURE 19.25 Electrochemical (EC) detection of DNA: GD1 (1 nM), GD2 (100 pM), GD3 (10 pM), and GD4 (1 pM).

Micropumps are critical components in the microfluidic biochip design. The biochip requires a transport of a number of liquid solutions in a volume of tens to hundreds of μL. Most conventional pressure-driven, membrane-actuated micropumps suffer from complicated designs, complicated fabrication, or high cost. In this device, electrochemical pumps are used that rely on electrolysis of water between two stainless steel electrodes in a saline solution to generate gases when a DC current is applied. The gas generates a pressure that, in turn, moves liquid solutions in the biochip. This pumping mechanism does not require a membrane and/or check valves in the design. As a result, the fabrication and operation are much simpler than most conventional micropumps.

The microfluidic biochip device was tested for detection of different concentrations of biotinylated oligomers. The microarray chip with an electrode density of ~ 1000 electrodes per sq cm was used. For this redox-based enzyme amplification electrochemical system, horseradish peroxidase (HRP) is used. HRP is an oxidoreductase that catalyzes the oxidation of a substrate (such as OPD) while using peroxide as the electron acceptor. The on-chip analysis starts from DNA hybridization. Similary, the system can also be used for immunoassays. A sample solution containing the $5'$-biotinylated 15-mer complementary sequences to GD1, GD2, GD3, and GD4 in varying concentrations (1 pM to 1 nM) was incubated in the array chamber for one hour. The array is washed with $2\times$PBST buffer and then filled with a solution containing the streptavidin-HRP conjugate for thirty minutes. Finally, the array is washed with $2\times$PBST and loaded with hydrogen peroxide and ortho-phenylenediamine before taking readings. Biological sample and reagent solutions are loaded into the device beforehand; electrochemical signals corresponding to pathogenic or genetic information are the primary output. The results are shown in Figure 19.25; four sections of the chip had different 15-mers synthesized (cGD1, cGD2, cGD3, and cGD4P-6) on the matrix covering the chip. The sensitivity of the system is less than 10 pM in oligonucleotide concentration. Similarly, the microfluidic biochip can also be used to perform on-chip immunoassay (in progress).

19.6   CONCLUSIONS

The chapter describes advances in microarray technology and clearly shows that microarrays may be performed on CMOS chips. The capture antibody is self-assembled on the CMOS chip based upon the complementary DNA label that has been covalently attached to the antibody. The analyte and signal antibody are then added in a sequential manner. Detection is based upon redox amplified electrochemical detection. The results are robust; detection of several log units of analyte concentration is possible and is far superior to traditional fluorescence-based detection.

REFERENCES

1. Dill, K., Montgomery, D. M., Oleinikov, A. V., Ghindilis, A. L., and Swarzkopf, K. R., Immunoassays based on electrochemical detection using microelectrode arrays, *Biosens. Bioelectron.*, 20, 736, 2004.
2. Oleinikov, A. V., Gray, M. D., Zhao, J., Montgomery, D. D., Ghindilis, A. L., and Dill, K., Self-assembling protein arrays using electronic semiconductor microchips and in vitro translation, *J. Proteome Res.*, 2, 313, 2003.
3. Dill, K., Montgomery, D. D., Wang, W., and Tsai, J. C., Antigen detection using microelectrode array microchips, *Anal. Chim. Acta*, 444, 69, 2001.
4. Tefu, E., Maurer, K., Ragsdale, S. R., and Moeller, K. D., Building addressable libraries: the use of electrochemistry in generating Pd(II) reagents at preselected sites on a chip, *J. Am. Chem. Soc.*, 126, 6212, 2004.
5. Dill, K., Montgomery, D. D., Ghindilis, A. L., and Schwarzkopf, K. R., Immunoassays and sequence-specific DNA detection on a microchip using enzyme amplified electrochemical detection, *J. Biochem. Biophys. Methods*, 59, 181, 2004.
6. Aguilar, Z., Vandaveer, W. R., and Fritsch, I., Self-contained microelectrochemical immunoassay for small volumes using mouse IgG as a model system, *Anal. Chem.*, 74, 3321, 2002.
7. Aguilar, Z. and Fritsch, I., Immobilized enzyme-linked Dan hybridization assay with electrochemical detection for *Cryptosporidium parvum* hsp70 mRNA, *Anal. Chem.*, 75, 3890, 2003.
8. Meyerhoff, M. E., Duan, C., and Meusel, M., Novel nonseparation sandwich-type electrochemical enzyme immunoassay system for detecting marker proteins in undiluted blood, *Clin. Chem.*, 41, 1378, 1995.
9. Rossier, J. S. and Girault, H. H., Enzyme linked immunosorbent assay on a microchip with electrochemical detection, *Lab Chip.*, 1, 153, 2001.
10. Dill, K., Lin, M., Poteras, C., Hafeman, D. G., Owicki, J. C., and Olson, J. D., Antibody-antigen binding constants determined in solution-phase with the threshold membrane-capture system: binding constants for anti-fluorescein, anti-saxitoxin, and anti-ricin antibodies, *Anal. Chem.*, 217, 128, 1994.
11. Dill, K. and Bearden, D. W., Detection of human asialo-alpha(1)-acid glycoprtein using a hetero-sandwich immunoassay in conjunction with the light addressable potentiometric sensor, *Glycoconj. J.*, 13, 637, 1996.
12. Yuen, P. K., Kricka, L. J., Fortina, P., Panaro, N. J., Sakazume, T., and Wilding, P., Microchip module for blood sample preparation and nucleic acid amplification reactions, *Genome Res.*, 11, 405, 2001.
13. Burns, M. A., Johnson, B. N., Brahmasandra, S. N., Handique, K., Webster, J. R., Krishnan, M., Sammarco, T. S. et al., An integrated nanloiter DNA analysis device, *Science*, 282, 484, 1998.
14. Thorsen, T., Maerkl, S. J., and Quake, S. R., Microfluidic large-scale integration, *Science*, 298, 580, 2002.
15. Dodson, J. M., Feldstein, M. J., Leatzow, D. M., Flack, L. K., Golden, J. P., and Ligler, F. S., Fluidics cube for biosensor miniaturization, *Anal. Chem.*, 73, 3776, 2001.
16. Bohm, S., Pijanowska, D., Olthuis, W., and Bergveld, P., A flow-through amperometric sensor based on dialysis tubing in free enzyme reactors, *Biosens. Bioelectron.*, 77, 223, 1999.

20 Bioconjugated Quantum Dots for Sensitive and Multiplexed Immunoassays

Xiaohu Gao, Maksym Yezhelyev, Yun Xing, Ruth M. O'Regan, and Shuming Nie

CONTENTS

20.1 INTRODUCTION

Immunoassay using fluorophore-labeled antibodies is a popular technique for medical diagnostics and bioanalytical chemistry because of its high detection sensitivity, non-isotopic safety, multiplexing, and quantitative capabilities. However, these promising features are still limited by the use of traditional organic dyes with less than optimal absorption and emission properties. For example, organic dyes often undergo rapid photobleaching, making accurate quantification difficult; their broad and asymmetric emission profiles result in significant spectral overlap in multicolor applications. As a result, it is difficult or nearly impossible to detect more than 3 biomolecular targets using a single light source. Therefore, there is a need to develop new detection labels that can overcome the aforementioned problems.

Recent advances in nanotechnology has produced a new class of fluorescent labels based on semiconductor quantum dots (QDs) for a broad range of applications, including single molecule biophysics, biomolecular profiling, optical barcoding, and in vivo imaging [1–20]. In comparison with organic dyes and fluorescent proteins, QDs have unique optical and electronic properties such as size-tunable light emission, improved signal brightness, resistance against photobleaching, and simultaneous excitation of multiple fluorescence colors. Surface-passivated QDs are highly stable against photobleaching and have narrow, symmetric emission peaks (as narrow as 14 nm full width at half maximum or FWHM). It has been estimated that CdSe QDs are about 20–40 times brighter, depending on particle sizes and quantum yields and 1000 times more stable against photobleaching than single dye molecules. These properties render QDs the ideal fluorophore for ultrasensitive, multiplexed, and quantitative fluoroimmunoassays with important applications in clinical diagnostics.

This chapter discusses the immunoassays and bioanalytical applications of semiconductor QDs. Specifically, it will describe the unique spectral properties of QDs, bioconjugation chemistry, multicolor and quantitative protein staining, and optical barcoding for homogeneous immunoassay. The synthesis of high-quality QDs and their surface modifications will be briefly reviewed. This chapter concludes with a short discussion of future prospects and challenges in QD immunoassays.

20.2 OPTICAL PROPERTIES

Unlike conventional dyes and biological samples made from organic materials, QDs are made from inorganic semiconductors and emit photons via a process known as excitonic fluorescence [1]. This different light emitting mechanism offers unique spectral properties and leads to a better signal to noise ratio (S/N) in fluorescence detection. One feature is the large extinction coefficients of QDs that make them an intrinsically brighter probe under most fluorescence imaging conditions. The molar extinction coefficients of QDs typically have a value of $0.5-2 \times 10^6 \, M^{-1} \, cm^{-1}$ that is about 10–50 times larger than that of organic dyes [2]. Because of an increased rate of photon absorption, single QDs appear dozens of times brighter than organic dyes (Figure 20.1a). The brightness of QDs has been experimentally confirmed by this group and other researchers [3] and is important for detection of low abundance targets.

Second, the differential fluorescence lifetime provides another dimension for separation of QD fluorescence from the background, a technique also known as time-domain or time-gated imaging [4–6]. Fluorescence lifetime is an intrinsic characteristic of a fluorophore and describes the duration of its stay in the excited state before returning to the ground state. A simplified model for fluorescence signal decay is shown in Figure 20.1b for QDs and organic dyes. Assuming the initial fluorescence intensities of QDs and dyes after a pulse excitation are the same and the fluorescence lifetime of QDs is one order of magnitude longer than that of the dye, the QD/dye intensity ratio quickly increases with time (that is, $I_{QD}/I_{dye} = 1$ at time $t = 0$, and $I_{QD}/I_{dye} = \sim 100$ at $t = 10$ ns). Therefore, an imaging time window can be selected to improve the S/N ratio.

The sensitivity can be further enhanced by the large Stokes shift of QDs. For organic dyes, the Stokes shift is measured as the distance between the excitation and emission peaks, and it typically has a small value from a few nanometers to 10 s of nanometers. The Stokes shift of semiconductor QDs, in contrast, can reach up to 300–400 nm, as shown in Figure 20.1c. In some high-background materials such as cells and tissues, fluorescence with a small Stokes shift is embedded within an overwhelming autofluorescence background, whereas fluorescence with a large Stokes shift is clearly differentiated from the background because of a color or wavelength contrast. It is also worth noting that for most biological samples, the autofluorescence is very low, above 800 nm. Recent development of alloy and Type II QDs has shown great promise in making high-quality near-infrared nanoparticles for biomolecular imaging and detection [7,8].

In addition to sensitivity, it is also important to have multiplexing capabilities so that multiple molecular targets can be simultaneously measured. Figure 20.1d shows a comparison of fluorescence absorption and emission spectra between red QDs and a color-matched organic dye X-Rhodamine. The emission spectral width of QDs is mainly determined by the particle size distribution. For CdSe QDs, the emission line-width (or FWHM) is typically 30 nm or less, much narrower than that of most organic fluorophores. Recent development in alloy nanoparticles synthesis has produced QDs with FWHM as narrow as 13 nm at room temperature [9,10]. This will allow simultaneous use of multi-color QDs without serious spectral overlap. Even with overlapping spectra, multicolor QDs can be resolved by spectral deconvolution because their emission spectra are symmetric (Gaussian distribution). In addition, QDs absorb excitation light over a broad range

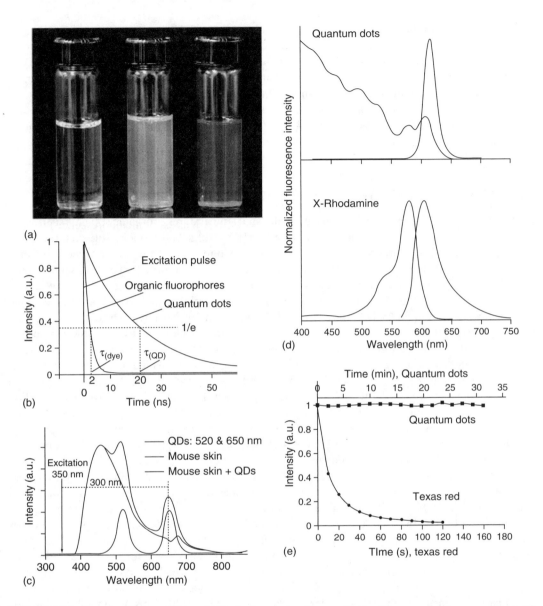

FIGURE 20.1 Novel optical properties of QDs for improving the sensitivity of fluorescence immunoassays. (a) Comparison of fluorescence light emission from organic dye tetramethylrhodamine isothiocyanate (TRITC; left vial), green QDs (middle vial), and red QDs (right vial) under normal room light illumination and at the same molar concentration (1.0 μM). Bright fluorescence emission is observed from the QDs but not from the dye, owing to the large absorption cross-sections of QDs. (b) Comparison of the excited state decay curves (mono-exponential model) between QDs and common organic dyes. The longer excited state lifetimes of QD probes allow the use of time domain imaging to discriminate against the background fluorescence (short lifetimes). $t_{(dye)}$ and $t_{(QD)}$ are the delay times for the fluorescence signals to decrease to $1/e$ of their original values, where e is the natural log constant and is equal to 2.718. (c) Comparison of mouse skin and QD emission spectra obtained under the same excitation conditions, demonstrating that the QD signals can be shifted to a spectral region where autofluorescence is reduced. (d) Fluorescence absorption and emission spectra of a representative organic dye, X-Rhodamine, and QDs. (e) Photobleaching curves showing that QDs are several thousand times more photostable than organic dyes (e.g., Texas red) under the same excitation conditions. (From Xiaohu Gao, Lily Yang, John A. Petros, Fray F. Marshall, Jonathan W. Simons, Leland Chung, and Shuming Nie, *Curr. Opin. Biotechnol.*, 16, 63–72, 2005. With permission.)

of wavelengths, and their molar extinction coefficient gradually increases toward shorter wavelengths that make it possible to simultaneously excite multiple colors with a single light source.

Another important parameter in fluoroimmunoassays is quantification. Rapid photobleaching of organic fluorophores and fluorescent proteins makes accurate quantification problematic. In contrast, QDs are dramatically (more than 1000 times) more stable against photobleaching and can tolerate extended illumination for repeated measurements (Figure 20.1e). Based on this group's calculations, single QDs are able to emit 4–5 orders of magnitude more photons before photobleaching than organic dyes.

20.3 SURFACE CHEMISTRY AND ANTIBODY CONJUGATION

Highly fluorescent QDs are often synthesized at elevated temperatures in organic solvents such as tri-n-octylphosphine oxide (TOPO) and hexadecylamine (HDA) (high boiling-point solvents containing long alkyl chains). The hydrophobic organic molecules not only serve as the reaction media, but they also coordinate with unsaturated atoms on the QD surface to prevent formation of bulk semiconductors. As a result, the nanoparticles are coated with a monolayer of the organic ligands and are soluble only in non-polar solvents. To overcome this problem, two simple methods based on surface ligand exchange were developed by Dr. Alivisatos and this group in 1998 [11,12]. These two pioneering reports demonstrated the use of QDs for in vitro cell labeling and endocytosis studies, but the bioconjugated probes had low quantum yields, stability problems, and cellular toxicity [13,14].

Intense interest in using QDs for live cell and animal imaging has led to the development of improved QD solublization and bioconjugation procedures (Figure 20.2). QDs are first solubilized with an octylamine-modified low molecular weight poly(acrylic acid) [15]. The hydrophobic alkyl side chains strongly interact with TOPO on the QD surface, whereas the hydrophilic carboxylic acid groups face outward and render QDs water soluble. After linking to antibodies or streptavidin, the QD bioconjugates have been used in both high resolution and high sensitivity cellular imaging studies [15,16]. The results revealed detailed cell skeleton structures as well as the diffusion dynamics of single receptors in living cells although fluorescence and binding affinity loss were also observed over extended incubation or storage, possibly as a result of polymer or ligand detachment [17,18]. For in vivo lineage-tracing of embryogenesis, Dubertret and coworkers encapsulated single QDs with PEG-derivatized phospholipid micelles and injected individual embryos with up to 2 billion particles [19]. Despite their relatively large sizes that may reduce the kinetics of biomolecular binding and nanoparticle extravagation from blood vessels for in vivo targeting, the QDs possess excellent stability and biocompatibility, leading to normal embryo development for up to four days. Most recently, Gao and Nie have used a tri-block amphiphilic copolymer that is specifically designed for in vivo molecular imaging and targeting [20]. This polymer is able to self-assemble on the surface of QDs, leading to well-dispersed nanoparticles that are encapsulated by a tight, hydrophobic layer. This block copolymer is another type of nanomaterial building block that is used in self-assembly and soft nanolithography [21,22]. A key feature shared by these amphiphilic polymer coating procedures is that QDs are solubilized in aqueous solution without replacing the coordinating organic ligands that are believed to be important for maintaining the optical properties of QDs and for shielding the core from contact with the outside environment [23].

Water-soluble QDs have little or no selectivity to target molecules except for some genetically engineered viruses or special type of cells [24–27], and they need to be attached to bio-affinity ligands such as antibodies. Because of the dimensional similarity of QDs and biomacromolecules, this can be achieved via a variety of conjugation chemistries such as adsorption, chelating, and covalent linkage. Two of the most popular crosslinking reactions are carbodiimide mediated amide formation and active ester–maleimide mediated amine and sulfhydryl coupling (Figure 20.3a and Figure 20.3b). The advantages of carboxylate-amine condensation using carbodiimides is that most biomolecules such as IgGs contain many primary amines and carboxylic acids and do not need

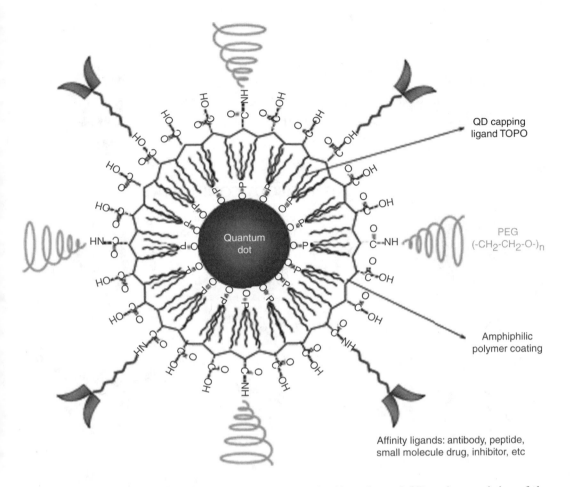

FIGURE 20.2 Schematic diagram showing the structure of a bioconjugated QD probe, consisting of the capping ligand TOPO, an encapsulating amphiphilic polymer layer, biomolecule-targeting ligands (such as antibodies, peptides, or small-molecule inhibitors), and polyethylene glycol (PEG). (From Xiaohu Gao, Lily Yang, John A. Petros, Fray F. Marshall, Jonathan W. Simons, Leland Chung, and Shuming Nie, *Curr. Opin. Biotechnol.*, 16, 63–72, 2005. With permission.)

chemical modification prior to coupling with nanoparticles. In comparison, sulfhydryl groups are less common in native biomolecules and are unstable in the presence of oxygen. QDs can also be functionalized with amines and carboxylic acids with a variety of amphiphilic polymers. But the abundant reactive groups could cause aggregation and render biomolecules randomly oriented on the QD surface that is detrimental to probe bio-activity (Figure 20.3a). Other conjugation reactions are possible, depending on the available chemical groups. For example, Pellegrino et al. recently used a pre-activated amphiphilic polymer for QD solublization [28]. This polymer contains multiple anhydride units and is highly reactive to primary amines without the need of an external coupling reagent. It is worth noting that polyanhydride is also under intense study and clinical examination for sustained drug delivery and tissue engineering [29,30].

The antibody orientation on the surface of QDs is critical to specific binding and can be controlled by several procedures as shown in Figure 20.3. Mattousssi and coworkers first explored the use of a fusion protein as an adaptor for IgG antibody coupling [31,32]. The adaptor protein has a positively charged leucine zipper domain for electrostatic interaction with QDs and a protein G domain that binds to the Fc region of antibodies. As a result, the Fc end of the antibody is connected to the QD surface and the target-specific $F(ab')_2$ end faces outward (Figure 20.3c). Direct

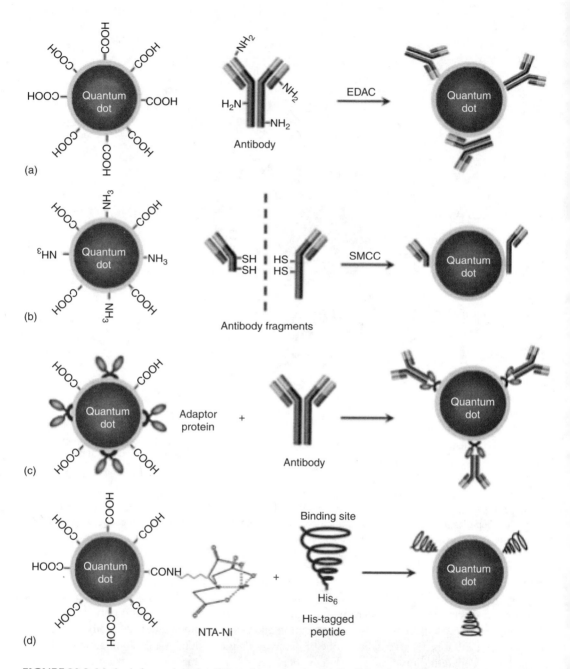

FIGURE 20.3 Methods for conjugating QDs to biomolecules. (a) Traditional covalent cross-linking chemistry using EDAC (ethyl-3-dimethyl amino propyl carbodiimide) as a catalyst. (b) Conjugation of antibody fragments to QDs via reduced sulfhydryl-amine coupling. SMCC, succinimidyl-4-N-maleimidomethyl-cyclohexane carboxylate. (c) Conjugation of antibodies to QDs via an adaptor protein. (d) Conjugation of histidine-tagged peptides and proteins to Ni-NTA-modified QDs with potential control of the attachment site and QD:ligand molar ratios. (From Xiaohu Gao, Lily Yang, John A. Petros, Fray F. Marshall, Jonathan W. Simons, Leland Chung, and Shuming Nie, *Curr. Opin. Biotechnol.*, 16, 63–72, 2005. With permission.)

conjugation via covalent bonds has also been achieved using chemical, enzymatic, or genetically modified biomolecules. For example, selective chemical reduction of disulfide bonds at the IgG hinge region and enzymatic digestion can produce antibody fragments with free sulfhydryl groups that are reactive to maleimide coated QDs (Figure 20.3b). It appears as though this antibody fragment conjugation leads to less aggregation and better target binding. Although the binding affinity of each antibody fragment to its target is reduced, this can be compensated by the increased avidity of multivalent binding (that is, multiple antibody fragments in the QD surface bind to multiple target receptors on cells or tissue specimen samples) [33]. More elaborate biomolecule modifications are also available through site-specific labeling or mutagenesis that could further improve the affinity and specificity of QD-antibody conjugates [34].

20.4 IMMUNOASSAY AND IMMUNOHISTOCHEMICAL STAINING

Detection of target molecules immobilized on the surface of a solid support is the most common format of QD immunoassay. For example, Kodadek and coworkers have reported the use of streptavidin-QDs as a fluorescent label to quickly identify peptide sequences from bead-bound combinatorial libraries [35]. Because of the strong autofluorescence of polymer resins, it is much easier to distinguish positive binding events using QDs, taking advantage of their large Stokes shift, than using traditional organic dyes. Similarly, using QD-antibody bioconjugates, Mattoussi et al. have successfully detected chemical compounds such as explosive TNT, and they also achieved a multiplexed toxin analysis [31,36]. Although multicolor QDs used in the experiments only have spectral separations as small as 20 nm, four fluorescence components can be readily resolved as a result of the well-defined QD emission peak shape (Gaussian distribution). Similar experiments can also be extended to gels and membranes (such as Western blotting) or even homogeneous assays in microfluidic channels [37,38].

Another emerging area of QD applications is in situ molecular analysis of cells and tissue specimens because no technologies are currently available to handle intact cells in a highly multiplexed and quantitative manner (Figure 20.4). Wu et al. linked polymer-protected QDs to streptavidin and showed detailed cell skeleton structures using confocal microscopy [15]. The improved photostability of QDs allowed acquisition of many consecutive focal-plane images and their reconstruction into high-resolution 3-D projections. The high-electron density of QDs also allowed correlated optical and electron microscopy studies of cellular samples [39]. Furthermore, Dahan, Jovin, and coworkers achieved real-time visualization of single molecule movement in single living cells [16,18], a task that is extremely difficult or impossible with organic dyes. This single-molecule imaging sensitivity has opened a new avenue for studying receptor diffusion dynamics, ligand–receptor interactions, biomolecular transport, enzyme activity, and molecular

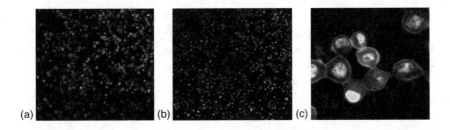

(a) (b) (c)

FIGURE 20.4 Multicolor fluorescence images of SKBR-3 cells stained with QD-antibody bioconjugates. (a) Red QD stained Her-2/neu receptor; image taken with a 10X objective. Note the uniform fluorescence distribution on the cell surface. (b) Counter staining of cell nucleus with 4',6-diamidino-2-phenylindole (DAPI); image taken with a 10X objective. (c) Enlarged cell image of the two colors overlaid; image taken with a 100X objective.

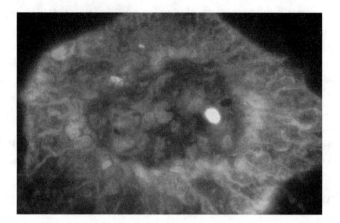

FIGURE 20.5 Immunohistochemical QD staining of formalin-fixed paraffin-embedded tissue sections of human prostate cancer. Mutated p53 phosphoprotein over expressed in the nuclei of androgen-independent prostate cancer cells is labeled with red color QDs (antibody from DAKO, DO-7). The Stokes shifted fluorescence signal is clearly distinguishable from the tissue autofluorescence. Quantitative spectroscopic analysis can also be done by using a spectrometer or an automated laser scanning microscope.

motors. This technology has been pushed a step farther toward clinical and translational research by analyzing formalin-fixed and paraffin-embedded (FFPE) patient tissue specimens (Figure 20.5). Combining the unique optical properties of multicolor QDs and the power of spectral imaging microscopy, the molecular signature of heterogeneous tumor specimens for individually tailored treatment was able to be probed [40]. This could become the first clinical application of semiconductor QDs.

20.5 SPECTRAL BARCODING AND BEAD-BASED IMMUNOASSAYS

Another immuno-application of QDs is bead-based barcoding for rapid screening of genes, proteins, and small molecules. Optical barcodes are prepared by embedding multicolor QDs into porous microspheres such as polystyrene or silica at precisely controlled intensity ratios [41–43]. Figure 20.6 illustrates the principles of multiplexed optical coding based on multicolor semiconductor QDs. The use of 10 intensity levels (1, 2,..., 10) at a single wavelength (color) gives 10 unique codes (10^1), and the number of codes exponentially increases when multiple wavelengths and multiple intensities are used at the same time. For example, a three-color/ten-intensity scheme yields 1000 codes (10^3), wheras a six-color/ten-intensity scheme has a theoretical coding capacity of approximately one million. In general, n intensity levels with m colors generate n^m unique codes. For each individual immunoassay, an antibody is linked to a distinct type of barcoded microsphere for capturing target molecules of interest. These encoded beads are then pooled together into a single vial for rapid screening or multiplexed analysis of solution samples. In a standard sandwich assay, target molecules are captured by the molecular probes on the bead surface, and QD-labeled secondary IgG is then applied as the reporter. Please note that the color of the reporter should be different (spectrally separated) from the colors used in the optical barcodes. The beads can be decoded with single bead spectroscopy or automated flow cytometry where fluorescence intensity of the reporter QD indicates the presence and abundance of the target, and the fluorescence barcode reveals the identity of the target (Figure 20.7). A unique feature of using QDs as both barcoding and reporter fluorophore is that all the colors can be excited by a single light source. In comparison with planar biochips, the QD-encoded beads are expected to be more flexible in target selection, faster in

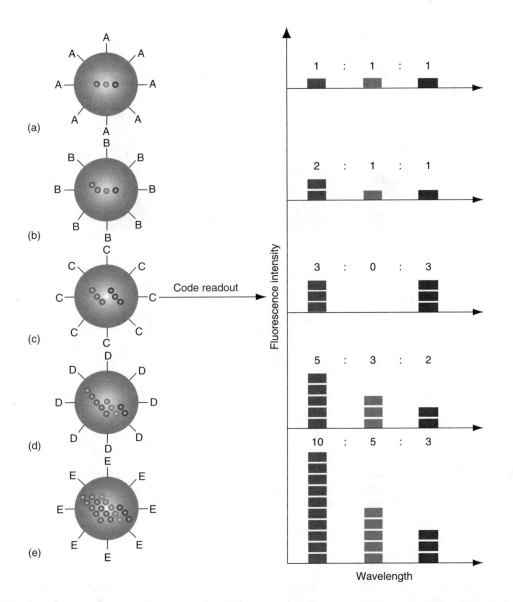

FIGURE 20.6 Schematic illustration of optical barcoding based on wavelength and intensity multiplexing. Large spheres represent polymer microbeads where small, colored spheres (multicolor quantum dots (QDs)) are embedded according to predetermined intensity ratios. Molecular probes (a–e) are attached to the bead surface for biological binding and recognition such as antibody–antigen interactions. The numbers of colored spheres (red, green, and blue) do not represent individual QDs, but they are used to illustrate the fluorescence intensity levels. Optical readout is accomplished by measuring the fluorescence spectra of single beads. Both absolute intensities and relative intensity ratios at different wavelengths are used for coding purposes. (From Mingyong Han, Xiaohu Gao, Jack Su, and Shuming Nie, *Nat. Biotechnol.*, 19, 631–635, 2001.)

binding kinetics, and cheaper in production. Besides the standard bench top flow cytometer for bead decoding, miniaturized devices composed of a diode laser, a microfluidic channel, and multiple PMTs (photomultiplier tubes) are currently under development. This kind of portable bead reader is not only useful for biomedical immunoassays, but it is also important for military applications such as multiplexed pathogen detection.

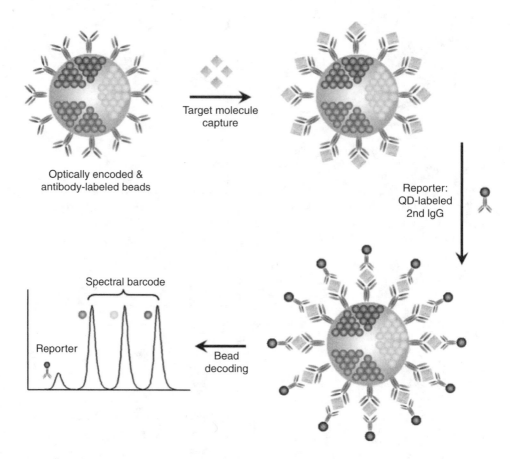

FIGURE 20.7 Schematic diagram showing how QD-barcoded beads can be used in suspension or homogeneous immunoassays. Note that the embedded QDs are not segregated, but they are randomly mixed inside the beads. This diagram is only for illustration purposes. See main text for details.

20.6 CONCLUSION

Development of highly sensitive and multiplexed immunoassays will play an important role in drug discovery, medical diagnostics, and personalized medicine. In the near future, novel platforms based on the combination of multicolor QDs, QD-barcode beads, and immunological reagents will be developed for biosensing and biodetection. Along with technology development, the next science wave and impact of QD nanotechnology are likely to happen in their applications to biological discoveries and clinical diagnostics that will directly benefit patient healthcare. In this context, some immediate directions include, but are not limited to, developments of QD-antibody sensors, ultrasensitive detection and quantification for protein arrays, multiplexed intracellular imaging, molecular profiling for clinical tissue specimens, high-throughput drug screening, and spectral coding and multicolor instrumentation.

ACKNOWLEDGMENTS

This work was supported by grants from the National Institutes of Health (R01 GM60562, P20 GM072069, and R01 CA108468 to SN), the Georgia Cancer Coalition (Distinguished Cancer

Scholar Award to SN), and the Coulter Translational Research Program at Georgia Tech (SN) and University of Washington (XG).

.

REFERENCES

1. Alivisatos, A. P., Semiconductor clusters, nanocrystals, and quantum dots, *Science*, 271, 933–937, 1996.
2. Leatherdale, C. A., Woo, W. -K., Mikulec, F. V., and Bawendi, G., On the absorption cross section of CdSe nanocrystal quantum dots, *J. Phys. Chem. Part B*, 106, 7619–7622, 2002.
3. Gao, X., Yang, L., Petros, J. A., Marshall, F. F., Simons, J. W., and Nie, S., In vivo molecular and cellular imaging with quantum dots, *Curr. Opin. Biotechnol.*, 16 (1), 63–72, 2005.
4. Jakobs, S., Subramaniam, V., Schonle, A., Jovin, T. M., and Hell, S. W., EGFP and DsRed expressing cultures of *Escherichia coli* imaged by confocal, two-photon and fluorescence lifetime microscopy, *Febs Lett.*, 479 (3), 131–135, 2000.
5. Pepperkok, R., Squire, A., Geley, S., and Bastiaens, I., Simultaneous detection of multiple green fluorescent proteins in live cells by fluorescence lifetime imaging microscopy, *Curr. Biol.*, 9 (5), 269–272, 1999.
6. Dahan, M., Laurence, T., Pinaud, F., Chemla, D. S., Alivisatos, A. P., Sauer, M., and Weiss, S., Time-gated biological imaging by use of colloidal quantum dots, *Optics Lett.*, 26 (11), 825–827, 2001.
7. Bailey, R. E. and Nie, M., Alloyed semiconductor quantum dots: tuning the optical properties without changing the particle size, *J. Am. Chem. Soc.*, 125, 7100–7106, 2003.
8. Kim, S., Fisher, B., Eisler, H. J., and Bawendi, M., Type-II quantum dots: CdTe/CdSe(core/shell) and CdSe/ZinTe(core/shell) heterostructures, *J. Am. Chem. Soc.*, 125, 11466–11467, 2003.
9. Zhong, X. H., Feng, Y. Y., Knoll, W., and Han, M. Y., Alloyed $Zn_xCd_{1-x}S$ nanocrystals with highly narrow luminescence spectral width, *J. Am. Chem. Soc.*, 125 (44), 13559–13563, 2003.
10. Zhong, X. H., Han, M. Y., Dong, Z. L., White, T. J., and Knoll, W., Composition-tunable $Zn_xCd_{1-x}Se$ nanocrystals with high luminescence and stability, *J. Am. Chem. Soc.*, 125, 8589–8594, 2003.
11. Bruchez, J. M., Moronne, M., Gin, P., Weiss, S., and Alivisatos, P., Semiconductor nanocrystals as fluorescent biological labels, *Science*, 281, 2013–2015, 1998.
12. Chan, W. C. W. and Nie, S. M., Quantum dot bioconjugates for ultrasensitive nonisotopic detection, *Science*, 281, 2016–2018, 1998.
13. Gao, X., Chan, W.C.W., and Nie, S., Quantum-dot nanocrystals for ultrasensitive biological labeling and multicolor optical encoding, *J. Biomed. Opt.*, 7, 532–537, 2002.
14. Derfus, A. M., Chan, C. W., and Bhatia, S. N., Probing the cytotoxicity of semiconductor quantum dots, *Nano Lett.*, 4, 11–18, 2004.
15. Wu, X. Y., Liu, H. J., Liu, J. Q., Haley, K. N., Treadway, J. A., Larson, J. P., Ge, N. F., Peale, F., and Bruchez, M. P., Immunofluorescent labeling of cancer marker Her2 and other cellular targets with semiconductor quantum dots, *Nat. Biotechnol.*, 21, 41–46, 2003.
16. Dahan, M., Levi, S., Luccardini, C., Rostaing, P., Riveau, B., and Triller, A., Diffusion dynamics of glycine receptors revealed by single-quantum dot tracking, *Science*, 302, 442–445, 2003.
17. Ness, J. M., Akhtar, R. S., Latham, C. B., and Roth, K. A., Combined tyramide signal amplification and quantum dots for sensitive and photostable immunofluorescence detection, *J. Histochem. Cytochem.*, 51, 981–987, 2003.
18. Lidke, D. S., Nagy, P., Heintzmann, R., Arndt-Jovin, D. J., Post, J. N., Grecco, H. E., Jares-Erijman, E. A., and Jovin, M., Quantum dot ligands provide new insights into erbB/HER receptor-mediated signal transduction, *Nat. Biotechnol.*, 22, 198–203, 2004.
19. Dubertret, B., Skourides, P., Norris, D. J., Noireaux, V., Brivanlou, A. H., and Libchaber, A., In vivo imaging of quantum dots encapsulated in phospholipid micelles, *Science*, 298, 1759–1762, 2002.
20. Gao, X., Cui, Y. Y., Levenson, R. M., Chung, L.W.K., and Nie, S., In vivo cancer targeting and imaging with semiconductor quantum dots, *Nat. Biotechnol.*, 22, 969–976, 2004.
21. Allen, C., Maysinger, D., and Eisenberg, A., Nano-engineering block copolymer aggregates for drug delivery, *Colloids Surf. B Biointerfaces*, 16, 3–27, 1999.
22. Ludwigs, S., Boker, A., Voronov, A., Rehse, N., Magerle, R., and Krausch, G., Self-assembly of functional nanostructures from ABC triblock copolymers, *Nat. Mater.*, 2, 744–747, 2003.

23. Gao, X. and Nie, S., Molecular profiling of single cells and tissue specimens with quantum dots, *Trends Biotechnol.*, 21 (9), 371–373, 2003.
24. Mao, C. B., Solis, D. J., Reiss, B. D., Kottmann, S. T., Sweeney, R. Y., Hayhurst, A., Georgiou, G., Iverson, B., and Belcher, M., Virus-based toolkit for the directed synthesis of magnetic and semi-conducting nanowires, *Science*, 303, 213–217, 2004.
25. Lee, S. W., Mao, C., Flynn, C. E., and Belcher, M., Ordering of quantum dots using genetically engineered viruses, *Science*, 296, 892–895, 2002.
26. Whaley, S. R., English, D. S., Hu, E. L., Barbara, P. F., and Belcher, M., Selection of peptides with semiconductor binding specificity for directed nanocrystal assembly, *Nature*, 405 (6787), 665–668, 2000.
27. Schellenberger, E. A., Reynolds, F., Weissleder, R., and Josephson, L., Surface-functionalized nano-particle library yields probes for apoptotic cells, *Chembiochem.*, 5 (3), 275–279, 2004.
28. Pellegrino, T., Manna, L., Kudera, S., Liedl, T., Koktysh, D., Rogach, A. L., Keller, S., Rädler, J., Natile, G., and Parak, J., Hydrophobic nanocrystals coated with an amphiphilic polymer shell: a general route to water soluble nanocrystals, *Nano Lett.*, 4 (4), 703–707, 2004.
29. Ron, E., Turek, T., Mathiowitz, E., Chasin, M., Hageman, M., and Langer, R., Controlled-release of polypeptides from polyanhydrides, *Proc. Natl Acad. Sci. U.S.A.*, 90 (9), 4176–4180, 1993.
30. Anseth, K. S., Shastri, V. R., and Langer, R., Photopolymerizable degradable polyanhydrides with osteocompatibility, *Nat. Biotechnol.*, 17 (2), 156–159, 1999.
31. Goldman, E. R., Anderson, G. P., Tran, P. T., Mattoussi, H., Charles, P. T., and Mauro, J. M., Conjugation of luminescent quantum dots with antibodies using an engineered adaptor protein to provide new reagents for fluoroimmunoassays, *Anal. Chem.*, 74 (4), 841–847, 2002.
32. Mattoussi, H., Mauro, J. M., Goldman, E. R., Anderson, G. P., Sundar, V. C., Mikulec, F. V., and Bawendi, G., Self-assembly of CdSe–ZnS quantum dot bioconjugates using an engineered recombinant protein, *J. Am. Chem. Soc.*, 122 (49), 12142–12150, 2000.
33. Mammen, M., Choi, S. K., and Whitesides, G. M., Polyvalent interactions in biological systems: Implications for design and use of multivalent ligands and inhibitors, *Angew. Chem. Int. Ed. Engl.*, 37, 2754–2794, 1998.
34. Weiss, S., Fluorescence spectroscopy of single biomolecules, *Science*, 283 (5408), 1676–1683, 1999.
35. Olivos, H. J., Bacchawat-Sikder, K., and Kodadek, T., Quantum dots as a visual aid for screening bead-bound combinatorial libraries, *Chembiochem.*, 4 (11), 1242–1245, 2003.
36. Goldman, E. R., Clapp, A. R., Anderson, G. P., Uyeda, H. T., Mauro, J. M., Medintz, I. L., and Mattoussi, H., Multiplexed toxin analysis using four colors of quantum dot fluororeagents, *Anal. Chem.*, 76 (3), 684–688, 2004.
37. Stavis, S. M., Edel, J. B., Samiee, K. T., and Craighead, H. G., Single molecule studies of quantum dot conjugates in a submicrometer fluidic channel, *Lab. Chip.*, 3, 337–343, 2005.
38. Wang, T. H., Peng, Y., Zhang, C., Wong, P. K., and Ho, C. M., Single-molecule tracing on a fluidic microchip for quantitative detection of low-abundance nucleic acids, *J. Am. Chem. Soc.*, 127, 5354–5359, 2005.
39. Nisman, R., Dellaire, D., Ren, Y., Li, R., and Bazett-Jones, D. P., Application of quantum dots as probes for correlative fluorescence, conventional, and energy-filtered transmission electron microscopy, *J. Histochem. Cytochem.*, 52, 13–18, 2004.
40. Maksym V. Yezhelyev, Xiaohu Gao, Yun Xing, Ahmad Al-Hajj, Shuming Nie and Ruth M O'Regan, Emerging use of nanoparticles in diagnosis and treatment of breast cancer, *The Lancet Oncol.*, 7, 657–667, 2006.
41. Han, M., Gao, X., Su, J. Z., and Nie, S., Quantum-dot-tagged microbeads for multiplexed optical coding of biomolecules, *Nat. Biotechnol.*, 19 (7), 631–635, 2001.
42. Gao, X. and Nie, S., Quantum dot-encoded mesoporous beads with high brightness and uniformity: Rapid readout using flow cytometry, *Anal. Chem.*, 76 (8), 2406–2410, 2004.
43. Gao, X. and Nie, S., Doping mesoporous materials with multicolor quantum dots, *J. Phys. Chem.*, 107 (42), 11575–11578, 2003.

21 Nanotechnology and the Future of Bioanalytical Methods

Lon A. Porter Jr.

CONTENTS

21.1 AN INTRODUCTION TO THE BIG FUTURE OF THE REMARKABLY SMALL

The astonishing idea of functional devices and structures 10,000–100,000 times smaller than the width of a single human hair (Figure 21.1) excites the imagination and fears of scientists, engineers, media, politicians, science fiction authors, Hollywood, and the general public. Over the past decade, the notion of nanotechnology has escaped the laboratories and disseminated into the popular culture. Browse the newspaper, surf the web, or look to recent sci-fi books and movies because nanotechnology seems to be everywhere. The study and manipulation of matter on the nanometer scale (measured in one-billionths of a meter) has been termed *nanoscience* or *nanotechnology*, an exploding field still in its infancy [1,2]. Unique, unpredictable, and highly intriguing physical, chemical, optical, and electrical phenomena can result from the confinement of matter into nanoscale features [3]. As a result, the promise of amazing discoveries relating to the understanding and preparation of structures exhibiting such interesting and unusual phenomena has attracted the focus of scientists and engineers from across the globe. Much of the driving force for building infinitesimal devices and features on the nanoscale derives its potential importance in existing and emerging technologies such as microelectronics, sensors, catalysis, and medical diagnostics, among countless others [1–3].

Although it is increasingly invoked as a stimulating plot device, nanotechnology is not science fiction. It provides a prime example for learning to appreciate the amazing opportunities that arise when various fields of science and engineering intermingle [4]. Although the stereotypical vision of nanotechnology where items are fabricated in a ridiculously inexpensive manner with negligible material or energy waste through direct atom-by-atom precision has yet to be realized, modern research in nanotechnology is progressing at an astonishing rate. Nanotechnology is one of the largest and most rapidly growing interdisciplinary research fields in modern science and engineering. Thousands of papers are published in the area every year, new journals dedicated to the

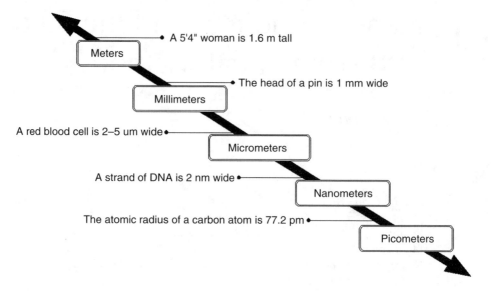

FIGURE 21.1 Putting the nanoscale into perspective: nanotechnology is focused on the study of matter containing at least one dimension within 1–100 nm. (Adapted from National Nanotechnology Initiative (NNI) Home Page, http://www.nano.gov, accessed April 2005.)

subject are in their early years of development, and new titles are emerging in both print and virtual formats. Research institutes dedicated to the area continue to be founded worldwide, and federal funding initiatives fuel this burgeoning field in the United States [5]. Whereas many of its proponents foresee nanotechnology ushering in the next industrial revolution [1], the field is principally still in the early stages of developing the tools and methods for manipulating and characterizing structures on the nanoscale.

As new developments in nanotechnology continue to generate an increasing amount of excitement and fill the headlines of research press releases and journals, it is critical to realize that nanotechnology is not as new as one might think. Explorations aimed at elucidating the fundamental behavior and stimulating possibilities offered through the direct control of matter at the nanoscopic level are encompassed within many of the various subdiciplines of chemistry, biology, physics, and engineering [6]. Traditional organic synthesis has always strived toward the systematic and efficient construction of complex molecular architectures through an elegant combination of established chemical transformations. The current pharmaceutical industry, along with many others, owes much of their success to the nanoscopic precision derived through synthetic organic and inorganic chemistry [7]. Similarly, cellular biologists have studied intricate nanoscopic assembly devices known as ribosomes that employ instructions from messenger RNA to manufacture the complex proteins required to foster life on this planet [8]. The ubiquitous personal computer, portable music devices, and video game consoles are all made possible by the multibillion-dollar microelectronics and semiconductor industry that now routinely pursues functional electronic structures smaller than 100 nm [9]. This all stems from an intellectual foundation made possible through the great progress made by pioneers in condensed matter physics. These few examples illustrate the point that research into nanoscience and nanotechnology has been pursued for some time, although it was not until recently that it enjoyed the fashionable nomenclature, increased funding, and popularization through print and visual media sources. It simply remains that for some nebulous, impalpable reason, the same science just would not be as evocative without the "nano" [10].

The formal beginnings of nanotechnology are generally attributed to the visionary physicist, Richard Feynman [11]. In December of 1959, during an address to the American Physical Society at the California Institute of Technology, Feynman concisely enunciated the central issues, possibilities, and motivations for turning the world's collective scientific focus onto the landscape of matter measured in micro/nanometers [12]. He spoke about the great rewards to be gained by working on the nanoscale as well as the difficulties that plagued such pursuits. While Feynman asserted that, "The principles of physics, as far as I can see, do not speak against the possibility of maneuvering things atom by atom," he respected the fundamental laws of science that would ultimately determine the success or failure of such endeavors. Although the idea of fabrication on the atomic level remains quite appealing, Feynman noted that, "You can't put them so that they are chemically unstable, for example."

It is important to remember that Feynman's address followed the Soviet launch of Sputnik by only two years. The fledgling National Aeronautics and Space Administration (NASA) was little more than a year old, and the resources of the nation were focused toward winning the Space Race. The research spotlight of the time was clearly fixated upon the remarkably large and Feynman attempted to entice the scientific community to begin to explore the unlimited possibilities offered by the extremely small. He remarked that, "We are not doing it because we haven't gotten around to it," and felt this area of investigation had been overlooked and was ripe for exploration. This invitation was clearly referenced in the title of his now famous speech, *Plenty of Room at the Bottom* [12].

One of Feynman's key observations was that nanotechnology could not progress forward at an accelerated pace until suitably powerful instrumentation for both the fabrication and characterization of nanostructures were developed and made widely available. He cited the need for a more powerful electron microscope (EM), first developed in 1931, as a central example [12]. The decades following Feynman's landmark address eventually saw the development and distribution of such instrumentation. Foremost on this list is the invention of the scanning tunneling microscope (STM) and its congener, the atomic force microscope (AFM). Created in 1981 by IBM researchers Binning and Rohrer, the STM allows researchers to obtain images at the atomic level [13]. Within 10 years of the STM's creation, Donald Eigler of IBM used the instrument to directly manipulate 35 xenon atoms into a pattern to write the company's name [13]. Not only did this window to the atomic realm truly open the way to nanotechnology as a viable research field, it captured the imaginations of scientists as well as many outside the scientific community. Science fiction author Greg Bear was one of the first to pioneer the use of nanotechnology as a central element in his 1983 short story *Blood Music* that warned of the tragic consequences of unchecked biological nanotechnology [14]. Many stories, television programs, and motion pictures quickly followed [15,16].

First published in 1986, K. Eric Drexler's *Engines of Creation: The Coming Era of Nanotechnology* served to predict, justify, and caution both researchers and the general public about the possibilities of molecular nanotechnology [17]. Although often criticized as overly speculative, this work is arguably most responsible for awakening public awareness of nanotechnology. Scores of similarly themed introductory and reference texts have followed in a response to peaked scientific and public interest. In 2000, President Clinton announced the National Nanotechnology Initiative (NNI) that served to promote research in nanotechnology through increased research funding [5]. The NNI budget has continued to grow each year as research has flourished. The following chapter will attempt to spotlight prevalent methods of structure and device fabrication on the nanoscale and explore the ways these techniques are beginning to impact bioanalytical methods.

21.2 EVOLVING METHODS OF FABRICATION ON THE NANOSCALE

21.2.1 PHOTOLITHOGRAPHY

Gordon E. Moore, co-founder and Chairman Emeritus of Intel Corporation, published an article in 1965 that predicted that the number of transistors the semiconductor industry would be able to place on a computer chip would double every couple of years [18]. Whereas this bold prediction was

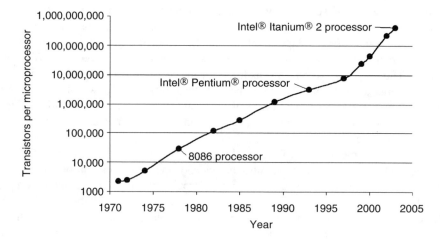

FIGURE 21.2 Moore's Law.

made only four years after the first planar integrated circuit was devised, the doubling of transistors approximately every 18 months has been maintained, and this prediction still holds true today (Figure 21.2). This trend, more commonly referred to as Moore's Law, is expected to persist through at least the end of this decade [19]. Although many enabling technologies have contributed to the pursuit of smaller, faster, and cheaper microelectronics, innovations and technological advances in photolithography have paved the way for the semiconductor industry to keep pace with Moore's Law.

   Photolithography is a process that utilizes electromagnetic radiation, typically in the form of ultraviolet light, to define patterns in various light-sensitive materials [20]. This method of micro/ nanopatterning is currently the cornerstone of modern, silicon-based microelectronics fabrication. As modern photolithography continues to push the boundaries of modern optics, anyone who has ever projected shadow puppets onto a wall using their hand and a flashlight will quickly grasp the foundations of this powerful, high-throughput fabrication technique. Once a pattern is designed, a photomask, or mask, must be created that contains a precise image of the design to be duplicated. A photomask is similar to a negative in print photography or equivalent to a person's hand in the shadow puppet analogy. The photomask is used as a master to optically transfer these images onto photosensitive layers of material [21]. Although inexpensive photomasks may be prepared with inkjet printer transparencies, metal grids/meshes, or photographic emulsion on soda lime glass, chrome on quartz glass is required for high-resolution deep UV lithography required by the microelectronics industry [22]. The devices contained on the silicon wafers used by the microelectronics industry are manufactured layer by layer, each requiring a unique photomask. The current generation of advanced chips has 25 or more layers [23].

   Once a quality photomask of the desired pattern is obtained, an alignment and exposure system must be employed to transfer the pattern onto photosensitive material. Employing the shadow puppet analogy once again, the sharpest, most well-defined figures are projected when the hand is positioned closest to the viewing wall. This limits the distortions resulting from diffraction. The same holds true in photolithography. In many instances, high-resolution pattern transfer can be accomplished using proximity and contact exposure/printing configurations (Figure 21.3). The proximity exposure method involves positioning the photomask so that a small gap (10–25 μm wide) is maintained between the silicon wafer and the mask during UV exposure [24]. This method reduces the pattern distortion because of diffraction, yet it prevents damage or contamination of the photomask by preventing direct contact with the material to be patterned. In contact exposure

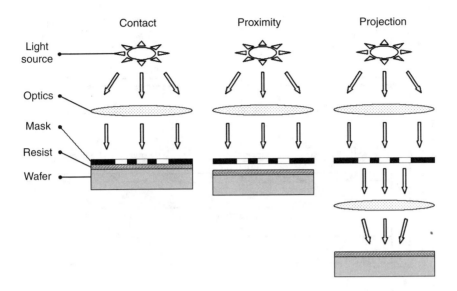

FIGURE 21.3 The three most commonly employed photolithography configurations: contact, proximity, and projection. (Adapted from Campbell, S. A., *The Science and Engineering of Microelectronic Fabrication*, Oxford University Press, New York, 2001.)

methods, the photomask is brought into direct contact with the surface to be patterned so that distortions in pattern transfer brought about by diffraction are minimized to the greatest extent. This method allows very high-resolution pattern transfer at the cost of higher potential for photomask contamination or damage.

Whereas both proximity and contact exposure methods offer enhanced resolution through reduction of diffraction-based distortions in pattern transfer, these techniques exhibit a few significant drawbacks [25]. First, the photomask pattern is limited to a one-to-one transfer scaling. Because both techniques leave no room for optics to scale the pattern, the size of the transferred pattern is restricted to the size of the pattern exhibited by the photomask master. This causes greater difficulty and cost in photomask preparation for applications aimed at the submicron/nanometer size regime where many photomasks are typically required for complex electronic architectures. This is a significant consideration because the possibility of contamination and/or damage is high for both of these exposure techniques. Furthermore, the photomask is typically prepared so that the entire wafer, not just a single device, is exposed in one illumination. This requires a photomask large enough to encompass the entire silicon wafer surface that also increases photomask production costs and introduces distortions resulting from planar deviations of the mask, wafer, or photosensitive material targeted for pattern transfer.

In order to circumvent difficulties associated with proximity and contact exposure methods, projection exposure/printing was established and quickly became the exposure method of choice for high-resolution photolithography requirements of the microelectronics industry [26]. Similar in nature to the optics bench experiments used to teach students in introductory physics courses, the projection method increases the distance between the silicon wafer and the photomask so that a set of optics may be introduced. Projection printing entirely avoids mask damage. An image of the mask pattern is projected onto the wafer that is many centimeters away. In order to achieve high resolution, only a small portion of the mask is imaged. This small image field is scanned or stepped over the entire wafer surface. As overall exposure speed is reduced for the entire wafer, the gains in scalability, resolution, and photomask preservation are considered to be well worth the sacrifice.

The exposure system transfers the photomask pattern onto a photosensitive material known as a photoresist. Widely employed by the printing and printed circuit board industries, the

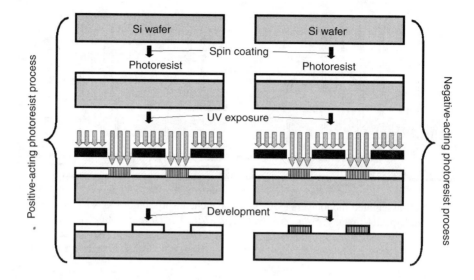

FIGURE 21.4 Outline of positive- and negative-acting photoresist processes utilized in photolithography. (Adapted from Van Zant, P., *Microchip Fabrication*, McGraw Hill, New York, 2000.)

semiconductor industry was quick to adapt photoresists for its own purposes [27]. Modern photoresists are typically light-sensitive polymers dissolved in a solvent system. The photoresist solution is applied to the surface of a silicon wafer or other substrate where high-speed centrifugal rotation is the standard method for applying photoresist coatings in microelectronics fabrication. This technique, known as spin coating, produces a thin, uniform layer of photoresist on the wafer surface after the prescribed amount of solvent is removed through evaporation with heating [28]. There are two main classifications of photoresist (Figure 21.4): negative and positive [29]. From a fundamental point of view, negative resists are composed of relatively small molecules that can be easily washed away with a simple solvent rinse. Exposure to the UV light causes the negative resist to become polymerized and more difficult to dissolve. The polymerized negative resist remains on the surface in regions exposed to light, and the developer solution removes only the unexposed, or masked, portions. Photomasks employed when using negative photoresists, therefore, contain the inverse, or photographic negative, of the pattern to be transferred. Negative resists enjoyed great popularity within the semiconductor industry up until the mid-1970s. By this time, device features had quickly outpaced resolution limits of negative resists. Other problems such as their susceptibility to oxygenation and the inherent difficulties associated with clear-field photomasks eventually led to positive resists becoming the photoresists of choice by the 1980s. Positive resists behave in the opposite manner. For positive resists, the resist is exposed with UV light wherever the underlying material is to be removed. In these resists, exposure to UV light changes their chemical structure to render the photoresist more soluble in a solvent wash known as a developer. Once exposed, the resist is washed away by the developer solution, leaving windows of the bare underlying material. The photomask, therefore, contains an exact copy of the pattern that is to be transferred onto the wafer or surface of interest.

Although polymeric photoresists have dominated traditional photolithography, a recent thrust has seen the abandonment of these materials for specialized applications demanding ultimate resolution in favor of utilizing molecular resists [30,31]. Molecular resists involve laying down a single layer of molecules, a self-assembled monolayer (SAM) that replaces the thicker layers of polymeric photoresists. These molecular resists are often chemically oriented and attached to the surface of interest through chemical bonds as opposed to traditional photoresists (Figure 21.5). Thiols on gold, siloxanes on glass, and a variety of other systems are currently under intense study

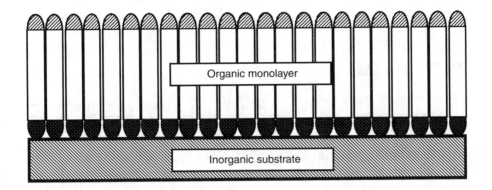

FIGURE 21.5 Self-assembled monolayers (SAMs) involve chemisorption of molecular species onto a solid substrate. (Adapted from Ulman, A., *An Introduction to Ultrathin Organic Films*, Academic, New York, 1991.)

by a multitude of research groups [32]. Molecular resists also operate in both negative and positive modes. For negative resists, a solution of free molecules is placed on the sample surface and exposed to UV light. The radiation facilitates a reaction between the surface and resist molecules, chemically bonding the two [33]. Therefore, the transferred pattern appears as the inverse, or photographic negative, of the photomask pattern. Positive molecular resists behave in the opposite manner. In the positive mode, a molecular resist is applied and chemically bound to the entire surface of the substrate, and harsh, high energy UV radiation is used to decompose or cleave the molecules from the surface. Because this removes the molecular resist in regions exposed to light, the transferred pattern is an exact copy of the photomask.

Negative molecular resists of various organic molecules on silicon and germanium (Ge) have received increasing interest as a result of the obvious technological focus on semiconductor-based sensors and microelectronics. Chidsey and coworkers at Stanford University

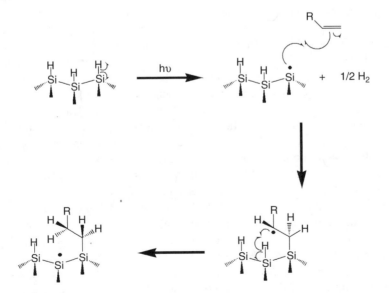

FIGURE 21.6 Mechanism of photochemical hydrosilylation on hydride-terminated porous silicon. (Adapted from Cicero, R. L., Linford, M. R., and Chidsey, C. E. D., *Langmuir*, 16, 5688, 2000; Linford, M. A. and Chidsey, C. E. D., *J. Am. Chem. Soc.*, 115, 12631, 1993.)

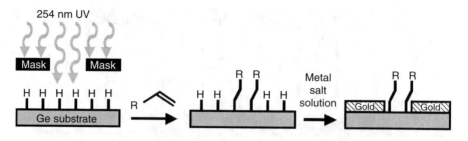

FIGURE 21.7 Hydrogermylation of alkenes by the germanium-hydride surface mediated by UV light in the unmasked areas results in areas of alkyl termination that prevent oxidation and subsequent electroless metal deposition (Adapted from Porter, L. A., Jr., Choi, H. C., Schmeltzer, J. M., Ribbe, A. E., Elliott, L. C. C., and Buriak, J. M., *Nano Lett.*, 2, 1369, 2002.)

pioneered UV-promoted surface reactions on flat silicon, resulting in molecular resists bound through robust, silicon–carbon bonds (Figure 21.6) [34]. A variety of functional surfaces/devices have been prepared employing these or related techniques [35]. Researchers at Purdue University recently employed a similar method in the preparation of metallic nanoelectrodes and nanocontacts on germanium (Figure 21.7) [36]. These nanoscale gold contacts serve as convenient attachment points for interfacing a wide variety of biologically important materials onto a semiconductor chip. First, a hydride-terminated Ge(100) surface was exposed to 254 nm ultraviolet light through a metal grid photomask in deoxygenated olefins. The illuminated regions react in the presence of 1-dodecene, forming a direct germanium–carbon bond, thereby anchoring the resist molecules in place. This leads to micro/nano-sized features of dodecyl for illuminated regions with hydride terminations remaining in masked regions. Metal deposition via electroless plating occurred preferentially in the hydride areas because the alkyl monolayer functions as an effective dielectric barrier. The hydride surface oxidizes in situ and subsequently dissolves in the aqueous medium. Metal salt reduction and deposition then occur via galvanic displacement, leading to metallization between the alkylated domains. As a result, photopatterning of noble metals can be achieved through tailoring of the germanium surface chemistry employing molecular resists [37].

Buriak and coworkers have also demonstrated a reaction between alkenes and alkynes on hydride-terminated porous silicon utilizing a simple white light source to facilitate the process [38]. Porous silicon has recently attracted considerable interest primarily because of its light emitting properties that make it a promising candidate for use in optoelectronic, sensor array, light harvesting, and/or biocomposite/bioanalytical applications [39]. Photoluminescent samples of porous silicon are coated with the neat alkene/alkyne or a methylene chloride solution of the substrates and illuminated. Sample illumination is maintained under inert conditions at a moderate intensity of 22–44 mW/cm^2 (essentially, an overhead transparency projector bulb) and results in efficient hydrosilylation at room temperature. Exposure times as short as only 15 min result in the formation of an alkyl surface termination for alkenes and an alkenyl surface termination when alkynes are utilized. These results were confirmed through Fourier transform infrared (FTIR) and Secondary ionization mass spectrometry (SIMS) analysis, and reaction conditions proved mild enough for the incorporation of a variety of functionalities. Through simple masking procedures, the hydrosilylation can be carried out in a regiospecific fashion on the surface because only the illuminated areas react as shown schematically in Figure 21.8 [40].

This approach toward silicon–carbon bond formation proved so mild that the underlying nanocrystallite subarchitecture remained largely intact. This is supported by the observation that alkyl functionalized porous silicon samples prepared in this manner maintain all of their intrinsic photoluminescence. These monolayers prove to be quite robust in serving to passivate the underlying porous silicon toward chemical attack and oxidation under ambient conditions.

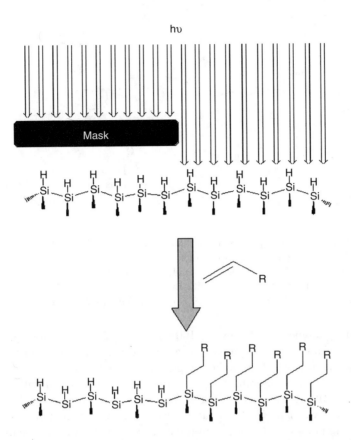

FIGURE 21.8 White light promoted hydrosilylation of alkenes on hydride-terminated porous silicon, utilizing a photomask to limit light exposure to predetermined resist domains (Adapted from Stewart, M. P. and Buriak, J. M., *Angew. Chem. Int. Ed.*, 37, 3257, 1998.)

Functionalized surfaces easily withstood immersion into hydrofluoric acid, organic solvents, and boiling alkaline solutions [40]. This method may serve as an excellent candidate for surface functionalization of photoluminescent samples of porous silicon for implementation into optoelectronic devices or bioanalytical devices.

By combining the chemical versatility and stability, subsequent organic transformations involving surface-bound reactive groups of similar molecular resists have been carried out to prepare functional interfaces. For example, Horrocks and Houlton recently reported the successful immobilization of single-stranded DNA (ssDNA) onto the surface of porous silicon (Figure 21.9) [41]. Surface-immobilized DNA is of interest for gene chip and biosensor applications. The ssDNA could not be prepared though direct surface functionalization, and instead, a molecular scaffold was employed as a chemical linkage. First, 4,4′-dimethoxytrityl-protected (DMT) ω-undecynol was functionalized onto the surface of porous silicon through thermal hydrosilylation. The protecting group was removed and a 17-mer oligonucleotide was synthesized at the surface, utilizing a specially modified DNA synthesizer/column setup. In order to assess the process efficiency, cleavable linkages were inserted so that gel electrophoresis could be utilized. Double-stranded DNA was also prepared from the single-stranded surface upon hybridization [41]. This serves to show the myriad surfaces yet to be explored through rational synthetic design of functional interfaces.

As photolithography is currently employed to mass-produce transistors at a rate of nearly three billion chips per second, this technique faces substantial fundamental and engineering challenges as

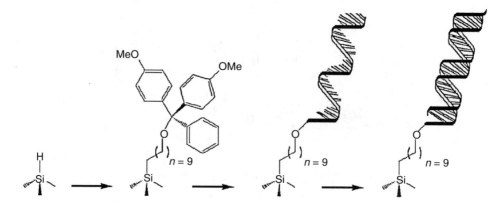

FIGURE 21.9 Hydride-terminated porous silicon functionalized with 4,4′-dimethoxytrityl-protected (DMT) ω-undecanol was deprotected and a 17-mer oligonucleotide was synthesized at the surface, utilizing a specially modified DNA synthesizer/column setup. The double-stranded DNA was prepared through a subsequent hybridization reaction (Adapted from Pike, A. R., Lie, L. H., Eagling, R. A., Ryder, L. C., Patole, S. N., Connolly, B. A., Horrocks, B. R., and Houlton, A., *Angew. Chem. Int. Ed.*, 41, 615, 2002.)

device dimensions plummet well below the 100 nm mark [42]. In photolithography, resolution is limited by the wavelength of light employed in a process as well as the optical elements of the system [43]. Optical lithographic methodologies are currently limited to mass production of features on the order of ~100 nm. The semiconductor industry's significant investments in extreme ultraviolet lithography are aimed at increasing the ultimate resolution by employing significantly smaller wavelengths of radiation. Using xenon plasma, soft x-ray radiation with a wavelength of 15.4 nm is obtained [44]. Unfortunately, all of the optical elements of a system using this radiation are limited to specially constructed, multilayered mirrors because no lenses transparent to this high-energy radiation are known at present. Whereas systems of this kind as well as systems employing even shorter wavelength x-ray sources may eventually find more widespread use, at present, they remain prohibitively expensive and highly complex [45].

Electron and ion beam technologies offer promising alternatives to the problematic issues associated with x-ray sources and optics. These charged particle beams are focused and rastered across a target surface using electromagnetic lenses [46]. Borrowing from electron microscopy, electron beams may be focused to a spot as small as 0.5 nm, offering ultimate resolutions limited only by the quality of the focusing system and the electron-resist scattering. Ion beams may be focused to spot sizes approaching 5 nm, but they are less prone to scattering because of their increased particle masses [47]. Both processes, however, are serial in nature. Each individual feature must be sequentially projected, taking hours to complete a single chip. Although both electron-beam and ion-beam lithography permit routine patterning on the sub-100 nm scale, vacuum requirements, extremely high voltages, equipment costs, and limited throughput currently restrict the utility of these methods. In order to truly realize the possibilities of mass production on the nanoscale, new, unconventional patterning technologies are required.

21.2.2 MICROCONTACT PRINTING

Surface chemistry is arguably one of the most developed and intensely studied realms of nanoscience. Self-assembled monolayers (SAMs) generated by the adsorption of organic molecules onto flat surfaces (Figure 21.3) represent an ever expanding field of research as a result of their potential use in applications such as chemical sensing, corrosion prevention, lubrication, electrical conduction, and adhesion [48]. The monolayer photoresists described in the previous section are examples of SAMs. Although many systems such as siloxanes on glass attract the focus of intense research,

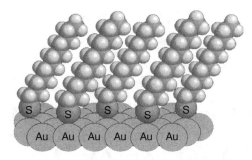

FIGURE 21.10 Alkylthiolate self-assembled monolayer (SAM) on gold. (Adapted from Ulman, A., *An Introduction to Ultrathin Organic Films*, Academic, New York, 1991.)

perhaps the most thoroughly investigated SAMs are those generated by the adsorption of thiols or disulfides onto flat surfaces of gold (Figure 21.10) [49]. Nuzzo and Allara reported the initial investigations in the formation of SAMs prepared by the adsorption of disulfides on flat substrates of gold [50]. This research served as the impetus for a multitude of similar studies. These systems have drawn special attention because the methodology is simple, and the resultant films are densely packed, highly ordered, and largely free of defects. Before long, it had been discovered that alkanethiols and disulfides have a great affinity for the surfaces of not only gold, but also silver, copper, and platinum as well [51]. Although these metals have been investigated, gold remains the most thoroughly studied of the flat metal substrates for one main reason. Gold does not form a stable oxide surface under ambient conditions. The surface is easier to preserve and protect from atmospheric contaminants. This allows simplicity and ease in manipulation under regular laboratory conditions.

In SAMs prepared from the adsorption of alkanethiols and disulfides, it is important to understand the chemical nature of the adsorbate. In both alkanethiols and disulfides, the sulfur atoms exhibit a great affinity for the gold or silver substrate. The surface active, sulfur head group is responsible for the most exothermic step in the adsorption process. The remainder of the molecule often consists of a relatively long alkyl chain of a derivatized alkyl group (two in the case of the disulfide). One theory provides that the adsorption process occurs in two steps. First, the sulfur head group chemisorbs onto the metal substrate. Electron transfer occurs from the metal surface to the sulfur head group. Some studies suggest that this process weakens the S–H bond, enough so that dissociation results. On the Au(111) surfaces, it is believed that the remaining thiolate then adjusts so that it equilibrates into one of the three-fold hollow sites of the lattice (Figure 21.11). The final S–Au/Ag interaction is believed to result in a strong covalent attachment ($\sim$44 kcal/mol) [49]. Helium diffraction and atomic force microscopy (AFM) studies report that the result of the chemisorption process is a monolayer, yielding hexagonal coverage scheme with an S–S distance of nearly 0.5 nm [52]. The monolayer is commensurate with the underlying Au(111) surface and is a simple ($\sqrt{3} \times \sqrt{3}$)R30° overlayer (the 2D lateral spacing of the S groups on the gold surface). This arrangement is shown in Figure 21.11 where the open circles represent gold atoms and the shaded circles represent sulfur atoms. After the initial chemisorption, it is believed that two-dimensional ordering initiates. The molecules are now close enough to allow for London-type van der Waals forces to become important. These forces result in a chain tilt that has been determined to be on the order of 27° for gold surfaces [49]. These forces combined with the van der Waals forces involved in packing, result in crystalline, well-ordered SAMs.

Self-assembled monolayers on flat gold substrates are typically prepared by soaking a clean gold substrate in a dilute ($\sim$1.0 mM) ethanolic solution of the selected thiol or disulfide adsorbate molecule [49]. As the gold substrate is exposed to the adsorbate molecule, gold–sulfur bonds are formed. This anchors the adsorbate molecules to the gold surface, providing a template for

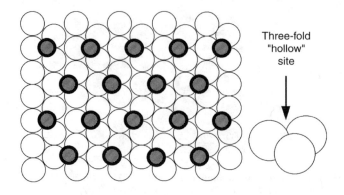

FIGURE 21.11 On the Au(111) surface (bright circles), thiolate ligands (dark circles) adjust to equilibrate into one of the three-fold hollow sites of the lattice. The result of the chemisorption process is a monolayer, yielding hexagonal coverage scheme with an S–S distance of nearly 0.5 nm. (Adapted from Ulman, A., *An Introduction to Ultrathin Organic Films*, Academic, New York, 1991.)

self-assembly of the organic monolayer. The hydrocarbon tails of the adsorbate molecules arrange themselves into an all-trans conformation because of van der Waals interactions (Figure 21.10). The resulting organic monolayer is highly organized and stable under ambient conditions. The energy of the gold–sulfur bond, the well-organized packing of the hydrocarbon tails, and the hydrophobic nature of the organic monolayer provide surfaces coverage that remains unchanged for extended periods of time [49].

Whereas certain technologically relevant applications utilize SAMs that completely cover a surface of interest, a growing number of devices require surfaces decorated with spatially defined patterns of SAMs [53]. Moreover, these patterns must be highly resolved with the proper orientation required by the application of interest. Microcontact printing is a technique that involves the use of an elastomeric stamp to apply adsorbate molecules to a surface with great precision, forming SAMs in reproducible patterns with features in the tens of nanometers [54]. To a first approximation, the technique differs little from its macroscopic congener used to stamp patters of ink onto paper. First developed in the early 1990s by the Whitesides group at Harvard University [55], this method has quickly become a principal method for the efficient mass production of nanoscale fabrication.

The elastomeric stamps are produced through a simple molding process that typically employs a hard silicon master. Because only a limited number of potentially expensive silicon masters are required to prepare a virtually limitless number of inexpensive, disposable stamps, electron beam lithography or other high resolution, low-throughput techniques are typically employed to produce masters exhibiting extremely small feature sizes (down to a few nanometers) [56]. A viscous liquid mixture of silicone elastomer and curing agent is poured onto the master and allowed to cure forming a rubbery polydimethylsiloxane (PDMS) stamp that is peeled off and removed from the master form (Figure 21.12). PDMS is the silicone elastomer utilized in microfluidic lab-on-a-chip applications and is found in many household materials such as waterproof sealants, caulks, and pressure-sensitive adhesives [57]. The resulting stamp contains a faithful reproduction of the surface patterns contained on the silicon master. This molding process may be repeated to produce numerous, identical stamps as long as care is taken to avoid damage to the master.

Once a stamp is removed from the master, it is ready for use and must be inked with an appropriate thiol reagent that will be used in the formation of the desired SAM [58]. A thiol solution is applied by simply pouring or spraying a mist of the solution onto the surface of the stamp. Alternatively, a lint-free cloth or paper may be saturated with a thiol solution, thereby forming an inkpad. Once the solution has been applied, the stamp is brought into contact with a gold substrate, or paper, in order to transfer the thiol solution with the desired arrangement. Self-assembled

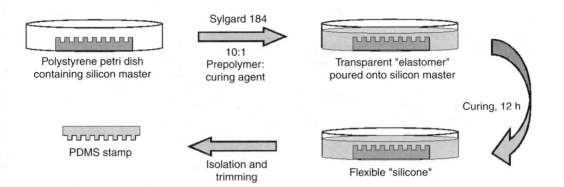

FIGURE 21.12 Schematic for the preparation of an elastomeric stamp from a silicon master for use in microcontact printing (Adapted from Dubois, L. H., Zegarski, B. R., and Nuzzo, R. G., *J. Chem. Phys.* 98, 678, 1993; Chidsey, C. E. D., Liu, G., Rowntree, Y. P., and Scoles, G, *J. Chem. Phys.* 91, 4421, 1989.)

monolayers are formed where the thiol solution was applied to the gold substrate, and the areas void of thiol remain bare (Figure 21.13) [59]. Because of capillary forces and stamp deformation among other variables, some spreading of this molecular ink is observed, yet patterns as small as 50 nm may be obtained in a routine fashion using this technique. More importantly, this technique opens the opportunity for extremely high-resolution pattern transfer on a massive scale with minimal cost and significantly higher throughput when compared to other fabrication techniques capable of such fine resolution [60,61].

Proteins, containing a significant number of thiol functional groups, are excellent candidates for patterning via microcontact printing for use in a variety of bioanalytical applications [62]. Self-assembled protein monolayers were first patterned via microcontact printing and utilized in a variety of immunoassays by Biebuyck and coworkers at IBM [63]. Using a PDMS stamp, a range of individual proteins (IgG, phosphatase, cytochrome c, streptavidin, peroxidase, etc.) were applied onto a surface in complex, submicron patterns. Efficient transfer of biomolecules from the stamp to the surface was accomplished with contact times of only a few seconds. The high resolution offered by this technique allowed the printing of fine features containing less than 1000 protein molecules. Fluorescent labeling, ellipsometry, and atomic force microscopy confirmed patterning and placement, whereas surface immunoassays confirmed retention of protein structure and activity. In most cases, the observed activity of the surface immunoassay was indistinguishable from its solution phase analog as monitored by fluorescent staining.

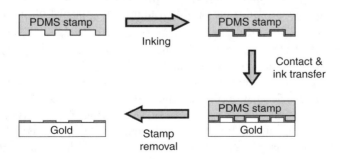

FIGURE 21.13 Microcontact printing. (Adapted from Love, J. C., Estroff, L. A., Kriebel, J. K., Nuzzo, R. G., and Whitesides, G. M., *Chem. Rev.*, 105, 1103, 2005.)

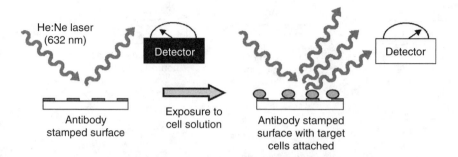

FIGURE 21.14 Diffraction based bacteria detection microarray fabricated via microcontact printing. (Adapted from St. John, P. M., Davis, R., Cady, N., Czajka, J., Batt, C. A., and Craighead, H. G., *Anal. Chem.*, 70, 1108, 1998.)

Microcontact printing has also been used in the preparation of patterned arrays of protein antibodies for use in bacterial detection applications. In one of the first such devices to implement microcontact printing, a diffraction-based, optical detection scheme was employed by researchers at Cornell University [64]. A PDMS stamp, subjected to a brief plasma exposure to increase its hydrophilicity, was used to apply a periodic pattern of antibody molecules onto a silicon substrate. Although the resulting antibody grating failed to produce a diffraction pattern, a significant signal was observed upon immunocapture of *Escherichia coli* cells (Figure 21.14). This inexpensive and efficient manner of antibody array fabrication was later expanded by groups at Purdue University to test the specificity of their antibody arrays against a solution containing a range of nonimmunologically targeted bacteria [65]. Utilizing atomic force microscopy as a detection method, the patterned antibody microarray demonstrated a definite selectivity for targeted bacteria over unspecified bacteria and an enhanced preference for cell binding to the antibody microarray than to unfunctionalized regions of the substrate. These studies demonstrate the great potential offered by microcontact printing in the quick and cost-effective patterning of protein monolayers with extremely high resolution.

21.2.3 NANOPARTICLE FUNCTIONALIZATION

Although the preparation of SAMs on flat surfaces has attracted considerable attention, recent years have witnessed an explosive growth in studies focusing on the adsorption of thiols and other molecular species onto colloidal particles of gold, silver, copper, palladium, and platinum [66]. These monolayer-protected colloids/clusters (MPCs) have found use in a wide range of technologies. The sol particle immunoassay (SPIA), for example, exploits colloid aggregation to monitor the interaction between an antibody or antigen bound to a colloid and its binding partner [67]. Surface-passivated metal colloids (Figure 21.15) also serve as useful materials for the manufacture of nanoelectronics, colloid-modified electrodes, labeling agents for electron and visible microscopy, and quantum-confinement materials [68].

The development of the Schiffin/Brust two-phase liquid/liquid route [69] to *n*-alkanethiol-derivatized metal nanoparticles has further advanced this area of research. This method made possible the facile preparation of bulk quantities of MPCs from inexpensive starting materials with no need for intricate apparatus. Functionalized nanoparticles generated in this manner exist as dark-brown/black solids that are easy to isolate, stable in air, and soluble in a wide range of organic solvents. MPCs can be repeatedly dissolved and precipitated from solution without decomposition. These features permit facile characterization using a variety of techniques [70] such as x-ray photoelectron spectroscopy (XPS), nuclear magnetic resonance (NMR) spectroscopy,

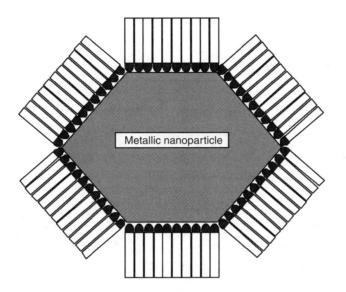

FIGURE 21.15 Self-assembled monolayers (SAMs) may also involve chemisorption of molecular species onto a variety of nanoparticle materials. (Adapted from Brust, M., Walker, M., Bethell, D., Schiffrin, D. J., and Whyman, R., *J. Chem. Soc., Chem. Commun.*, 801, 1994; Brust, M., Fink, J., Bethell, D., Schiffrin, D. J., and Kiely, C. J., *J. Chem. Soc., Chem. Commun.*, 1655, 1995.)

ultraviolet–visible (UV–vis) spectroscopy, transmission electron microscopy (TEM), surface-enhanced Raman spectroscopy (SERS), and FTIR spectroscopy.

The extremely large surface-to-volume ratios offered by these nanomaterials, coupled with their solubility in a wide range of solvent systems, makes them excellent candidates for use in a variety of important bioanalytical applications [71]. Mirkin and coworkers at Northwestern University have developed a sequence-specific method of DNA detection using gold nanoparticles functionalized with alkanethiol-capped oligonucleotides [72]. Whereas many of the traditional DNA detection methods are based on analyte hybridization with chemilumenescent, fluorescent, or radioactive probes, this nanoparticle method is colorimetric. First, citrate reduction of hydrogen tetrachloroaurate ($HAuCl_4$) yielded monodisperse, 15 nm diameter gold nanoparticles. A ligand exchange reaction was then carried out at room temperature to prepare $3'$- or $5'$-alkanethiol 12-base oligonucleotide-functionalized gold nanoparticles. These nanoparticle probes proved remarkable stable, exhibiting a shelf life in excess of three months with no detectable decomposition. The oligonucleotide-functionalized gold nanoparticles yielded a surface plasmon band in the visible spectrum ($\lambda_{max} = 524$ nm), serving as a signal of interest in this colorimetric detection scheme. Upon addition of the complementary target oligonucleotide, the red nanoparticle solution transitioned to purple within a period of five minutes. The oligonucleotide-functionalized gold nanoparticles align tail-to-tail onto a polynucleotide strand, resulting in the formation of an aggregated, three-dimensional network (Figure 21.16), thereby red-shifting the plasmon resonance from 524 to 576 nm. The color change of the solution is attributed to this shift that is due to a change in the interparticle distance of the gold nanoparticles [72].

Valuable information regarding the specificity of this detection scheme can be gleaned from dissociation or melting analysis of the resultant DNA-linked nanoparticle aggregate. At room temperature, the aggregate was observed to precipitate out of solution upon sitting for times in excess of 2 h, but this analysis was performed on the resuspended aggregates. The solution temperature was slowly increased from 25°C to 75°C, and a sharp absorption increase was observed at 260 nm, indicating the dissociation or melting temperature. This change could also be monitored as

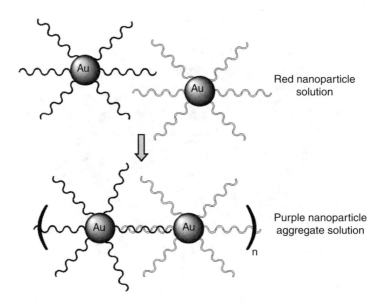

FIGURE 21.16 Sequence-specific DNA detection system employing functionalized gold nanoparticles (Adapted from Daniel, M.-C., and Astruc, D., *Chem. Rev.*, 104, 293, 2004; Rosi, N. L. and Mirkin, C. A., *Chem. Rev.*, 105, 1547, 2005.)

a purple to red shift in solution color. Imperfect targets were readily identified by a depression of the dissociation temperature. Once the perfect complement was identified, a water bath was maintained at this temperature, and any imperfect targets would dissociate and transition to red, while the aggregated, perfect target, remained purple [72]. This method allows for a quick and facile detection scheme that alleviates the need for radioactive or fluorescent reagents and requires little in the way of expensive instrumentation. Furthermore, it requires very little formal training and can be easily adapted for fieldwork.

21.2.4 SCANNING PROBE LITHOGRAPHY

Although the scanning probe techniques such as STM and AFM remain some of the most versatile characterization tools in nanoscience research, they have recently been employed as nanofabrication implements with increasing frequency for a variety of applications [73]. Scanning probe lithography involves the utilization of an ultra fine stylus to either deliver a reagent to, or remove material from, a surface with ultra-high lateral resolution. The production of the letters "IBM" referred to in the introduction of this chapter demonstrates the power of scanning probe techniques to deliver true atomic manipulation under certain conditions. Although still in early stages of development and commercialization, scanning probe lithography has attracted immense attention.

One of the most widespread and efficient manifestations of scanning probe lithography is dippen nanolithography (DPN). Analogous to its ancient, macroscopic inspiration, the fountain pen, DPN employs the tip of an AFM to deliver chemical reagents to spatially defined regions on a surface with nanometer resolution [74,75]. The process begins by immersing the tip of an AFM ($\sim$5 nm tip radius) into a solution of a predetermined chemical species. The tip is removed. The solvent then evaporates, leaving the solute as a thin coating on the tip. The tip is then brought into contact with a substrate where under conditions of high humidity, a water meniscus forms. The solute is then transferred from the tip to the substrate via this water layer (Figure 21.17). Typically, organic molecules are employed in this technique in order to prepare spatially defined organic

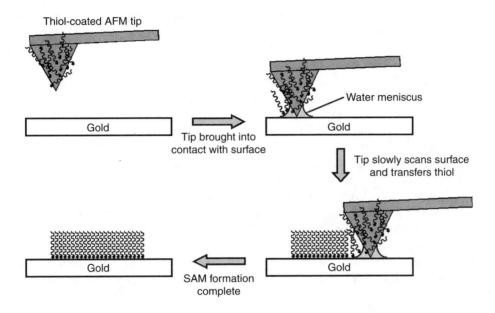

FIGURE 21.17 Dip-pen nanolithography (DPN) using an atomic force microscope (AFM) tip. (Adapted from Piner, R. D., Zhu, J., Xu, F., Hong, S., and Mirkin, C. A., *Science*, 283, 661, 1999.)

monolayers on various substrates [76]. Through careful modulation of parameters such as humidity and tip scan speed, lateral resolutions beyond 30 nm are routinely obtainable [77]. Vector scanning techniques are also widely employed to deliver reagents to produce SAMs in a variety of complex designs and patterns. As in the case of electron beam lithography, this resolution comes at the price of speed. This direct-write technique is serial in nature, and current studies aim to incorporate numerous tips, working in parallel to increase the throughput of DPN [78].

Mirkin and coworkers, pioneers of DPN, have employed the technique in high resolution patterning of mixed protein nanoarrays [79]. In order to enhance the protein adsorption onto the AFM tip, the silicon nitride tip was coated with gold and immersed into a dilute ethanolic solution of thiotic acid. The resulting hydrophilic SAM on the AFM tip facilitates more effective loading of the protein ink onto the AFM tip pen for patterning. Protein loading was accomplished by immersion of the AFM tip in a dilute, buffered solution of lysozyme (Lyz) or rabbit immunoglobulin-gamma (IgG). First, the Lyz protein nanoarray was prepared by vectoring the tip to prescribed regions of the surface, making contact under conditions of high humidity, thereby facilitating protein delivery. After loading the second protein onto a fresh AFM tip, an IgG nanoarray was patterned in the free space within the Lyz nanoarray. Finally, the remaining bare gold surface was passivated with a monolayer of 11-mercapto-undecylpenta(ethylene glycol) disulfide to avoid nonspecific protein adsorption onto regions of gold absent of any Lyz or IgG. The resultant mixed protein nanoarray was confirmed via atomic force microscopy height profiles both before and after exposure to anti-IgG coated gold nanoparticles. This effective approach to a mixed protein nanoarray successfully demonstrates the advantages offered by DPN in the preparation of the increasingly complex surfaces demanded by future generations of bioanalytical devices. Furthermore, DPN provides access to nanodot arrays with features ranging from 45 nm to a few microns in size, tunable by modulation of AFM-surface contact times, owing to larger feature sizes corresponding to longer contact times as a result of protein diffusion across the surface. Similar studies have focused on employing DPN in the patterning of nucleic acids and other pertinent biomolecules for diagnostic applications [80].

Dip-pen nanolithography in conjunction with electroless deposition has also been employed to form nanoelectrodes and other metallic structures on semiconductor surfaces [81]. At Purdue University, researchers immersed an AFM tip in a 1:10 (v/v) mixture of aqueous 20 mM $HAuCl_4$ and acetonitrile for 5 min and allowed the tip to dry under ambient conditions for approximately five minutes. The tip was subsequently used to perform galvanic displacement [82] with an extremely high resolution. An environmental chamber was used to maintain an atmosphere of 50% humidity, and the tip was rastered across the surface at a rate of 200 nm/s [36]. The result was a gold nanowire 500 nm long, 8 nm tall, and 30 nm wide (Figure 21.18). This technique employed vector-scanning software to obtain this simple structure, yet with sufficient programming, a variety of geometric configurations may be obtained for use in solid-state sensors or in the preparation of anchoring points for thiol containing biomolecules.

As an alternative to delivering chemical reagents with an AFM tip, mechanical plowing lithography utilizes an AFM tip to mechanically remove material from a surface [73]. This process can also be employed in the direct nanografting of various molecular species and metal clusters into SAM systems [83]. First developed by Liu and coworkers [84], the process begins by forming a SAM onto a gold surface that is subsequently immersed into a dilute solution of thols that differ from the ones contained in the initial SAM. An AFM tip is applied to the surface with a force on the

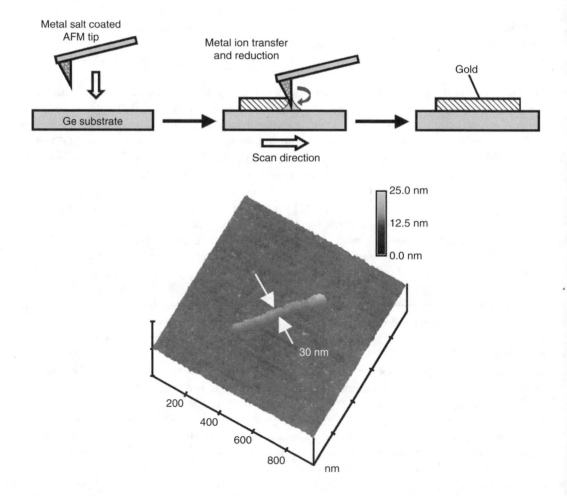

FIGURE 21.18 Gold nanoelectrode fabricated via dip-pen nanolithography (DPN). (Adapted from Porter, L. A., Jr., Choi, H. C., Schmeltzer, J. M., Ribbe, A. E., Elliott, L. C. C., and Buriak, J. M., *Nano Lett.*, 2, 1369, 2002.)

order of 5 nN and slowly scanned across the surface. The mechanical shear force is enough to shave away regions of the monolayer that come into contact with the tip, producing gaps within the SAM. The thiols from solution then quickly chemisorb onto the exposed gold, thereby replacing the molecules cleaved from the surface in the wake of the AFM tip. Recently, this technique has been employed in the nanografting of a variety of molecules of biological interest into SAMs on gold [85]. Liu and coworkers have demonstrated the ability to nanograft thiolated ssDNA into alkylthiolate SAMs, producing patterns of ssDNA down to approximately 10 nm in width (Figure 21.19) [86]. Hexanethiol and decanethiol SAMs on gold were first prepare by soaking thin gold substrates in a 1 mM ethanolic solution of the respective thiol for approximately 24 h. After washing away any excess thiol solution, the gold sample was mounted onto an AFM sample stage and immersed in a mixed solvent solution composed of 2-butanol/water/ethanol with a ratio of 6:1:1(v/v/v) containing 40 μM of thiolated ssDNA ($5'$-HS-$(CH_2)_6$CTAGCTCTAATCTGCTAG-$3'$). Nanografting was accomplished by initiating AFM tip contact with the surface using an applied force of 20 nN while scanning across the surface at a speed of nearly 800 nm/s. The samples were washed to remove any excess thiolated ssDNA following the nanografting procedure and characterized via AFM to confirm incorporation of the thiolated ssDNA within the original SAM. The thiolated ssDNA regions exhibited a height of approximately 8 nm, and the remaining hexanethiol or decanethiol regions were considerable shorter (<2 nm).

Stemming from fundamental scanning probe experiments focused on exploring polymer tribology [87], static plowing lithography has recently been utilized as a viable fabrication process. In this facile, yet effective process, an AFM probe is employed to mechanically plow away spatially defined regions of a target substrate with extremely high lateral resolution. Similar to DPN, vector scan software provides access to a variety of patterns and geometries. This technique has been utilized to produce etch and evaporation masks for semiconductor device fabrication [88] as well as in the preparation of nanoelectrodes and binding sites for the direct interfacing of organic and bio-organic molecules onto a semiconductor surface using a sequence of four primary operations (Figure 21.20) [89]: (1) application of a thin polymer resist onto a Ge(111) substrate; (2) utilization of an AFM tip to plow away patterns/domains of the resist, thereby exposing defined regions of the underlying Ge(111) substrate; (3) immersion of the substrate into a dilute metal salt

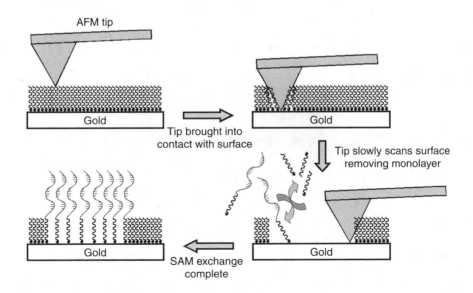

FIGURE 21.19 Schematic for the preparation of mixed monolayers using the nanografting technique. (Adapted from Liu, M., Amro, N. A., Chow, C. S., and Liu, G., *Nano Lett.*, 2, 863, 2002.)

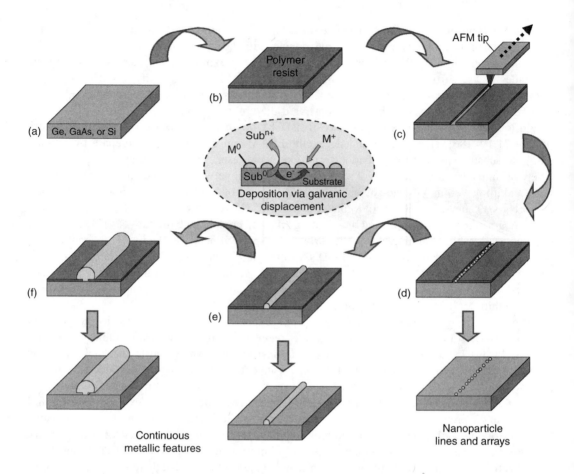

FIGURE 21.20 Metallic nanostructures via static plowing lithography: a Ge(111) substrate (a) is coated with a thin polymer resist layer (b) that is subsequently plowed away utilizing the tip of an atomic force microscope (AFM) (c). Immersion of the substrate into a dilute, aqueous $HAuCl_4$ solution for a brief time provides for the deposition of discrete nanoparticles onto the exposed Ge(111) surface via galvanic displacement (d). As deposition proceeds with longer immersion times, grain growth and nanoparticle coalescence eventually yields continuous metallic structures (d) that continue to increase in size with extended periods of electroless plating (f). Once the desired features are realized, the resist is completely removed with a solvent rinse. The inset represents a simplified schematic of the galvanic displacement deposition mechanism. (Adapted from Porter, L. A., Jr., Ribbe, A. E., and Buriak, J. M., *Nano Lett.*, 3, 1043, 2003.)

(such as $HAuCl_4(aq)$) where electroless deposition proceeds onto the areas of Ge(111) no longer concealed by the resist; and (4) final removal of all resist via a solvent rinse. Shorter plating times (≤ 30 s) resulted in the formation of discrete gold nanoparticles aligned in a linear formation consistent with the geometry of the resist furrow. Mean nanoparticle diameter and height were determined to be ~30 and ~5 nm, respectively. More lengthy immersion times (> 30 s) led to continuous metallic structures through a Volmer–Weber (3D island growth) concluding with nano-particle agglomeration as a consequence of Ostwald ripening. The resultant continuous gold nanostructures (approximated as semielliptical cylinders) were observed to increase in line width from 50 nm for a 1 min immersion to 200 nm for 30 min of deposition from 25 mM $HAuCl_4$ (aq). Analogously, the line height increased from 5 to 100 nm for the corresponding plating intervals. Both nanoparticle formations as well as continuous metal structures are obtained via galvanic displacement [82] from aqueous metal salt solutions (Figure 21.21). Vector scan operation

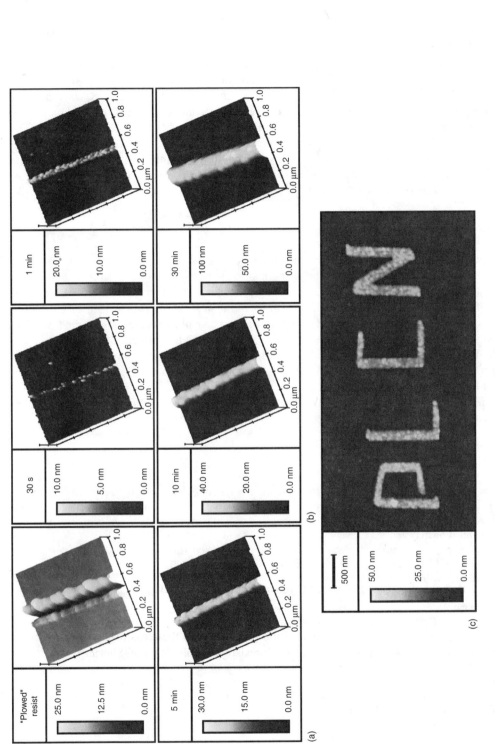

FIGURE 21.21 Intermittent contact (tapping) mode atomic force micrographs illustrating a resist furrow produced by static plowing (a) and the gold nanostructures (b) on Ge(111), resulting from increasing immersion times in 25 mM HAuCl$_4$ (aq) at 25°C following resist removal. Intermittent contact (tapping) mode atomic force micrographs illustrating a series of resist furrows (PLCN—Purdue Laboratory for Chemical Nanotechnology) produced by static plowing (c). (Adapted from Porter, L. A., Jr., Ribbe, A. E., and Buriak, J. M., *Nano Lett.*, 3, 1043, 2003.)

allows for the AFM tip to be driven along paths of arbitrary length scales at predefined angles while also providing for control over z-axis displacement and applied force. Figure 21.21 illustrates one vector profile utilized to prepare a series of resist furrows forming the letters, "PLCN" (Purdue Laboratory for Chemical Nanotechnology), employing static plowing lithography [89]. Realization of this pattern demonstrates the ability of this technique to successfully cope with various tip scanning directions. Additionally, solid features were generated by plowing away large domains of resist material through the use of multiple, closely spaced scan lines. These metallic nanostructures provide convenient bonding sites for the formation of SAMs as previous discussed. This method is compatible with photoresist coating processes ubiquitous to industrial and academic fabrication environments. Because the AFM is used solely to pattern the resist layer and not in the chemical transfer via the tip, this technique is not restricted by many of the limitations and complexities faced by other scanning probe lithographic methodologies.

21.3 CONCLUSION

This chapter has attempted to spotlight the major strategies and methodologies involved in functional structure and device fabrication at the nanoscale with an emphasis on their application to future developments in biodiagnostic technologies. Potentially, nanofabrication methods will have a profound impact on the bioanalytical methods of the near future. Biorecognition events that serve as the core to any screening method are processes that take place on the nanoscopic level [71]. Enhanced specificity, sensitivity, and practicality are but a few of the attractive features offered by nanoscale-based devices. However, a great deal of work remains before many of these and other significant developments mature beyond the research laboratory and make their way to compete with established, conventional methods in use today. At this stage, nanofabrication techniques continue to emerge, and only time will tell which will evolve into truly effective and efficient means for mass production on the nanoscale and which will fall by the wayside.

REFERENCES

1. National Nanotechnology Initiative (NNI) Home Page, http://www.nano.gov (accessed April 2005).
2. Timp, G., Ed., *Nanotechnology*, Springer, New York, 1999. (For more recent accomplishments, the reader is invited to peruse the most recent issues of *Nano Letters* (http://pubs.acs.org). The American Chemical Society Journal of Nanotechnology)
3. Wilson, M., Kannangara, K., Smith, G., Simmons, M., and Raguse, B., *Nanotechnology: Basic Science and Emerging Technologies*, CRC Press, Boca Raton, 2002.
4. Porter, L. A. Jr., *J. Chem. Educ.*, in press.
5. Fritz, S., Ed., *Understanding Nanotechnology*, Warner Books, New York, 2002.
6. Bhushan, B., Ed., *Springer Handbook of Nanotechnology*, Springer, New York, 2004.
7. Pirrung, M. C., Morehead, A. T., and Young, B. G., Eds., *The Total Synthesis of Natural Products*, Wiley, New York, 1999.
8. Garrett, R., Douthwaite, S. R., Liljas, A., Matheson, A. T., Moore, P. B., and Noller, H. F., Eds., *The Ribosome: Structure, Function, Antibiotics, and Cellular Interactions*, American Society Microbiology, Washington, DC, 2000.
9. Ghandhi, S. K., *VLSI Fabrication Principles: Silicon and Gallium Arsenide*, Wiley, New York, 1994.
10. Stix, G., *Sci. Am.*, 285, 32, 2001.
11. Roukes, M., *Sci. Am.*, 285, 48, 2001.
12. Feynman, R. P., *J. MEMS*, 1, 60, 1992. Feynman, R. P., *J. MEMS*, 2, 4, 1992. (The previous citations refer to reprints of Feynman's speeches entitled *Plenty of Room at the Bottom* (1959) and *Infinitesimal Machinery* (1983), respectively)
13. Binnig, G., *Phys. Rev. Lett.*, 56, 930, 1986. Binnig, G., *IBM J. Res. Dev.*, 30, 355, 1986. Binnig, G. and Rohrer, H., *Rev. Mod. Phys.*, 59, 612, 1987.
14. Bear, G., *Analog*, 103, 12, 1983.

15. Elliott, E., Ed., Baen Books, Riverdale, 1998. Crichton, M., Prey,, Avon, New York, 1998.
16. Nanotechnology in Science Fiction. http://www.geocities.com/asnapier/nano/n-sf (accessed April 2004).
17. Drexler, K. E., *Engines of Creation: The Coming Era of Nanotechnology*, Anchor Books, New York, 1986.
18. Moore, G., *Elec. Mag.*, 8, 114, 1965.
19. Intel Corporation. http://www.intel.com (accessed May 2005).
20. Whitesides, G. M. and Love, C. J., *Sci. Am.*, 285, 39, 2001.
21. Van Zant, P., *Microchip Fabrication*, McGraw Hill, New York, 2000.
22. Fay, B., *Microelec. Eng.*, 61, 11, 2002.
23. Advanced Micro Devices (AMD) Corporation. http://www.amd.com/us-en (accessed May 2005).
24. Campbell, S. A., *The Science and Engineering of Microelectronic Fabrication*, Oxford University Press, New York, 2001.
25. Levinson, H. J., *Principles of Lithography*, SPIE, New York, 2005.
26. Rothschild, M., *Mater. Today*, 8, 18, 2005.
27. Willson, C. G. and Trinque, B. C., *J. Photopolym. Sci. Technol.*, 16, 621, 2003.
28. May, G. S. and Sze, S. M., *Fundamentals of Semiconductor Fabrication*, Wiley, New York, 2003.
29. Deforest, W. S., *Photoresist: Materials and Processes*, McGraw Hill, New York, 1975.
30. Reichmanis, E., Nalamasu, O., and Houlihan, F. M., *Acc. Chem. Res.*, 32, 659, 1999.
31. Kadota, T., Kageyama, H., Wakaya, F., Gamo, K., and Shirato, Y., *Chem. Lett.*, 33, 706, 2004; Sugimura, H., Hayashi, K., Saito, N., Hong, L., Takai, O., Hozumi, A., Nakagiri, N., and Okada, M., *Mater. Res. Soc. Jpn*, 27, 545, 2002; Xia, Y., Zhao, X.-M., and Whitesides, G. M., *Microelec. Eng.*, 32, 255, 1996; Chan, K. C., Kim, T., Schoer, J. K., and Crooks, R. M., *J. Am. Chem. Soc.*, 117, 5875, 1995.
32. Ulman, A., *An Introduction to Ultrathin Organic Films*, Academic, New York, 1991.
33. Stewart, M. P. and Buriak, J. M., *J. Am. Chem. Soc.*, 12, 7821, 2001.
34. Cicero, R. L., Linford, M. R., and Chidsey, C. E. D., *Langmuir*, 16, 5688, 2000; Linford, M. A. and Chidsey, C. E. D., *J. Am. Chem. Soc.*, 115, 12631, 1993.
35. Buriak, J. M., *Chem. Rev.*, 102, 1271, 2002. Buriak, J. M., *Adv. Mater.*, 11, 265, 1999.
36. Porter, L. A. Jr, Choi, H. C., Schmeltzer, J. M., Ribbe, A. E., Elliott, L. C. C., and Buriak, J. M., *Nano Lett.*, 2, 1369, 2002.
37. Choi, K. and Buriak, J. M., *Langmuir*, 16, 7737, 2000.
38. Stewart, M. P. and Buriak, J. M., *Angew. Chem. Int. Ed.*, 37, 3257, 1998.
39. Canham, L. T., Ed., INSPEC, London, 1997; Sailor, M. J., Heinrich, J. L., and Lauerhaas, J. M., In *Semiconductor Nanoclusters*, Karmat, P. V. and Meisel, D., Eds., Elsevier, Amsterdam, 1997. Stewart, M. P. and Buriak, J. M., *Adv. Mater.* 12, 859, 2000.
40. Stewart, M. P. and Buriak, J. M., *Angew. Chem. Int. Ed.*, 37, 3257, 1998.
41. Pike, A. R., Lie, L. H., Eagling, R. A., Ryder, L. C., Patole, S. N., Connolly, B. A., Horrocks, B. R., and Houlton, A., *Angew. Chem. Int. Ed.*, 41, 615, 2002.
42. Benschop, J. and Kurz, P., *Microlithography World*, 10, 4, 2001.
43. Sheats, J. R. and Smith, B. W., Eds., *Microlithography Science and Technology*, Marcel Dekker, New York, 1988.
44. Hawryluk, A. M., Ceglio, N. M., and Markle, D. A., *Solid State Technol.*, 40, 151, 1997.
45. Chen, Y., Vieu, C., and Launois, H., *Condens. Matter News*, 6, 22, 1998.
46. Harriott, L. and Liddle, A., *Phys. World*, 10, 41, 1887.
47. Gamo, K. and Namba, S., *Ultramicroscopy*, 15, 261, 1984.
48. Adamson, A. W., *Physical Chemistry of Surfaces*, Wiley, New York, 1976; Kuhn, H. and Mobius, D., *Techniques of Organic Chemistry*, Wiley, New York, 1976; Polymeropoulos, E. E. and Sagiv, J., *J. Chem. Phys.*, 69, 1836; Sugi, M., Fukui, T., and Lizima, S., *Phys. Rev. B*, 18, 725; Furtlehner, J. P. and Messier, J., *Thin Solid Films*, 68, 233; Waldbillig, R. C., Robertson, J. D., and McIntosh, T., *J. Biochim. Phiophys. Acta*, 4481; Kornberg, R. D. and McConnell, H. M., *Biochemistry*, 10, 1111.
49. Dubois, L. H. and Nuzzo, R. G., *Annu. Rev. Phys. Chem.*, 43, 437, 1992; Bain, C. D., Troughton, E. B., Tao, Y.-T., Evall, J., Whitesides, G. M., and Nuzzo, R. G., *J. Am. Chem. Soc.*, 111, 321, 1989; Bain, C. D. and Whitesides, G. M., *Angew. Chem. Int. Ed. Engl.*, 28, 506, 1989; Whitesides, G. M. and Laibinis, P. E., *Langmuir*, 6, 87, 1990; Ulman, A., *Chem. Rev.*, 96, 1533, 1996.
50. Nuzzo, R. G. and Allara, D. L., *J. Am. Chem. Soc.*, 105, 4481, 1983.

51. Laibinis, P. E., Whitesides, G. M., Allara, D. L., Tao, Y.-T., Parikh, A. N., and Nuzzo, R. G., *J. Am. Chem. Soc.*, 113, 7152, 1991. (and references therein)

52. Dubois, L. H., Zegarski, B. R., and Nuzzo, R. G., *J. Chem. Phys.*, 98, 678, 1993; Chidsey, C. E. D., Liu, G., Rowntree, Y. P., and Scoles, G., *J. Chem. Phys.*, 91, 4421, 1989.

53. Love, J. C., Estroff, L. A., Kriebel, J. K., Nuzzo, R. G., and Whitesides, G. M., *Chem. Rev.*, 105, 1103, 2005.

54. Xia, Y. and Whitesides, G. M., *Angew. Chem. Int. Ed.*, 37, 550, 1998; Brittain, S., Paul, K., Zhao, X.-M., and Whitesides, G. M., *Phys. World*, 11, 31, 1998; Xia, Y., Rogers, J. A., Paul, K. E., and Whitesides, G. M., *Chem. Rev.*, 99, 1823, 1999.

55. Kumar, A. and Whitesides, G. M., *Appl. Phys. Lett.*, 63, 2002, 1993.

56. Gorman, C. B., Biebuyck, H. A., and Whitesides, G. M., *Chem. Mater.*, 7, 252, 1995.

57. Tomanek, A., *Silicones and Industry: A Compendium for Practical Use, Instruction and Reference*, Hanser Gardner, Cincinnati, 1993.

58. Xia, Y. and Whitesides, G. M., *J. Am. Chem. Soc.*, 117, 3274, 1995.

59. Larsen, N. B., Biebuyck, H., Delamarche, E., and Michel, B., *J. Am. Chem. Soc.*, 119, 3017, 1997.

60. Kumar, A., Abbott, N. L., Biebuyck, H. A., Kim, E., and Whitesides, G. M., *Acc. Chem. Res.*, 28, 219, 1995.

61. Gates, B. D., *Mater. Today*, 8, 44, 2005.

62. Whitesides, G. M., Ostuni, E., Takayama, S., Jiang, X., and Ingber, D. E., *Ann. Rev. Biomed. Eng.*, 3, 335, 2001.

63. Bernard, A., Delamarche, E., Schmid, H., Michel, B., Bosshard, H. R., and Biebuyck, H., *Langmuir*, 14, 2225, 1998.

64. St. John, P. M., Davis, R., Cady, N., Czajka, J., Batt, C. A., and Craighead, H. G., *Anal. Chem.*, 70, 1108, 1998.

65. Howell, S. W., Inerowicz, H. D., Regnier, F. E., and Reifenberger, R., *Langmuir*, 19, 436, 2003; Inerowicz, H. D., Howell, S., Regnier, F. E., and Reifenberger, R., *Langmuir*, 18, 5263, 2002.

66. Lee, P. C. and Meisel, D., *J. Phys. Chem.*, 86, 3391, 1982; Xu, H., Tseng, C.-H., Vickers, T. J., Mann, C. K., and Schlenoff, J. B., *Surf. Sci.*, 311, L707, 1994; Grabar, K. C., Allison, K. J., Baker, B. E., Bright, R. M., Brown, K. R., Freeman, R. G., Fox, A. P., Keating, C. D., Musick, M. D., and Natan, M. J., *Langmuir*, 12, 2353, 1996; Weisbecker, C S., Meritt, M. V., and Whitesides, G. M., *Langmuir*, 12, 3763, 1996; Leff, D. V., Brandt, L., and Heath, J. R., *Langmuir*, 12, 4723, 1996; Sarathy, K. V., Raina, G., Yadav, R. T., Kulkarni, G. U., and Rao, C. N. R., *J. Phys. Chem. B*, 101, 9876, 1997; Brown, K. R. and Natan, M. J., *Langmuir*, 14, 726, 1998; Ingram, R. S., Hostetler, M. J., Murray, R. W., Schaaff, T. G., Khoury, J. T., Whetten, R. L., Bigioni, T. P., Guthrie, D. K., and First, P. N., *J. Am. Chem. Soc.*, 119, 9279, 1997; Green, S. J., Stokes, J. J., Hostetler, M. J., Pietron, J., and Murray, R. W., *J. Phys. Chem. B*, 101, 2663, 1997.

67. Van Erp, R., Gribnau, T. C. J., Van Sommeren, A. P. G., and Bloemers, H. P. J., *J. Immunoassay*, 11, 31, 1990.

68. Schon, G. and Simon, U., *Colloid Polym. Sci.*, 273, 101. Dorn, A., Katz, E., and Willner, I., *Langmuir*, 11, 1313. Beesley, J. EW., *Colloidal Gold: A New Perspective for Cytochemical Marking, Royal Microscopical Society Microscopy Handbook*, 17, Oxford Press, Oxford, 1995; Hanna, A. E. and Tinkham, M., *Phys. Rev. B*, 445919.

69. Brust, M., Walker, M., Bethell, D., Schiffrin, D. J., and Whyman, R., *J. Chem. Soc., Chem. Commun.*, 801, 1994; Brust, M., Fink, J., Bethell, D., Schiffrin, D. J., and Kiely, C. J., *J. Chem. Soc., Chem. Commun.*, 1655, 1995.

70. Leff, D. V., Ohara, P. C., Heath, J. R., and Gelbart, W. M., *J. Phys. Chem.*, 99, 7036, 1995; Terrill, R. H., Postlethwaite, T. A., Chen, C.-H., Poon, C.-D., Terzis, A., Chen, A., Hutchinson, J. E., et al., *J. Am. Chem. Soc.*, 117, 12537, 1995; Hostetler, M. J., Green, S. J., Stokes, J. J., and Murray, R. W., *J. Am. Chem. Soc.*, 118, 4212, 1996; Badia, A., Demers, L., Dickinson, L., Morin, F. G., Lennox, R. B., and Reven, L., *J. Am. Chem. Soc.*, 119, 11104, 1997; Templeton, A. C., Hostetler, M. J., Kraft, C. T., and Murray, R. W., *J. Am. Chem. Soc.*, 120, 1906, 1998; Grabar, K. C., Brown, K. R., Keating, C. D., Stranick, S. J., Tang, S.-L., and Natan, M. L., *Anal. Chem.*, 69, 471, 1997; Hostetler, M. J., Stokes, J. J., and Murray, R. W., *Langmuir*, 12, 3604, 1996; Ingram, R. S., Hostetler, M. J., and Murray, R. W., *J. Am. Chem. Soc.*, 119, 9175, 1997; Badia, A., Cuccia, L., Demers, L., Morin, F., and Lennox, R. B.,

J. Am. Chem. Soc., 119, 2682, 1997; Hostetler, M. J., Wingate, J. E., Zhong, C.-J., Harris, J. E., Vachet, R. W., Clark, M. R., Londono, J. D., et al., *Langmuir*, 14, 17, 1998; Kang, S. Y. and Kim, K., *Langmuir*, 14, 22672, 1998.

71. Daniel, M.-C. and Astruc, D., *Chem. Rev.*, 104, 293, 2004; Rosi, N. L. and Mirkin, C. A., *Chem. Rev.*, 105, 1547, 2005.

72. Park, S.-J., Lazarides, A. A., Storhoff, J. J., Pesce, L., and Mirkin, C. A., *J. Phys. Chem. B*, 108, 12375, 2004; Nam, J.-M., Stoeva, S. I., and Mirkin, C. A., *J. Am. Chem. Soc.*, 126, 5932, 2004; Bailey, R. C., Nam, J.-M., Mirkin, C. A., and Hupp, J. T., *J. Am. Chem. Soc.*, 125, 13541, 2003; Jin, R., Wu, G., Li, Z., Mirkin, C. A., and Schatz, G. C., *J. Am. Chem. Soc.*, 125, 1643, 2003; Storhoff, J. J., Elghanian, R., Mirkin, C. A., and Letsinger, R. L., *Langmuir*, 18, 6666, 2002; Nam, J.-M., Park, S.-J., and Mirkin, C. A., *J. Am. Chem. Soc.*, 124, 3820, 2002; Storhoff, J. J., Lazarides, A. A., Mirkin, C. A., Letsinger, R. L., Mucic, R. C., and Schatz, G. C., *J. Am. Chem. Soc.*, 122, 4640, 2002; Reynolds, R. A. III, Mirkin, C. A., and Letsinger, R. L., *J. Am. Chem. Soc.*, 122, 3795, 2000; Mucic, R. C., Storhoff, J. J., Mirkin, C. A., and Letsinger, R. L., *J. Am. Chem. Soc.*, 120, 12674, 1998; Storhoff, J. J., Elghanian, R., Mucic, R. C., Mirkin, C. A., and Letsinger, R. L., *J. Am. Chem. Soc.*, 120, 1959, 1998; Elghanian, R., Storhoff, J. J., Mucic, R. C., Letsinger, R. L., and Mirkin, C. A., *Science*, 277, 1078, 1997.

73. Soh, H. T., Guarini, K. W., and Quate, C. F., *Scanning Probe Lithography*, Springer, New York, 2001; Nyffenegger, R. M. and Penner, R. M., *Chem. Rev.*, 97, 1195; Tang, Q., Shi, S.-Q., and Zhou, L., *J. Nanosci. Nanotechnol.*, 4948.

74. Piner, R. D., Zhu, J., Xu, F., Hong, S., and Mirkin, C. A., *Science*, 283, 661, 1999.

75. Hong, S., Zhu, J., and Mirkin, C. A., *Science*, 286, 523, 1999.

76. Rozhok, S., Piner, R., and Mirkin, C. A., *J. Phys. Chem. B*, 107, 751, 2003.

77. Mirkin, C. A., Hong, S., and Demers, L., *Chem. Phys. Chem.*, 2, 37, 2001.

78. Hong, S. and Mirkin, C. A., *Science*, 288, 1808, 2000; Bullen, D. A., Wang, X., Zou, J., Chung, S.-W., Liu, C., and Mirkin, C. A., *Mater. Res. Soc. Symp. Proc.*, 758, 141, 2003; Ryu, K. S., Wang, X., Shaikh, K., Bullen, D., Goluch, E., Zou, J., Liu, C., and Mirkin, C. A., *Appl. Phys. Lett.*, 85, 136, 2004;

79. Lee, K.-B., Park, S.-J., Mirkin, C. A., Smith, J. C., and Mrksich, M., *Science*, 295, 1702, 2002; Lee, K.-B., Lim, J.-H., and Mirkin, C. A., *J. Am. Chem. Soc.*, 125, 5588, 2003.

80. Zhang, H., Li, Z., and Mirkin, C. A., *Adv. Mater.*, 14, 1472, 2002; Smith, J. C., Lee, K.-B., Wang, Q., Finn, M. G., Johnson, J. E., Mrksich, M., and Mirkin, C. A., *Nano Lett.*, 3, 883, 2003; Wilson, D. L., Martin, R., Hong, S., Cronin-Golomb, M., Mirkin, C. A., and Kaplan, D. L., *Proc. Natl. Acad Sci. U.S.A.*, 98, 13660, 2001.

81. Maynor, B. W., Li, Y., and Liu, J., *Langmuir*, 17, 2575, 2001; Su, M., Liu, X., Li, S.-Y., Dravid, V. P., and Mirkin, C. A., *J. Am. Chem. Soc.*, 124, 1560, 2002.

82. Porter, L. A. Jr., Choi, H. C., Schmeltzer, J. M., Ribbe, A. E., Elliott, L. C. C., and Buriak, J. M., *Nano Lett.*, 2, 1369, 2002; Magagnin, L., Maboudian, R., and Carraro, C., *J. Phys. Chem. B*, 106, 401, 2002.

83. Xu, S., Miller, S., Laibinis, P. E., and Liu, G., *Langmuir*, 15, 7244, 1999; Liu, G.-Y., Xu, S., and Qian, Y., *Acc. Chem. Res.*, 33, 457, 2000.

84. Xu, S. and Liu, G., *Langmuir*, 13, 127, 1997.

85. Wadu-Mesthrige, K., Xu, S., Amro, N. A., and Liu, G., *Langmuir*, 15, 8580, 1999; Kenseth, J. R., Harnisch, J. A., Jones, V. W., and Porter, M. D., *Langmuir*, 17, 4105, 2001; Zhou, D., Wang, X., Birch, L., Rayment, T., and Abell, C., *Langmuir*, 19, 10557, 2003.

86. Liu, M., Amro, N. A., Chow, C. S., and Liu, G., *Nano Lett.*, 2, 863, 2002.

87. Jin, X. and Uertl, W. N., *Appl. Phys. Lett.*, 61, 657, 1992; Balta-Calleja, F. J., Santa Cruz, C., Bayer, R. K., and Kilian, H. G., *Colloid Polym. Sci.*, 268, 440, 1990.

88. Klehn, B. and Kunze, U., *J. Appl. Phys.*, 85, 3897, 1999; Bouchiat, V. and Esteve, D., *Appl. Phys. Lett.*, 69, 3098, 1996; Sohn, L. L. and Willett, R. L., *Appl. Phys. Lett.*, 67, 1552, 1995.

89. Porter, L. A. Jr., Ribbe, A. E., and Buriak, J. M., *Nano Lett.*, 3, 1043, 2003.

Index